768425

H6

$7495

11 25

D1224026

BOTANY
FIFTH EDITION

CARL L. WILSON
Dartmouth College

WALTER E. LOOMIS
Iowa State University

TAYLOR A. STEEVES
University of Saskatchewan

Line illustrations based on drawings by
HANNAH T. CROASDALE, *Dartmouth College*

HOLT, RINEHART and WINSTON
New York • Chicago • San Francisco • Atlanta • Dallas
Montreal • Toronto • London • Sydney

BOTANY

FIFTH EDITION

Preface to the Fifth Edition

THIS EDITION, although somewhat larger than the last, departs from it in no basic way with respect to organization and objectives. Emphasis has again been placed upon plant diversity and the relation of plants to human affairs. We have attempted to provide a text that includes many of the newer advances in cell biology as well as the more traditional information on structure, function, development, reproduction, evolution, and ways of living of both nonvascular and vascular plants. In other words, we have tried to write a balanced, general text, containing subject matter useful to instructors, whatever the nature of the introductory course they teach—whether it contributes to general education, prepares students to enter a major program, or is part of teacher or other preprofessional training. The text is designed for the two-semester course, but, as with previous editions, can be successfully adapted to shorter courses.

Careful attention has been given to every section of the text to increase its teaching value. Some parts have been rewritten for greater clarity, many have been expanded, and new topics have been added. The illustrations have been carefully scrutinized; a number have been replaced and approximately 40 new illustrations have been added to the text proper. This edition also contains two sections of four-color photographs.

It is impossible to review completely in the Preface all of the text modifications, but the changes mentioned here will serve as examples. Chapter 3, on the cell, now includes a description of the origin and development of the cell plate, together with additional information on cell organelles. In Chapter 4 the discussion of active transport has been revised, and a treatment of DNA in chloroplasts and mitochondria has been added. In Chapter 5 the discussion of light and dark reactions in photosynthesis has been reduced and simplified. In Chapter 9 the description of the mechanism of mineral absorption by roots has been largely rewritten. In Chapter 10 the material on the water and carbon cycles has been revised and the discussion of the nitrogen cycle has been expanded. Chapter 14 (formerly Chapter 11) has been largely rewritten to give better coverage to growth hormones and modern concepts of morphogenesis. In Chapter 17 the section on evolution and adaptation has been revised and enlarged. Chapter 19 ("The Algae") now contains a section on eutrophication. Chapter 25, on the early land plants, has been thoroughly revised, and to it has been added a consideration of the origin of life. The material on bacteria and viruses in Chapter 20 is more extensive than is usually found in a text devoted to plants. The bacteria, however, whether classified as plants, as protista, or otherwise, are ecologically intimately related to plants through their effects on soils, in natural cycles, and in other ways. They are also related to man through their roles as agents of disease in man, his domesticated animals, and cultivated plants. Similarly, viruses are of direct economic importance to man. Moreover, contemporary molecular genetics, which includes the study of the genetics of bacteria and viruses, is now considered essential to a full understanding of heredity in higher organisms. This chapter has therefore been expanded to include the types of recombination mechanisms in bacteria (conjugation, transformation, and transduction), the role of bacteriophage in the genetics of bacteria, and lysogeny.

The authors acknowledge with gratitude the assistance received from many individuals. Some read portions of the manuscript or assisted in other, major ways. Others patiently answered questions or contributed in less important ways. Professor J. P. Poole, Dartmouth, read large portions of the manuscript and contributed to organization and clarity of expression. Others to whom acknowledgments are due are H. P. Banks and Gene Likens, Cornell University; Dennis T. Brown, University of Maryland School of Medicine and Clarke Gray, Dartmouth Medical School; Roderic B. Park, University of California, Berkeley; R. W. Barratt, Humboldt State College; D. J. McDonald, Middlebury College; R. H. Wetmore, Harvard University; A. P. Nelson, Vermont Environmental Center, and A. E. DeMaggio, W. T. Jackson, D. A. Stetler, and H. L. Allen, Dartmouth College. Finally, particular acknowledgment is due our editor, Peggy Park, for the patience and skill with which she has guided the manuscript of this edition to publication. Any errors found in the text are, of course, our own.

C.L.W.

W.E.L.

November 1970   T.A.S.

Credits for Illustrations

The authors express their sincere thanks to the individuals and organizations who have made available the illustrations cited in the following list.

FIGS. 1-2 AND 1-3 (RIGHT) U.S. Forest Service

FIG. 1-9 Dr. B. M. Shaub, Northampton, Massachusetts

FIG. 1-11 Weyerhaeuser Company

FIG. 1-12 Dr. Harry T. Horner Jr., Iowa State University

FIG. 1-13 Paul Popper

FIG. 2-3 (CENTER) Dr. A. J. Eames

FIG. 2-12 U.S. Forest Service

FIG. 2-14 National Park Service

FIG. 2-20 Tucson Chamber of Commerce, Tucson, Arizona

FIG. 2-24 Dr. A. E. de la Rue. Permission Arnold Arboretum

FIG. 2-27 U.S. Department of Agriculture

FIG. 2-30 Dr. Donald B. Lawrence, University of Minnesota

FIG. 3-3 Dr. M. S. Fuller, from Fuller and Reichle, in *Mycologia*

FIG. 3-4 Dr. John L. Blum, University of Wisconsin, Milwaukee

FIG. 3-7 Dr. William T. Jackson, Dartmouth College

FIG. 3-8 Dr. P. K. Hepler, from Hepler and Jackson, in *Jour. of Cell Biology*

FIG. 3-9 Modified from Ledbetter, in Warren, K. B. Editor. *Formation and Fate of Cell Organelles.* Permission Academic Press

FIG. 3-10 Dr. Paul C. Bailey, Birmingham-Southern College

FIG. 3-16 Dr. J. C. O'Kelley, University of Alabama

FIGS. 4-12 AND 4-15 Dr. Harry T. Horner Jr., Iowa State University

FIG. 4-22 Dr. Christopher L. F. Woodcock, Harvard University

Fig. 5-1 Hawaiian Sugar Planters Association Experiment Station

Fig. 5-13 Dr. Harry T. Horner Jr., Iowa State University

Fig. 5-16 Professor Roderic B. Park, University of California, Berkeley

Fig. 5-17 From Ting and Loomis in *Plant Physiology*

Fig. 5-22 Dr. C. L. Prior, Iowa State University

Fig. 5-24 Visual Aids Department, North Carolina State College

Fig. 6-3 General Mills, Inc.

Fig. 6-5 Dr. S. F. Conti, University of Kentucky

Fig. 6-8 Subtropical Research Station, University of Florida

Figs. 6-12 and 6-13 Dr. Walter H. Hodge

Figs. 7-2 and 7-3 From Sass, *Botanical Microtechnique*, Third Edition. Permission Iowa State University Press

Fig. 7-4 Modified from Esau, in *American Journal of Botany*

Fig. 7-5 Photomicrograph by James E. Hannaford, Dartmouth College

Fig. 7-12 Dr. R. F. Evert, in *Univ. California Publ. in Botany*

Fig. 7-15 Dr. E. F. Artschwager

Fig. 7-16 A,B Redrawn from C. C. Forsaith, in *New York State College of Forestry Tech. Bull. 18*

Figs. 7-18, 7-22, and 7-23 U.S. Forest Products Laboratory

Fig. 7-19 U.S. Forest Service

Fig. 7-25 U.S. Forest Service

Fig. 7-32 Dr. L. H. MacDaniels, Cornell University

Fig. 7-35 Negative by Dr. John E. Sass. From Magee in *Iowa State College Journal of Science*

Fig. 7-36 U.S. Forest Service

Fig. 7-38 From M. H. Zimmermann, in *Science* Vol. 133, No. 3446, cover and p. 76. Copyright © 1961, American Association for the Advancement of Science

Fig. 8-1 Mr. D. N. Simpson, University of California, Davis

Fig. 8-5 Dr. J. K. Wilson

Fig. 9-3 Redrawn from Jensen and Kavaljian, in *Amer. Jour. of Botany*

Fig. 9-10 Professor J. P. Poole, Dartmouth College

Fig. 9-14 U.S. Forest Service

Fig. 9-15 Mr. George B. Stevenson, Tavernier, Florida

Fig. 9-16 From McMurtey, in *U.S. Dept. Agric. Tech. Bull. 340*

Fig. 9-18 From Walker and Barber, in *Science* Vol. 133, No. 3456, pp. 881–882. Copyright © 1961 by American Association for the Advancement of Science

Fig. 9-19 From U.S. Dept. of Agr. Misc. Publ. 923

Fig. 9-21 Mr. Ralph Walter, Grand Ridge, Illinois

Fig. 9-22 From Hoagland and Arnon, in *Univ. Calif. Agr. Experiment Stat. Bull. 447*

Figs. 10-4 and 10-5 From *U.S. Dept. of Agr. Farmers Bull. 2003*

Fig. 10-6 Professor C. M. Franco, Campinos, Brazil

Fig. 10-7 From G. Bond, in *Symbiotic Associations, 13th Symposium Society General Biology*. Permission Cambridge University Press

Figs. 10-8 and 10-9 Soil Conservation Service, U.S. Department of Agriculture

Fig. 10-10 Alabama Agricultural Experiment Station. Courtesy Professor Knox W. Livingston

Fig. 10-12 Designed by Professor W. W. Bowen

Fig. 11-3 *U.S. Dept. of Agr. Tech. Bull. 1286*

Fig. 11-5 National Park Service

Fig. 11-7 Boyce Thompson Institute

Fig. 11-8 Pineapple Growers Association of Hawaii

Fig. 11-11 U.S. Department of Agriculture

Figs. 11-12 and 11-16 Mrs. M. C. Lincoln

Figs. 11-13, 11-14, and 11-15 United Fruit Company

Fig. 11-17 U. S. Forest Service

Fig. 12-2 Marion Seiler

Figs. 12-16 and 12-17 *U.S. Dept of Agr. Circular 728*

Fig. 12-19 Dr. John E. Sass in *Iowa State College Jour. of Science.*

Fig. 12-21 Electron micrograph by Dr. Jean M. Sanger

Fig. 12-24 Dr. John E. Sass

Fig. 12-27 Modified from Thompson, in *Jour Agr. Research*

Fig. 12-29 Mrs. M. C. Lincoln

Fig. 12-30 From MacDaniels and Heinicke in *Cornell Agr. Exper. Stat. Bull. 497*

Fig. 13-1 Boyce Thompson Institute

Fig. 13-2 From Dr. Frank C. Dennis in *Science* Vol. 156, No. 3771, pp. 71–73. Copyright © 1967, by American Association for the Advancement of Science

Fig. 13-4 From J. G. Woodruff, *Peanuts, Production, Processing, Products*. Permission Avi Publishing Company

Fig. 13-5 U.S. Department of Agriculture

Fig. 13-15 From J. P. Nitsch, in *Amer. Jour. of Botany*

Fig. 13-16 U.S. Department of Agriculture

Fig. 13-25 Dr. Lewis Knudsen

Fig. 14-1 From Dorothy M. Winter, in *Amer. Jour. of Botany*

Fig. 14-2 Photo by Ivan McArthur, University of Saskatchewan

Fig. 14-3 Photo by Douglas DesBrisay and John Waddington, University of Saskatchewan

Fig. 14-5 From R. H. Wetmore in *Brookhaven Symposia in Biology*

Fig. 14-6 Redrawn from C. W. Wardlaw, *Morphogenesis in Plants*, Methuen and Co., London

Fig. 14-7 From T. A. Steeves, in *Phytomorphology*

Fig. 14-8 From T. A. Steeves and I. M. Sussex, in *Amer. Jour. of Botany*

Fig. 14-11 Adapted from F. Jacob and J. Monod, in *Cold Spring Harbor Symp. Quant. Biology* 26: 193–212

Fig. 14-12 Modified from R. H. Wetmore and J. P. Rier, in *Amer. Jour. of Botany*

Fig. 14-13 A, B, D, E, from Steward, Mapes, and Holsten, *Science* 143: 18–27. Copyright© 1964, by American Association for the Advancement of Science. C, from Walter Halperin, in *Amer. Jour. of Botany*

Fig. 14-14 Adapted from Bartel and Martin, in *Jour. Agr. Research*

Fig. 14-17 From S. H. Wittwer and M. J. Bukovac, in *Economic Botany*

Fig. 14-18 From F. E. Denny, in *Contrib. Boyce Thompson* Institute

Fig. 14-20 Redrawn from Hanson, in *Amer. Jour. of Botany*

Fig. 14-21 U.S. Department of Agriculture

Fig. 14-22 Dr. H. A. Borthwick

Fig. 14-23 From Kenneth Post, in *Cornell Univ. Agr. Exper. Stat. Bull.* 787

Fig. 15-1 From Marloth, *Flora of South Africa*

Fig. 15-2 Arnold Arboretum

Fig. 15-3 U.S. Forest Service

Fig. 15-4A Mr. James E. Hannaford, Dartmouth College

Fig. 15-5 Dr. E. J. Kohl

Figs. 15-7, 15-8, and 15-9 U.S. Forest Service

Figs. 15-11 and 15-12 Dr. Donald B. Lawrence, University of Minnesota

Fig. 15-18 Map outline: Goode Base Map Series, copyright University of Chicago

Fig. 15-19 Dominion Forest Service, Canada. Photo by Dr. L. Robinson, Geographic Northwest Territories

Figs. 15-21, 15-22, and 15-23 U.S. Forest Service

Fig. 15-24 Colorado Agricultural Experiment Station

Fig. 15-25 Mr. Bernardo Maza

Fig. 15-26 Mr. Patric Fitzimmons

Fig. 16-7 Electron micrograph by Dr. Stephen Vogel

Fig. 16-8 U.S. Department of Agriculture

Figs. 16-9 and 16-11 Connecticut Agricultural Experiment Station

Fig. 16-12 Modified from Sprague, Iowa Agricultural Experiment Station Bulletin

Fig. 16-13 Asgrow Seed Company

Chapter 17 opening photograph Brookhaven National Laboratory

Fig. 17-3 Dr. Paul Weatherwax, in *Jour. Heredity*

Fig. 17-4 Dr. H. P. Olmo, in *Jour. Heredity*

Fig. 17-7 W. Atlee Burpee Company

Fig. 17-8 From P. C. Mangelsdorf, R. S. MacNeish, and W. C. Galinat, in *Science* Vol. 143, No. 3606, pp. 538–545. Copyright © 1964, by American Association for the Advancement of Science

Fig. 19-17 Drs. S. F. Conti and Elizabeth Gantt

Fig. 19-20 Bausch and Lomb Optical Company

Fig. 19-23 Dr. H. B. Bigelow, in *Bull. U.S. Bureau of Fisheries*

Fig. 19-26 Dr. Robert B. Wylie

Fig. 19-27 Negative by W. A. Setchell, print from Dr. E. Yale Dawson

Fig. 19-30 From Drs. E. Gantt and S. F. Conti, in *Jour. of Cell Biology*

Chapter 20 opening photograph Dr. Dennis T. Brown

Fig. 20-2 From Dr. H. J. Welshimer, in *Jour. Bacteriology*. Permission American Society for Microbiology

Fig. 20-3 Dr. S. F. Conti, University of Kentucky

Fig. 20-4 Negative from R.C.A. Laboratories. Print from the Upjohn Company

Fig. 20-7 Dr. Thomas F. Anderson in *Annales de L'Institut Pasteur*

Fig. 20-10 Modified from Lamanna, in *Jour. Bacteriology*

Fig. 20-12 Wilmot Castle Company

Fig. 20-13 Illinois Agricultural Experiment Station

Fig. 20-14 Dr. Albert Kellner and Society of American Bacteriologists

Fig. 20-15 Dr. M. T. M. Rizki, University of Michigan

Fig. 20-16 Modified from Stanley, in *Chemical and Engineering News*

Fig. 20-17 Research Laboratories, Parke, Davis and Company

Fig. 20-18 From Dr. M. K. Corbett, in *Phytopathology*

Fig. 20-19 From Dr. Roy Markham, Virus Research Unit, Agricultural Research Council, Cambridge, England

FIG. 20-20   From Dr. M. K. Corbett, in *Virology*. Permission Academic Press, Inc.

FIG. 20-21   Department of Plant Pathology, Cornell University

FIG. 20-23   From Herriott and Barlow, in *Jour. Gen. Physiology*

FIG. 20-24   From A. H. Kleinschmidt, D. Lang, D. Jackerts, and R. K. Zahn, in *Biochim. Biophys. Acta*. Permission Elsevier Publishing Company

FIG. 20-26   From Dr. Dennis T. Brown, in *Journal of Virology*. Permission of American Society for Microbiology

FIG. 21-1   Modified from Grant Smith, in *Botanical Gazette*

FIG. 21-7   Modified from Schwarze, in *Mycologia*

FIG. 21-8E   Redrawn from Cutter, in *Bull. Torrey Bot. Club*

FIG. 21-13   Dr. M. C. Richards, University of New Hampshire

FIG. 21-16   Dr. Thomas Sproston, University of Vermont

FIG. 21-20   Department of Plant Pathology, Cornell University

FIG. 21-31   Dr. S. F. Conti, University of Kentucky

FIG. 21-32   F. and M. Shaefer Brewing Company

FIG. 21-35   Professor John Sheard, University of Saskatchewan

FIG. 22-9   U.S. Department of Agriculture

FIGS. 22-18 AND 22-19   Illinois Agricultural Experiment Station

FIG. 22-21   U.S. Department of Agriculture

FIG. 22-23   Illinois Agricultural Experiment Station

FIG. 22-26   Modified from Buller, in *Nature*

FIG. 22-27   Modified from Buller, *Researches on Fungi*, Vol. 7. University of Toronto Press

FIG. 22-32   Hitchcock Foundation, Hanover, New Hampshire

FIG. 22-33   From Webster and Cetas, in *Plant Disease Reporter*, U.S. Department of Agriculture

FIG. 22-34   Old John Beam Company

FIG. 23-3   Redrawn from Schuster, in *American Midland Naturalist*

FIG. 23-10   Modified from Durand, in *Bull. Torrey Bot. Club*

FIG. 23-12   National Park Service

CHAPTER 24 OPENING PHOTOGRAPH   U.S. Forest Service

FIG. 24-21   Dr. Philip L. Johnson, University of Georgia

FIG. 25-1   From Delevoryas, *Morphology and Evolution of Fossil Plants*, copyright © 1962 by Holt, Rinehart and Winston, Inc.

FIG. 25-3   Dr. Henry N. Andrews and Missouri Botanical Garden

FIG. 25-7   From Zimmermann, in *Die Phylogenie der Pflanzen*.

FIG. 25-8   From Dr. Jiří Obrhel, in *Geologie*

FIG. 25-9   Photo by Dr. Donald B. Lawrence, University of Minnesota

FIG. 25-10   E, After A. S. Rouffa, in *Canad. Jour. of Botany*

FIG. 25-11   From Dr. David W. Bierhorst, in *Amer. Jour. of Botany*.

FIG. 25-12   From Lang and Cookson, in *Phil. Trans. Roy. Soc. London*.

FIG. 25-13   Modified from Fritsch, in *Annals of Botany*

FIGS. 25-14 AND 25-17.   Modified from Zimmermann, in *Geschichte der Pflanzen*

FIG. 25-15   From DeMaggio, Wetmore, and Morel, in *C. R. Acad. Sc. Paris*

FIG. 25-17   Modified from W. Zimmermann

FIG. 26-1   Professor L. H. Millener, University of Auckland

FIG. 26-3   Dr. E. J. Kohl

FIG. 26-5   From Eames, in *American Fern Journal*

FIG. 26-9   Modified from Schaffner

FIGS. 26-10 AND 26-12   Modified from Hirmer, in *Handbuch der* Palaobotanik

FIG. 26-13   From H. Delentre, in *Mem. de l'inst. Géol. de l'Univ. de Louvain*

FIG. 26-15 (RIGHT)   Photo by E. M. Kittredge

FIG. 26-18   E, after Sharp; F, after Walker; G, after Jeffrey

FIGS. 26-19 AND 27-1   Courtesy Carnegie Museum

FIG. 27-2   A, after Halle; B, after Grand'Eury; C, from Andrews, *Ancient Plants and the World They Lived In*, Permission Cornell University Press; D, after Kidston

FIGS. 27-3 AND 27-4   Brooklyn Botanic Garden

FIG. 27-6   U.S. Forest Service

FIGS. 27-10, 27-13, AND 27-14 (LEFT)   Modified from Ferguson, in *Proc. Wash. Acad. Science*

FIG. 27-17   Modified from Buchholtz

FIGS. 27-21, 27-27 AND 27-28   U.S. Forest Service

FIG. 27-22   Photo by Keith A. Trexler, National Park Service

FIG. 28-1   Paul Popper

FIG. 28-2   Marion Seiler

FIGS. 28-10 AND 28-23   U.S. Department of Agriculture

FIGS. 28-12 AND 28-21   National Park Service

Contents

BOTANY
FIFTH EDITION

Plants and Man

THE EARTH on which we live is a green earth. This green is caused by plants, and these plants are essential to the life and welfare of mankind. Man, other animals, and plants themselves could not exist on this planet without the green coloring matter found in plants, and without the activities of the leaves containing this green pigment.

The earth's green carpet is in large degree composed of plants with leaves, stems, and flowers. But there are also other kinds of plants—plants without leaves, stems, or flowers, many of them small and even microscopic in size. Some of them are green, some nongreen. They may occur in the soil and in the earth's waters, and many of these plants without leaves are important to man's welfare and even essential to his existence.

All of these plants are alive: they absorb and expend energy; they grow; they reproduce; they have evolved and are evolving. Their bodies are enormously diversified in size, shape, and form, and they are found in a wide variety of environments. The study of plants and of all facets of their structure and behavior is the *science of botany*. This study should lead to a better understanding of the plant life of your environment and to a greater appreciation and enjoyment of it.

PLANTS FEED THE WORLD

Of all the activities carried on by plants, we shall mention here only food manufacture, postponing a detailed discussion of this activity to Chapter 5. In this process, which is called PHO-TOSYNTHESIS, two common substances, carbon dioxide and water, are synthesized into sugars. Photosynthesis occurs in green cells in the presence of sunlight and has not been duplicated synthetically. By other processes, not a part of photosynthesis, sugars are utilized in green plants in the synthesis of starch, oils, proteins, and other complex compounds. These materials are utilized in the construction of new tissues and as a source of energy. Plant tissues, in turn, provide food for all forms of life—for nongreen plants, such as bacteria and fungi, and for all animals, including man.

In the absence of photosynthesis, all living things, with a few minor exceptions, would cease to exist and would disappear from the earth within a short time. All the food and other plant products we use—sugars and starches, seeds and fruits, woods, oils, drugs, and fibers—are produced by plants, largely from the simple products of photosynthesis. Flesh-eating animals that do not consume plant products directly depend upon plants indirectly through the plant-eating animals upon which they prey.

This dependence upon photosynthesis in the green cell prevails in the seas as upon the land. Directly or indirectly, the animal population of the oceans is supported by microscopic or near-microscopic forms of plant life that are able to convert inorganic substances into the organic compounds upon which life depends.

PLANTS AS LIVING THINGS

Plants, like animals, belong to the world of living things; they are alive. There is no general agreement, however, on definitions of life and living, not only because it is difficult to exclude certain features of nonliving objects from such definitions but also because life is not a substance or a material. It is not even a process, but is a state, or condition, in a system of highly organized and specific chemical compounds that interact among themselves and with the environment. Life is a property of such a system as a whole, not of any of its parts. It is best discussed, not in terms of what it is, but in terms of what it does. Nevertheless, the following characteristics are useful in distinguishing living organisms from lifeless materials.

Growth One characteristic of living things is growth or the potentialities for growth. A seed, for example, is alive, but it grows only when supplied with water, oxygen, and a suitable temperature. Growth has been defined as irreversible increase in volume. True growth is also associated with the synthesis of new biological materials by the growing organism. In animals and in some plants, such as fungi and bacteria, the materials from which these new compounds are synthesized are organic compounds previously formed by other organisms. In green plants they are synthesized by the plants themselves.

Energy Transformation Living systems may accumulate and transform energy, which is derived, directly or indirectly, from the sun and is stored in organic compounds. An organism transforms the energy in food during the process of RESPIRATION. It is this energy that is utilized in the synthesis of complex compounds from simple ones and in many other processes, including the maintenance of life itself. The term METABOLISM is usually applied to the over-all chemical processes of the living body. Some metabolic processes break down the materials of the body into simpler constituents, while others build them up. Growth occurs only when the building up predominates over the breaking down.

Reproduction The formation of a new individual, a reasonably accurate reproduction of the parent, is another characteristic associated with living things. One important feature of reproduction is the appearance of variability among the offspring. One kind of inherited variability, called MUTATION, is especially important. Without mutations relatively few kinds of living things would now occupy the earth's surface.

Response Living organisms react to the environment; that is, they respond to stimuli. Both plants and animals may respond to such factors as temperature, light intensity, gravity, length of day, touch, and chemical agents. In most animals the response to stimuli may be expressed in movement of the entire body, although in some, as the attached corals and sea anemone, a stimulus may result only in local responses. Higher plants, anchored to one place by their roots, also may exhibit limited movement in response to such stimuli as light, touch, and changes in temperature. Such plant movements, including the opening and closing of flowers, the twining of tendrils, and the movement of leaves, can easily be demonstrated by the time-lapse camera. Often the layman considers movement as the sole criterion of life, and if an object, or any of its parts, does not move, it is considered to be nonliving. But a tree, a petunia, or an onion is alive, just as much alive as a goldfish or a squirrel.

Organic Composition   A final aspect of all living things is that they are constructed of ORGANIC compounds. Organic compounds all contain the element carbon, which possesses the capacity to form long-chain or ring structures composed of many atoms. Carbon also unites readily with other elements, such as hydrogen, oxygen, and nitrogen. Examples of organic compounds are sugars, starch, cellulose, proteins, fats, and alcohols. The word "organic" is a survivor from a period in the history of chemistry when it was believed that compounds containing carbon were produced only by living plants and animals. Most such compounds do originate from the activities of living organisms, but a large number have now been prepared synthetically in the laboratory and the term organic chemistry has been expanded to include these derivatives.

Among the simplest organic compounds are those composed of carbon and hydrogen. Examples of these HYDROCARBONS are lubricating oils, rubber, and methane or marsh gas, the chief component of natural gas. Hydrocarbons that are obviously not of biological origin have been identified by the spectroscope in stars, in comets, and in the sun. Living things, however, tend to be composed of large and complex molecules

usually containing many carbon atoms. Many of the largest organic molecules are PROTEINS— composed of the elements carbon, hydrogen, oxygen, nitrogen, and sometimes sulfur. The thousands of reactions that occur in living cells are dependent upon protein catalysts called ENZYMES (Chapter 4).

THE CELL AS THE UNIT OF STRUCTURE

Among the outstanding developments of nineteenth-century biology was the formulation of the generalization that plant and animal bodies are composed of small structural units termed CELLS. These units contain a complex, semifluid, living material known as PROTOPLASM, which is composed of many organic and inorganic constituents. Protoplasm is able to reproduce itself, and respiration and many other complicated chemical processes go on within it.

Some kinds of plants and animals are composed of a single cell, but in the larger forms the number of cells that make up an organism runs to many billions. Fundamentally, cells are alike, but they commonly differ in form and size as they grow and differentiate within the body of which they are a part. Thus, some cells are spherical, some brick-shaped, some irregular, some spindle-shaped, and some cylindrical. Cells vary greatly in the roles they play in the processes of the organism as a whole; there is great division of labor (FIG. 1-1). In an oak tree, for example, some cells function in food manufacture, some in food storage, and some in the transportation of water or food. Others are concerned with reproduction, with support, or with the prevention of water loss; some are unspecialized and continue to give rise to other cells.

THE SIZE OF CELLS AND PLANTS

The smallest plant cells are certain kinds of bacteria—microscopic spheres with a diameter of

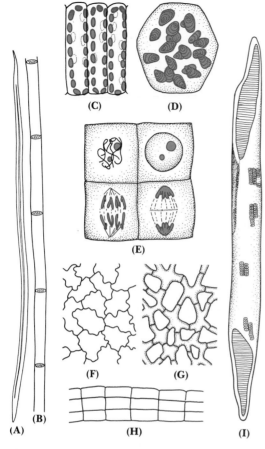

Fig. 1-1 Some kinds of plant cells. (A): a cell providing mechanical strength. (B): food-conducting cells. (C): food-manufacturing cells. (D): food-storage cell containing starch grains. (E): dividing cells. (F): epidermal cells. (G): cells from aerating tissue. (H): cork cells. (I): water-conducting cell.

about 0.00002 inch. Each of these cells is a single plant. Among the largest plant cells are the bark fibers of ramie, an important tropical and subtropical fiber plant. These cells attain a length of more than 20 inches.

This proportional range in the size of cells is about a million to one, but the range in the size of plants is much greater. It extends from the bacte-rial cells just mentioned to the redwoods, some of which attain a height of nearly 400 ft. The smallest flowering plant is the water-meal (*Wolffia*), a floating aquatic plant about 1/25 inch long. The largest and heaviest of all living plants are the giant sequoias of the Pacific Coast. The best-known of these, the General Sherman Tree, is 273 feet high and 83 feet 11 inches in circumference at 4½ feet above the ground. Its weight is estimated at more than a thousand tons.

WHERE AND HOW PLANTS LIVE

Plant life can generally tolerate more adverse conditions than animal life, and plants exist in virtually every environment in which living things can survive. Plants have become adapted to a wide range of moisture and drought, heat and cold. Some flowering plants are found in the hottest and driest sections of the deserts of Libya and Western Australia. Several kinds of lichens grow on bare desert rock moistened only by infrequent rain. At the other extreme, oceans, lakes, and streams may abound in plant life. Lichens and mosses have been found on antarctic rocks, and even the arctic tundra supports plant life on soil that is frozen for much of the year. Plants are quick to invade new environments—volcanic islands that have risen out of the Pacific have borne their first plant life in a year and developed a rich vegetation in a decade or so.

Not all plants require soil or water environments; many of them grow on other plants or animals. The alga *Protococcus*, for example, is found on the bark of trees. Some flowering plants, like the tangling, orange-yellow dodder, are parasites, taking food from their host. Some bacteria, also, are parasites that cause serious diseases in other plants and in animals. But many are useful; the intestinal tracts of animals, for example, swarm with bacteria that are frequently of direct value to the host. Other plants live on dead or decaying organic matter; still others live in associations that are mutually beneficial to the plant or animal organisms involved.

Classification	Nutritional relationships and methods of obtaining food	Examples
1. **Autotrophic**	able to manufacture all of their own food	
a. photo- (phototrophic)	using light as a source of energy in food manufacture	most green plants, as corn, oak, potatoes; a few kinds of bacteria
b. chemo- (chemotrophic)	using chemical energy for food manufacture	nitrifying bacteria, sulfur bacteria, iron bacteria
2. **Heterotrophic**	unable to manufacture their own food; require an external source of complex organic compounds. Most animals are heterotrophic	
a. parasitic	obtaining food in dissolved form from living tissues of plants or animals at expense of host	dodder, broom rape, squawroot, many bacteria, and fungi
b. saprophytic	obtaining food in dissolved form from non-living organic remains, plant or animal	puffballs, mushrooms, many molds, and most bacteria
c. phagotrophic	ingesting food in solid (particulate) form	animal-like flagellates, slime molds

3. **Symbiotic**
   a. parasitic
   b. mutualistic
	living in more or less intimate association with another organism to their mutual benefit	algae and fungi in lichens, legume-nodule bacteria, flowers with their insect pollinators

The term SYMBIOSIS is applied to any state in which two dissimilar organisms live together in a close relationship. If the relationship is advantageous to both participants, it is called MUTUALISM. If only one derives advantage, the condition is termed PARASITISM. Parasitism or mutualism may exist between plant and plant, plant and animal, or animal and animal.

One of many examples of a mutually beneficial relationship between plants and animals is found in the larvae of certain beetles, which can grow normally on a diet deficient in vitamins of the B group. They are able to grow because yeast cells, which live in the alimentary canal of the larvae, synthesize these vitamins. The yeast, in turn, is supplied with food from the environment. Even more interesting is the relation between a fungus and the leaf-cutting ants of the American tropics. The ants live in great nests, within which they grow the fungus, supplying it with food in the form of pieces of leaves cut from surrounding trees. The fungus flourishes on this nutriment, and bits of the fungus are cut off and used as food by the ants. This cultivation of a plant for food undoubtedly began long before the time of man, and we may therefore regard the ants as the world's first farmers.

Most of the nutritional relationships in the plant kingdom are summarized in the table at the top of this page.

A BRIEF SURVEY OF THE MAJOR PLANT GROUPS

I. Plants without supporting or conducting tissues. Plant body (with the exception of the seaweeds) usually small and inconspicuous.

ALGAE	Unicellular or multicellular pond scums and seaweeds of fresh or salt water; also found in damp places. Autotrophic.
BACTERIA	One-celled, usually colorless plants; usually saprophytes or parasites.
FUNGI	Molds, mushrooms, bracket fungi, mildews, and blights. Mostly multicellular plants; saprophytes or parasites.
LICHENS	"Plants" composed of an alga and a fungus; on tree trunks, rocks, and soil. Crustlike or erect and bushy. Mutualistic and autotrophic.
LIVERWORTS AND MOSSES	Liverworts: Delicate creeping or prostrate plants, often ribbonlike and flat on the soil, sometimes with stalked sex organs; or slightly differentiated into an axis with leaflike expansions. Autotrophic. Mosses: Plant body a short, erect, leafy stem, attached to the soil, and bearing stalked spore capsules. Autotrophic.

II. Plants with supporting and conducting tissues. Plant body usually large and conspicuous. Nearly all autotrophic.

A. Plants without Seeds

FERNS	Leaves commonly large and divided, bearing spores in minute spore cases on the lower surface.
CLUB MOSSES	Trailing or creeping plants with scalelike leaves. Spore cases usually grouped into slender clublike or conelike structures.
HORSETAILS	Plants with hollow, jointed stems and minute leaves. Spore cases in conelike structures.

B. Plants with Seeds

CONIFERS	Mostly evergreen trees: larch, fir, spruce, pine, hemlock, Douglas fir, cedar, redwood, etc. Mostly bearing seeds in cones.
FLOWERING PLANTS	Plants bearing flowers; seeds contained within a fruit. Generally subdivided into two groups, the monocotyledons and the dicotyledons. In the dicotyledons the embryo and seedling have two seed leaves (cotyledons). In the monocotyledons there is but one cotyledon. The groups also differ in the structure of the stem, leaf, and flower. Some dicotyledons are maple, oak, cotton, tomato, sugar beet, poppy, pea, aster, and sunflower. Some monocotyledons are lilies, irises, grasses, palms, orchids, cattails, sugar cane, and banana. (See Plate 1.)

THE PLANT KINGDOM

Botany was defined early in this chapter, but a satisfactory definition of a plant is more difficult.

The main distinctions between a typical plant and a typical animal are obvious. There are, however, numerous borderline organisms, microscopic or near-microscopic in size, that possess

Fig. 1-2 A flowering plant. Flowers and leaves of the great laurel (*Rhododendron maximum*).

Fig. 1-3 Conifers; seed cones and leaves. (Left): fir (*Abies balsamea*). (Right): longleaf pine (*Pinus australis*).

features of both groups of living things. This difficulty in definition emphasizes the complexity of the plant kingdom, which we now attempt briefly to explore.

The word "plant" may bring to mind the green leafy members of the plant kingdom, such as a potted geranium or a garden flower. These are conspicuous and familiar. If we pursue the common concept further, other plant organs in addition to leaves and flowers (FIG. 1-2) come to mind—roots, stems, fruits, and seeds. A survey of the plant kingdom reveals, however, that there are plants in which some or all of these organs are lacking. There are, for example, the conifers, or cone-bearing plants (FIG. 1-3), which have roots, stems, leaves, and seeds, but no flowers. A plant that bears flowers, then, is typical of only one segment of the plant kingdom, the flowering plants. Other groups—the club mosses (FIG. 1-4), horsetails (FIG. 1-5), and ferns (FIG. 1-6)—lack even seeds. Still other plants, the algae—the pond scums, seaweeds (FIG. 1-7), and their kin—lack the roots, stems, and leaves that mark the groups mentioned above. The algae are so constructed, however, that they can carry on the same basic

activities of plants, just as well as the structurally more complex plants. Another large group, the fungi (FIG. 1-8), lacks not only roots, stems, and leaves but even the ability to manufacture its own food. Still another group, the liverworts (FIG. 1-9) and mosses (FIG. 1-10), is set apart in many ways from all the others. All in all, about 350,000 kinds of plants have been described. About half of them are flowering plants; about one-third belong with the algae and fungi; the remainder belong to smaller groups.

Instead of attempting to define a plant, then, it would seem more useful to examine the great diversity of organisms that make up the plant kingdom. This survey will, moreover, supply a more accurate picture of the nature of plants than could be conveyed by any definition (see survey, page 8).

SCIENCE AND ITS METHODS

Botany can be described as the science of plants. Science, in turn, is viewed as a body of knowledge that is concerned with the world and its phenom-

Fig. 1-4 (Above, left): a club moss (*Lycopodium complanatum* var. *flabelliforme*).

Fig. 1-5 (Above, right): the common horsetail (*Equisetum arvense*).

Fig. 1-6 (Below): the maidenhair fern (*Adiantum pedatum*).

ena, and which is supported by data that can be tested by some sort of check or proof. Investigations directed to the answering of questions and the solving of problems are an integral part of science.

The Scientific Method The scientific method consists of the formulation of a problem or hypothesis, and observation or experiment, or both, to test it, as contrasted to speculation or the acceptance of authority. To be useful in solving a wide range of problems, the forms of scientific inquiry must themselves be variable. However, they have one characteristic in common: in all scientific work it is essential that another scientist be able to *repeat* and *confirm* what has been done or observed. Despite the authority, the belief, or the convictions upon which they may rest, descriptions and experiments are not acceptable unless they can be verified by further investigations.

The scientific method may be illustrated by the description of a typical pathway followed in the solution of a problem. The process begins with the accumulation of facts through observation or experimentation. A critical study of these facts permits the formulation of an *hypothesis*, often a tentative working hypothesis, that seems to explain these facts or establish relationships among them. This step is followed by further observations or experiments designed to verify or disprove the hypothesis. As evidence accumulates, hypotheses may assume the status of *theories*, and eventually, if widely accepted, of generalized principles, or *laws*. There is, however, no clear dividing line between hypotheses and theories or between theories and "laws." Thus we speak of Mendel's *laws*, Boyle's *law*, the *theory* of evolution, the germ *theory* of disease, the gene *theory* of inheritance, although all of these concepts have been generally accepted for many years.

Few, if any, problems in science, however, may be regarded as finally solved. A problem is regarded as solved only in the sense that the solution is highly probable—there is much evidence in its favor and little or none to the contrary. Sometimes hypotheses and theories fail to be reliably verified and must be discarded in favor of others that seem more probable. Even so-called laws have been accepted for a time and then have been replaced or modified. Some facts turn out to be nonfacts, and theories are contradicted by new evidence as new and more accurate methods of investigation are developed. The science of today is not that of yesterday and will not be that of tomorrow. Knowledge is never complete, and "truth" is always relative, never

Fig. 1-7 (Above, left): a brown alga (*Fucus*) on rocks at low tide.

Fig. 1-8 (Above, right): a fungus, the oyster mushroom (*Pleurotus ostreatus*), growing on a dead tree.

Fig. 1-9 (Below, left): liverworts (*Preissia commutata*). The caplike structures, borne on stalks 1–2 inches long, bear the female sex organs and, later, the spores.

Fig. 1-10 (Below, right): the pigeon-wheat moss (*Polytrichum*).

absolute. Science is constantly growing as erroneous concepts are discarded and more useful ones are substituted.

The wide dissemination of the facts and results attained through scientific investigation is necessary for progress, and the techniques of locating, digesting, and assimilating the mass of literature in his discipline are essential equipment for a scientist. In botany, for example, this training constitutes an important distinction between the botanist and the gardener, florist, or nature lover. The botanist has acquired enough fundamental knowledge of his science to understand its major problems and ways in which their solutions can be approached.

BOTANY AS A SCIENCE

Since botany is a science, its methodology must be that of science. To formulate his questions

Fig. 1-11 Plant scientists use radioactive isotopes in research on the movement of materials within the tree.

and solve his problems, the botanist must possess the knowledge and ability to choose the techniques most likely to lead to an answer. He may have to work in the field or the greenhouse; other problems may be attacked in the laboratory (FIG. 1-11). Whatever the techniques used, whether preparing microscope slides, using radioactive tracers, growing plants, or collecting them in a tropical rain forest, the investigator's methods are dictated by the problem at hand and the tools that are available.

Descriptive Methods Observation and description are the simplest, oldest, and best-known methods of science. In certain areas, including much of botany, direct observation and description are the chief applicable techniques. Despite newer procedures, description, identification, and classification are still essential, and such work proceeds actively today. Descriptive techniques

are used in such botanical problems as classifying a new species of plant, determining changes in plant life in a region after the introduction of sheep raising, or listing the characteristic plants of New England bogs. The facts needed to solve these problems and many others like them are obtained mainly by making ample and accurate observations and by keeping careful records of what is observed.

The validity of observational or descriptive studies may often be increased by the application of statistical techniques that help ensure that the observations include a representative sample of the plant population. The question "How do seasonal changes affect the kinds and numbers of free-floating microscopic plants in Burnett Pond?" obviously involves such subsidiary questions as "How many water samples should be taken for study, and how often? From what part or parts of the pond should they be taken? Should they be taken from the surface or at various depths?" Answers to questions on how the samples are to be taken, and hence on how the observations are to be made, will influence the data obtained. Since it is important that scientific studies be comparable—and essential that findings be verifiable—the development of techniques for sampling, observing, and recording has become important.

Experimental Methods Although botany originated as a descriptive science and some aspects of it remain so, other areas use controlled experiments to test hypotheses. In its simplest form an experiment involves a CHECK or CONTROL group compared with an EXPERIMENTAL or TEST group. The control group is held under constant conditions while the test group is exposed to the effects of various factors, one at a time. Any changes that occur in the test but not in the control group are assumed to be the result of the condition that is changed. Controlled experiments may be carried out in laboratory incubators, special growth chambers, greenhouses, or the open field. Each treatment, including the control, should be replicated, normally from three or four to as many as ten times, and the replicate cultures, potted plants, or field plots carefully distributed so that no plants being treated will be favored more than the others, as by light or temperature in the greenhouse, or soil variations in the field. Failure to observe these precautions makes valueless many experiments by amateurs. Experimentation is an exacting science, and only carefully designed, replicated, and recorded experiments can be expected to give dependable results.

BOTANY AND OTHER SCIENCES

A science as broad in scope as botany is obviously not isolated and independent. Botanical knowledge is closely related to many other sciences. Indeed, a course in general botany must include essential material from the field of physics—material on the physical nature of liquids and gases, on diffusion, and on the fundamental laws of energy. Some knowledge of the nature of chemical elements and compounds is involved. A grasp of some facts of geology is essential for understanding the plants of the past and their evolution. Recognition of these interrelationships among the biological and physical sciences has led to the development of such combined fields as biochemistry, biophysics, and paleobotany.

Facts uncovered in the field of chemistry have often provided the key to the solution of a whole group of plant problems. The knowledge, still rapidly accumulating, of the structure of the complicated molecules present in plant and animal cells makes possible a better understanding of cellular processes and of the mechanisms of heredity. The reverse situation is also true. Discoveries and problems in botany have stimulated advances in chemistry and other areas of science. The connections between botany and zoology, geology, meteorology, chemistry, and physics have become more and more apparent as these sciences have been studied in greater depth.

Most of the processes and tools of modern botanical science are also employed by other sciences. Among these methods and tools are the light microscope and its modifications; the phase, polarizing, and interference microscopes; the electron microscope (FIG. 1-12); various chromatographic techniques; electrophoresis; radioautographic techniques; the recording spectrophotometer; and radioisotope counters. The ultracentrifuge has made possible basic advances in the study of protoplasm. The microtome is used by all biologists for cutting thin sections. Com-

Fig. 1-12   An electron microscope. Sections of plant material 100 times thinner than those ordinarily used in the laboratory can be magnified hundreds of times more with a beam of electrons than they can with a light microscope.

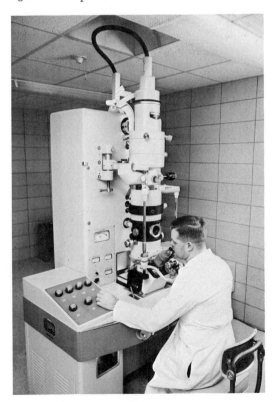

puters have greatly aided the analysis of data. These are only a few examples of the many instruments and methods that have directly furthered progress in botany.

DEVELOPMENT OF BOTANY

Before botany became scientific (about 1650), the collection, use, and cultivation of plants had been going on for thousands of years (Fig. 1-13). The beginning of agriculture ranks with the use of fire as a basic step in the development of civilization. We know that primitive man used plants for food, medicine, clothing, and shelter. Plants played an important role in many primitive religions. The fertility of the soil, the rebirth of plants in the spring, and the harvest were often the themes of tribal ceremonies. Primitive people used products of wild plants and later sowed seed of some of them to produce more easily harvested crops closer to their dwellings.

These utilitarian arts involving plants, so essential to man's existence, became the foundation of botany. The problems that arose were critical indeed, for then, as now, crop failure, famine, and death went hand in hand. The science of botany arose from the basic human needs for food, shelter, and the preservation of health. In spite of the tremendous technological progress made during the past centuries, these needs are still with us. The science of botany has, in fact, become more important than ever, for we face a rising tide of world population with no corresponding increase in arable land.

In its prescientific stage, much of what is now botany was descriptive. Early books described the forms and uses of plants that had real or fancied values as medicines or foods. Many of the botanical arts, such as grafting, cultivation, irrigation, and plant breeding, were known long ago, but there is little indication that the scientific basis for them was understood. Improvements in plants were due largely to chance because igno-rance of the nature of plants limited experimentation. Plant diseases, for example, were known at the time of Aristotle, but 2000 years passed before their causes and remedies were discovered. Botany became a science gradually. On the basis of newly acquired information, men began to question old authorities. Questioning and observation, checked by experience and experiment, eventually led to the development of botany as a science.

The historical periods during which inquiry and scholarship flourished were marked by progress in botany as well as in other sciences. Steps in this progress, for example, characterized the period of the Renaissance, the eighteenth-century period of liberalism, and the nineteenth-century period of Darwinism. The invention of the compound microscope, in 1590, opened broad new fields to the botanist as well as to other scientists. Bacteria and other minute forms of life could be studied with the aid of the microscope and, later, the cell was recognized as the structural unit of plant and animal tissues.

By the end of the 1700s, experimental methods had been introduced into the study of plant processes and activities. Photosynthesis had been discovered, and a beginning made in understanding sexual reproduction in plants. Botanical explorations had been carried out in many parts of the world, and progress made in plant classification. The 1800s saw studies on the geographical distribution of plants and the development of understanding of the relations between plants and climatic and other environmental factors. The universal presence of a cell nucleus was established, and the process by which it divides was discovered. Plant physiology became an established science, and many studies were carried out on the relations between plants and the soil. Immunity and the causal relationships between bacteria, fungi, and disease in animals and plants became known and subject to study.

Our own century, in turn, has seen immense advances in botany as in other fields of science.

Fig. 1-13 Winnowing grain. A Kurd threshes wheat with a method that was old 5000 years ago.

Not the least of these is the work on plants that has contributed to our knowledge of heredity and the formulation of the gene theory of inheritance. But although the accomplishments of the past several hundred years have been great, the sum of our knowledge is probably small in comparison with that which remains to be discovered. Fortunately, however, scientific knowledge seems to grow by geometric progression. The more man learns, the more tools he has for the solution of new problems.

THE SCOPE OF BOTANY

The "science of botany" is, in a sense, a misnomer, for botany is composed of a number of closely allied sciences. The rapid growth of botanical knowledge has made specialization inevitable. Certain focal areas developed into subdivisions of the science, maintaining a kind of local independence while remaining an essential and integral part of the whole.

The chief field of botanical activity, until about a century ago, was the classification and identification of plants—the science we now call PLANT TAXONOMY or SYSTEMATIC BOTANY. The importance of this subject has not decreased with the years. Modern taxonomy is essential in the fields of PLANT ECOLOGY, the study of the relationship of plants to their environment, and PLANT GEOGRAPHY, the study of plant distribution. Taxonomy is also an essential tool in the study of certain kinds of allergy; in weed control; in the control of plants poisonous to domestic animals; and in conservation, wildlife management, and the study of renewable natural resources. But classification and identification are only part of the activities of systematists. They are also students of organic diversity—they study variation and the nature of species, and are in fact sometimes more concerned with the processes of evolution than with classification itself.

With the development of new observational and experimental techniques and with advances in other sciences, other fields of botany have arisen and developed rapidly, especially in the present century. Prominent among these is PLANT GENETICS, the study of heredity. Other subdivisions include PLANT ANATOMY, the study of development and structure, mostly of seed plants; PLANT MORPHOLOGY, which stresses comparative studies on form and reproduction; PLANT CYTOLOGY, which deals with the structure and behavior of cells; PHYCOLOGY and MYCOLOGY, concerned respectively with algae and fungi; and PLANT PHYSIOLOGY, the study of the mechanisms and processes in plants and the interpretation of plant behavior in terms of physical and chemical laws. MICROBIOLOGY, which deals with microscopic forms of life, and includes the science sometimes called BACTERIOLOGY, may in large part be considered a subdivision of bot-

any, since most of the organisms with which it deals may be classified in the plant kingdom. A number of the fields cited—for example, genetics, ecology, and cytology—are equally botanical and zoological—that is, biological. This is also true of MOLECULAR BIOLOGY, a subject concerned with the ultimate physiochemical composition and reactions of living matter.

The fields of applied botany are numerous. PLANT PATHOLOGY is concerned with the control of plant diseases caused by fungi, bacteria, and viruses; ECONOMIC BOTANY is the study of the kinds and manner of utilization of the plants directly useful to man. HORTICULTURE, FLORICULTURE, FORESTRY, and AGRONOMY, the sciences of crop production, since they deal with plants, are subdivisions of botany. PHARMACOGNOSY, concerned with drugs and their uses, is in part a branch of botany, for many drugs are of plant origin.

SOME CONTRIBUTIONS OF BOTANY

Botanical investigations and discoveries have resulted in the improvement of crop growth and in the development and use of new and better plants. These results have come about in a number of ways. The environments of crop plants have been improved; weeds and plant diseases have been controlled; new varieties of plants have been produced by breeding; and new species and varieties of plants have been introduced. These changes, and others, have had profound and far-reaching effects upon the United States and the rest of the world.

Plant improvement by breeding has improved virtually every agricultural crop. An outstanding example is hybrid corn (maize). Before the introduction of hybrid corn, the average yield per acre of corn in the United States was around 26 bushels. At present the average yield per acre is about 70 bushels. Yields of over 100 bushels per acre are common, and more than 300 bushels per acre have been produced. Not all of this increase, of course, is due to the new hybrids, for much of it is the result of improved methods of production. In the developing tropical and subtropical countries, new high-yielding varieties of rice, wheat, corn, sorghum, and other grains have been produced, and these new strains are helping to feed a hungry world. Particularly important are the new dwarf, stiff-strawed, high-yielding, and disease-resistant wheats from Mexico and rice from the Philippines, which have been widely adopted in tropical countries. The stiffer stems prevent lodging (falling over) of heavy crops on heavily fertilized and irrigated soil. These new varieties can produce three and more times the yields of the old strains.

Plant breeding may be used to improve crop plants by means other than direct increase in yield. The development of hardy varieties enables certain crops to be grown in areas where soil and climate might otherwise be unfavorable. New varieties of cotton have extended the cotton belt westward. New varieties of soybeans have been developed that flower under varying daylengths, thus making it possible to crop them in latitudes in which the older varieties failed to blossom under the prevailing length of day. Drought-resistant varieties of corn, wheat, grain sorghums, and alfalfa have brought many acres of semiarid land under cultivation.

Botanists have also developed methods for the control of many diseases caused by viruses, bacteria, molds, rusts, and other fungi that limit crop yield. The development of such control measures has been vital to the production of cotton, potatoes, bananas, and many other crops. Plant diseases are also controlled by the development of disease-resistant crop plants—for example, varieties of oats have been produced that are resistant to stem rust, crown rust, loose smut, and halo blight. Plant breeders have developed varieties of sugar beet resistant to the virus causing curlytop, and varieties of mosaic-resistant beans and sugar cane, wilt-resistant peas, tomatoes, and sweet corn, and yellows-resistant cabbage.

Numerous other resistant varieties of plants used by man for food or ornamental purposes have been produced.

Plant Introduction With the exception of a few crops that are native to North America, all important fruits, vegetables, and grains in the United States have come from other parts of the world or have been derived by selection and breeding of such importations. Many of these economically valuable plants were introduced in early Colonial times, and the search for new crop plants has continued to the present. Beginning about 1897, the U.S. Department of Agriculture, through what is now known as the Section of Plant Exploration and Introduction, initiated a systematic program for locating plant materials for the diversification and improvement of American agriculture.

In these activities, botanists have taken a leading part. More than 180,000 introductions have been made, and many of these have become important crops. These include new strains of cottons; new legumes and grasses, such as Peruvian alfalfa, crested wheat grass, bromegrass, and Sudan grass; and new strains of cereals, including rice from Japan, wheat from Russia, and barley from Europe and Africa. Ornamentals include the Chinese elm and other shade trees. Among fruits are new varieties of dates, mangoes, avocadoes, prunes, peaches, apricots, cherries, and grapes. Numerous varieties of soybeans have been introduced, and from some have been bred new and superior strains. In the search for new crop plants, botanical explorers have traveled into the rain forests of the tropics, the mountains of central Asia, and African deserts so that new or better plants might become available to American agriculture.

Plant Improvement, Then and Now The development of crop plants is not something that plant scientists of today may claim as strictly their own contribution. They realize that virtually every major food plant we use today has been known and cultivated for at least 2000 years—some perhaps for 10,000 years or more. The earliest agricultural discoveries and the first steps in the improvement of plants, unconscious or fortuitous as they may have been, were made by unknown, semicivilized men in the highlands of Mexico, Peru, Bolivia, and Chile, and certain parts of China, northern India, Central Asia, Asia Minor, and possibly Ethiopia. The improvements that botanists have added to the basic advances made in these regions more than 20 centuries ago have been great. But authorities agree that all that has been done in improving corn, for example, is not yet equal to the development that this plant had undergone before Columbus saw it.

Other Botanical Activities Botanists are engaged in a wide variety of professional endeavors. They participate in weed research, including herbicides, life cycles, and cultural control, and in the study of plants poisonous to man and his domestic animals.

In the United States, weeds increase the cost of crop production by $4 billion annually. Botanists investigate the effects of growth regulators on plant development. They seek new ways of using fertilizers to increase crop production and to improve the soil. They have contributed to the solution of problems of storage and transportation of vegetables, fruits, and grains. Botanists identify fossil pollens, and use them in interpreting climates of the past. Botanical contributions to medicine are many: Botanists aid in the search for new drug plants and antibiotics and study the pollens that cause human disease. Botanists are called upon also in the search for new sources of fibers, tannins, essential oils, gums, and resins. And they play an increasingly prominent role in some of the most pressing problems of our times, such as environmental pollution and the proper utilization and conservation of soil, water, grasslands, and forests in the face of the accelerated growth of the world's population.

Finally, like other scientists, botanists carry on pure research, with the result that the frontiers of knowledge are extended and our cultural heritage is increased. The facts discovered in the pursuit of truth for its own sake, however, are often later applied to the furtherance of man's control over his natural environment and advancement of his material welfare.

THE INTERPRETATION OF PLANT ACTIVITIES

Our knowledge of botany, like that of all other sciences, has been built up slowly by repeated observations and carefully controlled experiments. This accumulated knowledge replaces opinions based upon popular beliefs and traditional authority. Scientific methods have established that plants are living machines whose operation is governed by physical and chemical laws and whose structures and processes are the result of the interaction between heredity and the environment.

Quite the contrary point of view is taken by many people, who commonly interpret the activities of plants in terms of human behavior. Because people usually act with a specific purpose and realize that certain acts usually are followed by predictable results, they tend to assume that plants behave in the same manner. This attitude is covered by the term TELEOLOGY, the doctrine that holds that the form and behavior of organisms are the result of conscious purpose or design. The implication is that a given structure has been developed to serve a particular purpose, or that plant activities are caused by the needs and requirements of the plant and result in certain anticipated goals. Such statements as "Trees in a forest grow tall in order to reach the light," "Plants adapt themselves to new environments," "In dry climates roots grow deeply in search of water," "Plants produce spines and thorns to protect themselves from grazing animals," are all teleological. In general, any statement that implies that structures or processes have developed "in order to" or "for the purpose of" achieving a certain result are teleological and unscientific.

Statements such as those quoted above are commonly made in connection with problems of adaptation—that is, changes in the composition of a group of organisms or in the physiology or structure of a single organism which result in the individual's or group's being better fitted to the environment. Structural adaptations include such modifications as air spaces in the tissues of aquatic or marsh plants. Examples of physiological modifications include the ability of certain arctic seaweeds to grow and reproduce at temperatures of 0°C (32°F) or lower, and the capacity of some plants to grow in the shade of other plants but not in open sun.

Plants are necessarily adapted to their environments; the fact that they survive and reproduce is evidence of this. But plants are not able to direct their own destinies or foresee the result of their activities. They cannot purposefully and intelligently develop structures that will enable them to survive. It follows, then, that any interpretation of plant structures or processes on a teleological basis tends to impede inquiry and search for biological truths. The student who interprets plant structures and behavior teleologically only confuses himself and hinders his understanding of the nature of plant life.

ECONOMIC USES OF PLANTS

Some of the economic uses of plants have been indicated earlier. The species that man has used are few compared with the third of a million kinds of plants known to exist.

The number of species discussed in one book on economic botany is over 1000. Only 40 of these are algae, fungi, lichens, and ferns. The remaining species are seed plants. Although such a compilation is somewhat arbitrary (only a few useful bacteria and molds are included), it undoubtedly contains most plants of major economic importance. The list might easily be increased tenfold if it included plants of occasional, local, or minor use. But even so limited a list is impressive because of the range in the kinds and parts of plants and because of the varied ways in which plants are used. Only a few of these are mentioned in the following paragraphs.

Of primary importance to man are the plants we use as food. The most important of these are the cereals, or grains. These are all grasses, and most of them have been cultivated since prehistoric times. Wheat, corn, rice, barley, rye, oats, millets, and sorghum are the principal cereals. Each includes many varieties adapted to specific soil and climatic conditions. There are more than 500 varieties of corn and nearly 200 varieties of wheat grown in the United States. Buckwheat, although not a grass, is commonly grouped with the cereals. Next to the cereals in use and importance are the legumes—peas, beans, soybeans, lentils, and peanuts. The legumes include many hay and forage plants, such as alfalfa and the clovers, as well as those used directly by man. The vegetables include a wide variety of plants with edible parts. Some, such as the potato, sweet potato, taro, cassava, turnip, and carrot, grow underground. Others, such as cabbage, lettuce, chard, and celery, are called leafy vegetables and grow above ground. Many kinds of fruit, from both temperate and tropical climates form an important part of man's diet. The seeds of nuts are valued both for food and for industrial use.

Three essential groups of foodstuffs are the carbohydrates (sugars and starches), fats and oils, and proteins. The carbohydrates are obtained directly from plants; the fats and proteins are derived from plants or from animals that have fed upon plants. Sugars are prepared from corn as well as from sugar cane and sugar beet. Honey, although prepared by bees, is essentially natural sugar collected from flowers. Much of the starch we use is obtained from corn and wheat, although some comes from potatoes, arrowroot, sago palm, rice, and cassava. Edible oils are extracted from cottonseed, olives, corn, coconut, soybeans, and peanuts. Other plant oils, such as tung, linseed, castor, and safflower, are widely used in industry. Plant fats include chocolate, from the seeds of cacao (Fig. 1-14).

Although not strictly foods, all our spices come from plants. Among the more familiar spices are pepper, allspice, ginger, nutmeg (Fig. 1-15), mustard, cloves, and cinnamon. Many herbs, such as sage, thyme, chives, and garlic, are used for seasoning.

The use of plants for medicinal purposes goes back to very early times, and many drugs, such as ergot, opium, and quinine, were used by primitive people. Other medicinal plants or plant products are digitalis, pilocarpine, morphine and codeine (from opium), ephedrine, scopolamine, ipecac, mustard, agar, physostigmine, reserpine, stramonium, and curare. The important antibiotics (Chapter 20) of modern medicine are produced by a number of bacteria and fungi. Finally, man uses a number of plant products for their stimulating or narcotic effect. These include opium (and its derivative, heroin), cola seeds, betel nuts, coca leaves, marijuana, tobacco, coffee, and tea.

Additional plant products of great importance are the fibers obtained from cotton, kapok, flax, jute, sisal, hemp, and ramie. Few plant products are more useful to man than cellulose, obtained chiefly from cotton and wood. We recognize cellulose most commonly in the form of paper, cotton cloth, and wood, but this substance has been transformed by chemistry into rayon fibers, explosives, plastics, lacquers and lacquered fabrics, synthetic sponges, cellophane, camera film, sausage casings, eyeglass frames—the list is endless. Many, also, are the direct products of the forest, from which we obtain not only lumber but also turpentine, gums, resins, waxes, rubber, chicle, and cork.

Finally, the esthetic values of plants cannot be ignored in any consideration of their usefulness to man. The fields and forests constitute retreats where peace and contentment are enjoyed by millions of people. And the beauty of plants has been responsible for a vast industry involving seedsmen, horticulturists, nurserymen, land-

Fig. 1-14 Cacao (*Theobroma cacao*). Cacao is a small, widely cultivated tropical tree. The pods, 5–13 inches long, are borne directly upon the wood of the stem or branches, and contain 20–60 seeds called beans, the source of the chocolate and cocoa of commerce.

scape gardeners, and florists. The care of rock gardens, herb gardens, and gardens of irises, gladioli, delphiniums, roses, and many other kinds of plants provides pleasure to many people, who eagerly await the offerings of new varieties as the annual seed catalogs appear. The numerous organizations associated with gardening further attest to the recreational values of plants. Pleasure is provided also by the plants of lawns, parks, streets, botanical gardens, conservatories, golf courses, and even cemeteries. Aside from the millions of dollars spent each year on ornamental plants, there are values in seeing and growing them that cannot be measured in monetary terms.

PLANTS, NOW AND IN THE FUTURE

The over-all picture of the values of plants is clear. Plants of the past—buried, altered, and preserved, and yielding coal, oil, and natural gas—and plants of the present—available in many forms for many uses—provide essential materials in our world of today. Mankind is dependent upon plants, directly or indirectly, for all food, most clothing, the fuels that warm our homes and operate our industries, and the lumber with which our houses are built. There is little possibility that we can free ourselves from this dependence upon plants. Despite all our technological advances, agriculture will continue to be basic among man's activities. Nuclear energy is now being used on an increasing scale, and it is assumed that it will become increasingly important in the future. As such energy becomes more widely available it will at best free us from dependence upon plants of the past in the form of fossil fuels. That chemists can free us from dependence upon plants of the present is improbable, for the raw materials used in the production of synthetic foods are mostly compounds derived from organic matter. The fundamental source of energy for living processes will probably always be supplied by green plants.

But plants are an important and frequently an essential part of the human environment for many reasons aside from immediate use. Grasslands and forests hold the soil and control floods and erosion. They affect climate through increase in humidity and equalization of temperature. They provide habitats for native animals, which disappear when their natural environments have been destroyed. Although man's welfare is dependent upon plants, he has exploited the earth's vegetation, to its detriment, from the time of the stone ax to that of the tractor-drawn plow, the bulldozer, and the chain saw. In converting wild acres to food production, little respect has been shown to plants or to the land upon which they grow. Man misuses plants, yet without plants he would soon return to the dust from which he came.

Fig. 1-15 Fruit of nutmeg (*Myristica fragrans*). Native to Indonesia, nutmeg has been known to Europe since the twelfth century. The single large seed is the nutmeg of commerce; around the seed is a scarlet, netlike outgrowth, called mace, also used as a spice.

The Plant Plan and Its Modifications

THE CASUAL OBSERVER of plant life is often impressed and even bewildered by the variety and diversity that confront him. The plants of his backyard, of city streets, or of parks and botanical gardens exhibit great dissimilarity. A visit to a flower show or the reading of a nursery catalog will make him aware of dozens, if not hundreds, of kinds of plants that differ from one another in color, form, habit of growth, hardiness, and other characteristics.

Diverse as they are, all these plants carry on the same basic activities, and although the structures differ markedly from one kind of plant to another, there is, in spite of their seeming diversity, great similarity in their construction. Their wide variation, in fact, emphasizes an underlying plant plan. Although most obvious in the flowering plants, this plan may be discerned in all plants possessing specialized cells that transport food, water, and mineral compounds or that give mechanical support to the plant. Such plants are called VASCULAR plants. Despite their differences, all flowering plants—palms and cacti, celery and potatoes—conform to a similar pattern.

Moreover, the plant organs and parts are interrelated—the plant is more than the sum of all its parts. A plant is more than roots, stems, flowers, fruits, and seeds, for its activities are so integrated that it is enabled to live and grow as a unit.

THE PLANT PLAN

The plant body consists fundamentally of an axis, which at one end becomes the ROOT and at the other, the SHOOT (FIG. 2-1). The shoot, in turn, is made up of STEM and LEAVES. The root and shoot, each with specialized roles in the economy of the plant, form an integrated whole.

The root serves to anchor the plant and to absorb water and mineral salts from the soil. The shoot functions primarily in support, conduction, and food manufacture. Stem elongation follows a regular pattern, and the leaves are borne singly or in groups along it.

A young stem or twig is marked by the presence of NODES, the points on a stem where a leaf or leaves are attached. The intervals between the nodes are called INTERNODES. The oldest leaves are found at the base of the shoot, the youngest near the tip. BUDS are usually to be found at the bases of the leaves, in the angle between the leaf and stem. These buds may grow into branches that duplicate the structures of the

Fig. 2-1 **Principal parts of a vascular plant.**

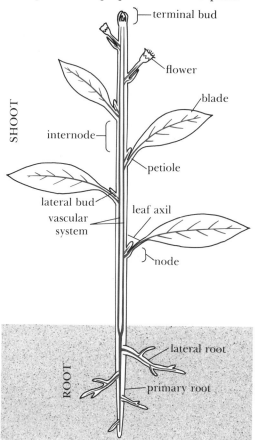

shoot upon which they were borne. A bud is found also at the apex of the stem where new tissues of the stem and leaf initials of the main axis are produced.

The leaves are arranged on the stem in a manner that is constant for a given species (FIG. 2-2). Each leaf is some distance removed from the one immediately below it. If only one leaf is attached at a node, the leaves are ALTERNATE. If two leaves are formed at the same node, on opposite sides of the twig, the leaves are termed OPPOSITE. In a relatively small number of plants more than two leaves arise from the same node in a WHORLED arrangement. These arrangements occur on herbaceous stems as well as on twigs.

Although the root has many structural features in common with the shoot, it shows conspicuous differences. The root bears no appendages comparable to leaves, and consequently has no nodes or internodes. There are, it is true, lateral roots, which result in a branching of the root system, but these arise within the tissues of the parent root rather than from externally visible buds.

One additional aspect of the plant body may be noted. An animal body is composed of organs such as the heart, stomach, and lungs, which perform definite functions. A plant body also is made up of organs in which one or more activities or processes are localized, but plant organs are not so clearly demarcated as those of animals. Along with the parts of the flower, the root, stem, and leaf are the organs most commonly recognized. Sometimes fruits and seeds are also considered plant organs. Organs are composed of tissues, and these, in turn, of cells. The various kinds of cells from which the tissues are formed are treated in detail in later chapters.

WOODY AND HERBACEOUS PLANTS

Plants may be classified according to the length of life of the entire plant or sometimes only of its aerial portion. Trees and shrubs have shoots that

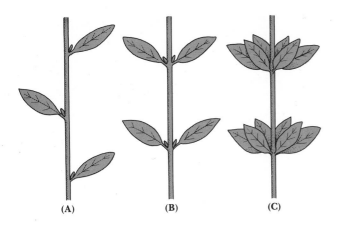

Fig. 2-2 Arrangement of leaves on a shoot. (A): alternate. (B): opposite. (C): whorled.

live for a number of years and are called WOODY plants. The stems and roots increase in diameter and in length each year, extending farther upward into the air and down into the soil. Because the tissues of the stem are dense and strong, the plant is capable of supporting a considerable bulk of branches, twigs, and leaves.

Plants in which the aerial portion is relatively short-lived and the tissues comparatively soft are called HERBS, or HERBACEOUS plants. In regions of temperate climates the aerial part in some species, or the entire herbaceous plant in others, lives for only one growing season. It is the commonly accepted view that in flowering plants the woody condition is primitive and the herbaceous type has been derived from it.

Among vines there are both woody and herbaceous forms. Grapes, for example, have woody stems, whereas the common morning-glory (*Ipomaea purpurea*), gourds, pole beans, and scarlet runner beans are herbaceous.

ANNUALS, BIENNIALS, PERENNIALS

A second convenient classification of plants is based upon seasonal growth. Plants are classified as ANNUALS, BIENNIALS, or PERENNIALS. An annual is a short-lived plant, the entire life cycle of which, from germination to seed production, takes place within one growing season. Among the annuals are many of the world's most useful plants, such as the cereal grains, peas, beans and soybeans, buckwheat, flax, jute, and tobacco. Biennials do not normally bloom until the second season after the seed is sown. The shoot of many biennials takes the form of a low ring of leaves (a ROSETTE) the first season. In its second year the plant sends up a leafy shoot bearing flowers and seeds. Many garden food plants—beet, celery, cabbage, carrot, and turnip—are biennials. Because they are harvested for food during the first season of growth, however, the average gardener seldom sees them flower and mature.

Perennials live from year to year, with varying blooming periods. Perennials are either woody (trees, shrubs, or vines) or herbaceous. In the temperate zones the aboveground shoots of herbaceous perennials die at the end of each growing season, but the plant persists by means of underground roots or stems. Perennial herbaceous food plants include asparagus and rhubarb; ornamental herbaceous perennials include peony, delphinium, primula, dahlia, and lupine. The spring flowers in our woodlands and meadows are mostly perennial herbs.

The distinction between annuals and perennials, and between woody and herbaceous plants, is sharper in temperate regions than in the tropics and subtropics. In temperate climates, for example, the castor bean is grown as an annual, dying with the frosts of autumn. In the tropics it grows from year to year and develops a considerable amount of woody tissue.

This classification of plants calls attention to the two phases of plant growth, *vegetative* and *sexually reproductive*. Vegetative growth increases the size of the shoot and root, the vegetative organs. In the sexually reproductive phase, flower, fruit, and seed are produced. In some flowering plants such as the annuals—corn and garden bean—the plant passes through a period of vigorous vegetative growth. It then passes into the reproductive phase and subsequently dies. Biennials pass their first year in the vegetative phase and complete the reproductive phase in the second. In trees and other perennials the vegetative phase continues, and reproductive phases occur periodically without exhausting the potentiality for further vegetative growth.

FORMS OF WOODY PLANTS

The fact that woody plants differ in size and form is obvious. Large woody plants with a single erect stem are TREES. Smaller woody plants with several stems emerging from the ground are SHRUBS. These categories are generally useful, but botanists know that plants that are clearly shrubs in one region or location are trees in another and that the climate is frequently the critical factor in producing shrublike or treelike growth in a species.

Among trees themselves there is great variation in HABIT—that is, in general appearance. Some, like larch, fir, and spruce, are spirelike. They have a single main trunk with smaller, lateral branches. This is the EXCURRENT growth pattern (FIG. 2-3A) and is due in part to the fact that the terminal shoot, or leader, grows faster than the lateral ones. Excurrent growth is characteristic of the conifers, but some deciduous trees show it also, especially when young.

Most broad-leaved trees follow the pattern that Longfellow noted in his line "Under the spreading chestnut tree." In these trees the rate of growth of the main axis, or trunk, declines with age and the lateral branches grow more vigorously. The result is a widely spreading system in which some of the branches are almost as large as the main axis. Such an axis may even be lacking above the point where the branching begins. Spreading trees have a DELIQUESCENT growth pattern (FIG. 2-3B). As might be expected, there are intermediate types. Some trees have an excurrent pattern when young and become deliquescent as they mature.

The coconut and other palms represent a third type of growth pattern. In this kind, lateral buds seldom develop, and as a result the stem is usually unbranched. Such growth is called COLUMNAR (FIG. 2-3C).

THE GROWTH PLAN

Growth is usually thought of as an increase in height. Whether this increase takes place at the rate of 6 inches or 20 feet during the growing season, whether it occurs in a corn plant or an oak tree, the basic plan is the same. The plant growth plan can best be approached by examining in detail a twig taken from a woody plant.

Growth of a Woody Shoot Twigs are present on most woody plants, but their structure is more easily studied in trees such as ash, maple, or poplar, which lose their leaves at the end of each growing season. Such trees are DECIDUOUS, in contrast to evergreen species, which shed their leaves gradually after one or more years and hence are never barren of foliage. Deciduous woody plants bear leaves only on that portion of the shoot produced during the current growing season. This young leaf-bearing part is, technically speaking, the TWIG. The older part of the shoot with the twig is called a BRANCH.

The twig also bears buds. A TERMINAL BUD occurs at the tip of the twig, and smaller LATERAL, or AXILLARY, buds are formed in each leaf AXIL, or angle where the leaf is attached to the twig

(A) (B) (C)

Fig. 2-3   Growth habit in trees. (A): excurrent growth of larch (*Larix laricina*). (B): deliquescent growth of white oak (*Quercus alba*) (C): columnar growth in the coconut palm (*Cocos nucifera*).

(FIGS. 2-1, 2-4). These lateral buds differ from the terminal buds only in their position on the twig. In addition to the terminal and lateral buds, ACCESSORY BUDS are present in some species. These are located at the nodal regions, either above the axillary bud or on either side of it. These accessory buds may not develop unless the axillary bud is destroyed.

Most buds develop only into leafy twigs, and such buds are termed LEAF BUDS. Other buds, called the MIXED BUDS, contain the rudiments of leafy twigs and embryonic flowers. Mixed buds are found in the apple and lilac. In FLOWER BUDS the bud scales enclose embryonic flowers only, as in the peach, cherry, red maple, and elm (FIG. 2-5). Flower and mixed buds may be accessory buds or may be terminal or axillary in position. They are often larger than leaf buds.

When leaves fall in autumn, a LEAF SCAR remains at that point on the twig where the break occurred. The axillary buds are located directly above such leaf scars (FIG. 2-4). On the surface of the leaf scar is a pattern of corky dots. These BUNDLE SCARS are the ends of strands of conducting tissue that linked the vascular tissues of the

stem with those of the leaf. The shape, color, and position of the buds, the shape of the leaf scar, and the number and arrangement of the bundle scars differ to such an extent that they may be used to identify different woody plants.

The terminal and lateral buds of most woody plants are covered by a number of overlapping BUD SCALES. In some species the scales remain for some time at the base of the developing twig; in others they are quickly shed. The bud scales are closely grouped, and when they fall, a zone of crowded, ringlike BUD-SCALE SCARS remains (FIG. 2-4). These scars persist for a number of years before they are finally obscured by the diameter growth of the branch. Since the distance between any two sets of bud-scale scars marks the growth in length the twig made during a season, the age of a small branch may be determined by counting the scars from the terminal bud down the branch.

The bud is composed largely of embryonic tissue, and this fact is probably the basis of the belief, common among laymen, that bud scales protect the delicate inner tissues from the cold of winter. But there is no evidence that the bud

THE GROWTH PLAN   27

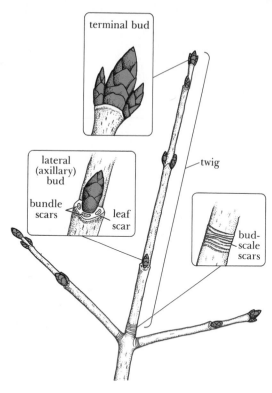

Fig. 2-4 The twig and its parts. Twig of sugar maple (*Acer saccharum*).

scales retain heat as does man's clothing. The temperature of the bud tissues is approximately that of the air. Moreover, bud scales occur on many plants in tropical and subtropical regions. Experimental and other evidence leads to the conclusion that bud scales are primarily effective in reducing the loss of the water from the delicate tissues within the bud and in protecting them from mechanical injury.

In the buds of certain woody temperate-zone plants, true bud scales are absent, and such buds are termed NAKED BUDS. In these buds, well-developed embryonic leaves, commonly covered by hairs or scales, enclose and protect the smaller young foliage leaves.

If a bud is cut lengthwise through the middle, several embryonic leaves surrounding a dome-shaped mass of tissue, the SHOOT APEX, may be observed (FIG. 2-6). The bud is really an embry-

onic twig. This fact becomes apparent in the spring when growth begins, for at this time the structures within the bud develop into a duplicate of the twig that it terminates (FIG. 2-7). Similarly, lateral buds develop lateral twigs. In some plants the full number of leaves that unfold in the spring already exist in embryonic form in the bud. In others the shoot apex produces additional leaves as growth proceeds.

The expansion of a bud in the spring thus reveals that a considerable amount of activity has taken place in the embryonic shoot during the previous growing season. The shoot apex is composed of young cells that at certain times undergo active division. This embryonic, or MERISTEMATIC, region initiates the tissues of the stem and also gives rise to leaves. Cells below the apex also divide and then enlarge, chiefly by an increase in length which accounts for the rapid elongation of internodes and the often spectacular growth of woody plants in the spring. In most woody plants of temperate climates, particularly in older trees, elongation is limited to the first weeks of the growing season. Following this elongation phase, the shoot apex continues its activity and produces another terminal bud ready for expansion after a period of winter dormancy.

To generalize from our study of the growth plan as shown by the twig: the activities of tissues in the terminal and lateral buds of woody plants result in the formation of new terminal and lateral twigs each season. Elongation of these twigs follows the opening of the buds. Other tissues, to be described later, bring about increase in diameter. A terminal bud forms at the apex of a new twig, and lateral buds develop in the axils of the leaves. During the succeeding growing season, these buds grow into twigs, which in turn produce lateral and terminal buds. In this manner terminal growth and branching continue to a potentially unlimited extent.

Growth of an Herbaceous Shoot Growth in herbaceous shoots takes place according to the same general plan. In such plants the terminal

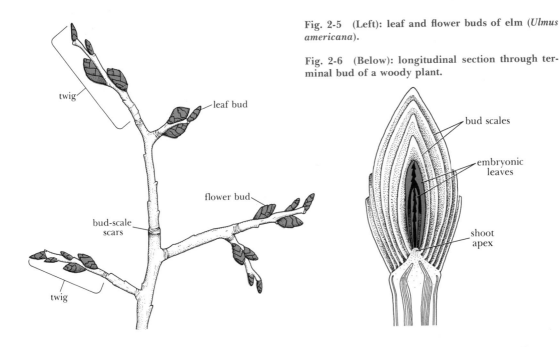

Fig. 2-5 (Left): leaf and flower buds of elm (*Ulmus americana*).

Fig. 2-6 (Below): longitudinal section through terminal bud of a woody plant.

Fig. 2-7 Unfolding of buds and development of leafy twigs in Norway maple (*Acer platanoides*). (A): terminal winter bud and several closely associated lateral buds. (B): early stage in growth. (C): the twigs emerge from the bud scales. (D): terminal and two lateral twigs, nearly mature.

(A)   (B)   (C)   (D)

bud is not a conspicuous feature. During the growth period as the shoot apex initiates new leaves and new stem tissues, the leaves expand without delay and internodal elongation proceeds continuously. Thus, although the apex is surrounded by embryonic leaves that it has produced, there is no buildup of young leaves in the formation of a massive terminal bud. There are also herbaceous plants in which there is no internodal elongation, and the shoot has the form of a rosette. Lateral buds are also commonly present in herbs, but these buds develop into lateral shoots during the season in which they are produced.

A significant modification of the growth scheme is found in the cereals and other grasses. In these plants the main axis of the shoot usually produces branches only from the basal nodes, those close to or just beneath the surface of the soil. These nodes are crowded together, for their internodes remain very short. Lateral buds in the axils of the leaves at the base of the main shoot, or primary axis, give rise to secondary aerial branches, which also have unexpanded basal internodes. The secondary shoots, or axes, in turn may produce branches of a third order from basal nodes (Fig. 2-8). This process may continue until a considerable number of branches are formed, the number depending upon heredity, closeness of planting, and environmental factors. This method of branching—from basal nodes—is called TILLERING, and the resulting branches are known as TILLERS. In many grasses only aerial shoots, or tillers, are formed; in other grasses some of the basal branches may become horizontal and grow in or on the surface of the soil, giving rise to aerial branches at the nodes, thus forming a sod.

Tillering has the important effect of enabling the plants in a sod, or thinly seeded cereals such as wheat or barley, to expand into any available growing space. Since the tillers, like the primary axis, usually produce grain, they may thus increase the yield. Tillering is especially important in sugar cane, where the single stalk arising from the node of a planted stem piece forms many sugar-producing stalks before harvest.

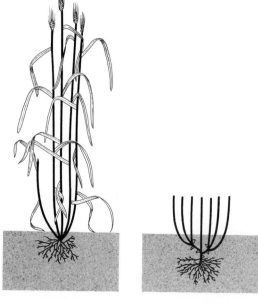

Fig. 2-8   Tillering. (Left): tillers in a cereal. (Right): the basal branching by which tillers are produced.

Significance of the Growth Plan Growth in length of the shoot is due to the activity of the meristematic shoot apex together with the elongation of the young internodes just behind the apex. The root apex, although it produces no leaves, is similarly meristematic, and immediately behind it is a region of elongation. In both shoot and root, terminal growth is a continuing process—it is potentially unlimited. In this phenomenon we see one of the most striking contrasts between vascular plants and vertebrate animals. An animal undergoes a period of development as an embryo, during which its basic organs and tissues are formed. Subsequently, after birth or hatching, it continues to grow, but what occurs is an enlargement and elaboration of the structures formed in the embryo. A dog, for example, is born with four legs and it will never have any more. The plant, too, has an important embryonic phase inside the developing seed during which a rudimentary plant body is formed, but after germination it continues embryonic activity, with its shoot and root apices continuing to initiate new organs and tissues. Moreover, this activity continues indefinitely, subject to certain limitations imposed by the environment, the natural life span of the plant, and its internal regulatory mechanisms. Thus, the examination of the embryo in a ripe seed will not disclose whether it will produce a roadside weed or a forest tree.

Plants also continue their growth by an increase in diameter, particularly in long-lived, woody perennials, but also to a limited extent in many herbaceous annuals. This increase in diameter is accomplished by growing layers near, but not at, the surface, which add to the conducting system and to the outer protective tissues.

In succeeding chapters it will be seen that the microscopic structure of the plant body is different from that of the animal body in that each protoplasmic unit (cell) in the plant is enclosed by a relatively rigid wall. This structural feature has resulted in an array of organisms that are essentially immobile. Under the condition of immobility, the scheme of unlimited growth becomes useful in allowing the plant to respond, at least in a limited way, to the environment. For example, a plant cannot roam about in search of new sources of water and mineral elements, but its ever-expanding root system can effectively invade untapped regions of the soil by growing into them. Similarly, the shoot system can respond to changing light conditions.

There is another aspect of unlimited growth, however, which may be even more important. It is well known that in the animal body there is an extensive turnover of cells, varying in degree in different tissues, but resulting in a constant replacement of worn-out elements. This seems to be possible in animal tissues, where dead cells can be resorbed and replaced by new cells resulting from cell division. Plant cells, on the other hand, encased in walls and cemented together, cannot be resorbed in this way, but are simply left in place as new ones are added. In the case of leaves, the whole organs are discarded and new ones are formed. The plant continually produces new, active cells that carry on its vital activities.

Control of plant growth Underlying the growth pattern of plants lie many unsolved problems. Potentially the growth is unlimited, but actually the potentialities are never realized although many species of woody plants exhibit great longevity. Why does an apple tree not attain the size and age of an oak? Why do growth patterns differ? How do heredity and environment affect the growth of plants? The answers to these difficult questions are far from complete.

Among the more significant advances in plant science has been the verification of earlier theories that plants produce organic chemical compounds that affect growth when present in very low concentrations. These growth-regulatory substances are termed HORMONES. A number of synthetic chemicals may produce some of the effects of plant hormones. These chemicals have been called GROWTH REGULATORS. Hormones are important in animals also—for example, those produced by the thyroid and pituitary glands and by the pancreas in man.

Prominent among the plant hormones are the (1) auxins, (2) gibberellins, and (3) cytokinins. Additional information on these and other hormones is given in Chapter 14. In general terms, the auxins and the gibberellins have the effect of promoting cell enlargement and the cytokinins incite cell division, but it is becoming increasingly clear that such a simple designation of function is not adequate to account for the diverse growth phenomena influenced by each of these hormones. Growth hormones are found in a number of plant organs, but are principally produced by actively growing tissues such as the shoot apex and young leaves and fruits. As the concentration of hormones increases, they may move away from the regions where they were formed, ordinarily through living cells.

Each of the categories of hormones is a group of chemically related substances rather than a specific compound, although indoleacetic acid (IAA) appears to be the common plant auxin. The action of a particular hormone often depends upon its concentration to the extent that it may stimulate growth at one concentration but inhibit it at another. Even more important is the fact that the different groups of hormones interact in the regulation of growth processes. Thus the balance among two or more hormones may be of greater importance than either the presence or absence or the concentration of a particular hormone.

Finally, there is increasing evidence of the important role of substances that are fundamentally growth-inhibiting rather than stimulating in their action, and these substances also interact with the other hormones. Considering such a system of chemical regulation, one can appreciate the complexities of the mechanisms by which varied growth patterns are produced.

Apical dominance The role of hormones in regulating the growth of plants is evident in the phenomenon of APICAL DOMINANCE, one of the most characteristic features of plant growth. In apical dominance the terminal bud of a shoot exerts an influence upon the growth of lateral buds, sometimes preventing their growth and sometimes retarding or modifying it. Thus, the frequency of branching is controlled, and lateral branches that do develop are often less vigorous than the parent axis, or are modified in their form. If the terminal bud is removed, axillary buds previously inhibited may grow out, or retarded laterals may be accelerated in their growth. Release from apical dominance may also result from a decrease in vigor of the terminal shoot. This may occur with age or if lateral buds are at a considerable distance from the terminal.

This inhibitory action of the terminal bud is effected mostly by its production of hormones of the auxin group. For instance, if a terminal bud is removed, its inhibitory effect may be simulated by applying auxin to the stump. The mechanism of the inhibition, however, is not fully understood. Some botanists believe that hormones from the tip of the twig may act directly to slow or prevent cell division and growth of the dormant buds. Others think that hormones from the tip act in some way upon the transport tissues to channel food materials to the tip and away from the lateral buds. Removal of the growing tip will,

of course, remove this directive influence. Moreover, it now seems clear that more than one hormone is involved, since the application of a substance of the cytokinin type directly to an inhibited bud can cause it to be released. Thus, the idea of hormonal balance is introduced, and it is possible to envisage the kind of refined control involved in the varied branching patterns of plants.

Apical dominance is important in the growing of tobacco (*Nicotiana tabacum*). This plant is a tall (4–6 feet in height) herb with a single stem bearing large leaves. The stem continues above the leaves as an axis with flowering branches. In most types of tobacco, as the plant begins to flower, the uppermost portion of the axis, including the young flower stalk and usually the youngest leaves, is broken or pinched off, a procedure known as "topping." The food materials that would have contributed to flower and seed production are now available to the leaves, especially the upper ones, which become larger and thicker. Topping, however, removes apical dominance, and axillary buds now grow out and produce lateral branches, or "suckers." The continued growth of these suckers would destroy the effect of the topping, and so they are removed by expensive hand labor. Research workers are experimenting with chemicals that will prevent growth of these axillary buds without reducing the quality of the tobacco leaves. Spraying the topped plants with a synthetic growth substance, maleic hydrazide, largely prevents suckering, but the tobacco from treated plants clogs cigarette-making machines and cigarette manufacturers generally refuse to buy it. Garden chrysanthemums (*Chrysanthemum morifolium*) may be topped several times in the spring to encourage branching and the production of more flowers; or the lateral shoots may be pinched from greenhouse chrysanthemums to produce a single, very large flower.

Apical dominance may be observed in potato tubers that have been stored for several months after harvest. A potato tuber is a modified stem, and the "eyes" are buds, comparable to buds on a

Fig. 2-9 Apical dominance in potato (*Solanum tuberosum*) tuber. (Top): a group of terminal buds have grown, inhibiting the growth of the lateral eyes. (Bottom): when a tuber is cut into pieces, each piece bearing an eye, each eye produces a sprout.

Fig. 2-10 Apical dominance in Norway spruce (*Picea abies*). The "leader" or terminal shoot (indicated by arrow) was injured by insects and one of the lateral branches has replaced the injured leader.

woody twig. If the tuber is allowed to sprout, the apical bud, or several buds near the apex, grow first, forming a vigorous sprout or group of sprouts (FIG. 2-9, top). The lateral buds develop slowly or not at all, indicating that the growth of the terminal bud has exercised an inhibitory influence. But if the potato is cut into pieces, each piece bearing an eye, as is usually done in planting, the influence of the apical buds is removed and all the eyes produce sprouts of approximately the same size (FIG. 2-9, bottom).

Excurrent and columnar types of growth may be explained, in part at least, on the basis of strong apical dominance of a special kind, such dominance being weak or lacking in the deliquescent type. Instead of their growth being prevented, the lateral buds are held to a subordinate, outward, or even downward growth. If storms or insects destroy the terminal shoot, one or more lateral branchlets grow upward and replace it (FIG. 2-10). In a short time a new leader will completely replace the original and exercise the same degree of apical dominance. If more than one new leader becomes established, a forked growth results, and the symmetry of the tree and its value for lumber are reduced. This dominance is more complicated than the inhibition of the growth of the lower buds in twigs and tubers, and no satisfactory explanation for it is available.

The elimination of apical dominance is the basis of many pruning practices. The removal of the terminal portions of twigs will check growth in height and bring about a more dense and compact type of growth. The natural habit of hedge plants and other ornamentals is thus altered to conform to horticultural requirements.

THE PLANT ORGANS

A description of the plant plan and the major organs of the body leads us to a more detailed consideration of the parts of the plant and of some of the modifications that these parts have undergone. In many plants the fundamental organs, root and shoot, have become greatly modified. Many of these modifications contribute to the ability of the plant to survive in an unfavorable environment.

As the plant grows, new buds and leaves develop upon the shoot, and the root branches, forming new roots. This general plan is modified in many species by the presence of ADVENTITIOUS ORGANS, which are structures produced in an unusual position or at an unusual time in the development of the plant. Among the adventitious organs are roots produced from stems, and shoots formed upon roots. Adventitious roots and shoots may arise during the course of normal growth or as a result of wounds or other stimuli. Such structures may be useful to the plant; for example, they are often an aid in reproduction. We take advantage of them in the propagation of food and ornamental plants.

The Leaf The leaf may be considered a flattened or expanded portion of the stem. It contains many of the kinds of cells and tissues found

in the stem, but it shows one characteristic difference. The stem follows a plan of unlimited growth, continuing to grow as long as it lives. The growth of the leaf, on the other hand, is limited. It soon ceases to grow; it matures; it functions for one or a few seasons; and it falls away.

The typical leaf has two parts: a thin, flat, expanded portion, the BLADE; and the stalk, or PETIOLE (FIGS. 2-1, 2-11A). When the petiole is absent, the leaf is said to be SESSILE. Leaflike or scalelike structures, the STIPULES, are sometimes found at the base of the petiole and are considered to be integral parts of the leaf. Stipules are usually small, but in some plants such as the garden pea, they are large and supplement the leaf as photosynthetic organs (FIG. 2-15A). Some stipules drop as the leaves expand; others persist throughout the growing season.

On the lower side of the leaf blade, and frequently on the upper surface as well, the VEINS appear as lines or ridges. Veins are composed of cells that serve in conduction of food and water and, to some degree, in support. The VENATION, or arrangement of the veins, may be used to distinguish the two great subdivisions of the flowering plants. In the monocotyledons, represented by iris, the palms, lilies, and corn and other grasses, the chief veins are parallel or nearly so. This arrangement is termed PARALLEL VENATION (FIG. 2-11C).

In the dicotyledons, represented by maple, geranium, apple, oak, and sunflower, the veins form a netlike pattern, and the leaves of these plants are said to have NET VENATION. Some net-veined leaves have a single, strong vein running the entire length of the leaf, with lateral veins like the barbs of a feather (FIG. 2-11A). Such leaves are said to be PINNATELY VEINED. In other leaves, several strong veins enter the base of a leaf from the petiole and spread through the blade like the fingers from the palm of the hand. These are termed PALMATELY VEINED leaves (FIG. 2-11B).

Simple and compound leaves Leaves may be classified as either SIMPLE or COMPOUND. In the simple leaf (FIG. 2-11) the blade is undivided,

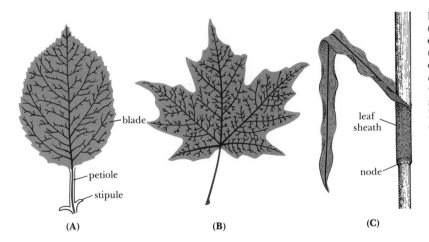

Fig. 2-12 Poison ivy (*Rhus radicans*). Compound leaves and fruit. Widely distributed in North America. In the west, the closely related Pacific poison oak (*Rhus diversiloba*), a species of similar characteristics, replaces *Rhus radicans*.

although the margin may be indented in various ways. In the compound leaf the blade is divided into several or many parts, termed LEAFLETS. Compound leaves are of two general kinds. In PALMATELY COMPOUND leaves, the leaflets are all attached at the tip of the petiole, as in horse chestnut, poison ivy (FIG. 2-12), red clover, lupine, and Virginia creeper (FIG. 2-13A). In PINNATELY COMPOUND leaves, the leaflets arise along the sides of a central stalk like the barbs of a feather, as in ash (FIG. 2-13B), potato, tree of heaven (*Ailanthus*), walnut (*Juglans*), and other woody and herbaceous plants.

The compound leaf is equivalent to the simple leaf, although it is sometimes difficult to distinguish a leaflet from a simple leaf. In woody plants the position of the winter bud, and in herbaceous plants the position of a bud or branch, will provide the proof. Buds or branches are found in the axil of a leaf, never in the axils of leaflets of a compound leaf.

Leaf modifications During their evolution, the leaves of many species of plants have become highly modified. Sometimes it is these modifications that make the plant useful to man. The thickened petioles of rhubarb and celery, and the fleshy leaves of cabbage and onion are examples of this. The upper parts of the tubular leaves of onion become dry at the end of the first season of

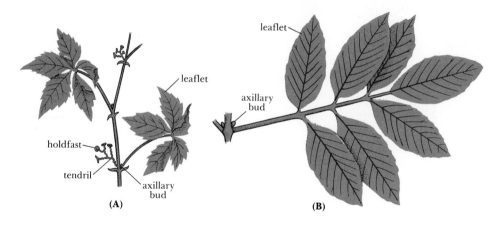

Fig. 2-13   Compound leaves. (A): palmately compound leaves of Virginia creeper (*Parthenocissus quinquefolia*). (B): pinnately compound leaf of ash (*Fraxinus americana*).

Fig. 2-14   Century plant (*Agave americana*) in Texas. The common name is applied because of the popular idea that the plant blooms only at the age of 100. Actually, it blooms at about ten years of age. The plant dies after blooming, but suckers around the base survive. Pulque is the fermented sap of this and certain other species of *Agave*.

growth. The basal portions remain as the concentric layers, or scales, that make up the onion bulb. These scales are rich in accumulated sugar and enable the plant to form leaves and flowers if the bulb is planted a second season. Many other plants (hyacinth, lily, narcissus) accumulate reserve foods in the modified leaves of bulbs.

Thickened and fleshy leaves in which water accumulates are also common, especially in plants of arid and semiarid regions. Many of these plants are cultivated in rock gardens and greenhouses as ornamentals, and are referred to as SUCCULENTS. Succulents also include plants in which water accumulates in a modified stem rather than in the leaves. Among the leaf succulents are the century plant (FIG. 2-14), ice plant (*Cryophytum*), and stonecrops (*Sedum*).

TENDRILS   Tendrils are, typically, modified leaves, which hold up the plant by coiling around a support. The petiole of a leaf, the entire leaf, the leaflets of a compound leaf, or even stipules may act as tendrils (FIG. 2-15). Leaf parts transformed into tendrils become long and slender and manufacture little or no food. Tendrils can be seen in vines such as pea, vetch, clematis, and greenbrier (*Smilax*). Although most tendrils are modified leaves or parts of leaves, some are modified branches.

SPINES Spines are leaves that have lost their green color and flat form and become slender, conical, and hard. They may be identified as modified leaves by their position just below an axillary bud or shoot. Stipules also may be transformed into spines, as in the black locust (*Robinia pseudo-acacia*). Spines are found in the barberry (FIG. 2-16), gooseberries (*Ribes*), and cacti. The terms *spine* and *thorn* are often applied indiscriminately. Technically speaking, THORNS are modified branches and usually arise from the axil of a leaf or just above the axil. Thorns are found in honey locust, Osage orange, and hawthorn. The spinelike structures of the rose and blackberry are termed PRICKLES, and are merely outgrowths of superficial tissues of the stem. Although their effect on anyone grasping the

plant is the same, they are botanically distinct from spines and thorns.

Some stem modifications

CLIMBING STEMS Climbing plants include such diverse forms as the great lianas or vines of the tropical rain forest, the wild grapes of our own river banks, and the pole bean and morning-glory of the garden. Some species merely clamber over or lean upon other plants, fences, or rocks; others climb by coiling around a support or by special modifications, such as tendrils. In the Boston ivy and Virginia creeper, the tips of the tendrils, which are modified branches, become enlarged and disklike and are termed

Fig. 2-15 Tendrils. (A): garden pea (*Pisum sativum*) leaflets modified into tendrils. (B): clematis (*Clematis virginiana*). The petiole acts as a tendril. (C): greenbriar (*Smilax rotundifolia*). The tendrils of *Smilax* have been considered to be stipules, but are structurally parts of the petiole. They split off from near the base of the petiole, one on either side, early in the growth of the petiolar region of the leaf.

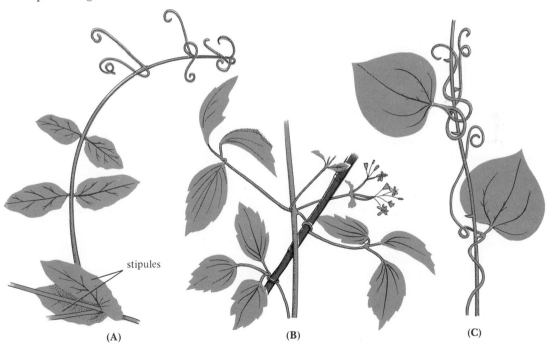

stipules

(A)   (B)   (C)

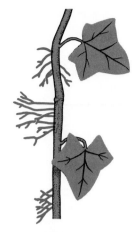

Fig. 2-16   Spine (modified leaf) of Japanese barberry (*Berberis thunbergii*).

Fig. 2-17   Aerial adventitious roots of the English ivy (*Hedera helix*).

HOLDFASTS (FIG. 2-13A). Adventitious roots are frequently found upon climbing stems and attach them to walls or trees (Fig. 2-17).

PROSTRATE BRANCHES   Many species that have upright shoots produce slender, horizontal branches from basal nodes of the stem. These structures, called RUNNERS or STOLONS, grow either over or under the soil surface, and are important in vegetative reproduction. It has been advocated that the term *runner* be restricted to aerial branches (FIG. 11-1), and the term *stolon* (FIG. 2-18) to horizontal branches in the soil. This usage will be followed here. In both runners and stolons, nodes and internodes are present, and adventitious roots, sometimes buds, are formed at the nodes. Leaves, also, are found at the nodes of stolons, but are usually minute and scalelike. Some species have both runners and stolons. Many of our most prolific weeds produce stolons. Many grasses, also, spread and are propagated by means of stolons. An example of the propagative potentialities of stolons is found in the spearmint (*Mentha spicata*) (FIG. 2-18). The tips of the horizontal stolons may bend upward and emerge from the soil as a leafy plant. Stolons arising from underground nodes spread through the soil and in turn may give rise to branches that immedi-

Fig. 2-18   Stolons of spearmint (*Mentha spicata*). (A): the tip of a stolon has grown upward, forming an aerial, leafy shoot. (B): stolons have developed from an underground node of a leafy shoot. (C): lateral branch of a stolon growing upward. (D): lateral branch growing through soil; the tip may eventually give rise to a leafy shoot.

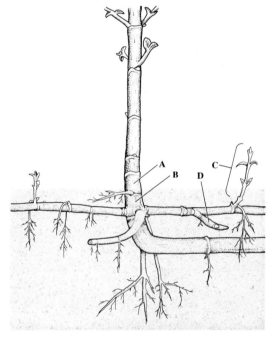

Fig. 2-19 Stem succulents. The modified stem accumulates water and manufactures food. (Left): *Stapelia* (carrion flower) a plant of the milkweed family. (Right): a cactus (*Neomammillaria*).

ately grow upward or continue to grow through the soil, the tip eventually giving rise to a leafy shoot.

STEM SUCCULENTS   In many plants the stems as well as the leaves are green and able to manufacture food. In a very considerable number of species this essential process is largely or entirely restricted to the stem because the leaves are much reduced or absent. Large quantities of water are accumulated in such fleshy and succulent stems. Examples of stem succulents are found in the milkweed, cactus, and spurge families (FIG. 2-19). Stem succulents and many leaf succulents are able to live for a considerable period of time without an external source of water. It is reported that the giant saguaro (*Carnegiea gigantea*), a large treelike cactus of the deserts of southwestern United States and northern Mexico (FIG. 2-20), accumulates enough water in the stem to last for several years.

Subterranean stems   In many plants, both wild and cultivated, the aerial parts, or branches, arise from underground stems. Such stems are significant to plants as regions of food accumulation, and some are utilized by man and other animals as important sources of food. Underground stems are of several kinds, each specialized in different ways.

RHIZOME   The rhizomes are perennial stems, usually horizontal in position. A rhizome has been defined as the main axis of the shoot, usually with relatively short and thickened internodes rich in accumulated food in the form of starch (FIG. 2-21). Stolons are also subterranean and, it will be recalled, are basal, relatively slender, prostrate branches of the erect shoot. Since rhizomes and stolons occur underground, they

Fig. 2-20 The giant cactus, or saguaro (*Carnegiea gigantea*). Arizona.

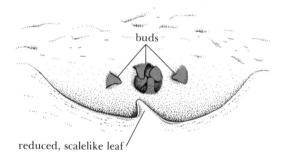

buds

reduced, scalelike leaf

Fig. 2-21 The large and fleshy rhizomes of the cow or yellow pond lily (*Nuphar variegatum*). These rhizomes have a high starch content. Roasted or boiled, they were used as food by North American Indians. They are eaten by muskrats, moose, and other herbivores.

Fig. 2-22 (Above): bud, or "eye" of potato tuber, greatly enlarged.

Fig. 2-23 (Right): corm of gladiolus.

may be confused with roots. Roots, however, never have nodes and internodes. In the rhizomes of some species, green leaves are borne at the nodes (Fig. 2-21), but more commonly the aerial leaves are produced on leafy branches, the result of the upward growth of lateral or terminal buds (Fig. 11-2). Examples of rhizome-bearing plants are sweet flag (*Acorus*), cattail (*Typha*), many violets, iris, cana, orchids, and water lilies.

TUBER A tuber is a much enlarged and swollen portion of an underground branch (stolon) of the vertical axis. The most familiar example is the potato tuber, which arises as an enlargement of the apex of a stolon (Fig. 11-12). The surface of the tuber is marked at one end by a scar, which was the point of attachment of the stolon, and by the "eyes" scattered over the surface. A single eye (Fig. 2-22) consists of a ridge bearing a minute, scalelike leaf, in the axil of which are several (usually three) minute buds. Since the tuber is a modified stem, this scale leaf and its axillary buds correspond to the leaf of a woody twig together with the bud in the axil of the leaf. The eye marks the position of a node, and the portion of the tuber between consecutive eyes represents the internode.

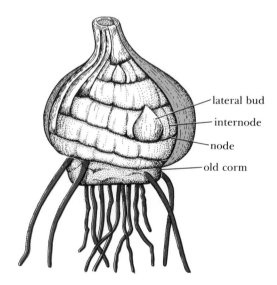

lateral bud

internode

node

old corm

CORMS The corm is an underground stem that is short, thickened, and fleshy. It usually grows erect and may be flattened vertically so that it is broader than high. Examples of corms are found in gladiolus, crocus, cyclamen, freesia, and tuberous begonias. The corms of some species exhibit no nodes, and in these the same corm normally lives from year to year throughout the life of the plant, producing new aerial shoots periodically. In other species the corm bears nodes and internodes, and in such forms the corm lasts only a single growing season. In gladiolus, for example (FIG. 2-23), as growth begins in the spring, one or more buds develop into shoots bearing leaves and flowers. The food used in the early growth of the shoot comes from the corm, which shrivels as growth proceeds. Later, after the leaves have matured, food is moved downward and one or more new corms are formed on top of the old, exhausted corm. The corms of the taro and dasheen, members of the calla lily family, are extensively used for food in parts of the tropics (Fig. 2-24).

BULBS A bulb is a very short stem wrapped in thickened, fleshy BULB SCALES, which are modified leaves. The scales are the chief sites of food accumulation. Bulbs of onions and hyacinth are said to be LAYERED (FIG. 2-25A), for the scales form a series of rings that can be seen when the bulb is cut across. The bulbs of most lilies are SCALY. The scales do not encircle the stem, but are small, fleshy, and rather loosely attached to the stem (FIG. 2-25B).

The Root The primary root system of a plant arises from the lower end (radicle) of the embryo during germination of the seed. The primary root grows downward into the soil and shortly produces lateral branches. If it persists and maintains its ascendancy, and so constitutes the chief root system of the plant, it is termed the TAP ROOT (FIG. 2-26A). This is the kind of root common in many herbaceous plants, such as the dandelion, carrot, and alfalfa. A number of

Fig. 2-24 Corms of taro (*Colocasia esculenta*) on a Pacific island.

Fig. 2-25 Bulbs. (A): section through a layered bulb of onion. (B): scaly bulb of lily (*Lilium tigrinum*).

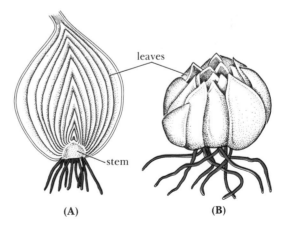

leaves

stem

(A)   (B)

forest trees, such as oak, hickory, and the conifers, have tap roots, at least when young.

In other species, including many herbaceous perennials and especially grasses, the growth of the primary root is soon equaled or surpassed by the growth of its branches, resulting in FIBROUS ROOTS (Fig. 2-26B).

Many plants produce adventitious roots in addition to the primary root system that arises from the embryo. The roots arising from the aerial portion of the stem, and from stolons, rhizomes, corms, and cuttings (pieces of plants used for propagation), are always adventitious.

Storage roots The roots of biennial and perennial plants are frequently rich in accumulated foods. Such roots, especially in cultivated varieties, become much enlarged. In the turnip, beet, radish, parsnip, and carrot, the tap root becomes fleshy as growth proceeds. In some of these plants the enlarged portions are derived partly

Fig. 2-27 Sweet potatoes at harvest time.

Fig. 2-26 Roots. (A): taproot of dandelion. (B): fibrous roots of a grass.

(A)   (B)

from the stem zone of the seedling and partly from the root zone. It is therefore difficult to determine the extent to which each organ is involved in food accumulation. In the carrot, for example, approximately the uppermost inch of the mature fleshy organ is derived from the seedling stem, the remainder being root tissue. In the dahlia and sweet potato, branch roots become swollen and function as storage organs (FIG. 2-27).

Aerial roots Numerous kinds of plants, both of the temperate zones and of the tropics, produce AERIAL ROOTS upon their stems above the ground level. Such adventitious roots are useful to the

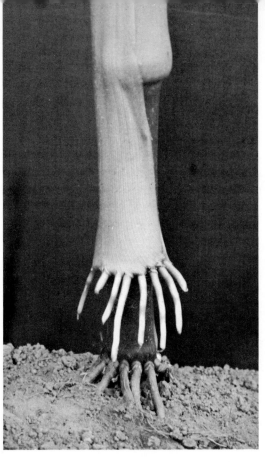

Fig. 2-28   Prop roots of corn.

Fig. 2-29   Prop roots of banyan (*Ficus bengalensis*).

plant in a number of ways. Some plants produce somewhat thickened aerial roots that absorb and probably store dew and rainwater. In a Malayan orchid found on tree bark, the stem is only an inch or so long, the leaves are reduced to brownish scales, and the roots are green and constitute the only photosynthetic organs of the plant. Roots that anchor or attach climbing stems are found in the poison ivy, English ivy (Fig. 2-17), and the trumpet vine (*Campsis radicans*). In some plants PROP ROOTS, which contribute support in addition to that provided by roots originating underground, develop from the stem and grow into the soil, where they branch profusely and absorb water and mineral compounds. The most famil-

iar example of prop roots is found in corn, in which adventitious roots are produced from several nodes above the soil (Fig. 2-28). Prop roots of notable size are found in tropical trees, such as certain palms, the mangroves, and the banyan, a tree of the fig family. The adventitious roots of the banyan grow downward many feet from aerial branches until they strike the soil (Fig. 2-29). They also grow in diameter to such an extent that they serve as additional trunks supporting the large limbs. The mangroves, which grow on ocean coasts or on the banks of tidal rivers, produce stiltlike prop roots from both the main trunk and the branches. These roots may form almost impenetrable thickets (Fig. 2-30).

THE PLANT ORGANS   43

Fig. 2-30 Edge of a mangrove swamp. Adventitious (prop) roots on the ground and growing downward from the branches.

SUMMARY

1. The body of a vascular plant is made up of a descending portion, the root, and an aerial portion, the shoot. The shoot is composed of stem and leaves.

2. Two phases of growth are found in flowering plants: a vegetative and a sexually reproductive phase. Annuals pass through both phases in a single season, biennials in two seasons. In perennials a sexually reproductive phase normally occurs annually. Perennials are woody or herbaceous, depending upon the longevity of the aerial part. In herbaceous perennials the aerial portion of the plant dies down annually, but the plant persists by underground parts. Three growth patterns in trees are recognized: excurrent, deliquescent, and columnar.

3. The twig is the growth of the current season in woody plants. It bears leaves and buds. Leaves are produced only on twigs, and a twig normally bears only one set of leaves.

4. Buds are classified as leaf, mixed, and flower buds. These are borne laterally or terminally. Lateral buds are located in the axils of leaves. After the leaf falls, the bud

may be seen just above the leaf scar.

5. A typical resting leaf bud consists of three parts: the shoot apex, embryonic leaves, and bud scales. Such a bud is an embryonic twig and is potentially capable of producing all the structures of a twig.

6. Plants are potentially unlimited in growth. New tissues are produced each growing season as long as the plant lives.

7. Hormones play an important role in the control of plant growth. The phenomenon of apical dominance illustrates this hormonal regulation.

8. The major organs of plants may be modified in various ways. These modifications may be useful to the plant and are frequently useful to man.

9. The plant body is highly plastic, and the organs of the body may give rise to other organs in an unusual position, not only during development but also after maturity. Such organs are termed adventitious.

10. The parts of a foliage leaf are petiole, blade, and stipule. Leaves are simple or compound. The venation of the leaf differs in the two great groups of flowering plants, the monocotyledons and the dicotyledons. The leaves of certain plants have become highly modified and store water or food. Leaves or leaf parts may also become converted into tendrils or spines.

11. Stems may be erect, climbing, or prostrate, may manufacture food and accumulate water, and may be aerial or underground. Aerial and underground basal branches of erect stems may be termed respectively runners and stolons. A tuber is the enlarged tip of a stolon. Rhizomes and corms are subterranean stems with relatively short, thickened internodes. Rhizomes are horizontal, corms erect. In bulbs the accumulated food is located in bulb scales, not the stem itself, which is reduced.

12. Roots, like stems and leaves, are sometimes greatly modified. Among these modifications are prop roots, roots that anchor climbing stems, and underground tap and side roots in which food accumulates.

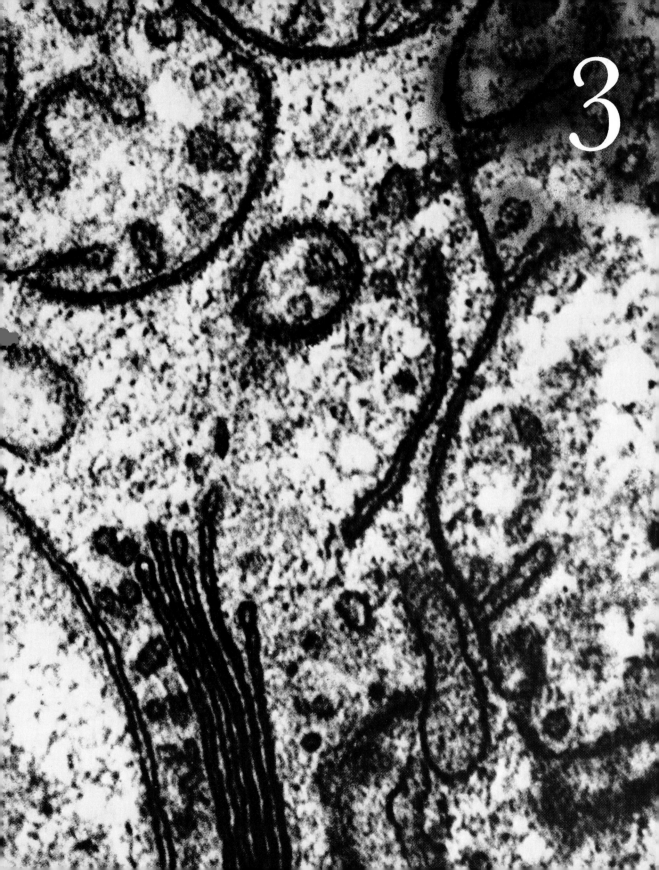

The Cell as the Basis of Plant Life

THE CONCEPT that the cell is the unit structure of plant and animal life is universally accepted. It has proved to be one of the most fruitful and penetrating ideas in the study of living things, for under its impact the basic problems on which scientists could previously only speculate were opened to direct investigation. The result of such research has been a tremendous increase in botanical knowledge, although there is still much to be learned about both the structure and the activities of living cells.

Speculation about cells dates back to the Greek philosophers, but the opening of the field of cellular biology had to await advances in other scientific areas—a clear example of the interdependence of the sciences. The microscope was invented in the late 1500s, but it was not until the middle of the sixteenth century that Anton van Leeuwenhoek and other investigators developed it into a practical instrument. The microscope rapidly became an important scientific tool, and many men of science turned their attention to using and perfecting it.

One of these men, Robert Hooke, an English scientist, published his observations in 1665, using the term *cell* for the first time. He saw that cells from fresh vegetable material had walls and were filled with "juices." A contemporary, Nehemiah Grew, continued these observations on the nature of plants and provided more detailed descriptions of cells in stems and other tissues.

Because plant cells are generally easier to observe and study than those of animals, it is not surprising that knowledge about them advanced more rapidly. It was not until the beginning of the nineteenth century, however, that Hooke's term *cell* came into common use. Even then, investigators did not regard individual cells as the units of structure and organization, but rather as globules in a continuous "membraneous" tissue. The emphasis was upon cellular tissue as a unit rather than upon cells themselves. It was not until 1824 that the French physiologist René Dutrochet clearly set forth the generalization that plants and animals are composed of cells and that these are the structural and physiological units of all living things. In 1833 Robert Brown recognized the NUCLEUS as an important organ of the cell—a further step in establishing the cell theory upon a firm foundation.

The life of any plant should be viewed not only in terms of the structure and activities of the cells that compose it, but also with respect to their interrelationships in the tissues of root, stem, leaf, and sexually reproductive parts. In the discussions that follow, the processes and activities of cells are related to specific structures wherever possible.

CELL STRUCTURE

Collectively, the contents of any particular cell are often referred to as the PROTOPLAST. In the plant cell (FIGS. 3-1, 3-2, 3-3) a CELL WALL surrounds the protoplast. This is typically a permeable, protective layer, analogous to the cover of a football. The protoplasm consists of the nucleus and the living material in which it is embedded, the CYTOPLASM. The nucleus is bounded by a NUCLEAR MEMBRANE. Studies with the electron microscope show that this double membrane

Fig. 3-1 Cell wall and organelles of a plant cell, highly diagrammatic: n, nucleus; nu, nucleolus; er, endoplasmic reticulum; m, mitochondrion; r, ribosomes; d, dictyosome; c, chloroplast; mt, microtubules; v, vacuole.

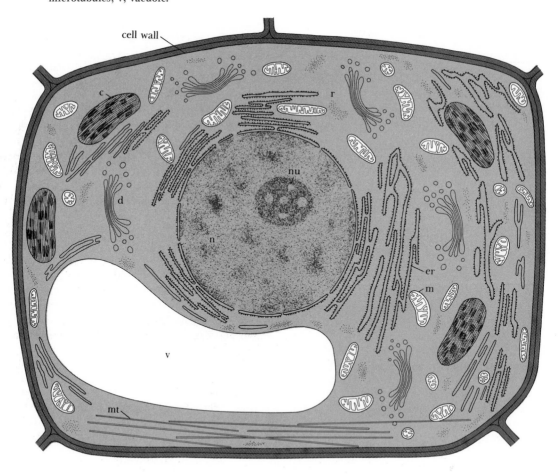

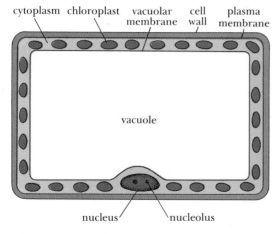

cytoplasm chloroplast vacuolar cell plasma
                        membrane wall membrane

vacuole

nucleus   nucleolus

Fig. 3-2 **General features of a green plant cell. Diagrammatic.**

commonly has pores in it (FIG. 3-1). The nucleus contains a NUCLEOLUS or several nucleoli, together with specialized structures that determine the hereditary characters of the cell and the organism. The outer layer of the cytoplasm, the PLASMA MEMBRANE, lying against the cell wall, is extremely thin. This membrane controls the passage of materials both into and out of the enclosed protoplasm. In addition to the nucleus, other bodies, readily observed under the light microscope, are the PLASTIDS of various kinds. In higher plants, the principal kinds are (1) CHLOROPLASTS (FIGS. 3-1, 3-2), which are green because of the presence of chlorophyll; (2) CHROMOPLASTS, colored other than green; and (3) AMYLOPLASTS, colorless and containing starch. The nucleus and nucleolus, the plastids, and other cytoplasmic bodies to be described, are termed ORGANELLES (little organs).

During the earlier decades of the twentieth century the cytoplasm was viewed as undifferentiated, and the emphasis was upon the nucleus and its role in heredity. Today, largely due to the development of the electron microscope, the cytoplasm is known to be packed with organelles. The MITOCHONDRIA (FIG. 3-1) are many times smaller than the plastids. By special staining methods they can be seen with the light microscope, but the electron microscope is necessary

for a study of the details of their structure (FIG. 3-3). They vary in shape from spherical to rodlike, and are the site of a number of enzymatic steps in respiration (Chapter 6). Other kinds of organelles, visible only with the electron microscope, include those discussed in the following paragraphs.

The ENDOPLASMIC RETICULUM is an extensive three-dimensional system of membranes. It may be tubular or composed of flattened sheets, which have been compared to a deflated inner tube. In sectional view the membranes appear in

Fig. 3-3 **Electron micrograph of a portion of a cell of a fungus (*Rhizidiomyces*): n, nucleus; nm, nuclear membrane; m, mitochondrion; er, endoplasmic reticulum; d, dictyosome. (44,000 ×)**

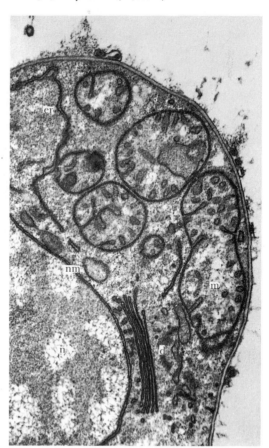

pairs (FIGS. 3-1, 3-3), enclosing spaces filled with a fluid containing synthetic products of the cell. The whole system is interconnected and may be viewed as an intracellular circulatory system. The DICTYOSOMES, also called GOLGI BODIES (FIGS. 3-1, 3-3), are organelles with some resemblances to endoplasmic reticulum. They consist of an aggregation of two to seven, commonly curved, disklike structures. Each disk tends to be enlarged at the margins and is composed of a membrane surrounding a space, as in the endoplasmic reticulum. The RIBOSOMES (FIG. 3-1) are minute bodies that are the sites of protein synthesis. They occur in the chloroplasts and lie free in the cytoplasm or are attached to the surfaces of portions of the endoplasmic reticulum. Organelles of still another type are the MICROTUBULES, rod-shaped bodies about 280 angstroms (see Glossary) in diameter and up to several microns in length (FIG. 3-1). They are relatively rigid, and vary in numbers and distribution within the cell. We shall return, in later pages, to a more detailed consideration of the organelles. It may be noted that, with the exception of the plastids, all the organelles mentioned above occur in animal as well as in plant cells.

In mature plant cells the protoplasm forms a layer just inside the cell wall, and a VACUOLE (FIGS. 3-1, 3-2), which may also be regarded as an organelle, develops in the interior of the cell. Vacuoles, several of which sometimes occur in a cell, are filled with a watery, nonliving CELL SAP, a fluid containing many substances in solution. The vacuole is bounded by the VACUOLAR MEMBRANE. In brief, a typical mature plant cell may be said to consist of a cell wall, cytoplasm with a nucleus, one or more vacuoles, membranes, mitochondria, and other organelles. It should be emphasized that cells are three-dimensional, having depth as well as length and width (FIG. 3-4). Their shape varies with the tissues in which they are formed and with the pressure and other forces that operate as they increase in size. Some cells are greatly elongated; others, such as those of apples or potatoes, are distorted spheres, flattened against the cells around them like soap bubbles in suds.

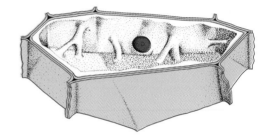

Fig. 3-4 Three-dimensional view of cell from epidermis of onion bulb scale.

THE NATURE OF PROTOPLASM

Every living cell contains a semiliquid material that is exceedingly complex in chemical composition and physical structure. This is the living material of the cell, the protoplasm. The term *protoplasm* is very old; it means the "first form," and was used in early Christian literature in describing Adam, the first form of man. In 1840 it was employed in describing very young animal embryos, and six years later Hugo von Mohl applied the term to the living contents of plant cells in more or less the modern meaning. Somewhat later (1861), it was pointed out that protoplasm is the living material of all cells, both plant and animal.

Protoplasm is a fluid, semitransparent, viscous, elastic material. It consists of 75–90 percent water together with solids that are largely protein. A chemical analysis of a funguslike organism (slime mold) that consists essentially of protoplasm reveals about 83 percent water. Of the solid materials, more than half are proteins; the remainder consists of fatty materials, carbohydrates, and other compounds, organic and inorganic. But an analysis of the chemical composition of protoplasm gives only a meager clue to its essential nature. The many chemical elements and compounds of which protoplasm is composed may be brought together, but the resulting mixture is not alive. Protoplasm, then, must be more than a random association of proteins and other materials. It is known, in fact, that protoplasm has a complex ultrastructure (FIG. 3-1)

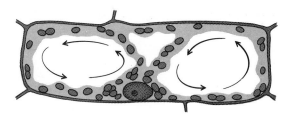

Fig. 3-5 Cytoplasmic streaming in leaf cell of water-weed (*Elodea*). Arrows indicate direction of movement.

ing NUCLEOPROTEINS, which are always present in living cells. The nucleic acids, like proteins, are large, threadlike molecules composed of many smaller units. The small units are composed of a chemical combination of a nitrogen base, a pentose sugar (ribose or deoxyribose), and a phosphate (FIG. 4-17). Nucleic acids are classified in two groups: (1) DEOXYRIBONUCLEIC ACID, called DNA for convenience; and (2) the closely related RIBONUCLEIC ACID, or RNA. As will be shown later, the nucleic acids are involved in both the storage and transfer of hereditary information and the synthesis of new protein.

and that innumerable chemical reactions and energy transfers occur within it. When these features are better understood, we shall have advanced our understanding of the nature of life itself.

Of all the components of protoplasm, the proteins have been most studied, for these compounds play a central role in the vital processes of the cell. Enzymes are protein catalysts that control the thousands of chemical reactions necessary to the functioning of the individual cell and of the organism as a whole. Protein molecules always contain the elements carbon, hydrogen, oxygen, and nitrogen, and are large—thousands of times larger than water molecules. Many proteins also contain a small percentage of sulfur. An approximate average composition of protein is carbon, 50–55 percent; hydrogen, 6–7 percent; oxygen, 21–23 percent; nitrogen, 15–19 percent; and sulfur, less than 2 percent. By suitable treatment, proteins may be broken down into AMINO ACIDS, the structural units, or building blocks, of which proteins are composed. About 20 different amino acids have been found in proteins. These are linked together in the long chains that form the protein molecule. Typical proteins are composed of about 300 amino acid molecules, although the number may be less or many times more. The kind of protein formed depends upon the number, kind, and arrangement of the amino acid molecules.

Some of the protein of the cell is combined with compounds known as NUCLEIC ACIDS, form-

Cytoplasmic Movement The cytoplasm of most cells, at some time during the life of the cell, exhibits motility, or cytoplasmic streaming. Under the microscope, cytoplasm may be seen flowing or streaming inside the cell in continuous movement, which may involve only local currents or may take the form of a general circulation. In aquatic plants such as *Elodea*, the chloroplasts are carried along by the cytoplasm like driftwood in a stream (FIG. 3-5). In terrestrial plants the chloroplasts also may change position, but so slowly that the movement is imperceptible. There is no generally accepted explanation for cytoplasmic streaming, but it should be noted that the movement requires energy; therefore the streaming of cytoplasm may be taken as a direct indication of life within the cell.

THE NUCLEUS

The nucleus in plants is usually spherical or elliptical, and its contents are more viscous than the cytoplasm. Its relative size varies considerably with the age and type of cell. In very young cells the nucleus is located centrally. In older cells, where the vacuole fills most of the cell cavity, the nucleus is located on one side, against the cell wall. The nucleoli (FIG. 3-1) are spherical or sometimes irregularly shaped bodies composed largely of RNA nucleoproteins. The nucleoli probably play several roles in the life of the cell. There is evidence, for example, that they are involved in the formation of ribosomes, particles composed of RNA and protein, found in great numbers in the cytoplasm. The RNA components of the ribosomes are known to be synthesized in the nucleoli. This RNA, together with protein, forms ribonucleoprotein precursor particles, which are later transferred to the cytoplasm. These precursor particles then add more protein and become functional ribosomes.

The role of the nucleus will be considered later, but it may be pointed out here that the nucleus is essential to the maintenance of the life of the cell. Interestingly enough, the cell does not die immediately if the nucleus is removed; the essential synthetic and respiratory activities can continue for a time. But eventually, in the absence of the nucleus, cell activities come to an end: the physiological mechanisms cease to operate, and the cell dies.

Every cell in the body of a plant or an animal owes its existence to the division of an older cell. Indeed, no new tissues of any kind can be formed except by cell division. The two daughter cells that result from cell division are genetically identical because of the way the nucleus divides. The importance of the intricate process of nuclear division, known as MITOSIS, is therefore apparent.

Mitosis The essential feature of mitosis is the behavior of the elongated, rodlike bodies that become visible in the nucleus. These bodies are termed CHROMOSOMES (color bodies) because they have a strong affinity for basic dyes. A chromosome possesses the ability to reproduce itself, utilizing materials in the protoplasm to construct a new chromosome that is its exact duplicate. One of these duplicates goes to each daughter cell during mitosis and gives each the same hereditary potentialities as the parent cell. This follows from the fact that the chromosomes carry hereditary units, called GENES, to be discussed later.

Mitosis is best observed in rapidly growing tissues such as those of the root tip. One can observe under the ordinary microscope the various stages in the cycle of chromosomes in the cell, shown diagrammatically in Fig. 3-6. Commonly, such studies are made on tissues prepared by specialized techniques, including the use of dyes, or stains, that emphasize the differences among cellular structures. In such preparations we do not see mitosis in progress but only stages in the process as it was interrupted by the action of killing fluids on the dividing cells. It is possible, however, to observe mitosis in living, unstained cells with the phase-contrast microscope. With this instrument, differences not visible to the eye are converted to intensity differences that are visible. Such a microscope was used in photographing the sequence of mitotic stages in the single living cell shown in Fig. 3-7.

The changes occurring in the nucleus during cell division are arbitrarily divided into INTERPHASE and the four stages of mitosis: PROPHASE, METAPHASE, ANAPHASE, and TELOPHASE (Figs. 3-6, 3-7). The interphase stage intervenes between one series of mitotic stages and the next.

Although the process from interphase to interphase is described as a series of stages, it is continuous, as pointed out above. Even in actively growing tissues, most of the nuclei will be in interphase, the contents appearing to be granular or as a network with no chromosomes individually distinguishable. Evidence indicates, however, that the chromosomes are present in the form of greatly extended, extremely slender threads, perhaps with lateral extensions, that extend at random throughout the nucleus (Fig. 3-6A). It is also during interphase that the chromosomes reproduce exact duplicates of themselves.

The chromosomes first become visible as individual structures in early prophase as elongated threads irregularly distributed within the nucleus. Under high magnification of the light microscope, each chromosome can be seen to be doubled longitudinally into two half-chromosomes, or CHROMATIDS, the two duplicate

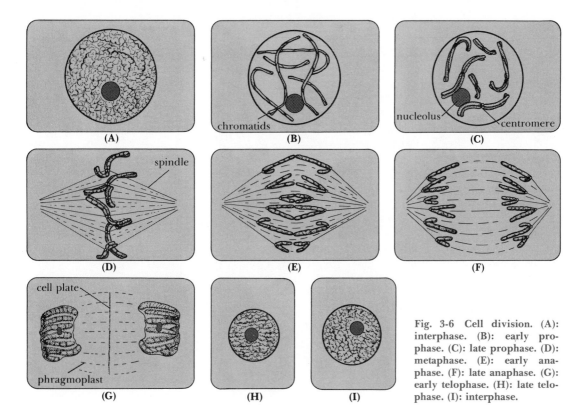

chromatids

nucleolus · centromere

spindle

cell plate

phragmoplast

(A)   (B)   (C)

(D)   (E)   (F)

(G)   (H)   (I)

Fig. 3-6 Cell division. (A): interphase. (B): early prophase. (C): late prophase. (D): metaphase. (E): early anaphase. (F): late anaphase. (G): early telophase. (H): late telophase. (I): interphase.

strands twisting about each other lengthwise (FIG. 3-6B). As prophase advances, the chromatids shorten and thicken, with a consequent untwisting that, by the end of prophase, leaves the chromatids of each pair lying side by side and nearly parallel (FIG. 3-6C). In late prophase there appears in each chromatid, at corresponding sites, a region of specialized function, the CENTROMERE. The location of the centromere varies among different chromosomes, but is constant for each chromosome. In ordinary preparations the centromeric region stains poorly, often appearing as no more than a colorless gap, or even a constriction, across both chromatids. The end of prophase is marked by the disappearance of the nuclear membrane and the nucleoli.

With the advance to metaphase (FIG. 3-6D) a structure called a SPINDLE is produced from protein material outside the chromosomes. It con-

sists of slender fibers converging at two poles, one at each end of the cell. The fibers are not unit structures, but each is composed, at least in large part, of an aggregation of microtubules so closely packed that they appear as a single strand. The scattered chromosomes now migrate and become arranged across the cell in the plane in which the cell will later divide. The centromeres of sister chromatids are directed toward opposite poles of the spindle, and a spindle fiber connects each centromere with the pole toward which it is directed. The fibers attached to the centromeres are known as CHROMOSOMAL SPINDLE FIBERS; other fibers extend from one pole of the spindle to the other, and are termed CONTINUOUS SPINDLE FIBERS.

As they pass into the anaphase (FIG. 3-6E, F), the centromeres of the sister chromatids separate, each chromatid becomes a chromosome,

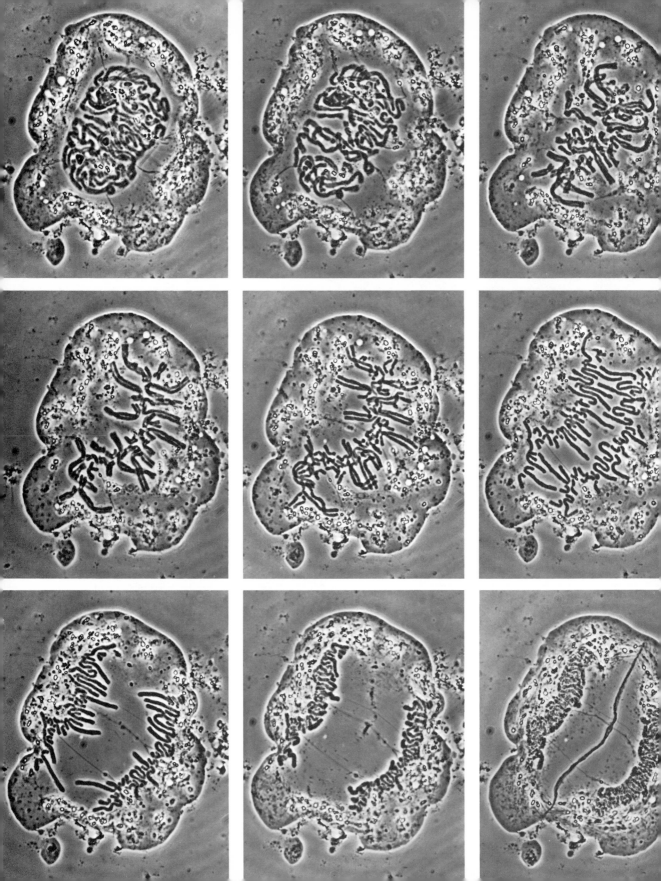

Fig. 3-7 (Facing)   Mitosis in a single living endosperm cell of the African blood lily (*Haemanthus katherinae*), as seen with the phase-contrast microscope. Since these cells do not possess a cell wall, they can be flattened in the same manner as animal cells. The chromosomes are very large in comparison with those of most other organisms, including man. The photographs, taken at times indicated, depict events from prophase through telophase. (Top row, l–r): 0 minutes, late prophase. 20 minutes, early prometaphase (nuclear membrane has just disappeared and chromosome ends are protruding into the cytoplasm). 60 minutes, late prometaphase. (Middle row, l–r): 100 minutes, metaphase. 120 minutes, early anaphase. 140 minutes, midanaphase. (Bottom row, l–r): 160 minutes, late anaphase. 170 minutes, early telophase (cell plate beginning to form). 201 minutes, telophase (cell plate complete).

and the two daughter chromosomes migrate along the spindle toward opposite poles. The mechanism of the movement of the chromosomes is not understood. Many investigators believe that the chromosomal spindle fibers contract and thus pull the attached chromosomes to the poles. Objections have been raised against this "traction" theory, however, and there is no generally accepted explanation of the anaphase movement of the chromosomes.

The end of the polar migration of the chromosomes marks the beginning of the telophase (FIG. 3-6G). The chromosomes gradually lose their compact structure, becoming indefinite in outline, and finally in late telophase (FIG. 3-6H) appear threadlike, twisting through the nucleus in an irregular manner, thus reversing their behavior in the prophase. A nuclear membrane is again formed, the nucleoli reappear, and the chromosomes can no longer be individually identified. The reappearance of nucleoli is brought about by nucleolus organizers, which are present on each member of certain pairs of chromosomes. If the chromosomes are adjacent, the nucleoli fuse into one as they grow. During the telophase, an extremely delicate structure, the CELL PLATE, is laid down across the equator of the spindle, and this gradually extends its margins to the lateral walls. At the end of the telophase the chromosomes have become reorganized into two new nuclei, and are once again in their synthetically active, interphase state (FIG. 3-6I).

The time required for a complete mitotic cycle varies with the organism and tissues concerned and with temperature and other factors. Es-

timates of the timetable of nuclear division in pea roots at two temperatures are shown in Table 3-1. Notice that active nuclear division, from prophase through telophase, was completed in 3 hours at the relatively cool temperature of 15°C (59°F) and in half that time at 25°C (77°F). This doubling of the rate with a temperature rise of 10°C (18°F) is what is expected of processes limited by enzymatically controlled chemical reactions. Note also that the nuclei completed one mitotic cycle, including interphase, in 26 hours at 15°C and in 16 hours at 25°C.

Even in these rapidly growing roots only 10 percent of the entire cycle, or generation time, was spent in active nuclear division; interphase accounted for 90 percent of the cycle. Generation times of from 3 or 4 hours to a day are characteristic of rapidly growing plant tissues. Cells of slowly growing organs like the potato tuber may have a generation time of a week or more, whereas bacteria have as little as 30 minutes.

TABLE 3-1
Duration of Mitotic Stages in Pea Roots (Minutes)

	15°C (59°F)	25°C (77°F)
Interphase	1356 (22.6 hours)	870 (14.5 hours)
Prophase	126	54
Metaphase	24	14
Anaphase	5	3
Telophase	22	11
Total	1533 (25.6 hours)	952 (15.9 hours)

In the cells of some plants, such as certain algae and fungi, the nucleus divides repeatedly without the formation of cell walls, forming multinucleate cells. Moreover, tissues are known in which cell walls are laid down only after nuclear division has proceeded for some time or is complete. It is apparent, then, that nuclear division and cell division are separate processes.

CELL DIVISION

In most tissues of higher plants, cell division is brought about by the formation of a cell plate. This separates the parent cell into two daughter cells, each with a daughter nucleus. The development of a cell plate is known to be a complicated process, and therefore only the main outlines, as currently conceived, are given here.

In the late anaphase or early telophase most of the continuous spindle fibers disappear. The cell plate is initiated in the middle of the cell by the appearance of large numbers of microtubules, arranged in clusters, that extend from the site of the future cell plate toward both poles (FIG. 3-8). The clusters meet and overlap in the region of the plate, but the tubules are seldom continuous through this region. These microtubules are collectively termed the PHRAGMOPLAST. In the plate region, where the ends meet, are now found large numbers of equatorially arranged vesicles—small, saclike, flattened, or rounded droplets with membranes (FIG. 3-8). These may enlarge irregularly, but eventually they fuse and form the cell plate. Additional vesicles are added, and the plate extends outward toward the wall of the mother cell. With continued growth, the microtubules begin to disappear from the older part of the cell plate and become more numerous along the margin. This results in the formation of a barrel-shaped phragmoplast (FIG. 3-9), clearly visible with a special light microscope in living cells or with the ordinary light microscope in stained sections. This barrel-shaped aspect of the phragmoplast persists until the plate reaches

Fig. 3-8 Curved portion of developing cell plate (endosperm of *Haemanthus*) showing microtubules and vescicles that will fuse to form the cell plate. (75,000 ×)

the wall of the mother cell, when the phragmoplast disappears.

The vesicles that fuse to form the cell plate are not believed to originate in that region. Rather, it is thought that they originate as budlike processes in either neighboring dictyosomes or endoplasmic reticulum, or both, and then migrate inward and become aligned among the microtubules in an equatorial position (FIG. 3-9). The membranes of the vesicles are believed to constitute the plasma membranes of the new cell surfaces, while their contents, presumably pectic in nature, constitute the material of the cell plate. This, in turn, as the cell matures, becomes the intercellular layer, the first layer of the cell wall (FIG. 3-14).

Chromosomes and Genes   During mitosis, each daughter cell receives the same number and kind of chromosomes from the parent cell. Located in these chromosomes are the hereditary units, the genes. Genes of many kinds are arranged in a definite linear order that is constant for each particular chromosome. Acting alone or in combination, they determine the basic nature and behavior of the organism. It should be noted that genes rather than chromosomes are regarded as the ultimate units of heredity, for in every organism studied, plant or animal, the number of inherited characters is greatly in excess of the number of chromosomes, normally hundreds of times greater.

It is generally accepted that a particular chromosome in all the cells of an individual bears the same genes in the same positions. It follows from this that the chromosomes must retain their individuality through successive nuclear divisions and that they are not re-formed in each mitotic cycle. Several kinds of evidence indicate this. Although it is usually not possible to distinguish chromosomes in the interphase, in a few organisms they have been observed as very slender threads. More important, perhaps, is the fact that chromosomes are recognizably different in size, shape, and structure when studied under high magnification (FIG. 3-10), and these differences are constant in successive mitotic cycles. Further, the number of chromosomes in the body cells of an organism is constant and usually characteristic of that particular species. This CHROMOSOME NUMBER varies from a few to a hundred or more. The lowest number in plants (four) is found in certain species of gill fungi and in goldenweed (*Haplopappus gracilis*), a small annual species of the aster family. In certain species of *Crocus* and a few other flowering plants there are six. Fourteen chromosomes are found in the garden pea, 16 in onion, 20 in corn, 32 in alfalfa, 48 in tobacco and tomato, 56 in American elm, and 80 in most varieties of sugar cane. A species of tropical sea grape (*Coccoloba diversifolia*) has been reported to have about 400 chromosomes. In flowering plants the most common numbers are 14, 16, and 18. The number in man was formerly thought to be 48, but the figure 46 is now generally accepted.

Chromosomes, then, are permanent structures of the cell. The chromosomes in each body cell are present in pairs. The members of each pair are similar in appearance (FIG. 3-10) and genic composition, and are said to be homologous to each other. Each parent contributes one member of each pair of chromosomes.

The significance of the mitotic process lies in the fact that during interphase the chromosomes

Fig. 3-9   Phragmoplast and cell plate. The fibers of the phragmoplast are composed of microtubules. Diagrammatic.

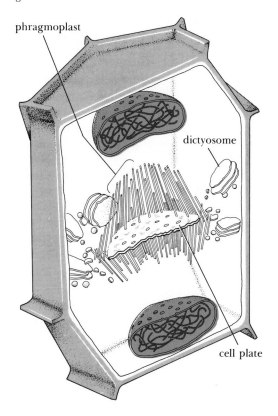

phragmoplast

dictyosome

cell plate

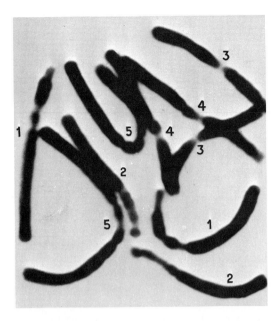

Fig. 3-10 Homologous chromosomes of purple trillium (*Trillium erectum*). The ten chromosomes consist of five pairs, each member of a pair derived from a different parent. The members of a pair are indicated by identical numbers.

have been duplicated lengthwise, forming daughter chromosomes that are identical throughout their length in the kind and linear arrangement of the genes they contain. Since the genes are different all along the chromosomes, only this kind of duplication provides for an orderly and exact distribution of genes to each new cell as the daughter chromosomes pass to opposite poles of the spindle toward the close of the mitotic process.

THE PLASTIDS

Plastids can be observed in considerable number in the cytoplasm of many cells, and they may be present in an undeveloped form in others. As listed earlier in this chapter, in higher plants the most conspicuous and best known kinds of plas-

tids are chloroplasts, chromoplasts, and amyloplasts.

Chloroplasts, usually disk shaped (Figs. 3-1, 3-2, 3-5), are green because of the presence of chlorophyll. The green color of leaves is due to 2 or 3 percent of chlorophyll in plastids that occupy perhaps 20 percent of the volume of the leaf cells. Although all pigments absorb light, chlorophyll is the only pigment that can use the energy of sunlight in the important process of photosynthesis (Chapter 5). Light is commonly necessary for certain final stages in the formation of chlorophyll; hence it does not develop in tissues that do not receive any light, such as roots, the inner leaves of lettuce or cabbage heads, or sprouts or seedlings grown in the dark. Chlorophyll may develop in exposed roots or tubers, or disappear from older stems as the dead outer cells of the developing bark cut off the light.

In addition to chlorophyll, the chloroplasts contain yellow or orange (occasionally red) pigments called CAROTENOIDS. These also occur independently of chlorophyll and are widely distributed. More than 70 have been isolated in the plant kingdom. Their function within the plant is not fully known, but there is evidence that they protect chlorophyll from destruction by sunlight and that some of them function indirectly in photosynthesis (Chapter 5). Although carotenoids are found also in animals, all evidence indicates that they are not synthesized by animal tissues but are obtained from plants. In the chloroplasts the carotenoids are masked by the chlorophyll and become evident only when chlorophyll is absent. When grass is covered by a board or stone, for example, the chlorophyll disappears and the yellowish tinge the blades assume is due to the remaining carotenoid pigments.

Two groups of yellow-to-orange carotenoids, the CAROTENES and XANTHOPHYLLS, are normally present in the chloroplasts. Our interest in the carotenoids results in large part from the relationship between certain members of these groups (chiefly the carotenes) and vitamin A, required by man and other animals for normal growth and development. When carotenes are taken into the body in foods, they may be converted into vitamin A, and they are therefore termed *pro-vitamin A*. Only animals can accomplish this transformation; its chief site in mammals and birds is believed to be the intestinal wall.

The most important carotene from the standpoint of vitamin A synthesis is known as BETA (β)-CAROTENE, with the formula $C_{40}H_{56}$. In many animals this molecule may be split in half and water added, forming two molecules of vitamin A:

$$C_{40}H_{56} + 2H_2O \longrightarrow 2C_{20}H_{29}OH$$
$$\text{β-carotene} \quad\quad \text{water} \quad\quad\quad \text{vitamin A}$$

Another carotene, ALPHA (α)-CAROTENE, has the same empirical formula as β-carotene, but a slight change in atom positions makes one half of the molecule unusable by animals. Thus, α-carotene has only half the vitamin value of the beta form. Other carotenes and xanthophylls are used, in part, by animals, but are of considerably less significance.

Animals that require vitamin A must obtain it either as the vitamin itself or as pro-vitamin A. If green plants should cease to synthesize carotenes, animal life could not long survive. Green and yellow vegetables or fruits are emphasized in the diet of man largely because of their pro-vitamin A value. Such dietary items include oranges, sweet potatoes, squash, carrot, rutabaga, and such green vegetables as kale, spinach, and broccoli. Grazing animals obtain carotene from grass or other forage plants. The preservation of the carotene of hay is a major problem, for it is readily destroyed during the curing of hay in the field. In alfalfa hay the loss of carotene may amount to 45–80 percent, and is known to be due to an enzyme whose action is favored by sunlight and by moisture from dew and from the plant tissues themselves.

The carotenoids of foods are in part excreted unchanged by animals, and in part retained and converted into vitamin A. In many animals they may also accumulate in various products. In the cow, for instance, the carotenes of the food are concentrated in the milk; hence the yellower color of cream when cattle feed on fresh grass. Butter and oleo normally contain added artificial color, which has no relation to the carotene content. The hen, on the other hand, tends to accumulate xanthophylls, which are chiefly responsible for the color of the yolk of the egg. Although eggs are a good source of vitamin A, a relatively small percentage of this is present as the pro-vitamin.

Chromoplasts owe their color to the presence of various carotenoids, which are the chief pigments responsible for the yellow, orange, and occasionally red shades of flowers, fruits, and seeds. Many yellow flowers, for instance, owe their color to the carotenoid pigments of their chromoplasts. These bodies are frequently disk shaped, but they may be long, spindle shaped, or angular (FIG. 3-11). Chromoplasts may be plastids from which the chlorophyll has disappeared, revealing the carotenoids, or they may be plastids that have never been green. During the ripening of the banana, for example, the chlorophyll breaks down and the skin of the fruit turns from green to yellow as the chloroplasts are converted into chromoplasts. The change in color during the ripening of yellow or even some red fruits may involve not only the disappearance of chlorophyll but also the formation of additional carotenoid pigments. This is the situation in tomatoes. Although these fruits contain carotene and are a source of pro-vitamin A, the color of the ripe fruits is due largely to lycopene, the orange-red carotenoid.

The chromoplast pigments are of interest because of the role they play in the coloration of autumn leaves. The colors of autumn may be divided into groups. One of these, which includes the reds, blues, and purples, is discussed later in this chapter. The yellow colors are due to carotenoids (Plate 2). As the leaves age, the chlorophylls and carotenoids tend to break

Fig. 3-11 Chromoplasts. (A): from petal of squash flower. (B): chromoplasts of squash flower, enlarged. (C): from petal of nasturtium. (D): from petal of jewelweed (*Impatiens*).

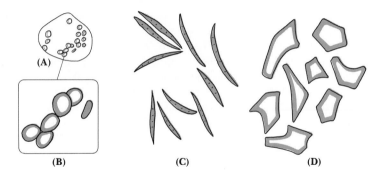

(A)

(B)　　　(C)　　　(D)

down, the carotenes more rapidly than the xanthophylls. These, combined with other substances, are thus the cause of the yellows of autumn leaves.

Amyloplasts, which contain no pigments, are important chiefly because they serve as centers of starch formation (FIG. 3-12). They are widely distributed in the storage cells of the nongreen parts of plants, such as seeds, roots, tubers, and portions of the shoot. Starch is the principal form in which carbohydrates are stored in higher-order green plants, where it is produced only in plastids and is not found in the vacuole or free in the cytoplasm. It is deposited as a starch grain within the plastid, successive layers of starch being laid down around a center, termed the HILUM. These layers, when brought into focus by adjustment of the microscope, appear much like the layers of a clamshell. When the starch grain is fully developed, the amyloplast may persist as a thin coating around it, visible only by special staining methods. An amyloplast usually encloses only one grain, but in some plants several grains of starch are produced within an amyloplast. The starch grains of various species differ to such an extent that it is possible to identify the various kinds by microscopic examination (FIG. 3-13). This characteristic of starch is frequently used in the detection of adulteration of foods and drugs.

Starch may also accumulate in the chloroplasts of the leaf when carbohydrates are formed faster than they are removed. Thus, a temporary starch accumulation occurs in the leaves of many plants during the day, when the rate of photosynthesis exceeds the rate at which its products are moved away. If the use of carbohydrates in growth is checked by deficiencies of water or minerals, this accumulation may become semipermanent.

Chloroplasts develop from PROPLASTIDS. In higher plants these are thought to be transmitted from one generation to another in the cytoplasm of the fertilized egg. At this stage they are extremely minute, probably globular, and internally devoid of structure. In very young dividing cells the proplastids increase to a size visible with the electron microscope, and begin the development of the internal membranes characteristic of the mature chloroplast.

Fig. 3-12 Amyloplasts in cell of young potato tuber. The amyloplasts enclose developing starch grains.

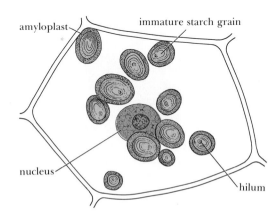

amyloplast

immature starch grain

nucleus

hilum

Chromoplasts and amyloplasts may also arise from proplastids, but sometimes plastids of one kind are converted into another kind. Thus, as previously noted, chloroplasts may be converted into chromoplasts. Chloroplasts, also, may be formed from amyloplasts. This occurs, for example, when potato tubers, exposed to bright light for a few days, turn green. The chloroplasts are formed in the outer tissues, just under the skin, and still contain very large starch grains, but they also develop the internal membranes characteristic of chloroplasts. Associated with this greening is the formation of toxic amounts of solanine, a poisonous alkaloid.

In actively dividing cells the proplastids become constricted and divide, their numbers increase, and they are distributed to daughter cells. Although there is evidence that mature chloroplasts divide in some higher order plants in general most of the divisions responsible for increase in plastid number appear to occur in the proplastid stage. Division of mature chromoplasts or amyloplasts has never been observed.

THE CELL WALL

The cell wall in plants is the most conspicuous structure of the cell as seen under the microscope. It is produced, of course, by the living protoplasm of the cell. The presence of a definite cell wall is a fundamental distinction between plants and animals, for a comparable wall is absent in animal cells. Both plant and animal cells, on the other hand, possess a plasma membrane, or outer limiting layer of the cytoplasm, which is distinct from the cell wall. In animal cells this layer is sometimes referred to as the cell wall, but it is a living part of the cell, whereas the mature plant cell wall is nonliving. Moreover, the wall of the plant cell is relatively rigid—that is, it does not change its shape to any extent—and as a result the mature plant body is usually constant in shape and fixed in position. If the cells of our bodies possessed walls like those of plants, we would literally be unable to move a muscle.

The structure of the cell wall and its chemical composition have much to do with the uses of plants by man. Plant tissues in which the cell wall is thin and soft, as those of fruits, leaves, and fleshy stems and roots, are commonly eaten by man or animals. Tissues in which the walls are thickened and strong or hard may be used as textiles, lumber, or other durable plant products.

Layers of the Cell Wall   In cells with thin walls, such as those from immature tissues and soft

Fig. 3-13   Starch grains. (A): banana (grains within a cell, and one grain enlarged). (B): garden bean. (C): potato. (D): sweet potato (E): rice (several grains within an amyloplast, sometimes called a compound grain).

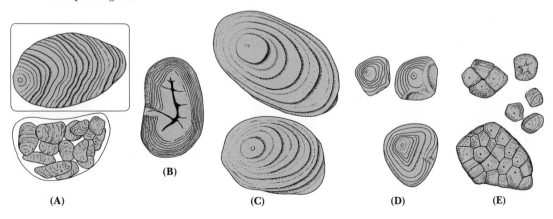

(A)   (B)   (C)   (D)   (E)

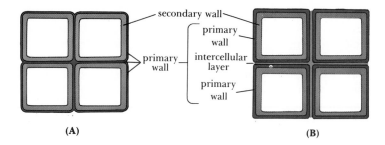

Fig. 3-14 The layers of the plant cell wall. (A): as seen under the ordinary powers of the microscope. (B): as determined by special methods.

secondary wall
primary wall
primary wall
intercellular layer
primary wall

(A)   (B)

parts of plants, the wall between two cells consists of three parts or layers. Each of two adjacent cells produces a PRIMARY WALL, and between them is found the INTERCELLULAR LAYER (also termed the *middle lamella*), which is common to both cells.

Under ordinary microscopic observation, these three layers cannot be distinguished, and appear as a single thin wall (FIG. 3-14A). The fact that they exist, however, may be demonstrated by removing with enzymes or other solvents the intercellular layer, which acts as a cementing material. The cells thus separated still retain cell walls. Where several cells come together so that their corners are adjacent, the intercellular layer may become more prominent, completely filling the intervening space. More commonly, however, an INTERCELLULAR SPACE (FIG. 3-15A) develops, which varies in size in different plant tissues. Such a space is usually not completely enclosed, but forms a part of a widely extended intercommunicating system, filled with varying proportions of atmospheric gases.

The primary walls and the intercellular layer are the only wall layers found in the cells composing the soft tissues of plants (FIG. 3-15). In fibrous and woody tissues a SECONDARY WALL is formed inside the primary one (FIG. 3-14B). The strength of fibers or the hardness of wood and nut shells is due to the relative thickness and composition of this secondary wall, which adds two more layers to the wall to make a total of five. The three layers between the secondary walls still appear as one, which may be termed the *primary wall*, although its triple nature should be kept in mind. Primary walls increase in area during the growth of the cell, but secondary walls are formed only after the cell has attained its

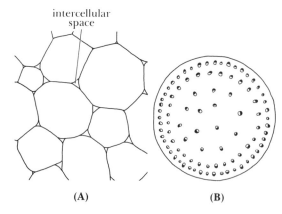

intercellular space

(A)   (B)

Fig. 3-15 (A): cells with primary walls only, from soft tissues of corn stem. (B): cross section of corn stem, showing units (vascular bundles) of conducting tissue. The cells shown at (A) occur between these units.

final size and shape. After the secondary wall has formed, the cell usually dies and the protoplast disappears.

Composition of the Cell Wall The chemical composition of the cell wall of higher plants varies with the different layers. The principal components are CELLULOSE, PECTIN, HEMICELLULOSE, and LIGNIN. Some cell walls contain, in addition, fatty substances such as SUBERIN and WAXES.

Cellulose and certain other wall components are CARBOHYDRATES, compounds of the elements carbon, hydrogen, and oxygen, in which the number of atoms of hydrogen is twice that of oxygen, as in water. Carbohydrates may be classified in two groups:

1. Simple carbohydrates, such as glucose and sucrose. The formula for glucose, or grape sugar, is $C_6H_{12}O_6$. This is classified as a MONOSACCHARIDE. Two such sugars linked together in a molecule with the elimination of a molecule of water constitute a DISACCHARIDE, such as sucrose, maltose, and lactose, with the formula $C_{12}H_{22}O_{11}$.

2. Polysaccharides. Starch, cellulose, and other carbohydrates are formed from the linking together of simple sugar units from which an oxygen and two hydrogen atoms have been removed. A generalized formula for many such compounds is $(C_6H_{10}O_5)_n$, the subscript indicating that a large number of these basic units—hundreds or thousands—are linked together in a single starch or cellulose molecule. Such compounds are classified as POLYSACCHARIDES. Pectin is related to the polysaccharides, but is composed of modified basic units. The term *hemicellulose* is confusing, for hemicellulose is not cellulose but refers to various polysaccharides of the cell wall exclusive of cellulose. Cellulose, hemicellulose, and other polysaccharides can be converted to sugars by chemical treatments and by many bacteria and fungi.

The intercellular layer in soft, nonwoody tissues is composed of pectin, a name given to a complex of pectic substances. It occurs typically in the form of calcium pectate, and is readily dissolved by various agents. For example, as fruits ripen, the insoluble calcium pectate of the intercellular layer is changed by enzymes to a more soluble form. The cells are thus less rigidly cemented together, and the fruit becomes softer in texture. Many bacteria and fungi that cause various soft rots of plant tissues produce the enzyme PECTINASE, which digests the pectin of the intercellular layers and reduces the tissues to a pulp from which the soft rot organisms obtain food. Fungi and bacteria are also active in the retting of flax, hemp, and jute. The stems of these plants are soaked in streams, ponds, or tanks, or allowed to lie in the field exposed to rain or dew. The enzymes of the organisms cause the bundles of fibers to separate from the surrounding tissues. The stalks are then dried, broken between corrugated rollers and combed to separate the valuable bark fibers from which linen or burlap are made.

The primary wall in soft, nonwoody tissues is chiefly composed of pectin, cellulose, and hemicellulose. The pectin content of the primary wall varies greatly and is much higher in some plants or plant parts than in others. It is higher, for example, in the cell walls of potatoes, beets, apples, and in the peel of citrus fruits. When jams or jellies are made from fruits low in pectin, or when it is desired to make an expensive fruit go further, commercial pectin preparations are added to the fruit or fruit juice to ensure the setting or jellying of the product. This is done in both home and industry. An important source of commercial pectin is the peel of citrus fruits, which has a pectin content of 20–50 percent. The pectin is extracted by hot dilute acids. Other uses of pectin are found in medicine and industry.

The chief constituents of the secondary wall are cellulose, hemicellulose, and lignin. In wood, the cellulose content is from 41–48 percent; hemicellulose, 25–40 percent; and lignin, 19–30 percent. In general, more than a third of the organic material produced by plants is cellulose. It is found not only in vascular plants but also in lower forms such as algae. Lignin, a complex organic compound (not a carbohydrate), occurs only in the cell walls of vascular plants, and only where secondary walls are present. But in such cells, lignin is found not only in the secondary wall but also in the other cell wall layers. Lignification actually begins in the intercellular layers and spreads inward so that the intercellular layer and primary walls contain more lignin than the secondary wall. Lignin contributes to the rigidity of the cell wall, although it does not increase the tensile strength. The flexible walls of the cotton fiber (about 94 percent cellulose) contain no lignin, and the flax fiber contains very little. Jute, sisal, manila hemp, and wood, however, contain a considerable amount, which is presumably responsible for the greater stiffness of these fibers.

Cellulose is the structural framework of the cell wall, and investigations, chiefly with the electron microscope, show that the cellulose is in the form of greatly elongated threadlike structures known as FIBRILS. The fibrils are built up of long, slender cellulose molecules. In the outer surface of the primary wall the fibrils occur in the form of an interwoven network (FIG. 3-16). In secondary walls, on the other hand, the fibrils are more or less parallel, but in successive layers they run in different directions in a crisscross pattern that contributes to the strength of the wall. The spaces between the fibrils contain pectin, hemicellulose, and lignin, depending upon whether secondary walls or only primary walls are present. It is possible, by the use of the proper solvents, to remove the filling substances and leave a coherent framework of cellulose (FIG. 3-16).

The cellulose of the cell wall is the most abundant organic material of the plant world, and has long been important to man in the form of plant fibers, wood, and lumber. Cellulose in industry is obtained from many sources, but the chief of these are cotton and wood. The United States produces some 15 million 500-pound bales of cotton annually and consumes more than 100 million tons of pulpwood. Chemical research laboratories are constantly seeking and finding new ways of converting cellulose into products useful in our daily lives; the cellulose molecule has transformed our world, and its consumption is an index of the standard of living of a country. One product alone, paper, may be cited as an example of the use of cellulose in everyday life. This is made mostly from wood—chiefly from softwoods such as spruce, fir, and pine, but also from hardwoods such as poplar and even oak. A remarkable development in the pulp industry is the rapid increase in the use of southern pines. These form quick-growing forests that can be harvested every 15–20 years. Their use in the production of wood pulp has become a major industry in our South.

In producing pulp, the wood fibers are separated either by mechanical means or by treatment of wood chips with dilute acids or alkalies that remove most of the noncellulosic components of the wood. From wood pulp are manufactured such products as newsprint; book, magazine, and other printing papers; building paper; wrapping paper and paper bags; milk and food containers; wall paper; and many other products such as rayon, cellophane, and celluloid. Products made from wood pulp are among the most widely manufactured commodities in the world today.

Until recent years, the lignin removed from wood in the production of paper pulp was entirely a waste product. Lignin is now utilized to some extent in the manufacture of synthetic rubber tires; in the ceramic, food, and drug indus-

Fig. 3-16 Electron micrograph of surface view of primary wall of a young cotton fiber. The non-cellulose components of the primary wall (chiefly pectin) have been dissolved, leaving the cellulose fibrils. Remnants of the external pectic layer persist as white patches. (12,600 ×)

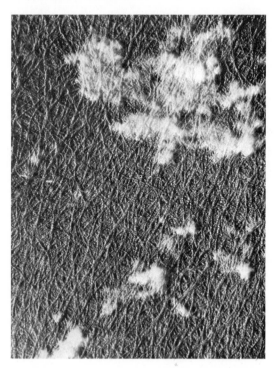

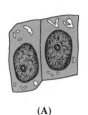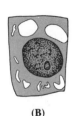

Fig. 3-17 Stages in the development of the vacuole of a plant cell. Diagrammatic.

(A)　　　(B)　　　(C)　　　(D)

tries; in the manufacture of pigments, synthetic resins, adhesives, and insecticides; as a tanning agent in the leather industry; and in the production of vanillin. Vanillin, the active principle of vanilla flavoring, was formerly obtained only from the immature beanlike fruit of a tropical orchid. Now much of our supply of vanillin is derived from the waste lignin of the paper pulp industry. Since lignin is available in large quantities at low cost, it is a raw material of great potentialities for industrial purposes.

THE VACUOLE

A vacuole within the cytoplasm, as pointed out earlier, is a region filled with cell sap and bounded by a vacuolar membrane (FIG. 3-2). Although this membrane appears to be a part of the vacuole, it is actually the inner limiting layer of the cytoplasm. The cell sap is composed of water (as much as 98 percent) containing dissolved or colloidally dispersed proteins, sugars, organic acids, water-soluble pigments, and other compounds. The reserve protein in the plant may be stored largely in the vacuole. The contents of the vacuole are nonliving.

Early stages in the development of vacuoles may be studied in the actively dividing cells of the tip of a growing stem or root (FIG. 3-17). Such a region, primarily concerned with the formation of new cells by division, is called a MERISTEM. The cells of apical meristems are generally uniform in diameter and are thin-walled, with a nucleus that is large in proportion to the volume of the cell. The vacuoles in such cells are usually numerous,

small, and distributed throughout the cytoplasm. The cells of the meristem will continue to divide. Other cells—behind the apex—enlarge, differentiate, and become mature cells of the stem or root. During the process of enlargement, water moves into the vacuoles, which increase in size and fuse with neighboring vacuoles. Eventually, a single large vacuole occupies most of the cell. The increase in size of cells as differentiation proceeds is more the result of water absorption than of the synthesis of new protoplasm, although protoplasm also increases in amount. For example, the water content of a root-tip cell may increase 20 times during enlargement while the dry matter increases only 5 times.

Most of the yellow and orange and some of the red colors of plants are due to plastid pigments. But tints of blue, violet, or purple, and most of the dark red or scarlet colors, are commonly due to pigments dissolved in the cell sap of the vacuole. These ANTHOCYANIN pigments are complex compounds composed of pigment and a sugar. Vacuolar pigments are soluble in water and will diffuse out of the cell if the cell membranes are destroyed by heating or by other means. These pigments may occur in any part of the plant. They are responsible for the color of the purple-topped turnip, the scarlet radish, and the garden beet; for the blues and reds of grapes, plums, and cherries; and for the purple, blue, and red of the aster, poinsettia, geranium, morning-glory, tulip, hyacinth, and a host of other flowers. The color of anthocyanins varies with acidity, so that the same pigment may be red in one flower and blue in another.

The anthocyanins, together with the carotenoids and perhaps the tannins, play an important role in the formation of autumn color in leaves. In many woody plants the anthocyanins develop so abundantly that they conceal the carotenoids, and thus are responsible for the shades of red and purple of many autumn leaves (Plate 2).

The production of anthocyanins in the autumn is determined in part by heredity and in part by environmental factors such as light and temperature. The accumulation of sugars generally and sometimes the direct effects of light stimulate the formation of anthocyanins, as seen in the brilliant colors of poison ivy leaves in strong light in contrast to the yellow color of the foliage growing in the shade, or the bright tints of the outermost leaves of a tree as compared to the yellow leaves deeper within the crown. Hereditary factors account for the fact that many woody plants such as privet and lilac develop no anthocyanins. Low temperatures commonly influence anthocyanin formation. Contrary to popular opinion, autumn colors may be at their best in the complete absence of frost, provided cool temperatures have prevailed.

CELLS AND TISSUES

As pointed out in Chapter 2, the cells of the plant are grouped into tissues. Tissues may be simple, composed of only one kind of cell, or complex, made up of several kinds of cells. Wood, for example, is a complex tissue. The most common kind of simple tissue is known as PARENCHYMA. Parenchyma is the chief tissue of fruits and flowers, and of the pith and other unspecialized parts of stems and roots. Parenchyma cells, when fully developed, usually have no secondary wall, but retain a nucleus and a thin layer of cytoplasm just within the primary wall (FIG. 3-18B, C). Their chief characteristic, in which they differ from most plant cells, is that they continue to live—sometimes for many years—after attaining maturity. Although they usually occur in the

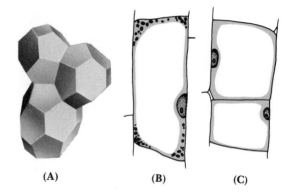

(A)   **(B)**   **(C)**

Fig. 3-18 Parenchyma cells. (A): highly diagrammatic. (B): cell containing starch grains, from stem of jack-in-the-pulpit (*Arisaema triphyllum*). (C): cells from pith of poplar (*Populus*).

form of a simple tissue, they may also be combined with other cells into a complex tissue. Parenchyma cells are useful to the plant in a number of ways: the parenchyma of the pith and other regions of the stem and root is frequently an important starch- or sugar-storage tissue, and in succulent plants such cells accumulate water. Parenchyma tissue is commonly penetrated by more or less conspicuous intercellular spaces.

Parenchyma cells illustrate, more clearly than most plant cells, the shapes assumed when cells are aggregated into masses and compressed on all sides. Although they may be somewhat elongated, they tend to be isodiametric—that is, of the same diameter in any direction. When examined in three dimensions, they are found to approximate a 14-sided figure, with 8 hexagonal and 6 square faces (FIG. 3-18A). However, an individual cell of this type occurs rarely, if ever, for the effects of contact with contiguous cells and other factors bring about wide departures from the ideal shape. Parenchyma cells may be viewed as representing a primitive type of cell from which, during evolutionary change, other kinds of cells of the plant body have arisen.

More complex, specialized tissues will be studied later in connection with their specific functions.

SUMMARY

1. Plants and all other organisms are composed of protoplasmic units termed *cells*. In plants, these units are usually enclosed within a cell wall.

2. Protoplasm is usually subdivided into nucleus and cytoplasm. The nucleus, nucleoli, and various cytoplasmic bodies are termed *organelles*. Some organelles, including the nucleus, nucleoli, plastids, and vacuoles, are readily visible under the light microscope. Mitochondria are just visible after special staining, and the endoplasmic reticulum, dictyosomes, ribosomes, and microtubules can be distinguished only with the electron microscope.

3. Protoplasm is composed of many substances, of which water and protein are the most abundant. Proteins are composed of amino acids. They may be linked with nucleic acids, forming nucleoproteins, present in both the nucleus and the cytoplasm. There are two general groups of these acids: deoxyribonucleic (DNA) and ribonucleic (RNA).

4. All cells arise by division of pre-existing cells. Cell division is preceded by nuclear division (mitosis) during which the daughter chromosomes are apportioned between the daughter nuclei, thus providing a uniform distribution of the hereditary materials.

5. Cell division is usually brought about by the formation of a cell plate in the cytoplasm. Associated with the cell plate is the phragmoplast, composed of aggregations of microtubules.

6. Genes are units of hereditary information, located on chromosomes. They determine the development, form, and functioning of an organism.

7. Chloroplasts contain chlorophyll and carotenoids. Chromoplasts contain only carotenoids. Amyloplasts, which contain no pigments, are centers of starch formation. Starch may form also in chloroplasts. Plastids of any kind may arise from proplastids, or mature plastids of one kind may be converted into another kind, as chloroplasts into chromoplasts.

8. Carotenoids are chiefly responsible for the yellow and orange colors of flowers, fruit, and seeds, and for the yellows of autumn leaves. A few carotenoids are red or orange. β-carotene is the chief precursor of vitamin A.

9. The presence of a true cell wall distinguishes plant from animal cells. The walls between cells in soft tissues are composed of three layers; those of cells in hard tissues, of five layers. The three layers comprise the primary wall. In hard tissues secondary walls are laid down.

10. The most abundant constituent of the cell wall is cellulose, but the wall also contains pectin, lignin, hemicellulose, and other compounds. In nonwoody cells (usually having no secondary wall) the intercellular substance is composed of pectin, and the primary wall of pectin, hemicellulose, and cellulose. The secondary wall is composed chiefly of cellulose, but commonly also contains a high percentage of hemicellulose and lignin. In woody cells the intercellular layer and primary walls also contain lignin. Cellulose in the cell wall is present in the form of elongated threads, called fibrils.

11. Vacuoles are spaces within the cytoplasm, filled with nonliving cell sap containing a wide variety of substances in solution, including the anthocyanin pigments. These pigments are responsible for most of the red, blue, and purple colors of flowers, fruits, and autumn leaves.

12. Aggregations of cells form simple and complex tissues. Parenchyma cells are nonspecialized and usually thin walled. They continue to live after maturity.

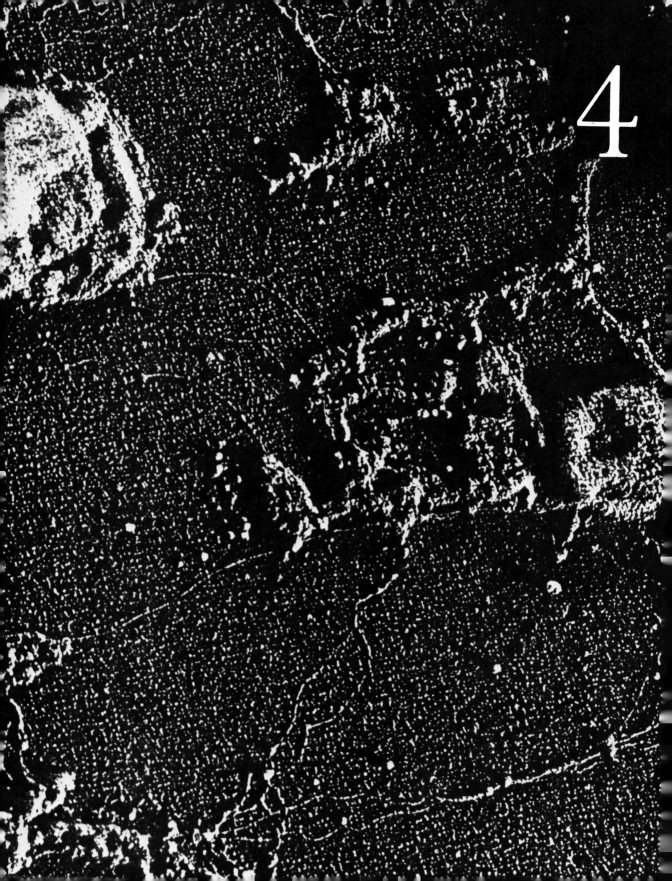

The Matter and Mechanics of Cells

CELLS ARE THE BASIC units of life, and all of the hundreds of life processes may occur in a single cell. Higher plants and animals, for example, begin life as a single cell, the fertilized egg. Many organisms—bacteria, algae, and protozoa—go through life as individual cells. Moreover, it is possible for single cells from a mature plant organ, such as a stem or root, to be multiplied in sterile cultures and then induced to form embryonic plants that develop to maturity. This is an indication that single cells contain the hereditary materials required for normal development and for all reactions of mature plants (Chapter 14).

Whether cells grow alone or as part of a complex organism, their development and activities are varied and complicated. Not many years ago they were considered too complicated to be understood. Molecular biology, essentially all of which has been developed in the twentieth century and much of it within the past few years, is rapidly uncovering the details of cellular reactions. They are complicated, but when divided into shorter steps can be understood and even duplicated in a laboratory test tube.

Modern biology has emphasized the long-held view that the reactions of living organisms are basically identical with the physical and chemical reactions of the nonliving world. The complexity arises from the number of simultaneous reactions that go on in a single cell and from the interactions among them. To understand these advances we need to be familiar with some of the simpler, basic concepts of physics and chemistry. Among other objectives, this chapter is intended to review some of these concepts and to show their connections with the life of plants.

FORMS OF MATTER

Atoms The concept that all matter is composed of invisibly small particles, of a specific kind for each substance, was postulated by the early Greeks more than 2000 years ago. These supposed particles were called ATOMS, meaning "indivisible." It was not until 1803, however, that an English chemist, John Dalton, established the atomic-molecular theory of matter on a sound basis. It is now accepted that the universe as we know it contains innumerable combinations of the 105 kinds of atoms that constitute the 105 known chemical elements, one particular kind of atom for each element.

Some chemical elements important in plant life or of other interest are listed in Table 4-1. For convenience the elements are commonly designated by their CHEMICAL SYMBOLS. These are abbreviations of the English, or sometimes the older Latin, names of the elements. Thus the symbol for aluminum is Al but that for gold is Au and for potassium is K, from the Latin names *aurum* and *kalium*. Atomic number refers to structural properties of the atoms, which will be discussed later. The last column shows the RELATIVE WEIGHT of the atoms when the weight of the commonest form of carbon is taken as 12. This weight is artificially set at a level that will avoid a figure of less than 1 for hydrogen (H), the lightest element.

It happens that 6×10^{23} (that is, 6 followed by 23 ciphers), multiplied by the actual weight of a hydrogen atom $(1.673 \times 10^{-24}$ gram$)$ gives a value of about 1 gram. The figure 6×10^{23} is therefore taken as the number of atoms in the GRAM-ATOMIC WEIGHT of any substance. This is the *relative* weight of atoms in grams and is the figure used in chemical calculations. Thus, if one atom of carbon (C) combines with two atoms of oxygen (O) to form carbon dioxide (CO_2), the combining weights of the two atoms would be $12C + 32(2 \times 16)O = 44$ units of CO_2. Carbon would thus constitute $^{12}/_{44}$ or 27.3 percent of the weight of the CO_2 molecule. Forty-four grams is the weight of 6×10^{23} molecules of CO_2. It is, therefore, the GRAM-MOLECULAR WEIGHT or, more commonly, the molecular weight of the compound. This quantity is also called a MOLE of CO_2.

While it is obvious that 10^{23} is a large number, it is difficult to realize how large it is. As an illustration, 1.008 grams of hydrogen contain 6×10^{23} atoms with a diameter of about 1 angstrom (Å) each. An angstrom is one hundred millionth of a centimeter. Assuming that the

TABLE 4-1
Chemical Elements Used by Plants or Otherwise of Interest

Element	Chemical symbol	Atomic number	Atomic weight
Aluminum	Al	13	26.98
Boron	B	5	10.81
Calcium	Ca	20	40.08
Carbon	C	6	12.01
Chlorine	Cl	17	35.45
Cobalt	Co	27	58.93
Copper	Cu	29	63.54
Gold	Au	79	196.97
Helium	He	2	4.003
Hydrogen	H	1	1.008
Iodine	I	53	126.90
Iron	Fe	26	55.85
Lawrencium	Lw	103	257.
Magnesium	Mg	12	24.31
Manganese	Mn	25	54.94
Molybdenum	Mo	42	95.94
Neon	Ne	10	20.18
Nitrogen	N	7	14.01
Oxygen	O	8	16.00
Phosphorus	P	15	30.97
Potassium	K	19	39.10
Silicon	Si	14	28.09
Sodium	Na	11	22.99
Sulfur	S	16	32.06
Uranium	U	92	238.03
Zinc	Zn	30	65.37

hydrogen atoms did not double up and form H_2 molecules, it would take 100,000,000, or 10^8, atoms side by side to equal 1 centimeter, $^2/_5$ of an inch. But 6×10^{23} hydrogen atoms in a row would form a line 37 billion miles long.

The Structure of the Atom   We know now that "indivisible" atoms are composed of smaller particles, but any division of atoms changes their chemical nature and properties. Of the subatomic particles that have been identified, three are of major importance in the structure and chemistry of atoms. These are the PROTON and the NEUTRON, which together form the NUCLEUS, and the much smaller but still highly important ELECTRON.

The proton is a relatively large particle with a positive electric charge. The number of protons in an atom is a specific character, called the ATOMIC NUMBER. Atoms are now known which have protons and atomic numbers starting at 1 (hydrogen) and going without a break to the 105th atom, hahnium—the latest of the artificial, radioactive elements, with 105 protons.

The positive electric charge of protons is balanced by the negative electric charges of an equal number of electrons. Thus hydrogen (FIG. 4-2) has one electron and lawrencium has 103 (Table 4-1). Electrons are some 1800 times lighter than the protons. The chemical properties of atoms are determined primarily by the number and arrangement of their electrons.

The third major particle of the atom is the neutron. These are bodies of very nearly the weight of protons but without an electric charge. They are thus an important part of the weight of the atom but have relatively little effect on its chemistry. It is believed that the neutrons contribute in some way to the stability of the nucleus. Of the known elements, only the most common form of hydrogen has no neutrons.

Isotopes Because of differences in the number of neutrons, many elements have atoms of varying atomic weight. Such forms are called ISOTOPES. Carbon, the element of charcoal and diamonds, may have atoms with relative or atomic weights of 10, 11, 12, 13, 14, 15, or 16. All of these have six protons and six electrons. Chemically, therefore, they are carbon. The differences in weight are due to the presence of from four to ten neutrons in the atom. Isotopes are commonly designated by the symbol of the element, with a superscript indicating the atomic weight of the isotope. Thus carbon of atomic weight 12 is ^{12}C, or common carbon, which forms

the base for the atomic weights. Carbon-13 (^{13}C), heavy carbon, constitutes about 1 percent and ^{12}C 99 percent of the earth's carbon. The atomic weight of natural carbon, 12.01, is the average weight of these two isotopes. Carbon-14 is a radioactive isotope present in the air in trace quantities, formed by high-energy cosmic or atom bomb radiation, and produced for laboratory use in an atomic reactor; ^{10}C, ^{11}C, ^{15}C, and ^{16}C are artificially produced isotopes that decay within a few minutes or seconds and are of little interest. Their behavior is assumed to be due to a seriously unbalanced nucleus with too few or too many neutrons.

Radioactive carbon, ^{14}C, with only two extra neutrons, decomposes very slowly, with a half-life of about 5700 years, forming nitrogen atoms and emitting electrons of relatively low energy. This reaction can be measured for accurate analyses of minute quantities of compounds containing ^{14}C. The isotope is thus a tool of major importance in biochemical studies of plants and animals.

The decay of ^{14}C consists of the separation of one of the neutrons into an electron and a proton. The electron is emitted and can be counted with a Geiger counter, while the proton remains in the nucleus. The carbon with six protons and eight neutrons has now become nitrogen with seven protons and seven neutrons, or $^{14}_{6}C \rightarrow {}^{14}_{7}N$. The missing electron is replaced from the environment.

Atoms and Molecules Atoms, then, consist normally of a nucleus containing a characteristic number of positively charged protons and a sometimes variable number of uncharged neutrons, the whole surrounded by very small, negatively charged electrons in a number normally equal to the number of protons. The electrons are grouped in layers or shells around the nucleus. Except for hydrogen and helium (He), which form a separate group with only one shell and one or two electrons, atoms have two to six or seven shells, each containing from one to as many as 32 electrons. The outer shell of most elements is "full" and stable with eight electrons and does not contain more than this. The so-called inert gases, such as helium and neon (Ne); have outer shells that are filled, and so they do not normally react to form molecules. Other atoms tend to combine into molecules of two or more atoms. Molecules may be as simple as the two-atom molecule of hydrogen gas (H_2) or as complex as the large protein molecules with thousands of carbon, hydrogen, oxygen, and sometimes other atoms in the molecular unit. The forces that hold atoms together in molecules are called CHEMICAL BONDS. Such bonds are primarily of two types, ionic and covalent, with various gradations between them, and, in addition a special weak bond, known as the hydrogen bond.

Ionic Bonds   In a typical IONIC BOND one or more electrons are transferred from one atom and become a part of the second atom. The atom from which the electrons are transferred becomes positively charged because of its excess of positively charged protons that are no longer balanced by negatively charged electrons. The atom to which the electrons are transferred becomes negatively charged because of the extra electrons. The electrostatic force between opposite charges then holds the atoms of the molecule together.

The sodium chloride (NaCl, or common salt) molecule, is a good example of ionic bonding (FIG. 4-1). The sodium atom has a single electron in its third (outer) shell, while the chlorine atom has seven. The transfer of the outer electron from a sodium to a chlorine atom leaves the sodium with a stable, eight-electron second, now the outer, shell and a net positive charge because of the loss of a negatively charged electron. This is the sodium ion, Na^+. The chlorine now has a stable, eight-electron, third shell and has a net negative charge because of its extra electron. This is the chloride ion, Cl^-. The two ions are held

together in the salt molecule by the electrostatic bond of their plus (+) and minus (−) charges. The ionization of sodium chloride in solution is sometimes discussed as though it happened in water. Actually the molecule itself is ionized, but the ions adhere to each other until they are dissolved in water. Water reduces the attraction between the sodium and chloride ions by nearly 99 percent so that they become largely separated in a water solution.

Covalent Bonds   In compounds of carbon, the organic molecules in which we are especially interested, and in other compounds, particularly of nonmetallic elements, COVALENT BONDS are formed by the sharing of electrons, which become in effect a part of both atoms. The covalent bonds of methane or natural gas (CH_4) are shown in FIG. 4-2. Carbon has only four elec-

Fig. 4-1   Planetary model of an ionic bond in the sodium chloride (NaCl) molecule. The single, outer, valence electron of sodium is captured by chlorine. The two ions are then held together by the electrostatic force of their plus (+) and minus (−) charges.

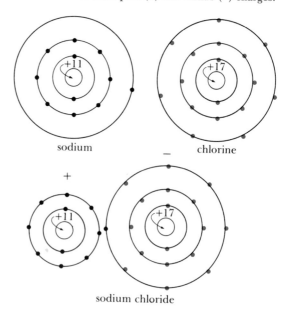

sodium   −   chlorine

+

sodium chloride

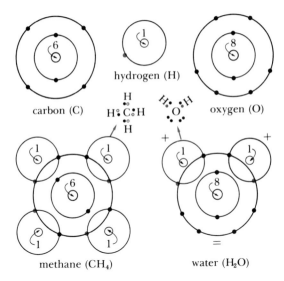

carbon (C)   hydrogen (H)   oxygen (O)

methane (CH₄)   water (H₂O)

Fig. 4-2 **Planetary models of covalent bonds in methane(CH₄) and water (H₂O).** Valence electrons are shared rather than transferred, and become a part of both atoms. Covalent bonding may be indicated by electron-pair diagrams, as in the small central figures.

trons in its outer shell and therefore four more are required to form a stable molecule. Four atoms of hydrogen, each with one electron, share these electrons and fill the outer shell of carbon. At the same time, each hydrogen atom shares one carbon electron and fills its own outer shell with two electrons. The shared electrons are considered to revolve about both atoms, thus binding them together in the molecule.

Covalent bonds hold the hydrogen and oxygen atoms together in water (H_2O, FIG. 4-2). Oxygen has six electrons in its second outer shell and it shares electrons with two hydrogen atoms, forming a stable combination. Water, however, is not a simple balanced molecule. Experimental evidence indicates that the water molecule has a V-shaped arrangement (FIG. 4-2). The large positive charge of the oxygen nucleus draws the hydrogen electrons toward the oxygen, thus giving the oxygen end of the molecule a small negative charge and leaving the hydrogen end positively charged. The molecule is said to be polar, and the hydrogens of one molecule are attracted to the oxygen of another, forming a hydrogen bond.

Hydrogen Bonds Hydrogen bonds are weak electrostatic bonds, only about 5 percent as strong as ionic or covalent bonds. Nevertheless they are very important in biology because of their effects on the structure of chromosomes and protoplasm and on the properties of water. As a consequence of their low energy, hydrogen bonds are easily formed or broken. They are ideal temporary or cross-bracing bonds in proteins and in the duplication of the chromosomes (FIG. 4-19). Hydrogen bonds in plants connect oxygen and nitrogen atoms through a hydrogen atom that is covalently bonded to one of these atoms. The pairing can be O—H··O, O··H—N, or N—H··N, in any combination. A hydrogen bond to oxygen in this case is indicated by dots. It is somewhat longer and much weaker than the other chemical bonds.

Hydrogen bonds are prevalent in water and result in many special properties for this liquid. These bonds, connecting the H of one molecule to the O of another, form unstable, rapidly changing polymers of H_2O molecules; compounds comparable, for example, to H_4O_2 and H_6O_3. This grouping of the simple H_2O molecules results in freezing and boiling temperatures that are many degrees higher than they would otherwise be, and so water is a liquid or a solid when it would be expected to be a gas. Hydrogen bonds increase the heat required to warm or to evaporate water, or that given off in cooling and condensation, and thus have major tempering effects on our environment. It is estimated that two-thirds of the large amount of heat necessary to evaporate boiling water is used to disrupt the hydrogen bonds in the liquid. When water freezes, its hydrogen bonds cause the molecules to form an open crystal structure in which the molecules are held farther apart than they are in the liquid. As a consequence, water expands on freezing, and ice is lighter than water. The expansion of ice bursts water pipes or radiators, and fractures rocks, breaking them down into soil. It also causes ice to float, in contrast to other solids, which normally sink in their liquid phase. Because ice floats, our lakes and oceans do not become solid blocks of ice in cold weather.

ENERGY TRANSFORMATION AND STORAGE

Oxidation and Reduction Energy transformations in living cells commonly involve oxidation and reduction. The term oxidation is derived from a common action of oxygen and we normally think of it in terms of this element. Thus wood, basically a carbohydrate with the general composition CH_2O, burns with the addition of oxygen in a reaction that may be indicated as

$$CH_2O + O_2 \longrightarrow CO_2 + H_2O + \text{heat energy}$$

Oxidation is not necessarily dependent upon oxygen, however, for chlorine and other elements can be strong oxidizing agents. Basically, OXIDATION is defined as the *loss* of electrons. Oxygen or chlorine oxidizes by taking up electrons that were previously in other atoms or molecules. A common oxidation in biological reactions consists of the removal of hydrogen atoms with their electrons. Thus the removal of hydrogen atoms from CH_2O would oxidize the compound to carbon monoxide (CO). Oxidation by the removal of hydrogen atoms is shown in Chapter 6 to be the basis of respiration.

Electrons do not remain free. They are taken up by some other atom or compound. This compound is then reduced; REDUCTION is defined as the *gain* of electrons. It follows that a reduction reaction occurs for every oxidation, and vice versa. Oxidation and reduction involve changes of energy as well as electrons. Two oxidation-reduction reactions transform the energy used by living cells. In one of these, photosynthesis (Chapter 5), the energy of sunlight is used to reduce carbon dioxide and oxidize water. Some of the radiant energy of sunlight is fixed in the

sugar that is formed. In the other type of reaction, respiration (Chapter 6), some of the energy of sugar is changed to available chemical energy and used, or stored in special phosphate compounds.

Energy Storage Compounds One of the important processes in living organisms is the formation and action of special compounds in which quickly available chemical energy is stored. A number of these high-energy compounds are known. Most of them are similar to ADENOSINE TRIPHOSPHATE, or ATP, the best-known and presumably the most important compound of the group. The structure of this compound is shown in FIG. 4-3. The significant part of the molecule is the No. 2 and No. 3 phosphates. The hydrolysis (decomposition in water) of these phosphates from a mole (506 grams) of ATP can yield nearly 15,000 calories of energy as heat. Adenosine triphosphate is changed first to ADENOSINE DIPHOSPHATE (ADP) and then to ADENOSINE MONOPHOSPHATE (AMP) and inorganic phosphate during the hydrolysis. It is sometimes stated that *high-energy bonds* are present in ATP or ADP, but *high-energy compounds* are considered to be a better description. ATP carries energy in small, quickly available packages that can be used by cellular enzymes in doing metabolic work in the cell. FIGURE 4-3 indicates the two high-energy phosphate groups that form ATP. If only one phosphate has been released, forming ADP, a new phosphorylation reaction can reconvert the ADP to ATP.

Adenosine monophosphate is a NUCLEOTIDE, one of an important group of compounds composed of a nitrogen base, a pentose (5-carbon) sugar, and a phosphate group (FIG. 4-3). Adenosine monophosphate, as it is shown in FIG. 4-3, is one of the four nucleotides that form the RNA molecule. The same compound, except that the starred oxygen in the ribose group is absent, is a nucleotide building block in DNA (FIG. 4-19). Other nucleotides may form high-energy phosphate compounds, but these seem to be limited in activity.

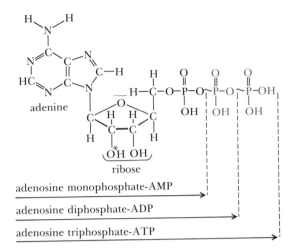

Fig. 4-3 **The structure of the adenosine phosphates: adenosine monophosphate (AMP), adenosine diphosphate (ADP), and adenosine triphosphate (ATP). ATP is considered to be a major source of available chemical energy in the cell.**

Adenosine triphosphate is formed in electron transfer systems in both the chloroplasts (Chapter 5) and the mitochondria (Chapter 6) of the cells. In these systems a high-energy electron, either from illuminated chlorophyll or from oxidized sugars, is moved through special enzyme systems and the energy of the electron is used to build ATP from AMP or ADP and inorganic phosphate. The result is the formation of what has been called the energy currency of the cell—legal tender for all of the hundreds of energy-requiring processes of life.

THE MATTER OF CELLS

The compounds present in cells range from those as simple as water and carbon dioxide to the highly complex pigments, hormones, carbohydrates, proteins, and other materials found in the various parts of the protoplast. These compounds may be present in various forms and in complex mixtures.

Physicists recognize three basic forms of matter: solids, liquids, and gases. At appropriate pressures and temperatures, matter may occur in any of these forms. Thus, water (at atmospheric pressure) is a solid at temperatures below 0°C (32°F), and a gas (vapor) at temperatures above 100°C (212°F). Carbon dioxide is a gas under ordinary conditions, a liquid in commercial cylinders under high pressures, and a solid (dry ice) at temperatures lower than −79°C (−110°F). Iron becomes a liquid (melts) at 1535°C (2795°F) and changes to vapor at about 3000°C (5432°F). We think of iron as a solid, water as a liquid, and carbon dioxide as a gas, but each of these substances can be changed to any of the three basic forms of matter by appropriate changes in temperature and pressure. Most liquids and some solids are volatile; that is, they tend to change to vapor at a rate that increases with the temperature. Thus water at atmospheric pressure evaporates slowly at room temperatures, but if superheated (raised to temperatures above its boiling point) it evaporates explosively.

In addition to the basic forms of pure substances, matter may exist in the mixed forms of solutions and colloids. In SOLUTIONS the *individual molecules or ions* of one substance, the *solute,* are scattered among the molecules of another substance; for example, sugar or salt dissolved in water. Theoretically, the dissolving substance, or *solvent,* may be a solid, a liquid, or a gas. In typical COLLOIDS, *groups of molecules* adhering together are suspended in a medium; for example, gold in ruby glass, clay in muddy water, or fog in air. Colloids will be discussed more fully later in this chapter.

THE KINETIC THEORY

During the 1800s the concept was developed that atoms and molecules are in continual motion, and that the rate of this movement is dependent upon temperature. This concept was proposed to explain the behavior of gases and was later extended to liquids and solids. In studying living organisms we are concerned primarily with the kinetic theory as it applies to gases, liquids, and solutions.

In these systems the motion of the molecules is random. The motion never ceases at any temperature above absolute zero, which is 273°C (459.6°F) below the freezing point of water. Motion is due to energy being transferred from the environment to the molecules, and to some of this energy being converted into KINETIC energy or energy of molecular motion.[1] At any one time a molecule is moving in a straight line with a uniform velocity, but the distance traveled will normally be very short, perhaps a millionth of a centimeter in air and much less in solutions because of collisions with other molecules. Such collisions may cause a change in both direction and velocity of the movement. The average velocity of all molecules will, however, remain the same if the temperature does not change.

Substantial support for the kinetic theory of matter is provided by what is known as Brownian movement. This movement was first observed by the British botanist Robert Brown in 1828 while he was studying a suspension of granular particles set free by crushing pollen grains. When observed under the microscope, the tiny particles were seen to be in constant zigzag motion. The same action can be seen in the laboratory by observing dilute India ink under a microscope. The slow movement of the carbon particles is due to their own kinetic energy. The irregular changes of direction are due to collisions between particles, or, more frequently, to collisions with water molecules. This movement may be observed in protoplasmic granules within living cells, or in bacteria and other very small bodies. When observed in bacteria, however, it should be distinguished from motility brought about by organs of locomotion (FIG. 20-4).

Diffusion An immediate consequence of the constant movement of molecules in gases and solutions is that these molecules tend to spread uniformly throughout any available space. If, for example, a teaspoonful of household ammonia is spilled on a table, the odor will soon be noticed throughout the room. The molecules of ammonia have spread by DIFFUSION, a process that may be defined as the net movement of a substance as a result of the independent motion of its individual molecules, ions, or colloidal particles. The molecules of ammonia gas, freed from the liquid, move off in random directions through the air. The inevitable collisions with one another and with air molecules send them in every direction. Some will move toward the spilled ammonia and others will move away from this region, but because of their greater concentration near the spilled ammonia—that is, greater numbers per unit volume—more molecules will move away from the ammonia than toward it. It follows that, until the concentration of ammonia is uniform throughout the room, there will be a net movement of particles away from the liquid. When the ammonia molecules have become evenly distributed throughout the room, diffusion is no longer proceeding, although molecular movements, of course, continue. A crystal of dye dropped into a beaker of water will show the same sort of spreading movement from high concentrations of the dye to lower concentrations.

As in the above examples, diffusion is commonly from a region of higher to one of lower concentration of the same kind of particles. As we shall see later, however, the direction in which diffusion occurs can be determined by factors other than concentration.

Diffusion is a process of great importance in the living cells of plants and animals. Water enters the roots of plants and moves through cells or is lost as vapor by diffusion. Gases, such as oxygen and carbon dioxide, and minerals and foods in solution, may move into or between cells by diffusion.

---

[1] Only a part of the energy of the molecules is considered here: the energy of motion of the entire molecule, or its speed (translational energy). Rotation and other motion within the molecule have other effects.

Evaporation and Condensation We have mentioned volatility as a common property of liquids.

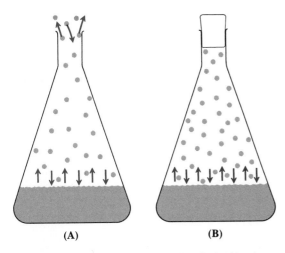

Fig. 4-4   Evaporation of water from a flask. (A): when the flask is open, more molecules escape from the liquid as vapor than return to it. (B): in a stoppered flask the concentration of water vapor molecules increases until molecules return to the liquid as fast as they leave.

Volatility, also, is a consequence of the motion of molecules. When moving molecules escape from the surface of a liquid, such as water, the process is called EVAPORATION.

Water molecules are thousands of times too small to be seen with a microscope, so we must attempt to visualize them. In a flask partly filled with water (FIG. 4-4A) the molecules of the liquid are in a state of active motion, moving at random in short paths, at an average speed that increases with the temperature. This motion is restricted by intermolecular or cohesive forces between the water molecules, and these forces tend to prevent the escape of molecules from the liquid. However, some molecules, moving with greater than average speed, possess sufficient kinetic energy to break away from these intermolecular forces and to escape, or evaporate, into the air of the flask. Since only the fastest molecules with the greatest kinetic energy escape, the average energy of the remaining molecules is reduced. The average kinetic energy of the molecules determines the temperature of the water, and a lower average energy results in a lower temperature. This is the cooling effect of evaporation. Evaporation thus causes a heat loss from the liquid, and this loss must be replaced from the environment if evaporation is to continue at the same rate.

At the same time that evaporation is occurring, water molecules (vapor) from the air, also in a state of active motion, may strike the water surface and become a part of the liquid water. This is CONDENSATION, the reverse of evaporation. If the rate at which water molecules leave the water is greater than the rate at which they enter it, the volume of the liquid is reduced. An interchange of molecules will also occur between the interior of the flask and the outer air. A greater number of water molecules will normally move out of an open flask than enter it in any given period of time. This diffusion reduces the concentration of water vapor within the flask and hence reduces the rate of condensation. Since evaporation continues at a constant rate if the temperature does not change, the volume of the liquid in the flask will decrease until no more remains.

If the flask is stoppered, however (FIG. 4-4B), the concentration of water-vapor molecules in the air of the flask soon becomes so great that the rate of condensation becomes equal to the rate of evaporation. The air within the flask is then saturated with water vapor. Since the rates of evaporation and condensation have reached a state of equilibrium, diffusion will cease and there will be no further change in the volume of the liquid.

The evaporation of water shows a special response to temperature that is of basic importance in the life of plants. Molecules diffuse faster at higher temperatures. In general, diffusion is increased about 3 percent by a temperature rise of 10°C (18°F). The rate of evaporation of water, however, is increased nearly 100 percent by the same temperature rise. This marked difference is due to the rapid decrease in the intermolecular forces of water as the temperature is increased. With a decrease in intermolecular forces the molecules can escape more easily, and the evaporation rate increases rapidly.

Evaporation of water from living plants is discussed in Chapter 8. In terms of quantities of materials concerned, this evaporation is the major process occurring in plants. Plants may lose hundreds of times their own weight of water during growth, and this loss is frequently the most important factor affecting their growth and survival.

Diffusion in Solutions   Concentration has been mentioned as a factor in diffusion. Since pure water has a concentration of 100 percent, differences in concentration of water must be obtained by dilution of the water. Water is used to dilute acids, alcohol, or solutions, but alcohol and sulfuric acid can dilute water as well as be diluted by it. Dissolved solids also dilute water. The increase in the volume of water when sugar is dissolved in it is illustrated in Fig. 4-5. The increase in the volume of the solution over that of pure water demonstrates that water is diluted by this added solute. The experiment, however, gives an exaggerated impression of the magnitude of the dilution because the water is diluted in proportion to the *number* of sugar molecules added, and not in proportion to their *size*, which is many times larger than that of water molecules.

The addition of sugar to water dilutes the water because some of the water molecules are replaced by sugar molecules. The water in a sugar solution is therefore less concentrated than in pure water, and there will be fewer water molecules striking any surface across which water is diffusing. As a result, water molecules will not evaporate so rapidly from diluted water or diffuse across a membrane so rapidly (Fig. 4-6) as will molecules from pure water. In Fig. 4-6D the direction of diffusion is determined by concentration; that is, water diffuses from a region of *high concentration of water* to one of *lower concentration of water*, that is, to water diluted by added sugar.

Diffusion Pressure   We have so far considered diffusion only in terms of movement from a higher to a lower concentration. Two other factors, *temperature* and *pressure*, may affect the direc-

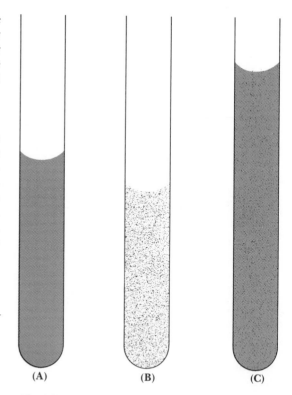

Fig. 4-5   Dilution of water by dissolved sugar. When the sugar in (B) is dissolved in the water in (A), the resulting solution (C) is spread through a larger volume and concentration of the water is reduced.

tion and rate of diffusion (Fig. 4-6). We may begin with a container separated into two parts by a *differentially permeable membrane*, considered to be inelastic, and indicated by a vertical dashed line in this figure. Such a membrane is permeable to one substance, but will retard or prevent the passage of others. The membranes of biological systems are readily permeable to water, but are impermeable or slowly permeable to solutes. If the concentration, temperature, and pressure are the same on both sides of the membrane, water molecules will move through the membrane in both directions at equal rates, and diffusion of water will not occur (Fig. 4-6A). If the cold water on one side of the membrane is warmed (Fig. 4-6B), the warm water will diffuse

Fig. 4-6 White arrows show that the direction of the diffusion of water across membranes may be determined by the concentration of water, the temperature, or the pressure. (A): no net movement, or diffusion. (B): the diffusion pressure of water increases with temperature, even though the water is diluted by expansion. (C): the diffusion pressure of water increases with mechanical pressure. (D): pure water diffuses into diluted water. (E): mechanical pressure can reverse the effect of dilution.

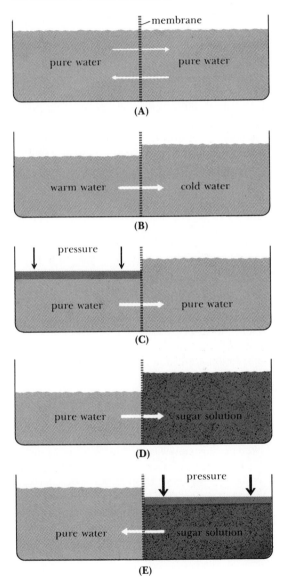

into the cold water. Since heated water expands and is therefore less concentrated, one might expect that cold water would diffuse into warm water. The increase of temperature, however, increases the energy of the water molecules. This energy effect is greater than the concentration effect, and the net movement is from the warmer (less concentrated) water into the cooler (more concentrated) water.

Pressure, also, may determine the direction of diffusion. Pressure concentrates a gas but has only a very slight effect on the concentration, or on the volume, of water. It will, however, increase the energy of the water molecules. If pressure is applied on one side of a membrane separating two bodies of pure water, as in FIG. 4-6C, water diffuses from the side to which pressure is applied. Such diffusion is due to the increased energy of the water molecules rather than to the negligible effect of pressure upon concentration. If sufficient pressure is applied on the solution side of a membrane separating a sugar solution from pure water (FIG. 4-6E), the pressure effect becomes greater than the concentration effect, and water will again diffuse from a region of lower to one of higher concentration of water.

We have now seen that the direction of diffusion may be determined not only by concentration but also by temperature and pressure. One, two, or all three of these may affect diffusion. They may also act in opposition to one another, as in FIG. 4-6E. The combined effect of these three factors is termed the DIFFUSION PRESSURE of a substance. Concentration is an important factor determining diffusion pressure, but, especially in water, temperature or pressure may be more important. We may now amplify our earlier definition of diffusion and call it the net movement of a substance from a region of higher diffusion pressure to a region of lower diffusion pressure of the substance. All gases and liquids have a diffusion pressure, although this may be evident only under certain conditions. That water has a diffusion pressure can be demonstrated when water is separated from a solution by a membrane. This is considered in the next section.

DIFFUSION IN CELLS

Osmosis   The diffusion process in which we are particularly interested in living systems is the diffusion of water and of dissolved substances into and between cells. This diffusion is complicated by the presence of differentially permeable membranes within the cells. The diffusion of water across plant and animal membranes occurs more rapidly than the diffusion of other common

Fig. 4-7   Demonstration of osmosis. The thistle tube was partly filled with a sugar solution and closed with a differentially permeable membrane fastened over the lower end. Water has diffused through the membrane and has pushed the sugar solution up the tube.

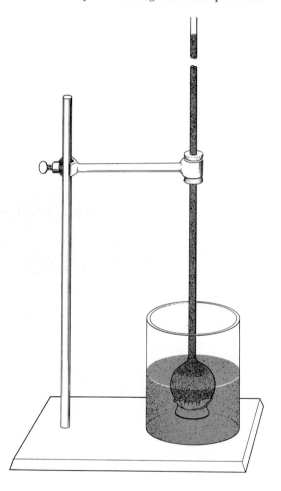

substances. As a result of the ready permeability of these membranes to water and their impermeability or slow permeability to substances in solution, there occurs a special type of diffusion called OSMOSIS. This is defined as the diffusion of *water* across a differentially permeable membrane.

The process may be demonstrated by means of a membrane separating two regions that differ in their concentration of water (FIG. 4-7). A thistle tube is filled with a 25 percent sugar solution. A membrane, previously soaked in water, is then tied tightly over the top of the tube and the apparatus inverted and placed in a beaker of water. The membrane is readily permeable to water but slowly permeable to sugar. Because of the diluting effect of the sugar molecules, the less concentrated water within the tube has a lower diffusion pressure than the water in the jar. Water will cross the membrane in both directions, but during any given period of time more water molecules will move through the membrane from the pure water outside than will move from the diluted water within the tube. As a result, the solution will gain in volume and will rise in the stem of the thistle tube. This gain in volume is the result of the greater diffusion pressure of the pure water outside the membrane. If the membrane were completely impermeable to sugar, the liquid would continue to rise until the added pressure inside the membrane had increased the diffusion pressure of the water in the solution until it was equal to the diffusion pressure of the water in the jar.

Turgor Pressure   If the stems of cut flowers are not immersed in water soon after cutting, the flowers and leaves will wilt. This wilting is due to the evaporation of water from the cells of the leaves and flowers, resulting in the partial collapse of the cells. In nonwilted plants the cells are said to be turgid, or to possess TURGOR. The cells of wilted leaves and flowers have no turgor. They may, however, regain it if supplied with water.

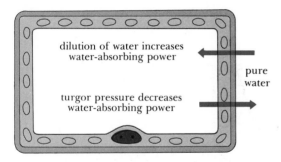

Fig. 4-8 Water-absorbing power. The diffusion of water into a cell because of the dilution of the cell sap is opposed by the effects of turgor pressure. When turgor pressure balances dilution, the cell is "full" and its water-absorbing power is zero.

To understand how cells gain or lose water, let us consider a plant cell such as that shown in FIG. 4-8. The cellulose cell wall, which offers little resistance to the passage of water or solutes, is lined within by a differentially permeable membrane, the outer layer of the protoplast. A similar layer surrounds the vacuole. The water in the cytoplasm and vacuole is diluted by sugars and various other organic and inorganic substances in solution. Because of this dilution, the diffusion pressure of the water is reduced, and water will tend to diffuse into the protoplast if the cell is in contact with water. The cellulose wall, like the cover of a football, resists extension, and a pressure (called TURGOR PRESSURE) is built up within the cell as water diffuses into it. Turgor pressure is the basis of mechanical support in most flowers and leaves, and in succulent young stems. If these tissues lose too much water, they collapse or wilt because of the absence of turgor pressure in their cells. In a woody twig, however, such support depends upon the thickened secondary walls of the cells; hence woody stems do not wilt.

It is important to recognize that the turgor pressure increases the diffusion pressure of water within a cell in the same way as would an externally applied pressure of the same magnitude. Turgor pressure thus increases the diffusion pressure of the water in the cell and reduces the rate of its inward diffusion. As osmosis continues, the turgor pressure builds up to a maximum, which is determined by the solute concentration—that is, by the dilution of water within the cell. At this maximum the lowered diffusion pressure within the cell, which resulted from the dilution of the water by solute particles, is balanced by an increased diffusion pressure due to turgor. The diffusion pressure of the diluted water inside the cell is now equal to that of the pure water outside, and the cell is "full."

Water-Absorbing Power Since all living cells contain solute particles, water will diffuse into wilted cells. That is, such cells have WATER-ABSORBING POWER[2] because the diffusion pressure of the water in the cell is less than that of pure water. But, as water diffuses into the cells, turgor pressure builds up and increases the diffusion pressure of the water within the cell. As the diffusion pressure of water *in the cell* approaches that of the water in the external environment, the rate of inward diffusion decreases and eventually stops. When the cell is full, its water-absorbing power is zero. This is shown diagrammatically in FIG. 4-8.

The above relationships can be expressed by the following equation:

Water-absorbing power of cell

= (dilution of cell sap) − (tugor pressure)

or $W = D - T$

Thus, when T is zero, $W = D$; when $T = D$, $W = 0$.

It should always be kept in mind that the movement of water from cell to cell, from the soil solution into roots, and from leaves to the external environment is dependent upon differences in diffusion pressure.

---

[2] Many plant physiologists use the terms *diffusion pressure deficit* for water-absorbing power and *osmotic concentration* for dilution of the cell sap.

Plasmolysis We have been considering the movement of water into cells whose sap consists of water diluted by dissolved materials. If, instead of water, a concentrated sugar solution is placed around the cell, water will diffuse out rather than in. The sugar solution penetrates the permeable cellulose wall of the cell and comes in contact with the differentially permeable plasma membrane. Inward diffusion of the sugar is stopped by the membrane, but water can move from what is now the higher concentration of *water* inside the cell to the more diluted water outside. The cytoplasm, which has been held against the wall by turgor pressure, now shrinks away from the wall (FIG. 4-9B). This shrinkage of the protoplast because of the loss of water to an external solution with a high solute concentration is called PLASMOLYSIS. If a plasmolyzed cell is transferred from the sugar solution to pure water, it will regain its form and turgor as water diffuses back into the protoplast, provided the membrane has not been damaged by excessive or long-continued plasmolysis.

The permeability of the cell wall to the sugar solution is indicated by the appearance of the plasmolyzed cells. If the clear space between the wall and the protoplast were air, it would show the bluish margins characteristic of air bubbles seen through a microscope. If it were pure water, the cell would not be plasmolyzed. The space must therefore be occupied by the plasmolyzing solution, and the wall is permeable to sugar.

Plasmolysis also demonstrates the location of the protoplasmic membrane and its differential permeability. Water, obviously, penetrates the membrane rapidly as it moves in or out. If the membrane were also permeable to sugar, the inward diffusion of this material would equalize its effect within and without the cell, and water would again be absorbed. If salts such as sodium chloride (NaCl) or potassium nitrate (KNO_3) are used in plasmolysis instead of sugar, a slow absorption does occur, and the cells tend to recover their turgor with time. The protoplasmic membranes are destroyed when a cell is killed, and a dead cell cannot be plasmolyzed. Extreme

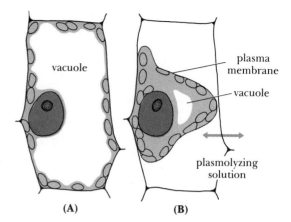

Fig. 4-9   Plasmolysis. (A): normal cell. (B): plasmolyzed cell. Water has diffused out of the protoplast into the plasmolyzing solution, which has penetrated the wall but not the membrane of the cell.

and continued plasmolysis will itself cause death. This principle is employed when common salt is used to kill plants. Heavy applications of salt or soluble fertilizers form concentrated solutions around plant roots, and the roots are then killed by plasmolysis (FIG. 4-10).

The behavior of a plasmolyzed cell should be distinguished from that of a cell in a wilted plant. In a plasmolyzed cell the protoplast collapses and detaches from the cell wall and the space between them is filled with plasmolyzing solution that has diffused through the wall. In the cells of a wilted plant, no external solution is present to fill the space formerly occupied by the protoplast. As a result, thin cell walls, as of parenchyma, will collapse with the protoplast. If the wall is sufficiently rigid so that it does not collapse, the tissue will not show wilting.

Imbibition In addition to osmosis, the process of IMBIBITION may be involved in the movement of water in plants. This term is applied to the absorption of water by absorbing surfaces such as those of cellulose and starch. The inward movement of water is accompanied by a swelling of the tissue and a gain in weight. An example of this

Fig. 4-10 Wilting in tomatoes caused by excessive applications of a soluble fertilizer. (Left to right): 1.0 gram, 4.9 grams, and 19 grams of ammonium sulfate applied to tomatoes in 4-inch pots.

effect is the swelling of starch or gelatin when water is added. The increase in volume of seeds in water or moist soil and the swelling of dry fruits are due in considerable part to imbibition. Imbition by wood causes doors and windows to stick in damp weather. A board on the ground tends to curl upward, or warp, because of imbibition by the cellulose in contact with moist soil. A dry wooden boat or tub will usually leak, but after soaking for some time, the wood swells and closes the cracks. Enormous forces can be developed by imbibition. Building stone has been quarried since prehistoric times by driving wooden pegs into rows of holes drilled in rock. The swelling of the pegs when water is poured over them can split off massive blocks of stone without shattering them as explosives would.

In osmotic absorption of water, the concentration and diffusion pressure of water within the cell are lowered because of dilution by the dissolved substances in the cell. In imbibition, no membranes are involved, as in osmosis. Imbibition occurs because the surfaces within the submicroscopic structure of cellulose, starch, protein, and other materials attract and hold water molecules by intermolecular forces. These surfaces therefore become coated with a film of adhering water molecules, held so strongly that they are no longer effective in diffusion, and more water diffuses in. This action of intermolecular forces at the surfaces of solids is called ADSORPTION. It results in an increased concentration but a decreased diffusion pressure of the adsorbed molecules. These layers of adsorbed water tend to push apart the surfaces of the absorbing particles and hence to cause swelling.

The cellulose cell wall can imbibe water whether the cell is alive or dead, and the wall must imbibe at least a certain quantity of water before water can pass through it. When dry seeds are placed in moist soil, they take in water by imbibition. This moisture activates enzymes, which begin to digest starch to sugar. It also brings about the reactivation of the protoplasmic membranes, and soon imbibition and osmotic absorption are acting together in the germination of the seed. When new protein is formed in meristematic cells, each protein molecule adsorbs hundreds of water molecules, thus contributing to the growth of the cell. This type of imbibition, frequently called *hydration,* is most important in very young cells, whereas osmotic absorption is more characteristic of vacuole formation in rapidly enlarging, older cells.

PERMEABILITY

No problem in cell physiology is more challenging than the determination of the nature of the processes involved in the movement of materials into and out of living cells. This subject has received the attention of many investigators, but many aspects of it are still poorly understood. Among other features, the problem involves the permeability of cell membranes. Various observations and experiments show that a membrane, the plasma membrane, is the outer limit of the protoplast (FIG. 4-9B), and that another membrane surrounds the vacuole. The latter is referred to as the vacuolar membrane, but it is a part of the protoplast and not of the vacuole.

Fig. 4-11 **The permeability of cell membranes. (Left): the membranes of living red beet cells hold the anthocyanin pigment within the cells. (Right): the membranes have been destroyed by heating and the pigment has diffused out.**

These membranes are fragile, extremely thin structures, each consisting of lipid substances and proteins. These lipids and proteins may be arranged either as three layers, the outer two being lipid and the middle layer protein, or as protein globules surrounded by lipids. The arrangement of the components varies with the kind of membrane. The term permeability refers to the properties of these membranes that determine the ease with which substances pass through them. They may be impermeable to the movement of some substances and freely or slowly permeable to others. The plasmolysis experiment cited above (Fig. 4-9) indicates that cell membranes are normally much more permeable to water than to sugar. Typically, the protoplast membranes of higher plants are readily permeable to water, slowly permeable to many salts (chiefly in the form of ions), and impermeable or nearly so to sugars, amino acids, and other organic compounds. Slices of red beets have been soaked in water for weeks without the loss of a detectable quantity of red anthocyanin pigment from the uninjured cells. If these slices are dropped in boiling water, however, so that the membranes are destroyed, the red color diffuses rapidly from the tissue (Fig. 4-11). In general, any injury to the protoplast, as by poisons, heating, freezing, or drying, increases the permeability of the membranes. The principle is used in extracting the sugar from sugar beets. Finely ground beet particles contain many intact living cells, the membranes of which still hold the sugar against extraction. The ground pulp is therefore heated just enough to destroy the membranes and sugar is then extracted with water. Heating must be carefully controlled to prevent pectins and similar materials from being solubilized. These substances interfere with the crystallization of sucrose from the extract and reduce the yield.

Plasmodesmata   If cell membranes are generally impermeable to foods, even in solution, the question arises at once of how foods move between cells. The answer seems to be that adjoining, living cells are connected by protoplasmic threads that pass through minute pores in the cell wall. These threads are known as PLASMODESMATA (singular, plasmodesma), a term derived from the Greek *desmos*, meaning a bond or connection. The plasmodesmata may be clustered into groups or generally distributed over the cell wall (Fig. 4-12). The threads of protoplasm connecting adjoining cells are so delicate that they cannot be seen at the magnification of the usual student microscope, and they are commonly visible at higher magnifications only after special treatments. The diameter of a thread is probably no greater than 0.5 micron (see Glossary) and it may be less than 0.1 micron. These small connections, however, are hundreds of times greater in diameter than typical molecules, so that sugars and even large protein molecules can diffuse through them. It should be noted that whereas the plasmodesmata connect the cytoplasm of adjoining cells, the vacuoles are isolated by the vacuolar membranes.

The presence of these threads allows the multicellular plant to be viewed as an organic whole rather than as a community of isolated cells. The role that plasmodesmata play in the life processes of the plant is not fully understood. It is generally held, however, that stimuli may be transmitted through plasmodesmata and that through them enzymes and foods in solution may move from cell to cell. The pores in the cell wall are, however, too minute to allow the passage of such relatively large cellular bodies as the nuclei and plastids. Plasmodesmata are probably most important as the normal pathway for the movement of organic substances between adjoining cells within the plant. Such movements could occur

Fig. 4-12 Plasmodesmata (P) connecting adjoining cells of shoot apex of *Selaginella emmeliana* (48,000×). (Lower right): longitudinal section showing continuity of plasma membranes through the pores that penetrate the cell wall. Upper left: surface view of the plasmodesmata in a fold of the wall.

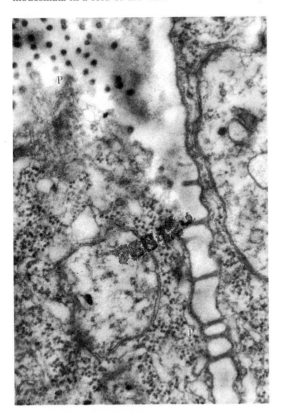

through plasmodesmata without the necessity of passage through surface membranes with their normally low permeability or impermeability to organic compounds.

Active Transport In plant cells the plasma and vacuolar membranes constitute barriers to the free movement of ions and molecules into the cell or vacuole. Yet these entities do move across membranes, and various mechanisms may be involved in this movement. One of these is diffusion; another, active transport.

Active transport involves the use of respiratory energy in accumulating unchanged ions or molecules into the cytoplasm or vacuole at concentrations several to many times higher than in the environment. A diagram illustrating the nature of the hypothetical process is shown in FIG. 4-13. The first step is believed to be the binding of the substance to be transported, called the substrate (S), at a site on the outer surface of the plasma membrane. At this point a carrier (C), presumably a protein specific for the material transported, combines with the substrate, forming a carrier-substrate complex CS. Because CS is formed on the outer surface of the membrane, its concentration builds up and CS diffuses across the membrane on a gradient of decreasing concentration. On the inner surface of the membrane, the substrate (for example, a potassium ion, K^+), is released into the cytoplasm of the cell, where it is held. The carrier then picks up new energy from an ATP molecule in the cytoplasm and moves back by diffusion to the outer surface, where it can bind a new potassium ion and repeat the process. It is presumed that the substrate cannot cross the membrane or that it crosses at a negligible rate unless linked to the carrier molecule. While ATP energy is required for active transport, its precise role is not known; it is possible that it functions in binding carrier and substrate.

Active transport has been studied extensively in animals and microorganisms. It is considered to be important in bacteria and in a variety of

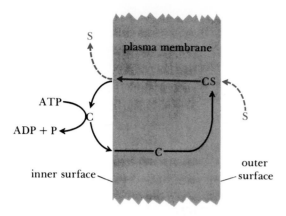

Fig. 4-13 Diagram of postulated mechanism of active transport. The transported substance S, combines with C and is carried across the membrane and released into the cytoplasm of the cell.

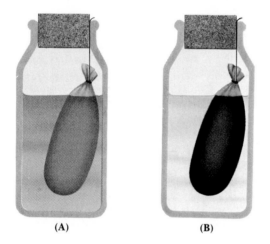

(A)   **(B)**

Fig. 4-14 Chemical accumulation. A membrane filled with water (A), and one filled with starch paste (B), were placed in dilute iodine solutions. Iodine diffused inward and reached equilibrium in (A). In (B) the iodine reacted with starch and was removed from solution. As a result, essentially all of the iodine diffused in and accumulated in membrane (B).

animal cells and tissues, such as those of the digestive tract and kidney. In plants, active transport may bring about accumulation within the vacuoles. Vacuolar accumulation involves a cytoplasmic layer and two membranes, the plasma membrane and the vacuolar membrane. In seaweeds, concentrations of potassium 30 times that of sea water are fairly common and iodine concentrations of 30,000 times are reported. This is largely accumulation within the vacuole. That ions accumulate in vacuoles of root cells is indicated by an experiment in which roots of barley plants were allowed to absorb radioactive bromine while the plants were in darkness and a high humidity. Under these conditions, quantities of the radioactive bromine accumulated in the roots, but very little of it moved to the tops of the plants. The roots were then rinsed, transferred to a solution containing ordinary bromine, and the plants moved to a dry location with bright light. Bromine then moved rapidly to the tops of the plants, but this movement was almost exclusively of ordinary bromine, while the radioactive bromine remained in the vacuoles of the root cells. This experiment suggests that active transport into roots is primarily into the vacuole and may result in storage rather than in normal transport through the roots to the shoot.

Chemical Binding Accumulation need not necessarily result from active transport. It may also follow chemical reactions in the cell or vacuole, and it is difficult to distinguish between this chemical accumulation and accumulation by active transport. In chemical binding, ions or molecules within the vacuole may be incorporated into insoluble substances or substances to which the membrane is impermeable. A reaction of this type may be demonstrated by the accumulation of iodine within a membrane filled with starch paste (FIG. 4-14). The membrane is placed in a stoppered bottle containing a dilute solution of iodine. The iodine, as it diffuses in, forms a dark blue starch-iodine complex that is held within the membrane. As a result the external solution may become almost free of iodine. An example of chemical accumulation is found in nature when nitrate nitrogen (NO_3^-) enters a cell and is used in the synthesis of amino acids. The concentration of nitrate is thus kept low and nitrate absorption continues. The amino acids are held or used within the cell and are accumulated.

COLLOIDS

Colloids, better termed *colloidal systems*, are of many different kinds, but all can be described in the same general terms—particles larger than typical molecules, called the *dispersed phase*, are scattered throughout a *dispersing medium*. Sugar in water forms a solution in which sugar molecules are scattered among the water molecules. Glue in water, however, constitutes a colloidal system in which microscopic clumps of glue molecules are scattered in the water. Glue may be considered a type of colloid. The word, in fact, is derived from the Greek *kolla*, meaning glue.

Because the molecules in a colloid are massed in groups that contain from a few hundred to many millions of molecules, they do not show the chemical reactions typical of solutions. Instead, they exhibit special reactions, known collectively as surface chemistry. The total amount of surface in a colloidal system can be astonishingly large. If, for example, a taw marble having a surface area of 1 square inch were ground into pieces the size of very small colloidal particles, its surface area would be increased to 70,000 square feet, or 10 million times its original size.

Colloidal particles are usually small. Typical particles range in size from 1 millimicron ($m\mu$) to 200 or 500 millimicrons in diameter. A millimicron is 1/25 millionth of an inch, and 500 millimicrons is 1/50 thousandth of an inch, a particle size barely visible in a laboratory microscope. A sphere 500 millimicrons in diameter could, however, contain millions of carbon atoms. A very small colloidal carbon particle might contain only 300 carbon atoms, but such a particle would have a relative weight of 3600, compared with that of a single carbon atom, taken as 12. In contrast, a single small protein molecule may have a relative weight of 36,000 and large protein molecules may have weights in the hundreds of thousands. Such molecules can be in true solu-

tion and still show pronounced colloidal properties because of their size and relatively great surface.

Colloidal particles may be of different sizes and shapes, and may be organic or inorganic. Water is an important dispersing medium, but both the colloidal particles and the medium may be solid, liquid, or gas. Among colloidal systems of biological interest are rubber latex and soil colloids. Much of the soil humus and many of the finer fractions of the soil clays are colloidal, and may exist as thin gelatinous films around the larger soil particles. The soil colloids greatly affect the physical and chemical properties of soils and their fertility. Emulsions, which consist of two immiscible liquids, such as oil in water or water in oil, are common. Examples of emulsions are homogenized milk, oleomargarine, mayonnaise, and pharmaceutical preparations such as lotions, creams, and ointments.

Stabilization of Colloids Small colloidal particles have enough motion as a result of their own kinetic energy to remain in suspension with little or no tendency to settle out. If these small particles, however, clump together and form large ones, the colloidal structure will be lost. Two general mechanisms prevent such clumping. In one, the colloidal particles become electrically charged and repel each other. In most biological colloids the dispersed particles adsorb layers of water that prevent direct contact with each other. Such colloids are called *hydrophilic*, or water loving. This water helps to keep protein particles from sticking together and coagulating. Its importance is illustrated by the effect of rapid drying on seed corn. If freshly harvested corn containing more than about 40 percent water is dried rapidly, it can be killed, even under moderate temperature. If, however, the grain is dried slowly, *all* of the water in the grain can be removed without reducing germinating capability. It is believed that rapid removal of the water that coats the active proteins of the immature grain causes the proteins to coagulate and so kills the seed. With slower drying, other coating materials typical of mature seeds can be deposited on the proteins and protect them when the water is removed.

Structure of Cytoplasm Cytoplasm was once regarded as a complex colloidal system. With the advent of the electron microscope and other advances in technique, this concept has been modified—cytoplasm cannot be regarded as a colloid such as gelatin in a dish. While it may exhibit the physical properties of colloids, superimposed upon these are other features, including structures, activities, and complicated interrelationships, all of which involve an *organization* not found in simple colloids. Cytoplasm may be considered under two heads: (1) the organelles, and (2) the ground cytoplasm, the substance that surrounds the organelles.

In considering the organelles we may amplify the account of these bodies presented in Chapter 3, with emphasis upon membranes. In the simplest kinds of cells, such as those of bacteria, there is only one membrane, the plasma membrane. The cells of most organisms, however, are characterized by a multiplicity of membranes, and much of the chemical machinery of the cell is localized in membrane-bounded structures. Most organelles are surrounded by a single membrane, but the nuclear, plastid, and mitochondrial membranes are composed of two parallel membranes. Many of the membranes undergo structural change, some to the extent that they may break down and be reconstituted. Notable here is the nuclear membrane, which in interphase of mitosis seems a rugged structure, but it disappears and reappears each time the cell divides.

The function of some organelles is clear, that of others not well understood. Some organelles may play several roles in the life of the cell. The plasma membrane, for example, not only regulates the passage of materials into and out of the cell, but also appears to be the site of numerous enzymatic processes. The endoplasmic reticulum (FIG. 4-15) is another organelle that may become disrupted and reassembled. In addition to serving as an intracellular transport system, it has been implicated in the formation of the nuclear membrane, the cell plate, and plasmodesmata. It has occasionally been observed to be continuous with the outer nuclear membrane.

The function of the dictyosomes (FIG. 3-3) is just beginning to be understood. These bodies, composed of varying numbers of flattened or curved plates, occur in higher plants in numbers from 30 to several hundred per cell. They are thought to transform and accumulate materials and to secrete them into the cellular environment. Individual dictyosomes are usually accompanied by budlike, membrane-bounded vesicles (droplets), and it is considered that these form, separate, and carry the products to other parts of the cell. Dictyosomes have sometimes been found continuous with the endoplasmic reticulum.

The mitochondria (FIG. 4-15) vary in number from one per cell (in certain unicellular organisms) to several thousand; in general perhaps 500–800 occur per cell. They are composed of protein (about 75 percent), lipids, and small quantities of nucleic acids. New mitochondria are formed by division of preexisting ones or by a budding process. These important organelles will be discussed further in Chapter 6.

Much of the RNA in the cytoplasm is located in minute particles, the ribosomes (FIG. 4-15). These particles, composed of RNA and protein, have no distinctive membranes, and are among the smallest of the known organelles. They may occur in aggregates, called POLYSOMES, held together by a single strand of ribonucleic acid. They are found not only in the cytoplasm but also in the chloroplasts and mitochrondria. The role of the ribosomes in the life of the cell is considered in the next section.

At interphase the microtubules are most abundant just within the plasma membrane of meristematic cells. It has been suggested that they determine the direction and path of movement in cytoplasmic streaming. They largely compose the mitotic spindle fibers, are involved in the formation of the cell plate, may play a role in the depo-

Fig. 4-15 Structure of the protoplasm in a cell of the apical region of *Selaginella emmeliana*. E: endoplasmic reticulum. M: a mitochondrion. N: nucleus. W: cell wall. R: ribosomes (black dots throughout the cytoplasm). (44,000×)

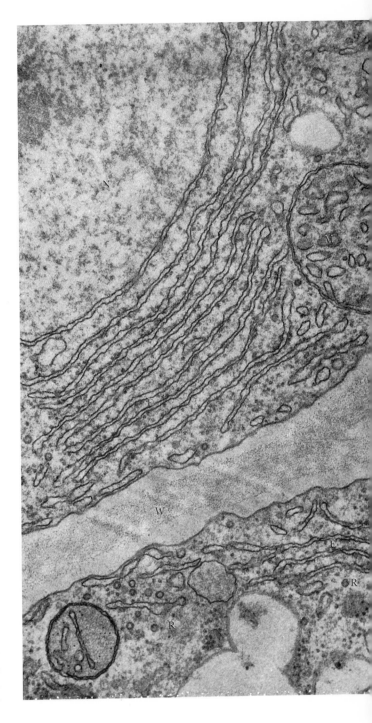

sition of the secondary cell wall, and no doubt have other functions.

Among other materials the ground cytoplasm contains water, solutes (both organic and inorganic), amino acids, ribonucleic acids, and granular materials that are probably chiefly protein in composition. There is no doubt a close interaction and exchange of materials between the ground cytoplasm and the organelles. Although ground cytoplasm under the light microscope appears to be homogeneous, it is really complex and heterogeneous, and is believed to possess a degree of organization much greater than could be found in a nonliving colloidal system.

Even though certain activities within the cell are localized within particular organelles or in the ground cytoplasm, nevertheless the components of the cell operate in harmony—their reactions are regulated and orderly. This is largely due to enzymes, present as particles or held on membranes. The cell, then, is a dynamic living system, an elaborate and complicated factory in which hundreds, perhaps thousands, of machines—the enzymes—perform exacting tasks in specific places, in a well-integrated assembly-line fashion.

PROTEINS

The general nature of proteins was discussed in Chapter 3. They are usually long chains of amino acids joined by peptide bonds (FIG. 4-16), and are folded into elongated (fibrillar) or globular molecules. They have molecular weights varying from about 13,000 to 500,000 or more, compared to 12 for the carbon atom. A PEPTIDE is a substance of lower molecular weight, composed of two to many amino acids linked by peptide bonds. A long peptide chain may be termed a POLYPEP-

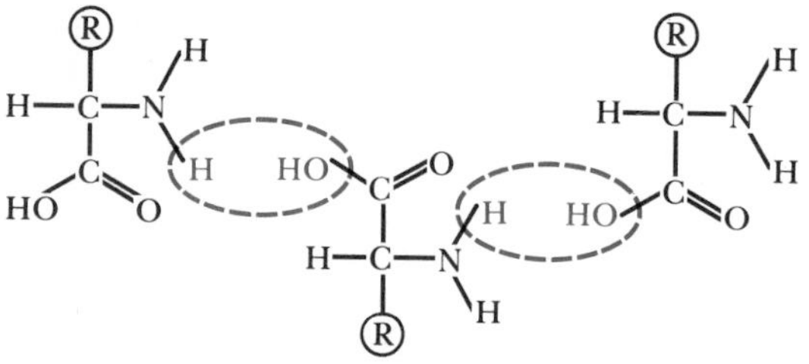

Fig. 4-16 Peptide bond formation. Amino acids are linked in peptide and protein molecules by the elimination of a molecule of water, (H_2O) in the circled areas. A covalent ("peptide") bond is then formed between the N of one amino acid and the terminal C of another. The circled R's represent the specialized side chains that distinguish one amino acid from another.

TIDE, and some of these are large enough to be considered proteins.

Among the various types of proteins we may distinguish (1) structural proteins, considered to be a general part of the protoplasm and especially concentrated in the multitude of membranes revealed by the electron microscope; (2) storage proteins, particularly abundant in seeds; (3) lipoproteins, combinations of lipid and proteins, characteristic of cell membranes; (4) nucleoproteins, combinations of nucleic acids with proteins. The chromosomes are composed of nucleoproteins and consist of double strands of nucleic acid, chiefly DNA, covered by proteins. These proteins, in addition to other functions, may protect the DNA molecules from enzymes that might break them down. (5) Enzyme proteins, the working systems of the cell, merit particular attention.

Enzyme Proteins   Enzymes are organic catalysts that accelerate chemical reactions by many times so that reactions that would otherwise be too slow to be biologically useful become a practicable part of cellular metabolism. The catalytic action of an enzyme operates by lowering the energy required for activation of the substrate. Energy of activation is the energy required by a molecule before it can enter into a reaction. If this required energy is lowered, which may occur when enzyme and substrate combine, the reaction can proceed. The addition of stored energy of adenosine triphosphate (ATP) to the enzyme-substrate complex can supplement the effect of the enzyme in lowering the required energy of activation.

The combination of enzyme with the substrate forms an intermediate complex, which breaks down and yields the reactive products and the enzyme. The velocity of enzymatic reactions is affected by, among other factors, temperature, pH, and the concentration of enzyme, substrate, and products. Enzymes can operate at very low concentrations, but the rate of reaction increases with enzyme concentration until the concentration of accumulated products or other factors becomes limiting. An increase in temperature can speed enzymatic reactions, but enzymes and other proteins may be destroyed by temperatures above 50°–60°C (122°–140°F). Conversely, enzymatic reactions are slowed by low temperatures, as are other chemical processes. The practical effects of low temperatures upon enzymatic reactions in ripening fruits and other foods and in decay organisms are considered in Chapter 6.

THE SYNTHESIS OF PROTEINS

It is now generally accepted that the kinds of proteins, particularly the enzymes, produced in the living cell determine in large degree the structure and activities of a cell or of a multicellular organism. How, then, are these proteins produced? How is their nature specified? There is convincing evidence that the instructions for these activities are carried by genes, constituting parts of DNA molecules, and that these operate by determining indirectly the arrangement, kind, and number of amino acids in a polypeptide chain. The fundamental aspect of gene activity, then, is the production of proteins, many of which are enzymes. The enzyme proteins, in turn, catalyze biosynthetic reactions, and the products of these reactions are used by the organism or become a part of it. Variations in gene activity can therefore result in changed proteins and subsequently in differences in the appearance of tissues or in the reactions that occur in them.

Structure and Function of DNA The DNA molecule is the information-storing structure of the chromosome. It is composed of units called nucleotides. Each nucleotide, in turn, is composed of one of the nitrogen bases ADENINE, THYMINE, CYTOSINE, or GUANINE (FIG. 4-17), combined with a molecule of deoxyribose sugar and a low-energy phosphate group, similar to AMP in FIG. 4-3. The DNA molecule usually consists of very long chains, two of which are twisted about each other in a regular double helix, or spiral (FIG. 4-18), with about ten nucleotides to each turn of the helix. Each chain consists of many nucleotides, the two chains being bound together

Fig. 4-17 The base pairs of DNA showing hydrogen bonding (colored lines). "Sugar" indicates the bond to deoxyribose. A nitrogen base plus a molecule of deoxyribose and a phosphate constitute a nucleotide (see Fig. 4-19).

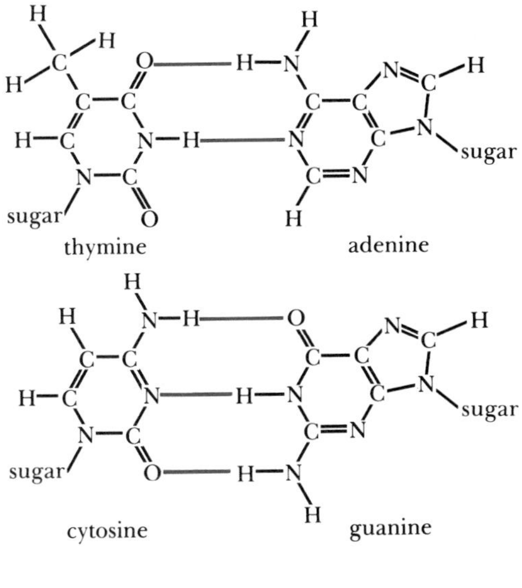

by hydrogen bonds between the pairs of bases (FIG. 4-19). Thymine (T) and adenine (A) are joined by two hydrogen bonds, while cytosine (C) and guanine (G) are joined by three. The bases are therefore always paired in the same way, A=T T=A and C≡G or G≡C. It is apparent from these arrangements that the two polynucleotide chains of the DNA strand are not identical. Instead, one is the complement, or reverse, of the other. The general structure of the DNA molecule and strand was revealed in 1953 by J. D. Watson, an American, F. H. C. Crick, an Englishman, and M. H. F. Wilkins, a New Zealander, who in 1962 shared a Nobel Prize for their work and began one of the most important scientific advances of modern times.

The DNA molecule appears to be responsible for only two operations. (1) During interphase it produces molecules that are duplicates of the two original chains. These duplicates, which are themselves arranged in double spirals, are then separated to different daughter cells at mitosis. Duplication of the DNA molecule is believed to involve separation of the hydrogen bonds of the double strand. As each nitrogen base is uncovered in this separation, it takes up a *complementary* base from the surrounding cytoplasm (FIG. 4-20). Thus, on one strand, A joins with T, or T with A, and C with G or G with C. The newly attached bases, with sugars and phosphates joined to form a "backbone," constitute a new chain that is a complement of the one on which it is formed and a duplicate of the other chain of the original structure. The same process on the second DNA chain results in the duplication of the double strand. (2) The DNA molecule acts as a template in the production of the very closely related ribonucleic acid, or RNA, which moves into the cytoplasm, carrying genetic information copied from the DNA (FIG. 4-21).

Structure and Function of RNA We come now to the mechanism by which the DNA controls the sequence of amino acids in proteins. As noted above, the genetic information of DNA is trans-

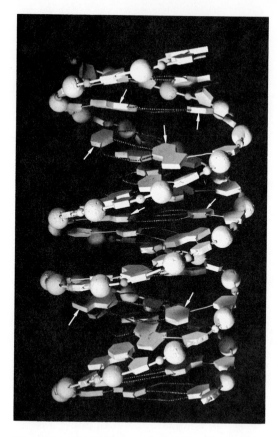

Fig. 4-18 A model of a small portion of a double-stranded spiral of DNA. The strands of the spiral are connected by hydrogen bonds between nitrogen bases, some of which are indicated by arrows.

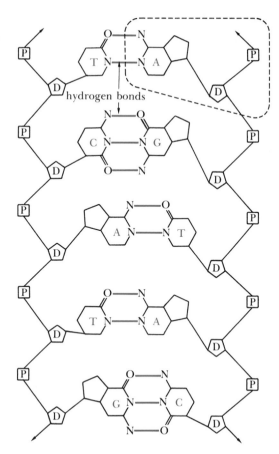

Fig. 4-19 Diagram of a short section of a DNA strand. Colored lines indicate hydrogen bonds that hold the two chains together, or separate when the strand is duplicated. D indicates the deoxyribose and P the phosphate molecules that join the bases in a DNA chain. The dashed lines in the upper right encircle an adenosine nucleotide, showing how the nitrogen base, adenosine (A), the sugar (D), and the phosphate (P), fit into the DNA strand. The diagram shows the reversed, complementary nature of the two chains.

ferred to another group of nucleic acids, ribonucleic acids or RNA. This group differs generally from DNA in three particulars; it has no thymine, but instead the nearly identical uracil (U); the sugar of RNA is ribose instead of deoxyribose; and RNA occurs and functions as a single strand instead of the double strand of DNA. There are three major types of RNA molecules: (1) messenger-RNA (mRNA), which carries the genetic information of DNA to the protein-building sites in the cytoplasm, the ribosomes; (2) ribosome-RNA (rRNA) molecules, which are important structural materials of the ribosomes; and (3) smaller, transfer-RNA

(tRNA) molecules, which carry specific amino acids to the correct positions on the growing polypeptide molecule.

Messenger-RNA The mRNA molecules are made on DNA, which serves as a template. It is visualized that, in local areas, the double strands of the DNA separate temporarily under the

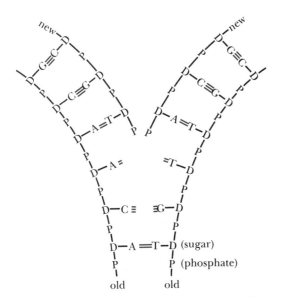

Fig. 4-20 Duplication of the DNA molecule. Coiling is not shown. As the strands separate with separation of the hydrogen bonds, the bases attract new complementary bases and new strands are formed with sugar-phosphate backbones. Each new strand is thus identical to a strand of the original helix.

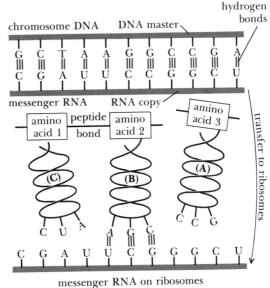

Fig. 4-21 Diagram of the formation and action of messenger-RNA; mRNA is believed to be formed in contact with DNA. It then separates, moves from the nucleus to the cytoplasm, and unites with a group of ribosomes. In this position the mRNA molecule serves as a template on which smaller, transfer-RNA molecules carrying specific amino acids become aligned. The matching of the triplet of exposed bases on the tRNA with the complementary triplet code of the mRNA ensures the proper order of amino acids, and a polypeptide results. (A): tRNA molecule approaching the mRNA. (B): tRNA bonded to the mRNA. (C): tRNA leaving the mRNA after amino acid release.

influence of an enzyme, and that in these regions a complementary RNA strand is synthesized, utilizing RNA nucleotides from the cytoplasm. Because the two chains of the DNA strand are complementary rather than identical, it is considered that only one chain of the pair acts as a template. In forming the complementary RNA single chain, the adenine of the DNA must pair with uracil, which takes the place of thymine (FIG. 4-21). Other complementary base pairs are formed as in the replication of the DNA molecule. (FIG. 4-20). The RNA molecules, once formed, separate and are moved to ribosomes in the cytoplasm. The genetic information now transcribed on the mRNA is the reverse of the DNA from which it was copied, just as the printing you are reading is the reverse of the printer's type but still carries the same information.

Within the cytoplasm the mRNA molecule combines with one or more ribosomes, and serves as a template on which the polypeptide molecule is formed. The ribosomes (FIG. 4-15), consisting of RNA and protein, are the sites of protein formation.

Transfer-RNA Although the mRNA, functioning as a template, provides for the arrangement of the amino acids in the polypeptide chain, there remains the problem of obtaining the proper amino acids from the cytoplasmic pool. This is accomplished by the tRNAs. These are short chains, believed, like the mRNA, to be formed on one chain of the DNA strand as it is opened up in local areas. The small tRNA molecule is probably folded back on itself with the ends coiled together. Where the molecule is folded, a three-base sequence is exposed (Fig. 4-21). This sequence can interlock with the complementary sequence on an mRNA molecule.

The joining of an amino acid with a tRNA molecule requires the action of a specific enzyme and ATP energy to activate the amino acid. The amino-acid tRNA complex then moves to a position on, and locks to, the mRNA ribosome group, this position being determined by the location of a base sequence on the mRNA that is complementary to the exposed bases of the tRNA (Fig. 4-21). As a ribosome moves along an mRNA molecule, a growing polypeptide chain is formed. The amino acid in a tRNA complex is joined to the exposed end of the polypeptide chain by enzyme action, the amino acid is then separated from the tRNA, and the tRNA is separated from the mRNA. The ribosome then moves along the mRNA until the next amino-acid tRNA complex is brought into a code position, and the process is repeated, adding the next amino acid to the polypeptide. The polypeptide molecule grows in this fashion until the end of the mRNA is reached. The polypeptide may be a small protein or it may combine with other polypeptides to form a large protein molecule.

THE GENE AND THE GENETIC CODE

We have seen that the sequences of bases on the DNA molecule constitute the ultimate key to the assemblage of amino acids into a protein. This sequence is thus a code that determines amino-acid sequence. The problem, then, is to spell out the thousands of enzymes that may be formed in cells with the four nitrogen bases A, T (or U), C, and G. If a four-letter alphabet seems restricted, consider that modern computers, with their enormous speed and capacity, use only two. Since about 20 different amino acids are present in typical proteins, it is obvious that one of these letters (bases) cannot specify a single amino acid.

Neither can two letters, for this provides for only 16 (4×4) combinations. With three bases, however, 64 ($4 \times 4 \times 4$) combinations may occur, and three bases are in fact believed to form the code for a given amino acid. Thus, although four bases are available, only three are used at any one time. With 20 amino acids and 64 base combinations, more than one combination may code for a single amino acid.

The gene is considered to be a part of a DNA molecule that contains a sufficient number of three-base sequences, or triplets, to code for the formation of a polypeptide chain. The length of a gene cannot be specified, for it varies from one gene to the next. It has been calculated, however, that an average gene is composed of about 300 coding triplets, that is, 900 nucleotides. The frequently used term GENETIC CODE may now be defined as the sequence of base triplets that specifies the kinds, number, and arrangement of the amino acids of a polypeptide. For example, GAU specifies the amino acid aspartic acid; UCG, the amino acid serine; and GGC, the amino acid glycine (Fig. 4-21).

It is considered that the code within a gene is always read in one direction and without interruption. This means that the substitution of one base for another in the gene could result in the placement of a different amino acid in the polypeptide chain, often changing the properties of the protein. The addition or deletion of a base could result in a completely changed reading within the gene and the formation of a grossly abnormal polypeptide chain. Such alterations of the genetic code are, at least in part, responsible for mutations, discussed in Chapter 17.

DNA in Chloroplasts and Mitochrondria It was once believed that all of the DNA of the cell is located in the nucleus, but it has now been established that double-stranded DNA also occurs in the plastids (Fig. 4-22) and mitochrondria. Here it is found in the form of fibrils, distributed within the matrix of the organelles. This DNA appears to replicate and be distributed to daughter proplastids or mitochondria following

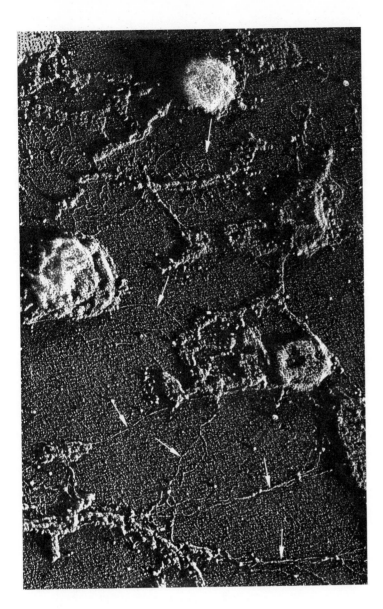

Fig. 4-22 Electron micrograph (30,000×), chloroplast of spinach disrupted by osmotic shock. Fibrils of DNA (arrows) are associated with the grana and the intergranum lamellae (Fig. 5-13).

division of the parent organelle. It is not, therefore, derived from the DNA of the nucleus, and there is some evidence that the DNA of both kinds of organelles may be synthesized in the absence of the nucleus. The DNA of both organelles is specific for these bodies; that is, it differs chemically and physically from the nuclear DNA of the cells in which the organelles occur.

Considerable evidence supports the belief that the plastids and mitochondria also contain mRNA, and that proteins are synthesized by the operation of the same kind of information transfer system that exists between nucleus and cytoplasm. This, at first sight, would seem to allow these organelles to be independent of the nucleus as far as protein synthesis and inheritance are concerned. There is evidence, however, that some of the enzymes and other proteins of the organelles are encoded by nuclear genes, and are probably synthesized in the cytoplasm and then transferred to the organelles. The plastids and mitochondria, then, are only semiautonomous with respect to protein synthesis, and the usefulness to the organism, in part at least, may be to increase the rate at which essential proteins are synthesized. Much remains to be learned concerning the integration between organelle-localized synthetic systems and those of the nucleus. Whatever this relationship may be, there is little question that genes of the organelles are inherited independently of nuclear genes, giving rise to what has been termed nonchromosomal or CYTOPLASMIC INHERITANCE.

The discovery of a protein-synthesizing mechanism in the plastids and mitochondria has been used to support a speculative concept regarding the evolutionary origin of these organelles. Briefly, according to this, chloroplasts and mitochrondria were once free-living cells, the plastids being comparable to certain living one-celled blue-green algae, and the mitochondria resembling living bacteria. Certain larger cells, perhaps amoeboid, ingested these free cells, which were not destroyed but entered into an obligate mutualistic relationship with the host. In the course of time, the invading entities lost a portion of their synthesizing capacities, which were taken over by the DNA of the host.

SUMMARY

1. The structure and reactions of plants are based upon atoms of many kinds. Atoms are the smallest units of a chemical element.

They are composed of a nucleus containing positively charged protons and uncharged neutrons. The outer part of the atom contains negatively charged electrons in a number equal to the number of protons.

2. Atoms combine to form molecules, which are held together by chemical bonds. These bonds may be ionic, in which electrons are transferred from one atom to another, or covalent, in which electrons are shared and become a part of both atoms. A special, weak bond, the hydrogen bond, is important in biology because of its effect on the properties of water, proteins, and the nucleotide chains of DNA.

3. Oxidation consists of the loss of electrons, typically, in biological reactions, by the removal of hydrogen atoms with their electrons, or by the addition of oxygen atoms, which are deficient in electrons. Reduction consists of the gain of electrons. The two reactions are always coupled.

4. Adenosine triphosphate is the best-known and most important carrier of available chemical energy in cells. The energy of ATP or comparable energy is used in many of the chemical reactions of cells.

5. Matter takes the basic forms of gases, liquids, or solids, and the mixed forms of solutions and colloidal systems. Solutions in water and hydrophilic, collidal systems are of major importance in biological materials.

6. Molecules or other particulate forms of matter tend to spread uniformly through available space because of their energy and constant motion. This spreading is called diffusion.

7. Water evaporates because some of the most rapidly moving molecules escape through the surface of the liquid into the air. Evaporation increases rapidly with temperature, approximately doubling with each temperature rise of 10°C (18°F). Water may be diluted by dissolved materials and such dilution reduces slowly the tendency of water to diffuse or to evaporate.

8. The diffusion of molecules is determined by their concentration and by the effects of temperature and pressure. The combined effect of these factors results in the diffusion pressure of any particular kind of molecule.

9. The diffusion of water across a differentially permeable membrane from a region of higher to one of lower diffusion pressure of water is called osmosis. If a wilted cell is placed in water, water will diffuse into the cell and build up a turgor pressure within the cell.

10. The dilution of water within a cell tends to cause water to diffuse into the cell. Turgor pressure tends to prevent this movement. The water-absorbing power of a cell is determined by the extent to which the dilution of water within it is not balanced by turgor pressure.

11. If a cell is placed in a solution with a high sugar content, water will diffuse out of the protoplast. As a consequence of this loss of water the protoplast will shrink away from the wall and the cell will be plasmolyzed.

12. Imbibition is due to the adsorption of water molecules onto the internal surfaces of wood and other substances. It is important in the wetting and swelling of wood and dry seeds.

13. The membrane that surrounds the protoplast is readily permeable to water, slowly permeable to dissolved salts, and nearly impermeable to sugars and other foods. If a cell is killed, as by heating, the membranes are destroyed and sugars, anthocyanins, and other substances will diffuse readily from the cell.

14. Foods may move between the protoplasts of adjoining cells through protoplasmic connections that pass through small pores in the walls. These connections are known as plasmodesmata. Movement into cells, and particularly into the vacuoles of cells, may be by active transport.

15. Colloids consist typically of a dispersed phase of small particles in a continuous phase, as the fat and proteins of milk dispersed in water.

16. Colloidal particles are very small but each may contain millions of molecules. Because of the small size of the particles, colloids may exhibit enormous surface effects. Protein molecules are so large that individual molecules show many of the properties of colloidal particles.

17. Proteins show combinations of solution chemistry and colloidal or surface chemistry that enable them to form the complex structures and the numerous enzyme systems of protoplasm.

18. Cytoplasm is a system of enormous complexity. It contains protein molecules, protein structures, fatty materials, sugars, and minerals, together with some 90 percent of water. The structure and reactions of this system constitute the qualities characteristic of living organisms.

19. Enzymes are specific proteins that speed up specific chemical reactions within a cell. It is estimated that there may be several thousands of kinds of enzymes in a single cell.

20. Chromosomes consist of long helical, double DNA chains coated with protein. Genes are sections along the DNA chains of the chromosomes. The information necessary to determine the sequence of amino acids in polypeptides is contained in the genetic code. The code consists of groups of three letters, each letter representing a nucleotide base located on a DNA molecule.

21. Protein formation starts with the production of messenger-RNA on a DNA chain. Messenger-RNA moves from the nucleus to the cytoplasm and becomes the template on which a specific protein is built. Transfer-RNA brings the amino acids to the template, where they are incorporated into the protein. The enzymatic proteins of the cell regulate the innumerable chemical reactions that occur in it.

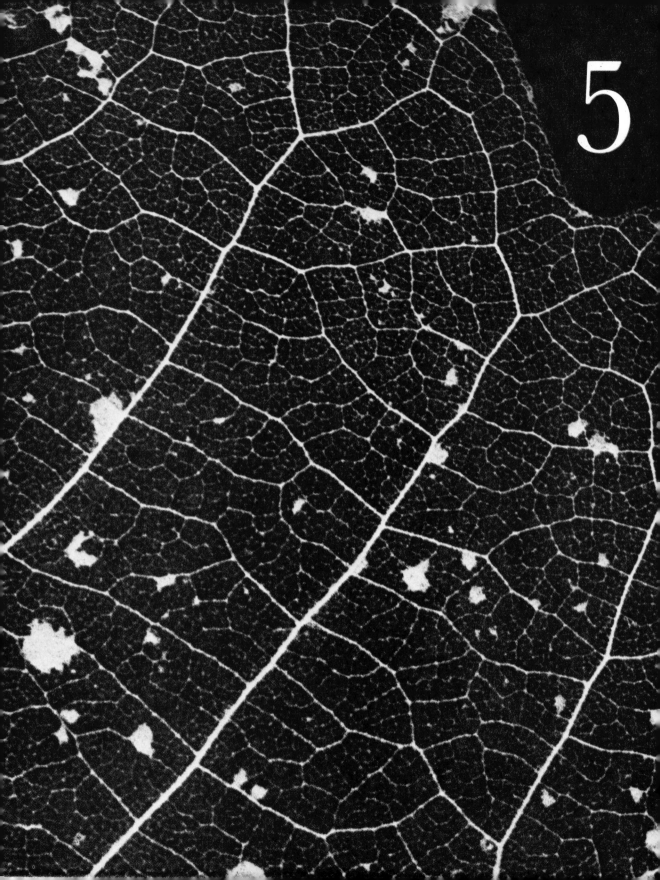

Photosynthesis and the Leaf

MANY PROCESSES OCCUR IN A LEAF, but the distinctive one and the most important is food manufacture. Green plants possess the ability to manufacture food from raw materials derived from the soil and the air, and upon this activity depends not only the life of plants but also the life of all animals, including man. Photosynthesis is the manufacture of sugar from two simple raw materials—carbon dioxide and water—in the presence of chlorophyll (the green coloring matter in plants), with sunlight as the source of energy. Since light is necessary, the name photosynthesis (*photos*, light, plus *synthesis*, a putting together) is apt.

SIGNIFICANCE OF PHOTOSYNTHESIS

All living things require energy, not only for growth and reproduction but also for the maintenance of life itself. This energy comes from the chemical energy in the food consumed; the food, in turn, has its origin in photosynthesis. Hence, through the agency of green plants, the radiant energy of the sun is trapped and made available to living things. Except for the activities of a few kinds of bacteria, photosynthesis is the only process by which organic compounds are constructed from inorganic substances, with a resultant storage of energy.

Photosynthesis has been called the most important chemical process known to man. It is also a major process in terms of mass of product. The total production of sugar by the world's plants, both on land and in the sea, has been estimated at about 100 billion tons a year. Unfortunately, much of this total is produced in the ocean and is largely unavailable. It is estimated that only about 2 percent of man's total food supply now comes from the ocean. The total production of sugar by land plants is on the order of 50 billion tons a year. This is many times the energy requirement of the nearly 4 billion people that comprise the earth's population. When, however,

99

we consider that a large percentage of the total goes into wood, straw, and other inedible materials, and that we recover as food only about 10 percent of the calories fed to animals, we can see why the earth will not be able to support a continually expanding population.

Although photosynthesis involves primarily the production of sugar or precursors of sugar, it also supplies materials for other important syntheses. Basic among these are the particular kinds of amino acids known to be essential. Only plants, including bacteria, are able to synthesize these compounds upon which life depends. Most of our vitamins come directly or indirectly from plants. Foods produced by plants may be used by the plant or they may accumulate as surplus—commonly starch. Foods accumulated in roots, stems, seeds, and fruits become available to the next generation of the plant or are used by it to initiate new growth after winter or other periods unfavorable for growth. Such accumulated foods are also harvested by man and other animals, whose food supply is obtained, directly or indirectly, from fruits, from seeds such as those of cereals and legumes, from stems such as sugar cane (Fig. 5-1) and potato, from roots such as cassava, turnip, and sweet potato, and from leaves.

Agriculture is the science of cultivating green plants whose products may be made available for the use of man or his domesticated animals. The quantity of plant products harvested for food reaches imposing totals. More than 6 billion bushels of wheat are produced in the world annually, and figures of the same magnitude may be cited for corn and rice. The world's annual production of sugar amounts to more than 50 million tons. The United States produces annually about 30 million bushels of dry beans, 400 million bushels of potatoes, 4 billion bushels of corn, and more than a billion bushels of wheat. The production of these foods depends upon photosynthesis in green leaves. Approximately 94 percent of the dry matter of a crop is derived from photosynthesis and only 6 percent from the soil.

The relationship of photosynthesis to man is important not only from the standpoint of food but also because man depends on this process for much of his economic existence. Many basic materials of industry are directly or indirectly of photosynthetic origin. The production of textile fibers, lumber, pulp products, vegetable lipids, gums, and resins depends upon the photosynthetic activities of green plants.

The extent of this dependence is even more apparent as we turn our attention from the present to the past. Our highly mechanized civilization is based largely upon the dead and partly decayed remains of plants of past ages. All coal was formed from plants. Petroleum and natural gas are produced by chemical changes in the remains of plants and perhaps of animals that fed upon plants. Although these materials are sometimes referred to as mineral fuels, they are organic in origin and are the products, direct or indirect, of plant activities. These fossil-plant fuels are largely the present basis of heat and power in our civilization, although nuclear energy may some day replace these vanishing materials.

Oxygen is liberated as a by-product of photosynthesis, and its significance should not be overlooked. Green plants and animals could not utilize the food produced by photosynthesis unless they also had available the oxygen released in the process. Studies indicate that the primitive atmosphere of the earth contained carbon dioxide, ammonia, methane, and other gases, but little or no oxygen. The present 21 percent by volume of oxygen in our atmosphere is due to photosynthesis and has accumulated slowly over millions of years. The oxygen in the air is being depleted by the respiration of living organisms, by the combustion of coal, oil, and other fuels, and by other processes. It is clear, then, that some factor must be operating to replenish the oxygen supply of the atmosphere. This factor is photosynthesis, and green plants constitute the only significant source of the free oxygen necessary to the existence of plant and animal life.

Fig. 5-1 Harvesting 100 tons of sugar cane per acre in Hawaii. Such a crop represents one of the world's highest rates of photosynthesis.

Early History Our present knowledge of photosynthesis, like that of many other plant processes, has developed over a long period of time. For centuries the concept of Aristotle (384–322 B.C.) prevailed—that plants obtained their food directly from the soil. This food was believed to consist of preformed particles of organic matter that were absorbed by the roots and utilized by plants without important chemical change.

The first experimental evidence against the theory of Aristotle was provided by Jan Baptista van Helmont (1577–1644), a Belgian philosopher, chemist, and physician. Van Helmont planted a small willow in a large pot of soil and grew it for 5 years, adding only rain water. In this time the willow gained 159 pounds and the soil lost 2 ounces of its dry weight.

Wherever the plant material came from, it had certainly not all come from the soil, and van Helmont concluded that the plant had been produced from rain water. His conclusions were wrong, primarily because the elementary concepts of chemistry were unknown at the time.

During the next three centuries many research contributions helped to piece together the story of how plants manufacture food by photosynthesis.

Of the scientists who contributed to the discovery of photosynthesis, only a few of the pioneers—Joseph Priestley (1733–1804), Jan Ingen-Housz (1730–1799), and Theodore de Saussure (1767–1845)—will be mentioned briefly. Priestley, English clergyman and self-trained chemist, is credited with the discovery of oxygen. In investigations of "air" (gases) he reported in 1772 that plants could "purify" air in which a mouse had suffocated. He did not realize that light was necessary for this process, and when he moved his experiments to a dark corner of his laboratory he was unable to repeat them. In 1779 Ingen-Housz showed that only the green parts of plants could purify air, and only in light. Not until 15 years later, however, did he recognize that plants as well as animals benefit from photosynthesis.

DeSaussure, in 1804, published the results of many chemical experiments showing that the carbon of plants comes from carbon dioxide, and the hydrogen from water. It was another 40 years before a German, Julius Mayer, showed that the energy released by burning plants or used in the metabolism of plants and animals is derived from the energy of sunlight, fixed in photosynthesis.

Little further important work on photosynthesis was done until about 1910, when two German chemists, Richard Willstätter and Otto Warburg, started modern research on chlorophyll and the utilization of light energy. The availability of radioactive carbon (^{14}C) in 1945 began the current intensive studies that are slowly revealing the complexities of photosynthesis and the many reactions that make up the over-all process.

THE STRUCTURE OF THE LEAF

Since the leaf is the chief site of the photosynthetic process, information on the structure of this organ is essential to a full understanding of its activities. The leaf is commonly studied in cross section (FIG. 5-2). The tissues of the leaf may be

classified as (1) EPIDERMIS, (2) MESOPHYLL, and (3) VASCULAR BUNDLES.

Epidermis The epidermis is composed of a single layer of interlocked cells that usually contain no chloroplasts. It is continuous over both surfaces of the leaf and is therefore distinguished as *upper* and *lower* epidermis. It is covered by the CUTICLE, a varnishlike layer, or film, that retards the movement of water and gases into and out of the leaf. In older leaves, and particularly in desert plants, waxes may be deposited beneath the cuticle, greatly decreasing its permeability. A heavy, cuticularized layer over the leaf may reduce loss of water through the epidermis to practically nothing.

The major water loss is normally through the STOMATA (FIG. 5-3). These are lens-shaped openings that perforate the epidermis and lead into intercellular spaces within the leaf. Stomata are also found in the epidermis of young twigs and herbaceous stems. The stoma is really an intercellular space formed by the splitting of the cell wall between two specialized epidermal cells

Fig. 5-2 Diagram of section view of a leaf, showing cells and tissues.

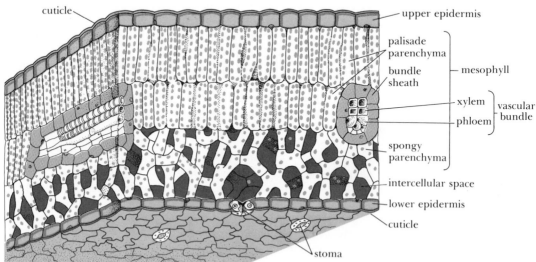

cuticle
upper epidermis
palisade parenchyma
bundle sheath
mesophyll
xylem
vascular bundle
phloem
spongy parenchyma
intercellular space
lower epidermis
cuticle
stoma

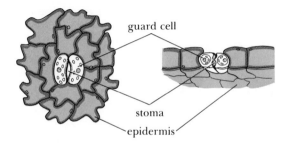

Fig. 5-3   The stoma. (Left): surface view. (Right): as seen in cross section.

known as GUARD CELLS. These, unlike other epidermal cells, contain chloroplasts. The guard cells, in surface view, are typically crescent-shaped. In cross section, the inner walls of the guard cells, those that border the opening, are seen to be much thickened. The upper and lower corners of the walls adjacent to the stoma are even thicker and frequently form projecting ridges.

As a result of these thickenings and the shape of the guard cells, changes in turgor within them cause the stoma to open and close. (The mechanism of this movement is discussed in Chapter 8.) Stomata are the chief path through which gases, such as water vapor, carbon dioxide, and oxygen, move from the leaf into the air, and vice versa. Stomata may be present in both the upper and lower epidermis, but the number in the lower epidermis is usually greater than in the upper. Many, perhaps most, woody plants have no stomata in the upper epidermis.

In many plants, the epidermis of both the leaf and stem bears hairs of various kinds (Fig. 5-4). A hair is the result of the outward growth either of a single epidermal cell or of a cluster of such cells. Unicellular hairs may later become multicellular by the formation of crosswalls. The presence of hairs frequently gives a silky, woolly, or felted appearance to leaves, especially to those of plants in dry environments and in bogs and alpine regions. Among the numerous modifications of hairs are glandular hairs, which may secrete a sticky substance, and the stinging hairs

of nettles. Stinging hairs are stiff and thick-walled, and end in a tiny swollen head. If this is broken by contact, a sharp point remains, which may penetrate the skin and inject a drop of an irritating, poisonous substance. Young stinging nettles (*Urtica*) have long been used as a cooked green vegetable. Cooking destroys the poison. For the most part, hairs are of doubtful value to the plant. We may, however, find them useful in identifying plants or dried plant parts, as in the detection of adulteration of drugs or foods.

Mesophyll   Within the leaf, photosynthesis is localized in the mesophyll. This tissue, with the exception of the vascular bundles, includes all cells between the upper and lower epidermis. The cells of the mesophyll are thin-walled and, even when mature, retain living protoplasm and a nucleus. Such thin-walled cells make up parenchyma tissue similar to the cells forming the softer parts of roots, stems, flowers, and fruits.

The mesophyll is usually divided into two parts. The cells toward the upper side of the leaf

Fig. 5-4   Plant hairs. (A): simple hair of geranium (*Pelargonium domesticum*). (B): glandular hair of geranium. (C): branched hair of mullein (*Verbascum thapsus*). (D): stinging hair of nettle (*Urtica dioica* var. *procera*).

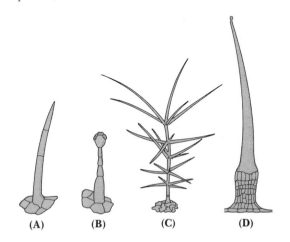

(A)   (B)   (C)   (D)

are elongated at right angles to the surface and form one to three compact layers. These cells make up the PALISADE PARENCHYMA, so called because of its resemblance to a palisade, or row of stakes.

Below the palisade cells, and extending to the lower epidermis, is a zone of irregularly shaped cells with large intercellular spaces. These cells form the SPONGY PARENCHYMA. The intercellular spaces of this tissue, together with similar but smaller spaces between the palisade cells, form a system of air passages extending throughout the leaf (FIG. 5-8), and connecting with the outer air through the stomata. Because of the presence of intercellular spaces, a considerable area of mesophyll cell wall is exposed to the intercellular air. This exposed surface is termed the *internal* surface of the leaf, and is usually considerably greater than the surface area of the leaf, or external surface. For example, in a lilac leaf the area of internal surface was found to be 13 times greater than the surface of the leaf. This internal surface is important in transpiration and in carbon dioxide absorption when the stomata are open.

All the mesophyll cells normally contain chloroplasts, embedded in the cytoplasm surrounding a large central vacuole. The chloroplasts are most numerous in the palisade cells, perhaps two to three times as numerous as in the spongy tissue. In a sunflower leaf, for example, the cells of the palisade layer averaged 77 chloroplasts per cell; those of the spongy mesophyll, only 27 per cell. Assuming about 400,000 chloroplasts to each square millimeter of an elm leaf—a conservative estimate—it has been calculated that the total surface area of the chloroplasts in the leaves of a mature elm tree may amount to as much as 140 square miles.

In some plants, such as corn and other grasses (FIG. 5-10), many aquatic plants, and conifers, the mesophyll is undifferentiated or only slightly differentiated into palisade and spongy tissue. Parenchyma cells containing chloroplasts are found in other parts of the plant, such as herbaceous stems, young twigs, and unripe fruits.

Fig. 5-5 Portion of a leaf of large-toothed aspen (*Populus grandidentata*) with mesophyll mostly removed, showing the larger veins and their branches.

The term CHLORENCHYMA is applied to such green cells, whether they are found in the leaf or in other organs.

Vascular Bundles Vascular bundles are specialized strands of tissue that function in both support and conduction. The larger bundles within the leaf can be seen on the surface as the main veins. The vascular bundles of the blade are branches of the vascular bundles of the petiole. These bundles, in turn, are continuous with vascular tissues of the stem. In the dicotyledonous leaf, the vascular bundles branch and rebranch, forming smaller and smaller strands that extend to all parts of the leaf. These smaller bundles are connected at frequent intervals and form a netlike system (FIG. 5-5). In most cross sections of such a leaf, therefore, vascular bundles may be seen in both end and side views. The vascular bundle system of the leaf is very ex-

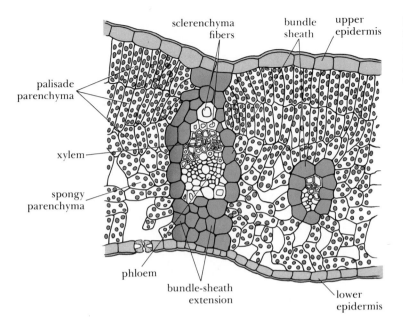

palisade parenchyma

sclerenchyma fibers

bundle sheath

upper epidermis

xylem

spongy parenchyma

phloem

bundle-sheath extension

lower epidermis

Fig. 5-6 Cross section of apple (*Pyrus malus*) leaf, showing bundle sheath and bundle-sheath extensions. The palisade tissue is composed of three layers.

tensive; it has been calculated, for example, that if the vascular bundles, large and small, in a single elm leaf were placed end to end, they would be more than 700 feet long.

The vascular bundles are usually located about halfway between the upper and lower epidermis (Figs. 5-2, 5-6, and 5-9). They are composed of two kinds of tissue of fundamental importance, XYLEM and PHLOEM. Together these constitute the VASCULAR TISSUES. The same tissues occur also in roots, stems, flowers, and fruits. The xylem and phloem are composed of elongated cells adapted to the movement of materials throughout the plant. The dead, woody cells of the xylem conduct water and supply mechanical support. Certain specialized living cells of the phloem are the channel through which foods are moved.

The xylem of flowering plants is composed of several kinds of cells. The most important, from the standpoint of conduction, is the VESSEL ELE-MENT (Fig. 5-7), a somewhat elongated cell with open ends. A continuous series of such cells is termed a VESSEL. In some vessel elements the secondary wall is incomplete and is laid down in

rings, spirals, or other patterns inside the primary wall (Figs. 5-7 and 5-8). The rigid, woody secondary walls of these cells enable them to function as noncollapsible tubes through which water may be moved under tension. When mature, the vessel element is nonliving and consists only of a permeable, rigid cell wall.

The most important conducting cells of the phloem are the SIEVE-TUBE ELEMENTS (Figs. 5-7 and 7-28), so named because of the presence of sievelike openings in the end and frequently in the side walls. Sieve-tube elements in the flowering plants are, like vessel elements, arranged end to end, forming SIEVE TUBES.

So numerous are the smaller vascular bundles that most cells of the mesophyll are only a few cells removed from the conducting tissue (Fig. 5-8). Water and mineral salts from the soil reach the mesophyll cells of the leaf through the xylem; food, manufactured in the mesophyll, moves in the opposite direction through the phloem. The smaller vascular bundles end blindly in the tissues of the leaf (Fig. 5-8). The bundle ends may be composed only of a few xylem cells or of elements of both xylem and phloem.

Fig. 5-7 Diagrams of principal cells of xylem and phloem of a medium-sized vein of oak (*Quercus rubra*) cells as seen in section view (left) and surface view (right).

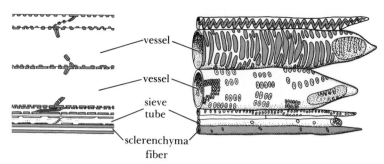

vessel

vessel

sieve tube

sclerenchyma fiber

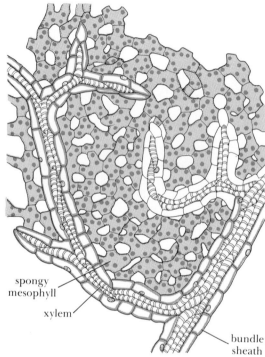

spongy mesophyll

xylem

bundle sheath

Fig. 5-8 Section of leaf cut parallel to the leaf surface and through the spongy mesophyll (in color). Branches of the smaller veins end in the spongy tissue.

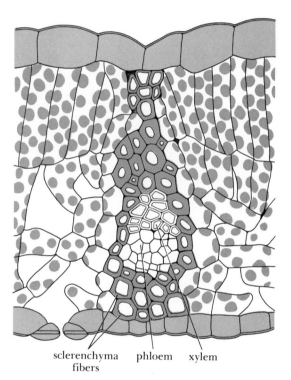

sclerenchyma fibers

phloem

xylem

Fig. 5-9 Cross section of a small portion of leaf of red oak (*Quercus rubra*), showing reinforcement of the leaf by sclerenchyma fibers surrounding the bundle and extending from the upper to the lower epidermis.

The smaller vascular bundles of the angiosperm leaf, and commonly those of intermediate size as well, are surrounded by a tight sheath of parenchyma cells, the BUNDLE SHEATH (FIGS. 5-2, 5-6, and 5-8). The cells of the sheath are

elongated parallel with the veins and have living protoplasm and thin walls, but usually have fewer chloroplasts than the mesophyll cells—in some species none at all. Intercellular spaces are lacking between the cells of the sheath and between these and the enclosed vascular bundle. The sheath cells are in intimate contact with both palisade and spongy tissue, and all materials moving from or into the vascular bundles must pass through this sheath. The sheath also extends over the bundle ends, and in this region food materials may enter the sheath and be transported through these cells to the nearest sieve tubes. The bundle sheath is especially conspicuous in the leaves of corn (Fig. 5-10).

In many dicotyledons the sheath surrounding bundles of intermediate size has extensions to one or, more commonly, to both epidermal layers (Fig. 5-6). The cells of these BUNDLE-SHEATH EXTENSIONS are without chloroplasts and may be somewhat larger and thicker-walled than those of the sheath itself. Because of their nature and position, the sheath extensions provide mechanical support for the blade. In addition, experiments have shown that dyes introduced into the leaf move quickly from the vascular bundles into the sheath extensions and through these into the upper and lower epidermis. Water lost from the leaf by evaporation may therefore be replaced by conduction through the sheath extensions and the epidermal layers.

In addition to the xylem and the sheath extensions that contribute to the mechanical support of the leaf, specialized strengthening cells are commonly present. The most conspicuous of these are the SCLERENCHYMA fibers, usually associated with the larger vascular bundles. These fibers are thick-walled and greatly elongated cells with tapering ends (Fig. 5-7). Like the vessel elements, the mature fibers are devoid of protoplasm. They are not conducting elements, but are strength and mechanical supports. They aid in preventing collapse of the leaf following loss of water, and they resist the very large stresses of wind. In many leaves the larger bundles contain fibers, as a surrounding sheath or in groups above and below the bundle. These fibers may extend to the upper and lower epidermis, thus forming a girderlike structure, very effective in strengthening the leaf (Figs. 5-9 and 5-10).

Sclerenchyma fibers are also found in the petiole, stem, and other parts of the plant.

Leaf Abscission   The falling of leaves is characteristic of woody and many nonwoody plants. The fall, or ABSCISSION, of the leaf is related to the formation of a SEPARATION LAYER, a zone of specialized parenchyma cells extending across the base of the petiole (Fig. 5-11). Toward the end of the growing season the pectins of the cell walls of this layer are dissolved by enzymatic action. The cells then separate and the petiole remains attached chiefly by the vascular bundles. The fall of the leaf is hastened by environmental factors such as the shrinkage of the petiole on warm, sunny days, wind, the impact of rain upon the leaf, or the formation of ice crystals within the separation layer. Before the fall of the leaf, or shortly afterward, a protective layer of cork forms just beneath the separation layer and protects the exposed stem tissues.

Leaf abscission is of great economic importance in cotton, where early leaf fall aids hand picking of the crop and is essential for machine picking. Many chemicals have been used to hasten leaf fall. A more or less normal abscission must be induced. Chemicals that kill the leaves on the plant are unsatisfactory. The dead leaves do not abscise and the crumbled tissue of the leaves mixes with the fiber and lowers its value.

PHOTOSYNTHESIS— THE PROCESS

The importance of photosynthesis as the basis of life, both in food production and in oxygen liberation, has been stressed. It is also one of the most complicated of biological processes, involving numerous physical and chemical reactions in the capture of radiant energy and its eventual conversion into the chemical energy of organic compounds. The study of photosynthesis involves the fields of radiation physics, photochemistry, bio-

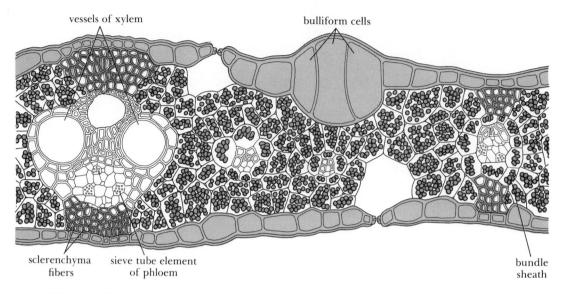

vessels of xylem

bulliform cells

sclerenchyma
fibers

sieve tube element
of phloem

bundle
sheath

Fig. 5-10 Cross section of portion of corn (*Zea mays*) leaf showing three types of vascular bundles. Loss of turgor in the bulliform cells is thought to contribute to the upward curling of wilted corn leaves.

chemistry, and the morphology, cytology, physiology, and ecology of green plants. It is the concern of physicists, chemists, botanists, and farmers, all interested in its reactions and its results.

The Over-all Reaction The chemical equation for photosynthesis has been written as

$$6CO_2 + 6H_2O + 672 \text{ kcal}[1] \longrightarrow$$

carbon   water   radiant
dioxide   energy

$$C_6H_{12}O_6 + 6O_2 \quad (1)$$

glucose   oxygen

This equation states that six molecules of carbon dioxide are combined with six molecules of water

to form one molecule of glucose and six of oxygen. Glucose, also known as dextrose and as grape or corn sugar, is a carbohydrate, made up of the elements carbon, hydrogen, and oxygen (FIG. 5-12). Glucose was once considered to be the direct product of photosynthesis. Modern research shows that fructose, a closely related sugar (FIG. 5-12), is formed first. The carbon dioxide used in photosynthesis is obtained from the air, the water from the soil, and the energy from light. It should be noted that 672 kilocalories is the energy that can be recovered by burning a gram-molecular weight (mole) of glucose. Some three times as much energy is required in glucose production. The oxygen produced is released into the air as an important by-product used by both plants and animals.

By the use of the heavy oxygen (^{18}O) isotope as a tracer, it has been shown that all the oxygen liberated in photosynthesis is derived from water (H_2O) instead of from carbon dioxide (CO_2), as was formerly believed. Equation (1) should now be modified to show this relationship:

[1] One kilocalorie (kcal) is the energy required to heat 1000 grams of water from 14.5° to 15.5°C, or, approximately, to heat 1 quart of water 2 degrees on the Fahrenheit scale. One can obtain 672 kcal by burning 1 gram-molecular weight (180 grams, or a little more than 6 ounces) of glucose. This is sufficient to heat the water for a hot shower.

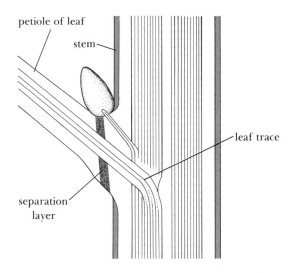

Fig. 5-11 Diagram showing position of separation layer before leaf fall.

petiole of leaf

stem

leaf trace

separation layer

$$6CO_2 + 12H_2O + \text{radiant energy} \longrightarrow$$

$$C_6H_{12}O_6 + 6H_2O + 6O_2 \quad (2)$$

Some of the oxygen atoms are marked in color in this equation to show their origin and final position. Notice that new water molecules are

Fig. 5-12 Structure of three important sugar molecules shown in the open-chain form. The 1-carbon is at the top.

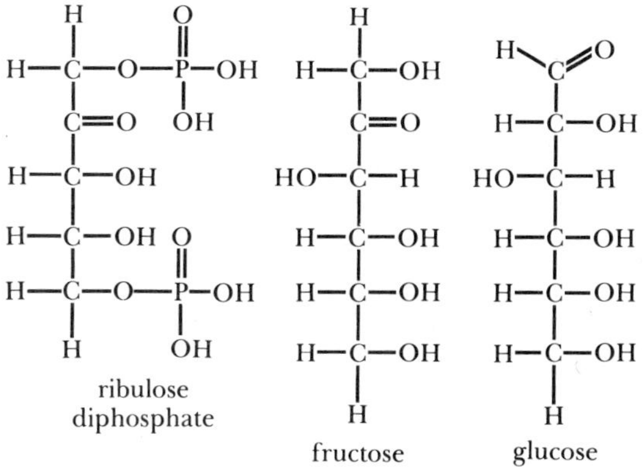

ribulose diphosphate

fructose

glucose

produced, as well as oxygen molecules. These water molecules become a very minor part of the total water supply of the cell. The fact that they have been produced, however, tells the chemist and plant physiologist a great deal about the nature of the process.

Structure of the Chloroplasts These organelles have an elaborate, complicated internal structure, which can be seen only with the electron microscope (FIG. 5-13). An outer, double membrane surrounds a finely granular mass, the STROMA. Embedded in the stroma is a system of layers, called lamellae, composed largely of protein and lipids. In higher plants, most of the lamellae are organized into DISKS, or THYLAKOIDS. Each disk is double, and closed at the edges, forming a hollow compartment. The disks, in turn, are aggregated into GRANA, cylindrical structures resembling a pile of hollow pancakes and showing pronounced layers (the disks) in vertical section. Some 5–25 disks may be found in each granum, and from 40–80 grana in a chloroplast. The grana are connected to adjacent grana by a system of flattened tubes, the intergranum lamellae. The chlorophylls and carotenoids are concentrated in the disks, associated with the protein. The stroma is by no means an unimportant part of the plastid; starch grains, when formed, are synthesized here, and the stroma contains the enzymes of the "dark" reactions of photosynthesis. The machinery of the "light" reactions, on the other hand, is localized in the grana.

Chlorophyll The chlorophyll molecules absorb the energy of light and start the reactions of photosynthesis. These molecules belong to a group of biologically important compounds that includes hemoglobin, the red pigment of blood. The higher plants contain two chlorophylls: a blue-green chlorophyll a, ($C_{55}H_{72}O_5N_4Mg$), and a yellow-green chlorophyll b ($C_{55}H_{70}O_6N_4Mg$). The structure of chlorophyll a is illustrated in FIG.

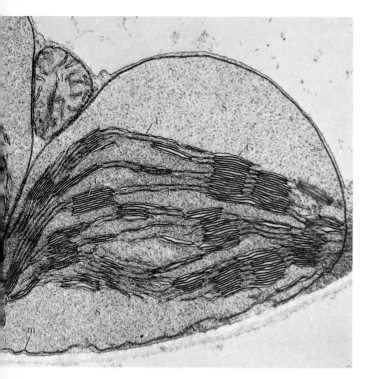

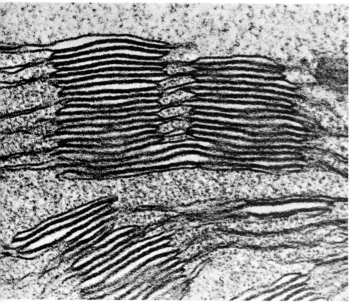

Fig. 5-13 Electron micrograph: chloroplast of sunflower (*Helianthus annuus*). (Top): entire chloroplast in section view (25,000×): m, limiting membrane; S, stroma; G, grana; L, intergranum lamellae; M, a mitochondrion from the same cell. (Bottom): a portion of the same chloroplast enlarged (93,000×) to show the lamellae and the disks.

5-14. Chlorophyll *b* differs from *a* only in the presence of a CHO group instead of CH₃ at the starred position in the upper right of the figure. The color of chlorophyll and its effectiveness in the absorption of sunlight are due to the alternating single (—) and double (=) bonds in the ring structure around the central molecule of magnesium. The long lipid tail, built of H—C—H units, is assumed to anchor and orient the chlorophyll molecule in the lipid and protein layers of the grana.

The relative proportions of the *a* and *b* pigments vary somewhat, but on the average they are present in flowering plants in the ratio of about three parts of chlorophyll *a* to one part of chlorophyll *b*. Chlorophyll *a* is always present in green plants above the bacteria. Photosynthetic bacteria contain a closely related pigment. Chlorophyll *b* is found with chlorophyll *a* in the green algae and all higher plants. Other modifications, known as chlorophyll *c*, *d*, and *e*, replace chlorophyll *b* in certain algae—chlorophyll *c* in the brown algae and diatoms, for example, and chlorophyll *d* in the red algae. Chlorophylls *b*, *c*, *d*, and *e* transfer the radiant energy they absorb to chlorophyll *a* rather than use it directly. Algae also contain nonchlorophyllous accessory pigments that are blue-green, brown, or red in color, and which transfer energy to chlorophyll *a*. Some of the yellow carotenoids (Chapter 3) also serve as accessory pigments. The movement of this absorbed energy, its transfer to chemically reactive sites, and its fixation as chemical energy are discussed later.

The green color of chlorophyll tells us that the molecule absorbs light. A more distinctive reaction may be observed by holding a solution of chlorophyll in a test tube against a source of light. The dark red color observed is due to FLUORES-CENCE. It results from light energy that has been absorbed and then *reradiated* within a millionth of a second in a wavelength characteristic of chlorophyll *a*. Fluorescence tells us that the chlorophyll *a* molecule has been raised to a high-energy level, however briefly, by the action of light. Energy transfer from other chlorophylls may also be tested by fluorescence. If one of these pigments is mixed with chlorophyll *a* and the mixture exposed to light of a wavelength that is absorbed primarily by the second chlorophyll, the color and intensity of the fluorescence are the same as though chlorophyll *a* had been tested alone with the same amount of energy, at a wavelength absorbed directly by chlorophyll *a*. We conclude, therefore, that the energy absorbed by the second chlorophyll has been transferred to chlorophyll *a*, and in this instance almost completely.

Fig. 5-14 Structure of chlorophyll *a*. Chlorophyll *b* has a CHO group instead of a CH_3 in the starred position in the upper right.

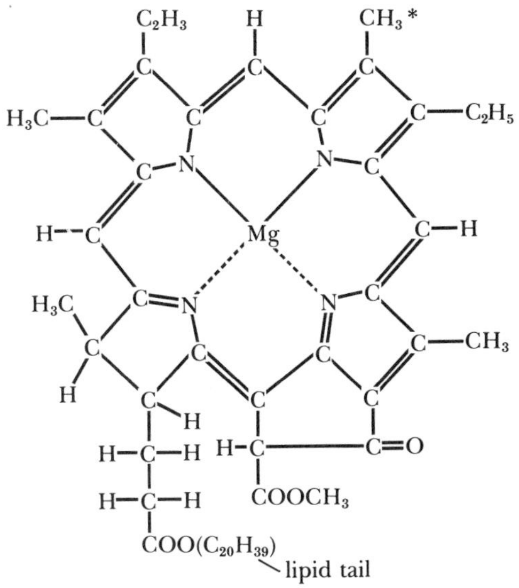

Absorption of Energy by Leaves The radiation from the sun is absorbed to a varying degree by the gases, vapors, and dusts of the atmosphere before it reaches the leaf. A part of the radiation that strikes a leaf is absorbed, a part reflected, and a part transmitted through it. The quantity absorbed by the pigmented parts of the leaf will depend upon the structure of the leaf and other factors, but it may amount to 50 percent of the sunlight energy that strikes the leaf. Only radia-tion that is absorbed can supply energy for pho-tosynthesis. However, the leaf is not very ef-ficient in utilizing even this energy, for it has been calculated that, in full sunlight, less than 3 percent and frequently less than 1 percent of the absorbed energy is converted into chemical energy. The percentage of absorbed energy uti-lized is higher in dim light, reaching about 10–15 percent.

The visible portion of the radiation from the sun can be separated by a simple prism into a number of colors, the *visible spectrum.* The color seen in each part of the spectrum is a function of the wavelength. The violet region, which lies at one end of the spectrum, passes, by impercepti-ble changes, into blue, then into green, yellow, and finally into orange and red. The wavelengths are measured in millimicrons (mμ)—one thou-sandth of a micron (μ) or $1/25$ millionth of an inch. The visible portion ranges from about 400 millimicrons at the violet end of the spectrum to about 760 millimicrons in the extreme red (FIG. 5-15).

A small percentage of the radiation from the sun is in the ultraviolet; of the remainder, about half is in the visible portion of the sun's spec-trum, which we call light, and about half is in the infrared. Chlorophyll absorbs little or none of the ultraviolet and infrared, and these wave-lengths are not important in the process of pho-tosynthesis.

Since only those wavelengths that are reflected from an object produce the sensation of color, the green color of leaves has been interpreted as meaning that radiation of these wavelengths is

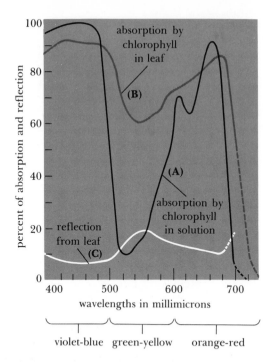

Fig. 5-15 Percentage of absorption and reflection of light by chlorophyll. (Curve A): absorption of light by chlorophyll in solution. (Curve B): absorption by the same quantity of chlorophyll in the leaf. Chlorophyll in the leaf absorbed three times more green light than the same chlorophyll in solution. (Curve C): the green color of a leaf is due to a slightly greater reflection of green than of other wavelengths.

unused, or little used, in photosynthesis. Such an impression may also be gained by studying the absorption of white light by a chlorophyll *solution;* here it is found that most of the absorption occurs in the blue and red wavelengths, and relatively little in the middle of the spectrum (FIG. 5-15A). However, the absorption spectrum of an intact leaf (FIG. 5-15B) differs markedly from that of a chlorophyll solution. In the experiment from which the curves of FIG. 5-15 were taken, a square centimeter of the flat absorption chamber containing the chlorophyll solution had the same quantity of chlorophyll as was present in a square centimeter of the leaf. The average absorption of light in the three general color regions by an

equal quantity of chlorophyll in solution and in the leaf is shown below.

	Violet-blue (400–500mμ)	Green-yellow (500–600mμ)	Orange-red (600–700mμ)
Pigments in solution	95%	23%	75%
Pigments in leaf	90%	68%	78%

This difference in the absorption of green light by chlorophyll in solutions and in the leaf is not always recognized. It is important because green wavelengths constitute a major portion of the energy of sunlight and because the energy of green wavelengths absorbed by chlorophyll in the leaves is just as effective in photosynthesis as the energy of the blue or the red. The greater absorption of green light by the leaf—in this experiment, three times as much as by the same quantity of chlorophyll in solution—is considered to be due to the multiple refraction of light by the cytoplasmic and chloroplast structures. Green is therefore an important color in photosynthesis. Although green light is strongly absorbed by leaves, some is also reflected (FIG. 5-15C). Therefore, although chlorophyll absorbs and uses radiation in all portions of the visible spectrum, leaves are green because chlorophyll absorbs somewhat less and reflects somewhat more of the green wavelengths.

Because any wavelength of the visible spectrum (except the longest red ones) can be absorbed and used by the leaves, plants can be grown under artificial light, which is usable in proportion to its visible radiation. Since, however, a light intensity more than 50 times that of a good reading light is required for satisfactory growth of most plants, artificially lighted basement gardens are expensive.

Energy Changes in Photosynthesis The work of changing carbon dioxide and water into sugar requires energy. Energy is neither created nor destroyed but is converted from one form into another. Some of the forms of energy are mechanical, electrical, thermal, radiant, and chemical. In photosynthesis, some of the radiant energy of light is converted into chemical energy, which is stored in the sugar molecule.

Although light shows wave properties, it is absorbed as "particles" called PHOTONS. Photons are minute packets of energy emitted or absorbed by atoms and molecules, for example, in a lamp filament or in the sun. When free, they travel at 3×10^{10} centimeters (186,000 miles) per second. The energy of a photon, called a quantum, is inversely proportional to the wavelength. Thus, red light, with long wavelengths, has less energy per photon than green or blue light. Quanta are the energy units of photosynthesis.

Fig. 5-16 **Photosynthetic units. The small quadrangular structures, located on the surface of the granum disks, are believed to be the sites where all the light reactions of photosynthesis are carried on (Fig. 5-18).**

The wavelengths that are used in photosynthesis are those in the absorption spectrum of chlorophyll (FIG. 5-15) and its accessory pigments. The longest wavelength absorbed by chlorophyll *a* and used by green plants is about 700 millimicrons. The photosynthetic bacteria have a different chlorophyll and carry on photosynthesis with wavelengths not detectable to the eye—890 millimicrons; these bacteria can carry on photosynthesis in the "dark."

The energy of a photon of red light is small, but compared to the kinetic energy of a chlorophyll molecule it is large. When a photon is absorbed by the chlorophyll molecule, the kinetic energy is not increased; instead, an electron within the molecule is raised to a much higher energy level than it occupied previously. The molecule then is said to be in an excited state. Such excited electrons may leave the chlorophyll entirely, or they may drop back to their normal energy levels in the molecule, losing energy as heat or as red light in fluorescence. It is this instability of the excited electron which puts a limit of a millionth of a second on the time from photon absorption by a chlorophyll molecule to its use in photosynthesis.

Photosynthetic Units One of the problems of understanding the mechanics of photosynthesis seems to have been solved by the discovery of the organization of chlorophyll molecules within the disks of the grana of the chloroplast. Elaborate studies of the light-reaction rates and of the ratios of different molecules indicate that chlorophyll works in closely packed units of perhaps 200 molecules each. This theory is supported by electron micrographs (FIG. 5-16) that show chlorophyll-containing structures of a size that could contain the calculated number of chlorophyll molecules. Each of these PHOTOSYNTHETIC UNITS has two REACTIVE SITES. Within these sites, single molecules of chlorophyll *a* are closely associated with enzymatic proteins, electron-acceptor molecules, electron-transfer systems, and essentially all of the complicated machinery of the light reactions of photosynthesis.

A packed sheet of chlorophyll molecules ensures that scattered units of radiant energy will be captured. Once captured, the energy can move through adjoining chlorophyll molecules, from higher to lower energy levels, until it reaches the reactive sites where it is used. The time required for this movement is estimated to be a fraction of a billionth of a second. The photosynthetic unit thus brings the energy of dim light to one spot where it can be stabilized and used in the energy-fixing steps of photosynthesis before it is lost. It is assumed that photosynthetic units make possible the rapid response of photosynthesis, at a low rate, to weak light. Photosynthesis can be demonstrated, for example, with the light of a single match.

Assembling of the Raw Materials Photosynthesis may now be visualized as a whole, from the raw materials to the final product. In land plants, water from the soil moves through the conducting cells of the root, stem, and leaf, and finally reaches the chlorophyll-bearing cells. Carbon dioxide comes from the atmosphere, where it is present to the extent of about 0.03 percent by volume, or 3 parts in 10,000 parts of air. Small as this quantity seems, the entire atmosphere contains more than 2000 billion tons of carbon dioxide—about 19 tons over each acre of the earth's surface. The carbon dioxide enters the leaf primarily through open stomata. It enters not because the leaf "needs" carbon dioxide to carry on photosynthesis but because, when photosynthesis is using carbon dioxide the concentration of this gas is lower within the leaf than in the air and the gas diffuses through the stomata, tens of thousands of which occur on 1 square inch of typical leaf surface (Table 8-1). Inside the leaf the carbon dioxide diffuses through the intercellular spaces and thus comes in contact with the wet walls of the cells that border upon these spaces (FIG. 5-8). Since carbon dioxide is soluble in water, it dissolves and diffuses through the wet cell walls into the protoplasm, where it is used by the chloroplasts. On the other hand, the oxygen liberated, because of its greater concentration within the intercellular spaces, diffuses outward through the stomata into the atmosphere.

The problem of the absorption of carbon dioxide through microscopic stomata has interested botanists for years. They have found that single, isolated, small pores are astonishingly effective in diffusion because molecules diffuse toward the pore from a large area around it. Leaves, however, may have 5000 to nearly a million stomata per square inch, each with diameters (when fully open) of 3 to about 20 microns. On the basis of the maximum diffusive capacity of each of the stomata, diffusion into or out of a leaf across a square inch of leaf epidermis, with a maximum of about 1 percent of the area in stomatal openings, could theoretically be 50 times the diffusion across a square inch of open space. Although such rates are not attained, due to the influence of adjacent stomata, the observed rates are still very high. Stomata are not isolated, and diffusion through each stoma is

Fig. 5-17 A "stomatal closure" curve. Smaller pores with the same spacing show relatively greater diffusion.

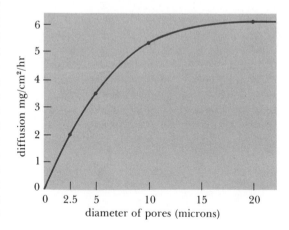

TABLE 5-1
"Stomatal Closing" Data of Fig. 5.17

Diameter of pores		Area of pores		Total diffusion
μ	%	μ^2	%	%
20	100	400	100	100
10	50	100	25	92
5	25	25	6.25	61
2.5	12.5	6.25	1.56	36

reduced by diffusion through the many stomata around it.

Actual diffusion through artificial membranes with pores 20, 10, 5, or 2.5 microns in diameter, all spaced 200 microns apart, is shown in FIG. 5-17 and Table 5-1. This experiment illustrates the effect of the gradual closing of rather large stomata. Reducing the area of each 20 μ pore by 75 percent reduced diffusion by only 8 percent. However, a pore 2.5 μ in diameter still has a diffusive capacity of 36 percent of the 20 μ pore. Thus stomata which are almost completely closed still have a high diffusive capacity.

Stomata are well adapted to the absorption of carbon dioxide. Unfortunately, as we shall see later, they are also well adapted to the loss of water vapor from the leaf.

REACTIONS OF PHOTOSYNTHESIS

The equations for photosynthesis shown earlier are greatly simplified, and indicate merely the over-all nature of the process. Actually the production of sugar from carbon dioxide and water involves many reactions and many intermediate products. Most of what is known about these reactions has been learned since 1940, and much of it since 1960. Radioactive carbon, ^{14}C, enables the chemist to determine the location of an individual carbon atom in a sugar molecule (FIG. 5-12), for example. Various chroma-

tographic procedures give rapid, clear-cut separation of closely related compounds in quantities measured in small fractions of a milligram—millionths of an ounce. In other areas of research, new techniques and new apparatus have made further progress possible. Electron-spin resonance spectrometers are now used to study the effect of light on individual molecules, with a time factor of a fraction of a billionth of a second. Electron microscopes show details of chloroplasts at magnifications of many thousands of times (FIGS. 5-13, 5-16). Many people working with these methods have learned much about photosynthesis, but a number of details are still incompletely known.

It has been known for many years that photosynthesis can be divided into reactions that use light directly and those that can proceed in the dark, but which are dependent upon energy previously obtained from light. These are called the LIGHT REACTIONS and the DARK REACTIONS of photosynthesis. Light reactions are very fast. The primary absorption of light by chlorophyll, for example, is completed in billionths of a second—practically instantaneous.

Because light reactions proceed so rapidly and are so hard to measure, our knowledge of them is fragmentary, and few of the steps believed to occur have been fully proven. Much of the detail of the biochemical dark reactions of photosynthesis, in contrast, is well established. These reactions are slow, thousandths of a second instead of billionths, and, more important, they involve carbon atoms that can be traced with ^{14}C, whereas photochemistry involves unmarked, very small, and incredibly fast electrons.

Light Reactions of Photosynthesis The light reactions of photosynthesis separate into three groups: (1) the splitting of water into electrons, hydrogen ions, and oxygen; (2) the production of ATP with light energy, called PHOTOPHOSPHORYLATION; and (3) the transfer of electrons from chlorophyll and of hydrogen ions split off from water to a hydrogen acceptor, NICO-

TINAMIDE ADENINE DINUCLEOTIDE PHOSPHATE, usually designated for convenience as NADP.

Splitting the water molecule The separation of water (H_2O) into hydrogen (H) and oxygen (O) is a distinctive reaction of photosynthesis. This is shown in FIG. 5-18 at the circled number *1*. Most of the other processes are duplicated in other systems and may be considered to be more or less general metabolic reactions. It is known that the oxygen liberated in photosynthesis comes from the water and that it is freed at the reaction site, P680 in FIG. 5-18. The label P680 represents a pigment group containing chlorophyll *a* (perhaps only one molecule) in combination with other molecules, the complex showing a maximum absorption of light in the red at about 680 millimicrons.

The activation of the electrons of chlorophyll by light is shown in fluorescence. With stronger light and in the photosynthetic units, electrons

Fig. 5-18   Diagram of the light and dark reactions of photosynthesis. ①: light is absorbed in the reaction centers P680 and P700. This light speeds up the electrons of chlorophyll until some of them are driven from the chlorophyll. If these electrons are held even briefly by the electron-acceptor molecules Y, they may be replaced by electrons from adjoining water molecules. ②: electrons trapped at Y are believed to escape through the enzyme C*f*, forming ATP from ADP as they pass on to P700. ③: here more electrons are driven off by light energy, and electrons escape to the acceptor X. They may then move through Fd and be taken up by NADP. These electrons give NADP a negative charge and it absorbs hydrogen ions (H^+) and forms $NADPH_2$, completing the light reactions. In the dark reactions, ATP and $NADPH_2$ move out from the photosynthetic units into the stroma of the chloroplast and react with ribulose, forming phosphoglyceric acid (PGA). These molecules are then reduced by $NADPH_2$ to the 3-carbon sugar phosphoglyceraldehyde.

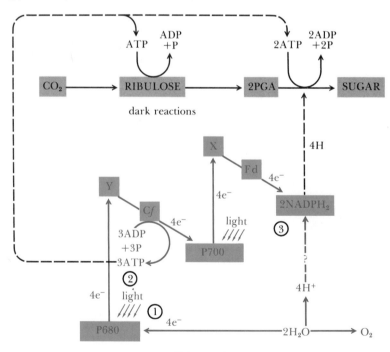

may be so activated that they escape from the molecule. Normally, these electrons would be expected to fall back quickly, but they may be caught and held temporarily by the electron-acceptor molecule Y in FIG. 5-18. Electrons then move into the chlorophyll of the reactive site from adjoining water molecules. As a result of the removal of electrons, water tends to break down into oxygen (O_2), hydrogen ions (H^+), and the electrons held in Y, thus completing the reactions labeled 1 in FIG. 5-18.

Photophosphorylation In the reaction labeled 2 in FIG. 5-18, the high-energy electrons held in Y move downward toward the nearby lower-charged photosynthetic site, P700. On the way, they move past the enzyme cytochrome f (Cf), where some of their energy is used in changing ADP to ATP (photophosphorylation), thus forming readily available energy for the dark reactions of photosynthesis.

At the point labeled 3 in FIG. 5-18, the free electrons, still containing some of the energy acquired at P680, now receive a second charge of light energy at P700 and move up to the electron acceptor X. The height of X above Y in FIG. 5-18 is an indication of the higher energy here. From X the electrons (e^-) move down through ferredoxin (Fd), an iron-containing protein, and combine with hydrogen ions (H^+) and NADP, forming $NADPH_2$ as shown by

$$NADP + 2 \text{ electrons } (e^-) + 2H^+ \rightarrow$$
$$NADPH_2 \quad (3)$$

The formation of $NADPH_2$ completes the light reactions. The energy of light is stored in ATP and $NADPH_2$, and these compounds can now be used in the dark reactions that lead to the formation of sugars. In fact, it is possible to separate the enzymes of photosynthesis from the chloroplasts, add ATP, $NADPH_2$, and CO_2, and get sugar production in a test tube in the dark.

Dark Reactions of Photosynthesis FIGURE 5-18 shows not only the light reactions of pho-

tosynthesis but also the dark reactions and the relationships among them. Two high-energy compounds, ATP and $NADPH_2$, formed in the light reactions, are transferred from the reaction sites of the disks to the stroma that surrounds the grana. ATP from the light reactions now phosphorylates a 5-carbon sugar (ribulose) found in the stroma, converting it to ribulose diphosphate. In a fairly simple reaction, one molecule of carbon dioxide, one of water, and one of the ribulose diphosphate[2] unite and split immediately, forming two molecules of phosphoglyceric acid:

$$P \sim C_5H_{10}O_5 \sim P + CO_2 + H_2O \longrightarrow$$

ribulose   carbon   water
diphosphate   dioxide

$$2P \sim C_3H_6O_4 \quad (4)$$

phosphoglyceric
acid

The phosphoglyceric acid is now converted into a 3-carbon sugar, phosphoglyceraldehyde. More ATP from the light reactions participates here, and the hydrogen that was taken up by NADPH in the light reactions is used to reduce the phosphoglyceric acid by removing one of the oxygen atoms:

$$2P \sim C_3H_6O_4 \sim P + 2NADPH_2 \longrightarrow$$

diphosphoglyceric   reduced NADP
acid

$$2NADP + H_2O + 2P \sim C_3H_6O_3 \quad (5)$$

oxidized   water   phospho-
NADP   glyceraldehyde

Two molecules of the 3-carbon sugar glyceraldehyde can be combined to form fructose or glucose (FIG. 5-12). Essentially, one 6-carbon sugar has been formed from one molecule of CO_2 and a 5-carbon sugar. The regeneration of the 5-carbon sugar ribulose is obviously basic to the continuation of photosynthesis. This, however, is a long and complicated process and will not be discussed here.

---

[2] In Eq. (4), $\sim$ P or P $\sim$ are symbols used to indicate high-energy phosphate.

Measurements of Photosynthesis The photosynthetic equation indicates that the rate of the process may be measured either by the rate at which carbon dioxide, water, or energy is used, or by the rate at which oxygen, water, or sugar is produced. All of these have been used, but percentage changes in water or energy are very difficult to measure.

A demonstration of changes in the rate of photosynthesis may be made with a water plant such as *Elodea* (FIG. 5-19). When shoots are cut from this plant and immersed in water in strong light, bubbles are emitted from the cut ends. If the bubbles are collected as shown in this figure and a glowing wood splinter is thrust into the test tube, the splinter will burst into flame, indicating an increased oxygen concentration. Analyses of the gas show that the bubbles are oxygen given off in photosynthesis and mixed with air that was dissolved in the water. The rate at which the bubbles are given off is a rough indication of the rate of photosynthesis. This, however, is subject to errors because of changing composition or size of bubbles, or because some of the oxygen liberated dissolves in the water without forming bubbles. The bubbling method can be used to show that photosynthesis varies with light intensity. It can also be used to demonstrate the necessity of carbon dioxide for photosynthesis, for no bubbles are produced in water that has been boiled to remove dissolved gases and then aerated with air freed of carbon dioxide.

Most of the fundamental research on the mechanism of photosynthesis has been done with algae in special flasks in which the volume of oxygen evolved can be measured to the nearest ten-thousandth milliliter. Skillful manipulation of expensive equipment is required for such measurements. In a useful method of measuring photosynthesis in higher plants, a leaf, still attached to the plant, or a plant is enclosed in a glass or plastic chamber in light. Air is then moved rapidly through the chamber and the volume and CO_2 content of the incoming and outgoing air is measured to show CO_2 absorption by difference. This figure is known as *net* or *apparent* photosynthesis. It can be corrected for the CO_2 produced by respiration (FIG. 6-2) to give *total* photosynthesis. Temperature, light intensity, CO_2 concentration, or various plant factors can be varied to determine their effect on the rate of photosynthesis. Plant chambers can vary from an airtight greenhouse covering an apple tree to a plastic envelope covering a leaf.

Investigations of the rate of photosynthesis over short periods of time contribute to an understanding of the photosynthetic effectiveness of a particular plant or crop. Important, also, are measurements of the dry matter produced throughout a growing season, for some 94 percent of the dry weight of plants is derived from the products of photosynthesis. By harvesting and drying plants at weekly intervals, a picture of total photosynthesis can be obtained. It has been shown by this method that a field of corn that yielded 100 bushels an acre gained on the average 160 pounds of dry matter per acre per

Fig. 5-19 Release of oxygen during photosynthesis. Bubbles of air enriched with oxygen are given off from the green plant in the sunlight.

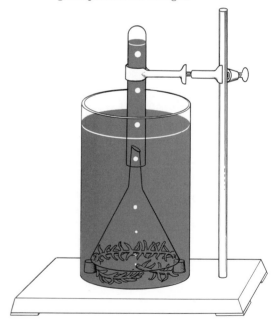

day from the time the plants were 2 feet high until grain production was approximately complete. Maximum rates of photosynthesis in corn and sugar cane (FIG. 5-1) are about 400 pounds of dry matter per acre per day, although a rate of nearly 600 pounds has been obtained under special conditions.

Factors Affecting Photosynthesis The rate at which photosynthesis proceeds is affected by a number of factors, both environmental and internal. Among the environmental factors that may limit the rate of photosynthesis are light intensity, carbon dioxide concentration, temperature, and water supply.

Light intensity Of all the limiting factors in photosynthesis, light intensity has probably received the greatest attention. Full sunlight on a clear summer day has an intensity of about 10,000 foot-candles. A foot-candle is the light cast by a *standard candle* at a distance of 1 foot. A good reading light has an intensity of 20 or 30 foot-candles.

In general, in leaves of plants whose normal habitat is bright sun, the rate of photosynthesis tends to be approximately proportional to the light intensity received by the leaves, up to a maximum of one third to one half of full sunlight. The point at which further increase in light intensity does not increase the rate of photosynthesis is called the LIGHT SATURATION POINT. At this point—which varies with the plant and the conditions, the rate of photosynthesis should not be affected by variations in light intensity during the day, as by clouds, as long as the intensity does not fall below the saturation point.

Variation in the light saturation point for Irish potatoes with varying CO_2 concentrations is shown in FIG. 5-20. Light saturation was at 3000 foot-candles with normal air containing 0.03 percent CO_2. When the CO_2 content of the air about the leaves was increased two and five times, light saturation occurred at 4200 and 5200 foot-candles. With more CO_2 the leaves used more

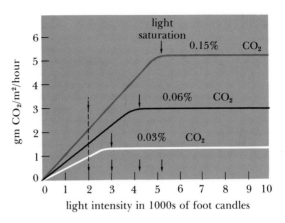

Fig. 5-20 **Effect of light intensity and CO_2 concentration of the air on photosynthesis in Irish potato (*Solanum tuberosum*). Photosynthesis increased with light up to light saturation (arrows) and with CO_2 concentration even when light was below saturation (vertical dashed line).**

light. Note, also that at 2000 foot-candles (dashed line), where light was limiting at all levels of CO_2, added CO_2 more than doubled the rate of photosynthesis. It is frequently but erroneously stated that when a process is conditioned as to its rapidity by a number of separate factors, the rate of the process is limited by the pace of the slowest factor. In other words, only the single, most limiting factor affects the rate of a process, and only additions of this factor—light, in this example—can be expected to increase the rate. This concept is called the LAW OF LIMITING FACTORS. The data of FIG. 5-20 illustrate the actual relationships. With light limiting (2000 foot-candles), added CO_2 increased photosynthesis. But CO_2 had much more effect when light was not limiting (more than 5000 foot-candles). More than one factor can be limiting at the same time. The greatest response can be expected from an addition of the "most limiting" factor, but other factors can still affect the over-all rate. Limiting factors do not act alone, but interact with each other.

Light saturation is a common condition for fully exposed leaves in full sunlight. For plants growing in the shade of other plants, and for the lower leaves of a plant that are shaded by the leaves above them, light is frequently a limiting factor in photosynthesis. A reduction in light intensity that would not affect the upper leaves could become seriously limiting for the lower leaves and thus reduce photosynthesis in the entire plant. As a result, any drop in light intensity tends to reduce total photosynthesis in trees or in plants growing thickly together in a forest or in a field.

The light under a low-growing maple may drop below 1 percent of that in the open—in other words, may reach 100 foot-candles as an approximate maximum. Mosses and a few shade plants may grow under such conditions, but grasses will not. The better lawn grasses are not shade-loving plants (Fig. 5-21).

Foresters estimate that they will obtain slow growth and satisfactory survival of shade-tolerant forest seedlings with 5–10 percent of full light. If light drops below about 2 percent, growth on the forest floor consists of ferns, mosses, and the most shade-tolerant weeds and shrubs. Timber species will then reproduce only if the canopy of the older trees is opened by the death of some of the trees or by cutting them.

The mosses, ferns, and various other plants that grow in heavy shade commonly have a low rate of growth and respiration, so that they use food slowly. They are also able to take advantage, even in low light (Fig. 5-21), of the higher carbon dioxide concentration resulting from the respiration of microorganisms in decaying forest litter. Many house plants are shade-tolerant; that is, they are able to survive and grow under light of low intensity, either daylight or artificial light. They normally benefit, also, from the extra carbon dioxide produced by human respiration. Even such plants, however, grow better if they receive good light at times.

It is completely uneconomical to increase the total light on a field scale. It is possible to catch more sunlight, however, by producing more and larger leaves on an acre. Thicker planting, irrigation, and fertilization result in increased yields primarily because they increase total leaf surface and thus total absorption of sunlight and carbon dioxide. It is necessary, however, to keep these factors in balance. Heavy planting, even with increased fertilization, can result in crop failure if the water supply is inadequate for the increased leaf surface.

In the search for higher grain yields in corn, the effect of close planting has received much emphasis. If planted closely, the upper leaves of corn with the usual horizontally oriented leaf blades shade the lower leaves, reducing the light received by them to a low value. New, upright-leaved varieties of corn are being developed (Fig. 5-22). In such varieties the erect leaves allow sunlight to penetrate to the lower leaves and so increase the total photosynthetic area of a thickly planted crop. The light-absorbing surface is increased by two to three times; the plants carry on more photosynthesis and produce more grain. Upright-leaved plants are also being developed in sorghum and other small grains.

Carbon dioxide When light, water, and other factors are optimum, the concentration of carbon dioxide is commonly a limiting factor in the rate of photosynthesis in both wild and crop plants. It should be possible, therefore, to increase photosynthesis by increasing the concentration of carbon dioxide in the air surrounding the leaves (Fig. 5-20). In short-time experiments in which the concentration of this gas was increased up to 13 times that of normal air, the increase in the rate of photosynthesis was 400–500 percent in potatoes, sugar beets, and alfalfa. In the few experiments in which plants have received carbon dioxide for longer periods, the results have been variable. More research is needed to determine the possible effects of continuously high levels of this gas. Actually, under average field conditions, the supply of carbon dioxide is relatively better than the supply of water and fertilizers. Adequate water supplies and proper fertilization will usually double or triple the average yield, even though the carbon dioxide supply is unchanged.

10,000 ft-c   300 ft-c

Fig. 5-21 Failure of blue grass (*Poa pratensis*) in heavy shade. Light intensities in foot-candles are shown above the stakes. No grass survived with less than 300 foot-candles of light.

In greenhouses where the CO_2 concentration can become very low—less than 0.01 percent—carbon dioxide may be added to increase photosynthesis. This can be done by releasing the gas slowly from high-pressure cylinders or, more economically, by burning natural gas (CH_4) in the house to produce both heat and CO_2. Propane gas ($CH_3CH_2CH_3$) can be used also. The gases from coal fires contain ethylene and other materials that are toxic to plants and, unfortunately, cannot be used.

Temperature The speed of enzymatic reactions, as is that of other chemical processes, is generally increased with a rise in temperature over a range from near freezing to above 40°C (104°F). It might be assumed, therefore, that a higher tem-

Fig. 5-22 A corn plant bred for stiffer, more upright leaves (left) compared to a normal plant on the right. The upright-leaved plants allow sunlight to penetrate to the lower leaves and so increases photosynthesis. The drooping leaves (right) shade out the lower leaves of the plant.

perature in this range would increase the rate of photosynthesis. In laboratory experiments with high concentrations of carbon dioxide, the rate of photosynthesis is commonly doubled by a temperature rise of about 10°C (18°F). Under field conditions, however, a number of plants have shown little or no effect of temperature on photosynthesis. The probable explanation of this difference is that normal carbon dioxide concentrations, some 100 times lower than those used in the laboratory, so limit photosynthesis under natural conditions that speeding up the enzymatic processes does not greatly increase the over-all rate.

Water supply The rate of photosynthesis in wilted leaves is reduced, sometimes to zero. This reduction is primarily due to reduced carbon dioxide absorption caused by the closure of the stomata with wilting. Only about 0.1 percent of the water absorbed by plants is used in photosynthesis as a source of hydrogen. It is improbable, therefore, that the water supply will be directly limiting for photosynthesis. In contrast, the indirect effects—closing of the stomata; reduced leaf area because of reduced growth; and possibly an interference with the functioning of enzymes because of a reduced water content of the protoplasm—may seriously reduce total photosynthesis.

Photosynthesis in Some Plants of Tropical Origin Beginning in the early 1960s, investigations have shown that certain flowering plants of tropical origin have a net photosynthetic rate almost twice that of other plants. These plants, mostly grasses, include corn, sugar cane, sorghum, and a number of forage grasses such as coastal bermuda grass (*Cynodon*), Bahia grass (*Paspalum*), and crab grass (*Digitaria*). It has been established that these "tropical grasses" combine CO_2 with a 3-carbon acceptor, forming stable, 4-carbon organic acids such as oxaloacetic. This replaces fixing CO_2 in the light by combining it with ribulose diphosphate, forming two mole-

cules of phosphoglyceric acid (Eq. 4). The acids of the oxaloacetic type formed in these grasses are then converted to simple sugars, probably by way of phosphoglyceric acid. The photosynthetic mechanism found in the "tropical grasses" is probably in part responsible for the higher rate of photosynthesis and consequent greater efficiency in the use of carbon dioxide. However, the greater photosynthetic efficiency of these tropical plants may be due more to the absence of minute particles, called *peroxisomes*, in the cytoplasm of their leaf cells. These particles are the site of a wasteful type of respiration. Other plants, possessing these bodies, waste a significant portion of the products of photosynthesis and thus have a lower net rate of photosynthesis than the tropical grasses. A possible contributory factor also is that the leaf cells that carry on photosynthesis at the highest rate are those of the bundle sheath (Fig. 5-10), thus allowing a more rapid rate of movement of photosynthetic products out of the leaf. If, however, sugars are produced more rapidly than they are moved out of the leaf, the excess is converted to starch, which accumulates in the bundle sheath.

Breeding programs, in which progeny are selected for the more efficient method of carbon incorporation and for lower rates of wasteful respiration in the light, could lead to doubling of photosynthesis in many of the crop plants that help to feed the world.

Photosynthesis, Food, and Population The number of people that can exist on this planet is directly dependent upon the capacity of photosynthesis, the only source of food for living creatures. In the face of an exploding world population, can we look forward to increasing the ability of the green plant to produce food so that the peoples of the world can always be fed? Can we anticipate constantly increasing yields from our cultivated food plants?

Yields of our major crops have, it is true, been increasing over the past hundred years, and this increase, for many crops at least, will continue.

Much has been accomplished by the development of improved varieties to give higher yields and to provide increased disease resistance. Examples of these are the new high-yielding varieties of wheat and rice now widely planted in Asiatic countries (Chapter 1). Improved yield has also been brought about by the use of hybrid corn, new varieties of soy beans, and many other crops. The protein content of crop plants can be increased and the shape and conformation of plants modified to allow more efficient use of sunlight. Even the efficiency of the photosynthetic process itself can be improved. Crop yields can also be increased by numerous agricultural practices such as increased use of fertilizers, new irrigation projects, and new soil-management methods. In spite of these and numerous other accomplishments and possibilities for increase in food production, there is a limit to the yield of the world's crop plants, but there appears to be no limit to population growth. Today, three-fourths of the world's people exist on a sub-standard diet. If the present rate of population growth continues until the end of this century, a tripling or even a quadrupling of the food supply of the world will be required to provide an adequate diet. It is generally held that such an increase in production is not possible, and it is certain that all efforts at increasing food supply will be futile unless there are concomitant measures to control population increase. Efforts at increasing food supply must continue, of course, but with the realization that they merely buy time pending the control of population size to a point where the nutritional needs of all mankind can be satisfied.

PLANT FOODS

As pointed out earlier in this chapter, it is considered that the simple sugar fructose is the typical direct product of photosynthesis. Conversion of carbohydrates from one into another takes place so rapidly within the plant that it makes little difference whether fructose, glucose, or some other sugar is the primary product.

The sugar manufactured in the green cell undergoes many transformations. Some of it is utilized in the process of respiration, either in the leaf or in other parts of the plant. The energy of the sugar is thus released to carry on the work of the plant. The sugar not used immediately is transformed into other carbohydrates, or is used in constructing lipids, proteins, or other compounds. These compounds are produced from sugar by transformations in which light is not necessary; the energy required is obtained from the chemical energy of ATP. This ATP is produced by oxidizing some of the sugar (Chapter 6).

Carbohydrates Among the important carbohydrates that may be constructed by the plant from simple sugars are sucrose, starch, and cellulose. Cellulose, although a carbohydrate, is not a food for higher plants but only a building material. The other carbohydrates are forms of stored food. In a number of plants, such as sugar beet, sugar cane, and onion, carbohydrate is stored in the form of sucrose, but in most species starch is the chief storage product.

During daylight hours, sugar may be synthesized in the leaf more rapidly than it is used or moved to other tissues. The excess sugar may then be transformed into starch, which appears as a temporary storage product in the chloroplasts of the mesophyll. This starch is later reconverted to a soluble sugar that is transported from the leaf. The movement of sugar out of the leaf is probably faster during the day than at night, but during the day the rate of outgo is normally overbalanced by the rate of production in photosynthesis.

Because of the rapid conversion of sugar into starch in many plants, a test that demonstrates the presence of starch in a leaf is considered to indicate that photosynthesis has occurred. The accumulation of starch in the leaf as a temporary reserve is readily demonstrated. If the chlorophyll is extracted with alcohol or acetone from the leaf of a starch-producing plant that has been in the dark for some hours and the leaf is then stained with iodine, only the brown color of the iodine will be apparent. But if a leaf that has been illuminated for a day is treated in the same way, it will stain blue or black, indicating the presence of starch (FIG. 5-23). This starch is sometimes termed the "first visible product" of photosynthesis, but the formation of starch is a reaction different from that in which sugar is produced. Starch may be formed from sugar in storage tissues not exposed to the light. Under experimental conditions, starch can also be produced in parts of leaves that ordinarily do not form starch by supplying them artificially with a high concentration of sugar.

Proteins Some of the sugar produced by the plant is utilized in constructing proteins—compounds essential to life and growth. As shown in Chapter 4, proteins are composed of simpler compounds, termed amino acids, that are linked together to form large, complex protein molecules.

Proteins accumulate in plants as protoplasmic protein or as storage proteins. Seeds, especially those of legumes, are rich in protein, as are corms, bulbs, rhizomes, buds, and storage roots. Plants use proteins in construction of new protoplasm, and these compounds are therefore required in large quantities during growth. In the germinating seed, stored proteins of the seed are converted to soluble forms and transported to the growing seedling. In the spring growth of woody plants, surplus protein is available from the living cells of the wood and, especially, from the living cells of the bark.

Apparently the manufacture of amino acids from carbohydrates and nitrogen can take place in many parts of the plant, although certain leaf and root tissues are more active in this respect. Once formed, the amino acids may be built into proteins immediately or they may be transported to other tissues and there transformed into protein and used or stored.

Fig. 5-23 The starch test on a leaflet of garden bean after exposure to sunlight. A stencil of the word "starch" was placed on the leaflet in the early morning. In the late afternoon the stencil was removed and the chlorophyll extracted with alcohol. The leaflet was then treated with iodine and photographed.

The amino acids are required in animal as well as in plant nutrition. While all amino acids are equally important in protein synthesis, feeding experiments have shown that the amino acids used by animals can be classed in two groups, the ESSENTIAL and NONESSENTIAL. The term "essential" means only that animals do not construct these from materials in the diet; rather, they must be supplied preformed and are derived originally from plant proteins. The nonessential amino acids are also required by animals, but they can be constructed by the transformation of one amino acid into another. Animals, therefore, require an adequate total of amino acids, including a supply of each of the essential forms.

Experiments have shown that ten amino acids are essential for the dog and rat, and eight are essential for adult human beings.[3] Most animal protein foods, such as eggs, meat, and milk con-

[3] The essential amino acids for adults are lysine, tryptophan, threonine, methionine, phenylalanine, leucine, valine, and isoleucine. Histidine is also required by infants.

tain all of the essential amino acids and are classed as high-quality proteins. A number of kinds of plant foods, however, such as potatoes, peas, beans, corn, wheat, and rice are deficient in at least one, generally more, of the essential amino acids. Most cereal grains (for example, wheat and rice) are deficient in lysine and threonine, and corn is deficient in lysine and tryptophan. By combining different plant foods it is possible to obtain a satisfactory amino acid balance. Such a balance is more easily obtained by the consumption of animal proteins in addition to the proteins from plants. Animal proteins are expensive, however. Grains now constitute the principal human foods—they annually contribute more than 40 million tons of protein to the human diet—and as the earth's population increases, their use must be increased. Two procedures, among others, may be used to increase the biologic value of cereal foods. One is the breeding of cereals with a higher content of essential amino acids. In this connection a great step forward was taken by the discovery, in the early 1960s, that either of two mutant genes in corn—opaque-2 and floury-2—will double the usual concentration of lysine in the grain. An infant-feeding formula, produced for use in developing countries, contains high-lysine corn as its main ingredient. The second procedure is the fortification of cereal products by the addition of small quantities of essential amino acids, produced synthetically or by fermentation. It is economically feasible to supplement bread and other cereal foods with essential amino acids, as they are now supplemented with vitamins and iron. It is estimated that 2 or 3 cents worth of lysine in a loaf of bread would make the proteins of this food equal in quality to those of milk. It is a commentary on scientific progress that poultry and stock feeds are already receiving important supplementary amino acids and other special nutrients.

Leaf proteins are usually of better quality than seed proteins, and many animals obtain their entire supply of essential amino acids from green leaves. Ruminant animals, such as the cow, sheep, and deer, not only feed upon leafy materials, but also obtain essential amino acids from bacterial synthesis in the rumen (Chapter 20), and are thus largely free of the need for specific amino acid content in their feed.

Lipids The lipids, or fatty substances, are widely distributed in plant tissues, primarily as droplets throughout the cytoplasm. Lipids include FATS, which are solid at room temperatures, and OILS, which have a lower melting point and are liquid at room temperatures. Lipids are found in animals mainly as fats, and in plants, chiefly as oils, although there are exceptions to both, such as the fat of the cacao bean and the oils of fish and seals. Oils, in general, can be converted to fats by being combined with hydrogen. This process, known as *hydrogenation,* is used in producing oleomargarine and cooking fats from plant oils. Some lipids, chiefly animal lipids but some plant lipids as well, are saturated with hydrogen in the natural state. It is believed by some physiologists that the consumption of saturated lipids is likely to result in an increased production of cholesterol in the blood stream and hence to lead to circulatory diseases.

Plant oils contain the same chemical elements as sugar, but the ratio of oxygen atoms to carbon atoms is much lower than in sugar. Oils are primarily sources of energy. They are stored chiefly in fruits and seeds, especially the latter. Among seeds rich in oil are the castor bean, Brazil nut, cotton, peanut, corn, palm, coconut, nutmeg, sunflower, sesame, safflower, tung and soybean. Fruits high in oil include the avocado and olive, about 50 percent of the dry weight of the latter being oil. Many seeds contain oil alone as a reserve food; in others the oil may be associated with starch. In animals, fats are the chief form in which reserve foods are accumulated. In contrast to their dependence upon plants—including rumen bacteria—for amino acids, animals are able to manufacture many fatty materials from carbohydrates. Most cereal grains are low in lipids, the content ranging from 2 to 5 percent. Farm animals that are fed upon such grains accumulate body fat greatly in excess of the fatty constituents of the diet. This fat is built up from the carbohydrates in the grain. Animals are dependent upon their feed, however, for small quantities of certain essential unsaturated fatty acids, which are derived from the oils of plants.

FOOD ACCUMULATION

The accumulation of reserve food materials in various plant organs is a very significant plant activity. In annuals, such foods accumulate chiefly in the seeds, but in perennials, reserve foods are deposited in woody stems, roots, corms, rhizomes, tubers, and bulbs. These reserves enable the plant to resume growth the next spring.

In flowering bulbs such as narcissus or hyacinth, the food reserves are sufficient for producing flowers and leaves. The replenishment of this food reserve in the bulb does not usually take place until after the flowering period; hence the leaves must be allowed to persist until they die. Another example of the importance of the green shoot in relation to storage is found in the growth and harvesting of asparagus. The edible shoots produced in the spring develop from food accumulated in the roots during the preceding growing season. The length of the asparagus harvest is restricted to 6 or 8 weeks in order that the new photosynthetic shoots that appear after this period may have time to accumulate food for the following year.

The continued removal of the aerial portions of growing perennials not only prevents the accumulation of food but also gradually weakens the subterranean structures through starvation. The overgrazing of pastures and ranges, for example, eventually causes the desirable forage grasses to die out, and these are replaced by unpalatable weedy plants. (FIG. 15-9). The principle of starving underground parts may be applied in the eradication of perennial weeds through frequent cultivation during the growing season. But the reserve food supply of certain weeds is so large that this treatment may have to continue for one or more seasons before the weeds are eradicated.

SOME NUTRITIONAL RELATIONSHIPS

Plants may be divided, with respect to their source of food, into two broad categories. In the first are the green, or autotrophic, plants, which manufacture all of their own food. This group includes, in general, the algae, liverworts and mosses, the ferns, conifers, and the flowering plants. The second class comprises the plants that must obtain a part or all of their food from external sources. Such plants are termed heterotrophic. The chief heterotrophic plants are the bacteria and fungi. The heterotrophs include both saprophytes and parasites, with many common and widely distributed species. Saprophytes obtain their food from nonliving materials such as decomposing plant and animal tissues. Parasites derive their food directly from other living organisms.

Although parasites are most commonly microorganisms, many species of flowering plants—as many as 2500 or more—also fall in this category. The degree of parasitism varies greatly. In some, such as certain mistletoes and root parasites, the shoot is well developed and photosynthesis occurs to such an extent that it is generally believed that the parasite obtains little but water and minerals from the host. Such plants have therefore been termed partial or HEMIPARASITES. At the other extreme are certain plants, such as members of the broomrape family, that lack chlorophyll and are completely dependent upon the host for water, minerals, and organic compounds.

The pernicious witchweed (*Striga asiatica*) long known in Africa, Asia, and Australia, and recently introduced into the United States, is an example of a root parasite. The aerial portion of this annual plant is 6–12 inches high, bright green, with small, narrow leaves and reddish

Fig. 5-24 Witchweed (*Striga asiatica*), a hemiparasite. The roots are attached to the roots of the corn host.

Fig. 5-25 Dodder (*Cuscuta gronovii*) twining about stem of jewelweed (*Impatiens*).

flowers (FIG. 5-24). The seeds are minute and numerous, and germinate only when near a susceptible root. Following germination, the tip of the emerging rootlet grows into contact with a host root and becomes swollen. From this swollen portion, projections grow into the conducting tissues of the host root. The original root then branches, and these secondary branches also attack the host. The plant does not appear above the ground for about a month after germination, and during this time it is a complete parasite. The aerial part, when it appears, is presumably hemiparasitic, since it seems amply provided with chlorophyll but is still dependent upon the host for water and minerals. Witchweed attacks a number of cultivated plants, including corn, sorghum, and sugar cane. The host plants, their roots damaged by the invading parasite, are stunted in growth and may be killed.

An even more specialized parasite is DODDER (FIG. 5-25), a slender, twining, leafless, yellow vine. It grows upon many kinds of wild and cultivated plants, and is especially destructive to al-

Fig. 5-26 Dodder (*Cuscuta*). (A): coil of dodder on red clover (*Trifolium pratense*), showing haustoria in contact with vascular tissues of host. (B): portion of host and parasite, greatly enlarged.

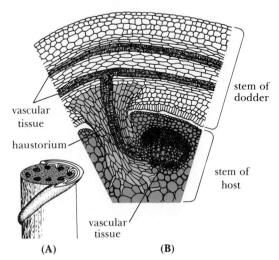

falfa and clover. Its water, minerals, and food are obtained through HAUSTORIA that invade the host tissues (FIG. 5-26). In dodder the haustoria are modified adventitious roots. Dodder is not an absolute parasite, for it still retains some chlorophyll in the seedling, buds, fruits, and even in the stem, but the quantity of food manufactured is of little significance in the economy of the plant. The presence of chlorophyll in dodder and some other parasites may be interpreted as evidence that the ancestors of such plants possessed normal chlorophyll and were autotrophic.

Heterotrophs among herbaceous flowering plants include certain nongreen species of orchids, members of the heath and gentian families, and others, commonly called saprophytes. These plants usually grow in humus, and their roots are covered by a web of fungus threads that invade the roots and extend into the soil (mycorrhizae, Chapter 9). It is through the

fungus that the flowering plant obtains carbon compounds, water, and minerals from the soil. The term saprophyte is unsuitable, however, for the flowering plants themselves do not decompose nonliving organic matter as do saprophytic decay bacteria and fungi. It is the fungus that is saprophytic, and the relation between higher plant and fungus may be mutualistic or even parasitic. The Indian Pipe, *Monotropa uniflora* (FIG. 5-27) is usually termed a saprophyte. Investigations in Sweden, however, on a species of the same genus, the pinesap (*Monotropa hypopitys*) indicate that the roots of the pinesap and spruce or pine trees are closely intermingled, that the fungus threads of the pinesap roots are continuous with those of the rootlets of the forest trees, and that the "saprophyte" receives all its nutrients, organic and inorganic, from the host trees through these threads. The pinesap is therefore a parasite, not a saprophyte.

Fig. 5-27 Indian pipe (*Monotropa uniflora*). The plants are about 7 inches high.

SUMMARY

1. The manufacture by green plants of food from inorganic materials is the process upon which life depends. This process is photosynthesis, the manufacture of sugar in the presence of chlorophylls and light, with carbon dioxide and water as raw materials. Oxygen is released from the water molecules as a by-product.

2. Photosynthesis takes place in the chloroplasts, which are located in the mesophyll of the leaf. The leaf tissues are epidermis, mesophyll, and vascular bundles. The mesophyll is composed of parenchyma cells and is commonly divided into palisade and spongy parenchyma. The vascular bundles are made up of xylem and phloem, collectively constituting vascular, or conducting, tissues. Strengthening sclerenchyma fibers are frequently associated with the vascular bundles. The epidermis contains openings, stomata, each with its surrounding guard

cells. The stomata are the chief paths of movement of gases into and out of the leaf.

3. Essentially all wavelengths of the visible spectrum can be used in photosynthesis. Leaves are green because they reflect somewhat more of the green wavelengths than of the red and blue. The difference in reflection is not great, however, and the green wavelengths are important in photosynthesis.

4. In photosynthesis, radiant energy is converted into chemical energy. In the light reactions, the energy of light is absorbed and used to split water into hydrogen and oxygen, and form ATP and $NADPH_2$. In the dark reactions, carbon dioxide and water are combined with a 5-carbon sugar (ribulose) and the 6-carbon molecule is then split into two molecules of 3-carbon glyceric acid. Hydrogen, derived from water through $NADPH_2$, is then used to reduce the glyceric acid, resulting in the formation of the 3-carbon sugar phosphoglyceraldehyde (PGA). Two molecules of PGA then combine, forming the hexose (6-carbon) sugars fructose or glucose.

5. The rate of photosynthesis may be measured by the amount of product formed or of materials used in a given period of time. It is affected by such factors as water supply, light intensity, and carbon dioxide concentration. Any one of these may be a limiting factor. Temperature may limit photosynthesis in laboratory experiments when high concentrations of carbon dioxide are used, but does not seem to be important under natural conditions.

6. The sugar made by green plants is used in various ways. Some of it is respired, some converted into other carbohydrates or lipids. Some is used in the manufacture of amino acids, the building stones of proteins. Certain amino acids, essential to the nutrition of man and other animals, are constructed only by plants, including bacteria. In general, animals are dependent upon plants not only for carbohydrates but also for essential amino acids and small quantities of certain fatty acids.

7. Surplus foods accumulate in seeds, stems, and roots. The surplus accumulated in stems and roots enables perennial plants to renew growth after a dormant period. The food used in this early growth is replenished later in the growing season. Underground parts of perennial plants may be gradually starved by repeated removal of the leaves. The surplus foods accumulated in plants constitute the chief food supply of man and other animals.

8. Plants may be classified, on the basis of their methods of obtaining food, as autotrophic or heterotrophic. Green plants are autotrophic, and nongreen plants are heterotrophic (saprophytic or parasitic). Heterotrophs include not only bacteria and fungi but also many species of flowering plants.

Respiration and Digestion

HUNDREDS, PROBABLY THOUSANDS, of diverse chemical reactions are constantly proceeding in every active living cell. The sum total of these chemical changes are referred to as METABOLISM. Some activities, such as photosynthesis, assimilation, and the synthesis of complex foodstuffs, are constructive—they build new compounds. Other reactions are degradative; the most important of these are digestion and respiration, in which complex food materials are broken down into simpler compounds. The term ANABOLISM is applied to the constructive processes, and CATABOLISM to the degradative. Anabolic and catabolic reactions are interrelated and interdependent, and commonly proceed at the same time in the same living cell. Both are essential to the survival and growth of the cell.

RESPIRATION

Plant and animal cells use energy in the building and maintenance of protoplasm, of protoplasmic membranes, and of cell walls. They obtain this energy, as does the automobile engine or the steam boiler, from the oxidation of organic compounds. When wood is burned, it must first be heated to a high temperature—on the order of 1600°F (870°C). At these temperatures the molecules are activated to the point where they combine spontaneously with oxygen of the air, and energy is liberated as heat. The burning of sugar in a living cell is comparable to the combustion of wood or coal, but spontaneous oxidation at high temperatures is impossible in a living cell. It is, rather, a slow process that can take place at room temperature and occurs in a number of steps under the control of numerous enzymes. This controlled burning within living cells is called RESPIRATION. Besides its slow speed, respiration differs from combustion in a bonfire in that a large percentage of the energy of respired sugars is converted into the energy of ATP, which can be used by the cell instead of being dissipated as heat.

In respiration, a sugar, such as glucose, is changed into simpler substances, with the transformation of energy. This may be represented by

$$C_6H_{12}O_6 + 6O_2 \longrightarrow$$

sugar   oxygen

$$6CO_2 + 6H_2O + energy\ (38\ ATP) \quad (1)$$

carbon   water
dioxide

Note that the reaction, from several standpoints, is the opposite of photosynthesis. The respiratory processes convert some of the energy captured in photosynthesis into a form that can be used by protoplasm.

The Use of Energy   It is clear that energy, the ability to do work, is involved in lifting a weight or in performing other mechanical work. It is not so clear that energy is used in the accomplishment of the internal work that allows a plant or animal to live and grow. The energy transformed in respiration appears partly in the form of heat, which is of no value to plants, and partly as the *useful energy* of ATP. It is the latter that is essential in the activities of the living cell.

A small amount of respiratory energy expresses itself as motion—in the streaming of protoplasm, the movement of chromosomes, and in other minor ways. Much of the energy is expended in ASSIMILATION, the process by which food is converted into cell walls and protoplasm. Energy is required in the chemical reactions by which proteins and lipids are synthesized from the products of photosynthesis. Energy may be used also in maintaining the working structure of the protoplasm, in the accumulation of solutes by cells, and in the movement of food materials through the plant.

In addition, the energy of respiration is used in the synthesis of a number of miscellaneous substances in plants. Some are of vital importance to the plant, but the significance of others is not yet known. These substances include pigments; vitamins; essential oils; gums; resins; organic acids such as malic, citric, oxalic, and tartaric; tannins; glucosides such as digitalin; and alkaloids such as caffeine, nicotine, and strychnine.

The internal work of the plant manifests itself externally in growth. Leaves, flowers, seeds, and fruits are formed. The plant increases in height and diameter. Roots extend through the soil. The utilization of energy by plants does not result in conspicuous motion, as it does in most animals—an obvious difference between most plants and most animals. But beneath this difference is found basic similarity, for energy is required to carry on the life processes within the cells of both. In the use of energy to carry on cellular processes, all life is essentially the same.

Evidences of Respiration   Respiration is usually accompanied by several phenomena that indicate the nature and rate of the process. A study of these should not, however, obscure the most significant aspect of respiration, the conversion of energy. Some of the more obvious aspects of respiration can be demonstrated directly.

Rise in temperature   Much of the energy of respiration may be released in the form of heat. In warm-blooded animals, this heat is important, for it may maintain the temperature of the body. In plants, it is commonly viewed as wasted energy, for it is quickly dispersed into the environment, and the temperature of the plant is therefore approximately that of the surrounding air. In this respect plants resemble cold-blooded animals, which possess no physiological mechanism for maintaining a constant body temperature. Actively growing tissues and organs, such as germinating seeds, young fruits, and opening buds, have a higher metabolic rate and respire at a higher rate than mature ones. If the heat generated by such tissues is confined, the temperature rises.

This temperature rise is readily demonstrated by placing a number of oat or other seeds, previously soaked in water, in a vacuum flask and inserting a thermometer through a loose cotton plug. After germination has begun, the rise in temperature may amount to 10° or 20°C (18° or 36°F), depending upon the kind of seed used. Bacteria and fungi that may be present can raise the temperature still higher. If the flask is stoppered to exclude oxygen, however, heating will drop to a low value as soon as the oxygen in the flask is used up.

When the outer layers of a mass of plant materials act as insulators, a temperature rise due to respiration can also be noted. For example, the interior of a crate of celery or green beans may register 6°C (11°F) higher than the outside air. Freshly harvested, damp bluegrass seed packed in bags to be taken to the drier may heat above the death point of the seed, about 50°C (122°F), within hours.

Use of oxygen and release of carbon dioxide The rates at which carbon dioxide is released or oxygen used by living plant tissue are generally accepted indexes of the rate of respiration. Qualitatively, the use of oxygen can be demonstrated by partly filling a large flask with germinating peas or other seeds. The flask is stoppered and allowed to stand overnight. If a burning splint is then lowered into the flask, the flame will be extinguished, indicating that oxygen has been used by the seeds.

A demonstration of carbon dioxide production makes use of the fact that calcium hydroxide solution (limewater) reacts with carbon dioxide to form insoluble calcium carbonate. Two flasks, one containing dry peas and the other an equal number of germinating peas, are stoppered and allowed to stand for several hours. The air in the flasks is then displaced by adding water through a funnel. The displaced air is conducted through a solution of calcium hydroxide. The air from the dry seeds contains little carbon dioxide and produces no observable change in the limewater. That from germinating seeds, on the other hand, brings about the rapid formation of calcium carbonate, as evidenced by a milky precipitate (FIG. 6-1).

Loss in weight That respiration results in a loss in weight is readily demonstrated. An experiment to show this in a green plant must, of course, be carried on in the dark, for in the light the plant may gain in weight as a result of photosynthesis. The loss in weight due to respiration is easily established with germinating seeds. A number of seeds of known moisture content are weighed and then germinated in the dark. When the seedlings are well developed and the stored food in the seeds is largely exhausted, the dry weight of the seedlings and seed remnants is obtained and found to be less than the calculated dry weight of the seeds with which the experiment started. The difference in weight represents the carbohydrate lost in respiration.

Respiration and Breathing Respiration must occur in every active, living cell if life is to be maintained. Not only plants but also all animals respire. In addition, some animals also breathe, and the distinction between respiration and breathing must be clearly understood—particularly because in common usage respiration is used incorrectly as synonymous with breathing. The more highly developed animals have a breathing mechanism, including lungs, by which air is passed over moist membranes filled with small blood vessels. Here oxygen and carbon dioxide are exchanged over a large membrane

Fig. 6-1 Demonstration of the production of carbon dioxide by germinating peas. The air from the flask at the left is displaced through limewater in the flask at the right. A milky precipitate of calcium carbonate indicates the presence of CO_2.

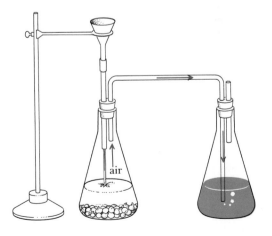

surface. The blood then carries the oxygen to the cells, where it is used in respiration. Plants, however, possess no such mechanism, and gaseous exchanges between the internal tissues and the outer air are brought about only by diffusion. Breathing in animals is a mechanical process that permits quantities of oxygen to reach active, deeply covered tissues. Respiration remains the important process, involving the actual conversion of energy. It is true, therefore, as frequently stated, that some animals breathe, whereas plants do not. But the respiratory processes in plants and animals differ in no fundamental way. Plants respire, whereas some animals both breathe and respire.

Gas Exchanges in Photosynthesis and Respiration A comparison between photosynthesis and respiration shows that the gaseous exchanges in the one are the opposite of those in the other. In photosynthesis, carbon dioxide is used and oxygen is released as a by-product. In respiration, oxygen is used and carbon dioxide is released. Since, in the green cells of plants in the light, both processes go on at the same time, the question arises as to the extent to which these two products are utilized by the plant.

The evidence is clear that the carbon dioxide released in respiration may be used in photosynthesis and that the oxygen released in photosynthesis may be used in respiration. Under light of low intensity, these two processes may exactly balance each other, and carbon dioxide and oxygen will move neither out of nor into the leaf. The light intensity at which this balance occurs in a leaf is called the COMPENSATION POINT.

The light intensity at the compensation point is low—usually less than 2 percent of full sunlight—but it varies in different plants. Temperature and other factors that affect either respiration or photosynthesis also affect the compensation point. At the compensation point there is neither a gain nor a loss in the dry weight of the leaves during the day. Under such conditions, however, a plant could not long survive, for during the day

the nongreen parts of the plant respire and use food, and at night respiration alone occurs. The plant as a whole, then, can survive at the compensation point only as long as stored food is available.

During the day, however, photosynthesis in adequately illuminated plants usually proceeds at a rate five or ten times that of respiration. All the carbon dioxide released in respiration is therefore used in photosynthesis, and additional carbon dioxide diffuses into the leaf from the air. Respiration uses some of the oxygen of photosynthesis; the excess diffuses from the leaf into the air. These gaseous exchanges are shown diagrammatically in FIG. 6-2. At night, of course, the plant utilizes only oxygen and liberates carbon dioxide. So rapid is the process of photosynthesis under favorable conditions, however, that, considering day and night together, the plant normally liberates much more oxygen than it uses, and it uses more carbon dioxide than it releases to the air.

Since photosynthesis normally exceeds respiration, the green plant manufactures more food than it uses in respiration. This excess food is available for immediate assimilation and for accumulation within storage cells. Such food makes it possible for growth to occur, since growth depends upon recently formed excess food or upon accumulated food reserves.

Effects of Respiration on Air of Rooms The amount of carbon dioxide liberated in plant respiration is sometimes overestimated by the layman; hence the belief that it is unhealthy to permit plants to remain in bedrooms during the night. In some hospitals potted plants and cut flowers are removed from rooms in which patients sleep. The practice was established initially in the mistaken belief that the plants would seriously vitiate the air of the room. It is sometimes continued because the life of flowers is prolonged if they are removed to a cooler room at night or because seriously ill persons may be unfavorably affected by the odors.

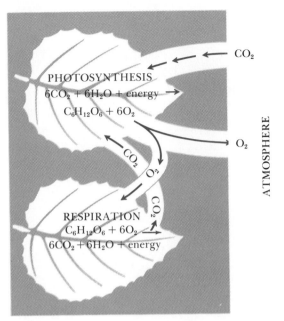

Fig. 6-2 Diagram showing gas exchanges due to photosynthesis and respiration in a green leaf in the light. Carbon dioxide is used five to ten times as fast as it is produced.

In the diagram:
PHOTOSYNTHESIS
$6CO_2 + 6H_2O + energy \rightarrow$
$C_6H_{12}O_6 + 6O_2$

RESPIRATION
$C_6H_{12}O_6 + 6O_2 \rightarrow$
$6CO_2 + 6H_2O + energy$

CO_2

O_2

ATMOSPHERE

Determinations made in the early morning in a greenhouse filled with plants have shown that the air contained about 0.05 percent carbon dioxide, or less than twice the normal amount. In contrast, carbon dioxide concentrations 30 times normal (0.9%) are maintained in atomic submarines and are considered desirable by the crews, partly, at least, because of the insulating effect of the gas in reducing body heat loss to the cold walls of the vessel. Concentrations of 3 percent, 100 times normal, are tolerable to man, although uncomfortable; 20 percent is anesthetic and 40 percent lethal. It seems clear that plant respiration is unlikely to have an adverse effect upon the air of rooms used for sleeping. In contrast, human respiration can raise the carbon dioxide concentration of crowded rooms above 1 percent.

Photosynthesis and Respiration Compared
The facts on photosynthesis and respiration can be presented in tabular form to contrast these two processes:

Photosynthesis	Respiration
1. Occurs only in the green cells of plants.	1. Occurs in every active, living cell of both plants and animals.
2. Takes place only in the presence of light.	2. Takes place during the life of the cell both in the light and the dark.
3. Uses water and carbon dioxide.	3. Uses food and oxygen.
4. Releases oxygen.	4. Releases water and carbon dioxide.
5. Solar (radiant) energy is converted into chemical energy.	5. Chemical energy is converted into heat and useful (ATP) energy.
6. Results in an increase in weight.	6. Results in a decrease in weight.
7. Food is produced.	7. Food is broken down.

Factors Affecting the Respiration Rate The rate of respiration in plant tissues is affected by temperature, moisture, injury, the age and kind of tissues concerned, the concentration of carbon dioxide and oxygen, the amount of available food, and other factors. Although all of these are of interest from the standpoint of understanding plant activities under both natural and experimental conditions, they become a matter of immediate and practical concern in connection with the transportation and storage of cereal grains, fruits, and vegetables.

The providing of food for our civilization, with its great concentration of population in cities, is a problem not only of production but also of storage and distribution. The problem is complicated because the activities of plant tissues do not cease with harvest. Citrus, apple, and other fruits, potatoes, and vegetables of all kinds continue to live and respire—sometimes for months—before they are used. Unless properly handled and stored, these foods will soon deteriorate, either through their own activities or by the action of bacteria and fungi. The handling of such products, involving what has been termed "after-harvest" botany, concerns both botanists and engineers and involves many problems of storage temperature and ventilation.

Immediately after harvesting, or during subsequent storage, the enzymes of fruits and vegetables act on various materials within their tissues and gradually bring about changes in color, texture, and chemical composition. These changes are dependent upon specific enzymes, and thus an important objective of proper storage is to slow down the rate of these processes. Up to a point, the rate of respiration and other enzymatic reactions increases rapidly with the temperature; hence the wide use of low temperatures or refrigeration in handling foods.

The effectiveness of low storage temperatures is due to slowing up the rate of action of respiratory and other enzymes of higher plant tissues and of bacteria and molds. The relation between temperature and respiration is similar to that between temperature and other chemical reactions: within certain limits the rate of respiration is doubled or tripled with each rise of 10°C (18°F) up to temperatures above 40°C (104°F). In one test the respiration of strawberries at 24°C (75°F) was ten times greater than at 0°C (32°F). A difference of this size is expected in enzymatic reactions. In the example here the increase was about 2.2 times for each 10°C. We might expect, then, that the rate of respiration and other enzymatic processes would increase ten or more times if the temperature were raised from 0° to 30°C (32° to 86°F), or decrease proportionally if the temperature were lowered over the same range—an adequate illustration of the value of refrigeration for fresh fruits and vegetables.

Respiration rates for some fresh fruits and vegetables are shown in Table 6-1. Since 1 milligram per kilogram is one part per million, the rates are all relatively low, many times lower than a typical rate of human respiration, for ex-

ample. The production of carbon dioxide from glucose involves a loss of dry weight.

TABLE 6-1
Respiration of Some Fresh Fruits and Vegetables*

Product	CO_2 Production (mg/kg fresh weight/hr)
Asparagus (*Asparagus officinalis*)	439
Strawberries (*Fragaria chiloensis*)	197
Green beans (*Phaseolus vulgaris*)	186
Carrots (*Daucus carota*)	171
Green onions (*Allium cepa*)	161
Lettuce (*Lactuca sativa*)	153
Bananas (*Musa sapientum*)	98
Pears (*Pyrus communis*)	73
Tomatoes (*Lycopersicum esculentum*)	63
Plums (*Prunus domesticum*)	52

* Recalculated from Benoy, *J. Agr. Res.*, Vol. 39, pp. 75–80.

But the loss of weight in stored fruit and vegetables also involves the loss of water by evaporation. The humidity of the air in storage rooms is therefore directly related to the keeping qualities of the products held in them. If the humidity is too low, the products dry and shrivel; if it is too high, they may mold and rot.

Grain stored in elevators (FIG. 6-3) loses *dry* weight at an appreciable rate because of its respiration, and this loss is an important factor in the cost of storage. Over the range of 10–18 percent moisture, the respiration rate of grain approximately *doubles* for each 1 percent increase in moisture content. For this reason elevators will accept grain with higher moisture content only at a sharply reduced price. Such moist grain may then be dried artificially or mixed with old, dry grain to reduce its average moisture and respiration rate. Respiration could be stopped by killing the grain and destroying its enzymes. Dead seeds mold readily, however, and chemical changes impart undesirable flavors to them.

In addition to these losses, other changes occur

Fig. 6-3 **Huge elevators for the storage of wheat. Why is respiration important here?**

after harvesting. Sugars are respired or converted into starch or into cell wall substances so that the products become tough, woody, and unpalatable. Such changes are particularly troublesome in asparagus, and are serious in peas and sweet corn. Because of such afterharvest changes, vegetables picked fresh from the garden are sweeter and more tender than those that are picked 2–4 days before they appear on the grocery counter. The conversion of sugar to starch in sweet corn, for example, is noticeable in

a few hours and is nearly complete in a day.

Poor storage conditions, such as high temperatures and lack of proper ventilation, give rise to physiological diseases, such as blackheart in potatoes, scald and browning in apples, and other conditions that reduce rapidly the food value and marketability of the products of farm and orchard.

RESPIRATORY REACTIONS

Equation (1) indicates that the over-all result of respiration is the oxidation of sugar and the conversion of the chemical energy that was fixed in the sugar in the process of photosynthesis. This equation might be interpreted to mean that the chemical energy of the sugar is changed in one step. Actually, however, the sugar is broken down stepwise through many enzymatically controlled reactions, in the course of which energy is made available to the cell, a small amount at a time. The identification of the various steps in the oxidation of sugar, and of the enzymes concerned with each, is one of the remarkable achievements of the past several decades. These reactions are so complicated that only their major aspects will be considered here.

The chemical energy of sugar cannot be used directly by the cell, but can be made available by its incorporation into high-energy compounds such as ATP. We saw in Chapter 5 that ATP can be formed in chloroplasts by the energy of light. The same compound may be formed with energy obtained from the oxidation of sugar and other organic compounds. This series of reactions is called OXIDATIVE PHOSPHORYLATION, in contrast to the photophosphorylation that occurs in the chloroplast during photosynthesis.

The ATP molecule can be visualized as a tiny storage battery that is charged with chemical energy at the time of its formation. When the energy of an ATP molecule has been used in doing cellular work, the molecule separates into adenosine diphosphate (ADP), or adenosine monophosphate (AMP) and inorganic phosphate. The freed phosphate then becomes part of the pool of inorganic phosphate in the cell. When energy of the required form is again available, the AMP or the ADP and the phosphate can be recharged and recombined to form ATP. Adenosine triphosphate is thus the fully charged and ADP or AMP the discharged or partially discharged form of the "battery" (FIG. 4-3).

Glycolysis The first series of reactions in either plant or animal respiration consists of changing glucose to pyruvic acid, and is called GLYCOLYSIS[1] (FIG. 6-4). These reactions occur in the cytoplasm, rather than in the chloroplasts, the mitochondria, or other organelles. Paradoxical as it may seem, the first step in obtaining energy from a glucose molecule consists of adding more energy to it. This energy is required in the activation of the glucose molecule before it can be changed by enzymes. Activation is accomplished by transferring some of the energy of ATP, derived from previous respiration, to the glucose molecule. Fortunately, ATP can be stored for long periods, as in a dry seed; it can thus be made available for renewed respiration at a later time. The addition of energy to sugar is accomplished by removing phosphate from ATP, together with its energy, combining it chemically with the glucose molecule and forming glucose phosphate. The glucose is now activated and ready for the next step. At the same time the ATP molecule is changed to the lower-energy compound, ADP.

The activated glucose phosphate molecule now goes through a series of enzymatic reactions that split it into two 3-carbon sugars. The glucose phosphate is first changed to fructose phosphate, and another high-energy phosphate is added to form fructose diphosphate. Each fructose diphosphate molecule is then split enzymatically

---

[1] For historical reasons, animal biologists usually include in glycolysis the additional steps of lactic acid fermentation, discussed later. Glycolysis as discussed here, however, constitutes the first part of respiration in animal as well as in plant cells.

into two 3-carbon sugar phosphate molecules [Eq. (2)].

$$P \sim C_6H_{12}O_6 \sim P \longrightarrow 2P \sim C_3H_6O_3 \qquad (2)$$

fructose   glyceraldehyde
diphosphate   phosphate

These molecules, interestingly, are the same glyceraldehydes that we found as the first sugar of photosynthesis. Note that P $\sim$ indicates a high-energy phosphate.

The glyceraldehyde phosphate molecules are now oxidized by the removal of two hydrogen atoms from each of them. This oxidation changes the 3-carbon sugar, glyceraldehyde, after some intermediate steps, into an important molecule, pyruvic acid [Eq. (3)].

$$2P \sim C_3H_6O_3 + 2NAD + 2PO_4 + 4ADP \longrightarrow$$

glyceraldehyde   oxidized   inorganic   adenosine
phosphate   NAD   phosphate   diphosphate

$$2CH_3COCOOH + 2NADH_2 + 4ATP \quad (3)$$

pyruvic   reduced   adenosine
acid   NAD   triphosphate

Oxidation by removal of hydrogen (dehydrogenation) rather than the addition of oxygen is the usual oxidative reaction in respiration. The hydrogen atoms removed from the glyceraldehydes are taken up by a hydrogen acceptor molecule, called nicotinamide adenine dinucleotide, or NAD. This molecule is identical with the hydrogen acceptor that functions in photosynthesis, except that it lacks the extra phosphate group of NADP.

At some time during these reactions each of the glyceraldehydes takes up a low-energy, inorganic phosphate molecule in addition to the high-energy phosphate carried from Eq. (2). The oxidation of glyceraldehyde to pyruvic acid increases the chemical energy of the molecule, and this energy serves to raise the energy level of the added inorganic phosphate to that of ATP. The four phosphates can now be transferred to ADP, forming four molecules of ATP. Two molecules of ATP are used for each molecule of glucose that goes through glycolysis and four are

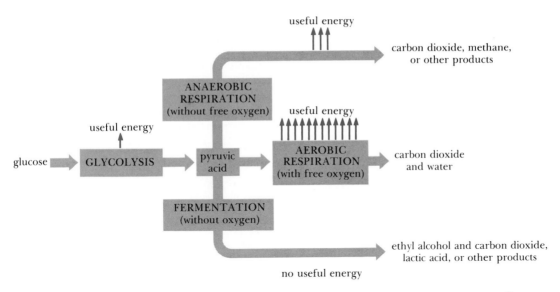

Fig. 6-4 Outline of the respiratory reactions. This diagram shows that glycolysis is common to all respiratory pathways and suggests the relative energy yield from each process.

recovered, for a net gain of two. This ATP is added to the useful energy supply of the cell. In some organisms, such as lactic acid bacteria, glycolysis constitutes the only energy source of the cells. It should be noted that glycolysis can occur either in the presence or absence of oxygen, but that it does not *use* oxygen.

As a result of the reactions of glycolysis, the relatively inert glucose molecules are changed into reactive pyruvic acid molecules and about 5 percent of the energy of glucose is recovered as ATP. Of the remainder of the energy, about 80 percent is left in the pyruvic acid and 20 percent is carried in the reduced hydrogen acceptor, $NADH_2$. Much of this remaining energy is recovered later in aerobic respiration, discussed below. Glycolysis has been presented here only in outline. An important end result of glycolysis is the clogging of the hydrogen acceptor, NAD. If this molecule could not be freed of hydrogen and reused, glycolysis would cease and death would result.

Types of Respiration The various kinds of respiration (FIG. 6-4) may be classified according to the method of disposing of the hydrogen that is split off during glycolysis. (1) If the hydrogen is recombined with pyruvic acid, or some organic compound derived from pyruvic acid, the process is called FERMENTATION. (2) If the hydrogen is united with the oxygen of some inorganic compound such as nitrate (NO_3) or carbon dioxide (CO_2), the process is called ANAEROBIC RESPIRATION. This type of respiration is established only in bacteria (Chapter 20). (3) If the hydrogen is combined eventually with free oxygen from the air, the process is called AEROBIC RESPIRATION. Aerobic respiration is the common process in cells of animals and plants. The term respiration has come to refer, unless qualified, to glycolysis plus aerobic respiration.

The diagram in FIG. 6-4 shows that glycolysis is common to all types of respiration. It also indicates, in a general way, the relative production of usable energy in each series of reactions. The steps in glycolysis and in subsequent respiratory reactions have been studied intensively by many workers. One fundamental result of such investigations is the finding that the respiratory reactions are essentially similar in man, other animals, and in all plants except bacteria. The reactions of bacteria are in part similar and in part diverse.

Fermentation The term fermentation is often used loosely by biologists as synonymous with anaerobic respiration, or as a special reaction of yeasts and bacteria. In many bacteria both fermentation and anaerobic respiration may be carried on, and it may be difficult to distinguish between them. In most plants and in animals any so-called anaerobic respiration is, in the strict sense, some type of fermentation. Of the numerous kinds of fermentations, two are typical and extremely common. One, lactic acid fermentation, is characteristic of animals and bacteria; the other, alcoholic fermentation, characteristically occurs in higher plants, some bacteria, and yeasts. The differences and similarities of these two fermentations are shown by Eqs. (4) and (5). In lactic acid fermentation the H_2 of $NADH_2$ is simply transferred to pyruvic acid. In alcoholic fermentation an additional reaction splits off CO_2.

$$2CH_3COCOOH + 2NADH_2 \longrightarrow$$
pyruvic acid   reduced NAD

$$2CH_3CHOHCOOH + 2NAD \quad (4)$$
lactic acid   oxidized NAD

$$2CH_3COCOOH + 2NADH_2 \longrightarrow$$
pyruvic acid   reduced NAD

$$2CH_3CH_2OH + 2CO_2 \quad + \quad 2NAD \quad (5)$$
alcohol   carbon dioxide   oxidized NAD

These final steps, following glycolysis, yield no energy. The process of glycolysis plus fermentation is poor in energy yield, and its end products are toxic to the cells carrying on fermentation. If these products accumulate, they will check the activities of the cells and may eventually kill them.

Lactic acid fermentation occurs in muscle and other animal tissues whenever glycolysis exceeds the subsequent steps of aerobic respiration, as in a straining athlete. This fermentation is a temporary, emergency process that permits the continuation of the glycolytic reactions with their small release of useful energy. Note also that this energy is released in the cytoplasm where it is immediately useful in muscular action. When the athlete stops "to catch his breath," the accumulated lactic acid is oxidized to carbon dioxide and water, with the formation of ATP, or is oxidized back to glucose. Among plants, lactic acid fermentation occurs chiefly in bacteria. The preservation of certain kinds of foods or feeds, such as sauerkraut and silage, results from the inhibition of bacterial growth by accumulated fermentation products, mainly lactic acid. This topic is discussed further in Chapter 20. Alcoholic fermentation, as pointed out above, differs from lactic acid fermentation in that carbon dioxide is split off during the reaction [Eq. (5)]. Yeasts are major producers of ethyl alcohol, but many bacteria also produce alcohol, as do most higher plants when deprived of atmospheric oxygen.

The ability of cells to survive under conditions of low energy supply and the accumulation of the products of fermentation varies greatly. Brain tissue is commonly irreparably damaged by more than 5 minutes without oxygen. Moist seeds, with a high rate of respiration, may be injured within a few hours, but dry seeds, with essentially no respiration, can survive for years without oxygen, and indeed may be injured by oxygen, particularly at higher temperatures. Differences in the ability of plant cells to survive with little or no free oxygen are probably due more to the ability of the cells to function with a low energy supply than to any extensive dependence on fermentation reactions. This effect applies to animals also. An active ground hog (*Marmota momax*) dies within 3–5 minutes without oxygen, but a hibernating animal, with a much lower rate of metabolism, may live for an hour or more under the same conditions, and some cold-blooded animals may live for a day or more. The roots of emergent plants, such as cattails or rice, in flooded soil usually obtain enough air through the stems of the plants to develop normally. The roots of upland plants that have developed in a well-aerated soil with a good oxygen supply will die within a short time if the soil is flooded and the oxygen supply cut off. The roots of wet-land trees, such as gums (*Nyssa aquatica*) or water oaks (*Quercus nigra*), may be flooded without injury during the winter when the trees are dormant and respiration is reduced by the low temperature. The same trees may be killed, however, by continued flooding during warm weather.

Anaerobic respiration In true anaerobic respiration the hydrogen is disposed of by combining it with oxygen or other atoms from some inorganic compound. For example, oxygen may be obtained from nitrate, liberating gaseous nitrogen. This process is known as DENITRIFICATION (Chapter 10). Other bacteria convert sulfate (SO_4) to sulfur or hydrogen sulfide (H_2S) in an analogous manner. Still others are able to combine hydrogen with carbon dioxide, producing methane (CH_4) and water. Some bacteria that carry on true anaerobic respiration are not capable of any other respiratory reaction, and their growth is prevented by free oxygen. Anaerobic respiration is considered to release more useful energy than fermentation but less than aerobic respiration.

A distinction should be made between an anaerobic mode of life—"life without air"—and anaerobic respiration. Fermentative bacteria and yeasts may be anaerobic, but they do not carry on anaerobic respiration in the sense defined here.

Aerobic respiration If free gaseous oxygen is available, the hydrogen that is split off in glycolysis and held in $NADH_2$ [Eq. (3)] is eventually combined with oxygen forming water. These reactions involve a complicated series of steps,

which are considered in the following discussion. During these reactions three molecules of ATP are produced from each molecule of $NADH_2$. The aerobic reactions of respiration thus yield three times as much ATP energy from the $NADH_2$ formed in glycolysis as is obtained in glycolysis itself. The pyruvic acid that is formed still contains the major portion of the energy of glucose. The processes by which pyruvic acid energy is changed to ATP fall into three groups: (1) the oxidation of pyruvic acid to CO_2 and $NADH_2$ in the Krebs or citric acid cycle; (2) production of ATP from the energy of the $NADH_2$ molecules; this is called oxidative phosphorylation; (3) terminal oxidation, in which hydrogen is finally disposed of by combination with oxygen from the air.

Mitochondria The enzymes that control the steps of the oxidation of pyruvic acid in the Krebs cycle (see below); the formation of ATP by oxidative phosphorylation, and the terminal oxidation of hydrogen to water are found in the mitochondria, although some of them occur in the cytoplasm also. Since oxidative phosphorylation is the major source of useful energy, the mitochondria have been called the powerhouses of the cell. They are minute structures, 0.5–2 microns in diameter and up to 7 microns in length, varying considerably in size and shape. They may be spheres, threads, rods, or oval bodies, and since they are barely visible under the highest powers of the light microscope, their internal structure can be studied only with the electron microscope (FIGS. 3-3 and 6-5). Among the other features of the mitochondria, it has been shown that the membrane is double, the inner membrane projecting inwardly in the form of plates or ridges that provide greater surface area for enzyme activity. Much of our knowledge of the detailed steps of respiration has been obtained by centrifuging mitochondria carefully from the debris of ground-up cells and studying their reactions. The addition of specific chemicals or enzyme inhibitors permits identification

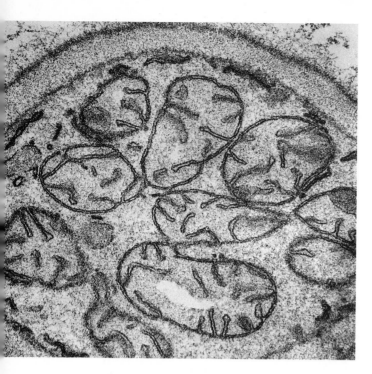

Fig. 6-5 Mitochondria in yeast cell (*Saccharomyces cerevisiae*). (21,000×)

of the individual reactions. Under suitable conditions, the complete oxidation of pyruvic acid to carbon dioxide and water, with the production of ATP molecules, can be demonstrated in these isolated structures.

Pyruvic Acid and the Krebs Cycle   Pyruvic acid is a key compound in plant and animal metabolism. It may be used in fermentation with the production of lactic acid or alcohol, converted into amino acids or lipids, or used as a source of respiratory energy. In respiration, pyruvic acid is oxidized in the mitochondria by the addition of water and the removal of hydrogen, in a complex series of reactions known collectively as the KREBS CYCLE.

In the Krebs cycle, important organic acids are formed that may be used in amino acid synthesis, for example; or the cycle may continue, adding H_2O, splitting off CO_2, and oxidizing by the removal of hydrogen. A considerable quantity of energy is freed in these oxidations and some of it may be used, although most of it is thought to be lost as heat. More energy is carried by the hydrogen atoms of $NADH_2$, which move directly, still in the mitochondria, to the next series of reactions.

The complete Krebs cycle, as it is believed to occur, is shown in FIG. 6-6. The reactions shown there can be summarized by Eq. (6).

$$2CH_3COCOOH + 6H_2O + 10NAD \longrightarrow$$

pyruvic   water   oxidized
acid   NAD

$$6CO_2 + 10NADH_2 \quad (6)$$

carbon   reduced
dioxide   NAD

Recall from Eqs. (2) and (3) that two pyruvic acid molecules are formed from one 6-carbon sugar; therefore Eq. (6) is shown doubled to indicate the reactions due to one glucose molecule. The CO_2 released in respiration is normally obtained from the Krebs cycle. The water used here as a source of hydrogen and oxygen is more than recovered in terminal oxidation, where $12H_2O$ are formed. The important product of the Krebs cycle, so far as respiration is concerned, is the $NADH_2$, which goes to oxidative phosphorylation.

Oxidative Phosphorylation   Ninety-five percent of the ATP produced in aerobic respiration is believed to be formed during oxidative phosphorylation. This is a complicated, poorly understood process, which is similar in its basic reactions to the photophosphorylation discussed in Chapter 5. In photophosphorylation (FIG. 5-18) electrons, raised to a high energy level by sunlight, are dropped down through a cytochrome-enzyme system (see below) in which some of the energy of the electrons is used to form ATP from ADP and inorganic phosphate.

In oxidative phosphorylation the electrons of the hydrogen atoms, taken up in $NADH_2$ during glycolysis and the Krebs cycle, carry much of the energy that was incorporated into sugar in photosynthesis. These electrons are then separated from the hydrogen, which is released into the cytoplasm as the hydrogen ion (H^+). You will recall from Chapter 4 that the plus sign here indicates a missing electron. These high-energy electrons are taken up by an electron-transport system built of enzymes, an electron-transfer molecule, riboflavin, and the special iron-containing respiratory proteins known as cytochromes. Here they are dropped down through riboflavin, cytochromes b, c, and a, and some of their energy is used to form ATP. Three ATP molecules are believed to be formed from ADP and inorganic phosphate for each pair of electrons that goes through the system. This is equivalent to three ATPs for one $NADH_2$.

Fig. 6-6 **The Krebs cycle. The hydrogens split off are combined with nicotinamide adenine dinucleotide (NAD) and moved to oxidative phosphorylation.**

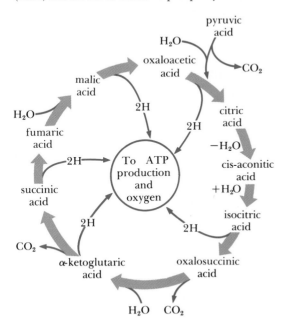

The process is analogous to the generation of electricity by use of a swift, mountain stream. The stream is diverted by one dam through one hydroelectric generator and the energy of the falling water is used to generate electricity. Then, as the stream falls farther, its remaining energy is diverted through a second and a third generator to produce more electricity.

Terminal Oxidation At the end of the electron-transport system of oxidative phosphorylation, the once high-energy electrons have lost their useful energy and are clogging the system. If they are not removed, respiration comes to a halt and the organism dies. If oxygen is present in a properly activated form, the electrons are transferred to it. The cytochromes are thus oxidized by the loss of electrons, and the oxygen that receives them is reduced. We think normally of oxygen as the primary factor in respiration. The fact that it comes last in time does not, however, make it any less important. The final removal of the electrons is an essential part of respiration and oxygen is thus vitally necessary to all forms of life except the relatively few species of anaerobic bacteria.

In its usual molecular form, (O_2), oxygen is nearly inert at ordinary temperatures. In some way this inert molecule is split and each atom can then take up two electrons to form the equivalent of oxygen ions (O^{2-}). These reactive ions then take up two hydrogen ions (H^+) each, perhaps the very hydrogen ions from which the electrons were separated in oxidative phosphorylation, and water (H_2O) is formed. An outline of the reaction is shown in Eq. (7).

$$O_2 \; + \; 4e^- \; + \; 4H^+ \longrightarrow 2H_2O \qquad (7)$$
oxygen   electrons   hydrogen   water
ions

We can summarize respiration by saying that some of the chemical energy of sugars has been transformed into the available chemical energy of ATP. Or, we can spread out the process in a simplified form, as in FIG. 6-7, which shows the

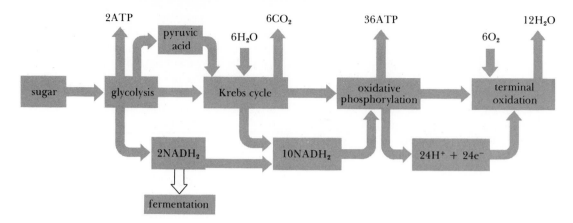

Fig. 6-7 A diagram of aerobic respiration. The main steps of the reactions are shown on the central line. 2 ATP and 36 ATP in the top line indicate the energy transferred to ATP during the reactions. The elimination of surplus electrons and H ions in terminal oxidation looks unimportant, but means life or death to a respiring organism.

relations among the four basic groups of reactions—glycolysis, the Krebs cycle, oxidative phosphorylation, and terminal oxidation. Fermentation is shown here in its proper position as a dead-end emergency reaction which permits a temporary continuation of glycolysis and its minor production of cytoplasmic ATP. Note that if the H_2O added in the Krebs cycle [Eq. (6)] is subtracted from that produced in terminal oxidation, the figures can be fitted into a simple, over-all equation [Eq. (1)].

Respiratory Efficiency If we summarize the ATP-yielding reactions of aerobic respiration, we find (see FIG. 6-7)

Glycolysis (net), 2 molecules
Oxidation of $NADH_2$ from glycolysis, 6 molecules
Oxidation of $NADH_2$ from the Krebs cycle, 30 molecules

The possible total of 38 ATP molecules is believed to represent a recovery of 40 to possibly 60 percent of the energy present in the sugar molecule. This is a good efficiency and better, for example, than the recovery of the energy of coal as electricity in the best modern power plants. The energy not recovered in ATP is lost primarily as heat.

DIGESTION

We have already discussed the nature of plant foods and the forms in which these are accumulated. Foods stored in an insoluble form, such as starch, lipids, or storage protein, can neither be used in the cells in which they are stored nor translocated to other parts of the plant until they have been changed to a soluble and diffusible form by enzymatic action, a process known as DIGESTION. After digestion the soluble foods can be used or transported to other cells and tissues, where they are assimilated or respired.

Digestion in green plants and in animals is fundamentally the same. There are similarities in the nature, kind, and behavior of the enzymes, the foods acted upon, and the substances formed. But one difference may be noted. In man and most other animals, digestion is, for the most part, *extracellular*, taking place in special digestive organs, where it is catalyzed by enzymes that carry on their activities outside the cells that produce them. In man, digestion proceeds in the mouth, stomach, and small intestine, into which enzymes are secreted from specialized cells or glands, such as the salivary glands, gastric glands, and pancreas. In autotrophic organisms, however, digestion is usually *intracellular*. That is, it occurs within any living cell in which food is stored, under the influence of enzymes produced largely by these cells.

Extracellular digestion is found occasionally in the higher plants, such as the carnivorous plants discussed at the end of this chapter. Another example occurs in the seeds of certain plants, some cells of which synthesize enzymes that are secreted by the cells where they are produced and digest foods in the surrounding tissues of the seed. In the lower organisms, bacteria and fungi, digestion is commonly extracellular: enzymes are secreted by the organism and act upon the insoluble food materials in the environment.

Digestion of Starch One of the most common digestive activities of plants is the transformation of starch to sugar. This may be accomplished by any of a group of enzymes known as AMYLASES. Amylases act upon starch, producing dextrins, maltose or malt sugar, and glucose in varying proportions. One amylase, known as β-amylase, produces nearly pure maltose from common starches. The amylase of saliva produces a mixture of maltose and glucose. Amylases and certain other digestive enzymes bring about changes by HYDROLYSIS, a process in which a large molecule is split into one or more kinds of smaller molecules with the addition of the elements of water. Hydrolysis may be illustrated with FIG. 4-16, where the removal of H and OH leads to the joining of two smaller molecules. In hydrolysis the H and OH of water are replaced and the smaller molecules are again separated. Hydrolysis of the peptide bonds of a protein forms

amino acids. Hydrolysis of common starch (amylose) by β-amylase leads to the formation of maltose [Eq. (8)].

$$2(C_6H_{10}O_5)n + n\text{-}1(H_2O) \xrightarrow{\beta\text{-amylase}}$$
$$\text{starch} \qquad\qquad \text{water}$$

$$n(C_{12}H_{22}O_{11}) \quad (8)$$
$$\text{maltose}$$

Maltose is rapidly changed to glucose by the enzyme MALTASE and does not normally accumulate in plant cells. The action of maltase is shown by Eq. (9):

$$C_{12}H_{22}O_{11} + H_2O \xrightarrow{\text{maltase}} 2C_6H_{12}O_6 \qquad (9)$$
$$\text{maltose} \quad \text{water} \qquad\qquad \text{glucose}$$

Glucose, as stated earlier, is one of the sugars commonly used in respiration. The digestion of starch in germinating cereal grains produces irregular channels or grooves on the surfaces of the starch grains so that they appear eroded or etched. Eventually the entire grain may be digested.

Maltose has been found in germinating seeds, particularly barley, and is one of the sugars present in the malt extract of the soda fountain. MALT, composed of partially germinated barley, is used in making beer and in the production of alcohol from grain. Malt is produced by soaking barley to moisten it and germinating the grain in layers, with frequent stirring to ensure an adequate air supply and prevent heating. As soon as germination is well started the malted grain is dried and ground for use. The important change during malting is the production of amylases and other enzymes. Germination is stopped before extensive digestion occurs. If the ground malt is mixed with water, as in brewing (Chapter 20), the starch and other reserve foods of the grain are digested rapidly by the enzymes present. Malt is the source of the enzymes used to digest the starch of other grains, as corn, in the production of grain alcohol and whiskey. The soaked, ground corn is mixed with barley malt and allowed to digest, with the formation of sugars that can be fermented by yeast.

Another starch-digesting enzyme is called STARCH PHOSPHORYLASE. The action of this enzyme is analogous to that of amylase, but here it is inorganic phosphate instead of water that enters into the reaction, and the product is the high-energy compound glucose phosphate rather than simple glucose. This is a phosphorylation in which the energy originally present in the starch molecule, rather than ATP energy, is used. When starch is digested to glucose by the amylase-maltase enzyme complex, the energy that holds the $C_6H_{10}O_5$ units together in the starch molecule is lost as heat. When starch is digested by starch phosphorylase, some of this energy is conserved in glucose phosphate. The glucose phosphate formed by starch phosphorylase is identical with the activated glucose formed by ATP, and its production saves valuable ATP energy. This activated glucose can then be used directly in respiration or in the synthesis of many cellular compounds, such as sucrose (table sugar), cellulose, and other products.

Starch phosphorylase was formerly thought to function in the synthesis of starch as well as in its digestion. Recent research indicates, however, that the synthesis of starch is catalyzed primarily by the enzyme STARCH SYNTHETASE. Both the amylases and starch synthetase are widely distributed in plant tissues. Starch is formed in higher plants only in plastids, either amyloplasts or chloroplasts, and starch-forming and digesting enzymes are assumed to be located within the plastids.

Digestion of Other Carbohydrates The sugars may be classified into simple sugars, as glucose, fructose, galactose, and ribose, and into complex sugars, as maltose, sucrose, raffinose, and others that can be hydrolyzed into two or more simple sugars. Fructose is abundant in fruits and is therefore termed fruit sugar. It has the same empirical formula as glucose, but differs in the arrangement of the atoms (FIG. 5-12). Ribose is present in RNA and ATP. Sucrose may be hydrolyzed by SUCRASE (also called invertase) into the simple sugars glucose and fructose, as shown by Eq. (10).

$$C_{12}H_{22}O_{11} + H_2O \xrightarrow{\text{sucrase}}$$

sucrose   water

$$C_6H_{12}O_6 + C_6H_{12}O_6 \quad (10)$$

glucose   fructose

Maltose, as shown in Eq. (9), is hydrolyzed into two molecules of glucose. Raffinose, found in low concentrations in sugar beets, forms glucose, fructose, and galactose. Sucrose has been shown to be the chief form in which carbohydrates are translocated in plants. It is also present in high percentages in sugar cane, sugar beets, and onions, and during the winter in the dormant tissues of many plants. A number of other sugars occur free or in combination in plants.

The digestion of cellullose is brought about by the enzyme CELLULASE, which converts cellulose into glucose in two stages. Cellulase is produced by many bacteria and fungi. An enzyme similar in properties to fungal cellulase has been reported to occur in very low concentrations in bean leaves, in the leaves and roots of tobacco, and in a few other plants. It seems to be present chiefly in young, growing cells.

The activity of cellulase shows that plant enzymes may be as dramatically powerful as the enzymes of such animals as owls and snakes, which swallow their prey whole. By the action of cellulase and other enzymes, fragile threadlike fungi are able to digest the cell walls of even the hardest woods. Cellulose-digesting bacteria and fungi are universally distributed and serve a natural function of great importance in the decay and disintegration of plant remains.

The enzyme cellulase is also produced by a few animals, such as some wood-eating insects, and shipworms. The latter, marine mollusks, obtain their carbohydrate by boring into ships' timbers, wharf pilings, and other submerged wood. Ever since man began to sail the seas, these mollusks have caused great damage to shipping. Termites, which also live on wood, produce no cellulase but depend upon microorganisms (protozoans) present in their digestive tract to produce this enzyme.

Cellulase is absent in mammals, including man. Herbivores, such as the cow, goat, and sheep, are able to digest grass, hay, and browse because of the activities of cellulose-digesting bacteria that live in their alimentary canals. Man thus indirectly benefits from cellulose when he dines on roast beef.

The rate of reaction of carbohydrate-digesting enzymes, like that of other kinds, is increased by a rise in temperature. It might be expected, therefore, that low temperatures would always retard the digestion of starch, but in many plants starch is converted into sugar at temperatures near freezing. For example, when potato tubers are stored for several weeks at temperatures below 5°C (41°F), much of the starch is converted into glucose. When cooked, the texture of such potatoes is soggy or watery, and they may be so sweet that they are unpalatable. Potato chips made from such potatoes are dark brown in color and may have a burned taste because the sugar is caramelized in hot cooking oil. Potatoes that have become sweet need not be discarded, however; if they are removed to room temperature, 20–25°C (68–77°F), the sugar is converted back into starch and the potatoes again become palatable. Parsnips, when subjected to low temperatures, also undergo changes similar to those of the potato, although in this case the increase in sugar content is desirable. The roots may be left in the ground in the autumn until freezing weather occurs, or harvested and stored at a temperature of about 1°C (34°F). The improvement in culinary quality is largely due to the conversion of starch to sucrose.

Digestion of Lipids   The lipids of plants are hydrolyzed by the enzyme LIPASE with the formation of glycerol (glycerine) and fatty acids. Glycerol is readily changed to sugars and other compounds that may then be respired or used in other processes. The utilization of fatty acids has been extensively studied in animals, where reactions that occur in the mitochondria of the liver yield useful energy and smaller molecules. The details of fatty acid utilization in plants are not so well known. Experiments show that the oil content of oily seeds decreases during germination while the sugar content rises. It seems possible that fatty acids may be used directly within the cells where they occur, but that if they are translocated to other tissues they are first changed to sugar.

The energy content of lipid is more than twice that of a corresponding quantity of sugar. Lipids constitute, therefore, a concentrated food reserve. They occupy little storage space and provide, besides, a high yield of energy to the plant or to some ultimate consumer of plant products.

Digestion of Proteins   Proteins are hydrolyzed into amino acids by enzymes known collectively as PROTEASES. The conversion of proteins to amino acids takes place in several stages, with the formation of intermediate products. Some enzymes are able to bring about the entire conversion; others carry through only part of the process. Some proteases found in plants do not occur in animals, and vice versa, but in general they are similar.

Proteases can be detected readily in germinating seeds, green leaves, fruits, bulbs, and roots. BROMELAIN and PAPAIN are prominent and interesting plant proteases. Bromelain, a component of fresh pineapple juice, has under experimental conditions digested live intestinal roundworms. The enzyme is inactivated or loses its power in the animal body, however, and is not used medicinally. If fresh pineapple is used in making gelatin salads or desserts, the gelatin loses its firm consistency on standing and becomes partly liquefied as a result of being digested by the bromelain in the pineapple. It has been discovered that the stems of the pineapple plant are an excellent and economical source of commercial protease preparations.

Papain is the commercial name applied to the

dried and purified latex, or milky juice, of the unripe fruit of the papaya, a small tropical or subtropical tree (FIG. 6-8). The ripe fruit is eaten like a melon. Papain contains several enzymes, the most prominent of which is a powerful protein digestant, also called papain. The enzyme has a number of uses, all dependent upon its ability to digest protein. It is used in medicine and surgery and is widely employed in the brewing industry to digest proteins that would otherwise precipitate when beer is cooled. It is used in the tanning industry and in the manufacture of chewing gum. Much of the papain imported into this country is used in the tenderizing of meat. Applied to meat before it is cooked, papain partly digests both connective tissues and muscle fibers. Tough cuts of meat are thus made tender.

Summary of Digestive Enzymes An enzyme acts as an organic catalyst upon a specific compound, called the SUBSTRATE. The name assigned to an enzyme consists, with some exceptions, of the name of the substance upon which it acts followed by the suffix "-ase."

TABLE 6-2
Some Enzymes and Their Action

	Enzyme	Product
Starch	β-Amylase	Maltose
	Starch phosphorylase	Glucose phosphate
Maltose	Maltase	Glucose
Sucrose	Sucrase	Glucose and fructose
Cellulose	Cellulase	Glucose
Lipids	Lipase	Fatty acids and glycerine
Proteins	Proteases	Intermediate products and amino acids

Plant Food On the basis of our understanding of photosynthesis, digestion, and plant metabolism in general, plant foods may be defined as organic compounds that, immediately or after a digestive process, can be respired, yielding energy, or can be used in the construction of tissues. As we have seen, plants manufacture their own foods from raw materials derived from the soil and air. Farmers, gardeners, and manufacturers of fertilizers, however, refer to inorganic fertilizers as "plant foods." The use of the term "plant foods" in this way carries the mistaken implication that the foods of plants and animals are different, when in fact they are the same—that is, proteins, carbohydrates, fats, and their derivatives. Water, carbon dioxide and inorganic solutes are not classed as "plant foods," although they are used by plants in the manufacture of food.

CARNIVOROUS FLOWERING PLANTS

Carnivorous, flowering plants have long attracted popular interest. About 500 species are known, classified in six unrelated families. A considerable number of these plants are found in North America. Carnivorous plants are adapted to the capture of insects and other small animals. The animal tissues are then digested by proteases and other enzymes secreted by the plants. This is an extracellular digestive process—an exception to the usual intracellular digestion of green plants.

A number of carnivorous plants are aquatic; some are aerial, growing upon other plants; and many are terrestrial, growing in sandy or boggy soils. There are various mechanisms by which animals are trapped. Several of the more common or better known kinds of carnivorous plants are described below.

Pitcher Plants Modified leaves are typical of all carnivorous plants. In the pitcher plants the leaves are flared into an upright or inclined tube, the upper end of which is further modified forming a lip (FIG. 6-9) or, in some species, a kind of

Fig. 6-8 Collecting latex from the papaya fruit. The unripe fruit is gashed with a sharp instrument, and the latex flows from the cuts. The enzyme papain is obtained from the latex.

overhanging hood. The leaves are clustered on a very short stem and may attain a height of 3 feet or more in some species. Several zones of downward-pointing hairs found at the mouth of the tube and along the inner sides prevent the escape of insects that fall into the cavity of the leaf. Glands within the tube, near the base of the leaf, secrete the digestive enzymes. A considerable area of the inner surface of the leaf is able to absorb the products of digestion. A single flower is produced at the end of a long stalk.

Sundews The carnivorous pitcher plants have a passive trapping mechanism. The sundews and some other carnivorous plants are more active in the capture of prey. The American species of sundews are small plants, growing in wet places (FIG. 6-10). Long stalks bearing small white flow-

Fig. 6-9 A pitcher plant, *Sarracenia purpurea.*

Fig. 6-10 Sundew (*Drosera intermedia*), a carnivorous plant.

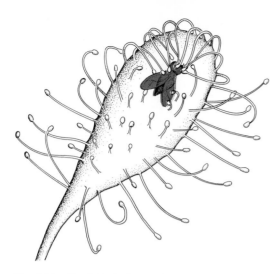

Fig. 6-11 Single leaf of a sundew. The tentacles are bending around a captured insect.

ers arise from a rosette of leaves. The blade of the leaf varies in shape. It may be slender and threadlike, somewhat expanded (Fig. 6-11), or almost circular, with numerous stalked glands or tentacles on the upper surface and margins. Each gland secretes a large mucilaginous drop, which, glistening in the sunlight, gives the plant its common name. Insects are caught and held by these viscid drops, which contain the digestive enzymes. The presence of an insect stimulates other tentacles to bend and further entrap the prey. If the stimulus is sufficiently great, the leaf blade itself may curve around the entrapped insect. The bending of the tentacle stalk appears to be brought about by differential growth on one side. Recovery is caused by growth upon the other side. The response of the tentacles is slow, and several hours are required to complete the bending movement.

Venus's-Flytrap The trapping mechanism of the Venus's-flytrap is even more active than that of the sundews, to which it is related. The plant consists of a rosette of leaves, varying in height

from 2 to 6 inches (Fig. 6-12). A long stalk, arising from the basal rosette, bears a cluster of flowers at the top. The leaf consists of two parts: a broad, expanded petiole, and a two-lobed blade, hinged down the middle. The blade is the trapping mechanism. From the margins grows a row of long, stiff, hairlike teeth. The upper surface of the blade bears sensitive hairs, together with glands containing a red pigment. Some of these glands are digestive, and others attract insects by a sugary secretion.

If the sensitive hairs are touched by an insect crawling over the blade, the two lobes quickly come together (sometimes in less than a second) and the teeth cross, thus forming a cage and preventing the escape of the animal (Fig. 6-13). This rapid movement is followed over several hours by a slower, tighter closing, bringing the upper surfaces even more closely together. The insect is digested and the products of digestion absorbed. The leaf opens again after a few days, ready to trap other prey. The mechanism of the trapping movement has not been explained adequately. Venus's-flytrap is native only in North and South Carolina, but is frequently cultivated in greenhouses.

Carnivorous plants have rather poorly developed root systems and usually grow in soils deficient in nitrogen. The additional supply of this and other elements derived from their animal food is therefore of direct advantage to the plant. On the other hand, carnivorous plants are amply supplied with chlorophyll and are able to mature and produce flowers and seeds even if deprived of animal food, although growth is stimulated when such food is available. The carnivorous habit, then, is useful to the plant but is not necessary to survival.

Carnivorous plants constitute a curious and interesting group with leaf modifications that are sometimes useful in supplementing nutrition. These modifications are probably the basis for the accounts of "man-eating trees" that occasionally appear in print. Such accounts have, of course, no basis in fact; the truth is remarkable enough.

Fig. 6-12 Plant of Venus's-fly-trap (*Dionaea muscipula*), showing the two-lobed leaf blades and expanded petioles.

Fig. 6-13 Venus's-flytrap. (Left): an ant has strayed upon the surface of the sensitive blade. (Center): the two lobes have come together with the teeth crossed, preventing the escape of the insect. (Right): a later stage, the lobes pressed tightly together.

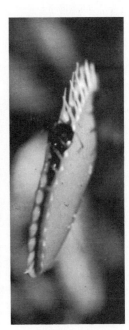

SUMMARY

1. Respiration is an intracellular process in which food is oxidized, with the conversion of energy. This energy is partly released as heat and is partly combined as the energy of ATP that is utilized in the work of the cell.

2. Respiration can be demonstrated by measuring the release of carbon dioxide, the loss in dry weight (in the dark), or the rise in temperature.

3. Respiration is an essential process that must occur in all active, living cells of both plants and animals. Breathing is a mechanical process that aids in gas exchange and may be associated with respiration. Both plants and animals respire, and some animals also breathe.

4. Under certain conditions, respiration and photosynthesis in the green leaf may exactly balance each other. The amount of food manufactured by photosynthesis, however, normally exceeds the total amount used in respiration during the day and night. This permits assimilation, growth, and the accumulation of food.

5. During a 24-hour period, more oxygen normally leaves the green plant than enters, and more carbon dioxide diffuses in than out. The explanation of this is found in the relative rates of photosynthesis and respiration.

6. Plant respiration has no important effect upon the carbon dioxide content of greenhouses, nor upon the health of persons in bedrooms at night.

7. Harvested vegetables and fruits deteriorate as a result of their own respiration as well as by the action of the microorganisms of decay. This deterioration is accelerated by high temperatures which hasten the activities of respiratory and other enzymes.

8. During respiration, some of the energy of sugars and other organic compounds is incorporated into high-energy phosphate compounds, primarily ATP. ATP is an energy carrier that picks up energy from reactions that yield it and transfers it as useful energy to reactions where it is used.

9. Respiratory reactions usually start with glycolysis, in which sugar is converted to pyruvic acid and a small amount of useful (ATP) energy is formed. Glycolysis occurs in either the presence or absence of oxygen, but does not use it. Oxidation by the removal of hydrogen occurs in glycolysis and is the basis of the energy conversion. The hydrogen is taken up by a hydrogen acceptor molecule, nicotinamide adenine dinucleotide (NAD).

10. Three types of respiration may be recognized, depending upon the disposition of the hydrogen removed in the oxidation of sugar; these are fermentative, anaerobic, and aerobic.

11. In fermentation the hydrogen that is split off in glycolysis is recombined with pyruvic acid to form lactic acid, alcohol, or other products. No additional energy beyond that of glycolysis is made available during fermentation.

12. In anaerobic respiration the hydrogen that is split off in glycolysis is combined with the bound oxygen of inorganic compounds, such as SO_4 and CO_2, and additional energy is obtained.

13. In aerobic respiration the hydrogen is moved through oxidative phosphorylation where ATP is formed in relatively large quantities, and eventually combined with oxygen, forming water.

14. In aerobic respiration the pyruvic acid as well as the $NADH_2$ formed in glycolysis may be oxidized and much of its energy used. The utilization of the energy of pyruvic acid starts with the Krebs cycle, in which water is added, and hydrogen and CO_2 split off until the pyruvic acid is oxidized to CO_2 and to hydrogen combined in $NADH_2$.

15. The NADH$_2$ from the Krebs cycle, as well as the small quantity from glycolysis, is then moved through oxidative phosphorylation and 36 molecules of ATP are formed per molecule of glucose, compared with 2 molecules in glycolysis plus fermentation.

16. As the final step in aerobic respiration, the hydrogen with its electrons, remaining after oxidative phosphorylation, is combined with oxygen, forming water.

17. Plant foods, properly, are the same sugars, starches, proteins, and lipids (fats or oils) that animals use. The difference is that green plants produce their own foods.

18. The digestion of foods by plants is similar to that in animals in the foods digested, the enzymes used, and the products formed. Digestion in green plants is primarily intracellular; animal digestion, and that of fungi and bacteria also, is extracellular.

19. A number of plants, called carnivorous, obtain amino acids and other foods by trapping and digesting small animals, primarily insects.

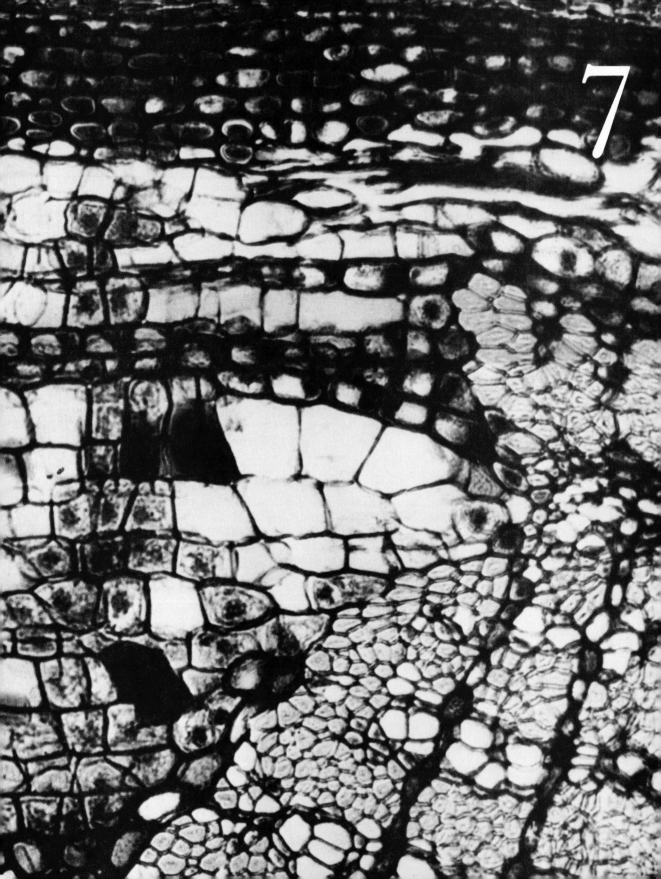

The Structure and Growth of Stems

SOONER OR LATER in a botany course the student becomes so concerned with the structural and functional details of plants that he may lose sight of the growing plant as a whole. This difficulty is not unique to the study of plants; there is a parallel dilemma in the study of animals. The reason lies in the structural complexity of the bodies of the higher plants and animals. The product of change over immense periods of time, these bodies are now composed of numerous kinds of cells, grouped together in many ways. The great variation thus resulting within every organism and groups of organisms presents enticing details that draw the attention of the botanist and zoologist. But the student, becoming involved in these structural details, may easily overlook the part they play as the plant lives and grows.

This dilemma of not seeing the forest for the trees is mentioned here because it is in the study of the structures of the stem that details can be most alluring. As we explore—largely by means of the microscope—the conducting tissues of the stem, for example, we must keep in mind that they are merely a part of a system of similar tissues found also in the roots and leaves. The vascular tissues—xylem and phloem—literally bind the entire plant together. There are many other kinds of tissues common to root, stem, and leaf, and their functions in the life of the plant are much the same wherever they are found.

The cells and tissues of the leaf have already been discussed. The stem, which is undoubtedly more familiar to us than the subterranean root, will be considered in this chapter, and the root will be dealt with in a later chapter.

SOME GENERAL FEATURES OF STEM STRUCTURE

Stems vary in form from the soft, spongy, or fragile stalks of aquatic plants to the towering trunks of forest trees 300 feet high. Internally, the tissues of the stem also vary greatly, in either the kinds or arrangement of the cells that compose them. But they also possess many features in common. Coniferous stems, and especially dicotyledonous woody and herbaceous stems, show many similarities in the arrangement of their tissues. In the stems of woody plants, the organization is relatively clear, and growth in height and diameter is easy to observe. In all groups of vascular plants the mechanisms by which increase in height is brought about are much the same, and there is a fundamental similarity in the kinds of tissues concerned in storage, support, and transportation.

A study of the internal tissues of a stem may begin with those of a woody dicotyledonous twig as seen in cross section (FIG. 7-1). This shows a circular arrangement of vascular tissue (xylem and phloem) located between the PITH and CORTEX. In longitudinal view the vascular tissues would, of course, appear as a cylinder. This general disposition of tissues is found in the stems of conifers and of flowering plants, with the exception of the monocotyledons.

The cylinder of vascular tissues is largely composed of vertically elongated cells. Between the

xylem and the phloem lies the VASCULAR CAMBIUM, a thin sheet of dividing cells responsible for increase in diameter of the stem. It is invisible to the naked eye, but is essential to the life and growth of the plant. Some of the kinds of cells in the xylem and phloem are already familiar. The xylem (wood) conducts water and dissolved salts upward from the roots. Wood also plays an essential role in the mechanical support of the trunk and branches of the tree and of the stem of herbaceous plants. The phloem conducts food—downward from the leaves or upward to unfolding buds and to fruits. The VASCULAR RAYS are bands of cells extending horizontally through both the xylem and the phloem, and are significant in food storage and in the lateral conduction of food and water.

If instead of a woody stem, an herbaceous dicotyledonous stem is examined in the same way, some differences appear. An important difference is the length of time during which the cambium functions in the short-lived herbaceous shoot. In many such plants the vascular tissue appears in cross section as vascular bundles arranged in a ring (FIGS. 7-2, 7-3). In longitudinal section they would appear as a cylinder (FIG. 7-4). The cambium produces additional xylem and phloem, which enlarge the bundles and may eventually connect them in a continuous ring of vascular tissue. In woody plants the cambium produces such a cylinder quickly. Nevertheless the original xylem portion of the bundles may be detected at the inside of the cylinder next to the pith (FIG. 7-1), and, in young twigs at least, the phloem portions may be seen at the exterior.

Supplementing the mechanical functions of the wood in most species, both woody and herbaceous, are sclerenchyma fibers. In stem cross sections these may be seen as caplike or circular masses composing the outermost part of the phloem, but they may also occur in clusters or bands throughout the soft parts of this tissue.

The pith and cortex are largely composed of parenchyma. Considerable quantities of food may accumulate in this tissue, in twigs, young woody branches, and the underground stems of

Fig. 7-1 Diagram showing important tissues of a young dicotyledonous woody stem.

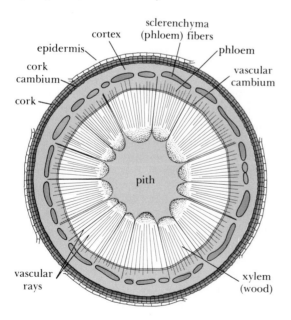

epidermis

cortex

sclerenchyma (phloem) fibers

phloem

cork cambium

vascular cambium

cork

pith

vascular rays

xylem (wood)

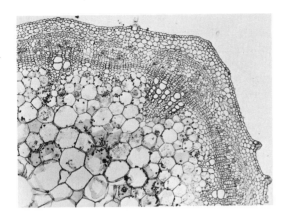

Fig. 7-2 Cross section, portion of stem of soy bean (*Glycine max*).

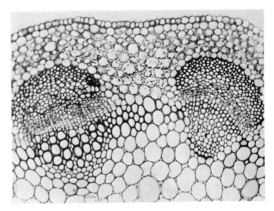

Fig. 7-3 Cross section, portion of stem of alsike clover (*Trifolium hybridum*).

herbaceous perennials. The outermost part of the cortex of twigs and herbaceous stems is frequently composed of chlorenchyma, which may give the stem a greenish color.

The outermost layer of the young stem is the epidermis, usually only one layer of cells thick. In woody plants the life of the stem epidermis is usually short, and it is soon replaced by a CORK layer, effective in reducing water loss from the tissues within. Cork is produced by the activities of a CORK CAMBIUM (FIGS. 7-1, 7-29).

The fully differentiated tissues of a vascular plant have long been classified in three tissue systems: (1) the DERMAL system, composed of epidermis or, later, of cork; (2) the VASCULAR system, composed of xylem and phloem; and (3) the GROUND TISSUE system. This includes all tissues exclusive of the dermal and vascular. By far the most common of the ground tissues is parenchyma, of which pith, cortex, leaves, and fleshy fruits are largely composed. But other more highly specialized cells, such as sclerenchyma, collenchyma (FIG. 7-10), resin ducts, oil ducts, and other secretory structures, may also occur in the ground tissue systems.

It was previously emphasized that the vascular tissues are continuous throughout the plant body in root, stem, and leaf. This is particularly evident in the nodal regions, where vascular bundles in the stem are seen to be continuous with bundles in the petiole. These bundles in the stem may be traced downward through varying distances until they merge with bundles associated with other leaves (FIG. 7-4). A bundle in the stem which extends out into a leaf from the point of departure from the vascular cylinder to the base of the petiole, is termed a LEAF TRACE. At each node, then, the trace or traces of a leaf may be seen diverging from a vertical course through the stem and extending out toward the leaf. In dicotyledons, one and three are common trace numbers for each leaf, but most monocotyledons have many. In a woody dicotyledonous twig, after the leaf has fallen, the leaf traces appear as bundle scars on the surface of the leaf scar (FIG. 2-4).

Where a leaf trace diverges from the stem vascular cylinder, it leaves above it a region filled with parenchyma tissue. Such a region is termed a LEAF GAP (FIG. 7-4). Above the node, adjacent bundles gradually fill the leaf gap, either by altering their course or by branching. The significance of the leaf gap in the economy of the plant is unknown.

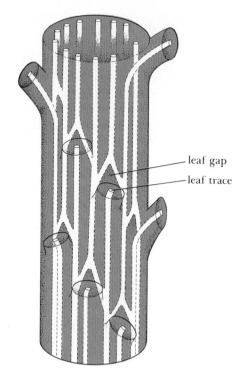

leaf gap
leaf trace

Fig. 7-4 Three-dimensional diagram showing the nature of the vascular system, leaf gaps, and leaf traces in an herbaceous dicotyledon. Based upon flax (*Linum perenne*).

GROWTH OF THE SHOOT

Growth of vascular plants includes (1) the formation of new cells and (2) the subsequent enlargement of these cells. Enlargement is commonly accompanied and followed by differentiation, which is the process of change during which cells may become different from one another. So closely related are these processes, growth and differentiation, that they may be considered together.

The formation of new cells is mostly localized in specific regions known as meristems. As previously stated, meristematic cells are those that are physiologically young and capable of continued division. Typically they possess rela-

tively large nuclei, abundant cytoplasm, and thin primary walls. Meristems may be classified in two groups: (1) APICAL MERISTEMS, found at the tips of the shoot and roots and (2) LATERAL MERISTEMS, which lie along the sides of stems and root. In addition, in elongating organs there are often zones that remain meristematic for a prolonged period and are referred to as INTERCALARY MERISTEMS. These, although important, are nearly always temporary, and therefore are commonly regarded as not of the same rank as apical and lateral meristems.

Terminal Growth In the discussion of the growth plan of the vascular plant in Chapter 2, the important role of the shoot apex, or shoot apical meristem, in initiating the tissues of the stem was mentioned. This group of meristematic cells is responsible for the continued or potentially unlimited growth of the shoot. The apical meristem (FIG. 7-5) not only produces new cells that are added to the tissues beneath, but it also produces bud scales, leaves, and frequently lateral buds. Simultaneously with the formation of new leaf primordia, new internodes are formed (FIG. 7-6). The apical meristem is thus primarily a region of *initiation* of tissues and lateral appendages and not, as is sometimes stated, the immediate source of all the cells of the mature stem (exclusive of those produced by cambial activity). Cell division is by no means confined to the shoot apex, nor is this the most important source of new cells. Cell division continues for a time behind the apical meristem, throughout each of the newly formed internodes.

If an actively growing stem is studied at increasingly greater distances from the apex, the results of progressive cell enlargement and differentiation may be observed (FIG. 7-6). In the top section of this diagram (A), which includes the shoot apex, cell division is dominant, but some cell enlargement and differentiation are also evident. It can be noted that some cells increase in size by enlargement in all diameters, while in other cells enlargement is much greater

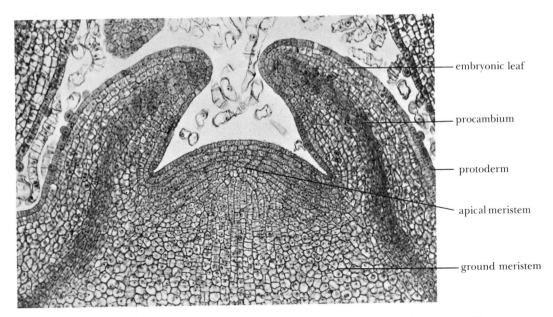

embryonic leaf

procambium

protoderm

apical meristem

ground meristem

Fig. 7-5 Longitudinal, approximately median section of shoot apex of white ash (*Fraxinus americana*), taken early in the growing season.

in a vertical plane. In the middle portion (B), cell division, even in vacuolated cells, continues, and cell enlargement and tissue differentiation are more conspicuous. In the lowest portion of the diagram (C), cell enlargement is again evident, but additional tissues have become differentiated. Some tissues, such as epidermis, pith, and cortex, are approaching physiological maturity, that is, they will shortly reach a state in which the cell will remain relatively unchanged in the absence of changes in the environment of the cell. Other cells such as the phloem fibers and certain xylem cells have lost their nuclei and cytoplasm, and are no longer alive. Such cells are certainly mature—no further differentiation is possible. The increase in size of cells from the top downward is due in part to the inward diffusion of water, which results in increase in size of the vacuoles, but the cytoplasm may also double in weight. The physiologically mature cell is typically 10 to 20 times the size of the meristem-

atic cell, although in some tissues such as those of fruits it may be hundreds of times larger.

We may now relate these processes to the elongation of the shoot, principally of a dicotyledonous woody plant, although the same principles apply to conifers and to dicotyledonous herbs and monocotyledons. It is only in the terminal part of the woody shoot—the twig—that increase in length occurs. Nearly all of this growth results from the division and enlargement of cells in the internodes behind the shoot apex, and the extension of these internodes pushes the shoot apex upward. This region of elongation extends for a considerable distance behind the apex, sometimes as much as 6 inches, and includes several internodes (FIG. 7-7). The dividing cells of each young internode constitute a temporary meristem, most of the cells of which will eventually become physiologically mature. It is chiefly the progeny of these dividing cells that enlarge and are responsible for growth in

GROWTH OF THE SHOOT **159**

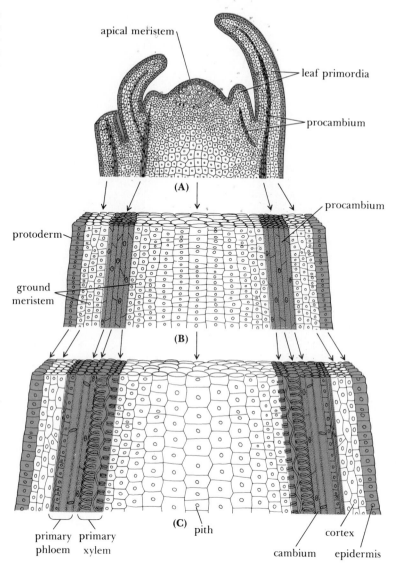

Fig. 7-6 Progressive changes (A,B,C) during growth in the terminal portion of a dicotyledonous stem. Longitudinal section, highly diagrammatic.

apical meristem

leaf primordia

procambium

(A)

procambium

protoderm

ground meristem

(B)

primary phloem   primary xylem

(C)   pith

cortex

cambium   epidermis

length. As a result of this enlargement, the process of elongation extends progressively upward from the base to the apex, passing through each internode in turn. Cell division ceases first at the base of each internode and later at the middle and top. Shortly thereafter, cell

enlargement also ceases at the base of the internode and later at the middle and top. Terminal growth of the stem, then, is largely the result of internodal elongation.

As pointed out in Chapter 2, in woody plants the bud is an undeveloped twig. Some—and in out much of the growing season. Some herbs continue to elongate until killed by frost; others, until the formation of flowers and fruits.

Origin and Development of the Dicotyledonous Leaf The leaf originates by the division of cells in a localized area along the side of the shoot apex. The first divisions usually take place beneath the outermost layer, but soon the cells of the surface layer also begin to divide. All of these cells—both the surface cells and those just beneath—divide more rapidly than the other cells of the apex and soon form a fingerlike or peglike protection, sometimes flattened on the side next to the apex (FIGS. 7-5 and 7-6). The lower part of the peg is the embryonic petiole; the upper part is the embryonic midrib, or axis of the blade. Two zones of cells, extending along the sides of the midrib portion, now begin to form the blade—the expanded portion of the leaf. Later growth is brought about by cell division and cell enlargement throughout the leaf. The rate at which these activities take place differs in the various tissues as the leaves unfold. The resulting unequal growth rates throughout the leaf bring about the formation of the numerous intercellular spaces as the leaf matures.

The dormant bud of a woody plant of temperate climates contains embryonic leaves in various stages of development, from those visible as relatively undifferentiated projections from the sides of the shoot apex to older leaves with the blade already outlined. A leaf originating during one season may thus have its growth interrupted and resume growth the following spring.

With some few exceptions, leaves are produced only by a shoot apex and, if injured or destroyed, cannot be regenerated. They can be replaced, but only by new leaves formed by an apical meristem. Since leaves are usually limited in growth and are short-lived, the adaptive advantages of continued production of new leaves in a scheme of unlimited growth are obvious.

(A)   (B)   (C)

Fig. 7-7 **Stages in growth of red oak (*Quercus rubra*) twigs. Increase in length results primarily from the lengthening of the internodes in the region of elongation.**

some cases all—of the rudimentary nodes and internodes of the new twig are present some months previous to bud expansion, and can be seen by dissecting a resting bud. In a woody plant with resting buds, terminal growth may be rapid, and the shoot may increase a few inches to several feet during a growing season. In woody plants of temperate climates, terminal growth is, in general, limited to the first half of the growing season, and is usually completed within a few weeks. In many temperate-zone herbaceous plants, however, such growth continues through-

Primary Growth We have now laid the foundation for an understanding of primary growth, one of the two general kinds of growth found in vascular plants. Primary growth is initiated in the apical meristems of roots and shoots. It is completed when derivatives of the apical meristems have differentiated into the mature tissues which collectively constitute the primary body. The tissues that compose the primary body are termed PRIMARY TISSUES. The phloem is primary phloem; the xylem, primary xylem. The cortex, pith, and epidermis are likewise primary tissues. The forerunners of these tissues can be distinguished just below the apical meristem as the initial stages of the three tissue systems: dermal, vascular, and ground tissue. These meristematic tissues (FIGS. 7-5, 7-6) are (1) the PROTODERM, which forms the epidermis; (2) the PROCAMBIUM, which gives rise to the primary vascular tissues; and (3) the GROUND MERISTEM, which produces the ground tissues, pith and cortex.

The procambium is composed of vertically elongated cells arranged, in dicotyledonous and coniferous stems, in a cylindrical pattern. In cross section (FIG. 7-8A) the procambial cells, small in diameter and rich in protoplasm, are seen to be in groups arranged in a circle and separated by tissue that remains, for a time, meristematic. These groups are actually immature vascular bundles, and they extend out into the leaf primordia that are developing at the same time. The first procambial strands are soon supplemented by additional procambial strands that differentiate from the meristematic cells between the original groups and are associated with more recently formed leaves. In some plants the procambial strands increase in width to the extent that they come into contact with their neighbors on either side.

The procambial strands then gradually differentiate and mature into vascular bundles, separated by interfascicular (interbundle) parenchyma, the result of the differentiation of the remaining meristematic cells between the procambial strands. The outer cells of a procambial strand differentiate into primary phloem, and the inner into primary xylem (FIG. 7-8B). Between the primary phloem and the primary xylem there remains, however, a single layer of cells, the vascular cambium, which does not become differentiated but remains meristematic. These cells retain indefinitely their capacity to divide and form new cells, thus bringing about increase in diameter of the stem.

The cells of the primary phloem are generally similar to the phloem cells that may be formed later by cambial activity. Most of the primary xylem cells, however, are characterized by the pattern of the secondary wall (FIG. 7-9). The first-formed (innermost) of the primary xylem cells are termed ANNULAR elements, for the secondary wall is laid down as rings, or annular, thickenings. The next in order of differentiation is the SPIRAL element, with spiral thickenings (FIGS. 7-6, 7-9). These two types of cells mature while the stem is still elongating, and a considerable amount of cell elongation may accompany organ elongation by the stretching of the primary walls between the annular and spiral secondary thickenings. Spiral elements are usually more conspicuous than annular elements, and are easily observed under the microscope as prominent components of the smaller veins of leaves in macerated or cleared preparations of soft leaves, such as those of the garden bean. The remaining kinds of cells of the primary xylem usually mature after elongation is completed. In order of differentiation these are the SCALARIFORM, RETICULATE, and PITTED elements. In scalariform (ladderlike) elements the individual coils of a spiral are interconnected. In the reticulate, the secondary wall is deposited in a netlike pattern. The pitted elements are very similar to the water-conducting cells of the secondary xylem.

All of the above types of primary xylem cells do not necessarily occur in a particular plant or plant organ, and they may intergrade with one another. For example, it is often difficult to distinguish between scalariform and reticulate, and such cells are best termed scalariform-reticulate.

The primary phloem often includes a con-

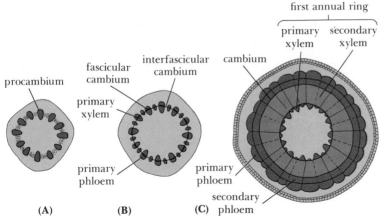

procambium

fascicular cambium

interfascicular cambium

cambium

first annual ring

primary xylem

secondary xylem

primary xylem

primary phloem

primary phloem

secondary phloem

(A)   (B)   (C)

Fig. 7-8 Diagrammatic cross section, showing arrangement of procambial tissue, origin of cambium, and primary and secondary tissues in young woody stems. (A): stem just below the apex. (B): formation of vascular bundles, origin of fascicular and interfascicular cambium. (C): end of first year's growth, showing relationship of primary and secondary tissues.

siderable amount of thick-walled, supporting sclerenchyma, usually in the form of elongated fibers. They also may occur in the cortex. In herbaceous and young woody stems the bulk of the tissue outside the phloem is the cortex, which is entirely or in large part composed of parenchyma cells. The storage of food in the cortex, and the occurrence of chlorenchyma, which manufactures a certain amount of food, have been mentioned previously. In herbaceous and young woody stems the cortex is covered by a single layer of cells, the epidermis, similar in most respects to the epidermis of the leaf. Like the leaf epidermis, the stem epidermis is protected from desiccation by a cuticle. Stomata commonly occur in the epidermis of the stem.

Many plants contain an important living primary tissue known as COLLENCHYMA. This tissue is composed of elongated cells with tapering (sometimes transverse) ends and thick primary walls of cellulose and pectin. It is a supporting tissue, supplementing the mechanical function of the vascular tissue. The thickening of the primary wall is typically restricted either to the angles where several cells come together (FIG. 7-10) or, less commonly, to the tangential walls. Intercellular spaces may be present or absent. Collenchyma is most characteristic of the herbaceous dicotyledons. It is usually found near the surface of an organ, such as the stem, and may occur all around the stem or only in vertical

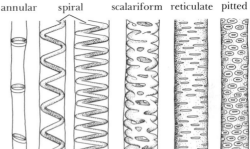

annular   spiral   scalariform   reticulate   pitted

Fig. 7-9 Types of cells of primary xylem in longitudinal view, showing the nature of the secondary wall. Pith to the left; outside of stem to the right.

ridges. It is also found in petioles, such as those of celery, where it is the chief component of the ribs or ridges visible on the surface. It is not infrequently found in large leaf blades, both woody and herbaceous, where it is usually associated with the midvein. During development, the collenchyma cells elongate to many times their original size. Wall thickening begins early, and the cells may elongate and form thick walls simultaneously, the tissue thus being well adapted to the support of growing organs.

Secondary Growth In contrast to primary growth, secondary growth results from the activity of a lateral meristem, chiefly the vascular cambium. A small amount of cork may also be produced by a cork cambium. Secondary growth is characterized by increase in thickness of stem and root, and the tissues formed during such growth are called SECONDARY TISSUES.

The cambium usually becomes active before the primary vascular tissues have become fully differentiated. It originates first in the developing primary vascular bundles as a layer of cells between the differentiating xylem and phloem. The cambium does not differentiate, but remains meristematic. The cambium within such bundles is termed the *fascicular* (bundle) *cambium.* Very soon, however, a cambium arises in the tissue between the bundles, thus connecting the fascicular cambium on either side. This cambium between the bundles is termed the *interfascicular cambium* (FIG. 7-8B), but it behaves like the fascicular cambium.

The cambium of the bundles thus becomes united and continuous and constitutes a cylinder that extends throughout the nodes and internodes of the stem and its branches. In some plants the procambial strands increase in width and come in contact with their neighbors. The cambium of the bundles is then continuous from the beginning, with no distinction between fascicular and interfascicular cambium. Since the cambium is located along the sides of the shoot and root, it constitutes a lateral meristem, in contrast to the apical meristems.

The cambium cells divide and produce secondary xylem on the inner side and secondary phloem on the outer side (FIG. 7-8C). It should be noted that the secondary xylem is laid down outside the primary xylem, and that secondary phloem is added to the inner side of the primary phloem. By the end of the growing season the first GROWTH, or ANNUAL, ring has been formed in the twig (FIG. 7-8C). This consists of a small amount of primary xylem next to the pith, the remainder of the ring being secondary xylem. In woody plants the cambium forms secondary xylem and secondary phloem, year after year, during the life of the plant. The older phloem is gradually crushed by the pressure of the new tissues formed within, but the secondary xylem is preserved and eventually forms the bulk of the plant body. It is seen, in cross section, as a series of growth rings (FIG. 7-11), only the innermost of which contains primary xylem.

The cambium The activities of the cambium, then, are responsible for the formation of the xylem and phloem of the tree and therefore for the wood and lumber used by man. Since the cambium is so significant to the plant and to man, it deserves intensive study.

The individual cells of which the cambium is composed are termed the CAMBIAL INITIALS. In cross section they are rectangular. If the bark is stripped from the tree so as to expose the cambium in face view, the initials appear under the microscope as in FIG. 7-12. It should be noted that the initials are of two kinds: (1) greatly elongated cells tapering at both ends, termed FUSIFORM INITIALS, and (2) much smaller cells, aggregated into lens-shaped clusters, the VASCULAR RAY INITIALS.

The fusiform initials give rise to the vertically elongated cells of the xylem and phloem. When these initials divide and form such cells, a tangential wall (parallel to the surface of the stem) is laid

Fig. 7-10 Collenchyma in outermost stem tissue of pigweed (*Chenopodium album*) as seen in cross section.

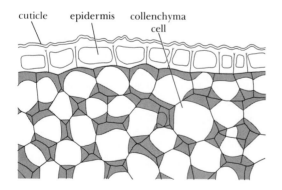

cuticle   epidermis   collenchyma cell

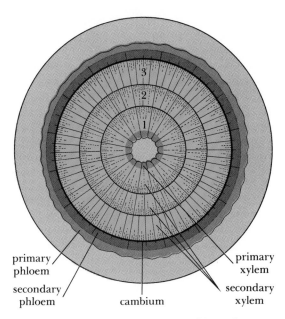

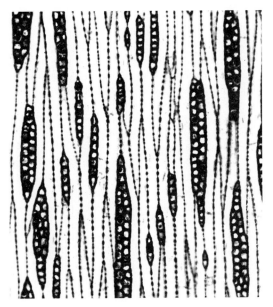

Fig. 7-11 Diagram of a three-year-old woody stem, showing position of cambium and primary and secondary vascular tissues 1, 2, 3: first, second, and third growth rings.

primary
phloem

secondary
phloem

cambium

primary
xylem

secondary
xylem

Fig. 7-12 Dormant cambium of pear (*Pyrus communis*), tangential view. The elongated cells are the fusiform initials; the rounded cells in clusters are the vascular ray initials.

down throughout the entire length of the cell, dividing it into two equal daughter cells (Fig. 7-13). One of these remains a cambial initial, enlarges to its former size, and divides again. The other daughter cell becomes a part either of the xylem (to the inside of the cambium) or of the phloem (to the outside). The recently formed cambial derivatives, however (at least in woody plants, where the cambium has been most studied), usually do not differentiate directly into mature xylem and phloem, but divide one or several times by tangential walls (Fig. 7-13). Such recently formed cambial derivatives, still dividing, are referred to as XYLEM and PHLOEM MOTHER CELLS. As division proceeds, the derivatives of the mother cells, beginning with those farthest from the cambium, gradually differentiate into xylem and phloem cells. Many more derivatives are produced on the xylem than on the phloem side—in fast-growing conifers, for example, more than ten xylem cells may be

produced for each phloem cell. As the trunk expands, new fusiform initials are produced in the cambium, which thus increases in circumference with the tree.

The vascular rays are built by the vascular ray initials in a way similar to that by which the vertically elongated cells of the xylem and phloem are formed. Cell walls are laid down in a tangential plane, dividing each ray initial into two daughter cells. One of these remains a ray initial, while the other, either directly or after one or more divisions, elongates somewhat in a horizontal direction and contributes to the radial length of the vascular ray. The ray initials add to the ray as long as the tree lives. As the trunk enlarges in diameter, the number of rays greatly increases. This is brought about by the formation of new groups of ray initials from fusiform initials in the cambium, and these give rise to new vascular rays at various points on the circumference. Some rays thus extend from a region near the pith, while others extend only through a few of the last-formed annual increments of the wood.

Consequently, during the growing season, when the cambium is active in the formation of new cells, a zone of thin-walled, immature cells may be seen in cross section between the wood and the phloem. This CAMBIAL ZONE (FIG. 7-14) is a number of cells in width and is composed of the cambial initials together with their derivatives that are still dividing or have not yet differentiated into xylem and phloem. It is difficult to identify the cambium itself, which is but one cell wide. As a result of the presence of these immature cells, a zone of weakness is present between the wood and the bark, and the bark is readily peeled from twig, branch, or trunk. The separation of the bark does not appear to take place through the cambium itself, but through the thin-walled cells of the xylem just inside the cambium, which have not yet developed secondary walls. When pulpwood is cut in the spring or early summer, the bark is readily removed by hand because of these soft-walled cells, but if cut at other seasons of the year when the cambium is inactive, it must be removed by other methods. Willow whistles, made by slipping the bark on a willow stick and returning it to place after shaping the wood of the stem, are easily made in the spring for the same reason.

Primary and Secondary Growth Compared
The contrast between the effect of primary and secondary growth upon the development of the plant cannot be overemphasized. Primary growth is responsible for increase in height; secondary growth, for increase in diameter. It is sometimes believed that limbs of a tree or initials carved in the bark are carried upward with the growth of the tree. This cannot happen because the region of elongation is the only region where upward growth occurs. When the cells of this region have differentiated, any additional growth results only from cambial activity and is growth in diameter.

Although terminal growth in woody plants is usually completed in the early part of the growing season, lateral growth continues much longer, sometimes until frost. Primary tissues make up a relatively small amount of the volume of woody stems, and after secondary growth has proceeded for some time, they play only an insignificant role in the economy of the plant. In many herbaceous plants, however, primary tissues constitute the bulk of the plant body. In monocotyledons a cambium is usually absent and the plant tissues are thus primary in origin.

SECONDARY TISSUES OF THE STEM

The Wood   The structure of the wood is rather complex. Wood includes a number of types of cells, most of them vertically elongated. Hardwoods include cells of different types than softwoods; hence it is convenient to discuss them separately. The words "softwood" and "hard-

Fig. 7-13 Diagram showing division of cambium initials with the formation of xylem and phloem mother cells. These in turn divide, and the derivatives enlarge and differentiate into cells of secondary xylem and phloem: ci, cambium initial; pmc, phloem mother cell; xmc, xylem mother cell; x, xylem; p, phloem.

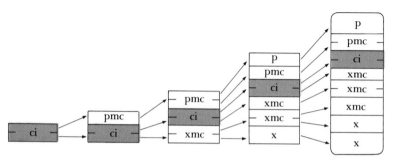

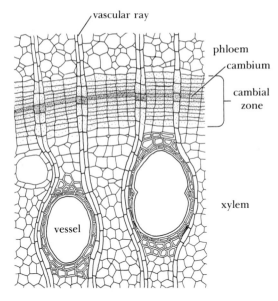

vascular ray

phloem

cambium

cambial zone

xylem

vessel

Fig. 7-14 **Portion of cross section of woody dicotyledonous stem, showing growth and differentiation following division of the cambium. Cells recently formed from the cambium (in the cambial zone) are just beginning to enlarge, and little differentiation has occurred. Earlier formed cells have become fully differentiated.**

wood" are popular names, and may cause some confusion. Lumber-producing trees are largely restricted to the conifers, or "evergreens," and to the dicotyledons among the flowering plants. Coniferous wood is called softwood, and wood from a dicotyledonous tree is known as hardwood, but some softwoods yield hard wood, and not all hardwoods produce hard lumber. For example, hemlock and yellow pines are softwoods, but their wood is harder than some hardwoods, such as poplar and basswood.

Soft, or coniferous, wood The wood of conifers is composed principally of cells known as TRACHEIDS. When viewed in cross section under the microscope, these appear as somewhat angular cells that fit tightly together, with few in-

tercellular spaces. They possess relatively thick secondary walls (FIG. 7-15). Entire tracheids are easily obtained for study by placing a small piece of pine or other coniferous wood in a mixture of chromic and nitric acids, which dissolves the intercellular layer, letting the tracheids separate.

Viewed from the side, these cells appear greatly elongated, with tapering ends (FIG. 7-16). Their average length is about 4 millimeters. They are not arranged end to end, but each tracheid overlaps the one above and below to about one-fourth of its length at each end.

Tracheids are characterized by the large size of their BORDERED PITS, which appear in longitudinal view as openings in the secondary wall. The primary wall in the pit area is continuous and its central portion is somewhat thickened. The secondary wall arches over an inner cavity. Water moves through these bordered pits from one tracheid to another, passing readily through the thinner areas of the primary wall. Other kinds of pits occur between tracheids and vascular ray cells, and between the cells of the vascular ray itself.

Microscopic examination shows that a tracheid consists only of cell walls. The cytoplasm and nucleus, present earlier in the life of the cell, have disappeared, and the nonliving cell is filled with water or air. Tracheids not only provide a series of tubes through which water may move from the roots to the leaves, but they also serve as the supporting or skeletal system of the coniferous tree. These thick-walled, nonliving cells enable huge pines, firs, and redwoods to support a tremendous weight of leaves and branches and to resist the added strains of snow and wind.

Tracheids are considered to be a primitive type of conducting cell. They occur not only in seed plants but also in the ferns, club mosses, and other lower vascular plants.

Another feature of softwoods is the vascular rays (FIG. 7-17), which extend in radiating lines from the vicinity of the pith through the wood and into the phloem. Most cells of the ray are parenchymatous; in the sapwood they are living

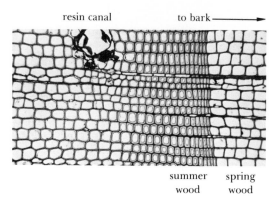

resin canal   to bark ⟶

summer   spring
wood   wood

Fig. 7-15 Cross section showing tracheids in the wood of red pine (*Pinus resinosa*).

and contain nuclei and cytoplasm. In coniferous wood the rays make up about 8 percent of the volume of the wood.

An understanding of the structure of the ray and of wood in general is gained by studying wood in three planes, or planes of section (FIGS. 7-17, 7-18, 7-22, 7-23). Wood is viewed not only in CROSS (transverse) section, but also longitudinally, both parallel to the rays and at right angles to them. In a cross section of pine wood (FIGS. 7-15 and 7-17) the ray is shown to be only one cell wide. A longitudinal cut made parallel to a ray reveals the RADIAL plane. Here the ray is seen to be several cells high, appearing under the microscope like a narrow ribbon passing through the wood. A longitudinal cut at right angles to the wood rays exposes the TANGENTIAL plane. In this the rays appear in end view as a group of three to ten or more cells arranged vertically in the wood. In general, softwood rays are only one cell wide, although in a few species the middle of the ray may be two cells wide.

Scattered through the wood and bark of some conifers—notably the pines, Douglas fir, larch, and spruce—are vertical and horizontal RESIN CANALS (FIGS. 7-15, 7-18). They are largely ab-

sent from other woods. The horizontal canals pass through some of the wood rays and form a connective system with the vertical canals. The canals are lined with living cells that secrete into them a resinous substance known as OLEORESIN. This is the material that yields the turpentine and rosin of commerce. Oleoresin for commercial use is obtained from southern pines by cutting through the bark and into the outer wood. A container is placed beneath the cut to catch the dripping resin (FIG. 7-19), which is then distilled to separate the turpentine.

Hard, or dicotyledonous, wood In the wood of conifers the tracheids serve for both conduction and support, but in hardwoods other types of cells carry on these functions. Tracheids occur in hardwoods, but WOOD FIBERS, cells of the same general form but considerably modified, are more common. Wood fibers have smaller cavities than tracheids, are longer and taper more at the ends, and have pits greatly reduced in size (FIG. 7-20A, B). The pits of wood fibers are so small, in fact, that little or no water moves through them. The function of the fiber is therefore confined to mechanical support.

Water conduction takes place in the vessels. A young vessel consists of a vertical series of cells, the vessel elements. Following cell enlargement, a secondary wall is laid down over the entire primary wall except at the ends of the cell, where perforations will appear (FIG. 7-21). As the individual elements approach maturity and the protoplast disappears, the end primary walls break down and a tubelike structure of considerable length is formed. This is the vessel. The perforations may be elongate and divided by transverse bars. Some vessel elements have oblique end walls, whereas in others the end wall is at right angles to the side walls (FIG. 7-20C, D, E). Vessel elements also bear pits on the side walls, and these connect the vessel elements with adjacent vessel elements and with ray and other cells of the wood.

The presence of vessels distinguishes hardwoods (FIGS. 7-22, 7-23) from softwoods. The

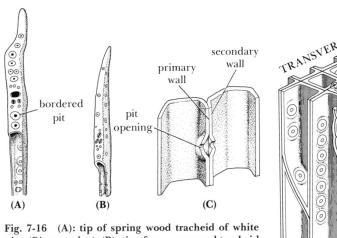

bordered pit

pit opening

primary wall

secondary wall

(A)   **(B)**   **(C)**

Fig. 7-16 (A): tip of spring wood tracheid of white pine (*Pinus strobus*). (B): tip of summer wood tracheid of white pine. (C): portion of two adjacent tracheids, and detailed structure of a bordered pit as seen in section view.

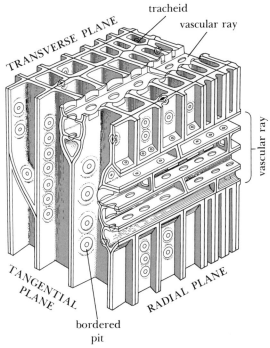

TRANSVERSE PLANE

tracheid

vascular ray

vascular ray

TANGENTIAL PLANE

RADIAL PLANE

bordered pit

Fig. 7-17 Portion of pine wood in three planes, greatly enlarged.

Fig. 7-18 Wood of white pine (*Pinus strobus*) in three planes. (Left): cross (transverse) section. (Center): radial section. (Right): tangential section.

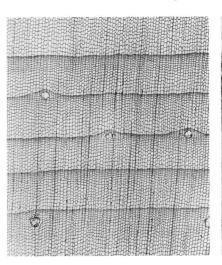
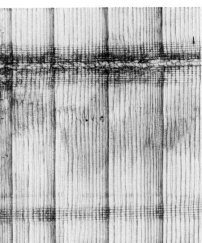
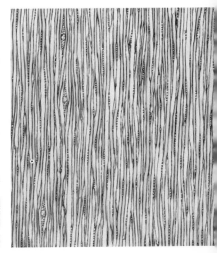

Fig. 7-19 Chipping southern pine for oleoresin, Georgia.

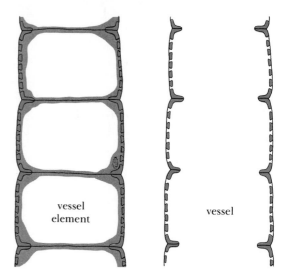

Fig. 7-21 Formation of a vessel. (Left): vessel elements arranged in a longitudinal series. The horizontal walls are primary only. (Right): the primary walls have disintegrated.

Fig. 7-20 Cells from dicotyledonous wood. (A): fiber of tulip poplar (*Liriodendron tulipifera*). (B): portion of fiber greatly enlarged, showing pits. (C): vessel element from wood of yellow poplar. (D): vessel element from early wood of oak. (E): vessel element from late wood of oak.

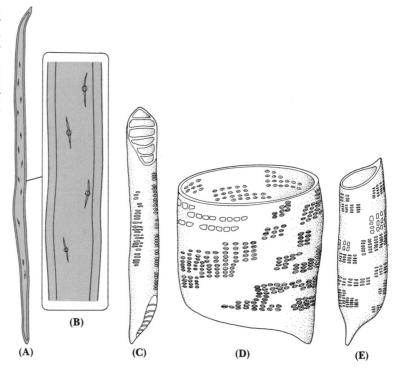

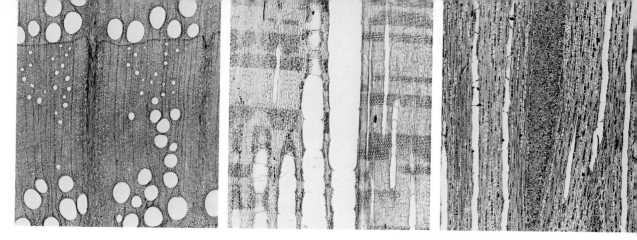

Fig. 7-22 Ring-porous wood of red oak (*Quercus rubra*) in three planes. (Left): cross section. (Center): radial section. (Right): tangential section. Note the single giant ray and the many smaller rays.

Fig. 7-23 Diffuse-porous wood of river birch (*Betula nigra*). (Left): cross section. (Center): radial section. (Right): tangential section.

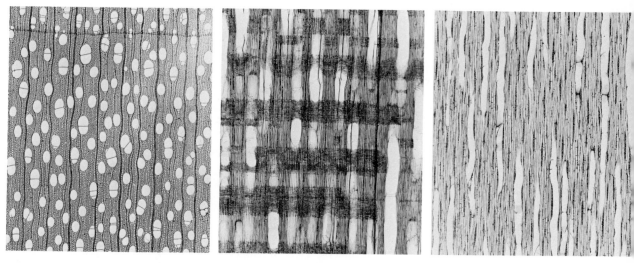

vessels, in cross section, may be seen with the naked eye or a hand lens as tiny openings, like pinpricks, in the wood. Fibers and vessels, like tracheids, die and lose their protoplasmic contents as they mature.

The rays of hardwoods usually differ from those of softwoods. In some hardwoods, as in softwoods, the rays are only one cell wide and comparatively few cells high, but more commonly they are several to many cells wide and extend vertically some distance in the wood. Both narrow and wide rays may occur in the wood of the same species. In some woods, such as oak, the wide rays are so large as to be conspicuous to the naked eye. The rays in hardwoods average about 17 percent of the volume of the wood.

Growth Rings The wood of trees of the temperate zone is characterized by the presence of layers, each usually representing wood formed during a single growing season. In a cross section of a branch, or on the end of a log, these layers appear as a series of concentric rings, termed growth, or annual, rings. Each growth ring consists of an inner layer, the EARLY, or SPRING, wood, and an outer layer, the LATE, or SUMMER, wood. The cells of the early wood, formed early in the growing season, may differ from the cells of the late wood in kind, arrangement, and size, and in thickness of the cell walls.

In conifers the early wood is composed of tracheids with larger cavities and thinner walls than those of the late wood (FIGS. 7-15, 7-18). The tracheids of the late wood, when seen in cross section, are somewhat flattened. As a result of these differences, the late wood appears more dense and darker than the early wood and is readily distinguished from it.

The hardwoods are classified into two large groups on the basis of the size of the vessels throughout the growth ring. In some hardwoods the vessels of the early wood are very large in comparison with those of the late wood, the transition from one type to the other often being abrupt. Such woods are called RING POROUS. Examples are oak (FIG. 7-22), ash, elm, black locust, hickory, and catalpa. Because of the difference in size between the vessels of the early and late wood, these two areas of the ring are usually clearly defined. In many ring-porous woods used for flooring or furniture, the vessels of the early wood are so large that a filler is used before a coat of wax or varnish is applied. In the second group, which includes many more species than the first, the vessels are approximately the same size throughout the growth ring, and such woods are known as DIFFUSE POROUS. Examples are birch (FIG. 7-23), maple, beech, yellow poplar, sycamore, and basswood. Since there is little or no size difference in the vessels throughout the ring, it is usually difficult to distinguish between early and late wood. In all hardwoods, the last-formed part of the late wood, a region only a few

cells wide, is composed chiefly of small, thick-walled cells that border upon the early wood of the next growth ring.

A number of theories have been advanced to explain the differences in cell size between the early and late wood. Early investigators believed it to be due to declining supplies of water or food to the cambium and immediate derivatives. A current concept is that cambial activity and xylem formation is regulated by downward-moving growth hormones produced by apical meristems and rapidly growing regions just behind them and by developing leaves. There is considerable evidence that the transition from the production of cells with large diameters in the early wood to those with narrow diameters in the late wood is correlated with a decline in the supply of growth hormones and an increase in the amount of sugar as the season advances and bud and leaf growth declines.

Annual growth increments are usually visualized as rings on the surface of a stump or the end of a log. In reality, each ring is a cross section of a hollow cone (FIG. 7-24) that encloses all the previously formed wood of the entire tree, not only of the trunk and branches, but of the roots as well. As the tree increases in height, it tapers from a considerable girth at the base to the diameter of the twigs at the top. The cones vary in height from the oldest increment, perhaps only a few inches high, representing the xylem formed in the seedling year, to the youngest, which extends over the entire tree.

The age of a tree may be determined, at least approximately, by counting the rings on the cut surface of the wood. To be reasonably accurate, this count must be made on the surface of a stump as close as possible to the soil level, for counts made higher up the tree will give only the age at the level at which the counting is done. The occurrence of *false* growth rings may cause the age of the tree to be overestimated. Such rings are produced by a temporary slowing of growth during the growing season. This produces a poorly defined zone of late wood within the limits of a growth ring. Among the factors responsible for false rings are drought and defoliation by insects.

The width of individual rings varies greatly, and much effort has been spent in attempts to correlate these differences with variations in the environment. Among the environmental factors are light, temperature, rainfall, soil moisture, length of growing season, and competition. The last of these commonly provides clear-cut effects on ring width. Such an effect may be observed in a tree closely surrounded by other trees, for the growth rate is slow and the rings relatively narrow. If the adjacent trees are cut, the growth rate increases, and the next rings formed may be several times wider (FIG. 7-25). This is partly because the greater intensity of light falling upon the leaves allows them to manufacture more food, and partly because the roots no longer compete with the roots of other trees for water and minerals from the soil.

In general, when the environment is favorable, the rings are wide; when unfavorable, the rings are narrow. In some regions a high degree of correlation has been found between rainfall and ring width. This is notably the case in the semiarid southwest United States, where rainfall is the most important climatic variable. In areas with more uniform rainfall, no such relationship has been found. Rainfall, especially in arid regions, tends to vary in cycles, and the pattern of growth rings tends to show the same variations. If the pattern of the inner rings of a living tree can be matched with that of the outer rings of a timber from an ancient building in the same area, the age of the building can be determined. By study of the growth rings of living trees and of beams from buildings of increasing age, the pueblos of Arizona and New Mexico have been dated back 1900 years to the time of Christ. Serious droughts in 840, 1067, 1379, and 1632

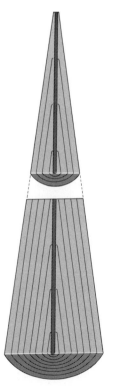

Fig. 7-24 Diagram showing the conelike shape of the annual increments of wood as seen in longitudinal view.

Fig. 7-25 Cross section of trunk of ponderosa pine, showing the increased width of the growth rings following the removal of adjacent trees. A count of the narrow rings shows that the growth rate was retarded for about 54 years.

are shown by very narrow growth rings. A calamitous drought, starting in 1276 and lasting for 23 years, brought starvation to the country and the abandonment of most of the ancient villages.

Heartwood and Sapwood   As the tree increases in age, certain changes take place in the inner part, resulting in the formation of HEARTWOOD. This region of the wood is frequently distinguished from the outer portion, or SAPWOOD, by its darker color (FIG. 7-26). The color of the heartwood results from chemical changes and from the infiltration of oils, gums, tannins, resins, and other complex organic compounds into the cell walls. These materials may also accumulate in the cavities of the cells. In some woods, such as black walnut, teak, red cedar, and redwood, the heartwood is conspicuous by the intensity of its color, whereas in others, such as fir,

Fig. 7-26 White oak (*Quercus alba*), heartwood and sapwood. The dark area in the center is the heartwood; the lighter part, the sapwood.

spruce, poplar, and basswood, color distinctions between heartwood and sapwood are small or absent.

As the tree ages, the inner portion of the sapwood becomes converted into heartwood; thus the radius of the heartwood increases with the age of the tree, whereas that of the sapwood remains approximately the same. The heartwood is usually durable and persists in the tree for many years, commonly for the life of the tree. Sometimes a part or all of the heartwood may decay as a result of the activities of fungi, and the tree becomes hollow. Such a tree may continue to thrive, however, since the cells of the heartwood do not function in conduction.

The Bark.   Bordering immediately upon the wood is the thin-walled cambium. The bark is the region of the stem outside the cambium. Like the wood, it includes several kinds of tissues.

Phloem   The phloem cells, of which there are several kinds, make up the inner portion of the bark, just outside the cambium. Prominent in the phloem are the vascular rays, which are continuations across the cambium of the vascular rays of the wood. The part of the ray in the phloem is usually the same width as the part in the adjoining wood. But in some trees, such as basswood, tulip poplar, and the hickories, certain of the phloem rays become V-shaped in cross section as a result of the radial division of the older cells of the rays (FIG. 7-27).

In flowering plants the most important feature of the phloem is the sieve tubes, for it is through them that vertical conduction of food occurs. The sieve tubes, as mentioned in Chapter 5, are composed of a number of sieve-tube elements, placed end to end (FIG. 7-28). The primary walls are relatively thick, and a secondary wall is usually absent. The ends of the sieve-tube elements are oblique or transverse. The most conspicuous feature of these cells, however, is the presence of SIEVE AREAS, which are somewhat sunken wall areas with clusters of pores. The best developed

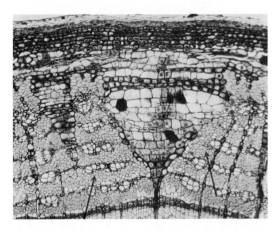

Fig. 7-27 Portion of cross section of basswood (*Tilia americana*) stem, showing the V-shaped rays in the phloem. Arrows indicate phloem fibers.

of the sieve areas are localized in the sloping or transverse end walls and are termed SIEVE PLATES (FIG. 7-28). If a sieve plate is composed of more than one sieve area, it is termed a COMPOUND SIEVE PLATE; if composed of but one sieve area, it is a SIMPLE SIEVE PLATE. Sieve areas are also found on the side walls of the sieve-tube elements of many plants, but these usually have smaller pores than the sieve plates. It is through the pores of the sieve plates that foods move from one sieve element to another.

In addition to the sieve tubes and vascular rays, phloem also contains COMPANION CELLS, PHLOEM PARENCHYMA, and sclerenchyma fibers (FIG. 7-29). One or more companion cells are associated with each sieve-tube element, connected to it by plasmodesmata passing through thin areas in adjacent walls. The companion cells usually arise from the same phloem mother cells as the sieve-tube element. The mother cell divides longitudinally once or several times, and one of the resulting cells becomes the sieve-tube element. Others, destined to give rise to companion cells, then divide transversely, producing a row of two or more elongated cells (FIG. 7-28). The parenchyma cells, which may contain starch and other organic materials, fit in among the sieve-

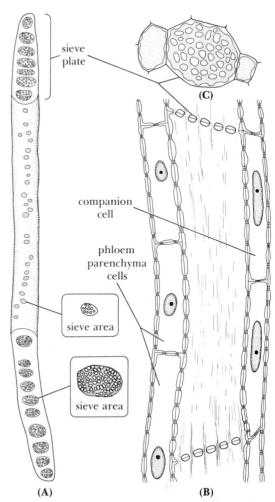

Fig. 7-28 Sieve-tube elements and associated cells. (A): from tulip poplar (*Liriodendron tulipifera*). The tapering end walls of the sieve-tube element bear compound sieve plates composed of several sieve areas. Small sieve areas also occur on the sides of the cell. (B): diagrammatic. Single sieve element with adjacent cells. A simple sieve plate (only one sieve area) is found at each end of the cell. (C): simple sieve plate and adjacent cells as seen in cross section.

tube elements, and are connected with other parenchyma cells and to ray cells and companion cells by plasmodesmata.

A unique feature of the sieve-tube element is

that the nucleus usually disintegrates during the late stages of differentiation. The cell remains alive, however, probably being maintained through its close physiological relationship with the nucleated companion cells. Other degenerative changes commonly also occur, such as the disappearance of the vacuole membrane, leaving only a thin film of cytoplasm just within the cell wall. The contents of the sieve-tube element include a proteinaceous slime, which in ordinary microscopic preparations may be seen condensed in a mass, the slime-plug, over the sieve plates and completely filling the pores. The slime does not occur in this form in the uninjured tissue, but as an extensive network within the cell. Some investigators believe that portions of this network, in the form of fine strands, extend through the pores of the functioning sieve tube, thus partially closing the pores. Others contend that the pores, during much of their functional life, are completely open.

Cells comparable to sieve-tube elements are present in conifers and other vascular plants. Such cells are elongated, with tapering ends that overlap, and are not arranged in a series, end to end. Sieve areas are present, but are not aggregated into sieve plates, which are characteristic only of angiosperms.

The functional life of the sieve-tube elements is short, commonly only a single growing season. The contents become disorganized, and in many plants the sieve tubes eventually collapse as the result of the pressure of surrounding cells. The phloem that functions in conduction is therefore confined to a narrow zone—usually less than a millimeter in width—immediately outside the cambium. In some trees, however, such as linden (*Tilia*), it has been shown that sieve-tube elements can function for five to ten years. Even more striking is the case of long-lived monocotyledons which, since they have no cambium that continually forms new secondary phloem, are dependent upon the original primary sieve tubes in the older portions of the stem. In one palm it has been shown that sieve-tube elements at least 50 years old were still functional, and in other species it has been estimated that the functional life span must be more than 100 years.

Sclerenchyma fibers, when they occur in the phloem, are termed PHLOEM FIBERS. They are very commonly seen in cross section as caps on the outer side of the phloem (FIG. 7-29). The fibers in this position originate from the procambial tissue and are therefore primary in origin. In many woody species, fibers occur also as groups or bands interspersed among the sieve tubes or phloem parenchyma (FIG. 7-27). These fibers are secondary in origin, for they are produced by the vascular cambium. Phloem fibers and certain other kinds of thick-walled cells are responsible for the hardness or toughness of the bark of many kinds of woody plants. The fibers from the bark of woody plants are very little used today, but in the past, fibrous barks were employed in the manufacture of clothing and writing materials. For example, the tapa or bark "paper" of the Pacific islands was prepared from the beaten bark of the paper mulberry, and the thin resulting sheets used as cloth. The Chinese and Koreans made a writing material from the bark of the same tree early in the Christian era. The ancient Romans wrote upon the inner bark, known as *liber,* of basswood and other trees. This word in time came to mean "book," and from it was derived the word "library."

In some woody plants annual growth increments may be distinguished in the phloem, but are not so clearly defined as in the xylem. They may be due to bands of fibers, to the banded appearance of tannin-containing cells, to the small size of the last-formed phloem cells, or to other causes. The zone in which such increments are distinguishable is small, however, since, as the tree ages, the older phloem is crushed or penetrated by layers of cork.

Cork Most woody and some herbaceous plants form a layer of cork on the outside of the stem (FIG. 7-29). Such a layer is usually to be found in the twigs of woody plants in the first season of growth. The cork is formed by a cork cambium, which arises in the epidermis or, more com-

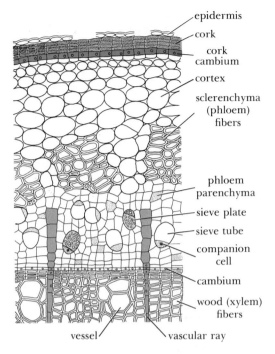

epidermis
cork
cork cambium
cortex
sclerenchyma (phloem) fibers
phloem parenchyma
sieve plate
sieve tube
companion cell
cambium
wood (xylem) fibers
vessel
vascular ray

Fig. 7-29   Important cells and tissues of the bark and adjoining regions of a dicotyledon, as seen in cross section. Diagrammatic.

Fig. 7-30   Cork in the potato tuber.

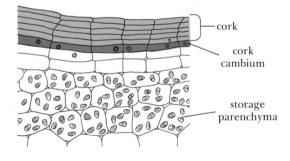

cork
cork cambium
storage parenchyma

monly, in the outermost layers of the cortex just beneath the epidermis. The cork cambium divides repeatedly and on its outer side produces a layer of cork cells arranged in radial rows. Oc-

casionally a few parenchymalike cells are formed on the inner side. As the outer cork cells mature, a fatty or waxy substance infiltrates into the cell walls, making them resistant to the passage of water and gases. The cork separates the epidermis from its supply of food and water, and so the epidermis dies and gradually scales away. Later, the cork cells die and become filled with air, with tannin, or, in the case of the paper birch, with a waxy material that gives the bark its whitish appearance and its value in lighting fires. Since cork is produced by a lateral meristem, it is classed as a secondary tissue.

Cork retards the loss of water from stem tissues. It is also formed during the healing of wounds, thus preventing the drying out of exposed tissues and the entrance of decay fungi. In addition, cork is an important insulator, and it is possible that the development of rough bark and thick cork in trees of temperate regions is related to the prevention of injury to the cambium by sudden temperature changes in the winter. Trees with a thick, corky bark may be uninjured by a ground fire that would destroy the cambium and thus kill a thin-barked species.

The potato tuber offers a familiar example of cork formation. A cork cambium forms early in the development of the tuber and produces a thin layer of cork. This thin layer, or "skin," is readily removed by brushing or scraping a freshly dug or "new" potato. If the tuber is stored, the cork layer increases in thickness, hardens, and can be removed only by peeling. This cork layer makes potato storage possible. The peel consists largely of cork cells (Fig. 7-30), and a person who eats the skin of a baked potato consumes cells similar in nature and origin to those of bottle cork.

Commercial cork is obtained from the cork oak, a native of the Mediterranean region. The cork cambium of this tree produces a thick layer of cork that is stripped from the tree at intervals of 8–10 years. The stripping destroys the cork cambium, but a new cambium, formed within the living tissues of the bark, develops another cork layer.

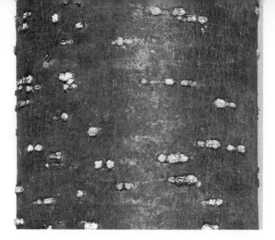

Fig. 7-31 Lenticels of alder (*Alnus rugosa*). This stem was 4 inches in diameter.

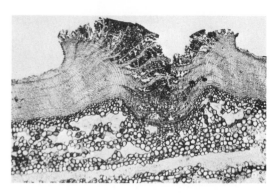

Fig. 7-32 Lenticel of black cherry (*Prunus serotina*), as seen in cross section of a young stem.

In a number of woody plants such as beech, certain cherries, and birch, the original cork cambium persists for the life of the plant and increases with the circumference of the tree. The cortex also enlarges by cell division, and the trunk, like the twigs and branches, remains relatively smooth. In most temperate-zone woody plants, however, the first cork cambium ceases to divide after a few years, and successive layers of cork are formed deep in the bark. The outermost layers of cork and included tissues soon die, and since these dead tissues are unable to increase with the increase in circumference of the trunk, cracks and fissures develop as the result of pressure from within. Such trees therefore develop a rough bark, in which the surface is broken by scales, ridges, and other markings so characteristic of the species as to be useful in identification. The outer layers of the bark remain on many trees for years and so may attain considerable thickness.

LENTICELS are structures in the bark that serve in aeration. They appear on stems and roots as raised, elongated, or circular areas of a corky appearance (FIG. 7-31). Lenticels are composed of masses of loosely arranged cells with numerous intercellular spaces through which oxygen may penetrate to the living tissues within. In some species the loosely arranged cells alternate with bands of more compactly grouped cells (FIG. 7-32). The lenticel is a product of the cork cambium, which locally gives rise to the cells of the lenticel instead of cork. A rather specialized kind of lenticel is found in bottle corks, in which the lenticels appear as brown channels or grooves traversing the surface at right angles to the long axis of the cork, or as pores along the sides.

DICOTYLEDONOUS HERBACEOUS STEMS

The development, structures, and processes of dicotyledonous herbaceous stems are in most respects identical with those of dicotyledonous woody stems. A cambium, with a few exceptions, is always present, but one difference lies in the length of time during which the cambium functions. Since the herbaceous stem, in temperate climates, lives for only the few months between spring and autumn, the cambium functions for only one growing season. In some herbs, such as sunflower, the stem becomes hard and woody. In others the amount of secondary tissue formed is small, and the stem remains relatively soft. In addition to the absence of growth rings, other, but minor, differences between woody and herbaceous dicotyledonous stems include the large size of the pith in many herbs and the frequent lack of cork development.

The internal anatomy of many herbaceous stems is very like that of the woody dicotyledonous twig. In some herbaceous stems the primary vascular tissues form an essentially continuous cylinder throughout the internode, and individual vascular bundles cannot be distinguished. The cambium also is continuous and forms a complete cylinder of secondary tissues. In other stems the primary vascular tissues, as seen in cross section, are composed of distinct vascular bundles, variable in number and size. The bundles develop a fascicular cambium and are subsequently united by an interfascicular cambium. As a result of cambial activity a complete cylinder of secondary tissues is formed.

In some herbaceous stems, evolutionary reduction has led to a further decrease in cambial activity. The interfascicular cambium between the bundles may produce only thick-walled parenchyma on the xylem side, as in alfalfa (*Medicago*), or cambial activity is limited to the vascular bundles, as in the clovers (*Trifolium*) (FIG. 7-3). In a few highly reduced types, chiefly the crowfoots (*Ranunculus*), a cambium is absent from the vascular bundles.

Phloem fibers in herbaceous stems are usually restricted to the outer side of the phloem, where they occur in cross section, as in woody stems, as cap-shaped masses. Phloem fibers from stems of herbaceous dicots are important as raw materials from which cord and textiles are manufactured. Among plants whose fibers are thus utilized are flax, jute, and hemp.

MONOCOTYLEDONOUS STEMS

The structure of the monocotyledonous stem differs from that of conifers and dicotyledons in two chief ways: (1) the vascular tissues are usually organized into separate bundles, and these, as seen in cross section, are scattered throughout the stem instead of in a cylindrical arrangement (FIG. 7-33). As a result of the scattered distribution of the vascular bundles, no distinction can be drawn between pith and cortex. In some monocotyledons such as bamboo, wheat (FIG. 7-33B), and other grasses, the internodal regions are hollow, but even in these plants the vascular bundles are irregularly arranged; (2) all the cells of the procambial strands mature into xylem and phloem. A cambium is therefore absent.[1] Lacking a lateral meristem, the tissues of a monocotyledonous stem are primary in origin, and so primary tissues are of much greater significance than in dicotyledons and conifers. Absence of secondary growth causes stems of monocotyledons, even of palms and large bamboos, to be columnar rather than tapering.

The arrangement of cells in the vascular bundles of monocotyledons may be illustrated by those of a bundle from the inner portion of the corn stem (FIG. 7-34). The first-formed xylem cells, those toward the center of the bundle, usually collapse as a result of lateral and vertical stresses during growth, leaving a large air space. The functioning xylem consists of two large vessels and several smaller vessels between them. The phloem consists only of sieve tubes and companion cells. The entire bundle is enclosed in a narrow sheath of thick-walled sclerenchyma fibers.

The sheath around the bundles in the interior of the corn stem contributes little to the mechanical strength of the stem, whose ability to resist breakage is largely the result of the structure of the "rind"—the outer part of the stem (FIG. 7-35), where the majority of bundles are found, each surrounded by a wide sheath of thick-walled fibers. The parenchyma cells among

---

[1] An exception to this statement is found in certain woody members of the lily and a few other monocotyledonous families. In these forms a cambium develops and may give rise to a considerable amount of secondary tissue. The cambium originates, however, in the parenchyma outside the primary vascular bundles, and does not produce phloem to the outside and xylem to the inside in the usual way. Most of the secondary tissue is formed to the inside of the cambium and consists of secondary vascular bundles embedded in secondary parenchyma. Some parenchyma of secondary origin is also formed on the outer side of the cambium.

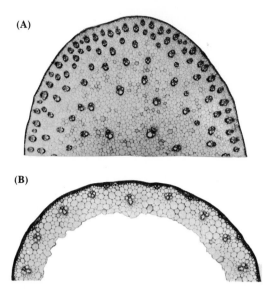

(A)

(B)

Fig. 7-33 Scattered distribution of vascular bundles of monocotyledonous stems. (A): corn (*Zea mays*). (B): wheat (*Triticum aestivum*). In addition to the larger bundles of the wheat stem, smaller bundles are located just under the epidermis.

Fig. 7-34 Cross section of vascular bundle of corn stem.

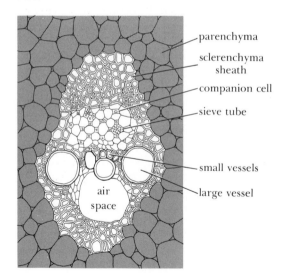

parenchyma

sclerenchyma sheath

companion cell

sieve tube

small vessels

large vessel

air space

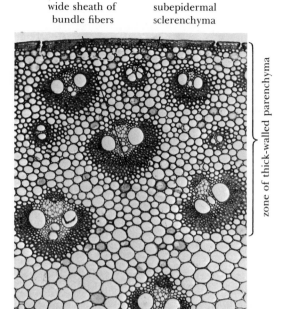

wide sheath of bundle fibers

subepidermal sclerenchyma

zone of thick-walled parenchyma

Fig. 7-35 Photomicrograph showing the structural features of the outermost part of the corn stem—the "rind"—which are responsible for the stiffness of the stalk.

which these bundles are embedded exhibit a feature that cannot be seen in the young stem usually studied in the laboratory, for in the mature stem they develop thick, woody walls. Just beneath the epidermis, also, is found a band of sclerenchyma cells, two or more cells thick (FIG. 7-35). These three parts of the rind—the subepidermal band of sclerenchyma fibers, the thick-walled parenchyma, and the sheath around each of the bundles—enable the mature corn stem to resist wind and the stresses produced by the weight of the leaves and ears.

Sclerenchyma fibers surround many of the bundles of monocotyledonous leaves and are the source of the commercial fibers derived from the leaf of sisal and the leaf sheath of Manila hemp.

Fibers of the same kind are also common in monocotyledonous stems as a circular band of varying width in the outer part of the stem. This band may be present just under the epidermis, as in corn, or several cells removed from the epidermis, as in asparagus. The upper part of the shoot, or spear, of asparagus is edible, largely because these fibers have not yet developed thickened walls. The woodiness of the basal part of the asparagus shoot, as cut for the table, is primarily the result of the maturation of these fibers. In preparing asparagus for cooking, the whitish basal part is frequently discarded. If it is peeled, however, thus removing the fibrous band, the tissue within is as soft and edible as the upper portion of the shoot.

TRANSLOCATION

The most superficial observer of plant activities realizes that materials of various kinds move throughout the plant body. It is commonly recognized that water enters the plant from the soil and that the application of commercial fertilizers and farm manures to the soil results in increased growth or yield of flower or fruit. Modern investigations have revealed a fairly clear picture of the movement of foods, water, and mineral salts within the plant, and of the tissues involved. The term TRANSLOCATION is applied to the movement of all such materials.

Translocation of Water The translocation of water takes place through the xylem, or wood. In the coniferous stem the water takes an irregular path through the bordered pits of the tracheids; in deciduous trees and herbaceous plants, the movement is chiefly through the perforations in the end walls of the vessel elements. If a trunk or branch is GIRDLED, or ringed by removal of a horizontal strip of bark extending inward to the wood, the leaves above such a ring do not wilt, indicating that the water supply to such areas has not been cut off. If, however, a deeper cut is made through the sapwood of the xylem, the top of the tree normally dies quickly from lack of water.

If the exposed xylem of the cut surface of a small branch is covered with some substance that obstructs the vessels, and the branch is placed in water, the leaves wilt, even though conduction in the phloem is not blocked. If an amputated branch is placed in a solution of a dye, a cross section of the stem shows that the wood cells become stained. In trees, the path of movement is still more localized, for here the wood of the trunk and larger branches is differentiated into heartwood and sapwood. Water conduction is usually confined to the outer part of the sapwood. A hollow tree, from which the heartwood has disappeared, may remain alive for many years (FIG. 7-36). In ring-porous species, water moves in only a few of the outermost growth rings, in some species perhaps only in the last-formed increment. In diffuse-porous trees and in conifers a number of growth increments are involved. The rate of water movement varies with the environment, but in general is most rapid in ring-porous trees, frequently several meters per hour. In other trees the rate is much lower—a meter or less per hour in conifers.

A discussion of the mechanism and forces concerned in water movement in plants is included in the next chapter.

Translocation of Foods and Mineral Salts The food materials from which new tissues are constructed and which are utilized in respiration are frequently synthesized or stored some distances from where they are used. The translocation of these organic compounds occurs normally in the sieve tubes of the phloem. Chemical analyses of the sieve-tube contents indicate that carbohydrates move generally as sucrose. The movement of organic nitrogen is probably in the form of amino acids and soluble proteins. Translocation follows the laws of diffusion in some instances, but it may be hundreds of times faster than can

Fig. 7-36 A hollow tree, still living and growing vigorously. Conduction of water takes place in the sapwood.

Fig. 7-37 Black cherry (*Prunus serotina*), end of second season of growth after girdling. Cambial growth was stopped below the ring but continued above it.

be accounted for by diffusion alone. Sometimes movement occurs against diffusion gradients, as when sugar moves from the leaves of the sugar beet to the root, or from the leaves of corn into the developing grain. Such movement is considered to involve active transport.

Evidence that foods are translocated in the phloem has been obtained chiefly by girdling woody stems. If a growing stem is girdled, activity in the cambium below the girdle soon stops. This change is thought to be due to the interruption of the polar, downward movement of growth hormones from the buds and leaves, but the area below the girdle will shortly also show the effects of starvation because of the interruption of the movement of sugar from the leaves. The cambium above the girdle will continue

growth at a normal or increased rate, and the stem above the girdle becomes greater in diameter than the portion below the ring (Fig. 7-37). Although a girdled tree may survive for some time, it will eventually die, even though translocation of water is still proceeding. As the foods present in the lower stem and roots become exhausted, the roots finally die. The lack of water absorption because of the death of the roots causes the leaves to wilt and die, and the plant dies.

The girdling of trees is of more than experimental interest. The girdling technique is useful for eradicating trees, such as some poplars, that sprout freely from the roots when the tree is cut. After girdling, the roots die slowly of starvation, and few or no sprouts are formed.

To clear fields for their crops, the early colonists in this country, pressed for time, girdled trees instead of felling them. Crops were then planted among the dying trees, which could later be cut and burned as time permitted. Girdling is still used in wooded areas to open up land for crops. Rabbits and mice frequently girdle orchard trees as they feed on the bark in winter. Such trees will normally die, but can be saved by bridge grafting them promptly—that is, by grafting both ends of a number of living twigs into the bark above and below the wound, thus bridging the wounded area. Finally, girdling is sometimes used by foresters as a means of improving woodlands. Stands of trees useful for lumber frequently contain worthless trees that shade younger trees around them. If these overtopping trees are felled, they damage the younger trees beneath. After the larger trees are girdled, however, they die and decay, and the gradual falling of their branches causes little damage.

Although the major movement of foods is downward in the stem, upward movement is also important. Food is translocated upward from storage regions to developing young shoots, from leaves to growing fruits, from seeds to seedlings, or, in herbaceous perennials, from underground storage organs to the young shoots. This movement, also, occurs primarily in the phloem.

Beginning about 1953 a new technique was developed that confirmed older ideas and revealed new information about the sieve tube and its contents. This is the aphid-stylet method, used with aphids that feed upon the younger branches of woody plants. In these aphids the mouth parts consist of three stylets and an enclosing basal sheath. When feeding, the stylets are protruded from the sheath and inserted into the bark until they penetrate a single sieve tube (Fig. 7-38). The contents of the sieve tube are then forced into the body of the insect by the turgor pressure of the tube. The exudate is high in sugar but low in amino acids, and the surplus sugar is discarded as honey dew that is commonly gathered by ants.

Fig. 7-38 (Bottom): mature aphid (*Longistigma caryae*) feeding on lower side of a branch of basswood (*Tilia americana*). The stylets are inserted in the bark and only the sheath can be seen. The aphid is about 6 millimeters long. (Top): stylet tips in an individual sieve element of basswood phloem.

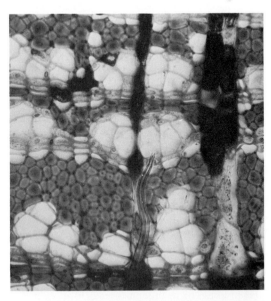

The unchanged sieve-tube exudate is obtained by anesthetizing a feeding insect with a stream of carbon dioxide, then cutting the insect away, leaving the stylets in place. The flow of sieve-tube sap may continue for a day or more, and can be collected with a fine capillary tube. For aphids feeding on willow (*Salix*), the rate of exudation was 1 or 2 cubic millimeters per hour, and the calculated rate of translocation through the sieve tubes was about 1 meter per hour. This involved movement through about 100 sieve-tube elements per minute. The major solute in the sieve-tube sap was sucrose, in concentrations that ranged from 5–15 percent by volume.

Both water and foods also move radially, especially in woody plants, and translocation in this direction is chiefly through the vascular rays. The rays are also important storage tissues of the tree, and at the end of the growing season the cells of the rays are commonly filled with starch. Sugars move downward through the phloem and radially into those parts of the rays lying in the sapwood of the xylem. Here they are converted into starch. This starch is later reconverted into sugar, and in the spring is translocated through the rays to the cambium and developing xylem and phloem, or to other tissues of the plant.

Many inorganic substances, including the essential elements (Chapter 9), are absorbed from the soil and translocated upward through the stem. Numerous experiments have been conducted to determine whether such substances are transported in the xylem or in the phloem. The evidence, derived in part from girdling experiments, is conflicting. The most probable answer is that inorganic substances such as potassium and calcium may move upward in either the xylem or the phloem. Usually, however, the main pathway of the movement of such elements is in the xylem. Thus, nitrogen in the form of inorganic nitrate or ammonia ions may be absorbed by the roots, diffuse into the xylem, and be carried to the leaves in the ascending water. Frequently, however, nitrogen is built into amino acids in the roots. When this occurs, the organic material thus formed normally moves upward through the phloem.

MISCELLANEOUS ASPECTS OF GROWTH AND STRUCTURE

Figure in Wood The figure, or pattern, found on sawed and planed wood surfaces is caused by a number of structural features, the most important being the growth increments and the vascular rays. Variations in color and pattern increase the value of many kinds of wood for use in furniture making, interior finish, and cabinet work.

The nature of the pattern or design varies greatly with the plane in which the wood is cut. This may be either tangential or radial (FIG. 7-39). Tangential cuts result in *plain-sawed*, also called flat-sawed, lumber. Such cuts are tangential to the circle of the log and at right angles to

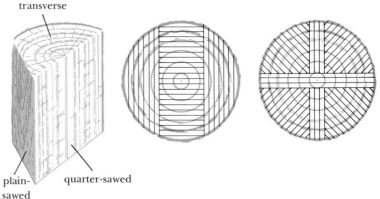

transverse

Fig. 7-39 Planes in which logs are cut to produce plain-sawed and quarter-sawed lumber. (Left): block of wood showing transverse, radial (quarter-sawed), and tangential (plain-sawed) surfaces. (Center): diagram of the end of a log, showing the planes in which the log is cut to produce plain-sawed lumber. (Right): diagram showing planes of sawing to produce quarter-sawed lumber.

plain-sawed

quarter-sawed

the longest dimensions of the vascular rays. The surface of such wood has a pattern or figure of stripes, concentric irregular parabolas, or ellipses caused by the differences in color or structure of the early and late wood of the growth increments (Fig. 7-40, left). These differences are especially prominent in certain conifers and hardwoods, in which the transition from early to late wood is conspicuous. Any difficulty in understanding the pattern on the surface of a plain-sawed board may be overcome by recalling that each growth increment is an irregular cone that tapers gradually from the base of the trunk to the top of the tree. A tangential cut through a log therefore exposes relatively broad surfaces of several growth layers. Since the cones are irregular in shape, there are many departures from a symmetrical geometric figure. The irregularities in the cones are the result of irregularities in the shape of the log and of differences in the width of the growth rings in various parts of the tree.

Wood is also cut in such a manner that *quarter-sawed* surfaces are produced. Quarter-sawed wood is cut approximately parallel to the vascular rays—that is, along a radial plane. In quarter-sawed wood, the rays, if large, appear as stripes or ribbons running across the surface. Oak is the chief hardwood used for quarter sawing, for many of the rays are large and conspicuous (Fig. 7-40, right). Quarter-sawed veneers are produced by cutting thin slices from steamed logs with a large, sharp blade. Other veneers may be cut by rotating the log against a blade, which cuts a continuous sheet in a tangential plane. Quarter-sawed softwoods, such as yellow pine and Douglas fir, are frequently used as flooring. The wearing quality of quarter-sawed softwood is much superior to that of plain-sawed lumber because the harder areas of late wood are closely spaced and exposed vertically to surface wear.

The Healing of Wounds When roots or stems are wounded, the exposed tissues become covered, after a lapse of time, with new tissues of bark and wood. The formation of such tissues is most easily observed in woody plants after the pruning of a large branch. A mass of undifferentiated parenchyma, known as CALLUS, is formed on the margins of the wound early in the growing season. The callus cells are primarily the result of proliferation from the cambium, but other tissues, such as the phloem and cortex, may also be involved.

As the callus forms, a cork cambium develops in the outer layers and a cambium differentiates within it in continuity with the original cambium of the tree. This extension of the cambium now produces swollen masses of tissues that advance from the margins and roll over the wound, a process sometimes referred to as "over-walling" of the cut surface (Fig. 7-41). When the opposite sides meet, the cambial layers unite. The continued activity of the cambium in forming new layers of bark and wood buries the wood more and more deeply each year.

 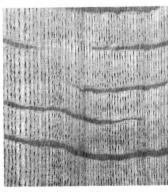

Fig. 7-40 Oak boards. (Left): plain-sawed (tangential surface). (Right): quarter-sawed (radial surface).

Fig. 7-41 This oak (*Quercus rubra*) limb was properly pruned close to the tree trunk, but the bark was injured on the left. After four years the right half of the wound is nearly healed, but healing has been retarded on the left.

Fig. 7-42 Diagram illustrating the burial of the base of a branch by secondary growth. The portion of the branch to the left of the line a—a was alive when buried, and its growth rings are continuous with those of the trunk. The portion to the right of a—a was dead when buried, and the growth rings of branch and trunk are not continuous. Pith shown in color, bark in dark tint.

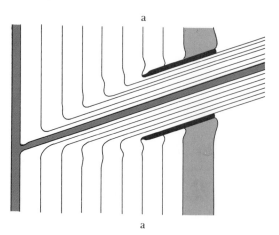

Knots A KNOT is a portion of the base of a branch that has been buried in the wood of the tree as the tree increases in circumference. The cambium of the trunk is continuous with that of the branch, and the growth rings formed over the branch are therefore uninterrupted extensions of those of the trunk. The growth rings merely change direction, more or less abruptly, at the junction of the branch and the trunk. As new growth layers are laid down over the trunk, the cambium moves outward, thus burying the base of the branch. The buried portion can no longer increase in diameter (FIG. 7-42). A plank sawed from a tree so that such a buried branch is exposed will contain a knot. The appearance of the knot is due to the difference in direction of the growth rings of the branch and the main trunk.

Under natural conditions in the forest the lowermost branches of a tree gradually die, chiefly from lack of light. When the branch dies, the cambium over it also dies, and the branch ceases to increase in diameter. But the cambium of the trunk continues to form new growth layers, and so the trunk increases in circumference. The dead branch thus gradually becomes buried in the form of a cylinder of nonliving tissue. As the trunk grows outward, the bark of the branch, if it has not already decayed, may be stripped away.

A dead branch or stub may persist for a number of years. The basal part of the branch, that nearest the pith of the tree, was alive when it became buried in the trunk (FIG. 7-42). Its growth rings are continuous with those of the tree. A knot in a board cut from that part of the tree is therefore a *tight* knot (FIG. 7-43, left). But a knot in a board cut from the outer part of the tree, where the branch has died and later been buried, is a *loose* knot. It is loose because the dead branch no longer forms growth rings continuous with those of the trunk (FIG. 7-43, right).

As long as dead branches remain on the tree, they form loose knots in the wood. These reduce the grade of lumber cut from the tree. If the dead branch breaks off or is pruned away, the

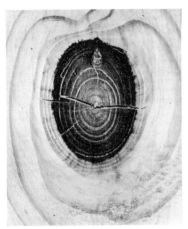 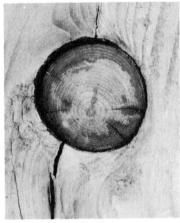

Fig. 7-43 Hard pine boards showing knots. (Left): tight knot. (Right): loose knot.

cambium of the trunk will heal over the stub, entirely burying it. It is therefore desirable to prune lower branches as they die, to obtain as much clear wood, free from knots, as possible. Moreover, decay organisms may enter the tree through dead branches and broken stubs.

Pruning of Shade or Fruit Trees The pruning of branches from trees is a common operation performed by the homeowner, and is frequently done incorrectly. A branch is often cut several inches away from the trunk, leaving a projecting stub. The branch should be removed close to the tree in such a way that the wound is flush with the surface. This reduces the possibility of decay, since wound tissue forms readily over such surfaces. A medium-sized or large limb should not be removed by a single saw cut because this may result in stripping the bark from the wood below the cut. The best method of preventing this stripping is first to remove the limb about a foot away from the trunk. This is done by making two cuts, the first by sawing upward from the lower side (FIG. 7-44A), and the second by a saw cut on the upper side a short distance outside the first (FIG. 7-44B). This is continued until the branch falls. The final cut is made to remove the stub by sawing flush with the trunk (FIG. 7-44C, D).

Fig. 7-44 Diagram showing proper method of removing a limb in pruning. (A): first cut. (B): second cut. (C–D): final cut. The much smaller cut made along the dashed colored line will not heal as quickly as a cut made closer to the trunk.

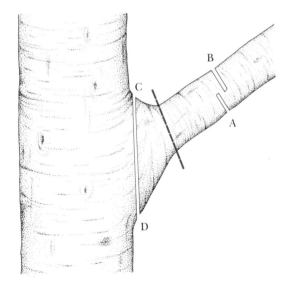

SUMMARY

1. Internally, the stem is composed of three tissue systems: dermal, vascular, and ground tissue, found also throughout the plant body. Initially, the conducting system of the stem is composed of vascular bundles, which also extend into the leaves as leaf traces. In dicotyledons the vascular bundles, as seen in cross section, are arranged in a ring; in monocotyledons they are scattered.

2. Growth in size is of two kinds, growth in length (height) and growth in diameter. Correspondingly there are two types of meristems: apical meristems that initiate growth in length, and lateral meristems, the cambium and cork cambium, which bring about growth in diameter.

3. With respect to origin, tissues derived from apical meristems are primary tissues; those derived from lateral meristems are secondary. Primary growth results in increase in height; secondary growth accounts for increase in diameter.

4. During growth in length, cell division occurs, not only in the apical meristem but also, and to an even greater extent, behind the shoot apex, in the young internodes. The derivatives of the temporary internodal meristems enlarge, chiefly by elongation, and bring about internodal elongation. Increase in length, then, which takes place only in the terminal portion of the stem, results from elongation of internodes just behind the shoot apex.

5. The mature primary tissues are the pith, primary xylem and phloem, cortex, and epidermis. Primary vascular tissues develop from procambial cells that originate from the apical meristem. These cells differentiate into primary xylem and phloem, and may also give rise to a cambium. The cambium forms secondary phloem to the outside and secondary xylem to the inside.

6. The cambium is composed of two kinds of initials. Fusiform initials give rise to the vertically elongated cells of the secondary xylem and phloem, ray initials to the vascular rays of the xylem and phloem.

7. The xylem of coniferous stems is composed chiefly of tracheids, which support the plant and conduct water. In hardwood stems, water conduction takes place through the vessels, and the chief supporting cell of the xylem is the fiber. Hardwoods may be distinguished from softwoods by the presence of vessels.

8. Vessel elements, like tracheids and fibers, bear pits on their lateral walls. Such pits are openings only in the secondary wall. Vessels are composed of vessel elements arranged end to end. These vessel elements are connected by true perforations in the end walls.

9. Tracheids, vessels, and fibers lose their nuclei and cytoplasm as they mature. The vascular rays remain alive (in the sapwood), but the bulk of the wood is nonliving. The vascular ray is a horizontal ribbon-shaped band of tissue extending radially through the wood and phloem.

10. A growth ring consists of the early and late wood of a single season. It is a cross section of a cone of woody tissue, the result of the activity of the cambium during a single growing season.

11. The tissues outside the cambium of a woody stem are collectively known as the bark. In the young shoot the bark is composed of phloem, cortex, and epidermis. In woody and in some herbaceous stems, the cortex is bordered on the outside by a layer of cork. In older woody stems, successive cork layers develop deep in the bark. The cork may be interrupted by lenticels, which serve in aeration.

12. Phloem includes sieve tubes, companion cells, phloem parenchyma, and sclerenchyma fibers. Certain of these may be ab-

sent in stems of different kinds. Sieve tubes are always present in angiosperms.

13. In woody plants the cambium functions for many years. Secondary growth is therefore prominent in such stems. In herbaceous stems the cambium functions for only a single growing season. Growth rings are absent, and secondary tissues are relatively small in amount. In monocotyledons, secondary tissues are usually absent.

14. Water is translocated through the tracheids and vessels of the wood, and food chiefly through the sieve tubes of the phloem. The xylem is the most important channel of mineral transport. The vascular rays store food and translocate food and water. Water conduction takes place primarily in the sapwood.

15. When logs are cut in a tangential plane, the result is plain-sawed lumber. The pattern on the surface of such boards is due chiefly to the growth rings. Wood may also be quarter-sawed, or cut in a radial plane. If the rays are large, they appear on the surfaces of the quarter-sawed boards as stripes or ribbons.

16. A knot is a section of the base of a branch buried in the wood of the trunk. A tight knot results if the portion of the branch through which the cut was made was alive when buried, with its growth rings continuous with those of the trunk. A loose knot results if the cut is made through a portion of the branch that was dead when buried, with no connection between the growth rings of the branch and trunk.

Plants and Water

O
F ALL THE ENVIRONMENTAL FACTORS that affect plant life, none is more important than water. Land plants, on an average, require about 400 pounds of water for each pound of dry plant substance produced, and this requirement frequently exceeds the supply. The physiological processes of the plant take place in solutions, or at least in the presence of water. Thus photosynthesis uses CO_2 from the cell solution, splits water as the source of hydrogen, and depends upon numerous enzymes functioning in a watery medium. Water may constitute more than 90 percent of the weight of young cells, where it serves as a support of the cell and tissue, and as the medium in which chemicals move and react. The minerals from the soil and the foods manufactured by the leaves move in solution to other parts of the plant. The percentage of water in physiologically active plant tissue ranges from about 40 in maturing seeds to 95 or more in rapidly growing cells.

It is estimated that two-thirds of the land surface of the earth lacks an adequate and dependable rainfall for satisfactory plant growth. Rainfall may be supplemented by irrigation when the water is available. Some 35 million acres, approximately the area of the State of Illinois, are irrigated in the western United States, and much of the agriculture of the world is partially or completely dependent upon irrigation (FIG. 8-1). It is sometimes assumed that increased irrigation, even with desalted sea water, will be used to increase the food supply of the earth's growing population. The economic returns of water used in agriculture are low, however, compared with the returns from industrial and domestic uses. These uses are even now competing with irrigation in our southwestern states. For the world as a whole, the availability of water

Fig. 8-1 Preplanting irrigation in California. Water is maintained at a uniform level by the canvas check dam at left and metered into the furrows with plastic siphon tubes.

as well as food will limit the number of people that can be supported at the level of the present, more developed countries.

TRANSPIRATION

In spite of the importance of water, the quantity actually used in plant processes is only a small fraction of the amount absorbed from the soil. Most of the water, about 99 percent, entering the plant escapes from the leaves and stem as water vapor. This process is termed TRANSPIRATION. Some water is lost from the stems of woody and herbaceous plants, but most water loss takes place from the leaves. Transpiration is of two kinds, STOMATAL and CUTICULAR. Most of the

water is lost through the stomata (FIGS. 5-2 and 5-3). Loss of water through the cuticle makes up only 5–10 percent of the total water loss from plants of temperate climates.

The loss of water by transpiration can be easily demonstrated with a small potted plant. Evaporation from the pot and soil is eliminated by covering tightly with metal foil or plastic film. The potted plant is then weighed at intervals and the loss of weight in unit time is recorded as the water lost in transpiration. If two plants are used, the effects of sunshine can be compared with shade, warm air with cool, and moist soil with dry.

Leaf Structure and Transpiration Most of the tissues of the leaf are directly or indirectly involved in transpiration. The structure of the leaf (Fig. 5-2) facilitates the transpiration of water. The leaf is usually broad and flat, with a large surface exposed to the air and sun. The presence of an epidermis covered with a cuticle on both sides of the leaf restricts the loss of water, but water evaporates from the surfaces of the wet mesophyll cells and escapes as vapor through the large numbers of stomata located on one or both surfaces of the leaf.

The number of stomata varies considerably from one species to another. The number per unit area also varies somewhat in the same species, depending upon the location of the leaf and upon environmental conditions during growth of the leaf. If the leaf grows rapidly after the stomata have been differentiated, they will be spread apart by the growth and show a smaller number per unit of leaf area. Table 8-1 shows typical numbers of stomata for several well-known plants. It should be noted that the leaves of the woody plants, oak, apple, and orange, have stomata only on the lower surface, while those of herbaceous plants have some, usually fewer, on the upper surface also. Such a distribution is commonly found.

TABLE 8-1
Number of Stomata per Square Inch of Leaf Surface

Plant	Upper epidermis	Lower epidermis
Apple (*Pyrus malus*)	None	250,000
Bean (*Phaseolus vulgaris*)	26,000	160,000
Corn (*Zea mays*)	39,000	64,000
Black oak (*Quercus velutina*)	None	375,000
Orange (*Citrus sinensis*)	None	290,000
Pumpkin (*Cucurbita pepo*)	18,000	175,000
Sunflower (*Helianthus annuus*)	55,000	100,000

Internally, the leaf is composed largely of thin-walled cells of the mesophyll, surrounded by a honeycomb of intercellular spaces that make up 15-40 percent of the volume of the leaf (Fig. 5-8). These cavities in the leaf form a branched system of air passages that connect with the larger spaces lying just behind the stomata. The exposure of portions of the surfaces of the mesophyll cells to the intercellular spaces greatly increases the surface from which water may evaporate into the air. The amount of internal surface thus exposed may be several times that of the epidermal surface.

During transpiration, the evaporation of water from the surfaces of the cells of the palisade and spongy mesophyll into the intercellular spaces normally maintains a high concentration of water vapor in these spaces. This vapor then diffuses through the stomata into the air whenever the stomata are open and when the concentration of water vapor in the air is less than that in the leaf. The water lost from the wet cell walls is replaced by water from the protoplast. The supply of water in the protoplast, in turn, is normally maintained by a movement of water from adjacent cells, and ultimately from the veins of the leaf (Fig. 5-8), which are parts of a conducting system extending to the water supply in the soil (Fig. 8-2).

The Mechanism of Stomatal Movement Since water escapes chiefly through the stomata, the size of the stomatal opening is related to the rate of transpiration, even though these openings usually constitute only a fraction of 1 percent of the leaf surface. It has long been known that the stomata of most plants open in the day and close at night. Under certain conditions they may close during the day also, particularly in the afternoon. The opening, closing, and changes in the size of the stoma are caused by changes in the shape of the guard cells, which are so constructed that an increase in their turgor causes the stoma to open and a decrease causes it to close. In a common type, the inner wall of each guard cell, next to the stoma, is thicker than the opposite, outer wall. The turgor of the guard cells increases when water moves into them from sur-

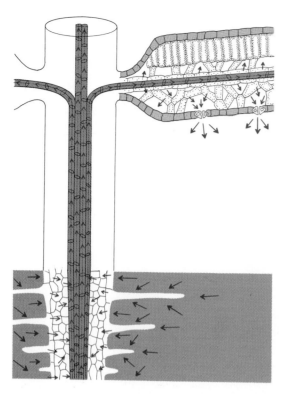

Fig. 8-2 Diagram showing the passage of water from the soil to the leaves and into the air. Water enters through the root hairs and other surfaces of the young root.

rounding cells. As the guard cells enlarge, the thin outer walls respond more readily than the thicker inner walls and expand into the surrounding epidermal cells. This change in the shape of the guard cells increases the size of the stomatal opening (Fig. 8-3A). A decrease in the turgor of the guard cells is caused by the loss of water, and the decreased volume allows the elastic inner walls to return to their first position, thus closing the stoma (Fig. 8-3B).

That stomatal movement is a turgor movement has been generally accepted since the first experiments on this mechanism, in 1856. Turgor changes can be influenced by several factors, including light, temperature, and humidity. Attention may be focused upon the first of these,

for in spite of numerous investigations the role of light in the opening and closing processes is still obscure. Numerous hypotheses have been proposed, of which the *starch-sugar* hypothesis is perhaps the best known. As pointed out earlier, during the night the surplus starch in the chloroplasts of the mesophyll cells of the leaf is digested to sugars that are transferred to other parts of the plant. In contrast, starch tends to accumulate at night in the chloroplasts of the guard cells. Also, starch-forming and starch-digesting enzymes of the guard cells seem to be particularly sensitive to changes in acidity in their environment.

According to the starch-sugar theory, then, stomata open because of the following sequence of events: (1) As light strikes the leaf, the chloroplasts of the guard cells carry on enough photosynthesis to reduce the concentration of the carbon dioxide that has accumulated from respiration during the night. The acidity of the guard cells is thus reduced. (2) This decrease is favorable to the enzymatic conversion of guard-cell starch to sugar. (3) The formation of sugars dilutes the water of the guard cells and causes water absorption and increased turgor, which result in stomatal opening. At night, the carbon dioxide produced in respiration accumulates, the guard cells become more acid, and the sugar is converted into starch. Water leaves the guard cells, and the stoma closes.

This is only one hypothesis explaining the action of light on stomata. Even if substantially correct, this mechanism might be operative only in some plant groups, other mechanisms prevailing in other groups.

The condition in which stomata are open in the day and closed at night does not always prevail. In a number of succulents, including members of the stonecrop family (Crassulaceae), the species of *Agave*, certain tropical orchids, cacti, and the cultivated pineapple, the stomata are open at night and closed, or nearly closed, during much of the day. Succulents in general have fewer stomata per unit of leaf surface than do nonsucculents, and in some (notably cacti and pineapple) the transpiration rate tends to be higher at night than during the day. The stomata respond to a combination of light and temperature, closing in light when the temperature is high, but not when it is relatively low. This response suggests that the water content of the guard cells is a controlling factor in opening, even though the plants show no other symptoms of water deficit.

Succulents exhibit an unusual form of organic acid metabolism, sometimes called "crassulacean" metabolism. This is characterized by the accumulation of organic acids, chiefly malic acid, in the chlorenchyma of the plant during the night. Sugars and CO_2 are required in the synthesis of these acids, the carbon dioxide being obtained from respiration and from the atmosphere. The organic acids break down in the light and release CO_2, which is then used in photosynthesis. This phenomenon of organic acid accumulation in the dark and breakdown in the light has been regarded as an adaptation that makes available a large amount of CO_2 for photosynthesis even though the stomata are closed during the day.

Rate of Transpiration Numerous estimates have been made of the quantity of water lost by transpiration. The rate at which water is lost from a plant varies greatly during the day and the season. It is governed by leaf structure and by environmental factors. The hourly rate of water loss in a number of flowering plants, during the middle hours of the day, averages 1.25 grams of water per 100 square centimeters (about ½ cup per square yard) of leaf surface. A single corn plant may transpire more than 2 quarts of water in a day, and an acre of corn may give off more than 400,000 gallons of water during the growing season. If a man used water at the same rate as a corn plant, he would drink 10–15 gallons of water a day.

The relative rate of transpiration is frequently expressed as the TRANSPIRATION RATIO. This is equal to the total weight of water transpired, divided by the weight of dry matter accumulated by the same plants during the same period. Ratios of 200 grams of water transpired per gram of dry matter accumulated are low, 300–600 being typical and 1000–2000 being high. It should be remembered that transpiration ratios depend upon both the water lost and the dry matter produced. This relation is illustrated by an experiment with corn on a poor soil. Corn grown in nonfertilized soil transpired 21,000 gallons of water per bushel of corn produced, or a transpiration ratio of 1590 pounds of water per pound of corn. An adjoining, well-fertilized plot produced much more corn and had a transpiration ratio of only 423. In general, plants that make good growth on fertilized soil produce a higher yield of dry matter per unit of available water. They also use more water per acre, so that fertilizers are of limited value in dry areas where the total water supply is limited.

The Mechanism of Transpiration Transpiration, except for certain differences resulting

Fig. 8-3 **The stoma in cross section showing the shape of the guard cells. (A): when the stoma is open. (B): when the stoma is closed.**

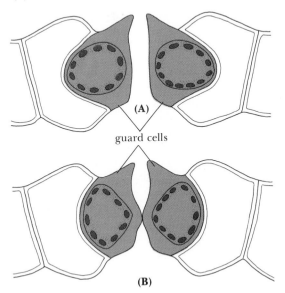

guard cells

(A)

(B)

from the structure of the leaf, is essentially identical with evaporation from a free water surface, and the physical processes involved are the same (Chapter 4). The wet walls of the mesophyll cells exposed to the intercellular spaces within the leaf provide a large evaporating surface (FIG. 5-8). The concentration of water vapor in the intercellular spaces is usually greater than in the outside air. When the stomata are open, the number of water molecules leaving the leaf through the stomata will be greater than the number entering during any given period, and water will be lost from the plant.

The most important environmental factors affecting the transpiration rate are temperature of the leaf, humidity of the air, light, wind, and the water content of the soil.

Temperature It was pointed out in Chapter 4 that the tendency of liquid water to evaporate is doubled by a temperature rise of 10° or 11°C (18–20°F). Because of this effect, the temperature of the leaf and of the water in it is the most important environmental factor affecting the transpiration of turgid leaves. The temperature of leaves in shade is approximately that of the air, but the temperature of leaves in the sunlight may be 5–10°C (9–18°F) or more above air temperature.

Humidity When leaves are well supplied with water and the stomata are open, the rate of transpiration depends upon the difference between the concentration of water-vapor molecules within the intercellular spaces of the leaf and that in the air near the leaf. Under natural conditions the air always contains water vapor, typically in a concentration of 1–3 percent. Some of these water-vapor molecules move in through the stomata of the leaf in a process that is the reverse of transpiration. The rate at which this inward movement occurs is proportional to the concentration of water vapor in the air—that is, to the humidity. The movement of water vapor from the air into the leaf will reduce the net rate at which water is lost. Thus, other things being equal, transpiration will decrease as the humidity of the air increases.

In order to understand the effects of temperature and humidity upon transpiration, we may consider a leaf at a temperature of 20°C (68°F). Assume now that the conditions are such that water molecules are escaping from the leaf at a rate of 100 molecules per second for a small surface area. At the same time, 50 molecules are entering the leaf because of the humidity of the air. The net loss, or transpiration rate, will therefore be 50 (100–50) molecules per second. If the temperatures of the leaf and air are increased to 31°C (88°F), the evaporation and escape of water from the leaf will be doubled to 200 molecules per second. The rate of return from the air will, however, be unchanged at 50 molecules per second. The small effect of the temperature rise on the speed of the vapor molecules in the air will be offset by the expansion, and therefore decreased concentration, in the unconfined air. The net loss will be 150 (200–50) molecules per second, and transpiration will have been increased three times. This response to temperature is due to its marked effect upon the evaporation of water and the buildup of water-vapor concentration in the leaf, with no direct effect upon the diffusion pressure of the vapor present in the air.

Light Light influences the rate of transpiration in two ways: (1) A leaf exposed to direct sunlight absorbs radiant energy. Only a small fraction of this energy is used in photosynthesis; the remainder is converted into heat energy, which raises the temperature of the leaf above that of the surrounding air. This heating normally increases transpiration. The fact that leaves in sunlight are warmer than the air makes rapid transpiration possible, even in vapor-saturated air. (2) Light—which need not be direct sunlight—also influences transpiration by its effect upon the opening of the stomata, by a mechanism previously described.

Wind Wind has two effects upon transpiration,

each of which tends to offset the other. First, wind moving over the leaf will sweep away any layer of water vapor accumulated near the surface as a result of transpiration. Wind will thus decrease the humidity of the air immediately over the stomata, and so decrease the return of vapor from the air into the leaf and increase the net loss of water. Under natural conditions, air is never still and therefore much of this effect is obtained without an appreciable wind velocity. When leaves are exposed to direct sunlight, however, a second effect becomes important. Such leaves are, as noted above, heated to temperatures that may be considerably above air temperature. Under these conditions, wind cools the heated leaves by bringing more air molecules into contact with them. The air-cooling[1] effect is thus increased, and the temperature of the leaves is reduced rapidly toward the average temperature of the air. Cooling leaves can greatly reduce the rate of transpiration. A study of the effect of wind on the transpiration of many plants in northern Italy showed that wind, on an average, neither increased nor decreased transpiration. In Hawaii, a moderate wind decreased the average transpiration of sugar cane rather than increased it. The results of an experiment with potted corn plants are shown in Table 8-2.

By way of summary, then, it may be stated that wind tends to increase the rate of transpiration, whether in the shade or in the sun, by sweeping away water vapor. In the sun, however, the effect of wind in cooling the heated leaf, and thus decreasing the rate of water loss, tends to become more important than its effect on the removal of water vapor.

Water content of the soil   The rate of transpiration will be influenced by the water content of the soil and by the rate at which roots can absorb water. During daylight hours, water is commonly transpired at a greater rate than it is absorbed from the soil. This causes a water deficit in the leaves. At night, this condition is reversed, since the temperature of the air and leaf is lower, and the stomata are normally closed. The effect of a water deficit in the leaf cells is to reduce the rate of evaporation, because there are fewer molecules of water present in the cells and the rate of replenishing the water evaporated from the surface of the cells is slower. As the water content of the soil decreases, because of absorption by the roots, the movement of water through the soil and into the plant becomes slower. This tends to increase the water deficit of the leaf and further reduce the transpiration rate. If the water content of the leaf continues to decrease, the turgor of the guard cells is reduced, resulting in almost complete closing of the stomata and greatly reduced stomatal transpiration.

TABLE 8-2
Some Effects of Sun and Wind on Transpiration of Corn

| | Grams Water Lost per 2-foot Plant per hour | | |
	No wind	4–6-Mile wind.	Change with wind
Plants in shade	13.0	17.5	+4.5
Plants in sun	29.1	22.0	−7.1

It would thus appear that water loss is, to a considerable degree, controlled by the plant. The stomata close, however, not in anticipation of water loss but only after considerable loss has already occurred; typically, complete closing occurs only after the leaves are visibly wilted. The control of water loss is not a purposeful act on the part of the plant but results from the structure of the leaf and the operation of environmental factors. Even partial closing of the

---

[1] Air cooling is a collective term for conduction and convection. In conduction, heat is transferred from the leaf to the air particles in contact with it. The heated air is carried away by wind, or expands and rises, carrying this heat with it. This is convection. The heat waves that can be seen over a meadow or along a highway on a still summer day are evidence of this air movement, or convection. The rising warm air is replaced by cooler air.

Fig. 8-4   Temporary wilting in squash (*Cucurbita maxima*) leaves. (Left): in the late afternoon of a hot day. (Right): early the next morning. The plants have absorbed water and recovered during the night.

stomata does not greatly affect the rate of water loss, for the rate does not drop markedly until the stomata are almost fully closed (see Fig. 5-17). In spite of these limitations, this passive regulation of water loss may be very effective during dry weather.

Injurious Effects of Transpiration   If the soil is well supplied with water, a high transpiration rate will cause no great harm to the plant, at least over short periods of time. But if water loss continues to exceed absorption, the injurious effects of transpiration soon become apparent in wilting of the leaves, a result of the loss of turgor. Thin leaves composed chiefly of thin-walled parenchyma cells become visibly wilted in a short time. Leaves of the jewelweed (*Impatiens*) may wilt minutes after picking. American holly (*Ilex*), on the other hand, may show no signs of wilting over long periods because the leathery leaves contain a great deal of supporting tissue.

Wilting of plants in soil is classified as *temporary* or *permanent*. Temporary wilting occurs when water is still available in the soil, owing to a temporary excess of transpiration over absorption.

Plants usually recover from temporary wilting as soon as the rate of transpiration decreases. Leaves that have wilted during the day will recover at night and appear fully expanded in the morning (Fig. 8-4). Leaves may also regain their turgor during the day if transpiration is reduced by clouds, a temperature drop, or a light rain, even though the water does not penetrate to the roots.

Permanent wilting, on the other hand, results from a serious deficiency of water in the soil. The roots are unable to absorb water and the plant eventually dies unless the water supply of the soil is renewed (see Chapter 9).

Repeated temporary wilting will have an adverse effect on the metabolism of the plant, and plants wilted repeatedly become stunted. An important reason for this is that water deficiency reduces the growth of young tissues, affecting particularly the processes of cell division and cell enlargement. This retardation of growth results in a decrease in the amount of food used by growing tissues, and carbohydrate accumulation almost invariably accompanies drought. Continued high carbohydrate levels may result in the permanent structural and physiological changes that are associated with stunting.

It is through such processes as those discussed above that drought affects crops. There are many areas in the world, including some in our own country, where rainfall is barely sufficient or is insufficient for the growth of crops, and where drought is a constant menace. The stunting of growth that results from a deficiency of water affects all parts of the plant. Growth in height and diameter is retarded, and older leaves wilt and die. The yield of fruit or seed is markedly reduced, and plants may die before they bear at all. It is no accident that drought and famine have gone hand in hand through the centuries.

Significance of Transpiration Although transpiration has been studied in detail, there are still a number of misconceptions about its significance to the plant. Evaporation uses heat energy, and it is sometimes believed that the cooling effect reduces the leaf temperature enough to prevent injury or death in hot weather. Even a high rate of transpiration, however, will normally account for a decrease in temperature of only 3–5°C (5–9°F), which will seldom have a significant effect upon plant survival. But if the absorbed radiant energy were not dissipated, the temperature of the leaf would rise so rapidly that it would shortly be killed. It has been calculated, in fact, that without some kind of cooling mechanism, the temperature of a leaf in direct sunlight could rise at the rate of 30°C (54°F) or more per minute. Starting at room temperature [20°C (68°F)], therefore, the temperature of the leaf could reach the boiling point of water in less than 3 minutes. Such overheating of the leaf is prevented, however, not by transpiration, but primarily by radiation[2] and air cooling. Most leaves are thin—less than 1/100 of an inch thick—

and their absorbed energy is quickly conducted to the air. The heated air is then carried away by air currents. Wilting and death of leaves during periods of drought therefore normally result from loss of water by the plant rather than from effects of temperature.

The fact that the cooling effect of transpiration is not important can be emphasized by considering the temperature of a dead leaf in the sun. Such a leaf does not transpire; yet it absorbs energy, and because of its lower water content and weight, its temperature rises two or three times as fast as that of a living leaf. Such leaves, in the sun, would burst into flame within minutes if it were not for radiation and air cooling.

It is sometimes claimed that transpiration is necessary for movement of water in the plant. The consumption of water in photosynthesis and growth tends to create a water deficit in the leaf cells, which, in turn, causes water to move upward through the stem. Transpiration may create a greater deficit and increase water movement, if the moisture is available in the soil, but this extra movement is of no benefit to the plant.

Transpiration has been claimed also to be important in the absorption of minerals—nitrates, phosphates, and other inorganic materials required for growth. Hence, the reasoning goes, the greater the rate of transpiration, the greater the absorption of soil solutes. But the movement of minerals in the soil, and particularly their absorption across the membranes of the root cells, is independent of water absorption. Solutes may move into the root more rapidly or less rapidly than water, depending upon various factors in the root and in the soil. There is no mass flow that sweeps soil solutes into the root with the absorbed water.

Under experimental conditions and for short periods it is sometimes found that a plant absorbs more minerals if the transpiration rate is high rather than low. But the quantity of solutes is not proportional to the amount of water transpired, and no consistent relation can be found between the rate of transpiration and that of mineral absorption. In general, absorption of minerals is proportional to their use in growth and not to

---

[2] Radiation is most familiar as sunlight. Light energy moves from the sun to the earth at a speed of 186,000 miles per second. Similar radiation, in long, invisible wavelengths (infrared radiation), is given off by leaves and other objects. It is commonly less important in the cooling of plants than is air cooling.

water absorption. This dependence is illustrated by an experiment in which one group of plants transpired 19.4 liters of water during growth and had an ash (mineral) content of 14.0 percent. Other plants in a moist atmosphere transpired only 4.7 liters, but had an ash content of 14.7 percent.

Since minerals may move to the leaves of plants in the transpiration stream—that is, in the water being transported through the xylem—it is sometimes assumed that the stream is necessary for the movement of minerals. Plants typically absorb 2 or 3 tons of water to each pound of soil minerals. It is probable that a fraction of this water—perhaps 2 percent—would be more than adequate for mineral movement, even if this occurred only in the transpiration stream. There is good evidence, however, as pointed out in Chapter 7, that minerals may be transported in the phloem also, and such transport would be independent of any transpiration. It is doubtful, therefore, that the plant would suffer from a deficiency of soil minerals if no transpiration occurred.

Transpiration is thus of questionable value in any way. Excessive transpiration may retard growth and eventually cause death of the plant. On the whole, transpiration may be regarded as a compromise. Open stomata permit the rapid absorption of carbon dioxide during photosynthesis. The system of a cuticle with small stomata covering leaves and young stems retards water loss while permitting near-maximum carbon dioxide absorption. As long, however, as the stomata are open and the concentration of water vapor in the leaf is greater than in the air, water loss is inevitable, and this loss constitutes an ever-present source of potential danger to the plant.

GUTTATION

Most water is lost from the plant as water vapor, but it may also leave in liquid form. This process is called GUTTATION. The water appears on the tips and margins of leaves as clear drops, most frequently seen in the early morning. As the temperature rises with the sun, the guttation water evaporates or is reabsorbed. Guttation commonly occurs when water absorption by the roots is relatively rapid and conditions are unfavorable for rapid transpiration. It is the result of active absorption of water by the roots and the development of root pressure (see below). Many plants show guttation, among them nasturtium, potato, cabbage, tomato (FIG. 8-5), strawberry, and the grasses. Guttation may be heavy in some tropical plants. A single leaf of elephant's ear (*Colocasia*) is reported to have lost half a cup of water by guttation in a single night. The water commonly exudes through specialized stomata, called water stomata, located close to the ends of the main veins of the leaf. Water stomata differ from other stomata in that they appear to remain open, day and night, during the life of the leaf.

In late spring and early summer, guttation water is especially noticeable on lawn grasses, appearing in the early morning as a single large drop at the tip of each blade. Guttation is readily demonstrated in the laboratory by covering a pot of well-watered seedlings of corn or barley (FIG. 8-6) with a bell jar.

Guttation water must be distinguished from dew. After a warm day, when the sky is clear, the exposed surfaces of most objects, such as leaves, soil, and rocks, cool by radiation to a temperature below that of the air. The air in contact with these cooler surfaces is also cooled and its moisture condenses as dew, just as moisture from the air condenses on a pitcher of ice water. Such condensed moisture is distinguished from guttation water by its position. Dew forms as a film or as droplets that cover the entire surface of the leaf. Guttation water is confined to the tips and margins of the leaves.

THE CONDUCTION OF WATER

No subject connected with plants is more commonly misunderstood than the upward movement of water, or sap, in plants. It is commonly said that "sap rises in the spring." This belief may

Fig. 8-5 Guttation in tomato (*Lycopersicum esculentum*) leaves.

have originated from the observation that if some trees are cut or wounded in the spring, water drips from the stems or branches. The erroneous idea of "rising sap" should be discarded, for the tree contains water in the winter as well as at other times of the year.

Speculation on the causes of water rise in plants goes back hundreds of years. One of the earliest explanations, first suggested by Nehemiah Grew in 1682, is that water moves by the "pumping" action of living cells surrounding the conducting tissues. Water, however, will rise through long sections of stems killed by heat or poisons, indicating that other mechanisms are operative. Two of them merit brief consideration:

1. Probably the most common explanation advanced by the layman credits capillary action with the necessary force. If the base of a glass capillary tube is placed in water, the liquid will rise in the tube by capillarity. The surface of the water partly filling a capillary tube is concave, and is termed the meniscus (FIG. 8-7). The sides of the tube above the meniscus become coated with a thin film of water, and since the water on the margin of the meniscus is continuous with the water film on the tube, the total surface of the

Fig. 8-6 Guttation in barley (*Hordeum vulgare*) leaves.

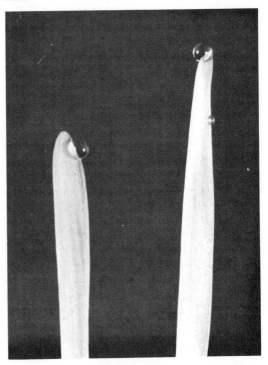

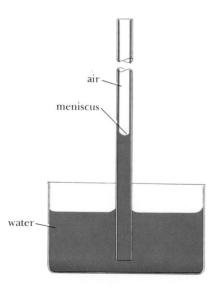

Fig. 8-7 Rise of water in a capillary tube.

air

meniscus

water

from osmotic forces developed in the roots. The combined effect of osmotic and perhaps other forces is the ROOT PRESSURE that causes bleeding and guttation.

Root pressure has been suggested as an important factor in the ascent of water, but a number of circumstances make this theory untenable. Not all plants bleed when the tops are removed. Further, bleeding is negligible or nonexistent in summer, when the transpiration rate is highest. If a manometer—an instrument for measuring pressure—is attached to the stumps, the readings are low, usually only a few pounds per square inch.[3] These pressures are not great enough to lift water to the tops of tall trees. Root pressure may occasionally be useful in forcing air out of vessels and reestablishing broken water columns, but it is certainly of no value in lifting quantities of water in transpiring plants.

meniscus is greater than it appears to the eye. This surface is affected by forces that cause it to contract to the smallest possible area, and thus it tends to become straightened. The contraction of the meniscus results in a pull at the top of the water column, and the column moves upward. The height to which it will rise is inversely proportional to the diameter of the tube; it may rise several feet in a very fine capillary.

According to this explanation, water-conducting cells of the tree would act as capillary tubes. But although small, the diameter of these cells is not small enough for water to rise more than a short distance in them. Further, the forces causing rise by capillarity are developed only in an open tube with air above the water column. The water-conducting cells in a stem, on the other hand, are normally filled with water and end in contact with living cells of the leaf, not the open air. Capillary rise is thus impossible.

2. If the tops of some plants are removed, water will exude, or "bleed," from the stumps (FIG. 8-8). The most conspicuous bleeding is seen when grapevines or various trees such as birch, box elder, sycamore, and butternut are cut in the spring. Bleeding results, in large part at least,

The Cohesion Theory The most adequate explanation of the lifting of water in plants is known as the COHESION THEORY. This theory is based upon the fact that water, although it changes shape readily, is very resistant to changes of volume under mechanical forces of either compression or tension. Because of the resistance to change in volume, water in a sealed, clean, glass tube, to which it adheres throughout, can be subjected to a strong pull or tension. Water in a closed tube, such as a xylem vessel, behaves much like a wire and literally can be pulled. This pull depends upon two factors: (1) the water molecules cohere because of forces of intermolecular attraction; and (2) they adhere to the walls of the vessel, and thus prevent bubbles of water vapor from forming and breaking the water column.

Another essential part of the cohesion theory is that water is pulled up by forces acting in the leaves. The leaf mesophyll cells contain sugar and other solutes in their cell sap, which dilute the water and reduce its concentration. Water

---

[3] A force of 14.7 pounds per square inch (1 atmosphere at sea level) will support a column of water about 34 feet high.

evaporates from the surfaces of these cells during transpiration so that turgor pressure does not build up and offset this dilution effect. The cells therefore develop a high water-absorbing force, and water moves from the xylem cells of the leaf veins into the mesophyll. The vessels of the xylem are closed at their ends in the roots and leaves by living cells, and they are strengthened against collapse by various types of woody reinforcement (FIG. 7-9). Also water adheres strongly to the wet walls of the tubes. In consequence, the loss of water from this closed, rigid system tends to develop a tension or pull on the water within the vessels. If the cohesive forces are adequate, this tension can be transmitted through the twigs, branches, and trunk down to the ends of the roots in the soil, and the water in the vessels can thus be lifted by the water-absorbing force of the leaf cells.

In view of the statement that the cohesion theory is the most successful attempt at explaining the rise of water in tall plants, it might be thought that this problem is entirely solved. Unfortunately, this is not true. No difficulties arise with respect to the water-absorbing power of the leaf cells—all measurements agree that this is sufficient to lift water to the tops of the tallest trees. The potential and actual values of cohesion in the stems of plants are, however, controversial. Some botanists consider that cohesion values of 300 or more atmospheres have been demonstrated experimentally. Such values are based upon improper techniques and calculations.

The method most frequently used for measuring cohesion employs water sealed into a glass tube under a vacuum. The tube is not quite filled, thus leaving a vacuum bubble. The tube must be chemically clean, and the water boiled to drive off dissolved air and other gases. When such tubes are heated slowly and carefully in a water bath, the water expands and fills the tube. If the water bath is then cooled slowly with constant stirring, nothing happens for some time. Then the vacuum bubble re-forms with a distinct click, indicating that the water in the cooling tube had been held in an expanded condition by cohesion. The difference between the tempera-

tures at which the bubble disappears and then re-forms is used to calculate the tension developed by cohesion. Such experiments yield repeatable values of 25–32 atmospheres of tension.

Other experiments by other methods have indicated that cohesion values of only about 30 atmospheres instead of 300 are due to cosmic rays, powerful and very short wavelength radiation that reaches the earth from space. These rays are so powerful that they penetrate inches of lead and many feet of wood or solid rock. When they strike water, including water in xylem vessels, they form submicroscopic cavities in the liquid. The surface tension of the water surfaces exposed in a cavity will normally draw the water together quickly and close the break. From the sealed-tube experiments we conclude that the cavities are not closed and the water column is broken if the tension is about 32 atmospheres.

The sealed-tube experiment has been repli-

Fig. 8-8 Demonstration of root pressure. A long glass tube is attached to the stump of an actively growing plant. The water has risen from the stump a considerable distance up the tube.

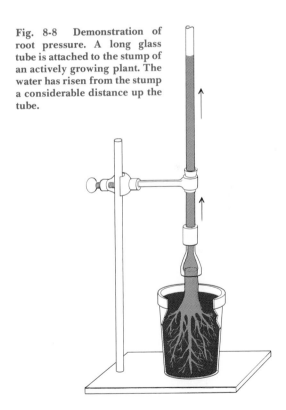

cated in at least five laboratories with comparable results, and two or three other methods also show 25 to 32 atmospheres of tension. We conclude, therefore, that 30 atmospheres is the approximate limit of cohesion in tall trees where the water columns are fully and continuously exposed to cosmic rays. It is calculated, however, that if cosmic rays could be eliminated, as in a deep mine, values in the hundreds of atmospheres could be obtained. But even 30 atmospheres is enough to lift water 1000 feet, and since our tallest trees are less than 400 feet high, it is a more than adequate force.

Under laboratory conditions, water columns under tension are easily broken by a slight jar. In contrast, the water columns within a tree are remarkably elastic and capable of adjustment to very severe conditions, such as rapid movement of branches in a high wind. Still other objections have been raised against the cohesion theory. Although it seems certain that cohesion is a factor in lifting water in plants, and is probably the major factor, much remains to be explained about the movement of water in stems.

Movement of Water into the Plant Water moves into the plant through the root hairs, which are hairlike outgrowths from the epidermal cells, and through the epidermis of the young roots (FIG. 8-2). It is believed that water is absorbed by two different mechanisms, which may be discussed under the heads of (1) ACTIVE ABSORPTION and (2) PASSIVE ABSORPTION. Active absorption of water should be distinguished from the active transport of dissolved substances across cellular membranes.

Active absorption occurs when soil moisture is high and the plant is transpiring slowly. Under these conditions, water absorption is considered to be primarily osmotic in nature. The inward movement of water depends upon a higher solute concentration in the xylem vessels than in the soil solution. It seems probable that sugars and other solutes tend to leak or to be secreted from the living cells of the root into the adjacent xylem cells. The exudation into the xylem may be

due to high turgor pressure in the living cells, causing their membranes to be abnormally permeable. When the solute concentration, and therefore the dilution of water, within the xylem becomes great enough, water will diffuse from the soil into the xylem, moving across the turgid cells of the cortex, which act as a complex, and differentially permeable membrane. The diffusion of water into the conducting cells of the root is responsible for root pressure and guttation, as discussed above. This movement is known as active absorption because it depends upon the solute content and permeability of living root cells. If common salt or an excessive quantity of soluble fertilizer is added to the soil, this movement can be reversed, and the plants may die for lack of water (FIG. 4-10).

When plants are transpiring rapidly, water intake is by passive absorption. Active absorption does not function under such conditions because the rapid movement of water through the roots washes away the xylem solutes upon which such absorption depends. Probably, also, rapid transpiration removes water and reduces turgor pressure in the living cells of the roots so that solute leakage is reduced. Under conditions of rapid transpiration, water tends to be lost more rapidly than it is taken in, and the water columns in the conducting cells pass into a state of tension. This tension is transmitted downward to the roots.

The simplest explanation of water movement into the plant under conditions of rapid transpiration involves differences in pressure inside and outside the plant. If a pressure of 5 atmospheres is applied to the water surrounding the roots, and at the same time the water inside the plant is under a pressure of 1 atmosphere, water moves into the plant because of the effect of this outside pressure in increasing the diffusion pressure of water. The pressure actually existing on the soil atmosphere is only 1 atmosphere, but that on the water in the xylem may be less than 1 atmosphere and may amount to several atmospheres of negative pressure because of the action of cohesion. The diffusion-pressure difference between the water within the plant and in the soil outside is thus the same as though a positive mechanical pressure had been applied to the soil moisture. Movement under these conditions is called passive absorption because the forces bringing it about arise at the top of the plant rather than in the roots, and because it will func- tion—for a time, at least—in dead roots. Passive absorption is dependent upon transpiration pull. It is far more important in volume than active absorption and may account for 98 percent of the total movement of water into roots.

SUMMARY

1. Virtually all plant processes depend upon water, and the water supply is commonly the most limiting factor in plant development.

2. Approximately 99 percent of the water absorbed by plants is lost to the air in transpiration. Most of this loss occurs through the stomata.

3. Stomata open and close with changes in the turgor pressure of guard cells. In general, they open in light and close in darkness, but they will also close if the leaves become wilted.

4. The rate of transpiration depends upon the difference between the rate at which water-vapor molecules leave the leaf and the rate at which they enter the leaf from the air. This difference is affected by the temperature of the leaf, by the humidity, and by light, wind, and the water supply of the plant.

5. Leaf temperature is the most important factor in the transpiration of turgid leaves. This importance is due to the rapid increase in the tendency of water to evaporate as the temperature rises.

6. Wind tends to increase the rate of transpiration, whether in the shade or in the sun, by sweeping away water vapor from near the surface of the leaf. In direct sunlight, however, where leaf temperatures are above air temperatures, increased air cooling with wind is more important than the removal of water vapor, and tends to decrease the rate of transpiration.

7. Light affects the rate of transpiration by bringing about the opening of the stomata and because absorbed sunlight increases the temperature of the leaf. Plants normally wilt or die, not because of high temperatures, but because of loss of water.

8. Wilting is temporary or permanent. A deficiency of water in the plant checks growth and reduces yield.

9. Transpiration is of little or no value to the plant. It does not cool the leaf to any marked extent, and it is probably not necessary for either the absorption or transport of minerals from the soil.

10. Guttation is the loss of liquid water from the leaf. It is caused by root pressure, the result of osmotic forces operating in the roots. Root pressure also becomes evident when water exudes from the stems of plants that have their tops cut away.

11. The cohesion theory is the best available explanation of how water rises in plants. The water-absorbing power of the leaf cells causes water to move into them from the xylem cells of the leaf. This creates a tension that is transmitted to the roots, and water rises by transpiration "pull."

12. The cohesion of water exposed to cosmic rays has been measured at about 25–30 atmospheres. This is sufficient to prevent breakage of water columns subjected to transpiration "pull."

13. Two mechanisms of water absorption function in roots. In active absorption, water diffuses into the root xylem because the xylem sap is diluted by solutes. Passive absorption is due to a difference in pressure. The transpiration "pull," exerted through cohesion, lowers the diffusion pressure of water in the xylem below that in the soil.

The Root and the Soil

THE PLANTS UPON WHICH WE DEPEND for the food we eat and for the oxygen we breathe depend in turn upon the soil. A good soil supplies the plants with the mineral elements they use. It conserves rainfall by rapid infiltration and by holding quantities of available water within the rooting zone. And it promotes the root growth without which both minerals and water are unavailable to the plant.

Soil is not a necessary requirement for plant growth. Many of the plants of the earth grow free or anchored to rocks (Fig. 19-26) in the ocean. Vigorous, highly productive plants can be grown in solutions of fertilizer minerals in the absence of soil (Fig. 9-22). Practical crop production, however, depends upon soil, even though water is applied in irrigation (Fig. 8-1) and heavy fertilization is practiced (Fig. 9-21).

In spite of its basic importance, the soil has been our most neglected and abused resource. Salt accumulation and wind and water erosion (Fig. 10-8) have ruined hundreds of millions of acres of land around the world, and the loss continues in areas that can least afford it. Soil conser-vation practices are now used extensively in the United States (Fig. 10-9). Even badly damaged land, such as that shown in Fig. 10-8, can be reclaimed if the basic soil material is good and erosion has not removed too much of it. Leveling, heavy fertilization, and growing grass-legume mixtures can make such areas productive again.

It is through the roots that the nutrient elements of the soil are made available to plants. Root-soil relations are intricate and poorly understood, and we need to know more about roots and their relations with the soil.

THE ROOT

Extent of Roots The nature of the root system of a plant and the depth to which it extends vertically and laterally depend upon such factors as the soil moisture, the soil air and temperature, and the physical nature of the soil. In general, the root systems of woody plants, such as trees, spread out laterally rather than penetrate deeply into the soil. The greatest number of roots is frequently found in the upper 4 feet of soil. The depth of penetration may vary from 2 to 30 feet. The lateral extent is commonly greater than the spread of the branches. An aspen tree 25 feet high was found to have a main lateral root 47 feet long. An oak tree 37 feet high had a tap root 14 feet long but lateral roots extended out as much as 60 feet from the base of the tree. The roots of a 35-foot elm may have a radial spread of over 60 feet. In their branching and rebranching, the lateral roots of both woody and herbaceous plants develop an amazing network. This cannot be appreciated by pulling a plant from the soil, for the finer roots are left behind. In research studies the soil is washed carefully from the roots, thus keeping them intact and revealing how extensive root systems may be.

In an intensive study of the fibrous roots of winter rye, an important grain and forage plant, H. J. Dittmer found that one plant, 20 inches high and consisting of a clump of 80 shoots, had the extraordinary root system shown in Table 9-1.

TABLE 9-1

Root System of a Plant of Rye (Secale Cereale)

Kind of root	Number	Total length, feet
Main roots	143	214
Branches of main roots (secondaries)	35,600	17,000
Branches of secondaries (tertiaries)	2,300,000	574,000
Branches of tertiaries (quarternaries)	11,500,000	1,450,000
Total	14,000,000	2,000,000 feet, or 380 miles

Enormous as these figures are, they are only part of the story. The total surface area of these roots amounted to more than 2500 square feet,

compared to an area of 51.4 square feet for 80 shoots with their 480 leaves. This spread of roots represented the growth, during a few weeks, of the adventitious root system that had developed from the bases of the stems.

Function of the Roots Roots play several roles in the life of the plant: (1) they anchor the plant in the soil; (2) they absorb water and minerals from the soil; (3) they transport these materials from the region of absorption to the base of the stem; and (4) they may serve as food-accumulating organs. In some plants the foods so accumulated are used only by the roots; in others a part of the food is translocated and used by above-ground parts, such as the developing shoots of herbaceous perennials. The structure of the root and its manner of growth are in a number of ways related to its penetration into the soil and to the absorption of minerals and water.

Fig. 9-1 Regions of growth in the terminal portion of the root.

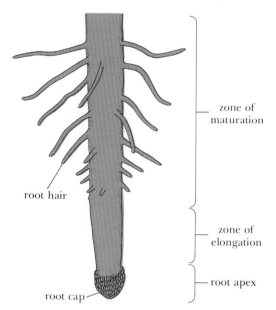

zone of maturation

root hair

zone of elongation

root apex

root cap

Terminal Growth of Roots Several general regions may be recognized in the terminal portion of a young root (FIG. 9-1). The first of these is the ROOT APEX, covered by a ROOT CAP. Behind this is a ZONE OF ELONGATION, varying in length from about 1000 microns (1 millimeter) in timothy (*Phleum pratense*) to as much as 10 millimeters in some kinds of corn. This region is thus much shorter than the comparable region in the stem. Following this is the ZONE OF MATURATION. Root hairs are produced here. The root cap and root hairs will be discussed in subsequent paragraphs.

The region of most rapid elongation in the terminal part of the root may be demonstrated by a method used by the German botanist Julius Sachs a century ago. He placed India ink marks 1 millimeter apart, beginning at the tip of the root.

Fig. 9-2 Region of elongation in a pea (*Pisum sativum*) root. (A): ink marks, 1 millimeter apart, were placed on a growing root. (B): the distribution of the marks several hours later shows that most of the growth occurred in what were orginally the second and third millimeters.

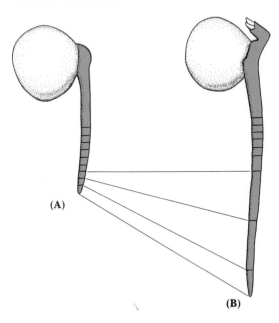

(A)

(B)

A few hours later the growth of the root had increased the distance between certain of the marks, as shown in FIG. 9-2. In FIG. 9-2B it should be noted that the second and third millimeters back from the tip made the most growth in the period after marking, with the first and fourth showing some elongation. If later observations were made on this root, it would be evident that the original third and most of the second millimeter had ceased to elongate. The original terminal millimeter, which had continued to expand slowly, would now include the region of most rapid elongation.

It is apparent that the processes of cell division and elongation are concentrated near the root apex, and do not, as in the stem, occur over a considerable distance behind it. The adaptive significance of this localization of growth in an organ growing through a compacted soil is obvious.

As in the stem, the root apex (apical meristem) is primarily a region of tissue initiation that establishes the pattern of the tissues behind it. The apical meristem includes a central group of relatively inactive cells bounded by more actively dividing cells that give rise to the primary meristems, the protoderm, ground meristem, and procambium (FIG. 9-3). While some cell division proceeds in the apical meristem, a much greater rate prevails behind it, in the subapical zone. The distance from the apical meristem at which new cells are formed at the greatest rate varies with different kinds of plants, the age of the root, and other factors. In one variety of hybrid corn the maximum rate of cell division was found at 1.25 millimeters behind the seedling root tip (including the root cap), and division ceased at 2.5 millimeters. In the onion root (FIG. 9-3) the cells that will differentiate into cortex show a maximum number of divisions at about 500 microns (0.5 millimeter) behind the apical meristem. Cells in the center of the root that will eventually become vascular tissue, on the other hand, show a maximum number of divisions at about 800 microns behind the apex.

In many respects, terminal growth in the onion

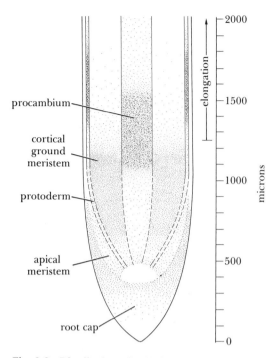

procambium

cortical
ground
meristem

protoderm

apical
meristem

root cap

elongation

2000

1500

1000

500

0

microns

Fig. 9-3 Distribution of cell divisions in the first 2 millimeters of the onion (*Allium cepa*) root tip at noon. The unshaded area (quiescent center) in the apical meristem showed no divisions, but they may occur at other times.

root is fairly typical of root growth in general. The meristematic zone not only includes the apical meristem but also extends more than 1000 microns behind this, into the region of elongation. But although maximum regions of cell division may be recognized in the meristematic zone, this is not a region of cell division alone. Division is accompanied by an increase in cross-sectional areas and by elongation. Divisions are still very common in the early stages of elongation, at the base of the meristem. There is thus a gradual transition from the meristematic zone to the region of elongation, over much of which only cell elongation occurs. Following the region of elongation is the zone of maturation. But some types of cells—xylem, phloem—begin or complete differentiation in the region of elongation,

or even close to the apical meristem. In the onion, for example, the first-formed phloem cells are physiologically mature at 1100 microns behind the apical meristem, where cell division is still proceeding.

The essential feature of the manner of growth described above is, of course, that extension of the root, like that of the stem, results from cell elongation.

The Root Cap The root apex is covered by a thimble-shaped or hoodlike mass of cells, found only in roots, and termed the root cap (FIG. 9-1). As the root cap is pushed forward between the soil particles by growth in the region of elongation, the root apex is protected from mechanical injury by the root cap. The cells of this tissue, loosely bound together in some species, more tightly in others, are continually worn away. The sloughing off of the root cap cells results in the formation of a slimy covering over the adjacent soil particles, thus facilitating the passage of the root through the soil.

The cells of the root cap are renewed from within by various methods, depending upon the species. In corn, for example, the cap is produced by a meristematic layer that is independent of the apical meristem of the root. In many dicotyledonous roots, both the root epidermis and the root cap originate from the outermost layer of cells of the apical meristem.

Primary Tissues In the zone of maturation most cells approach, or have attained, a condition of physiological maturity. Cell walls are thickened, and some cells lose their protoplasts and become nonliving. In this zone, and behind it, may be recognized highly specialized types of cells, distinct in structure and function.

Three regions of the young root can now be clearly distinguished, either in cross or longitudinal section: the vascular cylinder, the cortex, and the epidermis (FIGS. 9-4 and 9-5). The vascular cylinder at the center of the root is com-

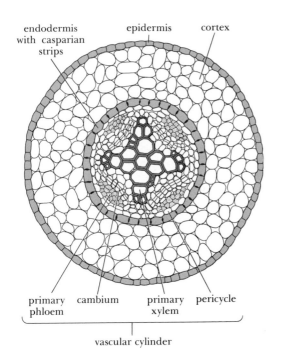

endodermis with casparian strips   epidermis   cortex

primary phloem   cambium   primary xylem   pericycle

vascular cylinder

Fig. 9-4   Primary tissues in a cross section of a young dicotyledonous root. Note the alternate arrangement of the primary xylem and phloem.

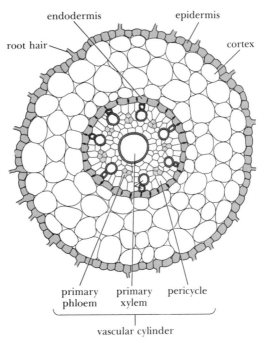

endodermis   epidermis

root hair   cortex

primary phloem   primary xylem   pericycle

vascular cylinder

Fig. 9-5   Cross section of a young root of barley (*Hordeum vulgare*), a monocotyledon.

posed of vascular tissues together with associated parenchyma cells. It functions primarily in the translocation of food, water, and minerals. In it are several types of cells. The most prominent are thick-walled xylem cells, which conduct water and minerals, and phloem, through which food is transported.

The xylem cells are so arranged that, in cross section, they form arms radiating from the center. The phloem cells lie in groups between the arms. Since the groups of primary xylem and primary phloem lie along different radii, these tissues are said to have *alternate* arrangement. This is in contrast to the disposition of these tissues in the stem, in which they lie on the same radius, with the phloem outside the xylem. The number of arms of xylem is fairly constant for a species, and is typically 2–4 in dicotyledons (FIG.

9-4). In monocotyledons (FIG. 9-5) the number is larger; in corn, 20–40. A pith may be present or absent in roots. In the conifers and most dicots the cells in the very center of the root mature into xylem, and pith is absent. In many monocots and some herbaceous dicots, on the other hand, xylem fails to differentiate in the center of the root, and some pith is present.

In many dicotyledonous roots a cambium arises late in primary growth, in the region between the primary phloem and the primary xylem (FIG. 9-4). The further development and activities of the root cambium are considered in a later section. Surrounding the conducting tissues, and constituting the outermost tissue of the vascular cylinder, is a narrow zone of cells, the PERICYCLE. The pericycle is composed of parenchyma cells, and as seen in cross section it is usually but one cell in width, although in some plants it is several cells wide. This is an extremely important tissue, for it is in the pericycle that branch roots, the cork cambium, and a portion of the vascular cambium arise. This is possible because the cells of the pericycle, like many other plant cells, retain their capacity for cell division and further growth even after physiological maturity.

Surrounding the vascular tissue is the cortex, a sheath of parenchyma tissue a number of cells in width. The cortex contains intercellular spaces and is important in the young root. Much of the bulk of the root at this stage is cortex, and these cells expose a large surface of plasma membranes across which mineral nutrients may be absorbed into the living cells, which are connected by plasmodesmata. This connected, protoplasmic system is known as the SYMPLAST.

The cortex is separated from the pericycle by the endodermis, a ring only one cell in width. The endodermis is distinguished by specialized thickenings of the cell wall. These thickenings usually take the form of thin, waxy strips on the radial and transverse walls. These strips are called CASPARIAN STRIPS. They form a seal between the protoplasts of the adjoining cells. This situation is shown in FIG. 9-6. Materials in solution may diffuse within the walls between the cortical cells, as shown by the red arrows, but the casparian strips prevent diffusion in the walls between the endodermal cells. In older monocot roots, diffusion through many of the cells of the endodermis is blocked by thickened inner walls. Some cells remain unchanged, however, and permit passage of materials between the cortex and vascular cyclinder.

The third and outermost region of the root is a single layer of cells, the epidermis. The outer walls of the epidermal cells usually lack a cuticle, and these cells are active in absorption.

The Root Hairs An examination of the terminal portion of the root of a radish seedling grown in moist air reveals a mass of short, whitish root hairs, the longest perhaps ½ inch in length (FIG. 9-7). Root hairs are specialized structures that increase absorption by increasing the root surface and the contact with soil particles. These hairs form only in cells behind the region of cell elongation; thus they are not destroyed by elongation of the root through the soil. They are simple, unbranched tubes with rounded ends, produced as outgrowths of the epidermal cells in the zone of maturation. They are a continuous part of the epidermal cells, and the cell nucleus may be located either in the epidermal cell or in these hairlike projections from it. Inside the cell wall of the root hair is a thin layer of cytoplasm that surrounds one or more large vacuoles (FIG. 9-8).

Root hairs are comparatively short-lived. As the root progresses through the soil, new root hairs are formed at the younger end of the zone of maturation; the older hairs, farther back on the root, collapse and slough away. The area of the root that bears the root hairs, therefore, does not change greatly in size or relative position as growth proceeds.

The formation of root hairs begins shortly after the seed germinates, and continues as long as the roots grow. In root hairs and certain other structures, such as pollen tubes, growth takes place only by elongation and is confined to the tip of the cell. Root hairs elongate very rapidly and may attain full size within hours.

Fig. 9-6 The endodermis and casparian strips. (A): diagram of endodermal cell in three dimensions. (B): diagram showing diffusion (colored arrows) within the walls of the cortical cells until this movement is stopped by the casparian strips.

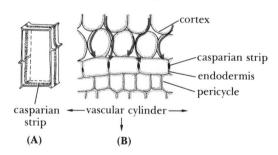

cortex

casparian strip

endodermis

pericycle

casparian strip

vascular cylinder

(A)   (B)

Fig. 9-7 Root hairs of young root of radish (*Raphanus sativus*).

Fig. 9-8 Stages in development of root hairs of timothy (*Phleum pratense*).

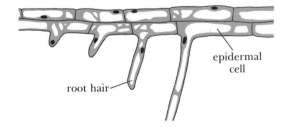

epidermal cell

root hair

Many land plants absorb water and solutes through the root hairs. Water and minerals are also absorbed by the young root through the epidermal cells. The root hairs are extremely numerous, and an idea of the number and extent of the roots and root hairs will help in understanding how plants can obtain large amounts of water from soil that seems nearly dry to the touch. The single rye plant previously described (Table 9-1) had more than 14 billion root hairs, with a total area of more than 4000 square feet—equal to the floor space in two or three good-sized houses. This, added to the area of the roots proper, gave this one grass plant a total root area of more than 6500 square feet, about 130 times the total area of the shoots and leaves. Yet the volume of soil penetrated by the roots totaled only about 2 cubic feet.

Root hairs increase the soil-contact surface of a root system as much as 20 times, and because they extend so widely through the soil, they make available a supply of water and minerals that the plant could not otherwise obtain. This intimate contact of root hairs and small roots with individual soil particles has a practical disavantage to the gardener and nurseryman, however. In the process of transplanting, these rootlets and root hairs are torn from the plant or lost by drying, greatly reducing the ability of the plant to absorb water. This difficulty can be minimized by moving a block of soil with the roots, and by watering and partially shading newly transplanted plants.

Some kinds of woody plants grown in nurseries are transplanted readily while dormant. The soil is removed and the roots are wrapped in damp peat moss or other material so that they will not dry out. But plants such as conifers, rhododendrons and azaleas, flowering dogwood, beech, birch, and some magnolias, are dug with a ball of earth around the roots. This ball is then wrapped firmly in burlap or placed in a basket to hold the soil in place. Another practice is to "root prune" some time before the tree is to be moved. If the spreading roots are cut at a short distance from the base of the tree, new root growth is stimulated near the trunk. These new

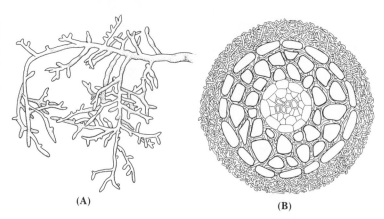

Fig. 9-9 Mycorrhizae. (A): swollen mycorrhizal rootlets of spruce (*Picea glauca*). (B): cross section of mycorrhizal rootlet, showing sheath of fungus threads around the root and threads between the cortical cells.

(A)

(B)

roots form a compact mass close to the stem, and months later, when the ball of earth is dug, it will contain a relatively large root system. If the transplanting is done with care and the reset plant given ample water, success is reasonably certain.

Mycorrhizae In numerous green plants the root hairs and other surfaces of the young root constitute the chief absorbing surfaces. In many other plants the young, active portions of the lateral or "feeding" roots are invaded by diverse kinds of soil fungi. The resulting condition is intimate and complex, and the combination of root and fungus is termed a MYCORRHIZA (literally "fungus root"). In mycorrhizae the root hairs are usually absent and the fungus replaces them, at least in part, in the absorption of water and minerals from the soil. In spite of more than 100 years of observations and investigations on mycorrhizae, active research continues, and there is much still to be learned. Two general types of mycorrhizae, sometimes intergrading, have been recognized; ECTOTROPHIC and ENDO-TROPHIC.

Ectotrophic mycorrhizae This type occurs in both deciduous and coniferous trees of the north temperate zone, such as the pines (*Pinus*), spruce (*Picea*), firs (*Abies*), beech (*Fagus*), oaks (*Quercus*),

birches (*Betula*), and poplars (*Populus*). It is found also in all species of *Eucalyptus* that have been investigated and in several tropical leguminous trees. Ectotrophic mycorrhizae are readily recognized, for the short, lateral, absorbing rootlets are much branched and swollen (FIG. 9-9A), the result of the presence of a SHEATH, or MANTLE, composed of tightly packed fungus threads. Threads extending from this sheath pass into the soil where they branch among the soil particles. In some mycorrhizae these extensions are abundant, in others, sparse. Threads from the sheath also extend into the root cortex, where they form a network, extending between the cells, through the middle lamella and intercellular spaces (FIG. 9-9B) and sometimes into the protoplasts of the cells.

The physiological relationship between fungus and root in this type is clearly mutualistic. The fungus obtains its necessary carbon compounds in the form of sugars and some amino acids from the host root. The root may also release to the fungus such vitamins as thiamin and biotin. The absorbing rootlets are so closely invested by the fungal sheath that they are dependent upon the fungal associate for mineral nutrients and water from the soil. This is especially important in soils deficient in available nutrients, where mycorrhizal plants absorb proportionately greater amounts of minerals from the soil than noninfected plants. This is due not only to a greater absorbing area, brought about by the branching of the rootlets and the extensions of their associated fungus into the soil, but also to the greater efficiency of the mycorrhizal threads in absorbing ions, especially phosphate, from the soil.

A knowledge of the nature and function of ectotrophic mycorrhizae is involved in certain forestry practices. For example, in the establishment of forest on land not previously forested, such as prairie, it is necessary to use seedlings grown on forest soils containing the mycorrhizal fungus. Similarly, in nurseries developed on previously nonforested land, mycorrhizal fungi may be introduced by the application of a thin layer of forest leaf mold.

Endotrophic mycorrhizae This type is more common than the ectotrophic, and more widely dispersed in the plant kingdom. It is found in great numbers of species of forest trees, both angiosperms and conifers; in shrubs; and in herbaceous plants, including most of the heterotrophic flowering plants called saprophytes (Chapter 5). Usually no fungus sheath is formed, and the lateral rootlets are not enlarged. These rootlets are associated with a loose web rather than a sheath of fungus threads. Some of these threads extend into the soil while others penetrate between and into the cortical cells of the root. Some species of fungi form coils (FIG. 9-10) within the cells, while others, by repeated dichotomous branching, form minute treelike outgrowths that almost fill the cells. The fungus obtains its organic compounds from the soil. Autotrophic vascular plants with this type of mycorrhiza derive a portion of their carbon compounds from the fungus, and heterotrophic flowering plants obtain all their carbon compounds from the associated fungus. This is accomplished by digestion of portions of the invading intracellular fungus. If the fungus forms coils, the threads tend to clump together, lose their contents, and a mass of undigested cell wall material accumulates (FIG. 9-10). In autotrophs, however, the most important role of the fungus is the disintegration of soil materials, releasing available nutrients, and serving as efficient absorbing structures.

In both types of mycorrhizae the relationship is usually termed mutualistic. If there is mutualism in endotrophic mycorrhizae, however, the contribution of the higher plant is not obvious; it may produce growth-stimulatory compounds useful to the fungus. Whatever this relationship may be, it seems certain that mycorrhizae of both types are ecologically important factors in growth and that they are among the most important and widespread symbiotic associations in the plant kingdom.

Fig. 9-10 Intracellular digestion of fungus threads in the endotrophic mycorrhiza of an orchid (*Goodyeara pubescens*).

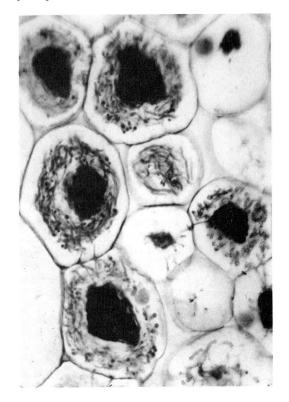

Mineral Absorption by Roots   The mechanisms involved in mineral absorption by roots are complex; many aspects are controversial, and the subject can be considered here only in general terms.

Mineral salts are obtained from the soil by diffusion into the cell walls of the root hairs and of the young tissues near the root tip. Having entered the root, the minerals move by diffusion within the cell walls of the cortex. These walls are readily permeable to water and dissolved substances. This permeable wall system, as distinct from the protoplasts of the cells with their plasma membranes, is called FREE SPACE; that is, it is free to movement of water and solutes. Movement within the free space brings the ions in contact with the large surface of plasma membranes throughout the cortical tissue, and through these surfaces they may pass slowly into the cytoplasm by diffusion or more rapidly by active transport. Once within the cytoplasm, further movement in the symplast is by diffusion, probably accelerated by cytoplasmic streaming, the ions passing from cell to cell through the plasmodesmata and bypassing the vacuoles. Continued radial movement brings the ions to the endodermis (FIG. 9-6). It has been held that the symplast ends at the endodermis, where movement through the free space is blocked by the casparian strips. There is evidence, however, that the endodermal walls are penetrated by plasmodesmata. Thus, even though the free space ends at the endodermis, the symplast may continue into the vascular cylinder, with its many conducting cells through which minerals as well as water can move to other parts of the plant.

The continued absorption of minerals by roots depends upon the removal of absorbed ions from the cortical cells. Three mechanisms may contribute to this removal. (1) The ions may be translocated to the tops of the plants, primarily through the xylem. (2) Ions may be changed chemically within the cytoplasm of the root cells. Nitrate ions, for example, may enter the cytoplasm and there be built into amino acids. These molecules do not normally penetrate back into the soil and do not interfere with the absorption of more nitrate. The amino acids may then be used in root growth or translocated through the plasmodesmata and the phloem to other parts of the plant. (3) Finally, ions may be accumulated into the vacuoles of the cortical cells by active transport. Experiments have shown that minerals that have accumulated in the vacuoles tend to remain there—they are not readily translocated to other parts of the plant. It is possible that if minerals become deficient in the root environment, ion transport to the xylem and the shoot may continue at the expense of ions accumulated in the vacuoles.

Secondary Growth in Roots   Secondary growth is characteristic of the roots of conifers and other gymnosperms, and of dicotyledons. It is characteristically absent from the roots of monocotyledons. When secondary growth begins, a cambium arises first in the parenchyma between the arms of primary xylem and inside the primary phloem. The cambium now forms secondary xylem to the inside and secondary phloem to the outside, as in the stem (FIG. 9-11). The cambium then is extended laterally, by differentiation of cambium initials in the pericycle, around the ends of the xylem arms and begins to form secondary tissues there also. At this stage the cambium is lobed, conforming to the outline of the primary xylem arms. But as growth continues, the lobed outline disappears, for secondary tissues are formed more rapidly between the xylem arms. The cambium then forms a circular zone, inside of which is a cylinder of secondary xylem that completely encloses the primary xylem (FIG. 9-11) Outside the cambium lies a zone of secondary phloem. The primary phloem and the endodermis are usually crushed by the pressure of the growing tissues within.

The secondary tissues of a root are not fundamentally different from those of the stem. In woody plants the bark may become similar to that of the stem. The first cork cambium appears in the pericycle, and as cork is formed, all tissues to the outside die and may be lost. Other cork layers

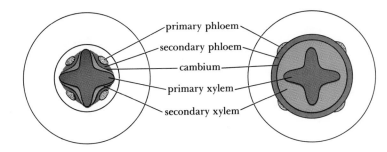

Fig. 9-11 Diagram showing secondary growth in roots. (Left): early stage in secondary growth; the cambium has developed and formed a small amount of secondary phloem and xylem. (Right): later stage, after considerable cambial activity has occurred.

primary phloem
secondary phloem
cambium
primary xylem
secondary xylem

may form at successively greater depths in the secondary phloem. Growth rings of secondary xylem are formed, as in the stem. Because of growing conditions, however, growth rings are commonly less distinct in roots than in stems. The wood cells formed by the cambium tend to be thinner-walled than those of the stem, and the proportion of vessels to other types of cells is greater. The vascular rays also are generally larger or more abundant in roots than in stems.

In plants with edible storage roots, such as beet, carrot, turnip, parsnip, horseradish, sweet potato, and cassava, the greater part of the storage organ is composed of secondary tissues. However, the cambium does not produce thick-walled xylem cells like those of the stem. It produces thin-walled vessels, usually in groups, together with masses of specialized xylem parenchyma. Such roots are edible because of the thin-walled xylem cells and abundant parenchyma (Fig. 9-12). If thick-walled vessels and fibers were produced, as they are in woody roots and stems and in wild carrots and parsnips, the roots would be hard and not usable as food.

Lateral Roots Branch, or lateral, roots commonly begin to grow before the cambium becomes active. A new, conical root apex is formed by the division of several cells of the pericycle at the end of a xylem arm. This apex then pushes outward as it grows, crushing or pushing aside the cells of the endodermis, cortex, and epidermis and emerging on the outside of the root (Fig. 9-13). It is frequently stated that the wounds made by emerging branch roots may serve as infection pathways for soil fungi. As the apex of the branch root grows, the cells behind mature and form a duplicate of the tissues of the main root. Cells at the base, in the pericycle region, differentiate and make connections between the vascular tissues of the main root and the branch root. At a later stage, the cambium of the main root becomes continuous with that of the branch. This origin of the lateral root from tissues lying deep within the parent organ is strikingly different from the origin of leaves and branches in superficial tissues of the stem.

THE SOIL

An understanding of roots and their activities requires a parallel study of the soil. An awareness of the interactions between roots and the soil is essential to an understanding of plants.

Soils vary in kind almost as much as the plants that live upon them. Many factors influence the nature of the soil: the nature of the rock from which soil was derived, the climate of the region, the plant and animal life in and on the soil, the exposure, and the slope. To these must be added the activities of man, who may change the soil for better or for worse through his methods of cultivation, through the grazing of his domestic animals, or through his use or misuse of forests.

Soil is popularly thought to be composed only

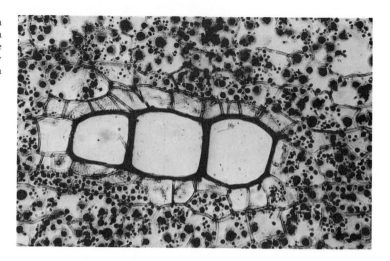

Fig. 9-12 Cross section of a portion of fleshy root of cassava (*Manihot esculenta*). Note the group of vessels surrounded by starch-filled xylem parenchyma cells.

of lifeless materials. It is, however, the abode of many organisms, for here live bacteria, fungi, algae, protozoa, insects, and worms of various kinds, from earthworms to microscopic nematodes. There may be 400 billion bacteria in a cubic foot of surface forest soil (over 100 times more bacteria in this volume than there are people on earth). It is through the activities of plants and other organisms, combined with the effect of chemical and physical weathering processes over ages of time, that the soils of the earth have been formed.

Soil Texture Soil contains materials in three states: solid, liquid, and gaseous. The solid portion is composed of both inorganic and organic materials, the latter alive or dead. The solid, inorganic portion is composed of mineral particles classified as sand, silt, and clay. Sand particles range in diameter from 2 to 0.02 millimeter, and silt from 0.02 to 0.002 millimeter. Particles less than 0.002 millimeter (about $1/10,000$ inch) in diameter are classed as clay. The varying proportions of these particles determine the *soil texture*, and on this basis soils may be divided into four large groups: SANDS, LOAMS, SILTS, and CLAYS.

A typical loam soil is one containing about 40 percent each of sand and silt with 20 percent of clay. Loamy soils vary widely in composition, however, as is indicated by the names sandy loam, sandy clay loam, clay loam, silt loam, and silty clay loam. Clay is the most important frac-

Fig. 9-13 Cross section of a young root of soybean (*Glycine max*), showing origin of lateral root.

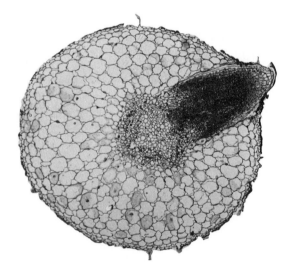

tion determining the characteristics of soils, and those containing more than about 40 percent of clay are classed as clays rather than loams. Many important agricultural soils are loams.

Soil Structure The elementary soil particles are generally grouped into larger, secondary particles, called AGGREGATES, of varying size and form. The particles in the aggregates are cemented together by compounds produced, in large part, by bacteria during the decay of organic material in the soil. The extent of the aggregation varies in different soils, depending on plant cover, organic matter content, texture, rainfall, temperature, and other factors. The quantity, size, shape, and arrangement of these aggregates determine the *soil structure*. In most soils the higher the content of organic matter and of the finer silt and clay particles, the greater is the degree of aggregation.

We have so far considered only the solid portion of the soil. No less important are the PORE SPACES between the soil particles and between the aggregates. The size and shape of the pore spaces are determined by the size and arrangement of the particles and aggregates of the soil. The pores contain air or water, or both, and make up between 30 and 60 percent of the soil volume. The ratio of air to water in the pore spaces greatly affects the suitability of the soil for plant growth.

The pores are of two general sizes, although there is no sharp line between them: the small, or micropores, usually filled with water, and the large, or macropores, containing water and air or air alone. In sandy soils, in which macropores dominate, aeration is good and water is absorbed rapidly, but retention is slight and the moisture drains away quickly. The addition of organic matter increases the water-holding capacity of such soils, in part because the organic matter itself contains micropores, and in part because the soil aggregates formed have a higher water-holding capacity than a sandy soil without such aggregates. Heavy soils, such as fine clays, have few macropores, and the movement of water and air is consequently slow. A clay soil may hold three to six times as much water as a sandy soil. Aeration in these clay soils, however, may be inadequate for respiration of roots, although it is greatly improved by the addition of organic matter. Organic matter increases aggregation and consequently favors the formation of macropores. The most desirable soil structure possesses a proper combination of large and small pores. The large pores allow infiltration of rainwater, rapid drainage, and proper aeration, while the small pores and the surfaces of the soil particles themselves retain water.

Air in the Soil The importance of air for plant roots is commonly underestimated. Roots respire as do other parts of the plant, and oxygen is as necessary for respiration of these organs as for that of the stem, leaves, flowers, and fruits. Roots, together with the plant and animal life of the soil, deplete the oxygen and increase the carbon dioxide concentration in the soil air. As biological processes proceed within the soil, however, diffusion is also operative. Carbon dioxide diffuses out of the soil, and oxygen diffuses in, thereby permitting continued aerobic respiration. This type of respiration prevails in well-drained soils during most of the growing season.

In some soils, however, the interchange of gases is impeded because of inadequate pore space. This may result from a puddled soil—that is, a heavy soil that has been compacted when wet, so that the soil aggregates are broken down and the pores largely closed. Continuous cropping with cultivated crops, as corn, cotton, tobacco, sugar beets, and many vegetables, depletes the organic matter, destroys the structure, and reduces the pore space of the soil. The concentration of oxygen reaches a minimum in inadequately drained or flooded soils. Crop plants grow poorly or die in such soils because of conditions unfavorable to root respiration and

Fig. 9-14   Stand of bald cypress (*Taxodium distichum*) in South Carolina, showing cypress "knees" in the foreground.

Fig. 9-15   Air roots (called pneumatophores) of the black mangrove (*Avicennia nitida*), Florida.

growth. Inadequate aeration reduces the growth of the root itself, curtails the absorption of minerals and water, and affects the activities of soil organisms. Upland plants vary greatly in their ability to withstand a reduced oxygen supply to their roots. Some species show injury in flooded soils within a few hours or days; others, including species of trees, can survive for many weeks or even months, particularly in cold weather when respiration rates are low.

On the other hand, many species, both woody and herbaceous, flourish in bogs and swamps, even though there is a very low oxygen concentration in the medium surrounding the roots. The ability of plants to grow under such conditions appears to depend upon structural or physiological adaptations. Herbaceous swamp plants commonly have large intercellular spaces in the shoot and roots through which oxygen diffuses downward. Lowland rice, for example, which is grown in 3–9 inches of water, possesses large air spaces in the leaf bases and roots.

Woody plants that grow normally in marshy soil or other areas deficient in oxygen include willows, the tupelo gum (*Nyssa aquatica*), several species of ash, the mangroves of sea coasts, and the bald cypress of southern forests. When the bald cypress (FIG. 9-14) grows in swamps, it produces upright conical growths, called cypress knees, from horizontal roots. Knees do not develop when the cypress grows in dry situations, nor are they formed in deep water. It has long been assumed that these structures are aerating organs, allowing gaseous exchange between the atmosphere and the submerged roots, but this has not been demonstrated and may be unimportant. Air roots are found in a number of tropical and subtropical plants, notably certain species of mangroves. Horizontal roots, buried in the mud, extend outward from the parent tree. The air roots, 10–15 inches long, extend upward into the air from the buried roots (FIG. 9-15). These aerial roots act like ventilating chimneys. They have a spongy structure with large intercellular spaces, which permits interchange of gases between the atmosphere and the submerged roots.

The raising of the soil level around shade trees frequently results in the injury or death of the tree. Such conditions may occur when new highways are built or where lawns or terraces are graded. Trees such as oak, hickory, beech, sugar maple, and certain conifers are most seriously injured, but other kinds also may be affected. The deeper the fill, the more likely that injury will result, especially if heavy subsoil is placed over the old soil level. It is assumed that such injury is due to reduced oxygen supplies to the roots and the base of the trunk.

Injury from soil fills may be reduced by constructing a dry well, extending down to the old soil level, around the base of the trunk. In addition, agricultural tile may be laid down in shallow trenches radiating from the well and extending out to points beneath the outer spread of the branches. The tile allows penetration of air and also drains excess water away from the base of the tree. In general, upland trees are most injured by fill, and flood plain species, such as willow and some poplars, ashes, and elms, are most resistant.

Water in the Soil   The water in the soil may be classified as (1) gravitational, (2) capillary, and (3) hygroscopic. Much of the water added to the soil by rainfall or by irrigation seeps into the subsoil and eventually becomes a part of the ground water or drains away into streams. This water, which the soil is unable to retain against the force of gravity, is known as gravitational water. In some soils the movement of gravitational water is rapid, and in others slow, and the rapidity with which it leaves the soil is of agricultural interest. Since gravitational water fills the larger pore spaces of the soil and thus interferes with aeration and root growth, it is frequently necessary to construct drainage systems of tile or open ditches to remove this water from the soil. The movement of gravitational water through soils also increases the difficulty of maintaining soil fertility, for it leaches out of the soil considerable quantities of soluble fertilizer materials, which are thus lost to plant life. Gravitational water can be absorbed by roots, but is of little value because it normally drains away within a few hours or days. If it saturates the soil for any length of time, its injurious effects are greater than the beneficial ones.

The water held in the soil against the force of gravity is capillary water. It fills the smaller pore spaces of the soil and is present as films around the soil particles and as wedge-shaped masses at the points of contact between the particles. This capillary water, in turn, may be classified as: (1) free capillary water, (2) available capillary water, and (3) unavailable capillary water.

Free capillary water   After drainage has ceased, following rain or irrigation, water may still continue to move by capillarity, much as water will rise in a glass capillary tube (FIG. 8-7). This movement may be downward, lateral, or upward into drying surface soil. Such movement of free capillary water may reduce the moisture percentage of a drained soil by one third or more, changing a muddy soil to a moist one suitable for cultivation. The upward movement tends to keep the surface moist. A dry soil surface is evidence, therefore, that rapid capillary movement to the surface has stopped. As the soil is dried by this movement, the moisture films become thinner until a point is reached in most soils where the moisture columns break and capillary movement becomes small or ceases.

Available capillary water   When the movement of free capillary water has ceased and the soil surface is dry, the soil just below the surface commonly contains an ideal level of moisture for cultivation or for the growth of plants. The percentage of moisture in the soil at this point has been termed the FIELD PERCENTAGE (also called the FIELD CAPACITY). The field percentage moisture level is a most important one for plants and agriculture, and represents the maximum moisture *storage capacity* of typical soils. Dry lands from Kansas to Saskatchewan are fallowed—that is, clean cultivated without crops—every second or third year to control weeds and to permit storage

of soil moisture for use in cropping years. This stored moisture is held, at or near the field percentage, between dry subsoil and a dry surface layer. In the absence of plants, much of such moisture may be held in the soil for months. Plants, however, whether crops or weeds, will absorb about half of this stored moisture. Absorption is rapid at first and then slows as the moisture films become thinner and the water is more strongly adsorbed on the surfaces of the soil particles.

When these adsorption forces reach the equivalent of 14 or 15 atmospheres—about 200 pounds per square inch—the remaining water in many soils becomes, for practical purposes, unavailable to plants. Growth stops and the plants become wilted, even in shade. The percentage of moisture in the soil at this point is called the PERMANENT WILTING PERCENTAGE. Wilted plants will stay alive in this soil for a period that is a measure of their drought resistance, but they will not recover from wilting and will eventually die without added water. Plant growth decreases as the soil becomes drier, and stops before permanent wilting is reached. Photosynthesis and increase in dry weight may continue, however, until the plants actually wilt.

It should be noted that a permanent wilting tension of 15 atmospheres is an approximate figure, found in many of the better agricultural soils. Drought-resistant plants on certain desert soils, in which moisture movement continues slowly at high tensions, may show much higher values. The permanent wilting point is related to a cessation of root growth at moisture tensions in the region of 15 atmospheres and to the near-zero movement of capillary moisture in most soils at this tension. If the roots do not grow to the water or the water does not move to the roots, there will be no absorption.

The cardinal points of soil moisture are (1) the field percentage, where moisture is readily available and the soil is not too wet for root respiration and growth or for cultivation, and (2) the permanent wilting percentage. The capillary water in the range between these points is termed the AVAILABLE CAPILLARY WATER. This is the most important and frequently almost the only source of moisture for plants.

The quantity of water held in a particular soil at the field percentage or the permanent wilting percentage is determined primarily by the size of the soil particles—that is, by the soil texture. The small particles of clay have more surface and produce a greater number of fine capillary spaces. A clay soil, therefore, will hold much more water than a sandy one. Typical soil moisture percentages in several soil types appear in Table 9-2.

TABLE 9-2
Typical Soil Moisture Percentages

Soil type	Field percentage	Permanent wilting percentage	Storage of available capillary water
Sand	6	3	3
Sandy loam	10	5	5
Silt loam	20	10	10
Clay	40	20	20

The clay shown in Table 9-2 holds four times as much water as the sandy loam but in a form that plants cannot use (permanent wilting percentage). It also stores four times as much as they can use (available capillary water). Heavier soils, from silt loams to silty clay loams, are preferred for dry-land farming because of their greater ability to store moisture for crop use. The lighter-textured sandy loams and loams are preferred where sufficient water is available because they are easier to work, absorb rainfall more rapidly, and permit better aeration for root growth.

Unavailable capillary water This is the water in the range between the permanent wilting percentage and the hygroscopic water. The latter is the water remaining in air-dry soil, in the form of molecular films around the soil particles. It moves only in the form of vapor and is of no significance to plant life.

Movement of Available Capillary Water It was long believed that capillary water moved rapidly and that as the water supply in the vicinity of the roots became exhausted, it was replenished by water moving from unexhausted areas. If, in the laboratory, a cylinder is filled with soil and one end of it placed in a vessel of water, the water rises rapidly through the soil by capillarity. This experiment may apply to the watering of house plants in pots, but does not represent conditions in the field, for the water in the vessel would correspond to the WATER TABLE, or upper limit of permanently saturated soil, and in most agricultural regions of the United States the water table lies so far beneath the surface that upward movement of water from it is not significant as a source of water for plants.

It has been shown that there is normally very little movement of available capillary water (water at or below the field percentage), with the exception of slow movements over short distances to nearby roots. During periods of high transpiration, the water in the immediate vicinity of the roots is removed so rapidly that each absorbing root becomes surrounded by a layer of nearly dry soil. Since water does not move appreciably in the moist soil in which plants typically grow, continued moisture absorption by roots depends upon rainfall, or upon the growth of roots into new areas of soil. The significance of the rate at which roots grow, therefore, becomes apparent. Apple roots may grow in length 1/8 to 3/8 inch a day. Native western prairie grasses average more than 1/2 inch a day. Potato roots may elongate at the rate of 1 inch a day over a period of several weeks, and the main vertical roots of corn sometimes penetrate downward at the rate of 2 to 2 1/2 inches a day over a period of 3 to 4 weeks. These roots grow, not in search of moisture, but because local conditions of moisture and aeration are favorable to their growth. Roots will not grow through a few inches of soil at the permanent wilting percentage, but will grow wherever available moisture, air and minerals occur.

Cultivation Cultivation of the soil by the hoe or by power cultivators has several objectives: (1) To kill weeds. This is the chief value of cultivation, for weeds compete with crops for soil nutrients, water, carbon dioxide, and, as they grow larger, for light. (2) To increase the porosity of the soil and thus bring about better aeration. (3) To increase the absorption of rain water. A compact, uncultivated soil surface generally permits less infiltration of rain water. It allows runoff and erosion, especially on slopes, during heavy rains. The last two reasons for cultivation are of minor importance in soils that are moderately sandy or in which the organic matter content is kept high by good soil management. They are of most value in heavy soils low in organic matter, especially if they have been compacted by excessive working with heavy machinery. Cultivation, however, cannot take the place of the good soil structure that results from a liberal supply of organic matter in the heavier soils.

Cultivation was at one time also carried on for the purpose of producing a loose, dry layer of soil, or mulch, over the surface. This, it was believed, would break the upward capillary movement of water from below and thus reduce loss by evaporation. Since rains would reestablish the capillary action and destroy the dust mulch, it was advocated that the soil receive shallow cultivation after every rain. This "dust-mulch" theory, advocated as early as 1895, was especially favored as a procedure for the prevention of failure of crops by drought in dry-land farming in the Great Plains area, where the precipitation is limited to an average of about 20 inches a year.

The practice of cultivating to produce a dust mulch has been generally abandoned. In dry-land farming, the finely pulverized soil surface increased runoff and was subject to serious wind erosion. Moreover, upward capillary movement in an uncultivated soil ceases as the surface dries, and a layer of dry soil, through which water does not move by capillarity, is thus produced without the aid of cultivation. Crops and weeds remove water that is stored in the deeper layers of the soil, beyond the reach of surface evaporation. Weed control is therefore vital in dry-land farming, both in the growing crop and during the fallow years when the fields are kept bare to allow the accumulation of water for the next year's crop.

In dry-land farming regions where soil moisture is the limiting factor in crop production, good agricultural practices involve the following: (1) the use of drought-resistant crops, (2) wider spacing of the plants to reduce total moisture use, (3) cultural practices, including the use of herbicides, that destroy weeds but at the same time maintain a rough rather than a dust surface, (4) keeping the land in clean fallow (without crops or weeds) every second or third year to accumulate extra moisture. The western Kansas or Montana farmer of today takes pride in the covering of stubble and large, durable clods that protect his fields from blowing and which allow rainfall to infiltrate the soil with maximum speed. Cultivation is thorough, but is done just often enough to control weeds and in such a way that a coarse covering over the soil is maintained.

Although dust mulches are no longer in general use, other protective coverings, or mulches, composed of hay, straw, sawdust, manure, leaves, peat moss, crop residues, and other organic materials may be used for special crops. These mulches retard water losses from the surface layers of the soil. They reduce or prevent weed growth and they protect the soil surface from the impact of raindrops, thus keeping it mellow and porous and increasing its capacity to absorb water. Straw mulches shade the soil and keep it cooler. Irish potatoes, particularly, may be benefited by this cooling when grown during the summer. Mulches of dark-colored paper absorb sunlight and warm the soil in early spring. In colder climates a winter mulch of coarse straw is desirable for bulbs, strawberries, and many evergreen and deciduous ornamentals. Such mulches protect the plants against the effects of low temperatures and particularly against alternate freezing and thawing.

THE ESSENTIAL ELEMENTS

Although the many kinds of green plants appear to exhibit bewildering complexity, they are surprisingly similar in chemical composition. The synthesis of plant foods and the utilization of these foods by plants require not only chemical elements contained in air and water but also a number of others obtained from the soil. Since these elements are necessary to growth and reproduction, they are termed the ESSENTIAL ELEMENTS. The necessity of these elements is taken for granted in agricultural practice today, but knowledge concerning them has resulted from extensive investigations by plant physiologists and soil chemists. The accumulation of this knowledge represents a major scientific accomplishment of the past hundred years.

After it had been established that mineral elements from the soil were essential, much research was still required to determine which elements are necessary to plant life, and which are not necessary, even though they are present in plants. The basic technique used in making these determinations is to grow plants in "water culture" (FIG. 9-16). In this method seedlings are grown in containers with their roots submerged in a dilute solution of known chemical composition. By a study of the behavior of plants growing in cultures from which certain elements are omitted, it is possible to determine which elements are essential and which nonessential. This method also indicates the symptoms that arise from deficiencies of the various elements.

Isotopes A new method for research with essential elements has become available in recent years. This involves the use of isotopes (Chapter 4). The radioisotopes of a number of essential elements are available. Among the more useful of these are calcium, carbon, and phosphorus. Usable isotopes of hydrogen, iron, manganese, and zinc are also available, and isotopes of the nonessential elements bromine and strontium are used for special purposes. Nitrogen forms four radioactive isotopes, but they decay too fast to be generally useful. However, ^{15}N, with one extra neutron, is a stable, natural isotope that can be used, but is much less convenient than a radioisotope would be.

If a plant is grown in soil, sand, or nutrient solutions containing radioisotopes, the rate of absorption, path of translocation, and concentration in various tissues and compounds can be studied (FIG. 9-17). In one study on tomato plants, for example, radioactive phosphorus was detected by the Geiger counter in the tips of 6-foot tomato plants 40 minutes after the roots had been placed in a nutrient solution containing this radioisotope. It was also found that if phosphorus is deficient in the plant, it may be withdrawn from the leaves and moved into the growing fruits, which thus develop at the expense of the vegetative organs.

A striking example of the use of a radioisotope to study mineral absorption by roots is shown in FIG. 9-18. Rubidium-86 was mixed uniformly with a small quantity of soil. This soil was then spread thinly over a plastic window in a special, plant-growing box. The remainder of the box was filled with ordinary soil, and corn was planted so that the roots would grow against the window and through the soil containing the isotope. After several days of growth, a negative was made in the dark by holding a photographic film tightly against the plastic window. The darker shades of the radio autograph are due to radiation from the rubidium-86. The blackness of the root tips indicates isotope accumulation, and the light color of the soil near the roots shows the areas of absorption.

The use of radioisotopes has made it possible to determine whether an element taken up by plants is derived from fertilizers added to the soil or from the supply of the same element already present in the soil. The effectiveness of various

Fig. 9-16 Tobacco (*Nicotiana tabacum*) plants in culture solutions, showing the effect of deficiency of various elements. 1: no nitrogen added. 2: no phosphorus added. 3: no potassium added. 4: no calcium added. 5: no magnesium added. 7: no boron added. 8: no sulfur added. 9: no manganese added. 10: no iron added. 6: all the above elements added.

Fig. 9-17 Radioactive phosphorus (^{32}P) in soybean (*Glycine max*). The isotope was added to a nutrient solution in which the roots were growing. After 18 hours, phosphorus has accumulated in the roots, nodes, and growing leaves. The radioautograph was made by pressing the plant against an x-ray film and exposing the film in darkness for four days to the electrons emitted from the ^{32}P. Note the near absence of phosphorus from the older leaves of the plant.

methods of adding fertilizers to the soil, and the response of various crops to high rates of fertilizer application, have also been studied by means of isotopic tracers.

The Elements Essential to Higher Plants An essential element is one that is necessary for the growth of plants and the production of flowers and seeds. These essential elements may be divided into three groups. The first of these includes those elements derived from the air and from soil water—carbon, hydrogen, and oxygen. Up to thousands of pounds an acre of these relatively free and plentiful elements are used by crops. The second group includes nitrogen (N), phosphorus (P), sulfur (S), potassium (K),

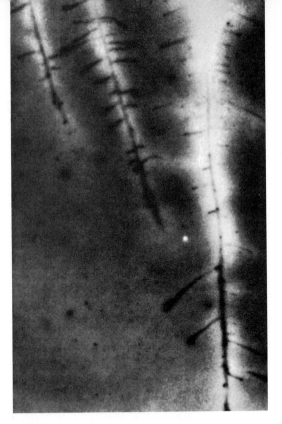

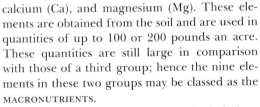

Fig. 9-18 The absorption of a radioisotope, rubidium-86 (^{86}Rb), from soil by roots of corn (*Zea mays*).

Fig. 9-19 Zinc deficiency in potatoes (*Solanum tuberosum*) in Washington State. The plants in the background received zinc and phosphorus fertilizers; the dying plants (arrow) in the foreground received phosphorus but no zinc.

calcium (Ca), and magnesium (Mg). These elements are obtained from the soil and are used in quantities of up to 100 or 200 pounds an acre. These quantities are still large in comparison with those of a third group; hence the nine elements in these two groups may be classed as the MACRONUTRIENTS.

The elements of the third group include iron (Fe), copper (Cu), manganese (Mn), zinc (Zn), (FIG. 9-19) molybdenum (Mo), boron (B), and chlorine (Cl). Iron has been known to be an essential element since 1847, but the necessity for the others has been established only since 1922. The elements of this group are required in very small quantities, but are just as essential for growth as nitrogen, phosphorus, or potassium. The concentration of boron required for normal growth, for example, is usually not more than one part per million parts of water, and the requirement for molybdenum can be satisfied by one part in one billion parts of water. Because only very small amounts of the elements of the third group are necessary for growth, they are termed *minor*, *trace*, or more commonly, MICRONUTRIENT elements. The necessity for the micronutrient elements, other than iron, was not discovered earlier because minute quantities were already present either in the seeds from which the experimental plants were grown or as impurities in the water or the chemicals used. With even greater refinements in experimental procedures, it is probable that still more elements will be added to the list.

Animals require most of the elements essential

to plants, together with others such as sodium, iodine, and cobalt. Most of the essential elements required by animals are obtained from the plant tissues upon which they feed, or from other animals that have fed upon plants. This constitutes another example of the importance of plants to the life of animals, including man.

Essential Elements and the Soil The inorganic materials of the soil that are insoluble in water are slowly brought into solution by acids produced during the decay of organic materials and by other agencies. The essential elements in nonliving organic matter are released by the activities of the bacteria and fungi of the soil. The soil is thus the site of slow transformations vital to green plants.

The essential elements are not found in elemental form in the soil water, nor do they occur in elemental form when they are added as fertilizer. They occur in the soil as compounds or as ions—defined in Chapter 4 as atoms or groups of atoms carrying positive or negative electric charges. For example, when a salt such as potassium nitrate dissolves in the soil water, it breaks up, or ionizes, into two ions, potassium (K^+), which carries a positive charge, and nitrate (NO_3^-), which carries a negative charge. The essential elements move into the plant primarily in the form of ions. The rate of movement of most ions through the soil is slow. The rate at which roots grow therefore determines, in considerable degree, the rate of absorption of essential elements as well as of water.

The kind and quantity of elements absorbed by roots are determined by such factors as the rate of root growth and the chemical composition of the soil and root. Essential elements present in high concentrations in the soil often move into the plant in excess of the amounts required for normal growth and may even damage tissues or kill the plant. Plants also absorb elements that are not essential for growth. In general, any mineral present in the soil solution may be expected to be absorbed by roots. The rate of absorption will vary, however, with the permeability of the root cells to the particular material and with the difference in the concentration of this material in the root and in the soil solution.

Differences in concentration depend upon the rate at which soluble, generally ionized, forms of a certain mineral are produced in the soil, and upon the rate at which the absorbed ions are changed to other forms within the plant. These changes typically consist of some chemical combination that removes the absorbed ion from solution. We could say that available ions are absorbed in proportion to their "use" in the plant. Thus the nitrate ion (NO_3^-) is built into amino acids or nitrogen bases, removing the ion from the plant solution; phosphate ions (H_2PO_4) are built into ATP, or into the DNA and RNA chains and various other organic compounds, and potassium ions (K^+) are held electrostatically by negative charges on organic acids and other molecules. Nonessential ions also may be bound within the plant and accumulated. For example, because silica is precipitated out of solution in the straw, it can continue to be absorbed by wheat. Also, the active concentration of mercury within a plant remains low because the poison is used up in reacting with the protoplasm; the plant may continue to absorb mercury and eventually be killed.

Most of the elements from soil that are present in the plant may be determined by an analysis of the ash obtained by burning the plant. Nitrogen, carbon, hydrogen, and oxygen are lost during the incineration process, but in the residue, which may consititute 5 percent or more of the dry weight, about 60 elements have been identified. Some of the presumably nonessential elements, such as silicon, may be present in considerable quantity. Other elements believed to be nonessential and usually found only in traces are aluminum, arsenic, lead, barium, mercury, bromine, tin, gold, nickel, and selenium.

The concentration of mineral elements present in the plant varies in part with the soil. Soils formed from granite are commonly high in potassium. Limestone soils should contain available calcium, but may be deficient in available iron and other micronutrients if their lime content is high. Sands are usually low in potassium and frequently in other elements. Hence the kinds and quantities of elements found in plant tissues vary to a limited extent with the soils in which the plants grow, and with the climate. It is surprisingly difficult, however, to determine from an analysis of plant tissue which fertilizer elements might profitably be added to a particular soil.

Absorption of nonessential elements frequently varies to a greater degree than that of the essential ones. The soils of certain areas of the western United States, for example, are rich in selenium, which may be absorbed in concentrations that are toxic to animals that eat the plants. Investigations have shown that the mineral content of the tissues of various species growing in soils above mineral deposits is higher than in the same species occurring in areas devoid of such deposits. Such accumulations in plants have been used in prospecting for metallic ores—a specialty termed geobotanical prospecting.

Role of Essential Elements Evidence from water and sand cultures and from field observations has revealed that a deficiency of a certain element may show as a recognizable deficiency symptom (FIGS. 9-16, 9-19). These tests show also that there is an interaction among the elements and that a balance of nutrient elements is necessary for satisfactory plant development.

In spite of a vast amount of research, the specific function of many of the essential elements is still imperfectly understood. The roles of carbon, hydrogen, and oxygen are clear. As building units of sugars, cellulose, protein, and other plant compounds, these elements form on an average about 94 percent of the dry weight of plants. Nitrogen is an essential component of the amino acids of which proteins are formed and of the nitrogen bases present in the nucleic acids and in related compounds such as ATP (FIG.

4-3). Phosphorus, as we have seen, is present in nucleic acids and functions in the transfer of usable energy. This element is also present in lipid combinations that are thought to be important in the structure and permeability of the protoplast. Sulfur is present in three amino acids and functions in essential respiratory and other enzymes.

So far, we can say that carbon, hydrogen, and oxygen are basic building elements, and that nitrogen, phosphorus, and sulfur are concerned in the formation and functioning of protoplasm. The role of the three remaining macronutrient elements, calcium, magnesium, and potassium, is not so clear. Calcium is present in the cell wall, and magnesium in the chlorophyll molecule, whereas potassium is not known to occur in any specific plant compound. These three base-forming elements seem to have in common the general properties of stabilizing the structure of the protoplasm and of aiding in the action of enzymes. Magnesium, in particular, must frequently be added to isolated plant enzyme systems to enable them to function. Definite roles in the functioning of specific enzymes can be assigned to the micronutrient metals iron, manganese, copper, zinc, and molybdenum. It is assumed that boron and chlorine function similarly. The essentiality of cobalt for higher plants has not been demonstrated. It is, however, essential, although in very low concentrations, to the syntheses of vitamin B_{12} compounds by nitrogen-fixing blue-green algae and the mutualistic nitrogen-fixing microorganisms of leguminous and nonleguminous higher plants (Chapter 10). These vitamin B_{12} compounds, in turn, are necessary to the nitrogen-fixing process—without cobalt, nitrogen fixation will not occur.

Many of the older ideas regarding the action of essential elements can be dismissed as oversimplifications. Among these are the generalizations that nitrogen makes leaf growth; that phosphorus makes grain; that potassium is especially important for root crops; that nitrogen delays maturity while phosphorus hastens it; and that calcium neutralizes toxic acids. Nitrogen, phosphorus, and potassium are required by all living cells. Either nitrogen or phosphorus, in balanced supply, can hasten growth and maturity of grain, vegetable, or other crops. An oversupply of nitrogen alone, however, may delay maturity by causing a relative deficiency of phosphorus. The acids, either within the plant or in the soil, that may be neutralized by lime are rarely toxic to plants in the concentrations at which they occur. Lime for acid soils, as discussed later, has a much more complex effect than the simple neutralization of soil acids.

Because the essential elements may have specific functions within the plant, it might be assumed that a deficiency of a particular element would lead to a specific symptom. Much attention has been given to this problem without outstanding success. Any nutrient deficiency will result in reduced plant growth, but this response has no diagnostic value. In some, but not all, plants a deficiency of phosphorus is indicated by the appearance of red or purple colors in the leaves. A nitrogen deficiency shows as yellow-green leaves on normally dark green plants, but other conditions can have the same effect. The lack of potassium may show as yellow areas on the leaves, which later become dead and brown. A shortage of magnesium will appear as an irregular *chlorosis* (yellowing) of the leaves. Typically, the veins remain green until the injury becomes extreme. Unfortunately, however, a lack of iron or manganese, or an excess of manganese, may produce similar symptoms. Leaves deficient in iron fail to develop chlorophyll and become chlorotic. It has been assumed, therefore, that iron has a specific function in chlorophyll formation. It may be, however, that the chlorosis is simply a visible symptom of the failure of the respiratory or other processes that are dependent upon iron-containing enzymes. A trained observer, familiar with a particular plant in a particular environment, can draw many conclusions from mineral-deficiency symptoms, but there are no simple and generally applicable rules that can be followed.

Soil Acidity and Liming   A large percentage of the soils of the more humid areas of the earth are classed as acid, or sour. This does not mean that they are acid enough to *taste* sour but only that they are more acid than pure water. A very small number of water molecules ionize into acidic hydrogen ions (H^+) and an equal number of basic hydroxyl ions (OH^-). Since there is an equal number of hydrogen and hydroxyl ions present in pure water, it is neutral. The ionization of water is so slight that 3 million gallons of pure water contain only about 1 gram of hydrogen ions. On the logarithmic pH scale used to express acidity, water is at pH 7, or neutral. The term pH 7 means that the water contains 10^{-7} gram of hydrogen ion per liter, or is 1 ten-millionth normal in H^+. Alkaline solutions or soils contain fewer hydrogen ions and more hydroxyl ions than water; acid soils contain fewer hydroxyl and more hydrogen ions. Because the pH units are negative logarithms, pH 6 is ten times more acid than pure water; pH 5 is 100 times more acid; and so on, as shown in Fig. 9-20.

Soil acidity develops primarily as the result of the removal of basic elements, especially calcium, from the soil by the leaching action of percolating rain water and by cropping. Lime, usually in the form of finely ground limestone (calcium carbonate, $CaCO_3$), is commonly added to acid soils. The effects of liming are complex. Where the soil is low in available calcium, or when crops such as peanuts (Fig. 13-4) which are particularly sensitive to a calcium deficiency, are grown, lime constitutes a calcium fertilizer. Liming may also have important effects on the availability, and sometimes the toxicity, of a number of other elements. The availability of phosphorus in acid soils may be increased markedly by liming the soil to near-neutrality. On a number of soils in Australia and New Zealand a major effect of lime is to increase the availability of molybdenum in the soil. If available molybdenum is added to these soils at the rate of a few ounces per acre, much less lime is necessary. In contrast to phosphorus and molybdenum, lime decreases the solubility of iron, manganese, copper, and zinc. Overliming may thus reduce the supply of these elements to plants. Such deficiencies occur naturally in the high-lime soils of arid or semiarid regions. In excess, however, all of these elements, manganese especially, may be toxic. Lime will reduce this toxicity and also that due to soluble aluminum and other metals.

Finally, the liming of acid soils favors the growth of soil bacteria at the expense of soil fungi. Liming may thus reduce or prevent such fungous diseases as clubroot of cabbage and the damping-off of legume and other seedlings. It will, however, increase the severity of diseases caused by soil bacteria. For example, Long Island farmers want limed soil to prevent clubroot of

Fig. 9-20 pH and soil acidity or alkalinity. Note that actual acidity or alkalinity changes ten times, or 1000 percent, for each unit of pH; also that the point of direct acid toxicity is below the level of strongly acid soils.

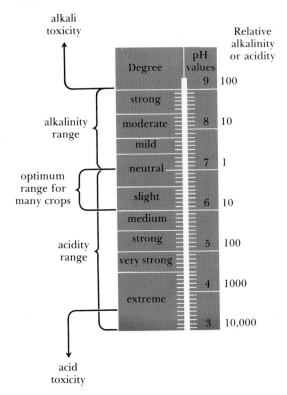

cauliflower but need strongly acid soils for Irish potatoes, to prevent the development of the bacteriumlike organisms that cause potato scab. Obviously, therefore, the two crops cannot be grown successfully in the same rotation. Lime is especially important for legume crops such as alfalfa and clover. These plants use large quantities of calcium and phosphorus. They are generally sensitive to soluble manganese and aluminum in the soil. They may be seriously affected by fungous diseases in the seedling stages, and they are dependent upon the acid-sensitive, nitrogen-fixing bacteria (Chapter 10) for their normal development. Lime with organic matter improves the physical properties of heavy, acid soils by favoring the bacterial growth that contributes to the formation of soil aggregates and the improvement of soil structure.

The liming of acid soils thus has many direct and indirect effects. It acts not only as a fertilizer in increasing the available supply of an essential element, calcium, but also as a soil amendment, and results in important chemical, biological, and structural changes in the soil. Some of these may increase and some decrease plant production, depending upon the conditions. It is necessary, therefore, that liming be done carefully and that excess lime be avoided.

Fertilizers and the Soil The use of fertilizers is almost as old as agriculture. Early farmers learned by observation that the use of animal manures resulted in larger and better crops. They learned to apply other materials in addition to animal manures: lime and marl, fish, wood ashes, and seaweed. An understanding of the essential elements and their use by plants makes it clear why these practices benefited the crops.

The Greeks and Romans thought that plants obtained their organic matter from the soil; hence their stress on the value of organic manures. In spite of the work of Ingen-Housz and others in the late 1700s (Chapter 5), this belief was prevalent until the middle of the nineteenth century. The teaching of Justus Liebig, a

German scientist and the "father of agricultural chemistry," was largely responsible for the general recognition of the fact that plants obtain their carbon from the air through photosynthesis, and only mineral salts and water from the soil. Liebig developed and patented an inorganic manure but, unfortunately, his fertilizer was insoluble and hence unavailable to plants. His work, however, and that of Lawes in England, who established the Rothamsted Experiment Station in 1843 to study the use of fertilizers, led gradually to the use of chemical fertilizers, which are a major factor in modern agriculture.

The normal objective in the use of fertilizers is not to supply all of the minerals used by a crop but to increase the quantities and availability of those elements in which a particular soil is most deficient. Nearly all the soils of the world will respond to nitrogen fertilization, provided the plants have enough moisture to support the increased growth. Most soils will be benefited by additions of phosphorus, particularly after their nitrogen content has been built up. Many soils require addition of potassium. The addition of calcium, in the form of lime, is beneficial on a large percentage of the soils of humid areas. As discussed earlier, lime has important effects on the chemistry and microbiology of soils, and also supplies an essential element. Of the other nine elements known to be obtained from the soil, magnesium is beneficial on some of the sandy soils of the Atlantic coast, and sulfur is used in certain areas of the Pacific Northwest. Organic, or peat, soils are frequently deficient in available copper and sometimes in manganese and zinc (Fig. 9-19). Molybdenum is important to considerable areas of Australia and New Zealand, and boron is used on soils of the eastern United States that have been heavily limed, as for the production of alfalfa. Iron may be unavailable in calcareous or other soils, and iron salts added as fertilizers to these soils are quickly changed to the unavailable forms. Leaf sprays of iron salts have some value, and good cultural practices, specifically nitrogen fertilization, increase the ability of the plants to absorb and use iron.

The fertilizers are applied as compounds; nitrogen, for example, may be applied as ammonium sulfate, ammonium nitrate, or as the anhydrous gas NH_3 (Fig. 9-21). The ammonium sulfate or ammonium nitrate may be used alone or else as components of a *mixed fertilizer*. Mixed fertilizers usually contain some combination of the elements nitrogen, phosphorus, and potassium, together with enough lime to offset any acid-forming properties of this mixture, and with other elements under special conditions. The analysis, or grade, of a mixed fertilizer is expressed as the percentage of nitrogen (N), phosphoric acid (P_2O_5), and potash (K_2O), in that order. A fertilizer labeled "4-12-4" contains 4 percent of nitrogen, 12 percent of phosphoric acid, and 4 percent of potash by weight. An 0-16-8 fertilizer might be purchased to supply phosphorus and potash for a nitrogen-fixing legume crop that does not require added nitrogen (Chapter 10), and a 10-20-0 fertilizer might be used for corn on soils naturally high in available potassium. The older fertilizers were mixed with relatively low percentages of nutrient elements to keep the per-ton price low. The fertilizer salts were diluted with a "filler," frequently soil. With increasing freight costs and heavier rates of fertilization, the trend is to high-analysis mixtures. Thus a grower who has used 500 pounds of a 4-8-6 fertilizer and wishes to double the rate will find 500 pounds of 8-16-12 fertilizer cheaper than 1000 pounds of the 4-8-6 fertilizer. An extreme example of a concentrated fertilizer is that of gaseous ammonia (NH_3), with a "grade" of 82-0-0, used as a source of nitrogen. The gas, transported in tanks at high pressure, may be dissolved in the water of an irrigation stream so that water and nitrogen are applied together, or it may be injected 12–18 inches beneath the surface of moist soil (Fig. 9-21), where it is adsorbed and held by the soil until it is used by plants.

Fertilizers are broadcast upon the soil, placed below the seed, or run in parallel bands along the row of seeds or plants. It is necessary to apply these chemical fertilizers carefully to avoid fertil-

Fig. 9-21 Applying anhydrous ammonia (NH₃) for corn in Illinois. The gas is forced into the soil under high pressure at a depth of 12–18 inches.

izer injury, which may result from the toxic effect of the salts or from a high solute concentration in the soil solution (FIG. 4-10). Properly used, commercial fertilizers may greatly increase the crop yield. In one experiment, corn was grown continuously for 70 years in an Illinois plot and the yield was down to 23 bushels an acre. Fertilizer alone increased this yield nearly six times.

Although fertilizers are usually added to the soil to be taken up through the roots, essential elements are sometimes supplied directly to the leaves, usually as a foliage spray. The best-known and one of the oldest of this method of applying essential elements is found in certain pineapple-growing regions of Hawaii, where iron is abundant in the soil but unavailable to plants. In such areas a weak solution of iron sulfate is sprayed at frequent intervals directly upon the pineapple leaves. The small amount absorbed is sufficient to maintain normal growth. Other examples of this method of applying fertilizer elements are found in the application of iron-containing spray solutions to the leaves of azaleas and pin oaks and of zinc sulfate to the leaves of citrus, tung, and pecan trees. Orange trees are sprayed with copper and manganese salts, and zinc-copper sprays are used on avocados.

Sprays containing essential elements have usually been applied as emergency treatments to overcome symptoms resulting from a deficiency of the micronutrient elements or others required in small amounts. In recent years, however, it has been found that nitrogen in the form of urea is absorbed by the leaves of some plants, especially apple trees, when applied as a spray, and this compound is now used in some apple orchards. Although the use of foliage sprays may increase, the application of fertilizers to the soil is likely to remain the preferred method of application.

Soil Testing Simple plant or soil tests that will indicate the need of specific fertilizer elements have been sought for many years, with only partial success. A fertile soil may be defined as one that will supply the essential elements, at the time and in the quantities needed by plants. The elements contained in the soil can be identified by chemical analyses, but the availability of these elements can be determined only by experiments with plants. In spite of this difficulty, much progress has been made in the development of special chemical tests that will indicate the fertilizer needs of a particular soil. These are tests for *available* nutrients rather than for their total content in the soil. In one type of test the soil is studied directly; in another, the mineral content of plants growing on the soil is examined carefully.

The development of soil tests for available nutrients starts with the determination of the responses of plants to the addition of fertilizers to a specific soil. After the plant responses to phosphorus or potassium or calcium fertilizers on this soil are known, the soil chemist develops analytical methods that will give measures of the availability of the individual elements. These methods must give results that are in agreement with the growth of plants on the same soil samples. His chemical procedures, in other words, are standardized on the basis of plant response.

Plant tests, sometimes called leaf analysis, must also be standardized by comparison of plant responses to particular soils. If it is found that plants respond to nitrogen fertilization of a certain soil, nitrogen is added to small plots in increasing quantities. The yields from the varying nitrogen applications and the total nitrogen content of selected leaves are then compared. It will normally be found that a certain low percentage of nitrogen indicates the need of nitrogen fertilization, and that a considerably higher percentage indicates no such need.

These methods are used by state experiment station workers who know the soils of their area and their responses in great detail. They test soils and recommend fertilizer programs for farmers who use thousands of dollars worth of fertilizers in large-scale production. The home gardener has a different problem. Normally he can get both recommendations and fertilizers from his local dealer in garden supplies, who can supply a fertilizer mixture adapted to local conditions and recommend an application rate. The fertilizer will usually be a complete N-P-K mixture plus other elements, as recommended for the area, and the rate of application will be heavy but safe. The gardener is more interested in results than in saving pennies on costs.

Fertilizers and World's Food Supply In the effort to increase the world's food supply until zero population growth is attained, the supply of fertilizers plays an essential role. Much is said about food for the surplus millions, but it is not sufficiently emphasized that increase in food production depends not only on high-yielding crops but also upon an increased supply of fertilizers and water. This is all the more important when it is realized that underconsumption of fertilizers, especially in the developing countries, is responsible for much of the malnutrition in the world today. The supply of the most easily depleted essential elements (nitrogen, phosphorus, and potassium) originally present in most arable soils has now been greatly reduced. To ensure future crops it is necessary that this supply be replenished by the use of fertilizers produced by the chemical industry. Millions of people depend upon this replenishment for survival. The essential elements, as components of food, are brought to large centers of population, where they end up in sewage and are lost into streams, lakes, and eventually into the oceans. Some recycling of the nitrogen and phosphorus in domestic and some kinds of industrial sewage is possible, and various methods have been proposed. The most promising of these involves chemical approaches to sewage treatment, such as precipitation and coagulation.

The supply of raw materials for the fertilizer industry seems adequate for the forseeable future—perhaps for centuries to come. We may anticipate a shortage of food long before there is a shortage of raw materials for fertilizers. Of the most heavily used essential elements, nitrogen is obtained by combining nitrogen from the atmosphere with hydrogen under high temperature and pressure to form ammonia. Immense reserves of phosphate rock, potassium salts, and limestones are available. Despite this abundant supply of raw materials and an expanding fertilizer industry, however, there are vast areas of agricultural lands, especially in developing countries, where crops remove more essential elements from the soil than are returned as fertilizers. The soil has not been replenished and perhaps never will be. The situation will become even more serious as the population increases in these countries. Increasing population will bring about increased demands for fertilizer in both developed and developing countries. This will require a greater consumption of energy by the chemical industry and will create additional problems in transportation. Increased quantities of fertilizer will also lead to increased pollution of water supplies by runoff of inorganic nutrients, notably nitrogen and phosphorus, into rivers, lakes, and coastal areas, enriching these waters and encouraging rapid algal growth (eutrophication, Chapter 19).

Organic Farming Beginning in the 1930s, an international cult—organic farmers and gardeners—has promoted the belief that mineral fertilizers exhaust the soil and that plants produced with them are deleterious to health. These same objections were raised a hundred years ago when Liebig and Lawes first proposed the use of inorganic manures and when Lawes started experiments to test them. Some of the fields at the Rothamsted Experiment Station in England have received only inorganic fertilizers for more than 100 years. These fields are highly fertile today and produce high-quality, nutritious crops.

The proponents of organic farming advocate the use of animal manures, composts, and crop residues to the exclusion of commercial fertilizers. The value of maintaining the organic matter content of soils has been discussed, and is considered again in the next chapter. A compost heap, in which leaves, old mulch material, stalks, and other plant residues are mixed with some high-nitrogen fertilizer and kept moist and compacted while the plant materials decay, is a valuable adjunct to any garden.

The organic farmers, however, overlook two important points: (1) Organic matter does not normally supply enough of the most needed elements, particularly nitrogen and phosphorus. The farmer or gardener who supplements the available organic matter with inorganic elements as required can expect the maximum returns. (2) The materials absorbed by the roots of the plants under organic farming have been changed by soil bacteria (Chapter 10) to the same inorganic ammonia, phosphate, or potassium ions that are absorbed by plants given these materials directly in inorganic forms. There is no reason to believe that the form in which the essential elements are supplied will affect the palatability or nutritional value of properly grown plants in any way. Given suitable growing conditions, the nutritional value of plants is a property of the plant and not of the medium in which they grow.

Hydroponics The growing of plants with their roots in a watery solution of fertilizer salts is an important laboratory and research technique. A number of years ago several adaptions of the

Fig. 9-22 Growth of tomato (*Lycopersicum esculentum*) in fertile soil, in nutrient solution, and in sand irrigated with a nutrient solution. All plants have made excellent growth and set large numbers of fruits. (A): fertilized soil. (B): nutrient solution. (C): pure sand with nutrient solution.

water-culture method were developed for commercial use, particularly in greenhouses, where the former practice of changing the soil in greenhouse benches each year was time-consuming and expensive. This method of growing plants was named HYDROPONICS, from a Greek word meaning water culture. Under favorable conditions, maximum yields of crops can be obtained by this method (FIG. 9-22).

Hydroponics as a commercial practice met two difficulties. The equipment required is expensive, and a very considerable knowledge of chemistry and plant physiology, together with a special skill in plant growing are required of the operator. The practice has now been abandoned in favor of a system in which the greenhouse benches are filled with a mixture of peat, soil, and sand. After each crop is removed, this mixture is pasteurized with steam, refertilized heavily, and re-used repeatedly.

SUMMARY

1. Water and mineral salts are absorbed by plants through roots. The extent of the root system is very great. In woody plants the lateral roots may extend well beyond the spread of the branches.

2. Viewed externally, the terminal portion of a young root is composed of the root apex covered by a root cap, a zone of elongation, and a zone of maturation. Increase in length of the root results chiefly from cell elongation in the region of elongation, which is much shorter in the root than in the stem.

3. The root cap protects the root apex. The cells of the root cap slough away and are renewed from within. The root hairs are extensions of the epidermal cells of the root, and constitute an important path along which materials from the soil move into the root.

4. In the root apex (apical meristem) as a whole, cell divisions are relatively infrequent. Most new cells are produced in a meristematic zone behind the apical meristem, and this zone extends to and overlaps with the zone of elongation. Not only cell division but also enlargement and some differentiation occur in this subapical meristematic tissue.

5. In the zone of maturation most cells have become physiologically mature and the chief tissue regions of the young root can be distinguished—epidermis, cortex, and vascular cylinder. In cross section, young roots may be distinguished from stems by the alternate arrangement of primary xylem and primary phloem.

6. In many plants, both woody and herbaceous, the young rootlets are associated with fungus threads that permeate the soil and extend into the cortical tissues. These mycorrhizae are of two types, ectotrophic and endotrophic. The fungus threads

replace the root hairs in the absorption of minerals and water from the soil and may have other functions.

7. In plants bearing root hairs the roots absorb minerals from the soil through the root hairs and other young tissues of the root. Ions and other soluble substances enter and permeate the cell walls (free space) of the cortex by diffusion. Movement within the free space is stopped by the casparian strips of the endodermis.

8. Permeation of the cortical cell walls brings minerals in contact with the plasma membranes of all cortical cells, which provide a large surface for the movement— by diffusion or by active transport—into the cytoplasm. Once within the symplast, minerals move by diffusion, accelerated by cytoplasmic streaming, across the endodermis and living cells of the vascular cylinder to the conducting cells of the root. Minerals may be accumulated, by active transport, from the cytoplasm into the vacuoles of the cortical cells.

9. Lateral roots originate within the parent root, in contrast to the superficial origin of leaves and branches.

10. Soils are composed of solids, liquids, and gases. Living organisms are important components of the soil solids. Soils may be classified according to their particle size and mineral composition.

11. The size of the pore spaces of the soil is determined by the size and aggregation of the soil particles. These pore spaces contain air or water, or both. Roots, as do other living parts of the plant, use oxygen in respiration, and the availability of oxygen is an important factor in root growth.

12. Soil water may be classified as gravitational, capillary, or hygroscopic. Most of the water absorbed by plants is available capillary water lying in the range between the field percentage and the permanent wilting percentage.

13. The movement of soil water by capillarity is slow or absent, and roots obtain most of their water by growing into zones where water is available.

14. The production of a dust mulch by cultivation is of little value in restricting water loss from soil. Weed control is the most important method of conserving soil moisture in cultivated crops or in fallow.

15. The elements essential to higher plants are obtained from the air and the soil. Green plants require at least 16 elements for normal growth and reproduction. Carbon, hydrogen, and oxygen are obtained from the air and water. The essential elements obtained from the soil are nitrogen, phosphorus, potassium, calcium, sulfur, and magnesium, together with certain micronutrient elements required only in minute amounts. These are iron, copper boron, manganese, zinc, molybdenum, and chlorine. Green plants also absorb many elements from the soil that are not essential to growth. The essential mineral elements have been identified by growing plants in purified, nutrient solutions.

16. Each essential element plays one or more roles in plant metabolism. Some enter directly into the composition of the plant; some function in other ways. Examples of the direct utilization of essential mineral elements are found in the chlorophyll molecule, which contains nitrogen and magnesium; in proteins, which contain nitrogen and sulfur; and in the intercellular layer, which contains calcium.

17. The essential elements most commonly added to the soil in the form of fertilizer to replace those used by crops are nitrogen, phosphorus, and potassium. These are the major nutrients of a mixed fertilizer.

18. Liming the soil provides available calcium fertilizer. It also brings about changes in the chemistry of the soil and in the kinds of microorganisms occurring in the soil. These changes are due to changes in the acidity of the soil and may be either beneficial or injurious to higher plants. A moderate soil acidity is desirable for many plants.

Natural Cycles and Plant Life

THE RELATIONSHIPS OF PLANTS to soil and water, as described in the two preceding chapters, indicate how closely the organic and inorganic worlds are bound together. Plant and animal life depends directly on about one-sixth of the naturally occurring chemical elements. The use of these elements and their compounds in metabolism and growth involves energy that came originally from the sun. In terms of the mass of the earth and the energy given off by the sun (in temperate zones about 500 kilocalories of solar energy strike every square foot of land or water each day), the amount of matter and energy involved in the life processes of plants and animals is small indeed. But because plants are so abundant, the flow of materials and energy through them is on a grand scale. More important is the fact that much of this flow of matter and energy involves interactions between living and nonliving matter.

The organic-inorganic relationships of matter have been termed CYCLES. Perhaps the earliest recognition of a cycle appears in Genesis III: "Dust thou art, and unto dust shalt thou return."

Numerous natural cycles are known: nitrogen, carbon, phosphorus, water, oxygen, and sulfur. Most of these involve changes in which the soil, plants, and animals, large and small, all play important roles. Although each of these cycles may be isolated for identification and analysis, all are markedly interdependent and operate continually and concurrently. The term *cycle* implies a return to a starting point, but this return may be a devious one in time and space. Although the broad outlines of the natural cycles are fairly clear, our understanding of the details of many of them is limited.

Some cycles involve only a physical change, as when liquid water is converted to water vapor and returns to the liquid state. This occurs during evaporation or transpiration, and condensation. Other cycles involve chemical as well as physical change. In all cycles, energy is necessary to produce the physical and chemical changes, and energy is either used or liberated as a result of them. Certain aspects of these energy exchanges will be considered later in this chapter.

239

THE WATER (HYDROLOGIC) CYCLE

The most obvious of the natural cycles involves the movement of water from the land and the seas into the atmosphere and back again (FIG. 10-1). The quantity of water of the earth is fixed and stable and is used over and over. It is present as (1) water in the oceans (97 percent of the total) and in fresh water lakes and streams; (2) water in polar ice and glaciers; (3) subsurface water, including soil and ground water; (4) water vapor in the atmosphere, and (5) water in living organisms.

Radiant energy, falling upon the earth's surface, brings about the daily evaporation of enormous quantities of water. Water evaporates from the oceans, and, on the land, from open bodies of water, such as lakes, reservoirs, and streams. Important sources of atmospheric moisture also are evaporation from moist soil surfaces and transpiration from vegetation. Water loss from these two sources is collectively termed EVAPOTRANSPIRATION. Wind currents, also caused by solar radiation, are important in the distribution of atmospheric moisture.

Some of the moisture evaporated from the continental surfaces falls back upon the land, but the direct return of this moisture to the land is much less than is generally assumed. Land air masses are relatively dry, even after absorbing evaporation from the land, and most of their water vapor does not precipitate upon the continents but is carried out over the oceans by air movements (FIG. 10-1). There, the air takes up more moisture from the ocean evaporation. These wet air masses may lose their water as precipitation over the oceans or, after being blown back over the land, their moisture is precipitated as rain or snow.

On the land masses the water that falls as annual precipitation may follow either of two pathways. In the United States about 70 percent (62 percent on a global scale) returns to the atmosphere by evapotranspiration and evaporation from open bodies of fresh water. This portion is not used directly by man, but much of it is beneficial, since water in the upper aerated soil layers is essential to the existence of the land vegetation, including forests, grasslands, and nonirrigated agricultural lands. Some 30 percent of the precipitation is diverted into stream flow, and

Fig. 10-1 The water cycle.

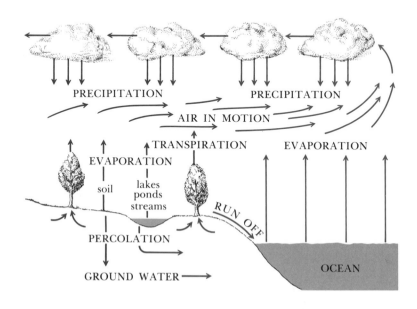

it is chiefly this water that is available directly for irrigation, for domestic uses, and for industry. Much of this stream water is polluted by use, but can be made available for repeated use by purification methods.

Streams receive water from several sources. One is runoff from saturated surface soil into lakes and streams. Another is from saturated subsurface soil which gradually percolates (seeps) downward. Some of this water may penetrate deeply to the water table, the top layer of the ground water. This ground water is a vast natural reservoir that maintains the dry weather flow of springs and streams, helps maintain the level of lakes and ponds, and is the source of water for dug and drilled wells. Ground water continually moves by seepage and slow flow to rivers and to the sea. In many parts of the world, however, ground water is tapped directly by deep wells and is an important source of water for irrigation and domestic and other uses.

The hydrologic cycle thus involves both the organic and the inorganic world and includes many subsidiary cycles and bypaths. To it are related such problems as soil erosion, water power, food supply, water pollution, industrialization, urbanization (water supply is a critical problem for most of the world's large cities); and flood control. In many parts of the world water requirements already exceed natural stream flow, and ground water is being tapped by deep wells faster than it can be replenished from precipitation. With increase in population more water will be required for all purposes, and many authorities believe that water is one of several factors that will limit the number of human beings that the earth can support.

LIFE IN THE SOIL

As we saw in the preceding chapter, soils consist not only of varying mixtures of sand, silt, and clay, but also of nonliving organic materials and living organisms. Most soils are inhabited by enormous numbers of bacteria and fungi. The bacteria alone may number as many as 5 billion per gram of soil, depending upon the amount of organic material, the temperature, the acidity, and other factors. Fungi are especially abundant in moderately acid forest soils of the temperate zones and in the soils of the humid tropics.

As a result of the activities of green plants, great quantities of essential elements are brought into organic combination. Since animals feed upon green plants, their bodies also contain these essential elements. The quantity of these elements, in forms which plant life can use, is limited. In the course of time they would become locked up in organic form and so become unavailable to future generations. But bacteria and fungi prevent this. They decompose the dead bodies of plants and animals and reduce them to water, carbon dioxide, ammonia, and simple mineral compounds. The essential elements are thus returned to the soil, or, in the case of carbon dioxide, to the air.

The reduction of the fresh organic materials to simpler end products proceeds gradually, and during the process many different compounds, known collectively as the SOIL HUMUS, are formed. Humus serves as a reservoir of mineral nutrients that are gradually liberated in a form available to plants as the humus undergoes further decomposition. Most of the decay processes take place in the upper layers of the soil—the TOPSOIL—which becomes enriched by the residues resulting from decay. Thus, the elements pass through a cycle in which they are a part successively of life, death, and decay. In the absence of this soil cycle, life on earth would eventually cease.

THE CARBON CYCLE

The carbon cycle, like the water cycle, seems relatively simple, but it involves many byways. Its main features are concerned with carbon dioxide in the atmosphere and with carbon incorporated in organic compounds. Although the proportion of carbon dioxide in the atmosphere is relatively small—usually stated as 0.03 percent by volume —because of the great mass of the atmosphere this small fraction amounts to approximately 2000 billion tons, equivalent to about 550 billion tons of carbon.

Carbon is continually being withdrawn from the air as carbon dioxide is used by green land plants in photosynthesis. The quantity of carbon dioxide withdrawn annually has been estimated at nearly 60 billion tons, equivalent to about 16 billion tons of carbon. This carbon becomes incorporated into the bodies of plants and of animals that feed upon plants. It is obvious that if it were not returned to the air and again become available for photosynthesis, the carbon dioxide content of the air would eventually be exhausted. On the basis of the quantities estimated above, exhaustion would occur in about 33 years. Fortunately, the carbon is returned to the air as carbon dioxide by various pathways. One of these is the respiration of animals. Another, more important, is respiration by green plants, including the nongreen parts of plants, such as root systems. The most important—indeed essential—source of renewal, in terms of quantities of carbon dioxide involved, results from the activities of saprophytic microorganisms of decay. As previously noted, at death the bodies of plants and animals are acted upon by huge numbers of bacteria and fungi, and carbon dioxide is released in the processes of aerobic respiration and fermentation. It is primarily the activities of these decay organisms that make enough carbon dioxide available for green plants to manufacture the food upon which life depends (FIG. 10-2). The carbon of the earth's atmosphere is thus used over and over, passing through a continuous cycle. For innumerable millenia the carbon in our own bodies has circulated throughout the world in a cycle of eternal change—from the air to the green plant, from plant to animal, from organic matter to organisms of decomposition in one eternal round. Any one of the atoms of carbon in the bread or meat that appears on the table today may have formed a part of a plant or an animal living millions of years ago, long before man, other mammals, or flowering plants appeared upon the earth.

The carbon cycle operates in the seas as well as on the land. The zone in the oceans equivalent to the zone occupied by phototrophic organisms on the land is the upper, lighted portion of the water, which extends downward from the surface as much as 600 feet, and comprises about 5 percent of the oceans. Except for some transfer to the atmosphere above and the depths below, most of the carbon is cycled in this zone. Most of the phototrophs here are minute floating algae (Chapter 19) whose bodies contain carbon as do the bodies of the animals that feed directly or indirectly upon them. Just as on the land, this carbon becomes available as carbon dioxide by the respiration of plants, animals, and organisms of decay. A tremendous quantity of carbon dioxide, estimated at some 60 times that of the atmosphere, is dissolved in the waters of the oceans. This reservoir is a great stabilizing factor in the maintenance of the concentration of carbon dioxide in the atmosphere at a roughly constant level. Over long periods of time the concentrations of carbon dioxide in the atmosphere and in the sea tend to come to equilibrium.

Increase in Atmospheric Carbon Dioxide Some of the carbon dioxide removed from the atmosphere by photosynthesis is locked up within tissues of plants under conditions that do not favor the activities of decay organisms. Extensive beds of peat and other organic materials in shallow lakes, bogs, and swamps are examples. The carbon content of our fossil fuels—coal, oil, and natural gas—resulted from comparable accumulations in plants that grew in past geologic ages. This carbon is now being restored as atmospheric carbon dioxide by combustion of fossil fuels in our modern industrial civilization, and measurements indicate that the atmospheric concentration of this gas is slowly increasing. Some of this combustion carbon dioxide will be taken up by the waters of the sea but much will

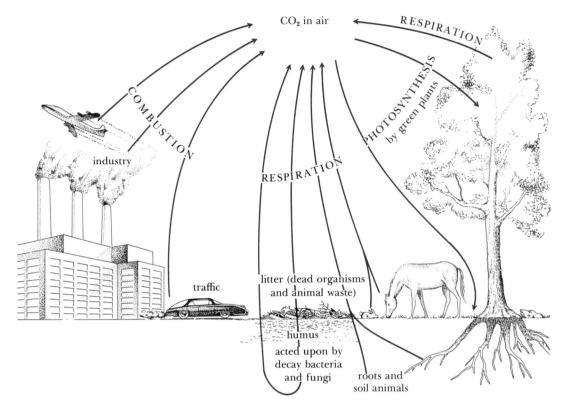

Fig. 10-2   The carbon cycle on the land.

remain in the earth's atmosphere. The possible effect of this increase in atmospheric carbon has been the subject of much discussion. Carbon dioxide transmits most of the visible radiation of sunlight, but is a strong absorber of the long wavelength infrared rays that normally radiate heat from the earth into space. Back radiation of infrared could cause an increase in the average temperature of the lower atmosphere. According to one estimate this so-called "greenhouse effect" could increase the average temperature of the earth's atmosphere by 1.1°C per century. In 300–400 years this could cause a melting of the polar ice caps, thereby raising the level of the oceans by several hundred feet, and thus destroying most of the world's major cities and flooding much of

the best agricultural land. On the other hand, there are factors that tend to cool the earth. Clearly, numerous difficulties arise in attempts to predict our future climate.

THE NITROGEN CYCLE

It has been stated that at least half the world's population suffers from malnutrition. This may result from a deficiency not only of energy-supplying foods, but also (perhaps even more so) of protein and other nitrogen-containing compounds, and some vitamins and hormones. It should be obvious that a deficiency of nitrogen leads to a deficiency of protein, which is composed of about 18 percent nitrogen. The importance of the nitrogen cycle, not only to man but to all living things, assumes greater significance when the relationships between nitrogen and protein are kept in mind.

Deficiencies of nitrogen and phosphorus are more commonly limiting factors in plant growth than are those of any other element. A deficiency of nitrogen is indeed strange when it is realized that nitrogen constitutes about 78 percent by volume of the earth's atmosphere. The air over a single acre of land contains some 35,000 tons of this element. But in an acre of soil the nitrogen may amount to only a few hundred pounds—rarely as much as 2 tons. Unfortunately, except for certain kinds of algae, the molecules of free atmospheric nitrogen are not available to green plants.

This dilemma of an abundance of an essential element that cannot be utilized directly was not generally recognized until the middle of the nineteenth century. The German chemist Liebig had assumed, in 1840, that plants use the free nitrogen of the air. Sir John Lawes in England, however, believed that this was not true, and set up field plots, some of which are still maintained, to test the point. From the results with these plots, and from careful laboratory experiments, Lawes was able to show that the free nitrogen of the air cannot be utilized directly by higher green plants.

The organic matter of the soil consists largely of plant remains and contains nitrogen in proteins or related compounds. This nitrogen, for the most part, is no more available to green plants than is the uncombined nitrogen of the atmosphere. The nitrogen supply of plants is derived from the soil, but from inorganic compounds in which the nitrogen is chemically combined with the elements hydrogen or oxygen. Such chemically combined nitrogen is produced during the decomposition of organic matter by the bacteria and fungi of the soil.

Mineralization of Nitrogen As a result of the processes of decay, the organic nitrogen of the soil normally goes through a series of changes that result first in the production of ammonia (NH_3), then in nitrite (NO_2), and finally in nitrate (NO_3). These are inorganic or "mineral" forms of nitrogen, and all are available to plants in varying degree, although nitrites are toxic when present in more than low concentrations. Most soil bacteria and fungi are able to decompose proteins with the formation of ammonia. The process is known as AMMONIFICATION. Various organic compounds are formed at the same time, from the proteins and from other organic materials present in the soil. The microorganisms utilize these decomposition products, including the ammonia, for their own growth.

If the materials being decomposed are high in nitrogen, the surplus of this element appears in the soil as ammonia. If, however, the nitrogen content of the material is low in relation to the percentage of carbon compounds, the decay organisms may use in their own growth not only the ammonia formed in decay but any other available nitrogen present in the soil. Thus, the decomposition of clover hay, which is high in nitrogen, results in a considerable addition of ammonia to the soil. Wheat straw, however, is very low in nitrogen, and the decomposition of this material in the soil commonly leads to a deficiency of available soil nitrogen because of bacterial use.

When ammonia is formed in well-aerated and not strongly acid soil, it is acted upon by nitrifying bacteria and changed first to nitrite by one kind of bacterium and then, very quickly, to nitrate by another kind. The combined process is called NITRIFICATION. Only a few kinds of nitrifying bacteria are known, but they are widely distributed in soils. The oxidation of ammonia to nitrite and nitrate is an energy-yielding process, and nitrifying bacteria use the energy obtained from nitrification in building carbon dioxide into organic compounds. Green plants change carbon dioxide to organic forms with the energy of sunlight, but nitrifying bacteria carry on similar reactions with the energy obtained by oxidizing ammonia. The organisms are thus chemoautotrophic (chemotrophic) as contrasted with photoautotrophic (phototrophic). The nitrifying bacteria are readily affected by adverse environmental conditions. Nitrification proceeds most vigorously if the soil is nonacid and well aerated, at temperatures of about 27°C (81°F), and if calcium and other essential elements are available in abundance.

Decaying organic matter was long a major source of nitrate used in the production of gunpowder. Black powder is made by grinding potassium nitrate (KNO_3), charcoal, and sulfur together and forming the mixture into granules. Nitrate was extracted from the soil of caves, where it is formed by nitrification of the ammonia produced from bat manure. It was also produced by allowing plant and animal remains, mixed with sand and soil, to decay in large compost heaps. These were wetted with blood and urine to add more nitrogen. After about two years the potassium nitrate formed in the compost as a result of nitrification was extracted and used in producing gunpowder.

The belief was once current that green plants could use only nitrate as a source of nitrogen. It has now been established that they can utilize the nitrogen from ammonia as well as nitrate. Some plants, particularly those that grow normally on acid soils, utilize ammonia in preference to nitrate. Others make a better growth with nitrate.

Nitrogen Lost to Plant Life Ammonification and nitrification do not increase the total amount of nitrogen in the soil. They merely convert the organic nitrogen in the tissues of dead plants and animals into forms that plants can reuse. The fundamental and most significant part of the nitrogen cycle is therefore the movement of nitrogen from the soil into the plant and from plant or animal tissues back again into the soil (Fig. 10-3).

There are, however, many leaks in the cycle. Nitrogen is continuously being lost from the soil. In cultivated soils the chief causes of this loss are erosion, crop removal (including grazing), and leaching. Because the nitrogen compounds present in the soil are concentrated on the surface or topsoil, erosion (Fig. 10-8) results in serious losses of nitrogen. In fact, nitrogen losses may be the most serious consequence of erosion. Estimates of total losses of nitrogen by erosion vary from an amount about equal to crop removal, to twice this quantity. Nitrogen is also removed from the land when crops are harvested, the quantity varying with the crop and the way in which it is handled. More than 100 pounds of nitrogen per acre may be removed in the grain of a good corn crop. Average losses with various crops are perhaps 25 pounds per acre per year. If the harvested crop is fed to animals and the manure handled carefully and returned to the soil, this loss is markedly reduced. Grazing animals remove only the nitrogen retained in their bodies in protein, but this may be a considerable amount. Nitrate nitrogen may be leached from the soil by percolating rain or irrigation water. Ammonia, in contrast, tends to be held in the soil and does not leach readily. If it is applied deeply (Fig. 9-21) in moderately heavy soils, reduced aeration at these levels retards nitrification of the ammonia and reduces leaching to a low value. Also, ammonia can be applied in the autumn in regions where low winter temperatures prevent the development of nitrifying bacteria. In warmer areas, where nitrification may occur in the winter, it is desirable to plant a winter crop to absorb nitrate nitrogen and prevent its loss by leaching.

Fire causes a complete loss of the nitrogen content of burned corn or cotton stalks, bean straw, and other plant materials. Nitrogen loss, in the form of oxides of nitrogen, such as nitrogen dioxide (NO_2) and nitric oxide (NO), is one of the most serious consequences of forest fires, even of ground fires that do little damage to trees. It is estimated that one such fire resulted in the loss, in the gases of the smoke, of 500 pounds of nitrogen per acre. The cost of replacing this much nitrogen can be more than the land is worth. The combustion of such organic materials as fossil fuels (coal, oil, and natural gas) and domestic refuse also releases, among other compounds, oxides of nitrogen into the air. These are of particular interest in large industrial areas,

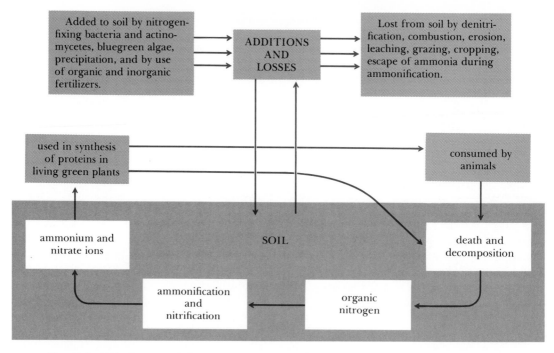

Fig. 10-3   Main features of nitrogen cycle on the land.

where they may constitute important atmospheric pollutants.

Finally we may consider the nitrogen lost to the air through the activities of microorganisms. A part of this is ammonia freed during the ammonification of proteins in surface residues, but more important is the nitrogen lost through the action of soil-inhabiting DENITRIFYING BACTERIA. These bacteria are normally aerobic, but are capable of growing anaerobically in the presence of soil nitrates and a much reduced oxygen supply. Under such conditions they employ a complicated respiratory process in which nitrates, rather than oxygen, function in terminal oxidation. Nitrogen is freed—chiefly as nitrous oxide (N_2O) and elemental nitrogen (N_2) and lost from the soil. Conditions favoring DENITRIFICATION may arise from flooding, prolonged wet weather, compaction of the soil, or depletion of

the soil oxygen by rapid bacterial growth in warm, moist soil high in organic matter. Denitrification can result in an expensive loss of nitrogenous fertilizers from good agricultural soils. It also proceeds in poorly drained, uncultivated soils, and the total amount of nitrogen losses by this process is probably much greater than is generally assumed.

The nitrogen supply is constantly being used, depleted, lost, and restored. The annual loss of nitrogen from the cultivated soils of the United States, through leaching, denitrification, cropping, grazing, and erosion, has been estimated at about 23 million tons. This loss is replaced to the extent of about 16 million tons, resulting in an annual net loss of about 7 million tons. The losses of other elements are even greater; the net annual loss of potassium from the soil is about 45 million tons, and of calcium over 55 million tons.

Nitrogen replacement It is evident that soils tend to decline in fertility and that every effort must be made to prevent and replace the loss of essential elements. The nitrogen losses are in part balanced by (1) the application of nitrogen fertilizers, crop residues, green manure, and farm manure; (2) nitrogen compounds, such as ammonia, brought down in solution in rain and snow; and (3) the activities of nitrogen-fixing bacteria and other microorganisms.

An estimated 35 million tons of commercial fertilizers, with an average nitrogen content of about 15 percent, are used in the United States each year. The atmosphere, as pointed out previously, contains ammonia and oxides of nitrogen. A considerable proportion of the oxides becomes converted by photochemical processes to nitrates, and these, together with ammonium ions, are returned to the soil in rainfall. The quantity of nitrogen returned varies from time to time and from one region to another, and may amount to 3–6 pounds per acre annually in temperate climates. This is of some significance in the maintenance of soil fertility. An important replacement of soil nitrogen is due to the activities of nitrogen-fixing bacteria that live on the roots of certain plants.

Mutualistic Nitrogen Fixation Three hundred years before Christ, Theophrastus wrote that the Greeks used crops of broad beans to enrich the soil. The Romans, according to Pliny, raised similar crops and turned them under as green manure. They recognized the value of lupines and alfalfa in the enrichment of the soil.

The crops so valuable in this way are the legumes, a large and important family of flowering plants, which includes the clovers, alfalfa, soybean, peanut, vetch, pea, bean, and many others. The value of these plants is due to the activities of specialized bacteria (genus *Rhizobium*) that live in swellings, called NODULES, or TUBERCLES, on the roots (FIG. 10-4). The bacteria penetrate the root hairs and, having gained entrance, stimulate the cells of the cortex to divide, perhaps through the action of growth hormones produced by the bacteria. The dividing cortex cells produce the nodules in which the bacteria live.

By reducing free, gaseous nitrogen obtained from the soil air, the legume bacteria are able to produce ammonia, using sugar or other compounds obtained from the legume plant as a source of energy. The process is known as NITROGEN FIXATION. No higher green plant can do this. Some of the ammonia formed is combined with more sugar from the plant and used by the bacteria to produce amino acids and proteins for their own use. Under normal conditions an excess of ammonia is formed, and this valuable compound is taken up and used by the legume plant. Both the bacteria and the legume

Fig. 10-4 Nodules of nitrogen-fixing bacteria on the roots of crimson clover (*Trifolium incarnatum*).

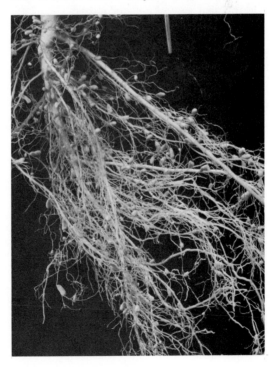

thus benefit from this mutualistic relationship.

Mutualistic nitrogen fixation proceeds in nature only when the two organisms are combined; it cannot be carried on by either the bacteria or the legume alone. If the legume possesses no tubercles, its entire nitrogen supply must come from the soil. Even when nodules are present, some of the nitrogen supply is commonly obtained from the soil.

The work of many men was involved in the series of discoveries that led to an understanding of nitrogen fixation by legumes. Lawes and others had shown that green plants could not use atmospheric nitrogen directly. However, it gradually became recognized that the legumes differed fundamentally from other green plants with respect to their source of nitrogen. Experiments showed that the legumes could acquire more nitrogen than was present in the soil, seed, and water. Finally, in 1888, two German botanists, Hellriegel and Wilfarth, reported the results of extensive experiments in which they showed that legumes obtain nitrogen from the air through the activities of microorganisms that live in nodules on the roots.

In one of their experiments they grew peas and lupines in sterile sand. To some pots they added an extract of fertile soil in which these legumes had previously been grown successfully. In such pots the legumes produced nodules, grew vigorously, and gained in nitrogen. This extract, of course, contained bacteria of the proper kind. Peas and lupines in sand to which the extract was not added grew poorly and soon died. The same experiment was carried out with oats, a nonlegume, but the addition of the soil extract produced no results. That the effect of the soil extract was not due to the combined nitrogen it contained was shown by heating it to kill the bacteria. Plants to which such a heated extract was added behaved in the same way as those in the sterile sand alone. Thus the significance of the nodules, the presence of which had long been known, was explained.

Legumes are important because of their own values and for improving the yields of other crops. Nonleguminous plants grown in combination with leguminous plants (timothy grown with clover or alfalfa, for example) may benefit from the association. It is probable that the nitrogen-rich roots and tubercles of the legume slough off and decay in the soil, eventually releasing nitrogen in inorganic compounds. The stubble and roots of many perennial legumes left behind after the plants are harvested for hay are sufficiently rich in nitrogen to aid in maintaining soil fertility and to increase the yield of later crops.

The greatest benefit to the soil and to succeeding crops results, however, from plowing under the entire leguminous crop as green manure. This practice adds appreciably to both the nitrogen and organic matter content of the soil. The organic nitrogen, following mineralization, becomes available to green plants. The stimulating effect of a leguminous green manure upon crop growth is usually apparent for several years. In many fertile soils of the temperate zone, however, the value of the lost grain or other crop is now so high that it is commonly cheaper to buy ammonia (FIG. 9-21) than to grow a green manure crop.

The amount of nitrogen fixed annually by legume bacteria varies with the nature of the soil, the species of legume, and other factors. A few nodule organisms have been found to fix as much as 250 pounds of nitrogen per acre annually, but the average is perhaps 80 pounds per acre. It is estimated that some $5\frac{1}{2}$ million tons of mutualistically fixed nitrogen from cultivated legumes are added annually to the soils of the United States.

Mutualistic nitrogen fixation will proceed only when certain bacteria are present. Hence, it is usually advantageous to introduce the bacteria when the leguminous crops are planted, for they may not be present in the soil. The nodule bacteria are added to the seeds at the time of planting. The bacteria are first grown in large numbers in a nutrient medium, or food supply, under conditions that exclude other bacteria or molds. Such a growth is termed a PURE CULTURE. Cultures of nitrogen-fixing bacteria, combined with some kind of carrier, are sold under various trade names by commercial seed firms. Mixing the bacteria with the seed just before planting, a procedure called INOCULATION, ensures their presence at or near the roots at the time of germination (FIG. 10-5). Research has disclosed several species and various strains or varieties of the nitrogen-fixing bacteria. A given strain will not infect all kinds of legumes but only those within particular groups. The groups that require different kinds of bacteria are seven in number: alfalfa, clover, pea, bean, lupine, soybean, and cowpea. The addition of lime to the soil favors bacterial growth by neutralizing soil acidity. Bacterial development and nodule formation are best at or near neutrality.

The use of legumes to improve the soil is important in the tropics and subtropics as well as in the temperate zone. The species employed, however, are generally different, and are adapted to the different climates and soils. In the tropics, tree legumes are frequently grown in plantations of cacao and coffee, where they provide light shade and add combined nitrogen to the soil (FIG. 10-6).

The preceding discussion has been concerned primarily with the importance of legumes in agriculture. However, only about 200 species of this large family (about 14,000 species) are cultivated, and we cannot ignore the significance of the great numbers of species and individuals scattered over the earth in varied habitats, such as seashores, grasslands, savannahs, equatorial forests, thorn woodlands, and semidesert scrub. The legumes of natural habitats include perennial herbs, woody vines, shrubs, and trees. Not all species are nodule-bearing, but probably 90 percent of them are, and they are believed to contribute greatly to the supply of soil nitrogen that, after mineralization, is available to green plants.

Although mutualistic nitrogen fixation is best known in the legumes, it is being found in an increasing number of nonleguminous flowering plants. Plants belonging to some 13 genera and more than 100 species—all woody dicotyledons—

are known to develop nodules in which nitrogen is fixed. The microorganisms involved are widely held to be actinomycetes, threadlike organisms intermediate in some respects between true fungi and true bacteria (Chapter 20). Prominent among these nonleguminous flowering plants are alder (*Alnus* sp.), Australian pine (*Casuarina* sp.), and sweet gale (*Myrica gale*), a shrub of northern North America and Eurasia. The growth of sweet gale, with and without inoculation with its nitrogen-fixing organism, is shown in FIG. 10-7. Because of their ability to add fixed nitrogen to the soil, the ecological importance of these nodule-bearing plants is thought to be considerable in localities low in combined nitrogen, such as recently deglaciated areas, sand dunes, and the margins of rivers and lakes. Planting for soil improvement has not been extensively practiced, but sweet gale is showing promise in increasing growth of young conifers when they are planted together in northern areas.

Nonmutualistic Nitrogen Fixation Nitrogen fixation is also carried on by free-living microorganisms—bacteria, blue-green algae, and possibly some actinomycetes and fungi. The free-living nitrogen-fixing bacteria comprise species belonging to some 15 genera—*Azotobacter, Clostridium,* and others—most of which are heterotrophic, obtaining their food from soil organic matter. In most cultivated soils of the temperate zone, however, these forms are considered to be of little importance. Their numbers may be small, and they require a large amount of energy for their nitrogen-fixing activities. Typically, they fix about 1 pound of nitrogen for each 100 pounds of carbohydrate consumed. The process may be more important in the tropics, but in general the significance of these forms is that they fix small quantities of atmospheric nitrogen over long periods of time, thus contributing on a world-wide basis to the supply of combined nitrogen in the soil and waters.

Some experimental evidence supports the occurrence of nitrogen fixation by certain free-

Fig. 10-5 Inoculated and uninoculated soybeans. The plant on the left was grown from uninoculated seed; that on the right from seed inoculated with soybean bacteria.

Fig. 10-6 Coffee (*Coffea arabica*) planted in the shade of a leguminous tree, *Albizzia endocarpa*.

Fig. 10-7 Plants of *Myrica gale* after one season's growth in a culture solution free of combined nitrogen. Plants on the left with nodules; those on right, without.

living actinomycetes and fungi, including some soil yeasts. The subject is controversial, however, and the significance of such fixation is unknown.

Nitrogen fixation by blue-green algae (*Anabaena, Nostoc,* and others Chapter 19), on the other hand, is thought to be among the more important ways by which molecular nitrogen is made available to both terrestrial and aquatic forms of life. The importance of these algae in nitrogen fixation has been much underestimated. The process is not found in all blue-greens, but seems restricted to about 14 genera with certain structural features. More than 40 species have been found, in pure culture studies, to fix nitrogen. These and other blue-greens are found in many habitats such as deserts and semideserts, cultivated soils, grasslands, rock surfaces, fresh-water lakes and swamps, salt marshes, and the sea. The nitrogen compounds produced are released into the environment, either during life or after death and mineralization. Under favorable conditions the rate of nitrogen fixation by blue-greens can be comparable to median or good rates of fixation by legume bacteria.

In many semidesert grasslands, notably in western North America, blue-green algae occur in crusts on the soil surface. These organisms are a source of organic matter, improve infiltration of rain, and add combined nitrogen to the soil. The algae are highly drought-resistant and become active in nitrogen fixation within hours after a light rain. For thousands of years blue-green algae have probably been the chief source of combined nitrogen in tropical rice fields. In east and southeast Asia, where rice is grown in flooded fields, production in many areas is maintained at a moderate level without fertilization. Nitrogen is removed from the soil by leaching, denitrification, and cropping, but is restored by the nitrogen-fixing activities of blue-greens that grow luxuriantly in the rice paddies or wet fields. Finally, nitrogen-fixing blue-greens are considered an important source of combined nitrogen in the oceans, although a full assessment of their significance is difficult. It is certain that a number of species of marine blue-greens fix nitrogen, and "blooms" of blue-greens have been observed over hundreds of square miles of oceanic surface in many latitudes.

An indication of the magnitude of the quantity of nitrogen restored to the soil by living organisms is found in the following data. An estimate for the world gives a total of 100 million tons of nitrogen added each year. Of this, 90 million tons is estimated to be from biological fixation, and 10 million tons from fertilizers. Nitrogen fixation by living organisms may be regarded as second only to photosynthesis in its importance to the maintenance of life on earth.

SOIL EROSION AND CONSERVATION

All land surfaces tend to erode away under the influences of weathering, water, and wind. Soil, as contrasted to rock or even to sand, is particularly susceptible to erosion by rainfall and running water. The splash of raindrops loosens the surface particles and fills the pores of the soil so that infiltration of the water is decreased and runoff is increased. This runoff carries the loosened soil particles with it. As the running water collects into small streams, these cut deeper, forming gullies (Fig. 10-8). Erosion removes the topsoil with its generally desirable texture and its reserve of nitrogen compounds; it leaves gullies that interfere with working the land; and it may result in the removal of all soil down to bedrock.

Soil erosion is basically dependent upon two factors, the amount of runoff and the protection of the surface of the soil by plants and plant materials. A dense grass sod protects the surface of the soil so that the pores remain open, more rainfall penetrates the soil, and less runs off. Also, the plants slow the runoff of unabsorbed water and protect the soil particles from the moving water. Plant roots are of some protection and aid in binding the soil, but a dense soil cover of living or dead plants is more important than

Fig. 10-8   Uncontrolled erosion. The vegetation in the foreground does little to prevent further erosion.

the roots. In forests the protection is due mainly to the cover of leaves and leaf mold on the soil surface rather than directly to the trees themselves. Paradoxically, desert soils frequently erode more than those of humid regions. Desert soils are largely unprotected by plants, the soils (not sands) generally absorb water very slowly, and desert rains, although infrequent, tend to be torrential, with most of the rainfall running off rapidly and causing serious erosion. Sands or very sandy soils absorb rainfall so rapidly that water erosion is usually not a problem. The lack of any binding structure, however, introduces serious problems of wind erosion. In the United

States, wind erosion on bare-fallowed dry-land soils in the 1930s forced the abandonment of the former practice of dust mulching in favor of stubble and clod mulches that resist wind action.

The undisturbed soils of humid regions are normally thoroughly covered with a forest or prairie vegetation that reduces erosion to a negligible level. Cultivation, burning, and overgrazing, however, can bring about a radical change. The early settlers in this country found dense forests, clear streams, and rich soil in the east, and mile after mile of grasslands covered with tall native grasses in the Middle West. There was little or no erosion. But as settlement and agriculture moved westward, accelerated erosion followed the ax and the plow. The covering of trees and grasses was removed and the soil was exposed to the destructive effects of wind and rain. Continual planting of soil-depleting crops robbed the topsoil of its organic matter and its fertility. In many areas, unprotected soil was carried away by water erosion (FIG. 10-8), thereby both thinning the slopes and the hillsides and burying fertile valley soils under deposits of eroded subsoil.

The importance of soil erosion, and the necessity for soil conservation, received little attention in the United States until the decade 1930–1940, when conservation agencies and public programs were first established. During this period the results of surveys of erosion by water and wind are impressive. It was estimated that 3 billion tons of soil were annually being lost from overgrazed pastures and from croplands. On 775,670,000 acres, or 41 percent of the total land area of the United States, erosion had been moderate; that is, from one-quarter to three-fourths of the original surface soil had been lost. On 12 percent, or 225,000,000 acres, erosion had been severe, with a loss of more than three-fourths of the surface soil. About 57,203,000 acres, or 3 percent, had been destroyed for cultivation. In many parts of the country, accelerated erosion had become an economic and a social menace.

Substantial progress has been made in soil conservation since the 1930s, but the prevention of soil erosion is still an important aspect of conservation programs. Soil conservation is of concern not only to rural dwellers but also to those who live in cities. Society as a whole, as well as the farmer, is affected by destructive floods, dust storms, muddy streams, silted reservoirs, and eroded hillsides.

Methods of Soil Conservation The conservation of the soil requires that soil fertility be maintained and that erosion be prevented or kept to a low level. Soil fertility depends in part upon management, as in preventing erosion and in

growing legumes. Today, however, the emphasis is shifting to the use of commercial fertilizers to maintain fertility. The prevention of erosion is more a matter of soil management, although good soil fertility is needed to grow an effective vegetative soil cover and to ensure the maintenance of the organic matter content of the soil.

Organic matter is not directly necessary for the growth of plants. Maximum growth of many plants has been obtained in solution or sand cultures (Chapter 9) where no organic matter is present. In soils, however, organic matter serves as a reservoir of plant nutrients, particularly nitrogen, sulfur, and phosphorus. These materials may be held in a stable form, but are made available by the decomposition of the organic matter as they are used by plants. In general, the greatest value of organic matter in soils is its action, discussed in Chapter 9, of improving the structure of fine-textured soils by its effect on the aggregation of the soil particles into moderate-sized groups. This aggregation leaves macropores through which rainfall can enter the soil, thus increasing infiltration and soil aeration and decreasing runoff. Aggregates also resist the hammering, soil-sealing action of raindrops and are not so easily moved by running water as the individual fine particles.

Cultivated crops—corn, tobacco, cotton, and potatoes—commonly known as row crops, can bring high returns per acre. Because of the value of these crops, relatively large quantities of fertilizers can be applied to them to maintain soil fertility. The production of row crops, however, leaves the soil greatly exposed to the direct action of rainfall and tends to deplete the organic matter of the soil, thus destroying soil structure and leaving the soil in a form to erode readily. Contour farming and strip cropping help to reduce erosion because the rows and strips on contours across the fields (FIG. 10-9) slow the runoff of rainfall and increase infiltration. Contour terraces have been extensively used in the southern United States to carry runoff on a gentle grade to a grassed waterway at one edge of the field. Completely effective erosion control on

Fig. 10-9 Contour strip cropping, South Carolina. Contoured rows and alternate strips of grass or grain increase infiltration of rain and decrease runoff and erosion.

any but very gently sloping land, however, requires an established, undisturbed forest or the dense cover of a vigorous grass or grass and legume sod. Grass forms a better cover than legumes on fertile soil, but a mixture of legumes with the grass is desirable because of the inexpensive nitrogen supplied to the vegetation by the bacteria in the nodules of the legume. Proper management of these hay or pasture lands will include the control of weeds or brush and the control of grazing so as not to weaken the sod and reduce the cover.

The per-acre returns from grassland are rela-

tively low. Fortunately, the large quantities of active organic matter remaining in the soil for one or two years after a sod is plowed up can maintain soil structure and infiltration and reduce erosion sufficiently so that a more valuable row crop may be grown. The farming system is then changed to one of CROP ROTATION. The production of different crops on the land in succeeding years helps in controlling weeds, plant diseases (Chapter 22), and insects, as well as in maintaining fertility, decreasing erosion, and increasing crop returns on cultivatable land. Land that is too steep or too rough for cultivation should be left in permanent pasture or converted into forest. Land that is somewhat better may be farmed in a four- or five-year rotation of a row crop, as corn or cotton, grain seeded with a grass-legume mixture, and two or three years of hay and pasture. Still better land may be planted in intensive row crops for two years, or be used to grow a crop of peas or soybeans before being returned to sod for a period of only one or two years. Finally, on the best, nearly level, corn-belt soils, the farmer may plant corn for a number of years in succession, depending upon heavy fertilization, chemical control of insects, and the return of the organic matter of his corn stalks to maintain yields.

When soil conservation is not practiced, the soil becomes depleted in essential minerals and organic matter. As depletion increases the plant cover becomes poorer and the soil is more exposed to erosion. This erosion removes the topsoil and the destruction is accelerated. Badly depleted and eroded land may still be reclaimed, however, if the subsoil has sufficient depth and a workable texture. Gullies may be leveled and the land fertilized and planted in grass and legumes or in forest. The recovery of an abandoned cotton field after planting in forest is shown in Fig. 10-10. Spectacular improvement can be obtained on some soils, but continuous care is required to maintain the land in a high state of productivity.

The loss of nutrients not only results in lower crop yields but also may affect the health of animals that feed upon plants grown on exhausted soils. Bone diseases of cattle are recognized in regions in which the soil is deficient in calcium and phosphorus, and other deficiency diseases result from an inadequate supply of iron, cobalt, manganese, copper, and iodine. The effect of soil-nutrient deficiencies upon human health is a subject of importance that is being studied extensively.

The soil supports man and all other animals and all plants that live upon the land. Without soil, the land areas of our earth would be barren wastes. The conservation of the soil, the establishment of a stable, scientific agriculture, and the efficient use of land are important international problems. The results of unchecked soil erosion are evident over great areas of Africa, China, the Middle East, the Mediterranean region, India, and Latin America. In many parts of the world rich farmlands have been replaced by deserts of rock and shifting sand. Uncontrolled soil erosion can lead to extinction of a people or a culture as certainly as the destruction of modern warfare.

THE ECOSYSTEM

Numerous examples have been noted of relationships among living organisms and their nonliving environment. These interconnections may be brought together by a brief consideration of ecological systems (ecosystems). An ECOSYSTEM is an arbitrary unit consisting of an assemblage of organisms—plant and animal—together with their nonliving environment. These organisms react with each other and with their environment.

This conceptual unit may have any size. For example, the BIOSPHERE, that part of our planet including air, water, and soil in which life can exist, could be considered a single ecosystem. For such an ecosystem we could examine such problems as the world-wide circulation of carbon or phosphorus. Two general kinds of ecosystems are recognized: terrestrial, such as a tundra, and aquatic, such as an ocean. Workable considerations, however, usually require that large ecosystems be broken down into smaller units for study, such as a forest, a pond, an expanse of native grassland, a weed-covered field, a salt- or fresh-water marsh, an aquarium with animals and green plants, a manned space capsule—any circumscribed area, natural or modified by man, that includes organisms and their nonliving environment.

A model of an ecosystem is shown in Fig. 10-11. It includes a BIOTIC portion or phase (phototrophs and heterotrophs) and the physical

Fig. 10-10 (Left): this eroded and abandoned Alabama cotton field was planted to loblolly pine (*Pinus taeda*) in 1927. (Right): picture taken from same spot 42 years later.

Fig. 10-11 Generalized diagram of an ecosystem.

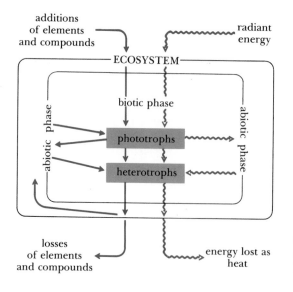

environment. The latter could be considered as composed of (1) physical factors, such as radiant energy, air and soil temperatures, wind, and fire; and (2) the ABIOTIC PHASE, which contains the nonliving substances, elements or compounds, that directly affect the biotic portion. These include such materials as water, minerals, atmospheric gases, and nonliving organic matter. The biotic phase contains all the living organisms. For example, in a mature deciduous forest the biotic phase would include the trees and shrubs, and the herbs, ferns, and mosses of the forest floor; the insect pollinators and feeders upon leaves, seeds, and fruits; the wood, bark, sap, and nectar feeders; the consumers of dung and carrion; the mites, sow-bugs, snails, beetles, ants, and other insects; the millipedes, worms, protozoa, and decay bacteria and fungi of the forest litter and humus; the plant and animal parasites; and the mosquitoes, wasps, centipedes, spiders, birds, and mammals. More than 3000 species of organisms might live together within this single ecosystem. All of these, together with the physical environment, may be regarded as a single interacting unit and studied as such.

Energy Flow through Ecosystems Essentially all of the radiant energy of the earth is derived from the sun. Some is released by the decay of radioactive elements, but this is negligible compared to the sun's energy. The radiant energy reaching the earth's outer atmosphere in 1 minute is equal to that contained in 375 million tons of coal. About 60 percent of this energy is reflected or absorbed by the atmosphere and the clouds, so that the amount of energy reaching the surface of the earth is considerably reduced. But only a part of the remaining energy is directly available to living things. Much of it warms the rocks and the soil, or is absorbed by or reflected from the seas. This portion of the solar energy is important in bringing about the evaporation that sets the water cycle in motion. Water is made available to land plants, to animals, and to cities in this way, and we benefit indirectly by obtaining water power for hydroelectric use and other purposes.

Of the total amount of solar energy that reaches the earth, nearly 1×10^{18} (one billion billion) kilocalories are combined annually in products of photosynthesis. This is about 0.2 percent of the total sunlight energy reaching the earth's outer atmosphere, but is approximately equivalent to that required in the production of some 200 billion tons of sugar. Even under favorable conditions, plants rarely use more than 2 percent of the energy of sunlight in photosynthesis. Radiant energy thus becomes converted into chemical energy in the bodies of green plants. Transformations of energy, however, unlike those of matter, do not occur in cycles. Following biological fixation, energy moves in a constant downhill flow, during which it is converted into thermal (heat) energy and ultimately dissipated into space. According to the law of conservation of energy, energy cannot be destroyed; it can, however, be lost from living systems in the form of thermal energy.

Food Chains The energy fixed in photosynthesis flows through the biotic portion of the ecosystem, from organism to organism, in a number of FOOD CHAINS, each composed of a sequence of organisms in which each kind of organism is food for a later member of the sequence. Phototrophic organisms, ranging in size from one-celled aquatic plants to giant forest trees, lie at the bases of food chains, and are the PRODUCER organisms. The energy-yielding and tissue-building materials of the producer organisms can then be utilized by CONSUMER organisms such as animals and heterotrophic plants. However, since the phototroph not only manufactures organic compounds but also consumes them, it is a consumer as well as a producer—a fact not always recognized. Herbivorous animals are important primary consumers of plants and plant products. These animals include all the hoofed animals such as the cow, sheep, and deer, and the rabbits, squirrels, and most other rodents. There are thousands of plant-eating insects and many seed- and fruit-eating birds. More than half of all the kinds of animals, vertebrate and invertebrate, obtain their food directly from plants. Some of these HERBIVORES, in turn, are fed upon by CARNIVORES (predators), consumers of animal food only, and OMNIVORES, like man, that feed on a mixed diet. Carnivores may be secondary or tertiary consumers. Thus the herbivorous aphid may be eaten by beetles, which may be consumed by small birds, and these in turn may be prey for larger predators, birds or mammals. Any organism in a food chain may be preyed upon by parasites, external or internal. To those mentioned above may be added SAPROVORES, small animals that feed upon dead organic matter such as plant and animal remains of the forest floor. Saprovores include earthworms, millipedes, small crustaceans, snails, termites, mites, thrips, and many other forms. These may lie at any consumer level between the first and the last.

Insects, birds, worms, rodents, reptiles, amphibians, fish, and all other animals have their place in food chains, as primary consumers, as intermediate links in a chain, or as more remote consumers of the products of the photosynthetic cell. At death, the bodies of all organisms, producers and consumers, become food for *decomposers* (saprophytic bacteria and fungi), which convert them into simpler compounds of the soil and air.

Food chains are even more complicated than indicated above. We might consider a specific food chain: perennial grasses and other herbs→meadow mouse→weasel→owl. The producers are fed upon not only by mice but also by insects and other invertebrates; the mouse may be preyed upon not only by the weasel but also by other predators, as the short-tailed shrew; the weasel may consume not only mice but also insects, and may be devoured by hawks as well as by owls. Because of these and other complexities, it is customary to refer to the totality of interconnecting relationships of a community as the FOOD WEB (FIG. 10-12).

From the standpoint of energy flow, the various parts of a given food chain are termed TROPHIC LEVELS. Thus, green plants represent the first trophic level, herbivorous animals the second trophic level, primary carnivores the third trophic level, and so on. At each trophic level a large part of the energy is dissipated in respiratory activity and passes out of the ecosystem as heat. Some of the stored energy, also, is not assimilated or even ingested.

As a consequence, a unit amount of food initially incorporated into plant tissue will support progressively smaller amounts of living protoplasm at higher trophic levels. About 10 calories in feed must be eaten by a cow or sheep, for example, to produce 1 calorie in meat. Because of this continued decrease of energy at each trophic level, ecosystems are restricted in the number of trophic levels they contain—rarely more than five. Eventually all of the energy fixed in photosynthesis is returned to the environment as heat energy at various trophic levels, and so all living things may be regarded as stages in the conversion of radiant energy of the sun into thermal energy. Whenever photosynthesis occurs, it is new energy—energy from the sun— that is stored. Carbon and other elements circulate and are used repeatedly, but energy is constantly lost as waste heat and cannot be recaptured in the cycles of matter. Millions of years from now, when the sun burns out, this constant supply of energy will fail and life will cease to exist on earth.

Fig. 10-12 Simplified diagram of a food web.

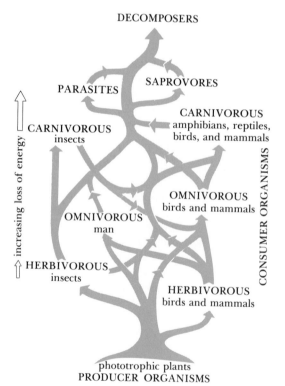

DECOMPOSERS

PARASITES

SAPROVORES

CARNIVOROUS insects

CARNIVOROUS amphibians, reptiles, birds, and mammals

increasing loss of energy

CONSUMER ORGANISMS

OMNIVOROUS birds and mammals

OMNIVOROUS man

HERBIVOROUS insects

HERBIVOROUS birds and mammals

phototrophic plants
PRODUCER ORGANISMS

Some Interrelationships: Biotic Phase and Nonliving Environment As a result of the activities of decomposers, the constituent elements of plant and animal matter are returned to the abiotic phase. The nongaseous mineral elements are taken up by the roots of the plants and recycled through food chains. This recycling neither adds to nor subtracts from the total supply of mineral elements in the ecosystem. In any ecosystem, however, the abiotic components are subject to additions and losses from outside the system. Additions to the abiotic phase include precipitation, nitrogen in rain and snow and from biological fixation, atomic fallout, minerals released by the weathering of underlying rocks, and, in man-modified ecosystems, irrigation water and fertilizers. Water is lost in streams and by underground drainage and evapotranspiration, while minerals are lost in dust carried away by wind, as dissolved substances, or in particulate matter in moving water, and through harvested crops. Essential elements with a prominent gaseous phase, such as carbon and oxygen, may not cycle in any given ecosystem but rather may be lost to the atmosphere and replaced from that source.

Although the components of an ecosystem vary and are in a continual state of change, most natural ecosystems are relatively stable in terms of maintaining levels of productivity and an equilibrium between organisms and the environment. Changes occur, but slowly. But man has modified ecosystems on a vast scale and within relatively short periods of time. In many man-managed ecosystems—lands under cultivation, forests properly harvested by selective cutting—stability and productivity have been maintained for long periods of time. But man often modifies ecosystems to his own detriment because he fails to consider the total and long-term effects of his actions. Clearing forest on steep, poor land, inept farming practices, and the indiscriminate use of fire affect natural cycles, causing accelerated losses of mineral nutrients and a lowering of productivity. For example, an undisturbed deciduous forest in New Hampshire gained nitrogen from rain and snow at the rate of about 4 pounds per acre per year over a period of years. Normally the forest loses very little nitrogen in drainage waters. However, when the forest was cut and the remaining vegetation sprayed with herbicides to prevent plant growth, the area lost 150 pounds of nitrogen per acre per year in drainage water.

The discharge of sewage into streams, lakes, or oceans not only causes drastic changes in aquatic ecosystems but also results in a continuing loss of minerals from terrestrial ecosystems. It is estimated that the loss of phosphorus alone into the seas, as sewage from great cities and in mineral-loaded runoff from large land areas, amounts to 3.5 million tons a year. Phosphorus tends to become buried in the bottom of the oceans and thus is unavailable. The draining or filling of swamps and marshes may produce arable farm land or room for urban expansion, but it also results in the loss of food and shelter for wildlife and the loss of important water-storage areas. The introduction of pesticides into ecosystems may kill not only unwanted insects but also insect predators and pollinators, and may result in outbreaks of new insect pests. Radioactive fallout or wastes added to ecosystems may be concentrated at some levels in food chains to the ultimate detriment of man. The introduction of foreign species into native ecosystems may result in drastic imbalances such as those that occurred following the introduction of the rabbit into Australia. And the destruction of predators, in a predator-control program, may result in herbivores overrunning the food supply of their habitat.

It has often been suggested that we may look to the soils of the tropics, particularly to those of the forests, as a source of food for the world's excess millions. But many tropical soils are deficient in minerals or badly eroded, and their forests are already exploited in large degree. The ill effects of man's interference with an established ecosystem are nowhere better illustrated than by the results of land clearance in many tropical regions of high rainfall. A tropical forest is delicately balanced; the forest products are returned to the soil and minerals are recycled—it is largely a closed system. When the forest is cut, the soil is exposed to the tropical sun and the high temperatures accelerate the decomposition of organic matter. The heavy tropical rains quickly wash away the minerals and after a very few years the soil becomes unproductive and cultivation is abandoned.

Man, like other organisms, is affected by the environment in which he lives, for no single living thing exists independently in nature. The components of his ecosystem—light, air, water, soil, plants, animals, and men—are closely linked, and any disturbance in one sets up a chain reaction that affects others. Man cannot afford deleterious chain reactions, not only because the resources of the earth are not inexhaustible, but also because the massive growth of the human population demands even more wisdom and skill in managing the resources we have. Man must pay careful attention to his ecosystem as a whole and hope to achieve an equilibrium that will allow his own survival.

SUMMARY

1. Many major series of transformations in the natural world, involving organic and inorganic matter, occur in cycles.

2. The water cycle involves the movement of water as water vapor from the land and the seas into the atmosphere and the return of this water as precipitation. Some two-thirds of the precipitation upon the land sustains the wild and cultivated plants of the earth's surface, and is returned to the atmosphere by evaporation and evapotranspiration. About one-third contributes to stream flow. Water not withdrawn from streams or evaporated goes to ocean storage before it is returned to the atmosphere.

3. Most of the essential elements taken into the plant are returned to the soil through the activities of decay bacteria and fungi. These elements therefore pass through a cycle in nature.

4. The concentration of carbon dioxide in the atmosphere is small, and is gradually withdrawn by green plants in photosynthesis. It is returned to the air by various pathways, the most important of which is the respiration of bacteria and fungi of decay.

5. The nitrogen cycle is more complex than the carbon cycle, for atmospheric nitrogen is not directly available to green plants except to some kinds of blue-green algae. The nitrogen in organic matter becomes available only after it has been acted upon by living organisms in the soil or water. The nitrogen used by plants comes in a form combined with hydrogen or oxygen.

6. Organic nitrogen in the soil is converted to ammonia by the ammonifying bacteria. Ammonia is oxidized to nitrates by the nitrifying bacteria. The energy obtained by the nitrifying bacteria is used to manufacture food. Ammonification and nitrification do not increase the supply of nitrogen in the soil but convert it into a form that green plants can use.

7. Nitrogen is lost from the soil by erosion, denitrification, cropping, grazing, leaching, and burning.

8. Nitrogen is restored to the soil by the addition of fertilizers and organic matter, by snow and rain, and by the activities of nitrogen-fixing bacteria, actinomycetes and blue-green algae. The most important of these bacteria are mutualistic, living in the roots of legumes. The actinomycetes are also mutualistic, living in the roots of nonleguminous flowering plants.

9. Accelerated erosion often follows the cultivation of the soil. Methods of preventing erosion and of maintaining soil fertility include planting to hay or pasture grasses, reforesting, crop rotation, fertilization, and contour farming.

10. The many interactions among living organisms and with their nonliving environment are all included in the ecosystem concept.

11. Nearly all the energy on earth is derived from the radiant energy of the sun. A fraction of 1 percent of the tremendous amount that reaches the earth's surface is converted by green plants into chemical energy and stored in carbon compounds.

12. This stored energy is gradually dissipated into space in the form of heat, following the respiration of living things—plants, animals, and decay organisms. The radiant energy of the sun thus flows through the bodies of living things in a constant "downhill" direction. Energy flows through food chains, the various parts of which, from the standpoint of energy, are called trophic levels. Because of the loss of usable energy at each trophic level, food chains are restricted in the number of trophic levels.

11

Vegetative Reproduction

PLANTS DISPLAY two prominent phases in their life cycles: a vegetative phase, composed of root and shoot growth, and a sexually reproductive phase. Sexual reproduction in seed plants involves the fusion of two cells to form a fertilized egg, which develops into an embryo within a seed. When the seed germinates, the embryo gives rise to a new plant. In the formation of the seed, however, the activities of the vegetative phase cannot be ignored. Seed formation is made possible only because of the numerous activities of the plant body. The manufacture of food, the absorption of water and minerals, and many other processes lead to seed formation and contribute to it.

Most higher plants produce seeds after fertilization, and many kinds are also able to reproduce vegetatively—that is, they can multiply by means other than seeds. Vegetative reproduction is also known as asexual reproduction and vegetative propagation or multiplication. Reproduction by vegetative means occurs widely in nature and is artificially carried out on plants by man. Our dependence upon vegetative propagation for the production of food and other economically important plants is indicated in Table 11-1 (page 264).

SOME GENERAL FEATURES OF VEGETATIVE REPRODUCTION

Production of Clones Artificial plant propagation commonly involves the separation of a portion from the parent plant. Almost any organ of the plant may be capable of vegetative reproduction. When plants are propagated by this method, adventitious roots or buds—that is, buds or roots produced in an unusual position— usually develop.

The term CLONE is used in horticulture to designate all the descendants of a single plant, produced by vegetative methods. All of the individual plants of a named variety of apple, such as the Delicious or McIntosh, comprise a clone. So do the individual plants of varieties of strawberries, peaches, roses, grapes, carnations, potatoes, and many other plants produced by grafting, budding, cutting, or other vegetative methods. An excellent example of a clone is the common orange day lily, a widely cultivated perennial known in Europe for more than four centuries. The plant is propagated by dividing the basal parts of old plants. None of the plants of this clone produce seed when pollinated with their own pollen, although it is possible to pollinate them with pollen from other day lilies, thus producing hybrid seed and a new variety. The great numbers of plants of the orange day lily, cultivated or escaped from cultivation, in this and other countries, are all branches of one individual plant and so are units of a clone. Still older clones are known in European varieties of grapes that have been propagated vegetatively for more than 2000 years. Since the individuals of a clone are all parts of the same original plant, grown by mitotic cell division, they may be expected to possess the same heritage.

TABLE 11-1
Methods of Propagating Some Important Cultivated Plants

By seed	By vegetative methods
Asparagus	Avocado
Buckwheat	Banana
Cereals (wheat, rice, corn, barley, sorghums)	Citrus fruits
	Date
	Fig
Coconut	Manila hemp
Coffee	Manioc (cassava)
Cotton	Nut crops (English walnut, pecan)
Flax	
Legumes (peas, beans, peanut, soybean)	Pineapple
	Pome fruits (apple, pear)
Opium poppy	Potato, white and sweet
Sugar beet	Sisal hemp
Tobacco	Small fruits (strawberry, grape, cranberry, blackberry)
Truck crops (tomato, lettuce, celery, carrot, turnip, cabbage)	Stone fruits (peach, plum, apricot, cherry, olive, almond)
	Sugar cane
	Vanilla

Natural Vegetative Reproduction The capacity to reproduce vegetatively is largely confined to herbaceous and woody perennials. Important among the structures that make possible this type of reproduction are prostrate, aerial or subterranean branches of aerial stems (runners and stolons), rhizomes, and horizontal roots. Prostrate aerial stems, where they touch the soil, may send up shoots and produce adventitious roots at the nodes (FIG. 11-1). The strawberry, many weeds, and other plants reproduce in this way.

Reproduction by means of underground stems—stolons, rhizomes—is even more common. Stolons may form axillary buds and adventitious roots at the nodes. The buds give rise to branches that eventually emerge as aerial shoots. Independent plants arise as older parts die. A number of wild plants, ornamentals, and weeds, such as lily-of-the-valley (*Convallaria majalis*), many mints, perennial grasses, and wild morning-glory (*Convolvulus sepium*) spread in this manner.

A detailed example will illustrate reproduction by rhizomes. The plants concerned are the familiar cattails, several species of which are widely distributed, often occupying great areas of marshlands and margins of lakes and streams. The rhizome, $1/2$ inch or more in diameter, grows through the mud, forming several nodes with vestigial leaves. Eventually the shoot apex curves upward, forming an emergent shoot (FIG. 11-2). As this shoot grows, lateral buds are produced at the base. These buds grow out horizontally as rhizomes, and each again develops an apical emergent shoot with accompanying basal buds.

The rate of vegetative growth of the common cattail (*Typha latifolia*) in the first growing season was demonstrated by investigators of the Agricultural Research Service, U.S. Department of Agriculture. In May a single seedling grew in a tank 6 feet in diameter containing soil covered with water (FIG. 11-3). By November the plant had produced 34 emergent shoots 18–48 inches tall, 29 smaller shoots 4–18 inches tall, and 35 nonemergent shoots 2–4 inches tall—nearly 100 potentially individual plants by vegetative reproduction in a single season.

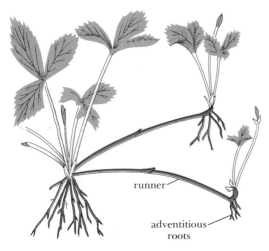

Fig. 11-1 Vegetative reproduction in wild straw-berry (*Fragaria virginiana*).

runner

adventitious roots

Fig. 11-2 Vegatative reproduction in cattail (*Typha*).

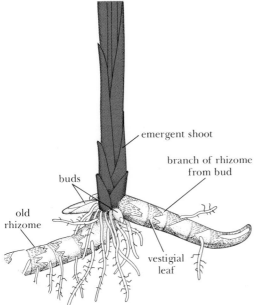

emergent shoot

branch of rhizome from bud

buds

old rhizome

vestigial leaf

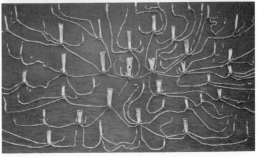

Fig. 11-3 (Top): cattail (*Typha latifolia*) plant grown from a single seed in 6 months. (Bottom): the same plant with leaves cut back and roots removed. The star in center indicates the original shoot.

An extremely rapid rate of reproduction is found in the water hyacinth (*Eichhornia crassipes*), a troublesome yet attractive weed in tropical and semitropical regions, including many parts of the southern United States. It may form seeds, but reproduces mainly by stolons. It is estimated that 10 of these plants could produce 655,360 plants during a growing season of 8 months. These plants would cover an acre of water surface and block all navigation. Also striking is the reproduction of our native blueberries. A species of blueberry native to the southeastern United States was found to produce annually new rhizomes averaging about a foot in length. A clone of this plant, radiating from the center, covered an area half a mile in diameter and was estimated to be 1000 years old.

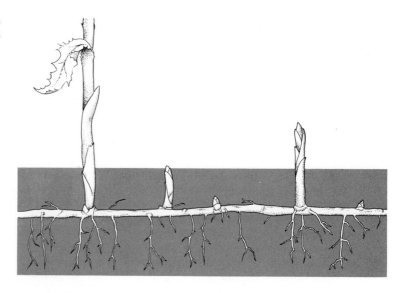

Fig. 11-4 Adventitious shoots from horizontal roots. Canada thistle (*Cirsium arvense*).

The production of adventitious aerial stems from horizontal roots is also common. Multiplication by adventitious buds on roots occurs in the Canada thistle (FIG. 11-4), bindweed, milkweed, and many other herbaceous weeds. Aerial stems also frequently arise from the horizontal roots of woody plants, such as Osage orange, poplars, sumac, beech, elderberry, and lilac. Thick groves or clumps of woody plants are often formed as a result of this method of reproduction.

Vegetative reproduction plays an important natural role. Many perennial plants, woody and herbaceous, are able to spread over wide areas because of vegetative multiplication. This ability enables grasses, for example, to become the dominant plant of meadow, prairie, and plains in many parts of the world. The rapidity with which some plants spread by vegetative reproduction enables us to use perennial grasses, legumes, and other plants in the control of erosion on sloping lands, in gullies and on sand dunes (FIG. 11-5). The masses of roots and subterranean stems, along with the aerial parts of the plants, stabilize the soil against the destructive action of water and wind (Chapter 10).

Intimately connected with vegetative reproduction is the capacity of a partially destroyed plant to regenerate—that is, restore, or partly restore—the lost parts. For example, a portion of a tap root of dandelion left in the soil following attempts to remove the plant by hand-digging forms adventitious buds that shortly produce leaves above the soil. Many hardwood forest trees, after cutting, produce sprouts from the base of the stumps. In the replacement of the forest in cut-over areas, this type of reproduction may be important.

Origin of Adventitious Roots and Buds As we pointed out earlier, adventitious roots or buds, or both, are formed in most kinds of vegetative reproduction. Roots are formed on the stems of many plants where they touch the soil. A piece of stem, inserted in the soil, may produce roots at the lower end. Roots may produce buds naturally, while still attached to the parent plant, or as a result of wounding.

Adventitious roots on stems generally start within or near the vascular tissues of the stem. Apical meristems are formed from various tis-

Fig. 11-5 Sea oats (*Uniola paniculata*) holding blowing sand. Cape Hatteras National Seashore, North Carolina.

sues, such as the outer part of the vascular ray or the parenchyma in the zone between the vascular bundles, the cambium itself, or parenchyma cells just outside the primary phloem. Buds formed from roots are commonly initiated in the pericycle. In these the shoot apex develops embryonic leaves while still embedded within the parental tissues (FIG. 11-6). Even physiologically mature plant cells, as long as they remain alive, may retain the capacity for renewed division and further growth under proper stimuli. Once a root or shoot apex is formed, it pushes through the overlying tissues and emerges into the soil or air.

Advantages of Vegetative Reproduction

Vegetative propagation is carried on by farmers, horticulturists, nurserymen, florists, and gardeners for a variety of reasons. A fully developed and frequently more robust plant may be obtained in much less time through vegetative reproduction than when seed is sown. Many lilies, for example, grow very slowly from seed, and four to seven years lapse between germination and flowering. When propagated vegeta-

tively, flowers may be produced in one or two seasons.

The number of kinds of flowers and other ornamental plants that are propagated vegetatively is very large. Among the more important are roses, carnations, chrysanthemums, poinsettias, geraniums, hydrangeas, gladiolus, narcissi, hyacinths, tulips, iris, dahlias, phlox, and delphiniums. Nearly all the flowering shrubs and ornamental evergreens are reproduced in this way.

Some varieties of plants lack sexual reproduction entirely, and can be propagated only vegetatively. Examples of these are the cultivated banana, the pineapple, some kinds of limes, the navel orange, the Marsh grapefruit, and certain varieties of grapes. Some varieties of sweet potatoes produce seeds, but many do not under usual conditions.

The most important advantage of vegetative reproduction of cultivated plants is the retention of desirable features of plants that do not "come

Fig. 11-6 Cross section of root of Canada thistle (*Cirsium arvense*), showing embryonic adventitious shoot. Compare branch root in Fig. 9-13.

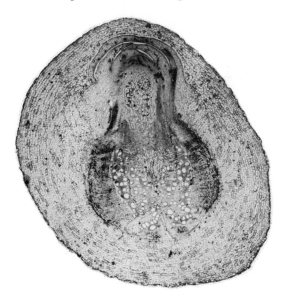

true" from seeds. Many useful plants, including most of our fruits, produce viable seeds but, because the plants are hybrids, they produce offspring that may be very different from the parents. If the seeds of a Delicious apple or a Bartlett pear are planted, it is very unlikely that the resulting trees and their fruit will resemble the parents sufficiently to deserve the varietal name or even to be of commercial value. Most fruits of the temperate zone in America are propagated asexually. The quality of fruits grown from seeds can be determined only after the plant has matured and fruited. With asexually propagated plants, on the other hand, the quality of flowers or fruit that each will produce is known in advance, for each generation will be the same as the preceding one.

PLANT ORGANS USED IN VEGETATIVE PROPAGATION

Propagation by Aerial Stems   One of the most widely employed methods of propagation is the use of CUTTINGS. These are vegetative parts of plants that have been removed and rooted in sand or some other suitable material, under proper conditions of warmth and moisture. Cuttings are made from stems, roots, and even leaves, but stem cuttings are most widely employed.

Stem cuttings are also often called SLIPS, especially if nonwoody. A stem cutting consists of a section of stem 3–12 inches or even more in length, bearing several nodes and lateral buds. When such cuttings are placed in soil, roots grow from the lower end, and the uppermost lateral buds grow into shoots. If the cutting is reversed, some species produce roots from the apical end and shoots from the basal end, but the growth is feeble.

A very large number of plants, woody and nonwoody, are readily propagated by stem cuttings. On the other hand, cuttings of a number of plants root poorly and are propagated in this manner only with difficulty or not at all. In 1935,

new techniques in plant propagation were made possible by the discovery that the rooting of cuttings may be stimulated by hormones (FIG. 11-7). Indoleacetic acid, obtained from organic sources, and later found to be produced by plants, stimulated the growth of roots when applied to the basal end of the cutting. Subsequently, other chemical compounds have been found that stimulate root production. Of these, indolebutyric acid and naphthaleneacetic acid are the best known.

These growth-promoting substances are used in dilute solution or, more commonly, in powdered form mixed with inert material. The basal end of the cutting is dipped in the hormone, then planted in the soil. Cuttings thus treated produce roots more rapidly and in greater numbers than untreated cuttings. Such plants as yew, holly, and camellia, difficult to root from cuttings, respond to this treatment. Commercial preparations, under a variety of trade names, are widely used by the amateur gardener as well as by the professional nurseryman.

Cacti are among the plants readily propagated by cuttings. A good-sized piece of the plant is removed with a sharp knife, placed in a dry atmosphere for a day or so to permit the cut surface to heal, and then rooted in sand. Many cacti spread naturally by vegetative propagation. Some, such as the opuntias, have jointed stems, which are readily detached. Others produce specialized side branches that break off at a touch. The joints or branches cling to the hair of a passing animal and may be carried considerable distances. When they fall upon the soil, they root readily and grow as independent plants. The spread of cacti by animals is particularly noticeable on overgrazed plains of the western United States. Some species of cacti are highly dependent upon vegetative propagation and rarely produce flowers or fruit.

Prominent among the food plants propagated by portions of the aerial stems are sugar cane, pineapple, and cassava. The varieties of sugar cane are grown by planting sections of stalks, each bearing one or more nodes and buds. Roots spring from the region just above the nodes, and buds soon grow into shoots.

The pineapple (Fig. 11-8) is propagated in several ways. On the flower stalk, just below the fruit, small shoots, known as slips, are produced. Other shoots, the suckers, are formed in the axils of the main leaves, close to the ground. Slips and suckers, removed from the plant and inserted in the soil, grow into new plants, and these are the parts generally used in propagation. The top of the fruit is prolonged in a vegetative shoot called the crown, and this also may be used in vegetative propagation. Suckers produce mature fruit in about 15 months, slips in 18–20 months, and crowns in about 22 months.

Cassava, a shrubby perennial plant 6–7 feet high, has been extensively cultivated for its edible roots in the American tropics since pre-Columbian times and has been introduced into other tropical regions. The plant is propagated by cuttings, about a foot long, of the semiwoody aerial stems. These are planted in rows, and the roots are ready for harvesting 9–12 months after planting. The roots (Fig. 11-9) are large, fleshy, tuberous, and rich in starch, which is the source of commercial tapioca. They are grated and washed, then boiled or roasted or made into bread. The roots contain hydrocyanic acid, sometimes in such quantities that they are highly poisonous when eaten raw. This poison is dissipated during washing and cooking.

The normally erect stems of many plants, if in contact with the soil, root readily at the base, at the nodes, or near the tips. This ability of many plants to produce adventitious roots from stems is utilized in the horticultural process of LAYERING. Several methods of layering are employed. Branches of raspberry, blackberry, rhododendron, rose, honeysuckle, and other ornamental or food plants are bent to the ground and covered with soil, with the tips exposed. After the roots have formed, the stems are cut off below the roots and the new plants set out. This is called SIMPLE LAYERING (Fig. 11-10, left).

An entire branch may also be laid down and covered with soil, either over the entire length of the branch or only at intervals. This is called COMPOUND LAYERING. Grapes, dogwood, hydrangea, and filberts may be reproduced by compound layering.

In MOUND LAYERING (Fig. 11-10, right) a plant is pruned back so as to cause it to produce new shoots close to the ground. The base of the plant is covered deeply with soil, and the new shoots develop roots. These shoots are separated from the parent and planted. Currants, gooseberries, and some varieties of apple and quince are propagated in this way.

With some plants it is impractical to bend a branch or the entire plant to the ground, or they may be difficult to root from cuttings. In such cases AIR LAYERING may be employed. In this method a branch is wounded by cutting part way through at a sharp angle, or, better, by girdling. The branch is then surrounded with a ball of damp moss at the site of the wound, which may be dusted with a root-inducing hormone. Frequent wetting of the moss is unnecessary if it is enclosed tightly in a sheet of waterproof plastic film. Roots grow into the moss after some weeks, and when this has occurred the branch may be cut off and planted in soil.

Greenhouse orchids may be propagated vegetatively by division of the rhizome, but a more rapid method, involving tissue culture, has been developed. Buds are removed from the bases of the leaves, and the apical meristem of each bud, including several leaf primordia, is dissected out under sterile conditions. The excised piece is then planted in sterile nutrient agar, where it grows into a small rounded body, several millimeters in diameter, called a PROTOCORM. If a protocorm is cut into several pieces, buds will develop on each piece and grow into new protocorms. Each protocorm is capable of forming a new plant. As long as the protocorms are cut into pieces they continue to yield new protocorms, but when cutting is stopped each protocorm produces a new plant. This technique could be regarded as a modification of the method of taking cuttings, but here only the apical meristem is involved.

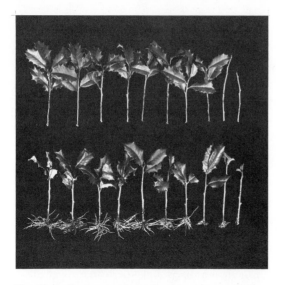

Fig. 11-7 Response to applications of growth hormones. Rooting of cuttings of American holly (*Ilex opaca*). The group at the bottom was treated; the top group was untreated.

Fig. 11-8 Fruiting pineapple (*Ananas comosus*) plant.

Fig. 11-9 Storage roots of cassava (*Manihot esculenta*), an important food plant in tropical countries. The pocket knife indicates relative size.

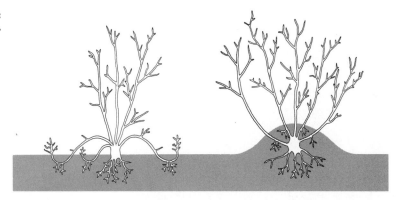

Fig. 11-10 Layering. (Left): simple, or common, layering. (Right): mound layering.

Propagation by Underground Stems Subterranean stems, modified in various ways, are probably more important in the vegetative propagation of plants than are aerial stems. Buds produced at the nodes of rhizomes give rise to branches that persist underground or develop into aerial shoots. Canna, iris, rhubarb, and peony are propagated by cutting the rhizomes into pieces, each piece bearing a well-developed bud.

The white, or Irish, potato is propagated by vegetative means. The white potato is the enlarged and swollen apex of a stolon. In preparation for planting (Fig. 11-11), the tuber is cut into pieces, each piece bearing at least one eye (Fig. 2-9). The central bud of each eye develops into a shoot, which shortly produces adventitious roots from the underground portion of the stem. Later, stolons develop from lateral buds located at underground nodes, and these stolons expand at the tips into new tubers (Fig. 11-12).

Plants with corms or bulbs are propagated in a variety of ways. Most bulbs increase naturally by the production of smaller bulbs (bulblets) in the axils of the outer scales. When mature, these are known as offsets and may be removed and planted to grow into flowering-size bulbs. Some lilies are propagated by separating the outer scales and rooting them in a mixture of sand and peat. Adventitious bulblets are formed at the base of each scale. Some lilies are propagated by stem cuttings. The production of bulblets in the hyacinth is hastened and greater numbers are produced by scooping out the stem and bases of the leaves from the underside. Adventitious bulblets are then formed from the leaf bases.

Corms, such as those of crocus and gladiolus, are the enlarged bases of herbaceous stems and differ from bulbs in being solid rather than composed of scales. When a corm of gladiolus (Fig. 2-23) is planted, from one to three new corms are produced at the bases of the new shoots that arise from the old corm. Small corms, or cormels, are produced near the base of each new corm. These may be used in propagation, although they require two growing seasons to reach flowering size.

The commercial varieties of banana produce no seeds and are propagated by the corm and by "suckers" or lateral shoots, which arise from buds on the corm. The "stem" of the banana plant arises from a huge, tuberlike corm as much as a foot in diameter. When a new plantation is started, the corm is cut into pieces, called BITS, each weighing 3 or 4 pounds and bearing at least one good eye, or bud. The bits are planted about a foot deep in rows. After a few weeks, a shoot appears above the ground and grows for 7–10 months before the flower stalk appears. During this time the base of the shoot enlarges into a corm and accumulates large quantities of starch.

Although the banana is called a tree, it is not

Fig. 11-11 Cutting and planting potatoes by hand for the production of certified "seed" potatoes. By cutting the potatoes in the field and planting the pieces together, all of the plants growing from a diseased tuber can be more easily identified and eradicated. The formation of tubers carrying diseases such as mosaic (Fig. 20-21) can thus be prevented.

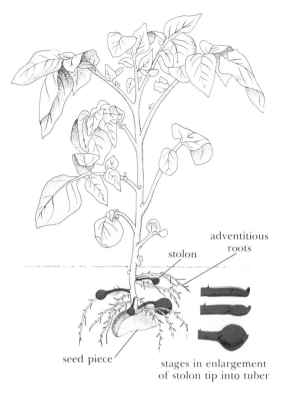

adventitious
roots

stolon

seed piece

stages in enlargement
of stolon tip into tuber

Fig. 11-12 Origin of potato (*Solanum tuberosum*) tubers.

Fig. 11-13   Cross section of banana (*Musa sapientum*) plant. The "stem" is composed of the sheathing bases of leaves.

Fig. 11-14 Root system of banana plant, showing "sword suckers" around the parent plant.

Fig. 11-15 Harvesting bananas.

woody and is really an herb. Until it flowers, the stem, or "trunk," is composed entirely of tightly wrapped leaf bases, or leaf sheaths, that support the plant (FIG. 11-13). As the plant grows, shoots ("suckers") grow from the buds of the corm and appear above the soil. These "sword suckers" (FIG. 11-14) may be detached and used in the propagation of new plants. Each sucker, when planted, develops a new corm.

Shortly before the banana plant attains full growth, the upwardly directed shoot apex of the corm at the base of the plant begins to elongate. It forms a thick shoot, which pushes up through the center of the stem, emerges at the top, and forms the flowers.

In 13–15 months after planting, the banana plant may have attained a height of 30 feet with a diameter at the base of 16 inches, and the fruit is ready for harvesting (FIG. 11-15). The parent plant is cut down, for the plant produces fruit but once. Sword suckers from the old corm are then allowed to grow and fruit in succession.

Propagation by Roots The sweet potato is one of the most important of the food plants propagated by roots. The plant is not related to the white, or Irish, potato, and the fleshy root is only superficially like the potato tuber, which is a stem. A sweet potato will sprout readily, and will grow as a house plant if it is partly submerged in a jar of water. For commercial propagation, sweet potatoes are set out in moist sand or soil. They soon produce adventitious shoots (FIG. 11-16) from which adventitious roots form at the basal end. These shoots are removed and planted in the field.

Many garden ornamentals, such as phlox, delphinium, columbine, oriental poppy, and dahlia, are propagated either by root cuttings or by the division of the roots. In the latter case, propagation may be by buds formed at the base of the old stem or on the roots themselves. Dahlias, widely grown for their attractive autumn flowers, form a clump of fleshy roots at the base of the stem. These produce adventitious buds at the junction of the root and stem. The clumps are stored over the winter and in the spring are divided into individual roots for planting.

Kudzu, a long-lived, perennial leguminous vine, grown in the southern states for pasturage, to control soil erosion, and to increase soil fertility, is prominent among the forage plants increased by vegetative means. It produces long runners that grow very rapidly and become woody with age. A single runner is known to attain a length of 50 feet in one season. The vine, once planted extensively, is becoming a nuisance (Fig. 11-17).

The plant spreads vegetatively, the runners rooting at the nodes as they grow over the soil. These roots enlarge, become branched and fleshy, and form buds at the top, which, when well grown, are called *crowns*. When a new area is to be planted, two-year-old crowns are dug during the dormant season and planted in rows. The plant produces seed sparingly, and germination is poor.

Propagation by Leaves The leaves of most kinds of plants do not ordinarily produce new plants, but there are a number of exceptions. The walking fern, which grows in shaded moist areas of the eastern United States, has long, slender, evergreen leaves. When the tips of the older fronds come in contact with damp soil, roots may form and a shoot develop. The new plant thus "walks away" from the parent plant and becomes independent when the old leaf dies. The leaves of a number of ferns, temperate and tropical, are able to produce new plants, sometimes in the absence of contact with the soil.

Among the dicotyledons, no group is more remarkable for its ability to reproduce from leaves than certain species of the genus *Kalanchoë*. In *Kalanchoë diagremontianum* (Fig. 11-18) the leaves, which are fleshy and notched along the margins, produce meristematic tissue in the notches, and this tissue gives rise to tiny plants while the leaf is still attached to the parent plant. These plantlets have tiny leaves, stems,

and sometimes roots. They readily fall away and continue their growth upon the soil. In the air plant or life plant (*Kalanchoë pinnata*), the plantlets are usually not produced while the leaf is still attached, but if a leaf is removed from the plant it will produce a crop of small plants along the margins when placed on moist soil (Fig. 11-19) or even suspended in air, hence the name "air plant." The kalanchoes are widely cultivated for their flowers and for the tiny plants that in certain species develop along the leaf margins.

Experiments have shown that the leaves of a considerable number of plants (24 percent of 1204 species tested) will produce roots and shoots under laboratory conditions. In general, however, only fleshy or leathery leaves are used for propagation on a commercial scale, since such leaves possess reserves of food and water sufficient to keep the leaf alive until a new plant is established. Among the plants regularly propagated by leaves are the bowstring hemp (*Sansevieria*), African violet (Fig. 11-20), gloxinia, and several species of begonia. In some cases the entire leaf is planted; in others the leaf is cut into pieces (leaf cuttings).

Fig. 11-16 Vegetative reproduction in sweet potato (*Ipomoea batatas*). (Left): adventitious shoots growing from swollen storage root. (Right): single shoot, called a "slip" or "draw," removed and ready for planting.

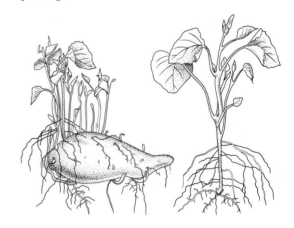

Fig. 11-17 A rank growth of kudzu (*Pueraria thunbergiana*) vine kills trees in Georgia.

Grafting and Budding Grafting is a horticultural process, known and practiced for centuries, in which a cutting from one plant is attached to a piece of the root or to the rooted stem of another plant. Unlike other methods of asexual propagation, the cutting does not regenerate new organs, but unites with and becomes an integral part of the plant onto which it is grafted. The root or rooted stem onto which the graft is made is called the STOCK, or ROOTSTOCK. The cutting that is attached to the stock is the SCION.

Grafts are commonly of three types: WHIP grafts (FIG. 11-21), WEDGE grafts, and CLEFT grafts. Whip grafting is generally restricted to apple and is done in the winter with dormant scions and roots. The grafts are then kept cold and moist until spring, when they are set in the

field to grow. The graft normally forms a close union during the storage period.

Cleft grafting is used to change the top of an older tree to a new variety or several varieties. A 1-inch or larger limb is sawed off and split or cleft for a short distance to allow the bases of two tapered scions to be inserted in the cleft. The cambium on the outer side of the scions is aligned with the cambium of the stock to ensure a quick growth union, and all cut surfaces are covered with a soft wax to prevent drying. Cleft grafting is done with fully dormant scions just before growth starts in the stock. It is essential to use dormant scions in grafting—cut and held in cold storage—so that a graft union can form before the scions leaf out.

In wedge grafting, a V-shaped cut is made in

Fig. 11-18 Upper and lower surfaces of leaves of *Kalanchoë diagremontianum*. The margins of the leaves bear young plantlets.

Fig. 11-19 Single leaf of *Kalanchoë pinnata*, removed from plant and placed upon moist soil, where it bears small plants along the margins.

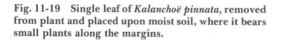

Fig. 11-20 Leaf of African violet (*Saintpaulia ionantha*) bearing adventitious plants at the base. These were produced when the petiole was set in soil.

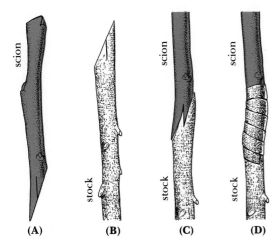

Fig. 11-21 Whip, or piece-root, grafting of apple (*Pyrus malus*). (A): scion of the desired variety. (B): stock; a piece of seedling apple root. (C): clefts in scion and stock forced together. (D): graft completed and wrapped.

the stock and a tapered scion is fitted carefully into the cut. The stock and scion should have approximately the same diameter. A new machine that makes both cuts accurately has been built in France to graft European grapes onto American rootstocks. The method promises to be adaptable to use with other plants.

BUDDING is a specialized form of grafting in which a single bud, with a surrounding small amount of bark, is used as a scion (FIG. 11-22). Budding may be done in the field in the spring, but is more commonly done in late summer when new dormant buds have been produced and can be used as scions. Peaches, plums, pears, citrus, and perhaps half of the commercial stock of apples for planting are produced by budding. The leaf that grows below the bud is cut away to reduce water loss from the bud. The piece is then slipped into a T-shaped cut made in the bark and wrapped snugly with a special elastic tape that holds the bud in place but does not prevent growth of the stock. The tape disintegrates after a few weeks and need not be removed. Summer buds should form a good union with the stock but remain dormant until the next spring. When they start growth, the stem of the seedling stock is cut away above the bud and the bearing part of the plant then develops from the bud and has the characteristics of the variety from which the bud was taken.

Grafting and budding are employed chiefly in the propagation of woody plants that do not come true from seeds and are not easily propagated by cuttings. They are also used to combine two kinds of plants, as when a weeping mulberry is grafted on an upright trunk to form an umbrella-type tree, or several varieties of apples are grafted on the same stock for use in gardens. European grapes, formerly propagated by cuttings, are now grafted on American stocks resistant to the *Phylloxera* root louse.

One use of grafting of general interest is the production of dwarf fruit trees for the home garden or commercial orchard. Dwarfing of fruit trees may be brought about by the use of rootstocks that have a dwarfing effect upon the scion. For example, dwarf pears may be produced by grafting on quince stocks, and dwarf peaches by grafting on apricot. The same method has been used in apples; that is, grafting a desired variety on an apple rootstock of a variety that has a dwarfing effect. This method has certain disadvantages in the case of apples and is being replaced by the use of dwarfing interstocks or interstems, which involves a standard seedling rootstock. In one procedure a 4-inch piece of the dwarfing variety is used as the scion and grafted onto the seedling stock. Later, the desired variety, such as Golden or Red Delicious, is budded on the scion, which then becomes the interstock and remains between the stock and the variety that will produce the fruit-bearing top. The short piece of dwarfing wood in the trunk makes the tree semidwarf and early bearing.

In making a graft it is important that the cambial region of the stock be closely associated with that of the scion. It is sometimes stated that the cambial layers themselves must be placed in actual contact, but in practice this is rarely achieved. Under the protection of the wrapping

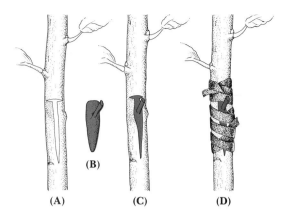

Fig. 11-22 Budding of peach (*Prunus persica*). (A): the growing, seedling stock is cut to receive a bud. (B): a bud with a shield of bark, cut from the variety to be propagated. (C): bud inserted in stock. (D): bud wrapped to protect and hold it in place.

or grafting wax, the cambium, the cells of the cambial zone, and sometimes other thin-walled cells of the inner bark form a mass of callus, or unspecialized wound tissue. Both stock and scion contribute to the formation of callus. The callus tissues meet and fuse, and a new cambium arises in this callus, uniting the cambium of the stock with that of the scion. The cambium now produces continuous layers of wood and bark, allowing translocation of food, water, and minerals to proceed without interruption between stock and scion.

It is usually stated that grafting is not possible with monocotyledonous plants, since cambial activity is essential to callus formation and the union of stock and scion. Monocotyledonous plants have been grafted, however, although only on an experimental scale. In one method, advantage is taken of the fact that in many monocots, chiefly grasses, a short zone just above the node remains meristematic for some time after the rest of the stem has matured. The upper portion of the stem is pulled out, thus rupturing the cells in the meristematic, nodal region. A scion similarly obtained, and of the same size, is inserted on the top of the stock, thus bringing the two meristematic zones into contact. The grafted region is supported by a waxed paper tube or by light splints tied around the stem. Union occurs and new vascular tissues are differentiated, uniting the vascular bundles of stock and scion. Grafts of this kind have been made in bamboo, sugar cane, and other monocots. The percentage of successful grafts is low, and although the method may be useful in certain types of investigation, it is improbable that it will ever be used on a commercial scale.

Grafting is most effective between related plants, usually those of the same or closely related species. Even closely related plants, however, may be uncongenial and no permanent union will occur—the graft will not "take." Grafts between different species of the same family, however, are not uncommon, as when varieties of pear are grafted on the roots of quince. Grafts between different species of plants within the sunflower, mustard, rose, and potato families have been made. Tomatoes may be grafted on potatoes, olives on lilac, and watercress on cabbage. Some grafts between plants of different families have succeeded.

Most such grafts as those mentioned have no economic value, but have supplied much information on the transfer of materials between stock and scion and on the organs concerned in the synthesis of certain plant products. For example, grafting experiments have shown that the nicotine of tobacco is synthesized in the roots and translocated through the stalk to the leaves, where it accumulates. Two types of grafts were made. In one, tomato scions were grafted upon tobacco stocks. The grafts were grown for about a month, after which it was determined by analysis that the leaves of the tomato scions contained large amounts of nicotine. On the other hand, when tobacco scions were grafted upon tomato stocks and grown for the same length of time, there was no appreciable accumulation of nicotine in the scions; the amount remained approximately the same as that found in scions of the same size as those used in making the graft.

SUMMARY

1. Vegetative reproduction is reproduction by means other than seeds. A clone is a group of plants propagated by vegetative or asexual means. All the individual plants of a clone have been derived by repeated vegetative propagation from a single individual.
2. Vegetative reproduction is common, and is important both to plants themselves and to man. Without this method of multiplication we would be unable to propagate many of our most useful plants in their present forms. Weeds, however, also spread by vegetative means. Plants that reproduce vegetatively may be used to control soil erosion.
3. Vegetative reproduction is most commonly brought about by aerial stems, rhizomes, or horizontal roots. Bulbs, corms, and leaves may also be used. Stems and rhizomes produce shoots from nodes and develop adventitious roots. Horizontal roots produce adventitious buds and roots.
4. Adventitious buds and roots usually arise deep within the parental tissues and push their way through the overlying tissue to the outside.
5. Among the advantages of vegetative reproduction are the production of mature plants more quickly than by seed, and the propagation of plants that do not produce seed or whose seed would give rise to plants differing from and less valuable than the parental stock.
6. Propagation by aerial stems is commonly practiced by means of cuttings or layering. Stem cuttings are portions containing nodes and buds; when cuttings are placed in moist sand or soil, roots grow from the lower end. Growth regulators are used to stimulate the rooting of cuttings. In contrast to cuttings, layering involves the formation of roots prior to separation from the parent plant.
7. Some kinds of plants are propagated by leaf cuttings. In some, new plants grow directly from notches in the leaf margin.
8. In grafting, the stem or bud of one plant (the scion) is attached to the rooted part or root of another (the stock). Callus tissue develops, and a new cambium is formed within the callus, uniting the cambium of the stock with that of the scion. Grafting and budding, for practical purposes, are carried on only with plants possessing a cambium, such as conifers and dicotyledons. Some monocots have been grafted experimentally, but such grafts have no economic value.

12

The Flower and Seed Production

THE PRECEDING CHAPTERS have been devoted largely to a survey of the structures, activities, and processes of the vegetative organs of seed plants, with particular attention to the flowering plants. Among other activities, vegetative organs may function in reproduction. But although reproduction may be brought about by roots, stems, or leaves, this process is commonly initiated within the flower. The flower is a part of a method of sexual reproduction that gives rise to seeds, from which new plants eventually arise.

Seed formation is essential to the survival of most kinds of seed plants, and for many species of animals that depend upon seeds for food. The cells and tissues within the flower that lead to seed and fruit production are as important, in many ways, to man and to other forms of animal life as are the activities of the green leaf. The steps in the production of the flower and the seed are the result of innumerable evolutionary changes extending over immense periods of time. It is a subject that the student may find perplexing but will certainly find intriguing.

An essential feature of sexual reproduction is fertilization, which is the union of male and female sex cells, or GAMETES, forming a ZYGOTE. This zygote, in the flowering plants, is the fertilized egg. It contains chromosomes of both parents and is the first cell of a new individual. It grows into an embryo within the seed, and this embryo, when the seed germinates, gives rise to the adult plant. Since the embryo contains the genetic contributions of both egg and sperm, inherited potentialities are transmitted through the seed from one generation to another.

The flower is usually a conspicuous stage in the process of seed formation. Commonly visualized as a unit, it is actually composed of a number of organs. But only two of these, the STAMENS and the PISTILS, are directly concerned in seed production. Stamens form POLLEN GRAINS, and these, in turn, give rise to male gametes. OVULES, each containing an egg, are found within the enlarged, lower part of the pistil. When the pollen grains reach the pistil during pollination, a POLLEN TUBE forms. The male gametes, or sperms, move in this tube through the tissues of the pistil to the ovule, where they are discharged. Fertilization follows, and the ovule develops into a seed.

An understanding of how seeds develop involves much more than this bare outline. It includes a complicated series of events within the stamen and ovule before fertilization, and a number of equally important and complex stages afterward. The casual observer of plant life will be unaware of these stages, since they involve cellular changes and interrelationships that can be seen only under the microscope.

THE FLORAL ORGANS

Flowers vary greatly in structure. The fundamental similarities among flowers of different kinds of plants are, however, greater than their differences, since all flowers have the same basic structural plan. Each individual flower is borne at the tip of a specialized stalk, the PEDICEL. The apex of the pedicel, where the floral parts are attached, is commonly more or less enlarged, and is termed the RECEPTACLE (FIG. 12-1).

This receptacular region in the young flower is similar to the shoot apex of a vegetative stem, but instead of producing leaves, it gives rise, typically, to four kinds of floral organs (FIG. 12-1) always arranged in the same order. The first of these floral organs, beginning at the base of the flower, are the SEPALS, collectively forming the CALYX. Next, proceeding toward the center, are the PETALS, collectively known as the COROLLA. The calyx and corolla together constitute the PERIANTH. Sepals are usually green and somewhat leaflike, and the petals are white or colored. Some flowers, such as fuchsia, delphinium, iris, and columbine, have sepals colored like the petals but distinguishable from them by size or form. In some plants, such as lilies and tulips, the sepals are colored like the petals and are also similar to them in size and form. In such cases the individual components of the perianth are termed TEPALS.

The corolla is usually the most conspicuous part of the flower and the only part many people

notice, but in some plants, such as clematis, hepatica, anemone, and marsh marigold, the sepals are the most prominent because the petals are small or absent (FIG. 12-7). In poinsettia and the dogwoods the perianth is highly reduced or inconspicuous, and the structures incorrectly called petals are modified leaves (bracts) that surround a group of small, inconspicuous flowers (FIG. 12-2).

Above the petals are located the stamens. Each stamen consists typically of a slender, elongated stalk—the FILAMENT—terminated by an enlarged ANTHER, which contains the pollen grains.

The pistil (or pistils) is located in the center of the flower. It is commonly flask-shaped, with a swollen, basal part, the OVARY, connected by a stalklike STYLE to a terminal, expanded portion, the STIGMA. The stigma, which receives the pollen, may be rough, smooth, sticky, branched, or feathery. The feathery type is particularly characteristic of plants pollinated by wind.

Pistil and Carpel Pistils are either *simple* or *compound*. A compound pistil is composed of two or more units, termed CARPELS—structures with certain resemblances to leaves. Ovules, which later develop into seeds, are borne on the inner surface of the carpels near the margins. The region or area where one or more ovules (or seeds) are attached is known as the PLACENTA, in analogy to the organ connecting the mammalian embryo to maternal tissues. A simple pistil is composed of but one carpel. A flower may contain only one simple pistil, as in the pea (FIG. 12-3) and bean. These have a row of ovules attached to the fused margins of the single carpel (FIG. 12-4A). In the flowers of plants such as buttercups, magnolia (FIG. 12-6), tulip tree, peony, columbine, and stonecrops (FIG. 12-10), there are a number of simple pistils. Usually, however, when a number of carpels occur, they are fused together, forming a compound pistil, and only one such pistil is produced in a flower (Plate 4).

Carpels are fused into a compound pistil in either one of two ways. In one, the carpels are

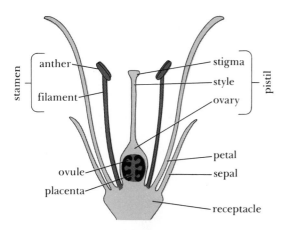

Fig. 12-1 Diagram showing floral organs.

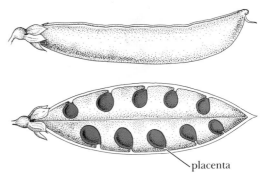

placenta

Fig. 12-3 (Top): nearly mature simple pistil of garden pea. The sepals appear at the point of attachment, and the shriveled remains of the style and stigma at the opposite end. (Bottom): pistil opened along the margin, showing attachment of seeds.

Fig. 12-2 "Flower" of flowering dogwood (*Cornus florida*). The "flower" is really a cluster of small flowers surrounded by four large petal-like bracts.

Fig. 12-4 Pistils in cross section. (A): simple pistil of bean. The placentation is marginal. (B): compound pistil of violet, carpels fused along their margins. The placentation is parietal. (C): compound pistil of lily, carpels fused along their sides. The placentation is axile. The shaded areas represent single carpels and their ovules.

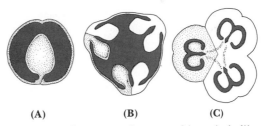

(A)   (B)   (C)

open or partly open and arranged in a circle like the staves of a barrel, thus surrounding a single cavity (locule). The margins of adjacent carpels are fused, and thus each placenta is formed by the union of two adjacent carpel margins. The arrangement of the ovules, or PLACENTATION, is thus PARIETAL, for they are now located in rows on the inner wall of the ovary (FIG. 12-4B). The number of placentae corresponds to the number of carpels. In the second type, the carpels are closed and fused along their sides, with the margins extending to the center, or axis. The placentation is now said to be AXILE (FIG. 12-4C). With axile placentation the number of loculi in the ovary corresponds to the number of carpels. The terms parietal and axile are applied only to placentation in compound pistils, for when the pistil is simple the placentation is said to be MARGINAL (FIG. 12-4A). From the standpoint of classification, these types of placentation are characters of considerable importance and usefulness.

MODIFICATIONS OF THE FLOWER

Although all flowers follow the same structural plan, they exhibit numerous modifications in design. Certain of these modifications increase the chances of pollination. Furthermore, they often reveal evolutionary relationships among plants, thus permitting botanists to place related forms together and to classify flowering plants according to their descent, from the most simple to the most advanced and specialized forms.

Number and Arrangement of Floral Parts The parts of the flower are usually arranged in whorls, or rings. These whorls are either four or five in number: one of sepals, one of petals, one or two of stamens, and one of carpels, united into a compound pistil. The number of parts in each whorl varies with the species, but is is usually constant and small.

The two large subclasses of angiosperms (flowering plants) are distinguished by the number of floral organs in each whorl. In the dicotyledons, the number is four or five or multiples of these numbers. A flower may have, for example, five sepals, five petals, ten stamens, and five carpels. However, the number of carpels in a compound pistil may be only two. In the

Fig. 12-5 Trillium (*Trillium grandiflorum*), a monocotyledon. The flower has three sepals, three petals, six stamens, and a compound pistil composed of three carpels.

monocotyledons, the number of organs in a whorl is three or multiples of three (FIG. 12-5). In the tulip, for example, there are six perianth parts, six stamens, and three carpels.

In both subclasses, however, certain families have numerous stamens and carpels—generally more than ten. Among the families in which this condition occurs are the magnolia (FIG. 12-6), buttercup (FIG. 12-7), rose, and water lily of the dicotyledons, and the water-plantain family of the monocotyledons. In some of these families the numerous stamens and carpels are attached to the receptacle spirally rather than in whorls. This combination of spiral arrangement and numerous stamens and carpels is considered indicative of a more primitive stage in evolutionary development than the whorled arrangement with parts reduced in number.

Fusion of Floral Organs Among the important modifications found in the flower are fusions of

simple pistil

stamen scars

perianth scars

Fig. 12-6   Mature fruit of magnolia (*Magnolia gran-diflora*). The ovaries of the simple pistils have opened along the back, allowing the seeds to fall away. Note the spiral arrangement of the pistils and of the scars left by the fall of the stamens.

Fig. 12-7   Flower of marsh marigold (*Caltha palustris*), a dicotyledon of the buttercup family. The flower has numerous stamens surrounding a cluster of simple pistils in the center. Petals are absent and the sepals are petal-like.

the organs of the same whorl and of different whorls. In many flowers, for example, the petals are fused (connate) along their margins into a COROLLA TUBE. This tube is commonly lobed (FIG. 12-8), the number of lobes normally corresponding to the number of petals. The sepals are usually distinct, but are sometimes fused at the base into a shallow tube. The stamens may be distinct, or may be fused by their filaments or by their anthers into one or several groups. Sometimes the filaments of the stamens are partially or completely fused (adnate) to the corolla tube (FIG. 12-8). The carpels, as we have seen, are sometimes separate, but more commonly united into a compound pistil.

In a flower with distinct parts, the sepals, petals, and stamens are clearly attached to the receptacle beneath the ovary. Since the ovary is situated above the zone of attachment, it is said to be SUPERIOR (FIG. 12-9, left). This is the condition in a very large number of flowering plants. In

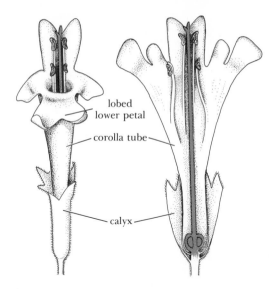

Fig. 12-8 Flower of a mint, ground ivy (*Glechoma hederacea*), illustrating fusion of floral parts and irregular symmetry of the flower.

Labels: lobed lower petal, corolla tube, calyx

plum, cherry, blackberry, and other species, the basal portions of the sepals, petals, and stamens are united, forming a cup-shaped FLORAL TUBE (also called *calyx tube*). The lobes of the sepals and petals extend from the margins of this tube together with the unfused portions of the stamens. The ovary is still superior in this type of flower (FIG. 12-9, center).

One other type of fusion is that of the floral tube with the ovary. In this type, the sepals, petals, and stamens apparently grow from the top of the ovary, and the ovary is said to be INFERIOR (FIG. 12-9, right). Actually, the pistil is still the uppermost structure of the flower, just as in flowers with superior ovaries. The ovary appears to be inferior because of its union with the floral tube. Inferior ovaries are found in such plant families as the carrot, bluebell, sunflower, evening primrose, iris, cactus, and orchid. The position of the ovary, whether inferior or superior, is significant in plant classification and also helps in interpreting the nature of many kinds of fruits.

The fusion of floral parts—of petals, forming a

tube, of carpels, forming a compound pistil, and of ovary wall and floral tube—is an evolutionary development. The various organs do not become fused during the development of the flower but are united from the very beginning of growth in the embryonic cells at the apex of the floral axis.

Symmetry of the Flower In many flowers, all the parts of each whorl are alike in size and shape, and the flower may be divided into two similar parts in more than one longitudinal plane (FIG. 12-10). Examples are tomato, apple, petunia, morning-glory, and many others. Such flowers are radially symmetrical, and are termed REGULAR or ACTINOMORPHIC (literally, "ray-formed"). The flowers of other species are bilaterally symmetrical, for they are capable of division into two identical parts along only one longitudinal plane. Bilaterally symmetrical flowers are termed IRREGULAR, or ZYGOMORPHIC (literally, "yoke-" or "pair-formed"). In such flowers, the corolla is commonly more modified than other floral parts. In the garden pea (FIG. 12-11) the single top petal, the STANDARD, is much enlarged. Two lateral ones, the WINGS, are much smaller, and the two remaining lower petals are united along one edge, forming the KEEL, which encloses the stamens and pistil. Nine of the ten stamens are united into a tube that surrounds the pistil. The tenth stamen is separate. Other examples of irregular flowers are the snapdragons, mints (FIG. 12-8), violets, and orchids. Irregular flowers are usually pollinated by insects, commonly bees.

Reduction in the Flower A flower in which four readily distinguishable kinds of floral organs occur—sepals, petals, stamens, and pistil —is called a COMPLETE flower. But, as previously noted, this is not the only kind. The flower may be reduced so that some organs are no longer evident, although they may be present in vestigial form that can be recognized by investigation. In many plants, for example, the petals have disap-

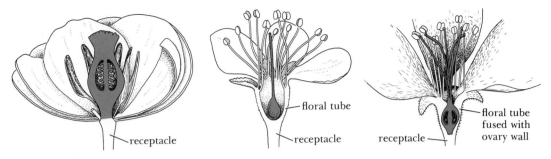

Fig. 12-9 Flowers with superior and inferior ovaries. (Left): flower of May apple (*Podophyllum peltatum*); ovary superior. (Center): flower of sour cherry (*Prunus cerasus*); ovary superior, floral parts fused, forming a floral tube. (Right): flower of apple (*Pyrus malus*); ovary inferior, the result of the fusion of the floral tube to the ovary wall.

Fig. 12-10 Regular (actinomorphic) flower of stone-crop (*Sedum*).

Fig. 12-11 Irregular flower of the garden pea (*Pisum sativum*). (Top): exterior view of the flower. (Bottom): section view.

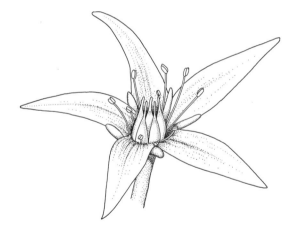

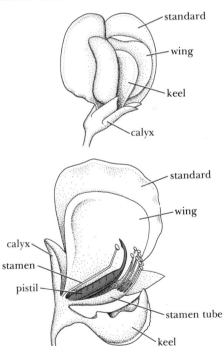

peared, and the sepals may be represented only by scales, bristles, teeth, or ridges. A flower lacking calyx, corolla, or both, is said to be IN-COMPLETE. Plants in which the perianth parts have become minute or inconspicuous include the grasses and many trees, such as the willows and poplars, a maple (*Acer negundo*) (FIG. 12-12),

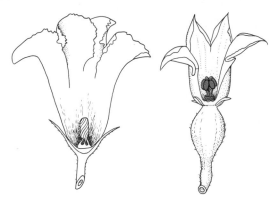

Fig. 12-13 The monoecious condition in squash (*Cucurbita maxima*). (Left): staminate flower. (Right): pistillate flower. Both kinds are borne on the same plant.

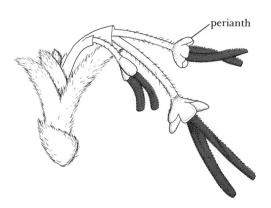

perianth

Fig. 12-12 Reduced unisexual flowers of box elder (*Acer negundo*). (Top): staminate flowers. (Bottom): pistillate flowers. The perianth is greatly reduced. This species is dioecious, for the two kinds of flowers are borne on different trees.

oaks, hickories, and elms. Contrary to popular belief, a conspicuous perianth is not an essential feature of a flower. Also, the use of the phrase "trees and flowers" is meaningless, for all angiosperm trees have flowers.

Reduction in the number of parts may also extend to the stamens and pistils. A flower in which both kinds of organs are present and functional is said to be BISEXUAL. If either stamens or pistil are lacking, or nonfunctional, the flower is said to be UNISEXUAL. Unisexual flowers are termed STAMINATE if they lack a pistil, or PISTILLATE if they lack stamens. A perianth may be present or absent in bisexual and unisexual flowers.

Both kinds of unisexual flowers may be present on the same plant, as in corn, most begonias, squash (FIG. 12-13), pumpkin and cucumber, cattail, calla lilies, and many woody plants such as oaks, beeches, birches, walnuts, and hickories. Such plants are called MONOECIOUS. In corn (FIG. 12-14), the staminate flowers are borne on the branched tassel at the top of the plant. The pistillate flowers are attached to the axis of the ear, or cob, and are surrounded by the husk. A single pistillate flower consists of an ovary to which is attached a threadlike silk 4–12 inches long. The silk bears numerous hairs that catch the pollen grains. The hairs are found even within the portion of the silk covered by the husks but are most numerous near the tip. Each thread of silk thus serves as a combined style and stigma (FIG. 12-15).

In DIOECIOUS plants the staminate and pistillate flowers occur on different plants of the same species. Examples of this condition are found in ashes and maples, in spinach, tree of heaven

Fig. 12-14 Corn (*Zea mays*), a monoecious plant. (Left): the tassel (staminate inflorescence), bearing staminate flowers. (Right): a young ear of corn (pistillate inflorescence). This consists of an axis, the cob, bearing pistillate flowers in rows. Each ovary bears a single "silk," the combined style and stigma.

Fig. 12-15 Germination of pollen grain of corn on a hair of the silk.

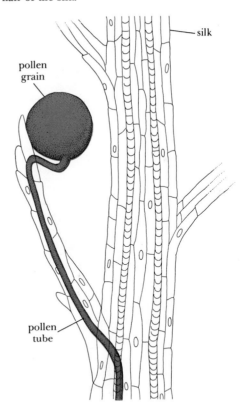

(*Ailanthus*), asparagus, hemp (*Cannabis*), the date palm (FIG. 12-16), hops, willows, holly, poplars, and American mistletoe.

The date palm probably represents the earliest recognized example of the dioecious condition, for records show that the tree was cultivated in Mesopotamia as early as 3500 B.C. The farmers of this region recognized that there were male and female trees and practiced artificial pollination by bringing clusters of staminate flowers in contact with the pistillate flowers. The methods of bringing about fruit production in the date palm today differ little from those of ancient times. Staminate trees to provide pollen are grown with the pistillate trees, which bear the fruit. Propagation is achieved vegetatively by planting offshoots from the base of the parent tree. The sex of the tree and the quality of the fruit that the pistillate plant will bear are thus determined in advance, for the offshoots always reproduce the parent type. Pollination is still done by hand, and several methods are used. A common procedure is to invert several of the long staminate clusters among the clusters of pistillate flowers (FIG. 12-17). The pollen is shed and drifts to the stigmas of the pistillate flowers.

Fig. 12-16 Date flowers on stalks removed from large flower clusters. (Left): pistillate flowers. (Right): staminate flowers.

Fig. 12-17 Artificial pollination of date flowers. (A): strands of male flowers being placed in center of the female cluster. (B): freshly opened cluster of female flowers ready for pollination. (C): female cluster after pollination. A string is tied around the strands to hold the male flowers in place.

INFLORESCENCES

Flowers are sometimes borne singly, as in tulip and cyclamen, but commonly they occur in a cluster or group on a floral axis or on branches of such an axis. Such a cluster is termed an INFLORESCENCE. The arrangement of the flowers in an inflorescence varies so greatly that many kinds of inflorescences have been recognized and given distinctive names. Since any particular kind may be constant for a species, genus, or even family, the kind of inflorescence is useful in identification of plants. In cauliflower and broccoli the edible portion is an inflorescence of undeveloped flowers, together with the condensed and fleshy branch system that bears them. Certain fruits, also, have developed from inflorescences (Chapter 13). Some of the more common and simple kinds of inflorescences are considered here.

In the RACEME (FIG. 12-18A) the stalked flowers are arranged along a single floral axis, as in lily-of-the-valley and blackberry (FIG. 28-7). The SPIKE (FIG. 12-18B) is like the raceme, but the flowers are sessile (without a pedicel), as in common plantain (*Plantago*), a widely distributed weed. The CATKIN (FIG. 12-34, left) resembles the spike, but the flowers are unisexual, without a perianth, and the staminate inflorescence is shed as a unit after flowering. The catkin is borne only on woody plants. In the UMBEL (FIG. 12-18C) the internodes of the floral axis are so short that the flower pedicels appear to arise from a common point, like the ribs of an umbrella. The umbel is commonly compound, as in the parsley family (FIG. 28-13). In the compound umbel the primary branches are branched again, the ultimate branches constituting the pedicels. In the last type to be considered, the HEAD (FIG. 12-18D), a large number of sessile or nearly sessile flowers are closely associated on a very short or vertically flattened floral branch. The whole forms a rounded (as in red clover) or flattened cluster of flowers, as in the composite family (FIGS. 28-18, 28-19). In the members of this family—for example, sunflower and chrysanthemum—the individual flowers are so tightly packed together that the head is popularly thought to be a single flower.

The grouping of flowers into a inflorescence is undoubtedly useful to many insect-pollinated plants, especially if the flowers are small, for they are thereby made more conspicuous to the pollinator.

DEVELOPMENT OF THE FLOWER

The flower, like the vegetative shoot, is composed of an axis with lateral appendages. These appendages constitute the floral organs. The axis is terminated by a floral apex, the organization of which differs in a number of respects—none of them really fundamental—from that of the apex of a leafy shoot. These differences are largely related to the limited growth of the floral axis, in contrast to the continued and potentially unlimited growth of the vegetative apex.

A shoot apex, after producing leaves, may be converted into a floral apex, or, more commonly, a flower primordium may arise as such and early exhibit differentiation into pedicel and floral apex. The apex becomes flattened, in some species domelike, and from it arise the floral organs. These first appear as protuberances, developing successively on the floral apex (FIG. 12-19). The sequence of development, from the outside toward the center, is commonly sepals, petals, stamens, and carpels. These protuberances, or primordia, increase in size by terminal and marginal growth in a manner similar to that of leaves. As the floral parts arise, the apex itself, at first broad, gradually becomes smaller, and meristematic activity becomes restricted to the developing primordia. The nodes of the floral axis (receptacle) are closely crowded and the internodes are largely or entirely suppressed. Growth of most parts of the flower is largely complete by the time the flower opens; some parts, however, most notably the carpels, continue to grow for some time as the ovary develops into the fruit.

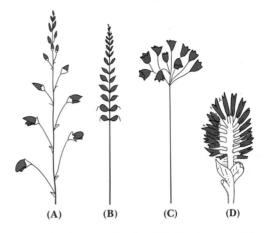

Fig. 12-18 **Common kinds of inflorescences. (A): raceme. (B): spike. (C): umbel. (D): head.**

Fig. 12-19 **Median longitudinal section through flower bud of tulip (*Tulipa gesneriana*). The carpels have not yet been initiated.**

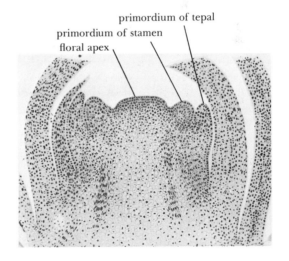

primordium of tepal
primordium of stamen
floral apex

FLORAL EVOLUTION

As indicated in the preceding section, the flower is a shoot of limited growth, comparable to a leaf-bearing shoot. The sepals may be regarded as modified leaves—that is, descended directly from structures that were primarily photosynthetic. In some primitive species the petals, also, are considered to be modified leaves, but in most angiosperms the petals are probably derived from stamens that became sterile, broadened, and leaflike. It is postulated that an ancestral flower was devoid of a perianth and that the perianth gradually evolved by the aggregation of leaflike bracts beneath the fertile organs. The problem of the origin of the stamen and carpel is more complex. These reproductive organs do show numerous resemblances to leaves—in position, arrangement, internal structure, and development. Based on an earlier concept that they represent modified leaves, the term SPOROPHYLL (literally "spore leaf") was applied to stamen and carpel, since certain structures essential to seed formation, known as *spores,* are located within ovule and anther. The stamen is termed the MICROSPOROPHYLL, and the carpel the MEGA-SPOROPHYLL. The flower then came to be defined as a group of sporophylls, usually accompanied by a perianth.

Although the preceding terms have been retained, and are in common use, the stamen and carpel are best regarded not as leaf modifications, but as homologous with leaves—that is, as having had a common ancestor. Thus, leaves and the fertile floral organs have evolved along similar lines, undergoing many transformations during their evolution. The carpel of the simple pistil is interpreted, then, as a modified, leaflike organ folded inward along the midrib and enclosing the ovules (FIG. 12-3). Its apical region became modified into the stigma, which receives the pollen.

Actually, the problem of the evolutionary origin of the flower is regarded as yet unsolved. There is, however, good evidence that the flowers of certain groups of angiosperms, such as the magnolias, water lilies, and buttercups, have retained many of the structures of the primitive flower. These groups, and others, are considered to be surviving remnants of ancient ancestral forms, and although much modified today, provide clues to the nature of ancestral angiosperms. They represent, in part at least, types from which other groups may have arisen, undergoing many modifications during evolutionary processes. Many families of flowering plants show characteristic modifications that are readily recognized by the layman. Examples of such families are the rose, pink, mustard, poppy, legume, carrot, gourd, potato, and heath, and aster family, the lilies, orchids, and others. Some of the more outstanding changes during floral evolution are considered further in Chapter 28.

DEVELOPMENT OF THE SEED

The seed itself is a complex structure, composed of a plant embryo, a seed coat, and a supply of stored food. In the seeds of many plants the food is stored within the embryo itself. In others it is stored in surrounding tissues. The full story of the seed must account for changes that take place in the stamen and pistil, for the process of pollination, for the development of the embryo, for the formation of the seed coat, and for the development of the supply of stored food used in the young plant when the seed germinates.

The essential parts of the flower, the pistils and stamens, are directly involved in the task of seed formation. The less essential and often more conspicuous parts of the flower may also contribute in various ways.

Anther and Pollen A close examination of a young anther (FIG. 12-20) shows that it is four-lobed and composed of four MICROSPORANGIA. These are merely chambers without walls of their own, enclosed within the sterile tissue of the anther. Two of these microsporangia lie on either side of a connecting tissue that is traversed by a single vascular bundle. A cross section of an anther from a very young flower bud reveals within each microsporangium a cluster of large cells with abundant cytoplasm and large nuclei. These are the MICROSPORE MOTHER CELLS, which extend in elongated masses throughout the anther. When first formed, the microspore mother cells are usually packed closely together, but later they separate and become spherical.

As the growth of the anther proceeds, the nucleus of each microspore mother cell divides. Then the daughter nuclei divide again. It should be noted that these are the special and important meiotic divisions, discussed below and in detail in Chapter 16. Cell walls form, giving rise to a group of four cells, known as MICROSPORES. Such a cluster of four spores is known as a TETRAD (FIG. 12-20). In some plants, such as the common cattail (*Typha latifolia*) and species of the heath family, the microspores remain together in the tetrad arrangement, but usually they separate and lie free within the microsporangium, surrounded by a nutritive liquid. The microspores then develop into pollen grains.

The conversion of a microspore into a pollen grain is brought about by the mitotic division of the microspore nucleus. The daughter nuclei move apart and with their surrounding cytoplasm form two adjacent cells, separated by a thin wall. One of these cells, the TUBE cell, is larger than the other, the GENERATIVE cell (FIGS. 12-20, 12-21). In a number of species the generative cell divides again, forming two male gametes, or sperms, before the pollen is shed. There are thus two kinds of pollen grains in the flowering plants: one containing only a tube and generative nucleus at the time of pollen shedding; the other containing a tube nucleus and two male nuclei.

The wall of the microspore becomes the wall of the pollen grain. It thickens and the outer surface becomes covered with spines, plates, ridges, or other markings characteristic of the pollen grains of different species (FIGS. 12-21, 12-22). At about this stage of development, the wall between the two microsporangia on each side of the anther usually disintegrates. This results in

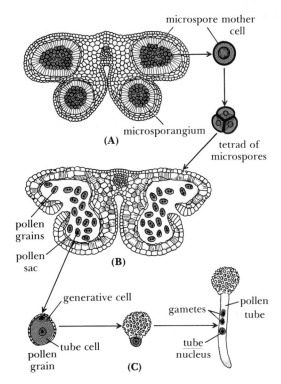

Fig. 12-20   The anther and pollen grain. Each microspore mother cell within a microsporangium divides, forming a tetrad of four microspores that shortly separate. The nucleus of each microspore then divides and a tube cell and a generative cell are formed within the wall of the microspore, which is thus converted into a pollen grain. Following pollination, the pollen grain germinates, producing a pollen tube, and the generative cell gives rise to two male gametes.

the formation of two POLLEN SACS, each filled with a powdery mass of pollen grains (FIG. 12-20). As the flower matures, each pollen sac opens, often by splitting along the length of the anther, and the pollen grains are liberated. If they reach the stigma of the same flower or another flower of the same kind, they germinate by forming a pollen tube that grows downward through the style.

Fig. 12-21 Electron micrograph of newly formed pollen grain of blood lily (*Haemanthus katherinae*): tc, nucleus of tube cell; gc, nucleus of generative cell; gcw, generative cell wall; v, vacuoles. (3500×)

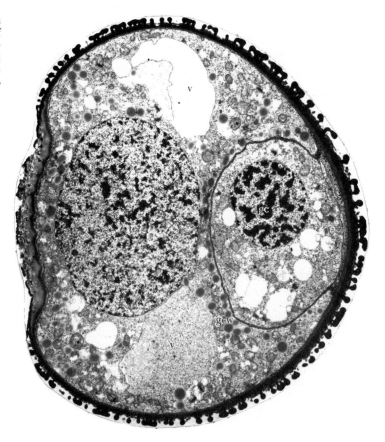

The Ovule and the Egg The events occurring within the ovule correspond to those leading to the formation of sperms. These steps lead to the production of the female gamete, or egg. The ovules, forerunners of the seeds, develop from the placental regions of the ovary walls. Each ovule first appears as a minute protuberance on the ovary wall (FIG. 12-23). This swelling consists of a jacket of tissue, one to several cells in thickness, termed the NUCELLUS, which typically encloses a single, large MEGASPORE MOTHER CELL (FIG. 12-24). This structure, in turn, is contained within a MEGASPORANGIUM. In seed plants other than angiosperms the megasporangium is walled. In the angiosperms the nucellus is usually considered to be the megasporangium wall. It is

more likely, however, that the nucellus is sterile, protective parental tissue comparable to the tissues of the anther that surround the spore-producing tissue, and that a megasporangium wall is absent. The megasporangium then would consist merely of a chamber, enclosed by the nucellus, within which the early events in the production of a seed occur.

The nucellus, as a result of basal growth, is soon elevated on a short stalk. Then, as development proceeds, one or two enveloping, protective layers, the INTEGUMENTS, grow up from the base of the nucellus and eventually completely surround it, except for a small opening, the MICROPYLE (FIGS. 12-23B–E, 12-24). The ovule may be erect, but in most flowering plants it

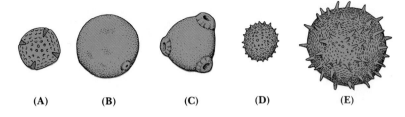

Fig. 12-22 Pollen grains. (A): of ash (*Fraxinus americana*). (B): orchard-grass (*Dactylis glomerata*). (C): fuchsia (*Fuchsia hybrida*). (D): ragweed (*Ambrosia artimisiifolia*). (E): mallow (*Malviscus arboreum*).

(A)   (B)   (C)   (D)   (E)

Fig. 12-23   (A–C): development of megaspore mother cell and megaspore. (D–F): stages in development of the embryo sac. (G): the mature ovule.

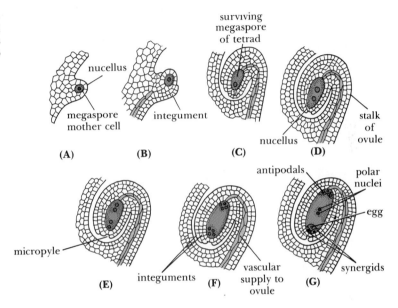

becomes inverted, with the micropyle pointing toward the placental region and the stalk fused or partly fused with the integument.

The megaspore mother cell soon divides twice, and four MEGASPORES are formed. These are meiotic divisions. The megaspores are arranged differently from the tetrad of microspores. They lie in a single line. Moreover, only one of the megaspores persists, usually the one farthest from the micropyle and nearest the food supply; the other three disintegrate (Fig. 12-23C).

The Embryo Sac   The single, large, surviving megaspore now gives rise to an oval or elongated EMBRYO SAC, so-called because the plant embryo forms here. During the development of the embryo sac, the megaspore nucleus divides three times by mitosis and forms eight genetically identical nuclei (Fig. 12-23D–G). Food and water are absorbed through the stalk of the ovule, and the sac enlarges along with the nucellus and integuments. Four of the nuclei become located at the micropylar end of the sac, and four at the op-

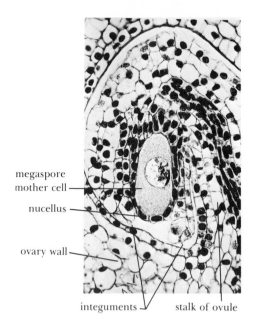

megaspore
mother cell

nucellus

ovary wall

integuments   stalk of ovule

Fig. 12-24 Photomicrograph of longitudinal section through young ovule of tiger lily (*Lilium tigrinum*), megaspore mother cell stage. The integuments have not yet completed growth.

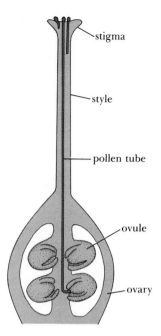

stigma

style

pollen tube

ovule

ovary

Fig. 12-25 Germination of pollen grains on the stigma and growth of a pollen tube to an ovule. Diagrammatic.

posite end. A single nucleus, the POLAR nucleus, now moves from each group of four to a central position, and thin membranes develop around each of the six remaining nuclei.

The three cells at the micropylar end are the EGG CELL and two neighboring cells partly surrounding the egg, known as SYNERGIDS. The three cells at the opposite end of the embryo sac are termed the ANTIPODALS. In some plants the antipodals continue to divide; in corn (*Zea mays*), for example, as many as 20 may develop. The synergids and antipodals are short-lived and are believed to be vestiges of structures in the ancestors of the flowering plants. Their function in modern angiosperms is not well understood, but the synergids, at least, appear to be metabolically active previous to fertilization and are rich in accumulated food materials. It has been

suggested that they function in the absorption and transfer of food from the surrounding parental tissue to the egg. The fully developed embryo sac, consisting usually of seven cells and eight nuclei, together with the surrounding nucellus, integuments, and stalk, constitutes the mature ovule (FIG. 12-23G).

Pollen Germination; The Pollen Tube A pollen grain germinates on the stigma (FIGS. 12-15, 12-25), and the pollen tube grows downward through the style to the ovule. If the generative cell has not already divided, forming the two male gametes, it divides in the pollen tube. The gametes (each consisting of a large nucleus surrounded by a cytoplasmic sheath) move down the pollen tube. The tip of the tube passes through

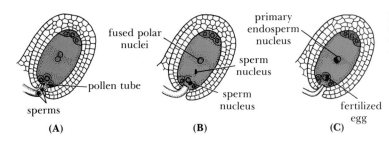

fused polar nuclei

pollen tube

sperms

(A)

primary endosperm nucleus

sperm nucleus

sperm nucleus

(B)

fertilized egg

(C)

Fig. 12-26 Double fertilization. One sperm nucleus fuses with the egg nucleus; the other with the polar nuclei.

the nucellus and enters the embryo sac, where it discharges the sperms (FIG. 12-26). The tube nucleus may move ahead of the sperms down the tube, and it was once believed that it directed the activities of the pollen tube during growth. But it may follow the gametes; moreover, numerous cases are known in which the tube nucleus degenerates, either previous to the germination of the pollen grain or in early stages of pollen-tube growth. Many investigators, therefore, view the tube nucleus as a vestigial structure that plays no role in the growth of the pollen tube.

Fertilization   The male gametes of most organisms are capable of active movement by means of specialized whiplike structures. No such structures are found in the male gametes of angiosperms, which apparently lack the power of independent movement. The mechanism of their movement down the pollen tube is unknown at present; cytoplasmic streaming as well as changes in the turgor of the pollen tube have been suggested.

In the embryo sac, one of the two sperm nuclei fuses with the egg nucleus, fertilizing it and forming the first cell of the new plant (FIG. 12-26). About the same time, a second fusion occurs, involving the two polar nuclei and the nucleus of the second sperm. The two polar nuclei may first fuse and then unite with the second sperm nucleus, or all three nuclei may come together simultaneously. The nucleus resulting from this fusion of three nuclei is called the PRIMARY

ENDOSPERM NUCLEUS, or TRIPLE FUSION NUCLEUS.

The union of egg and sperm nuclei, together with the fusion of the second sperm nucleus with the polar nuclei, is called DOUBLE FERTILIZATION. Double fertilization was discovered almost simultaneously by two botanists, one Russian and the other French, in the closing years of the nineteenth century. This remarkable phenomenon, found only in the angiosperms, has since been studied intensively in an attempt to understand its evolutionary history and its significance in seed formation. Double fertilization must occur within every ovule if seed formation is to follow. At least one pollen grain must germinate on the stigma for each ovule that develops into a seed. In the watermelon, for example, this means that hundreds of pollen grains are required to pollinate one flower.

The time elapsing between the germination of the pollen grain and fertilization is usually short, but it may be several days, weeks, or even months. In barley, it is less than 1 hour; in corn, about 24 hours; in tomato, approximately 50 hours; in cabbage, about 5 days. Witch hazel (*Hamamelis virginiana*), a widely distributed shrub of the eastern United States, flowers very late in the autumn. Pollination and pollen-tube formation take place then, but fertilization is delayed until spring, five to seven months later.

The Embryo Sac and the Seed   By the time of fertilization, or shortly afterward, the nucellus has become disorganized. Following fertilization, the

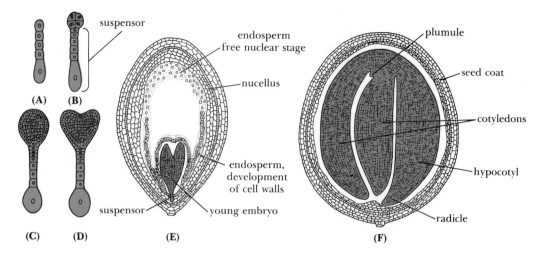

Fig. 12-27 Stages in the development of the embryo of cabbage (*Brassica oleracea* var. *capitata*). Diagrammatic. (A–D): differentiation of the embryo and suspensor from the proembryo. (E): young seed, embryo, and developing endosperm. (F): mature seed, consisting chiefly of the embryo and the seed coat.

synergids and antipodal cells also disintegrate. The fertilized egg grows into the embryo. The stages in the development of the embryo and seed of cabbage (FIG. 12-27) are typical of many dicotyledons. The zygote, after several divisions, produces a row of cells, the PROEMBRYO, which protrudes into the embryo sac. The uppermost cell of this row, the one farthest from the micropyle, divides by the formation of transverse and longitudinal walls to form a group of eight cells in two tiers of four cells each. This group of cells gives rise to most of the embryo; the remaining cells below constitute the SUSPENSOR. The growth of the suspensor pushes the growing embryo into the ENDOSPERM, where abundant food is available. The suspensor persists for a time, but eventually disintegrates (FIG. 12-27A–E).

The mature embryo (FIG. 12-27F) consists of an axis bearing two COTYLEDONS, or seed leaves. At the summit of the axis, above the cotyledonary node, is found the PLUMULE. This is the apex of the embryonic shoot, and in the cabbage and

some other plants it consists only of a small group of meristematic cells. In other plants, such as beans, the plumule is composed of an apical meristem together with several embryonic leaves. At germination the plumule gives rise to that portion of the shoot above the cotyledons. The basal, tapering end of the embryo, termed the RADICLE, develops into the primary root when the seed germinates. The region between the radicle and the cotyledons is the embryonic stem, or HYPOCOTYL. In the embryos of some seeds, such as corn, the hypocotyl is very short.

The general pattern of development of the embryo of monocotyledons is much the same as that of cabbage except that only one cotyledon is formed instead of two. In advanced monocotyledons, however, such as the grasses (including corn), the cotyledon has undergone extreme evolutionary modification. It is composed of two chief parts: (1) the SCUTELLUM, a food-absorbing organ; and (2) the COLEOPTILE, a protective cap over the plumule (FIG. 12-28B).

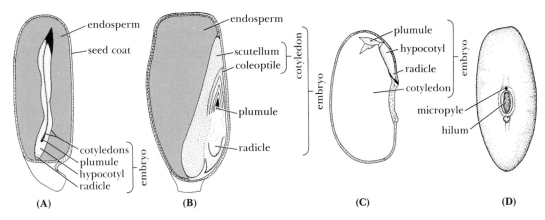

Fig. 12-28 Seeds. (A): longitudinal section of seed of castor bean (*Ricinus communis*). The plumule is minute. (B): longitudinal section of grain of corn (*Zea mays*). (C): bean (*Phaseolus vulgaris*) seed. One of the cotyledons is removed. (D): bean seed, showing micropyle and hilum.

Following fertilization, the primary endosperm nucleus soon begins to divide, eventually to give rise to a multicellular tissue, the endosperm. The fertilized egg grows into an embryo, but its growth lags behind that of the endosperm, for the zygote, following fertilization, usually enters a period of dormancy, which may last only a few hours or, in some species, several days. The endosperm grows by means of food supplied by the parent plant, and it, in turn, nourishes the embryo. In many species, in the early stages of endosperm formation, numerous free nuclei are formed (FIG. 12-27E). Later, walls develop around most of these nuclei. In other species, nuclear divisions are quickly followed by cell-wall formation. The endosperm grows faster than the embryo, and the young seed, supplied with abundant food, enlarges rapidly. In some seeds the embryo remains relatively small and is surrounded by endosperm. The endosperm persists, enlarges, and becomes a prominent tissue of the seed, rich in accumulated food in the form of oil or starch, and protein. The foods stored in the endosperm are utilized by the embryo when the seed germinates. Seeds in which the embryo is embedded in endosperm include castor bean (FIG. 12-28A), corn (FIG. 12-28B) and other cereals, buckwheat, date, and coconut. In the stony seed of the date the primary walls of the endosperm cells are much thickened, and composed chiefly of hemicellulose. This is the storage material of the seed, and is converted to sugars by enzymatic action at germination. The milk of the coconut is the free nuclear stage in the development of the endosperm and consists of approximately 1½ pints of fluid, containing great numbers of nuclei. As the seed matures, these nuclei migrate to the outside and cell walls form around them, producing the mature endosperm, or "meat."

In other seeds, perhaps in the majority, the embryo continues its growth until all the endosperm is absorbed. The embryo increases in size, and its cells become filled with accumulated food materials. Most of this food is accumulated in the cotyledons, which become greatly enlarged. Examples of seeds lacking endosperm are radish, cabbage (FIG. 12-27), sunflower, squash, acorns and walnuts, and the legumes, such as pea and bean (FIG. 12-28C, D).

The seed is surrounded by the SEED COAT, which has developed from the integuments of the ovule. The seed coat is usually thin as in the garden bean, and in the peanut, in which it forms the brown, papery layer around the embryo. But it may be thickened and hard or stony, as in the Brazil nut. The epidermis of the seed coat of the cotton seed produces the cotton fibers, which are spun into thread.

In many seeds the micropyle is still evident as a minute pore, and this is usually associated with a scar, the HILUM (FIG. 12-28D), which marks the position of the stalk that attached the seed to the placenta. As the seed ripens, the embryo gradually passes into the dormant state, in which it remains until it germinates.

The angiosperm seed is a complex structure, and its tissues are of varied origin. It may be composed of seed coat, endosperm, and embryo, as in corn, wheat, and rice; or of seed coat and embryo alone, as in bean, peanut, and soybean.

Besides being essential to plant reproduction, seeds are of great value to man. The production of edible seeds is probably man's most important agricultural activity. Such seeds as wheat, corn, rye, and rice are basic items in man's daily diet. Seeds are of equal importance as food for many other animals. Numerous birds and many rodents feed mainly on seeds. The food supply of a great proportion of land animals, including man, thus depends in whole or in part upon pollination and the intricate behavior of nuclei and cells within the anther and ovule.

ALTERNATION OF GENERATIONS

Alternation of generations is the occurrence of two phases, or generations, in the life cycle of an organism that reproduces sexually. One of these generations produces the spores and is called the SPOROPHYTE generation. The other, which produces the gametes, is known as the GAMETOPHYTE generation. The word GENERATION is used here in a different sense from the usual one,

which refers to the time interval between the birth of the parents and that of the progeny.

This alternation of generations corresponds with an alternation of the chromosomes number in the two phases of the life cycle of the plant. It will be recalled from the study of the cell (Chapter 3) that the number of chromosomes in each of the body cells of an individual of a given species is normally constant from one generation to another. When two gametes fuse to form a zygote, each gamete contributes a set of chromosomes to the fertilized egg, which thus acquires twice the number of chromosomes found in either gamete. The nucleus of the fertilized egg divides by mitosis, and so in each daughter cell half of the chromosomes originate from the sperm and half from the egg. All subsequent cells of the plant arising from repeated divisions of the fertilized egg also contain this double set of chromosomes.

Since the number of chromosomes varies with the species, it is convenient to designate the chromosome number of the fertilized egg, and hence of the cells of the plant body to which it gives rise, by the number $2n$. And since the number of chromosomes in each gamete is but half this number, the chromosome number of a gamete is designated as n. It is therefore apparent that the *doubling* of the chromosome number when the egg is fertilized is accompanied by a *reduction* in the chromosome number at some later stage in the life cycle. If this were not the case and the gametes contained the same number of chromosomes as the body cells—that is, $2n$—the fertilized egg and the cells of the resulting plant would contain $4n$ chromosomes. The next generation would have $8n$, and each cell would soon contain an impossibly large number of chromosomes.

In flowering plants a reduction in the chromosome number occurs, as has been pointed out, during the formation of microspores and megaspores from their respective mother cells. The two divisions, collectively called MEIOSIS, that lead to the formation of spore tetrads result in the halving of the chromosome number. The details of meiosis will be discussed in Chapter 16, where it will be shown that the process is significant from other standpoints in addition to the reduction in chromosome number.

The spore mother cells, then, possess the $2n$ chromosome number; the microspores and megaspores, the n number. All the structures formed directly from either the microspores or the megaspores also possess the n chromosome number. With the fusion of egg and sperm, the chromosome number is again doubled and becomes $2n$. All the cells of the plant are thus either n or $2n$, with the exception of those of the endosperm, which is formed following the fusion of three nuclei. Since each of these carries the n chromosome number, the endosperm is $3n$. The endosperm, however, is used up by the growth of the embryo, either before or during germination, and thus does not affect the chromosome number of other parts of the plant.

The limits of the two generations, sporophyte and gametophyte, are determined by the occurrence of meiosis and fertilization. The sporophyte generation is $2n$; the gametophyte is n. The first cell of the sporophyte generation is therefore the fertilized egg. This generation comes to an end with the formation of the two kinds of spore mother cells. Intervening between the fertilized egg and the spore mother cells is the plant body, such as that of a cotton plant or an elm tree, which is thus the most prominent part of the sporophyte generation. The gametophyte generation begins with the spores and ends with the gametes. Since, in the flowering plants, there are two kinds of spores, two kinds of gametophytes, males and female, are distinguished. The structures of each generation are summarized in Fig. 12-29. The endosperm has been omitted from the diagram, since, being neither n nor $2n$, it belongs to neither generation. Perhaps it is gametophytic tissue complicated by nuclear fusions.

Alternation of generations is not confined to the flowering plants but is general throughout the plant kingdom. Since the fusion of gametes and meiosis are universally associated in the same life cycle, not only fertilization but meiosis as well is regarded as an essential feature of sexual reproduction. Both the sporophyte generation, which produces spores, and the gametophyte generation, which forms gametes, are thus sexual generations. In certain dioecious plants, in fact, the $2n$ body cells contain chromosomes that are involved in sex determination.

In the vascular plants, such as the flowering plants, conifers, and ferns, the sporophyte generation is the prominent phase in the life cycle and the gametophyte generation is reduced to a relatively inconspicuous stage. But, as we shall see in later chapters, in other kinds of plants, such as the mosses and certain kinds of algae, the gametophyte generation is the more conspicuous.

POLLINATION

The fertilization of the egg and the subsequent development of the seed can occur only if pollen grains are deposited upon the stigma. POLLINATION, the transfer of pollen from the anther to the stigma, should be distinguished from fertilization, which is the fusion of the male and female gametes. Pollination is of two types. SELF-POLLINATION is the pollination of a stigma by pollen from the same flower or from another flower of the same plant. CROSS-POLLINATION is the transfer of pollen from the anther of a flower of one plant to the stigma of a flower of another plant of the same or related species.[1] The chief agents of pollination are wind and insects, but birds, other animals, and even water may carry pollen from one flower to another.

Self-pollination occurs in many plants. Among the economically important self-pollinated plants are oats, wheat, barley and rice, peas and beans, soybeans, peanut, flax, cotton, tomato, citrus fruits, peaches, and tobacco. Cross-pollination,

---

[1] The transfer of pollen between members of the same clone, however, is self-pollination.

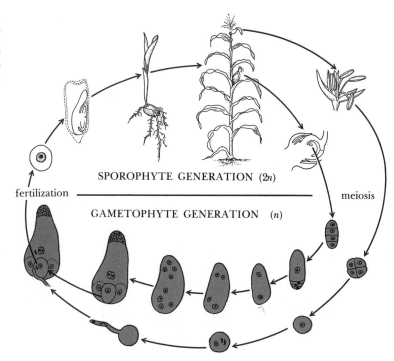

Fig. 12-29 Alternation of generations in an angiosperm (corn). Gametophyte generation in color. Note that more than three antipodals are found in corn and certain other angiosperms.

SPOROPHYTE GENERATION (2*n*)

fertilization

GAMETOPHYTE GENERATION (*n*)

meiosis

however, is more common than self-pollination and occurs, at least occasionally, in self-pollinated species. Cross-pollination brings about a more diverse combination of hereditary units of the two parents. This results in increased variability in the offspring and greater adaptability to new environments—conditions of evolutionary advantage to the species.

A more immediate effect of cross-pollination in many species is the production of more seeds or greater vigor in the offspring. A very large number of flowering plants have adaptations that prevent or reduce self-pollination. These many adaptations are usually related to pollination by insects. Prominent among them are modifications of the flower that make cross-pollination by insects possible and in some cases essential if seeds are to be formed. So numerous and varied are these adaptations to cross-pollination that ignorance of them has caused serious difficulties when new plants have been introduced. An example of this is afforded by the history of Smyrna

fig production in California. Large numbers of cuttings were introduced from Asia Minor about 1880. These grew well, but the fruits dropped from the tree before maturity. The realization that pollination is brought about by a tiny wasp resulted in the introduction of the insect, after which the industry flourished. Information on pollination is essential to the plant breeder when he attempts to produce varieties of greater usefulness to man.

Insect Pollination A large proportion of flowering plants are insect-pollinated. The insects are chiefly bees, wasps, butterflies, and moths, but beetles, flies, and other kinds also frequent flowers. Pollination is also brought about by birds and even mammals (bats). Insect-pollinated flowers are usually brightly colored or scented, sometimes both. Their pollen is heavy or sticky and is not readily scattered by wind.

Many such flowers contain NECTARIES,

specialized tissues or organs that secrete nectar, a fluid with a sugar content varying from 4 to 65 percent. This nectar is the raw material from which honey is made. Nectaries vary greatly in form and location and may be associated with any of the floral organs. Nectar may be secreted by the surface of the receptacle, by the walls of spurlike projections of the perianth, or by hairs on the petals or ovary. In many flowers the nectary consists of a ring around the base of the ovary. In some, the stamens or petals have become reduced and modified into nectaries.

Numerous experiments, especially on the honeybee, have confirmed the general belief that insects are attracted to flowers by color or scent. The sense of smell of the honeybee seems to be about as sensitive as that of man, but that of other insects, especially moths, is more powerful, and they may detect odors at considerable distances. Honeybees are partly colorblind. They are colorblind to pure red, which appears to them as black. They can perceive four main groups of colors, but do not distinguish between colors within each group: (1) yellow, including orange and yellow-green; (2) blue, including violet and purple; (3) blue-green, and (4) ultraviolet. Red flowers are usually pollinated by butterflies, which are not color-blind to red, or occasionally by humming birds.

Bees and other insects visit flowers and gather pollen or nectar as food for themselves or their progeny. Pollination is incidental to these activities. Pollen will usually be found adhering to the mouth parts, head, legs, and body hairs of a bee after it has visited a flower (FIG. 12-30). If the bee then visits another flower, some of the pollen may adhere to the stigma, resulting in cross-pollination.

Insect pollination is of the greatest importance from the standpoint of plant evolution and reproduction. It is of economic importance also, for many kinds of plants used directly by man or as food for domestic animals depend upon insect pollination. They include more than 50 kinds of crop plants in this country, among them apple and pear, the Smyrna fig, melon and cucumber, avocado, cabbage, buckwheat, alfalfa, (FIG. 28-8)

Fig. 12-30 Pollen grains of apple adhering to the hairy coat of a bee.

and many clovers. In the absence of certain insects, such plants yield neither seed nor fruit.

Floral Adaptations Favoring Cross-Pollination In many plants with perfect flowers, the stamens and pistils mature at different times. This, of course, favors cross-pollination. Such a flower is said to be DICHOGAMOUS. Dichogamy manifests itself in two ways. In the most common, the anthers ripen before the stigma is ready to receive pollen. In some plants the stigma matures before the pollen sacs of the stamens open. The anthers of Jacob's-ladder (*Palemonium caeruleum*), a garden ornamental, shed their pollen before the stigma of the same flower is ready to receive it;

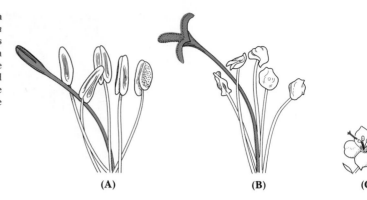

Fig. 12-31 Dichogamy in Jacob's-ladder (*Polemonium caeruleum*). (A): the stamens are beginning to shed pollen but the stigmatic lobes are closed. (B): the pollen is shed and the stigmatic lobes are receptive. (C): an entire flower.

(A)   (B)   (C)

the three lobes of the stigma are folded together at this stage (FIG. 12-31). After the pollen is shed, the lobes of the stigma spread apart and may be pollinated by pollen from younger flowers.

Flowers with pistils and stamens maturing at different times are extremely common, and most families of flowering plants have some species in which this condition is present. The interval between the maturation of the stamens and the time when the stigma is receptive varies from an hour or so to several days. Other examples of dichogamous plants are avocado, carrot, and the numerous species of eucalyptus.

Insects that rove or climb over the floral parts pollinate many kinds of flowers that have no special structural adaptations to insect visitors. These visits take place by chance, and the insects bring about cross-pollination only occasionally. However, the flowers of many plants are so constructed that they are visited by one kind or at most a few kinds of insects. These insects regularly visit flowers of the same species, so that pollen is commonly carried from the anthers of one flower to the stigmas of another. Such flowers have numerous adaptations that favor cross-pollination and that make self-pollination unlikely or impossible. The stigma is commonly placed so that it is brushed by an insect visiting the flower, but does not receive pollen from the anthers of the same flower. This condition commonly results when the elongated style projects well above the anthers.

In many flowers there is a close relationship between the depth of the corolla tube and the length of the sucking mouth parts of the insects that usually visit them. The nectar at the base of the corolla tube may be so deep that only a butterfly or moth with long tubular mouth parts can reach it. This is true of a number of strongly scented white or nearly white flowers that open in the evening. Soapwort (*Saponaria*), some species of tobacco, and the Jimson weed (*Datura*), which has a corolla tube 3 inches long, are pollinated only by night-flying hawkmoths. The red clover is pollinated by the wild bumblebee, whose mouth parts are of sufficient length to reach the nectary at the back of the flower. Honeybees, also, may pollinate red clover when collecting pollen, even though they cannot reach the nectar. Flowers with a short corolla tube or exposed nectar are likely to be visited by a variety of insects, including flies.

Cross-pollination is also promoted in plants in which the length of the style differs in individuals of the same species (HETEROSTYLY). Heterostyly is very common. It is found in a large number of plant families, including the pinks, St. John's worts, heaths, wood sorrels, gentians, poppies, pigweeds, and many others. On one kind of plant all the flowers have long styles and the anthers are located below the stigma. In the other kind the flowers have short styles and the anthers are above the stigma (FIG. 12-32). The mouth parts of an insect, thrust into a short-styled flower, come in contact with the anthers and become covered with pollen. If the insect should now enter a long-styled flower, the pollen is deposited upon the stigma, which is located at approximately the same level as the anthers in a short-styled flower. Similarly, the pollen adhering to the mouth parts of an insect that has visited a long-styled flower will be deposited upon the stigma of a short-styled flower, which is at the same level in the corolla as the anthers of the long-styled flower.

Self-pollination is not precluded by heterostyly, for an insect in withdrawing its mouth parts from the corolla of a long-styled flower may occasionally transfer pollen from the anther to the stigma of the same flower. Self-pollination will be even more likely to occur in the short-styled form, for the mouth parts can readily carry pollen downward and deposit it on the stigma. The extent to which self-pollination results in seed formation varies greatly, however, among heterostylic plants. Some are almost completely fertile when self-pollinated; others produce fewer seeds than when cross-pollinated; and some, such as *Houstonia* (Fig. 12-32), produce none. The reduction in fertility under self-pollination is due to the condition of self-incompatibility, discussed below.

Innumerable variations in pollination mechanisms and adaptation of flowers to insect visitors are known. A final example illustrates how complex the relation between flowers and insects has become.

Pollination in salvia The salvias (sages) have a pollination mechanism in which a sudden movement of the filaments results in the transfer of pollen to an insect visitor. This mechanism is best known in *Salvia pratensis*, a cultivated ornamental herb with blue flowers. The flower is irregular, with a strongly developed upper and lower lip (Fig. 12-33), the latter serving as a landing platform for the pollinating bees. Nectar is secreted at the base of the ovary. Only two stamens are produced, and the pollen sacs are concealed under the upper lip. What appears to be a filament is actually a downward prolongation of the connective tissue (the tissue connecting the pollen sacs on either side) and expanded

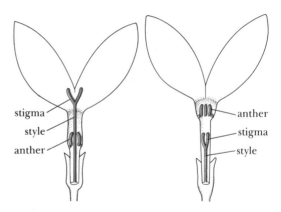

Fig. 12-32 Heterostyly in the bluet (*Houstonia caerulea*). (Left): long-styled flower. (Right): short-styled flower.

at the base. This expanded part is fused with a similar structure from the other anther, forming a plate that closes the mouth of the corolla tube. The true filaments are very short, and are located on either side of the connectives and attached to them.

The flower of salvia is also dichogamous. When it first opens, the pollen is ripe but the stigmatic lobes are closed and not receptive. If a bee enters the flower to obtain nectar, it encounters the plate, which is pushed backward and upward, causing the connectives and their anthers to tip forward. The anthers then come in contact with the back of the bee, which is dusted with pollen. When the insect withdraws, the connectives snap back to their former position. The stigma, located above the anthers, now matures; it grows downward and the lobes open. If the same bee then visits one of these older flowers, it brushes against the downwardly directed lobes, to which the pollen is transferred. Pollination of the stigma by pollen from the same flower would seldom occur.

Incompatibility In many plants with perfect flowers, fertilization and fruit and seed formation follow self-pollination. This condition is

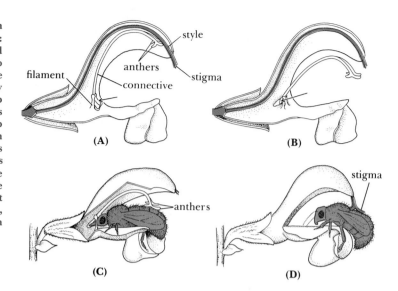

Fig. 12-33 Pollination in salvia (*Salvia pratensis*). (A): arrow points to plate formed by fusion of bases of the two connectives. (B): pressure against the plate (indicated by arrow) causes anthers to tip forward. (C): bee pushes against plate, anthers tip forward and deposit pollen on back of bee. (D): bee visits older flower. The stigma has now grown downward, the lobes open, and the bee brushes against the stigma that receives the pollen. (Stigma, style, and ovary are shown in color.)

known as SELF-COMPATIBILITY. In other plants with perfect flowers, fertilization does not occur when the stigma is pollinated by pollen of the same flower. Even pollen from another flower of the same plant or from flowers of other plants of like hereditary constitution will not result in fertilization. This condition is known as SELF-IN-COMPATIBILITY.

Many species that show the phenomenon of incompatibility can be arranged in groups of individuals that fail to set seed when pollinated by other individuals of the same group, but are fertile when pollinated by individuals of other groups. Many hundreds of such species, wild and cultivated, are known. One estimate, made some years ago, yielded a figure of more than 3000 species of angiosperms that are self-incompatible. It is probable that this is only a fraction of the actual number of incompatible species. Most cases of incompatibility are due to the slow rate of pollen-tube growth on the stigma or in the style when the flower is pollinated by pollen from flowers of like heredity. The rate of growth is governed by certain genes. If the pollen tubes carry the same sterility genes as the style, the rate of growth is so slow that the flower usually withers and falls away before fertilization takes

place. If the genes of the style are suitably different from those of the pollen tube, the tube grows rapidly and fertilization occurs.

In most ornamental plants, self-incompatibility is of little significance because these plants are grown for their flowers or foliage. Neither is incompatibility usually important in the commercial crop production of plants propagated by seeds. Red clover, for example, which is self-incompatible, can nevertheless produce seed because cross-pollination occurs between genetically different individuals in the field. But incompatibility in a plant propagated vegetatively and cultivated as a clone, such as apples, pears, cherries, plums, and almonds, is another matter. Some varieties of apples set fruits well if self-pollinated or pollinated by pollen from other trees of the same variety. But many are self-incompatible, and commercial orchards of these varieties are therefore planted with two varieties that have pollen which is effective in cross-pollination and blooming periods that overlap.

Certain combinations of varieties are unsuccessful, and the grower must have a list of the varieties that are compatible when crossed. The varieties are set out in the orchard so that several rows of one variety alternate with rows of another. Bees bring about cross-pollination. Frequently, hives are rented from beekeepers and placed in the orchard during the period of bloom. Even many of the self-pollinated varieties set a better crop when crossed with another variety.

Pears, and plums, like apples, include some self-incompatible varieties. Almonds and sweet cherries are self-incompatible, but sour cherries, nectarines, apricots, and peaches are not. A practical knowledge of the aspects of pollination may mean success or failure to the grower of fruit trees.

Wind Pollination   This is in many respects the simplest form of pollination. The flowers of wind-pollinated plants are usually small and inconspicuous. Devoid of bright colors and nectar, they are commonly grouped in dense clusters. The perianth is often reduced or lacking. The stamens may protrude prominently from the flower, suspended by long filaments. The stigmas are frequently large and branched. The pollen grains are small, light, and dry, and are commonly produced in larger quantities than in insect-pollinated flowers (FIGS. 12-14, 12-34).

Wind pollination is found in both woody plants and herbs. Among wind-pollinated woody plants are the conifers, the poplars, oaks, ashes, elms, birches, hickories, and sycamores. In many woody plants the staminate and sometimes the pistillate flowers are grouped into drooping clusters called catkins. The wind-pollinated herbaceous plants include pigweeds, sorrels, docks, plantains, nettles, meadow rue, hops, grasses, rushes, and sedges. Some species may be either wind- or insect-pollinated. The small flowers of willows, for example, are brightly colored and contain nectaries. But if the pollen is not removed by insects, it may be carried away by wind. Some maples, too, are both wind- and insect-pollinated.

Hay Fever   Large numbers of people are aware, to their sorrow, that numerous wild and cul-tivated plants are wind-pollinated. These people are sensitive to the pollen of such plants, a condition known as an *allergy*. The same pollen is harmless to other individuals. An allergy to plant pollens is popularly called "hay fever."

In order to cause hay fever, the pollen of any given plant must have three important characteristics: it must be produced in considerable quantities; it must be buoyant so that it floats readily in the air; and it must be irritating to allergic people. Thus, although nearly all hay fever is produced by wind-pollinated plants, all pollen carried by the wind does not cause hay fever. The cattails shed large quantities of pollen, as do the conifers, yet except for a few species of conifers these plants do not cause hay fever.

The goldenrods, cosmos, roses, and sunflowers are popularly believed to be important sources of hay-fever pollen. The pollen affects some people, but these plants are insect-pollinated. Their pollen is heavy and is not readily carried by the wind. Only intimate contact, such as handling roses or goldenrod, will produce symptoms of the disease in an allergic person. Usually these plants are only minor offenders in causing hay fever.

Pollination Seasons   Whether a sensitized person will develop hay fever depends upon the kind and amount of pollen in the air. This in turn depends, except for such atmospheric conditions as rain and wind, upon the plant species that occur in a given region. The time of year when the symptoms of hay fever appear is related to the flowering period of these plants. Three hay-fever seasons are recognized in many parts of the United States. But there is great local variation in the distribution of the plants that cause hay fever and in the pollination periods of these plants.

*Early spring (**March to May**)*   The usual cause of hay fever in this period is tree pollens. Among the trees concerned are the box elder, the elms, poplars, birches (FIG. 12-34), oaks, hickories, ashes, and the sycamores.

Fig. 12-34 Wind-pollinated plants that may cause hay fever. (Left): catkins of paper birch (*Betula papyrifera*). (Center): flowers of timothy (*Phleum pratense*). (Right): great ragweed (*Ambrosia trifida*), staminate flowers.

Late spring and early summer (May to July)
Grasses are the most important hay-fever plants during this period. Among the more important species are Kentucky bluegrass, redtop, orchard grass, timothy (FIG. 12-34), sweet vernal grass, and Bermuda grass. English plantain, a widely distributed weed, may also be a factor in hay fever during this season.

Late summer and fall (August to September) The ragweeds (FIG. 12-34) are outstanding pollen producers during this season. In certain western states, other plants such as Russian thistle (*Salsola*), saltbush (*Atriplex*), and sagebrush may be more important than the ragweeds.

The plants of significance in causing hay fever, the pollination periods of such plants, and the quantity of pollen in the air can be determined by pollen surveys for a specific region. Atmospheric pollen counts are made by exposing to the air microscope slides covered with glycerine jelly or Vaseline. The pollen that adheres to the slide is stained and studied under a high-powered microscope and is identified by surface features, which vary widely in pollen grains of different species (FIG. 12-22). Newspapers sometimes publish daily pollen counts for the information of hay-fever victims.

SUMMARY

1. Sexual reproduction in flowering plants involves the fusion of male and female gametes, resulting in a fertilized egg. The cells and tissues that give rise to the gametes are produced within the flower.
2. The essential parts of the flower are the stamens and pistil. The stamen is divided into filament and anther; the pistil, into stigma, style, and ovary. The pistil may be simple, composed of but a single carpel, or compound, composed of two or more carpels. A perianth, composed of sepals and petals, may surround the stamens and pistils, but this is not an essential feature of the flower. The dicotyledons and monocotyledons may usually be distinguished by the number of floral organs in each whorl.
3. The flower is usually considered to consist of an axis with appendages, the whole comparable to a vegetative axis and its leaves. The stamens and carpels, since they bear spores, are called sporophylls.
4. A generalized (primitive) type of flower is bisexual and regular, with a superior ovary. A perianth is present, the stamens and carpels are spirally arranged, and the parts of

the flowers are not fused. Flowers have undergone numerous modifications during evolution.

5. Among the modifications of flowers that have arisen are fusion of petals into a floral tube, the fusion of the floral tube with the ovary wall (inferior ovary), and irregular symmetry. Floral evolution seems marked by a reduction in number of parts and fusion of parts of the flower.

6. Some flowers are bisexual, with both pistils and stamens. Others are unisexual, with only one of these organs. A species is monoecious when both kinds of unisexual flowers occur on the same plant. It is dioecious if the staminate and pistillate flowers are borne on different individuals of the species.

7. Many aspects of seed formation are microscopic. The seed is preceded by one-celled microspores and megaspores contained within the anther and ovule.

8. A young anther is composed of four microsporangial cavities enclosed within the sterile tissue of the anther. In the microsporangia are found the microspore mother cells that give rise by meiosis to tetrads of microspores. The microspores in turn develop into pollen grains, each containing a tube cell and a generative cell.

9. Enclosed in the pistil are the megaspore mother cells, surrounded by the nucellus. A megasporangial wall is probably absent. One or two integuments develop from the base of the nucleus and grow over it, leaving the micropyle. A megaspore mother cell divides by meiosis, forming a linear tetrad of megaspores, only one of which develops further. The megaspore gives rise to an embryo sac with seven cells and eight nuclei. The mature ovule consists of the stalk, the nucellus (if not disorganized earlier), the embryo sac with its eight nuclei, and one or two integuments.

10. After pollination, a pollen tube grows through the style. Two sperms are formed; one sperm nucleus fuses with the egg nucleus, the other with the polar nuclei to form the primary endosperm nucleus. This is double fertilization.

11. The fertilized egg gives rise to an embryo, which consists of one or two cotyledons, a plumule, hypocotyl, and radicle. The primary endosperm nucleus initiates endosperm tissue, which nourishes the embryo. The endosperm may be consumed by the embryo during the ripening of the seed or may constitute a prominent food-accumulation tissue of the seed, which is used during germination. The ripe seed may consist of seed coat, endosperm, and embryo, or only of seed coat and embryo.

12. The life cycle of a flowering plant consists of two phases, a gametophyte and a sporophyte generation. The sporophyte generation is $2n$, begins with the fertilized egg, and ends with the spore mother cell. It includes the fertilized egg, the embryo, and mature plant; the stamen (microsporophyll), with its microsporangial cavities, and microspore mother cells; and carpel (megasporophyll), nucellus, and megaspore mother cell. The gametophyte generation is n, begins with the microspores and megaspores, and ends with the production of gametes. The female gametophyte generation includes the megaspore and the mature embryo sac with its contained egg, synergids, polar nuclei, and antipodals. The male gametophyte generation includes the microspore and the tube cell, generative cell, and sperms, enclosed within the pollen grain or pollen tube.

13. Pollination should be distinguished from fertilization. Pollination is of two kinds: self-pollination and cross-pollination. Conditions that prevent self-pollination or favor cross-pollination include structural features of the flower, maturation of stamens and pistils at different times (dichogamy), and incompatibility. The chief agents of cross-pollination are wind and insects. Wind-pollinated plants are almost exclusively the cause of hay fever.

13

The Fruit, Seed, and Seed Germination

T HE ACCOUNT of the flower and development of the seed presented in the preceding chapter forms only a part of the story of sexual reproduction in flowering plants. This story would be incomplete without a consideration of the fruit, of fruit and seed dispersal, and of seed germination. These and related topics are the subject of this chapter.

What is a fruit? After fertilization, the ovary, together with its seeds, develops into the fruit. The mature ovary wall, termed the PERICARP, encloses the seeds of flowering plants; hence the name "angiosperms," meaning covered seeds. Some fruits are dry at maturity; others are fleshy. Some dry fruits open at maturity. Some do not, and small, dry, one-seeded fruits such as those of sunflower and corn are popularly called seeds. Some fruits are composed only of the matured ovary and its seeds; others include parts of the flower and structures closely related to the flower. A FRUIT is defined, then, as a ripened ovary (or sometimes a group of ovaries) and its contents, together with any adjacent parts that may be fused with it. Since fruits arise only from floral organs, their production is limited to the flowering plants.

THE FRUIT

Pollination, Fertilization, and Fruit Development Although fertilization affects only the ovules directly, it normally determines indirectly the development of the entire fruit. If the stigma is not pollinated and fertilization does not occur within at least some of the ovules, the flower usually withers and drops without further growth. The pollen grain contains auxins and gibberellins, but further growth of the fruit is usually not the result of the limited quantity of hormones in the pollen grain or pollen tube. The greatest impetus to further growth is hormone production by the developing seeds, and this in turn stimulates hormone production in the young ovary wall. Perhaps all three of the growth hormones—auxins, gibberellins, and cytokinins—may be produced in varying quantities. The stimulus thus builds up until, within a few days after pollination and fertilization, the entire fruit is growing actively, and its ability to compete for food has increased from nearly nothing to the highest level attained in the plant. Flowers and very young fruits of apples, corn, and other plants are poor competitors for foods, and will continue growth only when these materials are readily available. In contrast, rapidly developing fruits on these same plants will compete strongly for foods over distances of several feet. It is assumed that this competitive ability is based upon hormone production by the developing seeds (FIG. 13-2) and other parts of the fruit, and that these hormones, in some way, channel the movement of foods to the young fruit.

Parthenocarpy Some kinds of plants produce fruits even though fertilization does not occur, a phenomenon known as PARTHENOCARPY. Parthenocarpic fruits are seedless, and the fruit apparently maintains a sufficiently high hormone content to continue its growth in the absence of seeds. Not all seedless fruits are parthenocarpic, however, for fertilization is required for some seedless fruits to develop, but the young ovules abort and pericarp development, once started, continues. Such a condition is found, for example, in some varieties of seedless grapes. Among parthenocarpic plants, with neither pollination nor fertilization, are the cultivated varieties of the banana and pineapple, the Washington navel orange, some kinds of figs, and some varieties of seedless grapes.

Parthenocarpic fruits have been produced artificially by the use of synthetic growth substances incorporated into a paste and applied to the cut surface of the style, injected directly into the ovarian cavity, placed on the soil to be taken up by the roots, sprayed upon the buds or flowers, or even volatilized in the air about the plant. By any of these procedures a number of plants, including watermelon, cucumber, summer squash, holly (FIG. 13-1), and tomato may be induced to form seedless fruits. Commercial preparations of synthetic auxins are now available for use as a spray by home or market gardeners to increase the yield of tomatoes. Such a spray is chiefly valuable in the production of an early crop, for the earliest flower clusters may not set fruit, especially in cool weather. Although such tomatoes are often seedless, the chief objec-

Fig. 13-1 Effect of hormones on fruit set in holly. This plant is sometimes grown as a house plant from cuttings. But holly is dioecious, and staminate plants must also be grown and pollen transferred by hand to the pistillate flowers. Fruit set can be induced with dichlorophenoxyacetic acid, a plant growth regulator. (Left): control, not pollinated, which lost its flowers. (Right): flowers and buds sprayed with a very dilute solution of the growth regulator.

tive of the treatment is to prevent flower drop and to stimulate the growth of the young fruit. Seedless tomatoes may be produced parthenocarpically in the greenhouse, where normal pollination is difficult to obtain, but complaints of poorly flavored fruit have caused the method to be abandoned commercially. The flavor of tomatoes is concentrated in the juice of the seed cavities, and these juices do not form normally in parthenocarpic fruits. Synthetic growth substances have also been used to replace pollination by wasps in the Calimyrna (Smyrna) fig, the edible and fleshy receptacle of which normally develops only after pollination and subsequent growth of seeds. The practice has been largely abandoned because of the pasty texture of the seedless fruits.

Sprays of the hormone gibberellic acid are used on some normally seedless grapes to increase the number and size of the fruits produced. The importance in fruit development of gibberellic acid from the seed is emphasized by an experiment in which a paste of embryonic apple seeds was placed on unpollinated flowers. The gibberellinlike activity of this extract was so high that parthenocarpic fruits developed, only slightly smaller than fruits produced following pollination (FIG. 13-2)

Kinds of Fruits   The fruit is important to the plant from the standpoint of seed dissemination. It is also vastly important as a source of food for man and animals. So numerous are the modifications of fruits that their study can yield valuable information concerning plant relationships and plant evolution. Fruits are so varied, in fact, that it is difficult to fit all fruits into even the most detailed schemes of classification. In any case, an understanding of the nature of the fruit can be obtained only by a study of the flower from which it arose. Often, also, it is necessary to study the fruit in various stages of development in order to classify it properly. Fruits are usually classified into three large groups: (1) simple, (2) aggregate, and (3) multiple. Only the more important or more familiar kinds of fruits in each of these groups are considered in the following paragraphs.

Simple fruits   The development of the ovary of one pistil, simple or compound, produces a simple fruit. The ovary may be either superior or inferior, and the fruit may therefore be composed of the ovary alone or an ovary with which other parts of the flower (usually the floral tube) are fused. Simple fruits may be *dry* or *fleshy*.

SIMPLE DRY FRUITS   In simple dry fruits the fruit wall, as it ripens, becomes leathery, papery, or woody. Simple dry fruits include such types as

follicle, legume, capsule, achene, grain, samara, and nut. The follicle, legume, grain, and samara always develop from a superior ovary. The FOLLICLE is probably the most primitive type of fruit. It is derived from a simple pistil and opens (dehisces) along one side only. The most familiar example is the fruit of the milkweed (FIG. 13-3A). Some flowers, such as peonies, larkspurs, and columbines, contain a number of simple pistils, each of which develops into a follicle.

The LEGUME (FIG. 13-3C) is a dry fruit similar to a follicle. It, too, develops from a simple pistil, but usually dehisces along two sides. This is the characteristic fruit of the pea family. In some members of the family, such as alfalfa and

Fig. 13-2   Seeded and parthenocarpic apple fruits in section view. (A): seeded fruits from pollinated flowers; seeds removed and placed on cut surface. (B): seedless fruits produced by treating unpollinated flowers with lanolin paste containing gibberellic acid. (C): seedless fruits produced by treating unpollinated flowers with a lanolin paste containing a paste of embryonic apple seeds.

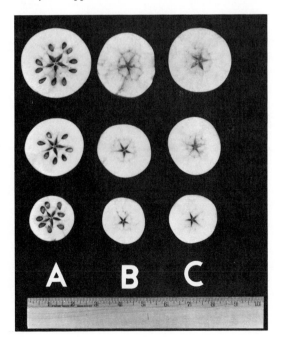

peanut, the fruit does not open at maturity (is indehiscent). A peanut is not (as it is popularly called) a nut, but an indehiscent legume. The legume is also called a pod, but this term is rather uncritically applied to the dry fruits of other families as well as the legumes.

The peanut deserves a further word, not only because of its economic importance, but also because of the unusual method of fruit production. The plant as cultivated is a low, herbaceous annual, 6–20 inches high. The small, yellow, pealike flowers are sessile and borne in clusters of three or more in the axils of leaves. Following self-pollination the flower withers, revealing a minute stalklike structure, the young fruit, that contains 1–3 minute, immature seeds at the tip. Within a few days the immature fruit exhibits a positive geotropic effect, and bends downward. The base of the young fruit is meristematic, and the activities of this zone produce an elongated stalk, or peg. The peg continues to grow downward, and finally penetrates the soil to a depth of 2–3 inches. After reaching this depth, the peg ceases to elongate, and the tip, containing the developing seeds, swells and matures into the familiar peanut fruit (FIG. 13-4). Thus, while the flower is aerial, the fruit is subterranean.

By the use of radioactive isotopes it has been found that the young fruits of the peanut are able to absorb certain minerals from the soil, to this extent behaving like roots. Calcium, especially, does not move from the plant into the

fruits in amounts adequate for normal development, and is supplemented by a supply taken up from the soil by the fruits. A deficiency of sulfur in the root area may also be compensated for by absorption by the fruits in amounts equal to about one-eighth that of root absorption.

The CAPSULE is a fruit that develops from a compound pistil with either a superior or an inferior ovary. It is one of the most common kinds of simple dry fruits and is characteristic of a number of plant families. The capsule may open in a variety of ways. It may split along the back of the individual carpels (FIG. 13-3D) or along the line where two carpels meet. The opening of the poppy capsule (FIG. 13-3B) is brought about by the formation of a circular row of pores near the summit of the fruit. When these capsules are shaken by the wind, the seeds sift out through the pores. The fruit of the horse chestnut is a three-valved capsule, the thick-shelled valves falling away at maturity and releasing the single large seed. The peculiar capsule of the Brazil nut (FIG. 13-5) is a thick, globular shell, sometimes 6 inches in diameter, weighing from 2–4 pounds and containing 12–20 seeds, each with a hard woody seed coat. A small lidlike cover develops at the upper end and separates when the fruit is ripe. The opening, however, is not large enough to allow the seeds to fall out. Germination of the seeds takes place in about three months, and the seedlings grow out through the opening in the capsule. The competition for light and space is so

Fig. 13-3 Simple dry fruits. (A): follicle of milkweed (*Asclepias syriaca*). (B): capsule of poppy (*Papaver*). When the unripe capsules of the opium poppy are gashed, a milky juice (latex) oozes out. When dried, this becomes the crude opium of commerce. (C): legume of lupine (*Lupinus*). (D): capsule of iris (*Iris versicolor*).

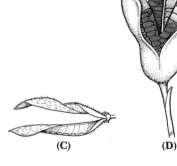

(A)   (B)   (C)   (D)

Fig. 13-4 Basal portion of a peanut plant (*Arachis hypogaea*), with mature fruits, immature fruits, and growing pegs that will produce fruit in the soil.

Fig. 13-5 Brazil nuts (*Bertholletia excelsa*). The globular shell is the fruit, or capsule. The structures within are seeds, not nuts.

great, however, that only a few survive, and these are finally released by the decay of the woody shell.

The ACHENE, GRAIN, SAMARA, and NUT, unlike the fruits just described, do not open at maturity; they are indehiscent. The achene (FIG. 13-6A) is a small dry fruit containing a single seed. The seed nearly fills the cavity of the fruit, but the seed coat does not adhere to the pericarp. The small mature achene, which may have developed from either a superior or an inferior ovary, is commonly mistaken for a seed. The achene is the most common type of fruit in the buttercup, buckwheat, composite, and other families. The familiar "seed" of a dandelion or sunflower is not a seed, but an achene, as are the "seeds" of strawberry.

The grain is a one-seeded fruit similar to the achene, except that the thin, usually colorless, seed coat is fused with the pericarp. The seed and fruit, therefore, merge into a single unit. The GRAIN (FIG. 13-6B) is the characteristic fruit of the grass family. Corn, wheat, rye, oats, barley, and rice are familiar examples. The last three of

Fig. 13-6 Simple dry fruits. (A): achenes of buttercup (*Ranunculus acris*). (B): grain of barley (*Hordeum vulgare*).

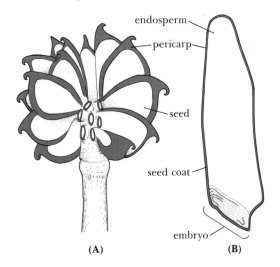

endosperm

pericarp

seed

seed coat

embryo

(A)   (B)

these, and most grass seeds, come from the thresher with attached bracts and must be hulled to uncover the fruits.

The SAMARA or key fruit (FIG. 13-18), is an indehiscent, usually one-seeded, winged fruit, the wing being formed by an extension of the ovary wall. Examples are the fruits of elm, ash, tree of heaven, and maple. In the maples the samaras are borne in pairs; in other genera, singly.

The term nut is applied to a one-seeded (exceptionally, two-seeded) rather large fruit, the wall of which becomes hard, stony, or woody upon ripening. Most nuts develop from pistils with inferior ovaries. Examples are the fruits of the filbert (hazelnut) and beech (FIG. 13-7A,B), oak, chestnut, hickory, and walnut. In these examples the nut is associated with structures formed from highly modified leaves (bracts) aris-

ing from the base of a flower or of several closely associated flowers. In the oak the bracts are fused, forming the cup of the acorn, which is essentially free, or separate, from the nut, merely surrounding it at the base. In the hazelnut the fused bracts form a loose leaflike bag around the nut. In the hickory, pecan, English walnut, beech, and chestnut, the bracts are more or less fused to the nut and do not split away from it until the fruit is ripe. In the black walnut and butternut, the husk of bracts becomes somewhat fleshy and remains permanently fused with the nut, the combination of fleshy husk and nut superficially resembling a drupe (see next page).

True nuts may appear on the market as the fruit or as the seed, separated from the fruit wall and thus only a part of the fruit. The term nut is often misapplied to the large seeds of fruits that are not nuts—for example, the seeds of the Brazil nut and cashew, neither of which are nuts; the edible part of the almond, which is the seed of a drupe; and peanuts, which are the seeds of an indehiscent legume. The horse chestnut, also, is not a nut but the single seed of a dehiscent capsule.

SIMPLE FLESHY FRUITS   Fruits in which a portion or all of the fruit wall is fleshy at maturity are classified as simple fleshy fruits. Important types falling in this category are the berry, drupe, false berry, and pome.

The true BERRY is a fruit in which the entire ovary wall ripens into a fleshy, often juicy and edible, pericarp. It is always derived from a superior ovary. In the botanist's definition of a berry are included more fruits than the layman calls by that name. The fruits of the date, tomato, grape, avocado, eggplant, persimmon, and red pepper are berries. In the date, an edible pericarp surrounds a single, elongated seed. In the tomato (FIG. 13-8A), the placentae become fleshy and greatly enlarged, bearing many seeds. Together with the side walls of the carpels, they form the interior of the fruit. Citrus fruits, such as orange, grapefruit, lime, and lemon, are modified berries, although the pericarp (rind or peel) is not

Fig. 13-7 Nuts. (A): beaked hazelnut (*Corylus cornuta*). (B): beech (*Fagus grandifolia*).

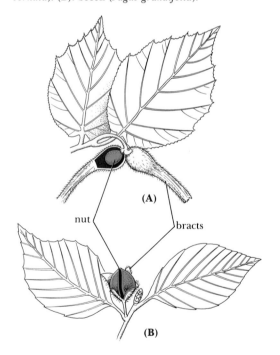

nut

(A)

bracts

(B)

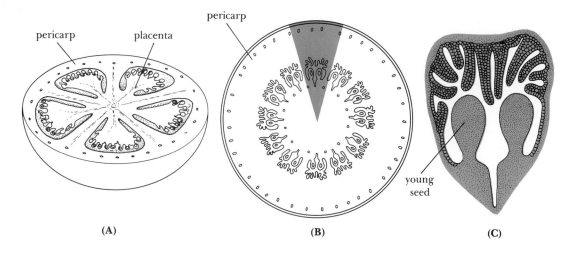

Fig. 13-8   Simple fleshy fruits. The berry. (A): section view of tomato (*Lycopersicum esculentum*) berry. (B): section of young citrus fruit; shaded portion, a single carpel. (C): enlarged interior of single carpel of citrus fruit showing the multicellular outgrowths that later constitute the pulp.

usually regarded as edible. Sections of very young citrus fruits show multicellular outgrowths developing from the surfaces of the carpel walls (FIG. 13-8B,C). These outgrowths gradually become much enlarged, juicy, and saclike, eventually forming the edible portion of the ripe fruit. The interior of an orange or grapefruit is readily separable into wedgelike segments, each representing a carpel filled with these multicellular outgrowths.

In the DRUPE, the pericarp is divided into three parts: an outer EXOCARP (commonly only a thin skin), the MESOCARP, which is a fleshy pulp, and the ENDOCARP, the stone or pit that encloses the seed. These terms are also applied to other kinds of fruits in which the pericarp is differentiated. Most drupes arise from flowers with superior ovaries. Examples of drupes are the olive (FIG. 13-9), plum, cherry, peach, and apricot. The almond, related to the peach, is a drupe also, but the exocarp and mesocarp are leathery and inedible and are removed when the fruit is harvested. The edible part of the almond is the seed. The shell is the endocarp. The husk of the coconut

(FIG. 13-10), another drupe, is the exocarp and mesocarp. Since the mesocarp is fibrous and not fleshy, the coconut is sometimes referred to as a dry or fibrous drupe. The coconut as purchased in the fruit store usually consists of a single large seed enclosed in the hard, woody endocarp. The stiff elastic fibers that largely compose the husk are known commercially as *coir*. These fibers are utilized in the manufacture of a number of products such as cordage, brushes, upholstery material, and the familiar coco door mat.

As pointed out earlier in the definition of a fruit, this organ may include not only the ripened ovary wall (pericarp), but also other parts of the flower or closely related structures. When such structures constitute an important or conspicuous part of the fruit, it is said to be ACCESSORY. Most accessory fruits are simple fruits that have developed from an inferior ovary and are derived, therefore, both from the ovary wall and from the floral tube, which is composed of the basal parts of the sepals, petals, and stamens. This tube is fused with the ovary wall, and becomes fleshy and ripens with it. Accessory

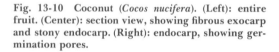

seed   exocarp

endocarp (pit)   mesocarp

Fig. 13-9   Simple fleshy fruits. Section view of an olive (*Olea europea*) fruit, a drupe.

Fig. 13-10   Coconut (*Cocos nucifera*). (Left): entire fruit. (Center): section view, showing fibrous exocarp and stony endocarp. (Right): endocarp, showing germination pores.

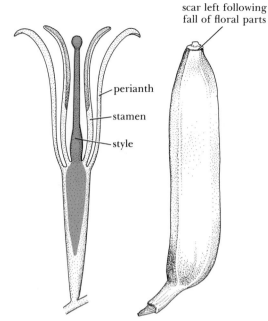

scar left following fall of floral parts

perianth

stamen

style

Fig. 13-11   Flower (left) and fruit (right) of banana (*Musa sapientum*). The ovary is inferior and the fruit accessory.

fruits in this category include the cucumber, squash, muskmelon, and other members of the gourd family, the cranberry, gooseberry, currant, and banana.

Since the entire fruit ripens fleshy, as in the true berry, such accessory fruits are termed *false* berries. False berries may usually be distinguished from true berries by the remnants of the flower that persist at the top of the fruit, opposite the stem. At the apical end of the cranberry, for

example, several small sepals can be seen. The tip of the banana fruit bears a large scar where the floral parts have fallen (Fig. 13-11).

The apple (Fig. 13-12) and pear are accessory fruits called POMES. These fruits have a pericarp that is differentiated into parts somewhat like a drupe. The outer part (exocarp and mesocarp) is fleshy, and the endocarp is leathery. In sectional views of an apple (Fig. 13-12C,D) the seeds appear in the center, surrounded by the leathery endocarp. A group of vascular bundles, sometimes called the "core line," marks the approximate position of the mesocarp. The tissue extending from the core line to the outside of the fruit is derived from the floral tube. That the apple is an accessory fruit is shown by the shriveled remains of the sepals and stamens, which can usually be seen at the tip of the mature fruit (Fig. 13-12C).

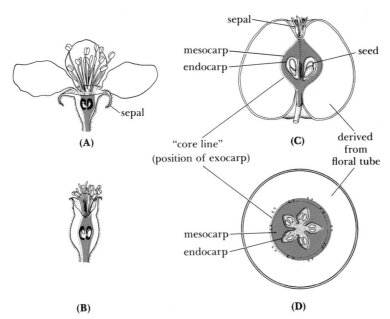

sepal

mesocarp

endocarp

seed

"core line"
(position of exocarp)

derived
from
floral tube

mesocarp

endocarp

(A)

(B)

(C)

(D)

Fig. 13-12 Development and structure of the apple (*Pyrus malus*), an accessory fruit. (A): flower of apple. (B): older flower, after petals have fallen; the unshaded regions in (A) and (B) represent the floral tube. (C, D): longitudinal and cross section of the mature fruit.

Aggregate Fruits An aggregate fruit develops from a flower that has a number of simple pistils. In the raspberry (Fig. 13-13, left), the simple fruits develop into tiny drupes (drupelets), which adhere to one another but which separate as a unit from the dry, dome-shaped receptacle. The blackberry (Fig. 13-13, right) is like the raspberry except that the elongated receptacle is also fleshy and forms a part of the fruit. The blackberry is therefore accessory as well as aggregate.

The strawberry is also aggregate-accessory. The numerous simple pistils of the flower develop into tiny achenes, which are seen on the outside of the fruit and which are commonly called the seeds. Each pistil contains a single ovule, which develops into a seed within the achene. The edible, accessory part of the fruit is the receptacle, which is fleshy and greatly enlarged. A longitudinal section through the fruit (Fig. 13-14) shows that the receptacle is a modified stem. In the center is the pith, separated from the cortex by the vascular cylinder. Vascular bundles or traces connect the main

Fig. 13-13 Aggregate fruits. (Left): raspberry (*Rubus strigosus*). (Right): blackberry, aggregate-accessory (*Rubus allegheniensis*).

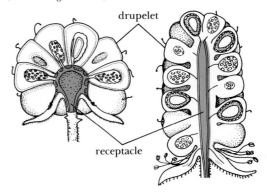

drupelet

receptacle

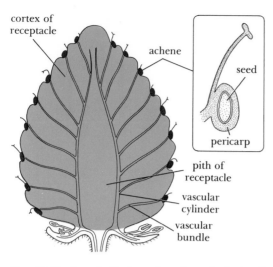

cortex of
receptacle

achene

seed

pericarp

pith of
receptacle

vascular
cylinder

vascular
bundle

Fig. 13-14 Longitudinal section through fruit of strawberry, an aggregate (aggregate-accessory) fruit. The enlarged and fleshy receptacle (in color), constituting the bulk of the fruit, is the edible part.

vascular system of the receptacle with the achenes on the surface.

It was stated earlier that developing seeds produce hormones that stimulate growth of the ovary wall. The growth of the strawberry receptacle is also stimulated by hormones from the seedlike fruits. This has been demonstrated by several interesting experiments. In one, all the achenes were removed from immature receptacles at intervals up to 21 days after pollination. No further growth occurred in any of the receptacles. In another experiment, pollen was placed only on the stigma of one pistil, after which it was observed that the immediately adjacent tissue of the receptacle became considerably enlarged. Similarly, the pollination of three pistils was followed by three areas of growth (FIG. 13-15A,B). In a third experiment, all the achenes were removed from the receptacles of developing fruits nine days after pollination. The receptacles of some fruits were then covered with a lanolin (wool fat) paste containing a synthetic growth hormone. Other receptacles were treated with lanolin alone. No further growth occurred in the receptacles receiving only lanolin, whereas

Fig. 13-15 Effect of auxin on growth of the strawberry (*Fragaria*). (A): growth of the receptacle immediately adjacent to a pollinated pistil. (B): three pistils were pollinated. (C, D): achenes removed from the receptacle nine days after pollination. (C): achenes replaced by lanolin alone. (D): achenes replaced by lanolin containing auxin.

(A)   (B)   (C)   (D)

the strawberries covered with lanolin containing auxin ripened in a normal fashion (Fig. 13-15C,D). From these and other experiments the evidence seems conclusive that the achenes produce hormones necessary in the growth of the receptacle.

Multiple Fruits A multiple fruit is formed from a cluster of flowers (inflorescence) grouped closely together, rather than from a single flower. A fruit is produced from each flower, and these fruits, at maturity, remain together in a single mass.

The pineapple (Fig. 11-8) is one of the best examples of a multiple fruit. The numerous (100–200) sessile flowers produce seedless fruits. These are attached to the elongated fleshy axis of the inflorescence, which is leafy at the top. Each flower is located in the axil of a bract, which is thick and fleshy below but tapers to a thin, papery tip. The flowers are fused with each other and with their bracts, and all ripen together, each of the units visible on the surface of the mature fruit representing a flower together with its bract. Since the ovary is inferior, each of these units is an accessory fruit.

Other examples of multiple fruits are the mulberry, Osage orange (*Maclura pomifera*), breadfruit (Fig. 13-16), and the fig. The edible portion of the fig is the mature axis of the inflorescence, which is hollow and encloses numerous tiny fruits, or achenes, each formed from a single pistillate flower.

Fruits and Vegetables. Confusion beclouds the use of the terms fruit and vegetable. Many fruits, such as the tomato, squash, cucumber, corn, and eggplant, are popularly called vegetables. From a botanical standpoint these are fruits, and they may be distinguished from vegetables if the definition of a fruit is kept in mind. A fruit always develops from a flower and is always composed of at least one ripened ovary with which may be fused other floral parts or structures associated with the flower. Any edible part of a plant that does not conform to this definition of a fruit should be classed as a vegetable.

The popular conception of a fruit as something eaten by itself or used as a dessert is of no botanical concern. The fact that some fruits are used as vegetables in the kitchen is also of no consequence. Although many fruits are popularly regarded as vegetables, few vegetables are considered to be fruits. One vegetable used as a fruit is rhubarb. Since the edible portion is the petiole of the leaf, it does not in any respect meet the botanical definition of a fruit.

Fig. 13-16 Breadfruit (*Artocarpus communis*), a multiple fruit; young fruit and staminate flower. The first attempt (1788) of the British government to introduce this plant into the West Indies from Tahiti failed because of a mutiny on H.M.S. Bounty. Breadfruit is now cosmopolitan in distribution in the tropics. The mature fruits weigh from 1 to 10 pounds.

DISPERSAL OF SEEDS AND FRUITS

As the seed develops, a new generation, in the form of an embryo, begins within it. This beginning is a limited one, for the growth of the embryo is soon arrested. The seed is then detached from the parent plant, and dispersal takes place. Germination eventually follows, commonly some time later. During this period the seed may have survived extremes of heat, cold, or dryness.

The seed may well have been the only structure disseminated in early angiosperms, and this is still true for many species. In others, however, the seed is retained within the indehiscent fruit and the seed and fruit dispersed as a unit. Examples are found in the achene, grain, samara, and nut. Other structures, such as the calyx and bracts, may also form a part of the *dispersal unit.* Further, in some plants, the inflorescence, in whole or in part, may be dispersed. It is convenient to apply the term dispersal unit to any detached part or organ of a plant involved in dispersal. This term is applied even to fragments of the vegetative plant body that give rise to new individuals.

The agents concerned in dispersal include wind, water, and such animals as birds, insects, and mammals, including man. Large numbers of plants have in their seeds and fruits structural modifications of value in dispersing the species, sometimes over great distances, and much has been written on this subject.

Some seeds can be dispersed by wind because they are small and light. The seeds of orchids are so small that several thousand to a million, depending on the species, may be found in a single capsule. In other dispersal units (seeds and some fruits), plumes or tufts of hair, effective in wind dispersal, are formed. These structures are outgrowths of the seed coat, as in milkweed (FIG. 13-17), cattail, and willowherb (*Epilobium*), or are modified floral parts, as in dandelion and thistle. Winged seeds (as in catalpa) and fruits (FIG. 13-18) are common. Dispersal by wind is also

Fig. 13-17 Hairy outgrowths of the seed coat of milkweed aid in dissemination by wind.

brought about, as in the tumbleweeds, by the movement of the detached inflorescences of the entire plant, which, when mature, breaks loose and rolls along the ground. An example of the latter is the Russian thistle (*Salsola*), which has spread over the dry plains of many parts of the world.

The action of water on dispersal units may be almost as important as that of wind. Seeds and fruits, large and small, are conveyed over short distances by the washing of rain or over long distances by streams and floods. Modifications useful in dispersal by water include air bladders, light spongy fruits, and relatively impervious seed coats. The dispersal units of coconut and other plants have been spread over great distances by ocean currents.

Dispersal by animals takes place in a variety of ways. Fruits may be eaten, and the seeds, their vitality unimpaired, passed through the body of the mammal or bird. It has been shown that the seeds of a number of species of angiosperms can remain viable in the intestinal tracts of some migratory shore birds long enough to be transported several thousand miles. Many fruits have hooks, bristles, or spines that adhere to the coat of animals or the clothing of man. Man, through commerce and travel, has played a major part in the distribution of plants. Most weeds have come from distant places through channels of transportation, and the seeds of crop plants are often contaminated with weed seeds. Seed analysts in federal and state seed-testing laboratories may condemn shipments of crop seeds that contain seeds of noxious weeds.

Finally, some plants possess mechanical devices that project the dispersal unit, usually a seed, away from the parent plant. In some species the mechanism is hygroscopic, as in a number of legumes, in which the simple pistil twists as it dries and then bursts open, scattering the seeds.

Fig. 13-18 Fruit dispersal. (Left): winged fruit (*samara*) of tree of heaven (*Ailanthus*). (Right): spiny fruit of beggar-ticks (*Bidens*).

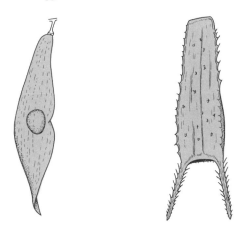

In other plants the seeds are expelled forcefully from the fruit as a result of sudden changes in internal pressure. In the dwarf mistletoe (*Arceuthobium*) it is recorded that the single seed (about 1/8 inch long) is expelled from the fruit to a distance of nearly 50 feet.

It should be noted that the possession of modifications that aid dispersal does not in itself ensure that a given species will be common and widely distributed. Many species with apparently effective dispersal mechanisms are limited in occurrence. Dispersal becomes effective only after a plant is in an environment in which it can grow and reproduce. Many species, conveyed to distant areas, are unable to compete with local plants under new conditions of soil and climate. Widely distributed and common weeds, for example, are not necessarily plants with the best dispersal mechanisms, but those that can persist under a great variety of environments, including poor soils, disturbed areas, and cultivated fields, where they successfully compete with crop plants. Some, at least, of the advantages of dispersal to the species are obvious. If all seeds were deposited in the immediate vicinity of the parent plant, the resulting seedlings would compete among themselves and with the parent for light, water, and minerals. Seeds dispersed at distances from the parent, on the other hand, may fall in habitats more favorable to survival and reproduction.

SEED GERMINATION

The story of the seed is completed only when the seed has germinated and the seedling has become established. Germination begins with resumption of growth of the embryo and concludes with the appearance of the radicle outside the seed coat. Establishment has been variously defined, but is here considered to be that period beginning with the end of germination and ending with a seedling independent of the accumulated food in the seed. Germination and establishment are crucial in the life of the plant, for it is in these stages of the life cycle of any species that the greatest number of individuals are lost. Environmental hazards are so great that some plants persist as a species only because seeds are produced in great numbers. The seed and seedling may be eaten by insects, birds, and rodents or other mammals. The seedling may fail to grow because of lack of water or light, or because of unfavorable temperatures.

The depth to which a seed is buried in the soil, either by accident or in planting, is an important factor in germination. Seeds on the surface may not have sufficient water to complete their germination. Those buried too deeply may fail to germinate, or may deplete their reserves of food before they break through the soil and reach the light. Large seeds, because they contain more food, can be planted deeper than small seeds, and can thus have the advantage of a more uniform moisture supply.

The various aspects of seed germination constitute problems in the fields of horticulture and agriculture, forestry, plant breeding, wildlife management, and weed and erosion control.

Environmental Factors and Germination
There are certain environmental conditions necessary for seed germination: (1) moisture, (2) oxygen, and (3) favorable temperature. In addition, the germination of the seeds of many species is favored by light, whereas that of others is retarded or inhibited by light. However, the effect of light or darkness may be modified by other factors, notably temperature.

The seeds of most plants have a low water content when ripe, and germination cannot occur until the seeds have imbibed water. The seed swells, and considerable pressure may develop if the seeds are confined.

The seeds of various species differ in their oxygen requirements, but oxygen is usually essential to germination. The oxygen concentration in the soil is affected by the amount of water present, and seeds may fail to germinate in a waterlogged soil. Some seeds, especially those of

weeds, remain dormant for several to many years of unfavorable growing conditions, as when buried in the soil. When brought to the surface by cultivation, these seeds may germinate because of better aeration and, sometimes, the effects of light.

Seeds differ greatly in their temperature requirements for germination, and many will germinate over a considerable temperature range. The lower limit is approximately 0°C (32°F), the upper, 45°C (113°F), but the percentage of germination is usually small at very low or very high temperatures. For most crop plants the optimum temperatures lie between 20 and 30°C (68 and 86°F), but certain seeds, such as peas, lettuce, radish, barley, and wheat, will germinate readily at 10°C (50°F), and hence can be planted earlier in the spring. In general, seeds of cool-weather crops can and should be planted early. Oats, hardy lawn grasses, and early vegetables should be planted as soon as the ground can be worked; corn and beans, after the trees begin to leaf out; and lima beans, squash, and melons only after the ground is warm and many early perennials are in flower.

The Germination Process Under favorable conditions, the imbibition of water by seeds is followed by a number of activities. The protoplasm is rehydrated and its enzymes begin to function. Starch is digested to sugar, lipids to soluble compounds, and storage proteins to amino acids. The availability of these materials permits the liberation of energy by respiration, the translocation of foods to the embryo, and the beginning of embryo growth.

Respiration in the dry, dormant seed is extremely slow. It is even possible that respiration stops in completely dry but still viable seeds. Moistening the seeds allows respiration to speed up rapidly, and by the time germination is well under way, the rate may have increased many times. This enormous effect of hydration on respiration is a major reason for the emphasis on low moisture percentages in stored grains and seeds. As a result of increased enzyme activity and available food and energy in the germinating seed, cell elongation begins in the embryo, and the development of the new plant that started with fertilization is under way again.

Methods of Germination In almost all seeds the first structure to emerge from the seed coat is the RADICLE, the embryonic root. The radicle grows out through the micropyle and produces the primary root. This, in turn, develops root hairs and, subsequently, secondary or lateral roots. The growth of the root somewhat prior to the growth of the other parts of the embryo enables the young plant to become anchored to the soil and to absorb water. The manner in which the seedling emerges and develops varies considerably in different species.

Germination of the bean and pea (dicots) The seeds of the bean and pea lack endosperm, and the food supply that nourishes the seedling is accumulated in the cotyledons. In the bean, after the emergence of the radicle, the hypocotyl elongates and becomes arched (FIG. 13-19). The apex of this arch is the first part of the seedling to appear above the soil surface. As the hypocotyl grows, it straightens and carries the two cotyledons upward into the air. Meanwhile, the plumule, lying between the cotyledons, has begun to grow and give rise to the true leaves and to that portion of the stem above the cotyledons.

The cotyledons and plumule of the bean are not pushed upward through the soil but are pulled out by the growth of the hypocotyl. Thus, injury to the shoot apex is prevented. This and other mechanisms that protect the apex of the stem are common in seed germination.

The food stored in the cotyledons is gradually digested and transferred to other regions of the rapidly growing seedling. The fleshy cotyledons of the bean turn green on exposure to light, but the amount of food they synthesize is negligible and, as the accumulated food is withdrawn, they wither and soon fall away. Other plants in which the cotyledons are carried above the soil at germination are radish, squash, sunflower, soybean, the clovers, cotton, ash, elm, and the maples. In some of these, the cotyledons are thinner and larger than in the bean and may function as leaves for a time, although they are unlike the true leaves that form later.

In the pea, the hypocotyl does not elongate, and the cotyledons, together with the seed coat, remain in the soil during germination (FIG. 13-20). The plumule produces a stem that arches until it reaches the soil surface. Then it straightens as it grows upward. The cotyledons in the soil finally disintegrate after the withdrawal of food by the seedling.

Germination of corn, onion, and coconut (monocots) The accumulated food of these seeds is largely located in the endosperm, and the chief activity in the cotyledon is the digestion and translocation of this reserve food to the growing parts of the seedling. In the germination of corn and other members of the grass family, the grain, containing the scutellum and the remains of the endosperm, remains within the soil (FIG. 13-21). The coleoptile, regarded as a portion of the cotyledon (FIG. 12-28B), covers and protects the plumule as it is pushed upward through the soil.

The primary root system, formed from the radicle, never becomes large and may be temporary. These primary roots are supplemented by a stronger, secondary root system, adventitious in origin, which develops from the lower nodes of the stem. These nodes are pushed upward through the soil in germination. If the secondary roots were to emerge during this movement, they would obviously be damaged or destroyed. Also, the young leaves of the corn plants would not be capable of pushing through the soil unless they remained covered by the coleoptile. The mechanism by which these various developments are controlled and correlated is a most interesting one.

Emergence of the corn seedling is brought about by elongation of the mesocotyl (FIG. 13-22B). This structure is regarded as a combination of the hypocotyl and of the cotyledonary tissues that originally connected the scutellum and the coleoptile (Fig. 12-28B). The elongation of the mesocotyl is dependent upon a liberal supply of growth hormone, which normally moves downward from a region of hormone formation in the coleoptile tip. The hormone concentration is at a level such that it stimulates the growth of the mesocotyl, but is so high that it inhibits the growth of the plumule and of the secondary roots.

When the tip of the coleoptile breaks through the soil surface, its rate of auxin production is sharply reduced by the action of light. The growth processes are then reversed. Elongation of the mesocotyl ceases, the plumule emerges from the coleoptile, and roots develop from the first node (Fig. 13-22A,B). The first adventitious roots thus form at approximately the same distance below the surface, in spite of considerable differences in the depth of planting, and other adventitious roots form on higher nodes as the plant develops. In shallow planting, the coleoptile reaches the surface and becomes affected by

the light so quickly that elongation of the mesocotyl is limited (Fig. 13-22A). With deep planting, the young leaves are protected by the coleoptile until this organ reaches the light at the soil surface (Fig. 13-22B).

The development of the onion seedling differs from that of corn in several particulars. After the emergence of the radicle, the single cotyledon emerges from the seed, increases in length, and arches above the ground (Fig. 13-21, right). As the cotyledon straightens, the seed is pulled from the ground and carried upward. The tip of the cotyledon remains within the endosperm until all the food material has been absorbed, after which the seed coat falls away. The cotyledon functions not only as an absorptive organ but also as a leaf, for it is green and manufactures food. As growth proceeds, the plumule, located at the junction of cotyledon and radicle, grows upward and finally bursts through the sheathlike cotyledon.

Germination and seedling development in the coconut illustrate a further and extreme development of the absorptive activities of the cotyledon. The meat of the coconut is the endosperm of a

Fig. 13-19 **Stages in germination of garden bean.**

Fig. 13-20 **Germination in pea.**

Fig. 13-21 **Seed germination. (Left): corn. (Right): onion.**

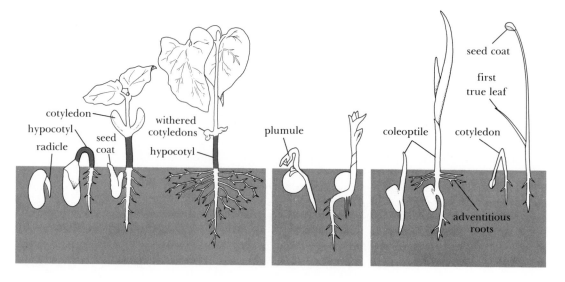

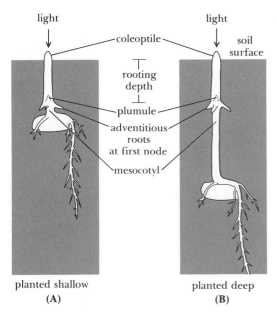

Fig. 13-22 The emergence of deeply planted corn and other grasses depends upon the elongation of the mesocotyl. This elongation is stopped when light strikes the coleoptile tip.

Fig. 13-23 Germination of coconut. (C): cotyledon. (E): endosperm.

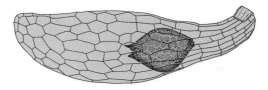

Fig. 13-24 Seed of showy lady's-slipper (*Cypripedium reginae*). The undifferentiated embryo is contained within the transparent seed coat.

single seed (FIG. 13-23). The seed is surrounded by a hard shell (the endocarp), which is a part of the mature ovary wall. The remainder of the ovary consists of a fibrous husk. The shell contains three shallow circular pits, or germination pores, at one end. Only one of these is functional, and beneath it, buried in the endosperm, is a short, peglike embryo. At germination the leaves and root grow out through the pore and through the fibrous husk into the air and soil. The single cotyledon grows into the cavity of the seed. Here it expands tremendously and develops into a pulpy mass that digests the endosperm and translocates the soluble food to the growing seedling. These processes continue over three or four months until the endosperm is exhausted and the roots and leaves of the seedling become established in the soil and air.

The orchid seed The seeds of orchids are strikingly different from most other kinds of seeds, both in structure and in the conditions necessary for germination. Orchid seeds are extremely minute. The seed has no endosperm, and the embryo is unspecialized, without cotyledon, axis, or radicle (FIG. 13-24). The very small amount of reserve food in the embryo does not permit seedling development, which occurs only when food and vitamins are obtained from outside the seed. Under natural conditions, following swelling of the seed, the embryo is infected by a mycorrhizal fungus (Chapter 9) which is able to convert insoluble carbohydrate and protein compounds of the soil organic matter into soluble form. In this manner the fungi obtain their own food, and the germinating orchid seeds, infected intracellularly, obtain their food by digestion of the fungal threads. The vascular plant could thus be regarded as a parasite upon the fungus. The seedlings are dependent upon the fungus until the first leaf appears, which may be several months or even years after germination. Most orchids are autotrophic in the later stages of development, but are heterotrophic in the earlier phases of growth.

Orchid plants were once collected in tropical forests and imported at great expense. Attempts to grow seeds of selected and improved orchid varieties on compost material inoculated with fungi obtained from tropical orchids were generally unsatisfactory. It was then discovered that these seeds can be disinfected and grown, as are bacteria (Chapter 20), on sterile agar containing sugar and mineral elements. After a period ranging from several months to a year, the sterile seedlings are large enough (FIG. 13-25) to be transplanted to soil for further development. This method is now extensively used by commercial orchid growers.

Dormancy of Seeds   The seeds of most wild and cultivated plants of temperate climates will not germinate immediately after ripening. At maturity they enter a state of dormancy that varies in length, depending upon the species, from a few weeks or months to several years. The seeds of some species germinate readily following an interval in dry storage, or in the spring after fall sowing; the seeds of other species may germinate irregularly over a period of two to many years. In the propagation of plants, it is frequently desirable to shorten the dormant periods of seeds. A number of methods, all depending upon a knowledge of the causes of dormancy, have been developed for accomplishing this. The two most common causes of dormancy are (1) impermeability of the seed coat, and (2) dormancy of the embryos. In some species, only one of these causes exists, but in a number of species both do.

Impermeability of seed coat   The impermeability of the seed coat to water and frequently to ox-

Fig. 13-25 Orchid seedlings growing on sterile nutrient agar. The seeds were sown on the agar and germinated there. The seedlings may now be transferred to soil.

ygen also is a common cause of delayed germination. The dormancy of many legumes, basswood, morning-glory, and other plants results from impermeability of the seed coat. Less commonly, seeds fail to germinate because the seed coat is mechanically resistant to the expansion of the embryo within. Under natural conditions, the permeability of the seed coat is gradually increased by freezing, thawing, and the action of microorganisms in the soil until germination finally occurs. The impermeable seeds of crop plants such as alfalfa are commonly SCARIFIED to hasten germination. The seeds are blown against finely abrasive surfaces that scratch the seed coats and thus permit water absorption and germination. On a smaller scale, seeds may be rubbed over fine sandpaper or abraded with a file.

Dormant embryos The seeds of many species fail to germinate even if the seed coat is completely removed and the seeds are placed under conditions that are favorable for germination. The seeds of many forest and fruit trees fall into this category. Failure to germinate is due to a physiological condition of the embryo, and such seeds must undergo a series of complex enzymatic and chemical changes collectively referred to as AFTER-RIPENING. For most seeds, low temperatures and moisture are necessary for the after-ripening process to take place. These conditions are supplied naturally during the winter in temperate climates. After-ripening is produced artificially by the process known as STRATIFICATION. This consists in placing the seeds in moist sand or peat moss at temperatures of 1–6°C (34–43°F) for various periods of times—often several months.

Dormancy can also be caused by various germination-inhibiting chemicals that may be present in the seeds. These inhibitors are commonly removed when the seeds are leached, as in wet soil. Inhibitors are considered to be advantageous to plants growing in regions with pronounced dry seasons. An occasional shower will not remove the inhibitors, and germination is thus delayed until the onset of the rainy season. The seeds are then leached and germinate at a time when their continued growth is possible.

Secondary dormancy Many weed seeds show secondary embryo dormancy. Primary dormancy may be broken by winter freezing, but if conditions are unfavorable for germination in the spring—if the seed is buried too deep, for example, and aeration is inadequate—the seeds revert to the dormant condition and must be afterripened a second time before they will germinate. Secondary dormancy is an important factor in the persistence of weed seeds in cultivated soil.

Nondormant seeds Dormancy in seeds is common but not universal. The seeds of some plants, especially those that mature early in the growing season, will germinate as soon as they have attained full size or when they drop from the parent plant. The seeds of willows, poplars, silver maple, and the white oaks germinate as soon as they fall upon the soil; willow seed is said to germinate within 12–24 hours following dispersal. In damp weather, the seeds of the garden pea may germinate within the pod. Under similar moist conditions, certain varieties of corn will germinate within the husk, while the ear is still attached to the parent plant. Some varieties of wheat and other small grains undergo after-harvest sprouting if moisture is available. Another example of lack of dormancy is found in the Spanish varieties of the cultivated peanut, the seeds of which may sprout in the ground if wet weather occurs after maturity and harvesting is delayed. The plant breeder, obviously, selects for enough dormancy to avoid these losses when possible.

The most extreme case of absence of dormancy is found when the embryo grows directly into the seedling without cessation of growth. This is known as VIVIPARY. Although not common, vivipary is known to occur both in monocotyledons, such as species of bamboo and other members of the grass family, and in dicotyledons. A striking example of vivipary in a dicotyledonous plant is found in the seeds of the red mangrove, a low tropical or subtropical tree growing on seacoasts or along tidal rivers. The fruit contains but a single seed, in which the embryo continues to grow while the fruit and seed are still attached to the parent tree (FIG. 13-26). The radicle grows downward, followed by the hypocotyl, which becomes peglike and greatly elongated. It may reach a length of 9 inches or more. The cotyledons remain within the seed coat and absorb food from the endosperm. The seedling remains attached for some months, but finally the hypocotyl and plumule become detached from the cotyledons and fall from the tree into the mud or shallow water beneath. Many seedlings fall vertically downward in such a way that they stand upright, thus planting themselves in the mud. Lateral roots are soon

Fig. 13-26 Stages in germination of viviparous seeds of mangrove (*Rhizophora mangle*).

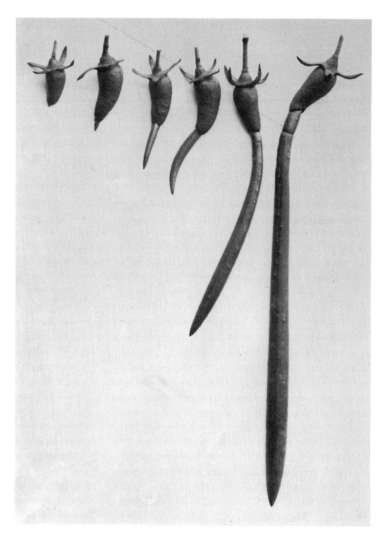

produced. Other seedlings may fall flat on the mud, where growth continues, or they may float away and be deposited in a distant area. Seedlings that lie horizontally gradually become erect by geotropic bending.

The absence of dormancy is probably a primitive condition, and dormancy is a later evolutionary development. It is presumed that a delay in germination is of survival advantage for many species, for immediate germination may subject the seedlings to unfavorable low or high temperatures or drought. Dormancy delays germination until conditions are more favorable for survival of the seedling.

Longevity of Seeds Seeds vary greatly in the length of time during which they will retain their vitality and ability to germinate. The seeds of some species remain viable for only a few weeks or months; those of others, for years. The longevity of most seeds is increased by dry storage at low temperatures. However, some moisture, with cool temperatures and reduced oxygen, favors longevity in many seeds, especially weeds. And the seeds of some tropical crops, such as cacao and coffee, must be kept moist and require special storage methods.

The periodic reports of the germination of wheat from Egyptian tombs, and of corn from Indian burial mounds, have never been verified and are considered to be untrue. A number of seeds of farm and garden plants are known to retain their viability for 10–25 years, the percentage of seeds germinating gradually decreasing with age. The longest record for wheat is 69 percent germination after 32 years in storage.

The longevity of weed seeds, especially when buried in the soil, is well recognized. It has been demonstrated by an experiment begun by Dr. W. J. Beal, of Michigan State University. In the autumn of 1879, Dr. Beal filled 20 pint bottles with sand holding 50 seeds of each of 20 kinds of herbaceous plants, mostly weeds—a total of 1000 seeds in each bottle. The bottles were buried 18 inches below the soil surface. A bottle was dug up at intervals and germination tests made. The seeds of only three species survived longer than 50 years. In 1950, 70 years after burial in the soil, the seeds of the evening primrose (*Oenothera biennis*) gave 14 percent germination, those of yellow dock (*Rumex crispus*) 8 percent, and those of moth mullein (*Verbascum blattaria*) 72 percent. In the 80-year test, made in 1960, the same three species germinated 10, 2, and 70 percent. The yellow dock will probably not survive another period.

There are, however, authentic records of the germination of seeds that have been stored for still longer periods of time. Seeds of a Chinese tree legume (*Albizzia*) germinated after 147 years in dry storage. The seeds of lotus (*Nelumbium*), of the water-lily family, are known for their longevity. One lotus seed, from a collection deposited in the British Museum in 1705, was germinated in 1942—still viable after 237 years.

Two other records are noteworthy. In Denmark, weed seeds excavated from archeologically dated sites were found to be viable after 100–600 years of burial, with two species germinating after 1700 years. The longevity of these seeds is attributed to the conditions existing at the considerable depths at which they were found—moist, cool, with a very low oxygen content. The age of viable lupine seed, found buried 10–20 feet in permanently frozen Alaskan tundra, is set at more than 10,000 years on the basis of the geological formation. The lupine seed had been stored by lemmings in burrows normally only a few inches beneath the soil surface. It is assumed that some catastrophic event, such as a landslide, trapped the rodents and buried their burrows in the permanently frozen zone. More data, including ^{14}C dating of similarly buried ancient and viable seeds, are needed to corroborate this very interesting record.

Experiments with storage of live seeds have shown that—in general—warm, humid storage not only increases the rate of degenerative changes within the seed but also encourages the attacks of insects and storage fungi. Corn, which has a normal life of five years under farm-storage conditions, remained viable more than 20 years when stored at low (5 percent) moisture and near-freezing conditions. Such storage is important to the plant breeder who wishes to repeat or run new experiments with the same seed or "germ plasm" at a later date.

The maintenance of viable supplies of seed has been a problem since the beginnings of agriculture. It was necessary to hold seeds from the time of harvest to planting, and desirable, when the harvest was abundant, to store excess seed for one or several seasons in case of crop failure. For example, plants bearing seed were suspended from rafters to dry; then the seed was placed in bags, baskets, jars, or pits in the ground. Some of these methods are still in use in developing countries, but modern storage methods control temperature and humidity more precisely. The proper storage of crop seeds is, of course, vital to the food supplies of mankind.

TABLE 13-1
Fruit Classifications

Kinds of Fruits		Examples
SIMPLE FRUITS	Fruits derived from a single pistil, simple or compound.	
Simple Dry Fruits	Fruit wall becoming papery, leathery, or woody at maturity.	
	Follicle Derived from a simple pistil; opens along one side only.	Milkweed
	Legume Derived from a simple pistil; opens along two sides.	Pea
	Capsule Derived from a compound pistil; opens variously.	Iris
	Achene Seed coat of single seed not fused with fruit wall.	Buttercup
	Grain Seed coat of single seed fused with the pericarp.	Wheat
	Samara A winged fruit.	Elm
	Nut A large one-seeded fruit with a woody or stony wall.	Hazelnut
Simple Fleshy Fruits	Fruit wall, or a portion of it, becoming fleshy.	
	Berry Entire ovary wall ripens into fleshy pericarp.	Tomato
	Drupe Pericarp divided into fleshy exocarp and mesocarp and stony or woody endocarp.	Peach
	False Berry An accessory fruit, the entire fruit wall fleshy, composed of pericarp and floral tube.	Cranberry
	Pome Like a false berry, but pericarp chiefly fleshy mesocarp and leathery endocarp.	Apple
AGGREGATE FRUITS	Fruit derived from a number of simple pistils, all from the same flower. Some aggregate fruits are aggregate only.	Raspberry
	Others are accessory as well as aggregate, for receptacle is a part of the mature fruit.	Strawberry
MULTIPLE FRUITS	Fruit composed of a number of closely associated fruits derived from different flowers, these fruits forming one body at maturity.	Pineapple

SUMMARY

1. The fruit develops from the pistil and is composed of at least one ripened ovary and its contents. The wall of such an ovary is termed the pericarp. The fruit may also include other parts of the flower—such as the floral tube—or structures closely associated with the flower—such as the receptacle.

2. Fruit development is controlled by hormones. These are produced following pollination and later by the developing seeds and other tissues of the fruit.

3. Fruits that have developed in the absence of fertilization are called parthenocarpic. Some plants form parthenocarpic fruits naturally. Such fruits have also been produced artificially, by the use of synthetic growth regulators.

4. Fruits may be classified as in Table 13-1.

5.	An edible part of a plant is a fruit if it consists of the mature ovary or the ovary and closely associated structures. Any other edible part of a plant is classed as a vegetable.

6.	In the development of the seed, the growth of the embryo usually is arrested while the seed matures and is dispersed. After a period of dormancy, germination takes place and growth continues. The agents of dispersal include wind, water, and man and other animals.

7.	Seed germination depends on external and internal factors. The chief external factors are moisture, temperature, and oxygen. Internal factors include impermeability of the seed coat and necessity for after-ripening. Treatments to induce germination include scarification and stratification.

8.	Some seeds will germinate as soon as they are fully formed. In some seeds, the embryo grows directly into the seedling without ceasing growth (vivipary).

9.	The radicle is the first structure to appear outside the seed coat at germination.

10.	In some seeds the cotyledon are raised above the soil during germination; in others they remain underground.

11.	The cotyledons carry on various activities during germination. They may be green and manufacture food; they may yield accumulated food or digest and translocate it. In some plants they have more than one of these functions.

12.	The viability of seeds varies greatly with age and storage conditions, but 10–20 years is a long life for most seeds.

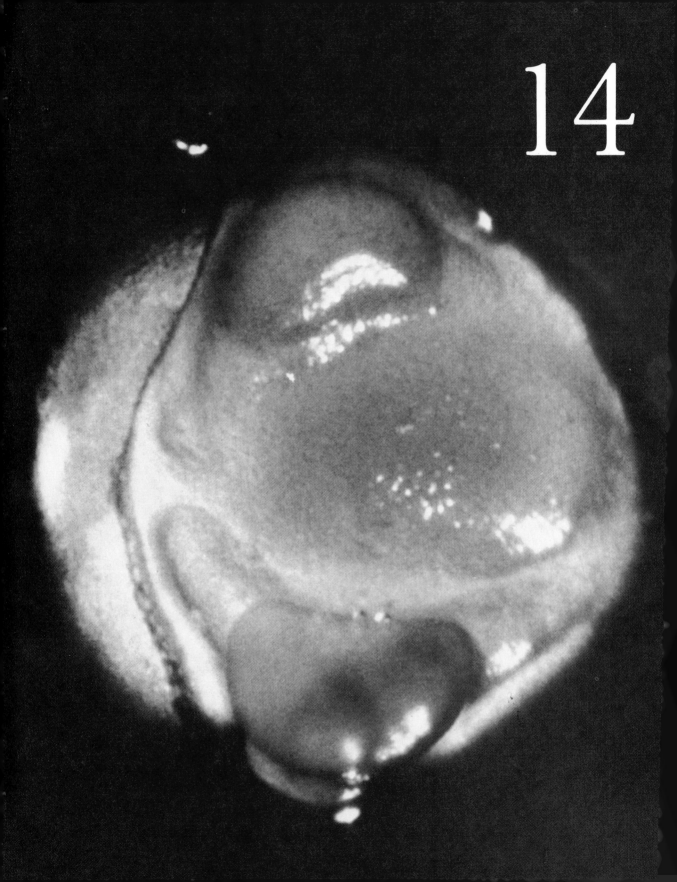

Growth and Differentiation

ROWTH IS a universal characteristic of living things. In the biology of plants it assumes particular importance because the plant characteristically continues to grow throughout its life, and much of the plant's metabolic activity is channeled into this process. Growth is usually defined as an irreversible increase in size, either of the organism as a whole or of any of its parts. At the level of the cell, growth can be resolved into two processes, *cell division* and *cell enlargement*. Increase in cell number by division does not itself result in growth, that is, an increase in size. It must be accompanied by enlargement of the divided cells. Many complex biochemical phenomena participate in plant growth, including the synthesis of new protoplasm and the elaboration of nonprotoplasmic components of cell walls. The uptake of water, which comes to occupy a large part of the volume of many cells, also plays an essential role.

Cell division and cell enlargement alone would be expected to produce formless masses rather than complex and integrated organisms. Within the plant, however, as cells divide and enlarge, they also undergo changes that result in the multitude of distinctive cell types found in the stem, root, and leaf. These changes are not haphazard; rather they occur in precise patterns that give rise to tissues with distinctive functions. The sequence of changes in cell form and function, whereby cells become different from their embryonic state and from one another, is referred to as *differentiation*. The term *development* is commonly applied to the combination of growth and differentiation, which produces the organized plant body.

Developmental biologists often employ the term *morphogenesis* to designate the development of the form or structure of an organism. In practice, however, the study of morphogenesis, one of the major fields of current biological research, is not limited to the study of structure, but is equally concerned with the physiological and biochemical mechanisms that control the origin of form.

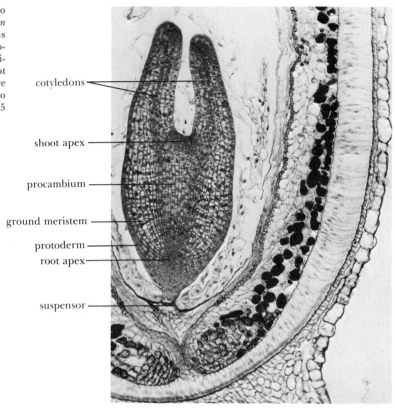

Fig. 14-1 Six-day-old embryo of Indian mallow (*Abutilon theophrasti*), a dicotyledonous annual. The protoderm, procambium, and ground meristem are delimited, and shoot and root apical meristems are established. The embryo attains full growth at about 15 days after fertilization.

cotyledons

shoot apex

procambium

ground meristem

protoderm

root apex

suspensor

EMBRYOGENESIS—THE BEGINNING OF THE PLANT BODY

In the development of a sexually reproducing plant, the all-important initial cell is the fertilized egg or zygote (Chapter 12), and it is with this cell that a study of development begins (Plate 5A). Although the zygote may not divide for several days, it is far from inactive. Important changes, including the development of endoplasmic reticulum, the synthesis and grouping of ribosomes, a reduction in the size of the vacuole, and other alterations in cell organelles seem to prepare this cell for intense metabolic activity. When it does divide, it does so in a regular fashion, with wall formation following each mitosis, producing in most species a short filament of cells attached at its base to the wall of the embryo sac (FIG. 12-27).

The filamentous embryo rather quickly becomes organized into two regions (Plate 5B). The suspensor at the base, that is, the end that is attached near the micropyle, consists of vacuolated cells and is probably concerned with the absorption of nutrients as well as with orienting the embryo in the endosperm. The apical end consists of cells that are more densely filled with cytoplasm and which constitute the embryo proper that will develop into a rudimentary plant. Repeated cell divisions give rise to a homogeneous mass of cells, the globular embryo.

Within a few days the beginning of a structural organization within the embryo can be detected.

Cell divisions at the surface become increasingly restricted to a plane at right angles to the surface, with the result that a recognizable protoderm, the forerunner of the epidermis, becomes apparent (Plate 5C and Fig. 14-1). At about the same time, a central core or cylinder of narrow, elongate cells, recognizable as procambium, marks the beginning of the vascular system. The remaining ground meristem cells enlarge and vacuolate as the ground tissue system. The similar initiation of the three tissue systems in the stem just behind the shoot apical meristem has been described in Chapter 7. Here their origin in the early embryo is seen.

Organ formation begins with the appearance of the cotyledons at the end of the embryo opposite the suspensor (Plate 5C). The embryo becomes heart-shaped (in dicotyledons) and has a polar organization, with its two ends differing (Fig. 14-2). The differentiation of tissues, such as epidermis or vascular tissue, ordinarily leads to a state of maturity in which cell multiplication ceases. If the whole embryo were committed to this kind of differentiation, its ultimate limitation would be established. But between the cotyledons, at what may be called the shoot pole of the embryo, a small group of cells becomes organized as the shoot apical meristem (Fig. 14-1). Near the opposite pole of the embryonic axis another group becomes the root apical meristem. These meristems are maintained in a state that permits their continued growth. The plant embryo thus remains an "open ended" system, which is capable of potentially unlimited growth. Were these growth centers not established in the embryo, the fundamental body plan of the plant, outlined in Chapter 2, could not exist.

Control of Embryo Development It has long been known that if the embryo is dissected from a mature or nearly mature seed and placed on a nutrient culture medium, it will develop into a plant. In fact this technique has been used by horticulturists and plant breeders as a way of bypassing seed dormancy, particularly in seeds in

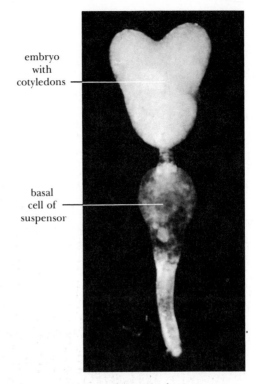

embryo with cotyledons —

basal cell of suspensor —

Fig. 14-2 Heart-shaped embryo of chickweed (*Stellaria media*) dissected from an ovule. The embryo proper is approximately 0.25 mm in length and is attached to a suspensor with a greatly enlarged basal cell.

which dormancy does not lie in the embryo itself. In this way, valuable time can often be saved. The technique is also used for the cultivation of hybrid embryos resulting from crosses between distantly related plants. In such hybrids the embryo may abort before reaching maturity. The embryos, however, may be dissected out and cultured before they abort, with the result that hybrid plants are produced which could not be obtained without such techniques. Procedures such as these, however, begin with embryos that are already well developed or even fully developed, and they tell us very little about the development of the embryo itself.

In some experiments, embryos as young as the globular stage have been cultured successfully. It has been found, however, that very young embryos have requirements for complex organic substances such as vitamins and amino acids, and for growth factors such as auxin and cytokinins for which older embryos are self-sufficient. The importance of a high osmotic concentration in the medium has also been discovered for young, undifferentiated embryos. It is interesting that these exacting requirements are similar to the factors that are provided by the enriched environment of the endosperm in which the embryo normally develops. In fact, the milk of coconuts, which is an endosperm, is often added to the medium in which the young embryos are grown. It is tempting to think that the complex components of the endosperm somehow control the activities of the zygote and impose an organization upon the cells which it produces. Another view, steadily gaining acceptance, argues that the control resides in the developing cells themselves and that the role of the embryonic environment is to provide the conditions and the substances necessary for the internal control to be expressed. This view in no way diminishes the importance of the embryonic environment, which is essential for development.

THE SHOOT APEX

Following germination, the apical meristems of root and shoot, organized during embryo development, are involved not only in the growth of existing structures but also in the initiation of new organs and tissues that were not present in the embryo.

A meaningful impression of the shoot apex may be obtained by dissection of the terminal bud of either a woody or herbaceous shoot under a stereoscopic microscope. The young leaves are pulled away from the outside in succession, the successive leaves being smaller and less well developed. The leaves are close together because the internodes at this level have not elongated.

The leaves are regularly spaced around the stem, and if the leaf arrangement is of the alternate type, the primordia are arranged in a helix ascending toward the apex. Ultimately, the shoot apex itself is exposed. This appears as a smooth, glistening area, usually from 0.1 millimeter to 0.25 millimeter in diameter (FIG. 14-3). It often has the form of a domelike mound or of a somewhat flattened cone, but it may be flat or even depressed like a saucer. The initiation of leaf primordia does not occur in the center of this area, but is restricted to its margins.

When the internal structure of the shoot apex is examined in longitudinal section (FIG. 7-5), it is clear that this region is highly organized. In the flowering plants, one or more layers of cells cover the surface. These layers result from the fact that above the level of leaf initiation, the cells at the surface divide in a plane at right angles to the surface, whereas divisions in the interior occur in various planes. Often it is possible to observe in the apex a central group of cells that are larger than those at the periphery and which stain less intensely with the dyes used in histological preparations. Mitotic figures are rarely seen in this central region, and some observers maintain that this is a region of low mitotic activity, in contrast to the peripheral region in which mitotic activity is high. If this is true, the peripheral region would be the major initiating portion of the apex.

The Unity of the Shoot Apex In normal development, the shoot apex functions as a unit, but experiments have shown that when it is subdivided artificially, it is capable of reorganizing several new apices from the parts. After the apex has been exposed by dissection, it is possible to perform surgery upon it by means of delicate knives fashioned from fragments of razor blades ground to an extremely fine edge. If the apex is divided into halves by a single, median cut (FIG. 14-4A), the two halves of the meristem continue their development and soon reorganize into two complete shoot apices, which produce two shoots

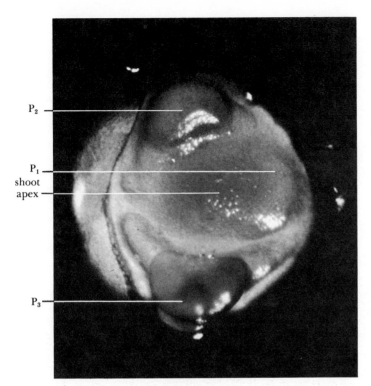

Fig. 14-3 An exposed shoot apex of lupine (*Lupinus albus*) viewed from above and magnified about 180 times. The apex itself is a domelike mound in the center and is surrounded by the three most recent leaf primordia it has produced; P₁ (youngest), P₂, and P₃.

(FIG. 14-4B). It is possible to subdivide the apex into even smaller pieces, many of which reorganize into complete apices. If the apex is not subdivided, it remains as a functional unit with no such independent development of its parts. Just why this is so and what sort of integrating mechanism of control causes the intact apex to function as a unit is unknown.

Autonomy of the Shoot Apex The delicate shoot apex is clearly dependent upon the shoot, from which it obtains its water, mineral elements, and sources of energy. However, it is one thing to be supplied with essential materials and quite possibly another to be controlled in the use of them, once supplied. Thus, in considering the organization of the shoot apex and its initiation of leaves and the structures of the stem, it is rea-

Fig. 14-4 Diagram illustrating the surgical subdivision of a shoot apex as seen in longitudinal section. (A): apex showing position of median cut. (B): the apex two to three weeks later, showing the apices reorganized from the halves in (A).

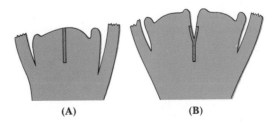

(A)   (B)

sonable to ask whether the apex is controlled by mechanisms within itself or whether control is imposed upon it by the organized shoot that it terminates. In other words, does the apex play a

Fig. 14-5 Maidenhair fern (*Adiantum pedatum*) grown in sterile culture, starting with an isolated shoot (rhizome) apex bearing no leaf primordia. Six months old.

directive role or does it merely respond to stimuli that emanate from the mature regions of the plant?

This question can be answered by removing the shoot apex from the plant and growing it on a nutrient culture medium, an experiment that has been performed on a number of species including ferns and other lower vascular plants as well as angiosperms. It has been shown by this technique that the apex, devoid of all leaf primordia, is able to give rise to an entire plant (FIG. 14-5). In some cases a shoot is formed first and later develops roots, but in others a root apex is quickly initiated opposite the shoot apex so that the result is much like a seedling plant developed from an embryo. In either case the shoot apex is evidently *totipotent,* that is, capable of forming an entire organism. It is worth noting that the nutritional requirements are simpler for the development of an isolated shoot apex of lower vascular plants than for an angiosperm apex.

The important role of the shoot apex as an organizing center in development is seen clearly when, instead of the apex, a segment of immature stem just behind the apex is isolated on a nutrient medium. Usually such segments grow, but they invariably give rise to unorganized masses of cells showing little or no differentiation (callus tissue) rather than to organized shoots or to whole plants. In other words, a piece of partially differentiated stem cannot provide organization, but the undifferentiated shoot apex can.

Morphogenesis of the Leaf Perhaps the most important characteristic of leaf growth is that it is limited, although the leaf primordium arises from a shoot apex that is potentially unlimited in its growth. Leaf growth, however, may be extensive, as in the case of some palms with leaves more than 50 feet in length. A second feature of importance is that the leaf, which arises from a radially symmetrical shoot apex, in most cases quickly becomes dorsiventral (bilateral) in its symmetry. This characteristic ultimately results in the flattened, expanded form of the leaf so important for effective photosynthesis. Finally, the leaf primordia arise in predictable locations around the shoot apex in a definite pattern. The question of leaf placement and what controls it will be considered first.

In many plants where leaves are attached singly at the nodes (alternate arrangement), each new primordium arises at an angular distance of approximately 138 degrees around the apex from the preceding one. In seeking to answer the question of why leaves are initiated where they are, experimental surgery has proved useful. A leaf primordium can be suppressed at an early stage by puncturing it with a needle or cutting it off with a knife, leaving a scar. In FIG. 14-6A a shoot apex is illustrated with its most recently formed leaf primordia clustered around it and numbered in the order of formation, beginning with the most recent. In FIG. 14-6B an apex is shown in which primordium 4 (P_4) was suppressed, its place being occupied by a scar.

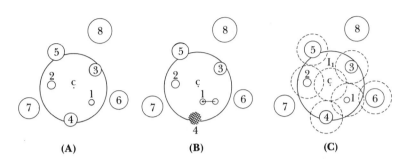

Fig. 14-6 Diagrams illustrating the placement of leaf primordia at the shoot apex. (A): a normal apex with the primordia numbered in succession, beginning with the most recently formed, grouped around the center of the apex c. (B): a comparable apex in which primordium 4(P_4) was suppressed shortly after its initiation. P_1 has arisen closer to the position of P_4 than in the normal apex. (C): the interpretation of leaf positioning based on the existence of inhibitory fields (indicated by dotted circles). The next incipient primordium (I_1) will arise where the inhibition is least.

Now we may compare the positions of succeeding leaves in the two apices. In the operated apex, the next two primordia to be formed, P_3 and P_2 arose in normal positions, but the next one, P_1, has been displaced toward the position at which P_4 should have developed. This does not mean that the primordium has moved, but rather that there was a shift in the position at which it arose.

This experiment, which has been repeated with various species, shows that when a leaf primordium is prevented from developing, subsequent primordia in the vicinity of its site arise closer to that site than they do when the primordium is present. This has led to the proposal that around each developing leaf primordium there is a field of inhibition, which prevents the development of new primordia. A similar field is thought to exist in the center of the apex. FIGURE 14-6C illustrates how such a system of fields could regulate the placement of new primordia. There is no information as to the nature of the fields, but they are often visualized as resulting from the production of an inhibiting substance by the leaf primordia and the central cells of the apex. The interaction of such inhibi-

tory fields also might explain why the shoot apex remains as an integrated unit, that is, why parts of it do not grow out at random unless artificially separated.

The field theory is capable of explaining how the positions of leaf initiation are controlled, but it does not suggest why the outgrowths become leaves and not shoots. Some experiments on the cinnamon fern (*Osmunda cinnamomea*) help to clarify this problem. If some of the youngest leaf primordia are removed from the apex and placed on a culture medium, many of them develop as shoots and ultimately, after initiating roots, as whole plants (FIG. 14-7). Older primordia, however, develop into small leaves only, and have lost the capacity to form an entire plant (FIG. 14-8). These experiments show that the capacity of a leaf primordium to develop into a leaf rather than a shoot is determined early in growth. The process of determination does not itself directly control the development of a leaf, but rather confers upon the primordium the capacity to control its own development. The nature of the stimuli that bring about the determination of a leaf primordium is unknown, but it is believed that they are complex.

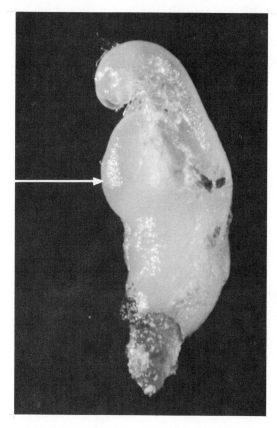

Fig. 14-7   Cinnamon fern (*Osmunda cinnamomea*). A very small, excised primordium in sterile culture may grow into a whole plant instead of a leaf; shoot apex (arrow) and first leaf.

Fig. 14-8   Leaf (frond) of cinnamon fern (*Osmunda cinnamomea*) grown in sterile culture from an older excised primordium.

The Development of Buds   The axillary buds that produce the branches of the shoot appear to have their origin in the shoot apex. In longitudinal sections of many shoot apices, it is possible to detect in the axils of recently formed leaf primordia small pockets of meristematic tissue termed *detached meristems* (FIG. 14-9). These detached meristems are evidently derived from the shoot apex. After they have been left behind by its further growth, they develop into small shoot apices and ultimately into axillary buds. Often

they develop only to a certain stage before being arrested by the influence of apical dominance. The exact mechanisms preventing the development of the detached meristems are not entirely understood, but it is this delay that allows them to produce buds. If they did develop with the leaf primordia, there is every reason to expect that they, too, would be determined as leaves.

Morphogenesis in the Root   Because of the presence of a cap the root apex cannot be exposed for external observation by simple dissection. In longitudinal section (FIG. 14-10) it is evident that the arrangement of cells is much more

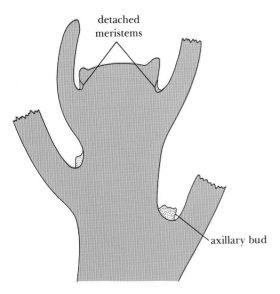

Fig. 14-9 The origin of lateral buds from detached meristems in the axils of leaf primordia.

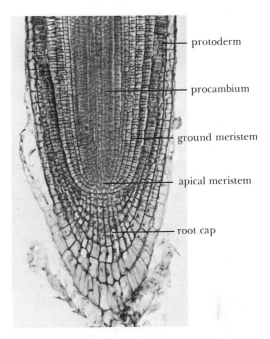

Fig. 14-10 The root apex of a radish (*Raphanus sativus*) seedling in longitudinal section.

regular than in most shoot apices, possibly because there are no appendages equivalent to the leaves and buds of the shoot apex which, in their development, disturb the regular sequence of cell divisions in the axis. As a result, the apex often consists of distinct layers that are continuous with specific tissues of the mature region of the root. The root apex, unlike the shoot apex, produces derivatives apically for the root cap as well as basally for the main body of the root. It is thus completely surrounded by its maturing derivatives and is not truly terminal in position.

In spite of these differences, there is a fundamental similarity between the root and shoot apices. Both have the potentiality for unlimited growth. In the root apex, as well as in the shoot, there is evidence for the existence of a central group of cells, which divide infrequently. The existence of this *quiescent center* (FIG. 9-3) has, in fact, been demonstrated more clearly in the root than in the shoot. If the root apex is divided by a vertical incision, the resulting parts reorganize into new root apices. Thus, in the root, as in the

shoot, there appears to be an integrating mechanism that causes the intact apex to function as a unit. The autonomy of the root apex has been demonstrated by experiments, particularly with pea roots, in which tips 0.5 millimeter in length, including the cap (about 0.2 millimeter of the root apex itself), have given rise to roots in nutrient culture. The isolated tips included, besides the root cap, little more than the apical meristem, which is the source of the organization of the root. Thus, the root apex, like the shoot apex, seems to control its own activities, but is capable of forming a root rather than a shoot or a whole plant.

DIFFERENTIATION AND GENE ACTIVITY

The phenomenon of differentiation has been mentioned in connection with growth, notably in the terminal region of an angiosperm shoot (Figs. 7-5, 7-6). The basis of this differentiation is change in individual cells, and it is remarkable that cells side by side behave as differently as they do. For example, within the procambium, xylem and phloem regions differ markedly in the later stages of their differentiation. Cells differ even within individual tissues. Yet, throughout this diversity, an organized system is maintained, and this is essential if the resulting mature stem is to function properly. It is important that vessel elements function as cells, but it just as important that they be formed in vertical rows in association with other vessel elements and in a channel that ultimately leads to a leaf.

We are a long way from understanding how the complex phenomena of differentiation are controlled, but some information has been obtained. One puzzling question arises as soon as we consider the great variety of cell types in the plant body. All of these cells are derived from a single original cell, the fertilized egg, and there is every reason to believe that, in a given plant, they are genetically identical. The development of buds on roots or leaves, and the initiation of roots on stem cuttings (Chapter 11), for example, indicate that differentiated cells have not undergone a fundamental genetic change, a mutation. If genes do determine structure, that is, if they do control development, how can cells carrying the same genes become radically different?

The development of a cell is due primarily to the chemical reactions that occur within it, and the individual steps of these reactions depend upon specific enzymes. The enzymes are produced by the action of genes. The important concept as far as differentiation is concerned is that, of the thousands of genes present in each cell, not all (perhaps only a small percentage) are active at any time and many never become active during the life of a particular cell.

It is, then, necessary to assume that individual genes may be activated or inhibited by conditions and reactions within the cell and to ask how this may be accomplished. No final answer is available, but a working concept, which may be termed the *inhibitor* or *repressor* hypothesis, has been widely accepted. In its simplest form, this concept is explained as follows: As we have seen (Chapter 4), protein synthesis is brought about by genes, which specify the kinds of amino acids in a polypeptide chain. These genes are termed STRUCTURAL GENES (Fig. 14-11). It is postulated that another set of genes, the CONTROLLING GENES, control the expression of the structural genes. Two important kinds of controlling genes are recognized, the OPERATOR GENES and the REGULATOR GENES. An operator gene may control the activity of a single structural gene or, frequently, of a series of such genes next to it on the chromosome (Fig. 14-11).

As long as the operator gene is not inhibited (*repressed*), the structural genes associated with it continue to produce messenger RNA for enzyme production. The regulator gene may be located some distance away in the chromosome complex of the cell. The regulator gene controls the operator. It does this by producing a *protein repressor*, which acts on the operator and prevents it from functioning, thereby blocking the formation of messenger RNA. The operator gene, however, may be activated (*de-repressed*). Some metabolic product, called an INDUCER, appears in the cytoplasm. The inducer may be a hormone or some simple food substance such as a sugar. The inducer combines with the repressor, inactivating it. The operator gene is now activated (de-repressed) allowing the structural genes which it controls to function in enzyme production. Ultimately, then, the regulator genes, acting alone or in conjunction with inducers, determine the initiation of gene activity and the period of time during which genes function, thus regulating, to a large extent, the form and function of the cells.

An illustration of gene control is shown by bacteria given a new carbohydrate food. If the necessary genes are present, these bacteria will quickly form the enzymes required to digest the new food. The food apparently acts as an inducer and leads to the de-repression of the appropriate genes. Similarly in certain experiments with both animals and plants, there is evi-

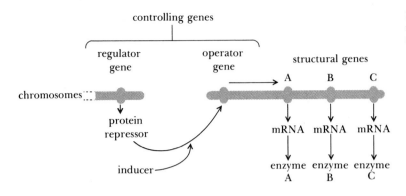

Fig. 14-11 Model for gene regulation of enzyme synthesis.

dence that hormones may activate specific genes. Messenger RNA is then formed and the enzymes produced are responsible for the physiological effects of the hormone.

A fully differentiated cell has thus become what it is, not only because of its inherited potentialities, but also because its position within the plant and because its specific environment have resulted in the selective activation of its genes. Environmental conditions could include the pressure of other cells; the supply of foods, such as sugar and proteins; the presence of growth hormones produced within the cell or transmitted through plasmodesmata from surrounding cells; the temperature; the concentration of carbon dioxide; the availability of oxygen; the pH of the cell sap; and the presence or absence of numerous organic and inorganic materials. These environmental variations are thought to influence the genes that control the enzymes, which control the chemical reactions, which control the development of the cell, by causing the production of specific inducers or, possibly in the case of chemical substances, by functioning themselves as inducers.

Although much of the evidence for the operation of such a system of genetic regulation has come from microorganisms, the existence of regulator genes has been demonstrated in higher plants. The attractiveness of this control system as a mechanism of differentiation lies in the fact that one regulator gene may influence a number

of structural genes and a single stimulus may thus have far-reaching effects on cellular development. Since the changes brought about in differentiation often tend to be of long term—as, for example, in a determined leaf primordium that continues to develop as a leaf after removal from the determining influence—there must be mechanisms that permit repression or de-repression to continue when the environmental stimulus has been removed. It has been suggested that if a particular active gene or group of genes causes the cytoplasm of a cell to produce a continuing stimulus (inducer), this kind of long duration could be accomplished. However, other possible mechanisms of gene regulation could be involved.

Experiments on Vascular Differentiation The role of genetic regulation in differentiation in plants has not as yet been demonstrated directly, but there is experimental evidence that specific substances of a hormonal nature can induce specific kinds of differentiation. For example, the involvement of auxin in the differentiation of vascular tissue has been indicated by grafting experiments in callus tissue derived from the cambium of lilac (*Syringa vulgaris*). The callus was devoid of vascular tissue at the beginning of the experiment and composed entirely of homogeneous parenchyma. The base of a bud of lilac, with bud scales removed, was cut into the shape of

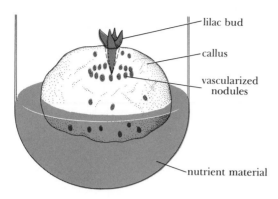

lilac bud

callus

vascularized
nodules

nutrient material

Fig. 14-12 Induction of vascular tissue by grafted bud of lilac (*Syringa vulgaris*) in callus from the same species.

a wedge and grafted into a V-shaped cut on the upper surface of the callus. A graft union occurred, and leaf expansion and internodal elongation began. After some 20–30 days, randomly distributed nests, or nodules, of actively dividing cells appeared in the callus along the sides of the graft. Other nodules appeared below the base of the graft, but these occurred in a circular pattern at right angles to the axis of the grafted bud (FIG. 14-12). Cells of all of these nodules gradually differentiated into vascular elements. In 50–55 days the nodules of the circular zone consisted of fully differentiated xylem on the inside and phloem on the outside, sometimes separated by a cambiumlike layer, which produced a few cells of xylem and phloem. Occasionally, when nodules occurred close together, the cambiumlike layer became continuous between them. It is assumed that these nodules were stimulated to form under the morphogenetic influence of auxins and perhaps other compounds moving downward from the growing bud. In grafted callus tissue about 50 days old, additional nodules, larger but randomly arranged, appeared below the original circle. These may have resulted from increased auxin production as the bud grew and the leaves enlarged.

In other pieces of lilac callus, instead of a bud, agar containing indoleacetic acid was placed in the incision. Again vascularized nodules developed

in a circle. With increase in auxin concentration in different experiments, the circle of nodules, as seen in cross section, increased in diameter, suggesting that an optimum concentration is required to promote the differentiation. An interesting aspect of these studies was that the addition of sucrose to the indoleacetic acid in the agar placed in the incision greatly enhanced the differentiation of vascular tissue. The effect of auxin was apparently dependent upon adequate supplies of sugar. Experiments such as these demonstrate the far-reaching consequences of specific stimuli of the sort to which cells in the intact plant may be exposed at certain phases of their development.

The Totipotency of Cells If the concept of selective gene action in differentiation is correct, there is no fundamental genetic change in differentiated cells. It then follows that differentiated cells, if they can be stimulated to grow again, ought to reveal their latent potentialities, perhaps even to the extent of producing whole plants. In other words, they ought to be totipotent. Cells derived from living vegetative cells in which division has ceased can, under certain conditions, resume division and produce structures that resemble embryos derived from fertilized eggs. This has been spectacularly demonstrated in the wild and cultivated carrot. In one set of experiments, bits of tissue were removed from the secondary phloem of the storage root of the cultivated carrot—physiologically mature cells that would normally undergo no further divisions. Growth of this tissue was possible if the culture medium contained growth-regulating substances that incite cell division. The most effective way of providing these substances was the addition of coconut milk, itself an endosperm that normally supports the growth of an embryo.

In a liquid nutrient medium of this type, the phloem tissue grew in an unorganized manner, forming callus, from which single cells floated free (FIG. 14-13A). These single cells grew into small nodular masses, from which roots emerged. Transferred to solid media, shoots formed in connection with the roots and plantlets resulted that could be planted in soil and give rise to the entire plant, producing flowers and seeds.

Other investigations have shown that cells produced in the manner just described can sometimes give rise to structures (embryoids) very like normal plant embryos. Cell suspensions derived from embryos of wild carrot (*Daucus carota*), when transferred to a solid medium, produce such embryoids in abundance. The phenomenon, however, is not restricted to cells of embryo origin. It has been observed in cultures derived from root and leaf petiole tissues from wild carrot plants and has been demonstrated for other plants as well. Many cells grow into globular colonies (FIG. 14-13B). Further growth results in elongated embryoids with the primordia of cotyledons (FIG. 14-13C) and finally in fully developed embryoids with well-formed cotyledons (FIG. 14-13D). A young carrot plant grown from such an embryoid is shown in FIG. 14-13E. These experiments provide support for one of the important tenets of the gene-regulation concepts of differentiation.

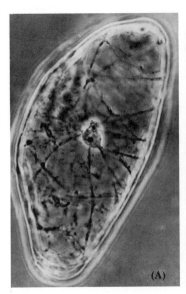

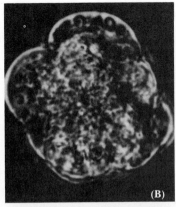

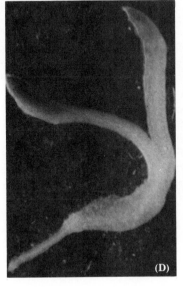

Fig. 14-13 The development of a carrot plant from a single cell. (A): freely suspended cell derived from surface of phloem callus. (B): globular embryoid resulting from growth of a single cell of a suspension of embryo origin. (C): embryoids showing cotyledon primordia from a suspension of leaf petiole origin. (D): fully developed embryoids. (E): carrot plant grown from an embryoid in a 1-liter flask.

MEASUREMENT OF GROWTH

Much of our knowledge of the control of development pertains to growth, and there is a substantial body of information relating to environmental effects on growth. A great deal of this information is of applied or practical value. Growth in plants can be measured in several ways.

One method is to measure increase in fresh or green weight. Such a measure of growth is reliable as long as plants are fresh and growing. It might be considered that growth could also be measured by increase in dry weight, that is, the material of the plant body remaining after all water has been removed. Such data could be misleading, since a plant that has completed growth (increase in size) may still increase in dry weight if photosynthesis continues and results in the accumulation of starch and other storage materials.

Measurements of length or height are frequently convenient and useful indicators of growth. Such measurements show that the rate of growth varies with the age of the organism. This is clear when the measurements are made on an annual plant such as corn or garden bean where the life span is limited. The growth of the plant is slow at first, then gradually becomes more rapid until a maximum is attained, after which the rate diminishes. When the increase in height of the stem, for example, is plotted over a period of time, a sigmoid (S-shaped) curve is obtained. The increase in fresh weight of the entire plant yields the same result (Fig. 14-14). This type of analysis is more difficult to apply to long-lived perennial plants such as trees, where the growth occurs in a long series of seasonal increments. That the result is probably the same is indicated by the fact that in all plants, perennials as well as annuals, the growth of individual parts or organs follows the same sigmoid curve. This has been recorded in many instances for elongation of individual internodes, increase in area of leaves, growth in diameter of fruits, and elongation of hypocotyls in seedlings. Measurements indicate that cells themselves conform to

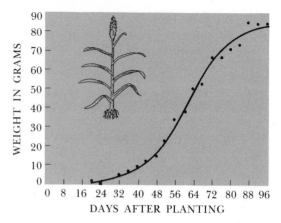

Fig. 14-14 Increase in weight of aerial portions of growing sorghum (*Sorghum vulgare*). The solid line represents the theoretical sigmoid curve; the circles, the actual weights of individual plant samples.

the typical growth curve. Thus, the over-all curve for a plant represents the summation of growth increments in its parts.

The general form of the growth curve appears to be constant for all plants investigated, but deviations occur as a result of variations in the environment. These deviations, however, concern the magnitude of the values that are plotted rather than the shape of the curve. The wide applicability of the growth curve suggests that its explanation lies in internal factors in the plant rather than in external or environmental factors. The nature of the causal factors is unknown.

GROWTH REGULATION

The growth of a plant gives evidence of being under rather precise control. In large part this control reflects the activity of chemical substances known as plant hormones. These may be defined as organic compounds produced by the plant, which are active in minute amounts and are transported through the plant body, influencing growth or other physiological processes either in other regions or in those in which they are produced. Hormones are used up during metabolic processes and must be renewed if their effects are to continue. Growth in one part of the plant may thus be dependent upon the cellular activities of another part. By means of hormones, the cells of a plant are changed from the status of independent units to correlated components of a unified organism.

Three major groups of plant hormones have a growth-promoting effect as their fundamental characteristic, although under particular conditions and in certain concentrations they may also inhibit or retard growth. These are the auxins, the gibberellins, and the cytokinins. Of course there are many substances that promote plant growth, including all nutritive substances and even water, but these are not to be confused with the hormones, which serve as regulating stimuli when present in minute amounts. The question of vitamins is somewhat more difficult to define, for these also regulate growth and function at extremely low concentrations. They are, however, considered to be essential factors in metabolism rather than growth-regulating hormones. Green plants, unlike animals, are able to synthesize the vitamins needed for growth, although isolated plant tissues and organs such as roots often require them in a culture medium if they are to be grown successfully. In addition to growth-promoting factors, it has long been known that plants contain substances that inhibit growth. Certain of these inhibitors, such as ETHYLENE and ABSCISIC ACID, have been demonstrated to interact with the growth-promoting hormones, and are themselves considered to be hormonal in nature.

The plant hormones, both promoting and inhibitory, produce a wide variety of effects, and this diversity makes it difficult to determine the precise manner in which hormones act. There is good evidence in some cases that certain hormones operate at the level of the gene, specifically at the point at which RNA synthesis occurs. It is by no means certain, however, that they always operate at this level, and it is equally possible that they act at various other sites in the machinery of the cell.

Auxins The plant hormones that have been most widely studied are the auxins. This term, from the Greek meaning "to increase," is applied to hormones that accelerate growth by stimulating cell enlargement. They may also act in conjunction with other hormones in stimulating cell division and in bringing about other changes. In different concentrations and circumstances, they can both stimulate and inhibit growth. Roots, for example, are generally inhibited by auxin concentrations which stimulate growth of the stem. Auxins are important in maintaining apical dominance (Chapter 2), in initiating roots on the bases of stem cuttings, and in the differentiation of vascular tissues. In these and other effects, auxins undoubtedly interact with other hormones. Where the auxin effect can be isolated to a certain extent, its role in promoting cell enlargement, especially elongation, appears as the fundamental one. In this promotion of cell growth, it has been shown that auxin causes an increase in extensibility of the cell wall so that turgor pressure brings about enlargement. The precise method by which the softening of the wall is accomplished is a subject of controversy.

The effect of auxins upon elongation has been studied chiefly by experimenting with the seedlings of grasses, especially oats. The sprout that arises from the germinating oat grain consists of a tapering sheath (the coleoptile), which encloses the bud and foliage leaves. This coleoptile is the "guinea pig" that has yielded much of the available information about growth hormones. It has long been known that the extreme tip of the coleoptile is necessary for elongation of the coleoptile and the stem below it. When the tip is cut off, growth is greatly retarded for a period of several hours, after which the apex of the stump may take over the physiological functions of the removed tip and renew the production of auxin. But if the severed tip is replaced immediately and cemented in position with a little warm

Fig. 14-15 Experiments on growth of the oat (*Avena sativa*) coleoptile. (A): the tip is removed and later replaced, bringing about resumption of growth. (B): auxin from the tip migrates into an agar block, which is then placed on a stump, bringing about renewal of growth. (C): an agar block containing auxin is placed on one edge of the cut surface of a stump, causing curvature.

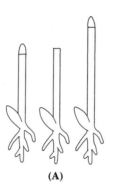

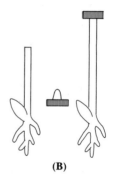

(A)

(B)

(C)

gelatin, growth does not stop (FIG. 14-15A).

The indication that soluble growth-regulating substances are produced by the apex is confirmed by removing the coleoptile tip and placing it upon a small block of agar gel for about 2 hours. If the cut tip is then discarded and the block of agar is placed upon the stump of a decapitated coleoptile, growth is again resumed (FIG. 14-15B). The use of a block of pure agar as a control shows no such effect. If an agar block containing auxin is placed upon one edge of a coleoptile stump, a curvature results, indicating that growth is more rapid along the side below the block (FIG. 14-15C). This procedure is used to test the amount of auxin present in the agar block, for in general the degree of curvature is proportional to the concentration of auxin in the block.

As long ago as 1880 it was suggested that migratory growth substances occur in plants. In 1910, and again in 1928, the existence of these substances was proved by experiments such as those just described. Since 1928, many fundamental investigations have been carried on. Auxin has been isolated from growing organs, such as germinating seeds, seedlings, and developing shoots, and, largely in an inactive or combined form, from resting seeds and pollen grains.

Several auxins have been isolated, one of which, indoleacetic acid (IAA) (FIG. 14-16A), is widely distributed in plants and is also produced

synthetically. In addition to the natural auxins, a large number of organic compounds have been found to possess similar growth-regulatory properties. These GROWTH REGULATORS, as contrasted to growth hormones that are formed by the plant, include indolepropionic, indolebutyric, and naphthaleneacetic (FIG. 14–16E) acids. Because the natural auxins can be obtained only in minute quantities, the discovery of synthetic growth substances has greatly stimulated investigations on plant growth. The synthetic compounds have also come into extensive use in agriculture and horticulture.

Gibberellins Much attention has been given to a group of growth substances discovered in Japan a number of years ago. These substances are produced by a fungus that causes the "foolish seedling" disease of rice, and are called gibberellins (FIG. 14-16B) after the fungus (*Gibberella fujikuroi*). The gibberellins are relatively inexpensive and nontoxic, and therefore can be used in many types of experiments, either as sprays or in droplets applied locally. They have been isolated as natural components of many plant parts. As do the auxins, they stimulate elongation growth in stem and leaf, and their effects can be dramatic in plants of the rosette type (FIG. 14-17). Equally remarkable is their ability to convert dwarf varieties of some plants, such as peas, into plants of normal size. Gibberellins are

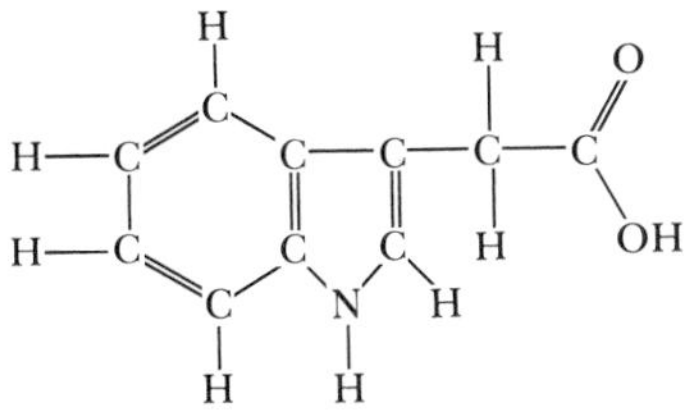

β—indoleacetic acid (IAA)

(A)

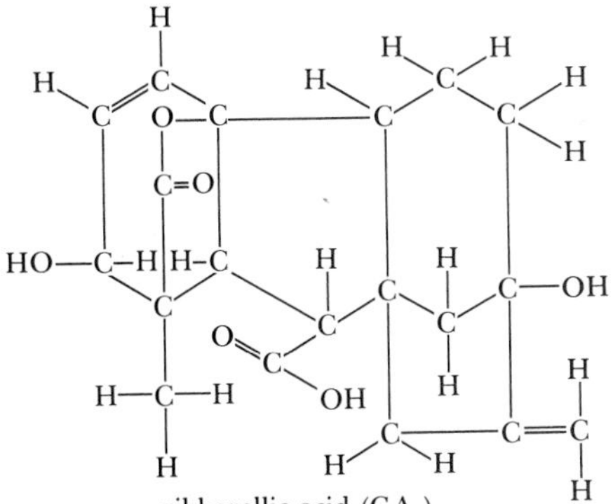

gibberellic acid (GA₃)

(B)

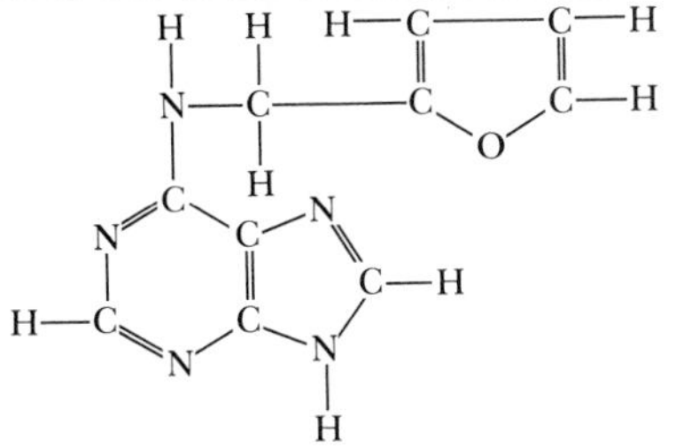

a cytokinin (6—furfurylamino purine)

(C)

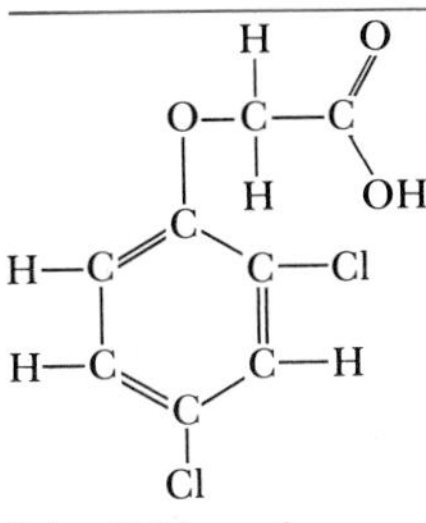

2,4—dichlorophenoxy-
acetic acid (2,4-D)

(D)

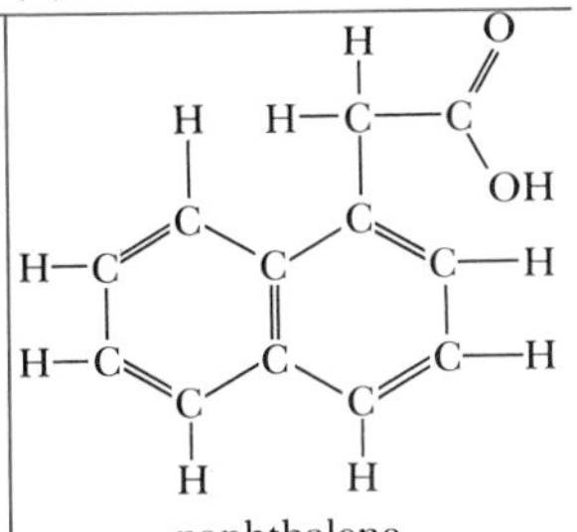

naphthalene-
acetic acid (NAA)

(E)

Fig. 14-16 (Left) Chemical structure of some plant hormones and growth regulators.

Fig. 14-17 Effect of gibberellic acid (GA) on growth of cabbage. (Left): untreated. (Right): sprayed several times with 0.1 percent GA.

usually classified as promoters of cell elongation, and in this they appear to function together with auxin. However, in stimulating the "bolting" (the premature production of a flowering stalk) of rosette plants, they cause an enormous increase in cell division as well. They have been implicated in the initiation of flowering, the growth of young fruits, the breaking of dormancy, and the differentiation of vascular tissues derived from the cambium. One of the most interesting chapters in the growth hormone story has been the discovery of the role of gibberellins in the germination of cereal seeds. In germinating barley grains, gibberellin released from the embryo induces the formation of a starch-hydrolyzing enzyme, α-amylase, by the outermost cells of the endosperm (aleurone layer). This in turn makes the starch of the endosperm available to the developing embryo. It is believed that the effect of the hormone is to stimulate the synthesis of the messenger RNA specific for the production of that enzyme.

Cytokinins The cytokinins (FIG. 14-16C), in contrast to the auxins and gibberellins, are considered to be primarily stimulators of cell division. In this action, however, they appear to interact with auxin. Cells in a tissue culture supplied with adequate nutrients and indoleacetic acid often enlarge greatly without dividing. If a cytokinin is also included in the culture medium, the cells both divide and enlarge, but the cytokinin without the auxin has little effect. Cytokinins have been detected as natural components of plant tissues, the richest sources being young fruits and endosperm tissues. It is significant that the endosperm, which provides the environment for the developing embryo, should contain plentiful supplies of these potent growth factors. The presence of cytokinins in coconut milk is undoubtedly an important factor in the ability of this material, when added to a culture medium, to incite cell division in mature, nondividing explants. This, it will be recalled, was an essential step in the experiments that revealed the totipotency of differentiated cells.

Inhibitors Interest in naturally occurring substances that have an inhibitory effect upon plant growth reached a peak with the discovery and characterization of abscisic acid. This hormone was independently discovered in young cotton fruits, where it eventually promotes abscission, and in the leaves of birch, where it induces dormancy in buds. Only after isolation and characterization was it realized that the same substance is involved in the two cases. It is now believed that abscisic acid is widely distributed in plants, and it has been shown to inhibit a variety of developmental processes, including elongation growth and seed germination. In different circumstances it will counteract the effects of auxins, gibberellins or cytokinins, and its inhibitory action may be reversed by the promoting hormones.

One of the most surprising of the inhibitory hormones is the gas ethylene. The effect of ethylene in promoting the ripening of fruits has been known for many years, but it has now been found to exert an inhibitory influence upon elongation. Auxin in excess of a certain concentration appears to stimulate the production of ethylene, and this in turn may act as a brake upon the auxin-stimulated growth. This is another example of the interaction of promoting and inhibiting hormones in the regulation of plant growth.

Some Uses of Growth Regulators The use of artificial, or synthetic, growth substances has greatly increased in recent years, particularly in the field of weed control. In high concentrations, these chemicals are toxic to many kinds of plants. A synthetic, hormonelike chemical, 2,4-dichlorophenoxyacetic acid, or more popularly, 2,4-D (FIG. 14-16D), is widely used as a weed killer. This and related compounds can be applied in such concentrations as to have selective effects. In general, broad-leaved plants, both monocots and dicots, are killed at the concentrations used, but grasses are not injured. The use of 2,4-D is especially effective in eliminating weeds other than grasses from lawns and from fields of corn, wheat, and other grains. Among

PLATE 1   Diversity in the plant kingdom: dicotyledons (the eucalyptus trees) and the monocotyledonous grass trees (*Xanthorrhea preisii*), remotely related to the lilies. Western Australia.

PLATE 2 Anthocyanins and carotenoids in autumn leaves of some North American woody plants. (A): white ash (*Fraxinus americana*); (B): red maple (*Acer rubrum*); (C): mountain maple (*Acer spicatum*); (D): sugar maple (*Acer saccharum*); (E): witch-hazel (*Hamamelis virginiana*); (F): blue beech (*Carpinus caroliniana*); (G): yellow birch (*Betula lutea*). (Adrian N. Bouchard.)

PLATE 3 Water hyacinth (*Eichhornia crassipes*), a troublesome perennial free-floating weed, which reproduces both by seed and by vegetative reproduction. A native of South America, it is now widely distributed over the warm regions of the earth, in streams, lakes, reservoirs, drainage ditches, and irrigation canals. The plants may block navigation, destroy fishing grounds and recreation areas, clog irrigation pumps, and interfere with the operation of hydroelectric power plants. Loss of water from the leaves may be 3 to 7 times greater than evaporation from a free water surface. Much effort is expended in control by herbicides. Florida. (Mary S. Shaub.)

PLATE 4 Flower and fruits of bloodwood (*Dillenia alata*). The fruits are in various stages of dehiscence, exposing the seeds and showing the carpels of which the compound pistil is composed. The species of this genus are mostly trees of the tropics of the Old World.

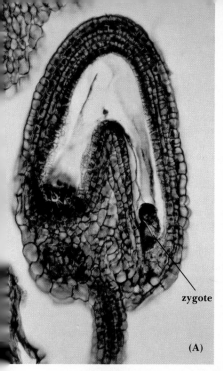

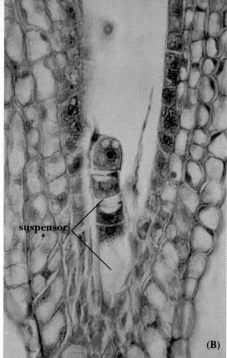

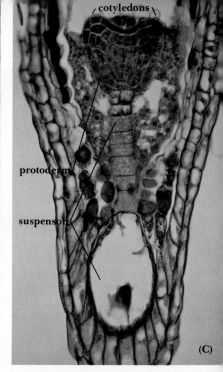

zygote (A)

suspensor (B)

cotyledons

protoderm

suspensor (C)

PLATE 5 (Above)   Development of embryo of shepherd's purse (*Capsella bursa-pastoris*). (A): the whole ovule in section view showing the zygote before its first division; (B): filamentous embryo with suspensor becoming recognizable at base; (C): embryo showing well-developed suspensor, the initiation of cotyledons, and the beginning of tissue organization. The basal cell of the suspensor is greatly enlarged. (Douglas DesBrisay and John Waddington, University of Saskatchewan.)

PLATE 6 (Left)   Inherited variations in seeds of Kahler "Baby Lima" beans. A bean with white seed coat was crossed with a bean with black seed coat. The progeny were then self-pollinated for eight generations. No two seeds of the eighth generation were alike, and some of this variability is shown by the illustration. The plants bearing these seeds also differed in other characters. (Adrian N. Bouchard.)

PLATE 7 Cultures of the pink soil mold, *Neurospora crassa,* showing the normal, or *wild* type (upper left), and five color mutants. All contain carotenoid pigments except the albino mutant (upper right) which lacks all such pigments. The color mutant at the lower right results from a mutation which causes a deficiency of an enzyme necessary in the biosynthesis of adenine (a component of nucleic acids). The alteration of this biosynthetic pathway causes the accumulation of a purple pigment. The other four mutants have abnormal carotenoids. This mutant fails to grow unless adenine is added to the culture medium. (Adrian N. Bouchard.)

PLATE 8 A population of *Eschscholtzia californica* (California poppy), native in western North America. Genetic variability is favored by cross-pollination by insects. The plants in this population are annuals, growing in an arid habitat. Perennial races of the species are found in more temperate habitats. (Mrs. Agnes Baptist, Modesto, California.)

the weeds controlled by the use of this chemical are dandelion, pigweed, ragweed, mustard, bindweed, cocklebur, and water hyacinth. Great caution must be exercised, however, in the use of these potent chemicals because when applied as sprays or dusts, they may be carried considerable distances by air currents, causing damage to other crops and vegetation. In the present period of growing concern about pollution of the environment, the possible effects of these substances in the soil and upon natural vegetation are being carefully evaluated.

Some of the important uses of plant growth regulators are discussed in other sections of this text. The stimulation of rooting in cuttings is considered in Chapter 11, and the use of growth regulators in the production of seedless fruit is covered in Chapter 13.

The inhibitory effects of growth-regulating substances have been used experimentally to prolong the rest period of potato tubers and thus prevent the heavy sprouting that occurs in potatoes stored for long periods. Naphthaleneacetic acid (FIG. 14-18) and chloroisopropyl phenylcarbamate (CIPC) have been used with some success. These chemicals are applied in the form of vapor to tubers in storage, or as a dip or dust on the tubers as they are being stored. Prolongation of dormancy and control of sprouting can also be obtained by spraying potato plants in the field with the compound maleic hydrazide while the tubers are still growing. The chemical, or some substance resulting from its action, is translocated to the tubers and delays sprouting after harvest. Maleic hydrazide has also been used to slow the growth of lawn grasses and reduce the number of mowings required.

Another and important application of the inhibitory effect of growth substances is the control of the preharvest drop of apples and pears. In a number of important varieties, the fruits fall from the tree before they are ready to be harvested. This results in considerable loss to the grower, since the fallen fruits are usually bruised and unsalable. The fall of fruits can be prevented by spraying the trees with synthetic growth-regulating substances shortly before harvest.

Fig. 14-18 Prevention of sprouting of stored potatoes (*Solanum tuberosum*) with a growth regulator, naphthaleneacetic acid. (Left): treated. (Right): untreated.

Naphthaleneacetic acid is most commonly used. Minute concentrations (5–10 parts per million) are effective, and usually only one application is necessary. The growth substance slows the formation of a separation layer at the base of the fruit stem and allows normal maturation of the fruit.

Hormone applications are also used to improve the quality and yield of fruits. In California, gibberellin sprays are widely used to produce enlarged fruits in the Thompson seedless variety of grape. If applied at the proper time (bloom stage), such sprays also result in looser clusters, which are more easily handled and are less susceptible to certain diseases than are compact clusters.

LIGHT AND GROWTH

Among the external factors known to affect plant growth are the availability of minerals, moisture and air in the soil, the humidity of the air, the light intensity and duration, and the temperature. Any one of these factors can be limiting for growth. Some of them have already been discussed in relation to other plant processes. The effects of light and temperature on growth are considered here in more detail.

The effect of light upon growth has been extensively investigated. Aside from its effect through photosynthesis and food supply, light influences the growth of individual organs or of the entire plant in less direct ways. The most striking effect can be seen by comparing a plant grown in normal light with the same kind of plant grown in total darkness (Fig. 14-19). The lack of light affects both the external form of the plant and the rate of elongation. The plant grown in the dark usually has a tall and spindling stem. The leaves fail to expand, and both leaves and stem, lacking chlorophyll, are pale yellow. Such a plant is said to be ETIOLATED. If plants that have been growing in the dark are exposed to light, the rate of elongation is quickly reduced. When grown in the light, the same kind of plant is sturdy in form, with fully expanded green leaves, spaced at shorter distances along the stem. Etiolation can most easily be demonstrated by planting potato tubers or seeds of garden beans or peas in pots kept in the dark. The tubers and seeds have large reserves of food, and the plants that result from sprouting or germination grow for some time before they are starved by lack of light for photosynthesis.

Plants growing in shade instead of darkness show a different response. Moderate shading tends to reduce transpiration more than it does photosynthesis. Hence, shaded plants may be taller and have larger leaves because the water supply within the growing tissues is better. With heavier shading, photosynthesis is reduced, and small, weak plants result. The greatest response to light and shade is shown by the leaf, which may vary markedly in structure with variations in light intensity (Fig. 14-20). Leaves growing in the sunlight commonly have more sugar and less water than those in the shade. As a result, the leaves in the sun become thicker and smaller, with a heavier cuticle and a greater amount of conducting and mechanical tissue. Shaded leaves have larger air cavities, and the palisade layer of the mesophyll may be reduced in size or may fail to develop altogether so that the interior of a shaded leaf consists entirely of spongy mesophyll.

In some tobacco-growing areas of the United States, and in Cuba, growers put to practical use

Fig. 14-19 Etiolated and normal potato (*Solanum tuberosum*) plants. The etiolated plants on the left were grown in the darkness; the normal plants on the right, in the light. The plants were the same age.

the effects of reduced light intensity on leaves. Because broad, thin leaves are required for cigar wrappers, and bring a premium price, certain tobacco varieties are grown in the shade under acres of thin cloth (Fig. 14-21). These coverings lower the light intensity, increase the humidity, and reduce the transpiration rate per unit of leaf surface. These factors contribute to the growth of leaves that are longer, broader, and thinner, with smaller veins. Because the leaves of shade-grown tobacco increase in size even though pho-

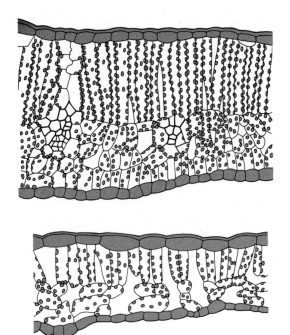

Fig. 14-20 Cross sections of sun and shade leaves of silver maple (*Acer saccharinum*). (Top): leaf from south periphery of an isolated tree. (Bottom): leaf from center of crown of an isolated tree.

Fig. 14-21 Tobacco (*Nicotiana tabacum*) grown under cloth shade, Florida.

tosynthesis is not increased, less material is available for cell-wall construction. Lignin, for example, which causes brittleness in dry tissues, is lower in shade-grown leaves; hence, they are more pliable and are especially valuable for cigar wrappers. The modern, less expensive substitute for these shade-grown leaves is a thin plastic sheet in which crumbled tobacco leaves are embedded.

Photoperiodism Both the intensity and duration of light may have different and characteristic influences upon plant growth. Duration of light, in particular, has been studied intensively, and it has been found that the relative lengths of the daylight and dark periods can have a striking ef-

fect upon the vegetative growth and reproductive activities of plants.

The discovery of this relationship forms an interesting story of botanical research. In 1906, a commercial variety of tobacco gave rise to a mutation called Maryland Mammoth. This variety grew to a height of 10–15 feet during the summer and produced a large number of leaves. But no matter how vigorous the growth, the plants did not flower and produce seed before cold weather set in. W. W. Garner and H. A. Allard, plant physiologists of the U.S. Department of Agriculture, set out to find out why.

They began with the observation that plants transferred to the greenhouse from the field at the end of the season eventually flowered and produced seed. The investigators then grew seedlings of the Mammoth variety in pots in the greenhouse during the winter and found that they blossomed and produced abundant seed, although attaining a height of only 3 feet. Finally they observed that although greenhouse plants blossomed during the winter, they ceased to bloom in late spring. In the words of Garner and Allard, "Naturally it became of interest from both a practical and scientific standpoint to determine the factor of the environment responsible for the remarkable winter effect in forcing blossoming."

In their search for the cause of this winter blooming, they studied not only the Mammoth tobacco but also soybeans and other plants. Experiments soon showed that such factors as temperature, fertilization, and intensity of light did not produce the explanation sought. They then considered whether the length of the daily exposure to light could be responsible for the winter blooming. In a number of experiments to test this factor, they obtained striking results.

In one set of experiments during the spring and early summer of 1918, they grew Maryland Mammoth seedlings in pots and also a variety of soybean that normally does not bloom until September in the latitude of Washington, D.C. These they kept in darkness from 4:00 P.M. until 9:00 A.M., giving the plants only 7 hours of daylight. All bloomed promptly. Further investigations showed that the effect of the duration of light upon plants is important and widespread. In 1920, Garner and Allard published important studies on the flowering of plants in relation to the length of day, a phenomenon that they later called PHOTOPERIODISM.

The length of day, the period from sunrise to sunset, varies greatly over the earth's surface, depending upon the latitude and the time of year. On March 21 and September 21 the days are 12 hours in length throughout the world. In the northern hemisphere, the days increase in length from December 21 to June 21, when at the latitude of Washington, D.C., they are 15 hours long. After June 21, the day length decreases to December 21, when the days are only 9½ hours long at the latitude of Washington. Farther north, this difference becomes more striking, with a maximum of 24 hours of daylight or darkness north of the Arctic and south of the Antarctic circles.

Responses to photoperiod Plants of temperate regions may be divided into three groups according to the effect of the photoperiod, or length of exposure to daylight, upon vegetative growth and reproduction. The first of these, the "short day" group, comprises plants that flower when the day length is short. If they are subjected to a long period of daily illumination, flowering is inhibited, although the plants continue to grow vegetatively (FIG. 14-22). Short-day plants grow vegetatively during the long days of summer and do not produce flowers until the days become shorter in the late summer and fall. Among short-day plants are poinsettias, most asters and goldenrods, the ragweeds, chrysanthemums, sorghum, and many others.

The plants of the second group, the "long-day" plants, come into flower only under extended periods of illumination and produce only vegetative growth when the photoperiod is short. Among long-day plants are hollyhock, radish, garden beet, spinach, iris, timothy, and red and sweet clover.

The dividing line between day lengths favorable to vegetative growth and those tending to cause flower and seed formation is called the CRITICAL LIGHT PERIOD for a particular plant. Short-day plants bloom when days are shorter than their critical day length, and long-day plants when days are longer than those that are critical for them. For many species, the critical day length lies between 12 and 14 hours of illumination, but small differences may be of major importance. One variety of soybean, for example, blossomed 15 days later when the photoperiod was increased from 14 to 14½ hours. This difference is equal to only a small difference in latitude, but for this plant it means that a crop grown near Charleston, S.C., may be successful, whereas one planted near Washington, D.C.,

Fig. 14-22 Photoperiodic response in chrysanthemum (*Chrysanthemum morifolium*), a short-day plant. The flowering plant on the left received 8-hour days and 16-hour nights. The vegetative plant on the right received 16-hour days and 8-hour nights.

may fail to produce mature seed before the fall frosts.

Many varieties of soybean, corn, and other short-day plants have been selected for "resistance" to long days until they may flower normally with 15- or 16-hour day lengths. These plants, however, will flower more rapidly with shorter days, showing their inherent short-day nature. Such varieties typically have a north and south zone of adaptation only 100 or 200 miles wide. If grown farther north they are too late to mature, and farther south they mature so rapidly that they make small vegetative growth and do not produce a full crop. In either situation a different, better-adapted variety must be used.

The plants of a third group are termed intermediate, or day-neutral, for they are insensitive to length of day. Included here are many varieties of garden beans, tomato, calendula, vetch, cyclamen, nasturtium, roses, snapdragon, carnation, and many common weeds.

The photoperiodism of plants is readily demonstrated by artificially increasing or decreasing the length of day. When the days are long, for example, the photoperiod may be shortened by covering the experimental plants in the morning and evening. If longer periods are desired, the prevailing photoperiod may be supplemented by the use of electric light. Increasing the photoperiod by artificial light during short days prevents the flowering of short-day plants. The additional light may be added at either end of the day, or it may be even more effective if given in the middle of the night. This last response calls attention to the fact that in short-day plants, the length of the dark period rather than that of the light period is of fundamental importance. Under natural conditions, of course, a long night means a short day, and vice versa. It has been concluded that flowering in short-day plants is induced by reactions occurring during the dark period, which must, therefore, be of sufficient length. Conversely, flowering in long-day plants is inhibited by reactions of the dark period, which must not be too long, and long-day plants generally will flower in continuous light with no dark period at all.

The light intensities used to lengthen the light exposure (after the plants have had several hours of good light for photosynthesis) can be less than one-thousandth of full sunlight—5–10 footcandles. The added light should be rich in red wavelengths, from incandescent rather than fluorescent lamps, since red light is most effective in photoperiod and related growth responses of plants.

The following example illustrates the effect of decreasing the photoperiod upon long-day plants. Several species of *Sedum* (stonecrops), perennial herbs that normally flower in the middle of the summer, were kept under a photoperiod of 12 hours or less for 9 years. During this time they continued to grow vegetatively, but they did not flower until they were returned to long summer days at the end of the experiment.

Certain short-day ornamentals, such as aster and chrysanthemum, may be forced into bloom at will if the plants are covered morning and evening with heavy black cloth, thus shortening the prevailing spring and summer photoperiods. This is now an established commercial practice. On the other hand, long-day plants, grown in the greenhouse in the winter, may be brought into bloom by the use of supplementary electric light. A knowledge of the photoperiodic responses of plants is particularly important to plant breeders. By the artificial control of the daily light period, they are able to bring about simultaneous blooming of long-day and short-day varieties that they desire to cross, or to produce an extra generation of long-day plants in a greenhouse during the winter.

Many other aspects of growth, in addition to flowering, may be affected by photoperiod. The formation of tubers by the Jerusalem artichoke (*Helianthus tuberous*) is a short-day response. The plant flowers on long days, but it does not produce tubers unless exposed to shorter days, such as those that occur in the autumn. Many biennial and perennial plants, such as sweet clover (*Melilotus alba*) and horse nettle (*Solanum carolinense*), form enlarged and thickened roots during the short days of September and October. These roots become filled with the accumulated foods that enable the plants to live through the winter and start growth the following spring. Many woody plants of temperate climates respond to short days by the cessation of vegetative growth, the formation of resting buds, and the onset of dormancy. With long days, dormancy is delayed or prevented, and dormant plants can be induced to resume growth if exposed to long days. For example, if actively growing plants of flowering dogwood (*Cornus florida*) are subjected to short-day conditions of 12 hours or less, elon-

gation of the stem and development of new leaves are arrested, bud scales are formed, and the plant becomes dormant. However, if the plants are maintained under 15-hour day or longer conditions, vegetative growth continues indefinitely. The adaptive significance of these responses in a region of alternating favorable and unfavorable conditions for growth is evident. There is also evidence that the seeds of some plants respond to photoperiod in their germination, and both long-day and short-day species are known.

Mechanism of photoperiodism The fact that photoperiodic responses may be experimentally altered by brief exposure to low light intensity—as, for example, in the interruption of the dark period in short-day plants—has made it possible to demonstrate that it is red light (wavelength about 660 millimicrons, FIG. 5-15) which is effective. Similarly, it has been found under laboratory conditions that far-red light (about 730 millimicrons) will reverse the red light effect, that is, if exposure to red light is followed by exposure to far-red, the result is as if there had been no exposure at all. If repeated alternating exposures to red and far-red are given, the final exposure determines the response. From these observations has come the discovery of a cytoplasmic pigment, PHYTOCHROME, now identified as a protein and partially purified, which is sensitive to red light. This pigment can exist in two forms, one that absorbs red light (Pr) and one that absorbs far-red light (Pfr). Red light converts Pr to Pfr, but this form is unstable and is gradually converted back to the Pr form. However, far red light causes a rapid conversion to the Pr form, and hence a reversal of the red light effect.

$$Pr\ 660 \xrightarrow{\text{red}} Pfr\ 730$$

$$Pr\ 660 \xleftarrow[\substack{\text{far-red (rapid)} \\ \text{or dark (slow)}}]{} Pfr\ 730$$

Phytochrome is believed to be active in the Pfr

form and thus the responses are produced by red light. In relation to flowering, reactions promoted by Pfr interfere with the dark reaction that promotes flowering in short-day plants and inhibits the flowering of long-day plants. Since it is the length of the dark period that is critical, it might be supposed that the reversion of Pfr to Pr determines the critical length of the dark period. The reversion, however, takes only about 4 hours in the dark, and this is much too brief a period to explain the timing. The timing mechanism or "clock" is as yet poorly understood. Plants appear to have natural, internal rhythms, different in different plants, in their metabolic processes, and it is suggested that phytochrome exerts its influence upon the natural rhythm of a plant. In effect, phytochrome, which receives the light stimulus, "sets the clock," but it is the natural rhythm of the plant that provides the clock mechanism and thus determines what the ultimate response will be.

Aside from photoperiodic phenomena, many other light effects upon plants are exerted through the phytochrome system. The germination of many seeds—lettuce (*Lactuca sativa*), for example—requires a light stimulus, and this is effected by red light and reversed by far-red. Stem elongation and leaf expansion are also influenced by red light, and the effects of etiolation are prevented by exposure to red light. The red light effect is reversed by far-red. Even the formation of anthocyanin pigments and the diurnal movement of leaves can be tied to the phytochrome system. The example of the sensitive mimosa plant (*Mimosa pudica*) is particularly instructive. In light, the leaflets are open, but 5 minutes after being placed in the dark, they begin to fold and the process is completed in about 30 minutes. A 2-minute exposure to red light at the beginning of the dark period causes the leaflets to stay open. This is reversed by a subsequent exposure to far-red, and a sequence of alternate red and far-red treatments may be carried out with the final treatment determining the response.

The precise way in which phytochrome func-tions is a matter of speculation. It has frequently been suggested that phytochrome may operate by influencing membrane permeability in the cells, not only of the plasma membrane but also of those of the nucleus and mitochrondria. In this way, profound influences upon the metabolism of cells could be exerted quickly.

Hormones and Flowering One widely held view of the control of flowering involves a flowering hormone called FLORIGEN. It is believed that this hormone is produced in the leaves under the proper photoperiodic conditions and is transported to the shoot apices, where it induces the flowering response. This idea arose from experiments in which only a part of a plant (such as a single leaf) was exposed to an inductive photoperiod; the entire plant responded, suggesting that a hormone moved out of the treated portions. In *Chrysanthemum indicum,* a short-day plant, the upper portion will flower if only the lower portion is exposed to short days. It is necessary, however, to defoliate the nontreated portion, probably because the noninduced leaves in some way inhibit the transmission of the stimulus. The stimulus apparently can cross an interspecific graft union, and this fact has been used to show that the hormone of one species can induce flowering in another. A branch of a short-day plant grafted to a long-day plant will often flower in long days, and vice versa. In spite of many attempts, however, the flowering hormone has not been isolated.

It has also been shown that other known hormones exert an effect upon flowering. In many instances, auxin has been shown to inhibit flowering or, more probable, to promote vegetative growth and thus to counteract stimuli to reproductive development. Substances that inhibit the action of auxin (antiauxins) can under certain conditions induce flowering. Soybean plants, for example, are vigorously vegetative in long days, but if they are sprayed with an antiauxin, they may be forced rapidly into flowering. Similar results can be obtained by removing the young expanding leaves at the top of the plant, which are the major auxin producers. Conversely, auxin sprays on plants growing in a day length short enough for flowering can prevent flowering. In fact, it has been suggested that florigen may act by reducing the activity of auxin to a level that permits the initiation of flowering.

Gibberellins, also, have frequently been implicated in the flowering process as promotors in certain types of plants, particularly in long-day plants that form rosettes. Thus, in lettuce (*Lactuca sativa var. Grand Rapids*), the stem will elongate and flower (bolt) on short days if treated with gibberellin. It is possible that the primary effect of the hormone is upon stem elongation and that the flowering follows from this. In short-day plants, gibberellin at most enhances the flowering response, but it cannot cause flowering under noninductive conditions.

The participation of the known growth hormones in the control of flowering has led to the suggestion that florigen may not exist as a single substance. Rather, it may represent a balance among several or many factors. This would certainly explain the difficulties in its attempted isolation.

The recognition of photoperiodism has helped botanists to understand better the northerly and southerly distribution of plants. It has also made possible the recognition of varieties adapted to growth in particular localities and the fixing of proper dates for planting. In addition, selection and hybridization have developed special varieties of a number of kinds of plants that are adapted to the photoperiods of the areas in which they are to be grown.

Man has long sought for greater crop yields by modifying the environment in which he grows his useful plants and by developing new varieties better adapted to specific environmental conditions. Each species and variety may have different requirements as to soil, humidity, temperature, and length of growing season. To these requirements, for a large number of plants, must be added length of day.

TEMPERATURE AND GROWTH

In general, growth is promoted when the temperature rises and is retarded when the temperature falls (FIG. 14-23). However, the growth rate does not continue to increase with rise in temperature, for injurious effects appear in time and the growth rate diminishes. High-temperature injury may result from desiccation and from a rate of respiration so high that the consumption of food materials tends to exceed production by photosynthesis. Temperature affects growth through its effect upon all metabolic activities: digestion, translocation, respiration, and the building of new protoplast and cell-wall material. Also, high temperatures increase transpiration (Chapter 8) and thus reduce turgor and growth, especially during the day.

Each species has a minimum temperature, below which it fails to grow; an optimum, at which the growth rate is highest; and a maximum, above which growth ceases. The optimum temperature may vary with each stage of development and with the length of time the temperature prevails. For example, the optimum temperature for elongation or increase in dry weight may not be the optimum for the production of flower and seed. Most plant growth occurs between 10 and 40°C (50 and 104°F).

Temperature can also affect the type as well as the rate of growth. An interesting relationship between temperature and photoperiodic effects has been found in the phenomenon known as VERNALIZATION. Certain plants, mostly long-day species but including some short-day plants, become sensitive to photoperiod only after an exposure to near-freezing temperatures. Some plants, such as winter annuals—which begin their growth in the autumn and complete it early in the next growing season—require exposure during or soon after germination, but others, such as many biennials, must undergo substantial vegetative growth before the cold exposure. Without both cold treatment and proper photoperiodic induction, flowering does not occur. In some biennials (for example, the carrot), the cold treat-

Fig. 14-23 Effect of temperature on the growth of a tropical ornamental plant (*Browallia americana*). (Left): grown at 10°C (50°F). (Right): grown at 18°C (65°F).

ment can be replaced by a gibberellin application.

Dormancy One of the characteristics of plants of temperate climates is their marked periodicity—a period of active growth alternating with one of rest, or DORMANCY. This is especially prominent in woody plants; the leaves drop in autumn and growth is not renewed until the following spring. Herbs that live more than a single growing season behave similarly, but here the dormant structure is a rhizome, corm, bulb, tuber, or root system rather than the entire plant. Numerous wild and cultivated perennials live through the winter and flower in early spring.

The aerial parts then die down and the plants enter a period of rest, even though temperature and light conditions are apparently favorable for continued growth. In most larger trees, terminal growth of the shoot usually comes to an end in early summer, although diameter and root growth may continue for some time longer.

The length of the dormant period varies, and for many plants a period of low temperatures is required to "break" the dormancy and permit growth to be resumed. Branches cut from some flowering trees or shrubs, such as dogwood, azalea, or lilac, fail to grow if brought indoors in the fall but come into bloom if cut in the late winter or early spring. Most deciduous fruit trees, such as the apple, peach, and cherry, require extended winter-rest periods and therefore can be grown only in temperate climates. Most varieties of peach must rest for 600–900 hours below 7°C (45°F) before the leaves and flowers will emerge from the buds. Pecans, English walnuts, almonds, grapes, and other crops also require low temperatures for periods of varying duration.

Some plants require freezing temperatures to break the rest period. In others the same effect is produced by low temperatures above freezing. Most bulbs, tubers, and other underground stems require at least a short rest period. The normal rest period of the potato tuber is not shortened by exposure to low temperatures, but freshly harvested tubers may be forced into sprouting by two months or more of warm, dry storage. It is possible and sometimes desirable to break dormancy or reduce its length to force early food or ornamental crops. Among the methods used to accomplish this are warm or cold storage or treatment with various chemical compounds. Dormant potato tubers, for example, will sprout in 7–10 days after harvesting if exposed to the vapor from the chemical ethylene chlorohydrin. Plant pathologists are thereby able to grow seed potatoes in the greenhouse during the winter to determine whether certain potato diseases are carried by the tuber. This same chemical may be used to break the dormancy of the gladiolus corm, which has a normal rest period of 6–7 months. By this treatment gladiolus flowers are made to bloom during the winter.

The participation of hormones in the onset and maintenance of dormancy in plants and their reactivation is well established. Abscisic acid, the inhibitory substance already mentioned, is particularly implicated in a variety of dormancy phenomena. Dormant buds of many trees, such as ash, peach, sycamore maple, and lilac, have been shown to contain appreciable quantities of inhibitor, and there is a progressive reduction in the amount present as the buds emerge from dormancy. In at least some cases the inhibitor has been identified as abscisic acid, and applications of this substance to leaves cause shoot growth to be arrested and induce the development of typical resting buds. Under natural dormancy-inducing conditions, such as a short photoperiod, abscisic acid is synthesized in the leaves and transported to the buds. This hormone is also involved in the dormancy of certain seeds, and applications can prolong seed dormancy in many species. In ash (*Fraxinus*) seeds the abscisic acid content of the embryo decreases during the stratification (cold treatment) required before they can germinate. Abscisic acid, as has been pointed out, appears to function by counteracting the growth-promoting activity of other hormones, notably the gibberellins.

PLANT MOVEMENTS

The larger and more complex plants are fixed in position, but many of their parts or organs carry on several kinds of movements. These movements are commonly overlooked because they are too slow to be detected by ordinary observation. Among them are curvatures, twining, leaf movements, and the opening and closing of flowers. Movement from place to place also occurs in simple motile plants and reproductive structures, such as spores and sex cells, but these will not be considered here. The basis of all plant movement is the fact that the protoplasm of plants, like that of animals, is sensitive and responds to external and internal stimuli.

Although long studied, the nature of plant movements and the causes and mechanisms underlying them are still poorly understood. Plant movements are of two types: (1) SPONTANEOUS MOVEMENTS, the result of internal stimuli and relatively independent of the environment; and (2) INDUCED MOVEMENTS, the result of response to stimuli coming from outside the plant. Among the external stimuli causing plant movements are gravity, contact, shock, and fluctuations in light and temperature.

The mechanisms responsible for movements include (1) growth movements, which result from an unequal rate of growth on different sides of the organ concerned and which are irreversible; (2) turgor movements, which result from an increase or decrease in the size of cells caused by a gain or loss of water, and which are temporary and reversible.

Spontaneous Movements The most common spontaneous movement is NUTATION. This consists of a movement of the tip of a growing stem or other organ, describing a more or less circular path in space (FIG. 14-24A). Nutation is most apparent in the stems and tendrils of climbing plants, but it occurs also in leaves, runners, flower stalks, and roots. Indeed, Darwin, who made extensive studies of nutation, declared that every growing part of every plant continually nutates, although the scale of movement of some parts, such as roots, is small. The mechanism of the movement is a spiral shift around the axis of the region of most active growth. Nutation is a growth movement and ceases as the tissues mature. The nutation of a slender, rapidly growing stem may lead to twining (FIG. 14-24B). The tip of a vine may swing in a circular path several inches across. If it strikes a slender object such as a twig, the nutation is restricted by the object, and the continued nutation of the stem tip results in the stem twining around the object as it grows.

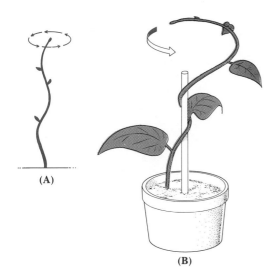

Fig. 14-24 Nutation. (A): diagram of nutation in a growing stem tip. (B): exaggerated nutation in a rapidly growing stem of morning-glory (*Ipomea purpurea*) results in twining.

Induced Movements Movements resulting from external stimuli are classed as TROPISMS and NASTIC MOVEMENTS. Tropisms are responses in which the direction of the movement is determined by the direction from which the stimulus comes. Tropisms are usually growth movements, and include such responses as PHOTOTROPISM, GEOTROPISM, and THIGMOTROPISM (a twining movement in response to contact). In nastic movements the response bears no relation to the direction from which the stimulus is applied. Examples of nastic movements are PHOTONASTY (response to light), THERMONASTY (temperature), and THIGMONASTY (touch). Of all induced movements, only phototropism and geotropism will receive further consideration.

Phototropism Phototropism is usually a positive response to light that comes in greater intensity from one direction than another. The stems and leaves (usually the petioles) curve or bend toward the source of light (FIG. 14-25). Roots, on the other hand, either do not respond to light at all or are negatively phototropic; that is, they bend away. Stems bend directly toward the light, but leaves become oriented in such a manner that the leaf blade is approximately at right angles to the light source. This is especially noticeable in vines on a wall or fence, where the leaf blades and petioles turn in such a way that few spaces are left unfilled. The leaves become so arranged that they present a maximum surface to the sun, with a minimum of overlapping, forming what is termed a LEAF MOSAIC (FIG. 14-26).

In spite of numerous investigations, the mechanism of the phototropic response is not completely understood. There is ample evidence, however, that an unequal distribution of auxin when the plant is illuminated from one side is involved in this response. Experiments with the oat coleoptile show that bending in an etiolated organ can result from exposure to a weak light source for no more than 1 second. This response appears to be caused by a migration of auxin from the lighted to the unlighted side of the coleoptile tip. This migration is demonstrated by placing an excised coleoptile tip upon two agar blocks. The blocks and the base of the tip are separated by a razor blade (FIG. 14-27). If the tip is now exposed briefly to light from one side, more auxin will diffuse into the block on the side of the tip away from the light. In an intact plant, the greater supply of auxin on the back side of the coleoptile causes increased cell elongation and a bending toward the side that has been lighted.

The type of bending shown in FIG. 14-25, in contrast, normally follows an exposure to hundreds of foot-candles of light for a number of hours. The common and ecologically important phototropic movements in plants are dependent upon such long-continued exposures to relatively high light intensities. The mechanism of this high-intensity phototropism is not understood, but it is believed that light inactivates some system involved in the growth process. In fact, auxin may not even be involved.

Fig. 14-25   Phototropism in coleus (*Coleus blumei*). The stem is inclined toward the light, and the leaf blades are approximately at right angles to the direction of the light.

Fig. 14-26 A leaf mosaic shown by the leaves of wild grape (*Vitis riparia*).

Fig. 14-27 Tip of oat seedling, showing migration of auxin from the lighted side to the opposite side.

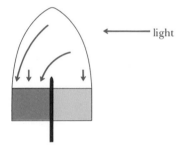

light

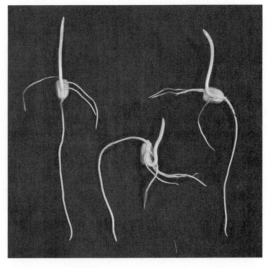

Fig. 14-28 Negative and positive geotropism in germinating corn (*Zea mays*). (Left): the grain was germinated in a vertical position. (Right): the grain was germinated in a horizontal position. (Center): the grain was inverted. In all, the shoot has grown upward (negative geotropism) while the root has grown down (positive geotropism).

Geotropism The effect of gravity upon the direction of growth is termed GEOTROPISM. Two main types of response exist: NEGATIVE GEO-TROPISM, as in the upward growth of the shoot of a plant, and POSITIVE GEOTROPISM, as in the downward growth of the root.

The curvatures that result from the stimulus of gravity on both root and shoot may be demonstrated by placing a young and actively growing plant in a horizontal position; the shoot will soon bend upward, the root downward (FIG. 14-28). These curvatures result, as in phototropism, from unequal growth rates in the zone of elongation. In the stem, this zone is long, and curvature can therefore occur at some distance behind the tip. In the root, where the zone of elongation is short, the region of curvature is located only a short distance behind the apex.

The explanation of geotropic bending is again found in the effect of auxin upon growth. When the stem is placed horizontally, auxin, under the influence of gravity, accumulates on the lower side in greater concentration than on the upper. The increased concentration on the lower side brings about a greater rate of growth on that side, and the stem bends upward. This explanation is well established and is supported by much experimental evidence, including the detection of lateral movement of radioactive auxin.

When a root is growing horizontally, it is believed that auxin, as in a stem or cleoptile in a similar position, accumulates on the lower side of the root in greater concentration than on the upper side. But the zone of elongation of the root is much more sensitive to auxin than the similar region of the stem, and the concentration on the lower side becomes too high and inhibits rather than stimulates growth. Growth continues on the upper side, and the root apex bends downward. It has been suggested that the inhibition of root growth on the lower side is not due to the direct action of auxin. Rather, it may result from the action of ethylene, the formation of which is induced by auxin.

SUMMARY

1. Growth, an irreversible increase in size, is accomplished by cell division and cell enlargement. In the development of the organized plant body it is accompanied by differentiation.

2. Development of the plant body begins with the zygote, which gives rise to the embryo with its suspensor. As differentiation begins in the embryo, the apical meristems of shoot and root are set off. The requirements for embryonic development may be investigated by removing the embryos at early stages and rearing them in culture.

3. The meristematic shoot apex is a highly organized structure. It is largely autonomous in the control of its activities and acts as an organizer in shoot development.

4. The placement of leaf primordia around the apex appears to be regulated by inhibitory fields. After initiation a leaf primordium undergoes a process of determination, which fixes its developmental pathway.

5. Lateral buds arise from detached meristems in the leaf axils.

6. The root apex produces no appendages and is surrounded by its differentiated derivatives. Like the shoot apex, it is largely autonomous in the control of its development.

7. In differentiation, cells that are genetically identical develop differently because of selective gene activity. A structural gene may be either repressed (inactive) or de-repressed (active) under the influence of controlling genes. Stimuli influence differentiation through their effect upon repression and de-repression. Differentiated cells may retain the capacity for further development, as has been shown by the production (in culture) of embryoids and ultimately whole plants from cells derived from differentiated tissues. The role of specific stimuli in differentiation is illustrated by experiments on auxin control of vascular tissue differentiation.

8. Growth may be measured as increase in size or weight, either of an organism or of its parts. When such measurements are plotted against time, a sigmoid curve—considered to be characteristic of growth—is obtained.

9. Growth in plants is regulated by hormones. Three groups of these, auxins, gibberellins, and cytokinins, in general have a promoting effect. Others, including abscisic acid and ethylene, are inhibitors. The interaction of promoting and inhibiting hormones is significant in the control of growth. Synthetic growth regulators have important commercial uses.

10. Both external form and internal structure of plants are affected by light. Plants deprived of all light become etiolated. Variations in light intensity may also affect form and structure through their effect upon photosynthesis and transpiration.

11. The length of day, or photoperiod, regulates the time of flowering of many plants. Short-day plants flower when the light period is shorter than their critical day length, whereas long-day plants flower when the light period is longer than their

critical day length. The flowering of day-neutral plants is not affected by day length. The length of the night is the important factor in the response of plants to day length. Photoperiod also influences the vegetative growth of many plants.

12. In photoperiodic phenomena it is red light that provokes the response, and this light acts upon the pigment phytochrome. The effect of red light is reversed by far-red. Phytochrome also brings about a variety of other light effects upon plant development. The plant hormones, especially auxins and gibberellins, are also important in the control of flowering.

13. Temperature affects the rate of all plant processes. Most plants grow best within an optimum temperature range, but this optimum varies with the plant and with time. In some plants the photoperiodic response depends upon prior exposure to low temperature.

14. Most perennial plants of temperate climates go through periods of rest or dormancy, which are known in some cases to be produced by inhibiting hormones. Periods of low temperature are often required before growth can be resumed.

15. Plant movements are either spontaneous or induced. Spontaneous movements such as nutation result from internal stimuli. Induced movements caused by external stimuli include tropisms and nastic movements.

16. Phototropic and geotropic responses are in large part the result of an unequal distribution of auxin. A lateral displacement of auxin may be caused either by light or by a gravitational stimulus.

Plant Ecology

MOUNTAIN CLIMBERS pausing to rest on rocky outcrops in the midst of widespread ice fields find plants (lichens) growing upon barren rocks. In Death Valley, California, the hottest and driest region of the United States, there are more than 300 species of native plants. The fact that plants are found growing in snow, in the salt-saturated waters of Great Salt Lake, in hot springs, on beach and desert sands, and even underground in caves and mines attests to the wide range of their adaptability.

These various examples illustrate the fact that plants exist in almost every known environment. The adaptations and relationships of plants within less extreme environments may be just as striking and significant. The study of the relationship between plants and their environment is known as ECOLOGY, or, more strictly speaking, PLANT ECOLOGY, since the same types of relationships also exist for animals and many of them for man as well. The study of plant ecology merges into plant geography, which treats of the distribution of plants over wide areas and attempts to account for the distribution and migration of species.

Plants usually live in COMMUNITIES, composed of one to many species living in a specific area under similar conditions. A community may consist of a mass of algae floating on a pond, lichens on a rock, a field of corn, a deciduous forest, the plants of a riverbank, a sand hill, a flood plain, a burned area, or an alpine meadow. The community, by definition, comprises only living organisms; if nonliving factors are also considered, we are dealing with an ecosystem (Chapter 10). The study of how plants and animals live together in ecosystems is an important aspect of ecology.

Closely related to the concept of environment is the HABITAT. This term is applied to the place where a particular kind of plant or community of plants usually lives. A habitat may be a south-facing, warm, dry slope; a mud flat; a damp ravine; a marsh; the floor of a moist woodland; or a sandy beach. It implies the presence of a group of particular environmental conditions and is therefore more specific than environment as a whole.

The growth and reproduction of plants are intimately connected with the chemical and physical aspects of such environmental features as water, light, soil, wind, and temperature. The great variations in these and other factors from region to region are in large part responsible for the differences among natural environments and for the kinds of plants found in them. The ecological factors responsible for these differences act as a whole and produce as a result of their combined effects a characteristic environment. Within this framework the contribution of each factor is fairly distinct, although its effect may vary over a wide range. It is possible to look at ecology in two ways: one may consider the ecological factors and their effects on habitats, or one may consider habitats in terms of their typical plant life and characteristics. Both points of view are essential and are included in this chapter.

ECOLOGICAL FACTORS

Ecological factors are customarily classified as: (1) CLIMATIC factors, such as precipitation, length of day, atmospheric humidity, temperature, wind, and intensity of light; (2) SOIL factors—the chemical and physical nature of the soil, the soil moisture, the soil temperature, and the soil aeration; (3) BIOTIC factors—those resulting from the activities of plants themselves and of man and other animals.

These categories overlap to a considerable degree. The amount of rainfall affects the moisture content of the soil, and both precipitation and temperature modify the chemical and physical nature of the soil. The intensity and duration of sunlight may be determined by the climate or by a biotic factor, such as shading by other plants. Moreover, plants are influenced by a host of agents, not by one factor alone. The effects of all must be integrated in order to obtain a comprehensive picture of the plant in relation to external influences. A clearer concept of these relationships is obtained if we distinguish between direct and indirect effects of factors on the plant. The forests of Ohio are deciduous rather than evergreen as a result of climatic conditions, past and present. It is easy to observe May apples, trillium, and Solomon's seal growing in open woods, and goldenrods, asters, and thistles growing in adjacent open fields. But achieving an understanding of the mechanisms that have segregated these plants into their special habitats is more difficult.

Ecological studies thus become extremely complex. But in spite of the difficulties, the study of plants in their environments has yielded a large body of ecological knowledge, contributing both to a better understanding of plants themselves and to a more effective use of plant, soil, and water resources. Every farmer or gardener is a practicing ecologist, since by such practices as cultivation, fertilization, irrigation, and spraying he affects plant behavior by changing the environment of the plant. The farmer who incorporates the best practices of soil conservation and agriculture in his work discovers that he is creating a specialized enviroment. In it, highly specialized, cultivated plants will produce a high crop yield while maintaining or even improving the soil.

Climatic Factors

Temperature Temperature and precipitation are the chief climatic factors that determine the occurrence of various plant species and the pattern of vegetation. They are the factors most closely related to plant survival. Temperature influences such plant activities as transpiration, respiration, germination, growth, and reproduction. Temperature extremes are important, but their duration and the length of the growing season, or frost-free period are also significant. The growing season is a major factor in determining the areas in which crop plants can be grown.

Cotton growing is restricted to the southern United States and southern California because this plant requires a warm growing period of 200 days or more. The mild, nearly frost-free winters of the Imperial Valley of California, or the lower Rio Grande Valley of Texas, and of southern Florida result in a 12-month growing period and make it possible for these regions to supply the northern states with out-of-season vegetables. In contrast to cotton, some varieties of corn and spring wheat will mature in less than 100 growing days. Potatoes, cabbages, peas, and other garden vegetables are adapted to cool weather and mature in a growing season of 75–90 days. They can be grown as far north as Alaska.

Low temperatures may affect plants both during the dormant period and in the stage of active growth. Winter injury may result in the death of the plant or in damage to the roots, bark, and buds. Many crop plants, both woody and herbaceous, are injured by frost in late spring. Damage to citrus, peach, and other crops is often reckoned in millions of dollars. The death of plants when they are exposed to freezing temperatures is not due to the direct effect of low temperature, but is the result of ice formation in the tissues. The formation of ice crystals within the protoplast normally results in the death of the frozen cell. In hardy plants, under natural conditions, ice crystals form between the cells rather than in them. At temperatures below freezing, water as ice has a lower diffusion pressure than it has in the liquid form. Water consequently tends to diffuse from the cells and to accumulate as intercellular ice. The removal of water from the cells causes dehydration of the protoplasm and may result, as the temperature drops, in coagulation of the protoplasm and death of the cell. Death may thus be due more to desiccation than directly to cold.

The ability to survive freezing temperatures varies widely among different species and is one of the important factors that determine northward and altitudinal distribution. The ability to survive freezing is known as frost resistance or frost hardiness. Hardiness is normally seasonal.

During the growing season, for example, conifers may be injured by a hard frost, but during the winter may withstand temperatures of 60°C (76°F) below zero or lower. The resistance of many plants to freezing may be increased by a process known as hardening. In nature, hardening takes place with decreasing temperatures as winter approaches. It is a common practice among growers of early cabbage to harden the seedlings by gradual exposure to low temperatures, sometimes also by reducing water, before transplanting them to the field. Similar practices may increase the resistance of plants to high temperatures or to drying. Hardening is brought about by the higher sugar concentrations in plants whose growth is reduced by low temperature or drought. This sugar results in chemical changes in the protoplasm, which increase its resistance to loss of water in freezing or drying.

Precipitation The annual rainfall is a major factor in determining the distribution of plants. Its influence, however, may be modified by the distribution of rainfall throughout the year, the amount falling at any one time, the slope of the land, the permeability and water-holding capacity of the soil, the wind velocity, and the temperature. In general, forests occupy areas where the rainfall is high; deserts occur where it is low; areas of intermediate rainfall are commonly grasslands or scrub vegetation.

The seasonal distribution of rainfall may be more important than the total amount. For agriculture, spring and early summer rains are generally preferable to those of fall and winter. If precipitation is concentrated in the winter, perhaps as snow, so that a deep layer of subsoil is soaked, forests may exist in regions of relatively low total rainfall. Summer rainfall under the same conditions favors shallow-rooted plants like the grasses.

Temperature is an important indirect factor in determining the effectiveness of rainfall. High temperatures increase the rate of evaporation from the soil, but more importantly, they increase the rate of transpiration. For example, nearly twice as much rainfall is needed to produce a crop of wheat in Oklahoma as in the cooler weather of central Saskatchewan.

In long-range agricultural planning, it is not enough to know the annual rainfall and its seasonal distribution. The rate of evaporation and degree of retention in the soil must be known so that the moisture actually available to crops can be calculated. Light rains in hot, dry weather will usually have no effect upon the soil-moisture content, for the water does not get down to the roots, but evaporates quickly from the surface of the plants and the soil. Heavy rains of short duration may also have little effect upon soil moisture, for the runoff may be great.

Water in relation to habitat and plant structure On the basis of the nature of the habitat to which they are adapted, plants may be classed as: (1) MESOPHYTES, (2) XEROPHYTES, and (3) HYDROPHYTES. Mesophytes grow in an environment that is neither very wet nor very dry, such as the meadows and forests of the temperate zones and corresponding regions of the tropics. Because most vascular plants are mesophytes, their structures and adaptations are already familiar to us. There is no hard and fast division between these plants and those adapted to drier environments on the one hand and to wetter ones on the other.

Broadly defined, a xerophyte may be regarded as a plant adapted to survival under conditions of a limited supply of water in its habitat. This deficiency may be persistent or periodic. Xerophytes are characteristic of desert and semi-desert regions. They also occur in mesophytic regions where available water is frequently deficient, as on sand hills, beaches, rocky ledges, or cliffs. Even where the annual precipitation is sufficient to permit a mesophytic vegetation, xerophytes may be found in place of mesophytes if the climate is characterized by sharply defined wet and long dry seasons. The xerophytes are better able to withstand the seasonal droughts.

Hydrophytes grow in or near the water, partly or entirely submersed. Some grow on land with their roots in saturated soil. Both xerophytes and hydrophytes are thought to have originated from mesophytes and by evolutionary change become adapted to new environments. In the course of this evolution, many changes in physiology, structure, and behavior related to the environment have arisen.

FORM AND STRUCTURE IN XEROPHYTES The xerophytes possess no common anatomical or physiological features. Some xerophytes are succulent, others are not. They are a heterogeneous group, and not all are "true" xerophytes.

The succulents are regarded as highly specialized xerophytes. Among the plant families containing succulents are the cactus, spurge, milkweed, stonecrop, lily, and amaryllis. The chief characteristic of the succulent type is that the bulk of the plant body is composed of water-storage cells that maintain the plant during periods of drought. The succulent organs are generally the stems or leaves, rarely the roots. If the leaves are succulent, the stem is generally much reduced; if the succulent organ is the stem, the leaves are reduced or absent. The volume of the shoot is great in proportion to the surface exposed, and this, combined with a thick cuticle and other features, retards water loss so that the transpiration rate is low. The succulents are said to resist, rather than endure, drought.

One of the most specialized of succulents is a little plant of the carpetweed family (FIG. 15-1) found in a sandy desert in South Africa. The club-shaped leaves, about 1½ inches long, grow in the sand, with only the flat or slightly convex, colorless or grayish apex of each leaf visible. The thin, green, outer shell of the leaf surrounds a mass of tightly packed water-storage cells. The exposed apex is covered by a thick epidermis and cuticle, and has a few stomata. It is through this windowlike portion that the leaf receives light, which penetrates to the chlorophyll along the inner surfaces. Each leaf is thus protected from desiccation, but at the same time exposes a considerable area to the light. During the blooming season the flower bud grows upward through

Fig. 15-1 A succulent (*Fenestraria rhopalophylla*) with window leaves, South Africa.

Fig. 15-2 A spurge (*Euphorbia ingens*) in the desert of southeast Africa. Note the resemblance to the cactus, a plant of another family, in Fig. 15-3.

the soil into the air; here pollination occurs and the seeds are disseminated.

Succulents illustrate clearly the principle that unrelated plants, under similar environmental conditions, may develop striking similarities in external form. For example, species of the spurge family growing in the deserts of Africa show such marked resemblance to some of the cacti of the American deserts that, in the absence of the flower (which is characteristic for each family), they might be considered close relatives (FIGS. 15-2, 15-3).

Most of the nonsucculent xerophytes, like the succulent types, are perennials, and include herbaceous forms, mostly grasses, as well as woody species. Many of the latter are more or less evergreen, or at least retain their leaves during periods of drought. The leaves of such plants commonly exhibit XEROMORPHIC modifications

that contribute to survival during dry periods. Thousands, perhaps millions, of square miles of the earth's surface support a vegetation markedly xeromorphic in character. Among xeromorphic modifications (FIG. 15-4) are (1) small leathery leaves; (2) a thick cuticle and thick epidermal cell walls; (3) sunken stomata; (4) strongly developed palisade tissue at the expense of spongy parenchyma; (5) increased mechanical tissue, such as fibers that accompany conducting tissue or are located beneath the epidermis; (6) cells small and compactly arranged; and (7) capacity for folding or curling of the leaf (FIG. 15-5). The latter is relatively common in certain xerophytic grasses. The stomata are located

Fig. 15-3 Organ-pipe cactus (*Lemairocereus thurberi*), Arizona. Compare with Fig. 15-2.

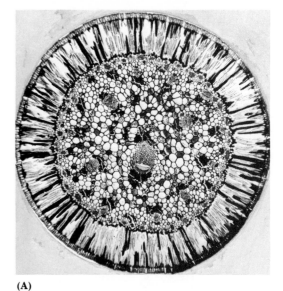

(A)

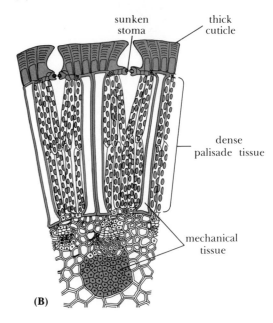

(B)

Fig. 15-4 Leaf structure of an Australian xerophyte (*Hakea platysperma*). The leaf is approximately cylindrical, several inches long, and tapering at both ends. (A): entire cross section (B): portion of cross section, enlarged to show details.

chiefly or entirely on the upper surface, and as the leaves curl or roll upward the rate of water loss is decreased.

It was long assumed that the modifications of the leaves of xerophytes are useful in reducing the rate of stomatal transpiration. Experiments show, on the contrary, that per unit of leaf area, the transpiration rate of such plants, when the stomata are open and soil moisture is available, is commonly as great as or even greater than that of mesophytes under similar conditions. These

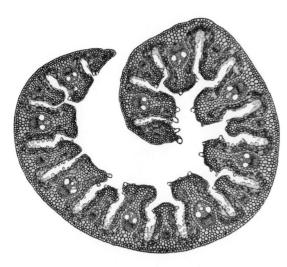

Fig. 15-5 Cross section of a rolled leaf of a xerophytic grass (*Ammophila arenaria*). The leaf is rolled upward. Stomata are absent from the lower surface, but are numerous along the sides of the furrows on the upper surface.

structural features are of value, however, under conditions of drought, for when the stomata close, the thick cuticle reduces cuticular transpiration to near zero. The mechanical tissues prevent the collapse of the leaf, which would otherwise result from dehydration. The effect of desiccation upon such xerophytes is thus much less than upon a mesophyte.

The capacity of many xerophytes for survival lies not only in structural features but also in the resistance of the hardened protoplasm to heating and drying, thus prolonging the life of the plant during periods of water scarcity.

Numerous controversies have arisen over xerophytes and their relations to their habitats, and many aspects of the subject are still not understood. Although a dry habitat is commonly regarded as characteristic of xerophytes, xeromorphic modifications are found in plants growing in habitats not usually regarded as dry. The EPIPHYTES (plants growing upon other plants) of

the canopy of many tropical rain forests, for example, include succulents, such as cacti, and nonsucculents with such xeromorphic modifications as thick, leathery leaves. Among these nonsucculents are heaths, orchids, and many other kinds of plants. Such epiphytes, however, may be regarded as xerophytes, for they are dependent upon rainfall for their entire water supply and between rains must endure or resist drought if they are to survive.

Xeromorphic modifications are found also in conifers and in evergreen flowering plants, especially those growing in alpine regions. Conifers may be subjected to drought when the soil is cold or frozen. Evergreen alpine plants are exposed to drying winds, periodic drought, and a cold or frozen soil from which water is obtained with difficulty.

Not all nonsucculent xerophytes have xeromorphic modifications. The leaves of many desert shrubs are thin, and the transpiration rate is high when water is available. They may have extensive root systems that penetrate the soil deeply over wide areas, and the leaves are frequently shed quickly when they begin to wilt; new leaves then develop with returning rains. Like xerophytes with xeromorphic modifications, they endure drought during rainless periods. A similar adaptation enables many perennial grasses to live in arid regions (FIG. 15-9). These plants grow quickly in spring or during a rainy season. The aerial shoots then mature and die as dry weather sets in. The underground parts may survive for a year or more without rain, while the dried shoots constitute the feed for sheep and cattle on arid ranges.

Other nonsucculents that possess no such modifications are the desert ephemeral annuals, which carry on active growth only during a brief rainy season. When the rains come, the seeds germinate; the plants grow rapidly and produce flowers and seeds within a period of a few weeks. As drought sets in, only the seeds survive. Such plants escape drought because they are able to complete their life cycles within a brief period, before the soil dries. The classification of the ephemeral annuals among the xerophytes may be questioned.

A true xerophyte is very likely to have one or more of the following physiological characteristics, independent of morphological differences: (1) a high rate of photosynthesis under favorable water supply and a high ratio of photosynthate to water used; (2) a quick cessation of photosynthesis and a change to an inactive condition under drought, but also a quick reversal under favorable conditions; (3) a very low rate of respiration and utilization of food materials during the inactive period.

FORM AND STRUCTURE OF HYDROPHYTES Some hydrophytes are free-floating plants; others grow submersed. Still others are emergent. Such plants grow up into the air, but the roots or rhizomes are anchored in the mud, perhaps beneath a foot or so of water. Internally, hydrophytes are commonly characterized by a well-developed aeration system (FIG. 15-6). This consists of greatly enlarged intercellular air chambers that extend throughout the plant. Mechanical and conducting tissues are often reduced in amount, and their cell walls are thinner than in most plants. In submersed plants the cuticle may be absent from leaves, and the roots may be reduced in size or absent. The leaves are commonly modified and may be thin, ribbonlike, or dissected. Some emergent plants, such as cattails, although possessing specialized air chambers, have retained well-developed mechanical and conducting tissues, particularly in their rhizomes.

Submersed plants obtain the oxygen used in respiration from that dissolved in water, as fish do, or from their own photosynthesis. The rhizomes, roots, and other submersed parts of emergent plants are in some degree independent of the oxygen in the surrounding water because free oxygen diffuses downward from the aerial organs of the plant through air passages in the leaves and stems. This oxygen is in part derived from the atmosphere and in part from photosynthesis. So much oxygen may move down

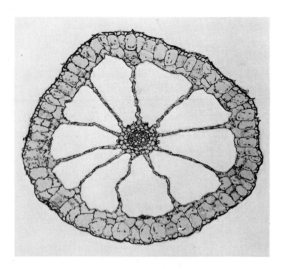

Fig. 15-6 Cross section of stem of a hydrophyte (*Ceratophyllum demersum*) showing air passages and reduced vascular system.

from the tops of emergent plants that the mud about the roots is partly oxygenated.

Light Light is the third important climatic factor that helps determine the distribution of plants and the formation of plant communities. The effect of light upon plant growth, flowering, and photosynthesis has already been discussed. One obvious effect of low light intensity is the poor growth of grasses when shaded by trees with dense foliage (FIG. 5-21). In the weak light of the forest floor, undergrowth may be entirely absent. When trees grow closely together, shade causes the death of the lower branches.

Many species have rather definite light requirements. Some species possess the capacity to survive and grow in light of low intensity, as on the floor of a forest. These are said to be SHADE TOLERANT. Other species, called SHADE INTOLERANT, require a high light intensity, in some cases 50 percent of full sunlight. These terms are usually applied to forest trees, but they may also be used with shrubs and herbs.

Tolerant and intolerant species may be iden-

Fig. 15-7 Shade tolerance. Engelmann spruce (*Picea engelmannii*), a shade-tolerant species, growing under a heavy canopy of aspen (*Populus tremuloides*), Colorado.

tified by a study of young trees in the shade of a forest canopy. Certain woody plants, such as sugar maple, beech, basswood, black gum, hemlock, fir, spruce (Fig. 15-7), and coffee are shade tolerant. Other species, such as aspen, paper birch, willows, box elder, bur oak, and some pines, are shade intolerant.

Shade tolerance is especially important in forest species where the seedlings must become established in the shade on the forest floor. The most shade-tolerant seedlings will survive with only 2 or 3 percent of full sunlight, but will die as they become older and require more food. A number of species will grow for some time with 4 or 5 percent of full light. Young plants of this

sort are ready to start rapid growth when any break in the canopy of older trees admits more light.

Although tolerance is important in determining the establishment of a species in a given area, competition for water and for essential mineral elements by roots is more significant than is generally realized. In some locations, light is the more important factor; in others, light and root competition operate together to influence the occurrence or predominance of a given species. A knowledge of the tolerance of the valuable timber trees is extremely important in the management of forests and woodlands.

The tropical crops coffee and cacao are shade

tolerant, and there has been much divergence of opinion regarding the importance of shade in the cultivation of these trees. In general, shade is not necessary in the cultivation of coffee. The highest yields are obtained without shade at altitudes of 3000 feet or more and with rich soils and high moisture content of soil and air. At lower altitudes and under less favorable conditions, coffee is also grown without shade, but when marginal conditions are present, shade may be advantageous. Shade reduces yields, but may extend the productive life of the tree. The use of shade trees is especially justified if woody legumes are used to contribute to soil fertility. Brazilian coffee is largely grown without shade because there is insufficient soil moisture for both coffee and shade trees.

Cacao behaves much like coffee. In the cultivation of this crop, shade is necessary when the trees are young; an important advantage is the protection of the organic matter of the surface layer of the soil from breakdown by exposure to high temperatures. Under optimum growing conditions, shade trees may later be removed, but on soils of low fertility the crop may be grown under permanent shade.

Soil Factors Some aspects of the nature of the soil have already been mentioned in connection with the study of roots (Chapter 9). Among the factors that affect plant distribution, growth, and survival are the temperature of the soil, its content of water, air, and organic matter, its structure (which affects its air and water capacity), its mineral composition, and its degree of acidity.

Temperature The temperature of the soil is a significant factor in plant growth, chiefly through its effect upon the absorption of water and minerals. At low temperatures the rate of root elongation is checked, resulting in a slower rate of penetration into new areas where water and minerals may be available. The rate of absorption of water and minerals in actual contact with the root is also decreased by low temperatures. Bacteria, too, are inactive in cold soil, so that minerals may not be made available to the roots. Growers of early crops frequently use nitrogen fertilizers that would not be needed if bacteria were active in a warmer soil.

Low soil and air temperatures, combined with strong winds, are responsible in large part for the dwarfed appearance of plants in alpine regions (FIG. 15-8). As pointed out earlier in this chapter, winterkilling is commonly due to desiccation of living cells following ice formation in the intercellular spaces. But winterkilling may also result from a relatively high rate of transpiration during mild sunny periods in midwinter or early spring. A bright sun increases the rate of transpiration at the same time that absorption is retarded or prevented by a cold or frozen soil. Moderately hardy evergreens and even hardy deciduous species are frequently killed under such conditions during the cold, dry winters of the northern Great Plains. The injury consists in the browning and dying of the foliage of evergreens, and is especially severe in trees with shallow root systems or in exposed situations. Liberal watering of evergreens in dry autumn weather is recommended to reduce this winter-drying injury.

Soil reaction The relation of soil acidity and lime to plant growth was discussed in Chapter 9. Many wild and cultivated plants grow well in soils that differ in pH over a wide range. Others are restricted to (or at least make better growth in) acid or alkaline soils, and the plants of such habitats are often very different. Among the plants that are especially sensitive to the effects of acidity and alkalinity are certain ferns, many orchids, and most legumes and heaths. Perhaps the most extreme examples of growth under conditions of low pH are the shrubby and herbaceous plants of peat bogs (FIG. 15-20). Many carnivorous plants are also restricted to acid habitats. Soils with a low pH contain relatively large quantities of soluble iron, and this is one factor that explains why certain heaths, such as blueberries, mountain laurel, and azalea, flourish in such

Fig. 15-8 Foxtail pine (*Pinus balfouriana*) at timberline. Growth is retarded by drying winds, wind-blown sand, and a cold soil, California.

soils, for they require little calcium and are inefficient in the absorption and use of iron. Homeowners may find that rhododendrons do not survive, or that they become chlorotic (yellow), around a brick or stone house where waste mortar, containing lime, has accumulated. In the cultivation of members of the heath family, it may be necessary to add acidifying materials, such as peat, oak or pine leaf mold, ammonium sulfate, ferrous sulfate, or powdered sulfur to the soil.

Biotic Factors Any study of the factors that affect plants must include a consideration of other living organisms, such as other plants, both green and nongreen, and all animals, including man. The activities of any of these may result in changes in the habitat and produce profound effects upon growth, structure, reproduction, and distribution of plants. The most important changes in the environment result from the competition or interference of plants with one another.

Competition As long as plants live as isolated individuals, growth is influenced only by such factors of the environment as climate and soil. But when plants live in groups, or communities, they compete with one another at all stages of development, from the seedling to maturity. This competition is for water, for light, and for essential soil elements. When plants grow close together, their roots invade the zones from which supplies of water and nutrients for other plants are obtained. The foliage of one plant may shade that of another. As a result of competition, certain more vigorous and better adapted plants become dominant. Others, less vigorous, are suppressed or eliminated. The invasion, or movement, of individual plants or species into areas already occupied by other plants depends upon the ability of the invaders to compete successfully with established species and individuals for essential growth factors.

The nature of the requirements for which plants compete varies in different habitats. In the forest, light is perhaps the most important, whereas in grasslands, deserts, and cultivated soils, competition is chiefly for water and minerals. Competition is a process of vital interest from the standpoint of food production. Most cultivated plants could not compete in the wild. Under cultivation, competition is, in large degree, under man's control. The destruction of weeds, for example, eliminates competition for minerals, water, and light between them and crop plants. The proper spacing of seeds at the time of planting controls competition among plants of the same kind and results in greater yields. If too many seeds are planted in a given area, the yield may be reduced as a result of competition for an inadequate supply of water and nutrients. On the other hand, thin planting, with the aim of

Fig. 15-9 Overgrazing (left) has allowed unpalatable brushy species to spread in a Nevada range. Protected, lightly grazed range (right) shows a vigorous growth of palatable rye grass (Elymus).

reducing competition, is not necessarily profitable, for the yield may fall short of the maximum because the available nutrients, water, and light have not been utilized fully. Competition should be permitted to operate to the extent of taking advantage of all the potentialities of the habitat. Farmers who once grew 7000 corn plants per acre are now growing 20,000 or more with heavy fertilization and perhaps irrigation.

Grazing by animals involves an effect on competition that is of both ecological and economic importance. Cattle and sheep feed on the more palatable grasses and legumes of pasture and range land. Competition is thus reduced for the unpalatable species, and they spread at the expense of the palatable plants. The effect is most serious when dry range lands are overgrazed by too many animals. In the area shown in FIG. 15-9, the range at the left has been so heavily grazed that grasses have been almost eliminated, leaving only unpalatable, brushy plants. In the protected and lightly grazed range on the right, valuable grasses constitute the dominant plant cover. Careful regulation of grazing is required to maintain range land in this condition

The most striking effects of the competition

for light are found in tropical rain forests. In regions of uniform temperatures, a long growing season, and precipitation of 80 or more inches evenly distributed throughout the year, the luxuriance of plant growth is greater than in any other regions of the earth's surface (FIG. 15-10). In such an environment plants are relieved in large measure from competition for water, and survival depends chiefly upon adequate exposure to light.

CLIMBERS and epiphytes are prominent adaptations related to light intensity. Climbers are either herbaceous or woody. The term LIANA is commonly applied to the woody climbers. The stems of these perennial plants may vary from about ¼ inch to more than a foot in diameter. Certain temperate-zone plants, as clematis, grape, ivy, and honeysuckle, may be termed lianas, but they are usually small compared to the lianas of the tropical rain forest. Here, rooted in the soil and supported by the trees around them, lianas extend upward until their leaves form a part of the forest canopy, sometimes 200 feet or more above the ground (FIG. 15-11). In this location the leaves are well exposed to the sunlight.

Epiphytes, or "air plants," grow upon other plants, usually trees. They are not parasitic but manufacture their own food. The support they obtain may place them more favorably in relation to light. Many kinds of plants may be epiphytic. Among them are ferns, lichens, mosses, orchids, and members of the arum, cactus, and pineapple families (FIG. 15-12). The Spanish "moss," a rootless flowering plant of the pineapple family, is an epiphyte that grows not only on trees but also on telephone poles and other objects. This plant (FIG. 15-13), found in parts of the southern United States and Central and South America, absorbs rain water and dew through minute scales that cover the leaves and greatly elongated slender stems. It is these scales that give the plant its characteristic gray appearance. It obtains its essential elements from rain water and dust. Most epiphytes obtain their essential elements from decaying organic matter that collects around their roots.

Fig. 15-10 Edge of a rain forest, Guatemala. The large-leaved plant is a monocotyledon (*Xanthosoma roseum*) related to the jack-in-the-pulpit (*Arisaema*) of eastern woodlands of North America.

Fig. 15-11 Lianas in a Jamaican rain forest.

Fig. 15-12 Epiphytes in a tropical rain forest; liverworts, lichens, filmy ferns, and bromeliads (related to the pineapple).

Fig. 15-13 Bald cypress (*Taxodium distichum*) trees covered with Spanish moss (*Tillandsia usneoides*), North Carolina.

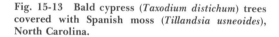

Fig. 15-14 Strangling fig (*Ficus*), Queensland, Australia.

Some epiphytes are found in the lower levels of the forest, but many others grow in situations exposed to full sunlight. These sun epiphytes are subject to periods of water shortage even in a climate with frequent rainfall, and, as mentioned previously, exhibit numerous xeromorphic features. In many species of bromeliads (pineapple family) the sheathing leaf bases form a cluster of reservoirs, which in large plants may hold a total of more than a gallon of rain water.

Some plants of tropical rain forests begin life as epiphytes, but later become independent of the trees that originally supported them. These are the STRANGLERS, belonging to several plant families, notably the fig. The seeds of the strangling figs germinate high on tall trees, often in a fork between the trunk and a large branch. The plants grow epiphytically for a time, then send down long aerial roots that penetrate and grow vigorously in the soil. Some roots spread downward along the trunk, where they enlarge, interjoin, and eventually form a network closely investing the tree (FIG. 15-14). After a time the host tree dies and rots away, leaving the fig as a hollow, independent tree whose "trunk" is composed of root, rather than stem tissue. The death of the tree is usually ascribed to the constrictive effect of the fig roots in crushing the bark of the support against the wood, thus reducing growth and translocation.

PLANT SUCCESSION

Competition in a plant community usually brings about changes in the conditions affecting plant growth. Plants react with one another and change the habitat in relation to light and supply of water and nutrients. Tall plants shade lower plants and bring about changes in the temperature of the air and soil, in relative humidity, and in force of the wind. A dry soil may become moist and more suitable for the growth of seedlings because humus accumulated from the decay of plant tissues has greater water-absorbing and, to some extent, greater water-retaining powers.

As the environment becomes modified, the kinds of plants that compose the plant population also undergo change. New species invade the area and, if they are better adapted to the changed environment, they become established and supplant the original inhabitants. In the course of time a given territory will be occupied successively by different plant communities, a process known as SUCCESSION.

As plants compete more and more with one another, certain species become DOMINANT; that is, they largely control the environmental conditions upon which the growth of plants depends. The course of a succession is marked by a sequence of dominant forms, for as new plants

invade, the nature of the competition changes until one kind of community is replaced by another. Eventually, the vegetation tends to become stabilized; that is, the kinds of plants occupying an area are those that are most successful in their reproduction and that make the greatest growth possible in the particular environment.

The end of a succession is characterized by a state of stability and is known as a CLIMAX. A climax area is also marked by the general aspect of the vegetation, and the constituent species. The principal types of terrestrial ecosystems, such as forest, grassland, desert, and tundra, are the climax of their respective areas. They represent the type of vegetation best adapted to the environmental conditions of the area. The dominant species of such climaxes, as the result of local climatic and other factors, are different from place to place, forming subdivisions or associations of the larger climax ecosystems.

The western forests of North America, for example, are composed of such subdivisions as western hemlock-western red cedar, spruce-fir, or western yellow pine-Douglas fir. Among the subdivisions of the Eastern Deciduous Forest (FIG. 15-22) are oak-hickory and beech-maple. Deciduous-coniferous subdivisions are prominent in the Hemlock-Northern Hardwood Forest (FIG. 15-21). The grasslands are composed of a number of subdivisions, each characterized by two or more species, while the tundra ecosystem may be characterized by lichen-moss in some regions, dwarf shrubs in others, and a mixture of grasses, sedges, and herbs in still others.

Successions are of various kinds, but all are composed of a number of stages, each characterized by different forms of plant life. The first plants to become established in any succession constitute the PIONEER stage. The nature of the pioneer plants is determined in large part by the nature of the soil or rock and by the water supply. Some successions, termed PRIMARY, begin in areas that have not previously been occupied by plants, such as open water, bare rock, or sand. SECONDARY successions begin wherever the existing vegetation has been destroyed without denuding the area of soil or covering it with new soil-building material. Secondary successions commonly develop after forest fires, lumbering operations, severe overgrazing, and on abandoned agricultural lands. Successions are so varied that only a few examples, illustrative of the general principle, can be described here. The first of these is a primary succession that begins on bare rock.

Primary Succession Succession on bare rock surfaces resulting from glaciation or from erosion by wind or water commonly begins with lichens as the pioneer stage (FIG. 15-15). These plants grow only when water is available, but in the intervals between rains they can remain alive even though desiccated. The mechanical action of the lichens on the underlying rock loosens particles, which, together with the decaying lichens, make up the first thin layer of soil. Simple, crustlike lichens may be succeeded by larger, leafy forms, which grow on the slight accumulation of soil and humus.

Lichens may be followed by mosses, which, like lichens, are able to survive during dry periods. The mosses shade the lichens and successfully compete with them for water and nutrients. The death and decay of the older mosses often produce a mat over the rock surface. As, over long periods of time, this mat becomes thicker and develops a greater water-holding capacity, a third stage in the succession follows. This consists of annual and perennial herbs, including grasses and, occasionally, ferns. The soil is increased in thickness by disintegration of the rock and by the decay of the roots, stems, and leaves of plants; more nutrients become available, and a fourth stage, dominated by shrubby plants, appears. The herbaceous plants of the preceding stage, now shaded, tend to disappear.

The shrubs, in turn, further enrich the soil and, many years later, may bring about conditions suitable for the development of the tree stage in the succession. This stage constitutes the climax in a forest climate. The first tree species to appear, however, are normally not the climax species, but forms adapted to relatively dry conditions. As conditions improve, such species will gradually be replaced by more mesophytic types. These may become dominant because their seedlings are shade tolerant in contrast to the shade-intolerant seedlings of the first tree invaders.

The species of the climax reproduce themselves, replacing the individuals that die, and remain the same unless the climate changes or the community is disturbed by fire, man, or other agencies. Succession on bare rock does not always follow the sequence just described. If small amounts of soil accumulate in crevices or if deposits of sand and gravel occur on the rock surface, the earlier stages in the succession may be bypassed, with herbs, shrubs, or even trees becoming established more rapidly (FIG. 15-16). Moreover, there are many rock outcrops which, after hundreds or even thousands of years, are still covered only with lichens, these constituting the climax. Periodic droughts, among other factors, prevent the establishment of other kinds of communities.

Note that the succession just described leads from xerophytic to mesophytic types of vegetation. This situation is common in regions of adequate rainfall where primary successions begin on areas devoid of soil. Another example of such a succession is found on sand dunes, where drought-resistant grasses are commonly the pioneers, followed by shrubs and trees.

Primary succession on land may lead to a grassland rather than to a forest climax, as evidenced by the grassland areas of North America and many other parts of the world. Three grassland subdivisions have been recognized in central North America, from east to west: (1) the tall-grass prairie, (2) the mixed prairie, and (3) the short-grass plains (FIG. 15-24).

Primary successions also proceed in open bodies of water, such as ponds and lakes (FIG. 15-17). The first step may consist of submerged

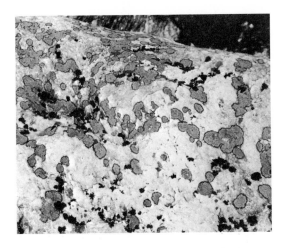

Fig. 15-15 Pioneer lichens (*Gyrophora parmelia*) on a bare rock surface.

Fig. 15-16 Succession on bare rock: lichens, mosses, grasses, and small shrubs. The small woody plants are becoming established in the crevices of the rock.

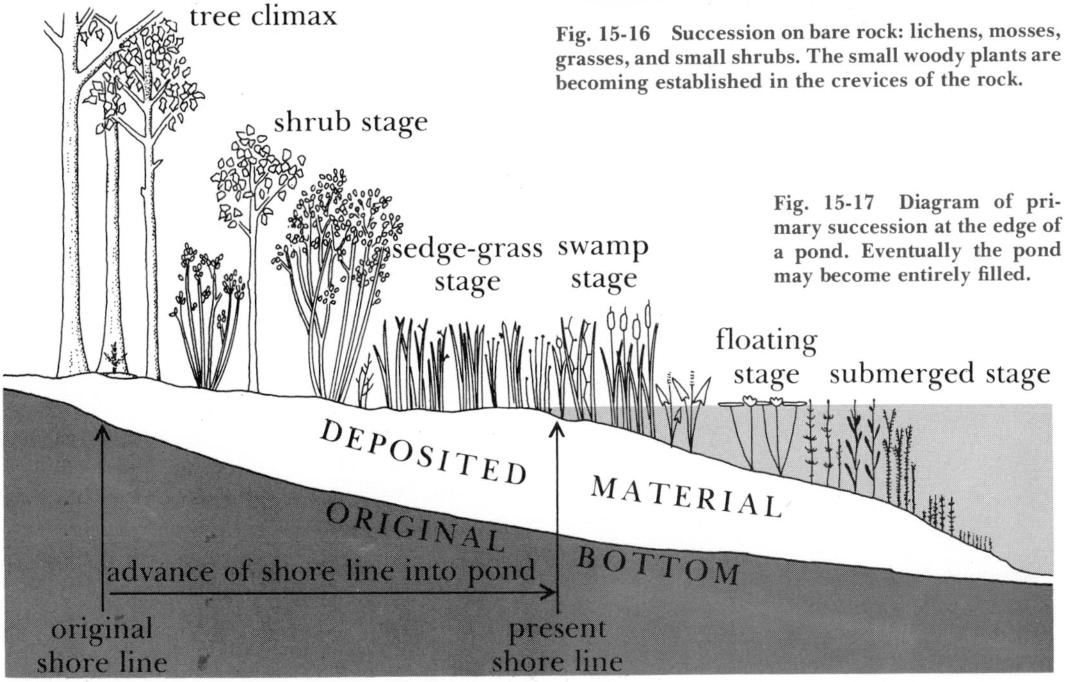

Fig. 15-17 Diagram of primary succession at the edge of a pond. Eventually the pond may become entirely filled.

tree climax

shrub stage

sedge-grass swamp
stage stage

floating
stage submerged stage

DEPOSITED MATERIAL

ORIGINAL BOTTOM

advance of shore line into pond

original
shore line

present
shore line

aquatic plants, which accumulate after their death and decay and, together with silt deposits, gradually raise the lake bottom. Such plants grow most luxuriantly in shallow water, and the accretion of organic and mineral materials therefore proceeds more rapidly near the shore than in deeper water. As the bottom is built up, a second, or floating, stage follows, characterized by water lilies, broad-leaved pondweeds, and other plants. These plants are rooted in the mud, and their leaves float upon the surface, shading the plants below them. With decreasing depth of water, a swamp stage is initiated. The area is invaded by emergent hydrophytes, such as cattails, bulrushes, arrowheads, and bur reeds. Eventually the soil becomes dry enough to afford a foothold for terrestrial species such as sedges and other low herbaceous plants. These may be shaded by a succeeding shrub stage, which is followed in turn by a climax of trees or prairie grasses. In grassland areas the shrub stage is replaced by lowland prairie and this by upland prairie, mixed prairie, or short-grass prairie.

Secondary Succession Secondary successions usually result from the activities of man and are therefore the most common type of succession in settled regions. They have fewer stages than primary successions, and the climax is reached more quickly.

The stages that follow the removal of the original vegetation vary widely in different regions, depending upon the climate, the soil, and the kinds of plants that invade the denuded area. Burned or lumbered lands may be colonized by grasses and other herbs, both annual and perennial. Fireweed is a typical example. These may be followed by shrubs, such as sumac, or by weed trees, such as box elder, aspen, and chokecherry or pin cherry. All these have seeds that are easily disseminated. If undisturbed by man, the area will again be covered, after a period of time, by a deciduous or coniferous forest similar to the original climax.

Succession on abandoned agricultural land may proceed from herbaceous annual weeds to perennial weeds and grasses, and then to prairie or woodland. A typical succession is revealed by studies of abandoned upland farms in the piedmont area of North Carolina. Scattered stands of undisturbed hardwoods, including trees 200–300 years old, show that oaks and hickories dominated in the forests before the advent of the white man. The early stages in the secondary succession follow each other rapidly. In fields abandoned for one year, the most conspicuous species were the annual crabgrass and horseweed. In the second year the dominant species were the common ragweed and a tall, bushy, white-flowered aster. By the third year, broom sedge (a grass) had formed massive clumps, with tops that shaded surrounding plants and extensive fibrous roots that competed with the roots of small plants for water and minerals. The broomsedge stage was succeeded by pines, seedlings of which became established between the clumps of grass. Within 10–15 years, the pines shaded out the grass and dominated for 70–80 years. The pines, however, do not reproduce in their own shade and are gradually replaced by oaks and hickories, which have developed in the lower strata of the forest. After about 150–200 years, the oak-hickory climax is again attained, with a few scattered pines remaining as relics of the preceding stage in the succession. Similar successions occur on western grasslands that have been cultivated for a time and then abandoned. The fields are invaded by early annual grasses and annual and perennial, broad-leaved weedy herbs. Unless recovery is hindered by overgrazing, these in turn are slowly supplanted by the climax grasses.

Still other kinds of plants may constitute the pioneer stage in both primary and secondary successions. These are the soil-inhabiting green, and especially blue-green, algae (Chapter 19) that live on the soil surface or just beneath it. They may form crusts, which stabilize the soil, fix nitrogen, add a small amount of organic matter, and allow germination of seeds and spores of higher plants. Where physical and chemical factors are favorable, they colonize rock surfaces, badly eroded lands, and soils denuded by drought.

Secondary successions are so common that the city dweller need not depart from his immediate environment to study them. They may be observed in vacant lots, along roadsides, and in bare spaces on lawns.

A knowledge of the principles of succession has been of the greatest value in the management of forests, ranges, agricultural lands, and wildlife areas. As a result of succession, our landscapes are in a process of continual change. Many of these changes may be observed in one's own lifetime; others are so slow that men living their whole lives in the same region scarcely notice them. Vegetation as we see it today, then, is not merely a haphazard covering of the land with species of plants. It represents an orderly process of colonization governed by the interaction of plants with their changing environments.

Other Biotic Factors   The emphasis in the preceding discussion has been chiefly upon higher plants as biotic factors. But lower plants also influence plant growth; soil fungi, for example, play an important role. Fungi are almost universally present in soils, and in many habitats may be essential to the growth of pine, beech, and other trees. Other unique and interesting relationships exist among plants.

Although plants are important as biotic factors, man and other animals are of equal, sometimes of greater, importance as agents that influence plant growth, reproduction, and distribution. Some of the effects of man's activities, such as lumbering, fires, and the grazing of domestic animals, have been indicated. The uncontrolled grazing of cattle, horses, sheep, and goats interferes with forest reproduction and destroys the native cover of grasslands (FIG. 15-9), thus causing soil depletion and erosion. Rodents use seeds as food and thus diminish the reproductive capacity of plants. They also consume leaves and underground stems and injure the underground parts by their burrowing activities. Locally, they may be important factors in range depletion.

Among the beneficial effects of animals is their role in the dissemination of seeds and fruits, the part played by insects in pollination, and the contribution of earthworms in increasing aeration, water absorption, and soil fertility. Many other examples of animals as biotic factors could be cited.

MAJOR TERRESTRIAL ECOSYSTEMS

The climatic and other ecological factors examined in this chapter have, over long periods of time, been responsible for varied patterns of vegetation over large geographic areas of the earth's surface. Regional differences in temperature and moisture, modified by soil and biotic factors, have produced major kinds of ecosystems. These constitute the highest category in any classification of terrestrial ecosystems. They are usually described on the basis of the characteristic type of vegetation, and the pattern may be repeated, under similar environmental conditions, on different continents.

Chief among the major types of plant cover of North America are the following: (1) tundra; (2) coniferous forest; (3) temperate deciduous forest; (4) tropical broad-leaved forest; (5) scrub (dense thickets of shrubs or small trees); (6) grasslands; and (7) deserts. In North America the precipitation over deserts is roughly 0–10 inches a year; over grasslands, 10–30 inches; and over forested areas, 30–90 inches. The general limits of the major vegetation types of North America are shown in FIG. 15-18. A brief description of the components of most of these vegetation regions is included in the subsequent paragraphs.

Tundra   The tundra zone extends from the northern limits of the forests to the limits of perpetual snow. Temperatures of the soil and air are low, and strong, drying winds prevail. The grow-

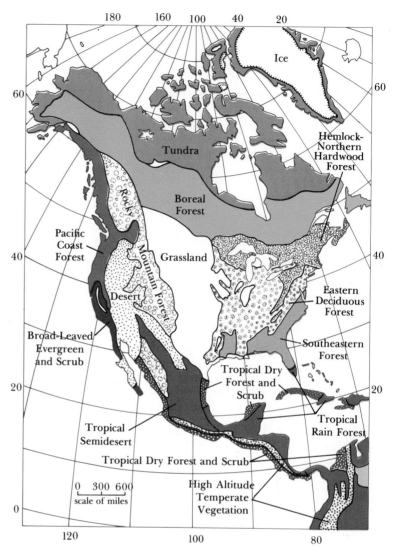

Fig. 15-18 Vegetation map of North America.

ing season is short, from a few weeks to two months. The summer days are long, but the light is of low intensity, owing to the inclination of the sun's rays and to fogs, clouds, and drizzling rain. As a consequence of the unfavorable environment, the region is largely treeless (Fig. 15-19). The vegetation consists chiefly of mosses and lichens, grasses, sedges, and dwarf shrubs. Simi-

lar conditions may prevail at high altitudes, such as mountaintops above the timberline. The appearance of the vegetation in these local areas is very similar to that of the arctic tundra, which extends for thousands of square miles across the northern parts of North America, Europe, and Asia.

The thin layer of vegetation on the tundra

Fig. 15-19 The tundra, Baker Lake, Keewatin District, Canada. Lat. 64½°N, 96°W.

Fig. 15-20 Boreal forest and succession in a boreal bog. A floating sphagnum-sedge zone is encroaching upon open water. Beyond this is an invasion of heath shrubs. The forest is composed of fir (*Abies*), spruce (*Picea*), and larch (*Larix*).

serves as insulation. If broken, as by the activities of man, the permafrost—the permanently frozen subsoil—thaws, and extensive erosion usually follows. In many regions the tundra still bears the scars of bulldozers and other heavy machines that were used heedlessly in the past.

Forest The forests of North America occur in three great belts—the Boreal, Eastern, and Western forests.

Boreal forest This most northern forest (FIG. 15-20), which extends across the continent, is primarily coniferous. Spruce and fir are the most important trees, but pine, cedar, and larch also occur. The hardwoods, or broad-leaved trees, are represented by aspens and birches. The climate in this region is rigorous, but the growing season is sufficiently long to permit tree growth. The Boreal Forest grows largely on thin-soiled, glaciated land, studded with lakes, ponds, and bogs. The plant succession in the ponds and bogs is characteristic (FIG. 15-20). Sphagnum mosses gradually fill the ponds and convert them into peat bogs, where sedges and acid-soil heaths, such as cranberries, blueberries, leatherleaf, and sheep laurel, take over, to be followed in succession by spruce and larch.

Eastern forest The Eastern Forest extends from the east coast across the Mississippi to the eastern prairie border of the grasslands. It, in turn, has been classified into a Hemlock-Northern Hardwood Forest, Eastern Deciduous Forest, and Southeastern Forest. These are the richest in the temperate zone in numbers of species of trees (approximately 600). The Hemlock-Northern Hardwood Forest (FIG. 15-21) includes hemlock and other coniferous trees such as red and white pine, spruce, and fir, and deciduous trees such as the birches, beech, maple, basswood, and the oaks. Among the important trees of the Eastern Deciduous Forest (FIG. 15-22) are the oaks, hickories, beech, maples, ashes, basswoods, black locust, yellow poplar, and walnut. The Southeastern Forest (FIG. 15-18) is composed of mixed

Fig. 15-21 Hemlock-Northern Hardwood Forest, Adirondacks.

Fig. 15-22 Eastern Deciduous Forest, Tennessee.

hardwoods and conifers. It includes several species of pine, among them those that yield the turpentine and rosin of commerce (Fig. 7-19) and, more recently, pulpwood for paper, red cedar (*Juniperus virginiana*), gums, oaks, hickories, bald cypress, and others.

Western forest The Great Basin, arid and hot, divides the Western forest into two sections—the Pacific Coast Forest and the Rocky Mountain Forest. Both sections are predominantly coniferous. Important trees of the Pacific Coast Forest (Fig. 15-23) are firs and pines, hemlock, cedar, redwood, Douglas fir (not a true fir at all), and the spruces. The western yellow, or ponderosa, pine is perhaps the most important species of the Rocky Mountain Forest but Douglas fir, western white pine, and other valuable conifers also occur in this forest.

To these forest types may be added the Tropical Forest, found only in scattered areas in the southern tip of Florida. The species found here, which include mangrove, palms, wild figs, and mahogany, are northward extensions from their ranges in the West Indies and Central and South America.

Grassland Extending westward from the Eastern Forest to the Rocky Mountains lies a vast, largely treeless area, variable in temperature and generally low in rainfall. Of this region, the part bordering upon the Eastern Forest is the prairie, once covered with tall grasses, but now the intensively cultivated western corn belt. This area merges into the Great Plains (Fig. 15-24), once covered with drought-enduring short grasses, such as grama and buffalo grass, now extensively planted to wheat. Grasslands are found also in Washington and California.

Over much of the Great Plains area, the 10–20 inches of precipitation is concentrated in the early part of the growing season. Because of the low precipitation and its distribution, and the high rate of evaporation, the climate does not favor the growth of trees. Recurrent drought

Fig. 15-23 Pacific Coast Forest, Washington. Mostly Douglas fir (*Pseudotsuga menziesii*).

Fig. 15-24 Short-grass plains, Colorado.

and misuse of the land by overcropping and overgrazing have led to wind erosion over much of the Great Plains area. Large parts of this region are suitable only for the carefully controlled grazing of cattle and sheep.

Desert Desert regions are markedly low in rainfall and have a high rate of evaporation and low relative humidity. Commonly, high day temperatures alternate with low night temperatures. In North America, the chief desert regions lie between the Rocky Mountains and the Sierra Nevada. They extend southward into the highlands of Mexico.

No sharp boundaries exist between desert and grasslands. The desert usually supports a meager yet interesting vegetation, for the most part clearly adapted to the environment. Low evergreen or deciduous, small-leaved shrubs are perhaps the predominant life forms (FIG. 15-25). Among other kinds of desert plants are the succulents, such as cacti; the yuccas, with their elongated narrow leaves (FIG. 15-26); wiry grasses; and ephemeral annuals, which germinate quickly, grow rapidly, and mature their seeds during the brief rainy periods.

This chapter has, in large part, dealt with adaptations—variations in structures or processes associated with some particular mode of living. It was at one time considered that adaptations arose in response to the requirements of organisms for survival; indeed, adaptations are frequently so closely correlated with environmental factors that it is tempting to conclude that plants have purposefully responded to environmental stimuli. Adaptations, however, rise by the selective action of environmental factors upon hereditary variations in the organism. A basis for understanding how these mechanisms operate is presented in Chapters 16 and 17.

Fig. 15-25 Desert shrub vegetation in the Mojave Desert, southern Nevada. Photo taken in summer when most shrubs are leafless. A plant of the strawberry cactus (*Echinocereus engelmannii*) appears in the center foreground.

Fig. 15-26 The Mojave yucca (*Yucca schidigera*).

SUMMARY

1. The many environments in which plants exist affect plants and are modified by plants. The study of the relationship of plants to their environment is ecology. The specific environment in which a plant or community of plants exists is the habitat.

2. The interacting environmental features that are responsible for different habitats and the plants found there are termed ecological factors. These may be classified as climatic, soil, and biotic.

3. Temperature and precipitation are the chief climatic factors. Temperature affects the length of the growing season; low temperatures limit plant growth by slowing down the rate of plant activities. High temperatures cause injury or death chiefly by

increasing transpiration. Precipitation affects plant distribution through its effect upon available soil moisture.

4. In relation to water, plants are classed as xerophytes, mesophytes, and hydrophytes. The plants of each of these groups usually possess external and internal features correlated with the environment.

5. Light intensity is generally uniform over wide areas but varies locally. With respect to light intensity, plants may be divided into two groups, shade tolerant and shade intolerant. The intensity of light reaching a plant is frequently a limiting factor in growth and reproduction.

6. Among important soil factors are texture, structure, fertility, acidity, and temperature. Low soil temperatures retard growth through their effect upon the activities of roots and of soil bacteria.

7. Competition is a major biotic factor. Plants compete with one another chiefly for water, minerals, and light. In the tropical rain forest, the most important competition is for light.

8. Plants tend to improve habitats by holding the soil and increasing its organic content and fertility. New species may then invade this improved habitat and crowd out the earlier plants.

9. The occupation of an area by a sequence of plant communities is known as succession. Each stage in a succession is characterized by certain dominant forms. Successions lead to a climax, and are either primary or secondary.

10. The principal terrestrial ecosystems of North America are coniferous, deciduous, and tropical forest, tundra, grassland, scrub, and desert. The type of vegetation of each of these is primarily determined by temperature and precipitation.

16

Inheritance in Plants

THE ABILITY OF LIVING THINGS to produce offspring like themselves is universally recognized. Beans grow into bean plants and acorns produce oaks, not apple trees. Because of inheritance, it is possible to classify plants and animals into groups of similar and presumably related individuals.

The bridge between parents and their offspring is a narrow one; for most organisms it consists of a single cell, the fertilized egg. This hereditary bridge is the link between the generations, and all hereditary potentialities are transmitted through this cell. It carries the essential materials, half derived from one parent and half from the other, that determine the kind of plant (or animal) that will develop.

The potentialities transmitted through the reproductive cells produce differences as well as similarities. They are responsible for the fact that variations exist to such an extent that no two individuals are identical (Plate 6). Individuals of similar heredity may also vary greatly, depending upon the environment in which they grow. Hence, it is necessary to distinguish between individual variations that are not inherited and differences that are inherited and transmitted from generation to generation.

The study of heredity and inherited variations—the science that seeks to understand the mechanisms governing the transmission of hereditary potentialities from parents to progeny—is known as GENETICS. The practical application of genetics to the breeding of useful plants has greatly modified agriculture and horticulture. Moreover, this same knowledge furnishes us with an approach to the problem of organic evolution. It has given us an insight into the mechanisms that through the ages have produced new and different forms of life.

The modern study of heredity dates from

1900, although prior to that time many experimenters had worked on the breeding of plants and animals. They had accumulated a considerable body of knowledge, but they were unable to formulate laws that could be used to predict the results of crossing plants having different hereditary characteristics (HYBRIDIZATION). In 1866, Gregor Mendel, an Austrian monk, published the results of eight years of hybridization experiments, chiefly with the garden pea. Although this work contained the solution to problems that had long baffled students of heredity, its importance was not recognized for many years. The results were finally rediscovered and confirmed by three independent investigators at the beginning of the present century.

One reason why others had failed to find laws governing heredity was that they observed simultaneously all of the many characters in which the parents differed. Mendel, on the contrary, ignored all characteristics except a few markedly contrasting ones, and these he studied individually. Another reason for the earlier failures was the belief that the hereditary characters of two organisms become blended in the progeny. Mendel, however, discovered that the nature of an organism is governed by the presence of unit factors in the reproductive cells and that these factors may be transmitted independently to the progeny.

LAWS OF MENDELIAN INHERITANCE

Mendel's experiments and the laws that he formulated are essential to an understanding of hereditary processes. The most significant of his findings are summarized in this chapter.

Dominant and Recessive Characters The several forms or varieties of the garden pea differ in a number of clearly defined features. Mendel selected seven pairs of readily observable characters, such as tallness and shortness, and cross-

pollinated plants showing opposite phases of these characters. Since the pea is normally self-pollinated, Mendel accomplished cross-pollination by opening the flower while it was still in the bud stage, before the pollen had matured. The stamens were removed and the stigma pollinated with pollen from the desired parent. Mendel's experiments dealt with all of the seven pairs of characters, but for the sake of simplicity only a few of them need be considered here.

In one of the experiments, a variety about 6 feet tall was crossed with one averaging 1 foot in height. When the seeds from this cross were planted, they produced plants not intermediate between the two parents, as might be expected, but all tall, like the 6-foot parent. Mendel made crosses to study the inheritance of six other pairs of characters and found that in every instance the hybrid resembled one of the parents with respect to this character. No trace of the other character in each pair was apparent in the hybrid.

The character of tallness, and the others that appeared unchanged in the hybrid, Mendel designated as DOMINANT. Those that did not appear in the hybrid he called RECESSIVE. In other crosses he found that round seeds are dominant over wrinkled seeds, that yellow color of the cotyledons is dominant over green, and that distribution of flowers along the stem is dominant over terminal flowers.

The hybrid plants, produced from seeds formed after the cross, are designated as the F_1 (pronounced *eff-one*), or first filial (daughter) generation.

Mendel now permitted the F_1 generation of each of his crosses to self-pollinate, forming seeds that in turn produced an F_2 (second filial) generation. When these seeds were planted, they produced not only tall plants but also dwarf plants—in a ratio of approximately three tall to one dwarf. The plants of the F_2 generation were permitted to self-pollinate, and an F_3 generation was raised. Mendel then found that the F_2 dwarf plants produced only dwarf plants. Of the F_2 tall plants, one-third produced only tall plants,

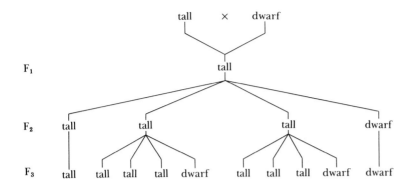

Fig. 16-1 Appearance of pea plants three generations after a cross between a tall and a dwarf plant. The F₁ and succeeding generations were self-pollinated.

whereas the remainder produced both tall and dwarf plants in the same 3:1 ratio as in the F_2 generation (FIG. 16-1). Since the progeny of the tall plants also included dwarf plants, it was apparent to Mendel that the hereditary constitution of a plant is not necessarily revealed by its external appearance. This was a highly significant discovery.

The Chromosome Basis of Mendelism The main facts about sexual reproduction in plants were known to Mendel, and he believed that his results were compatible with them. He was confident that his findings could be accounted for if factors influencing the appearance of the hybrids were carried in the egg cells and in the pollen of the parent plants. According to Mendel, plants that bred true when self-pollinated had originated in the fusion of egg cells and pollen of like heredity. Those that did not breed true—that is, produced hybrids—were formed from the fusion of germ cells that contained unlike factors. In all of these assumptions Mendel was entirely correct. Since 1900 it has been possible to proceed further and to explain Mendel's results in terms of structures and processes not known to him.

In order to understand Mendel's results in the light of modern knowledge, certain aspects of the life history of a flowering plant must be recalled. The nucleus of each gamete, sperm or egg, possesses only half as many chromosomes as the parent plant, a number designated by the symbol n. When the gametes fuse, the chromosome number is doubled and is then designated as $2n$, or the DIPLOID number of chromosomes. This $2n$ number is found throughout the cells of the plant body, but is again reduced to the n number (or HAPLOID number) when the spore mother cells divide to form microspores and megaspores. This reduction process occurs in both the anther and ovule, and the behavior of the chromosomes at this time accounts for Mendelian ratios.

Meiosis The method by which the chromosome number is reduced from $2n$ to n is essentially the same wherever it occurs. A description of the mechanism of this process as it occurs within the anther will therefore be sufficient. All the cells of the plant, up to and including the spore mother cells in the anther, possess the $2n$ chromosome number. All nuclear divisions up to this point are mitotic. Then four microspores are produced from each microspore mother cell after two divisions. These nuclear divisions, during which the chromosome number is reduced from the diploid to the haploid number, are known collectively as MEIOSIS.

Let us consider a specific microspore mother cell that possesses but four chromosomes, two coming originally from the sperm and two from the egg. During the process of meiosis, the four chromosomes of the nucleus of this spore mother

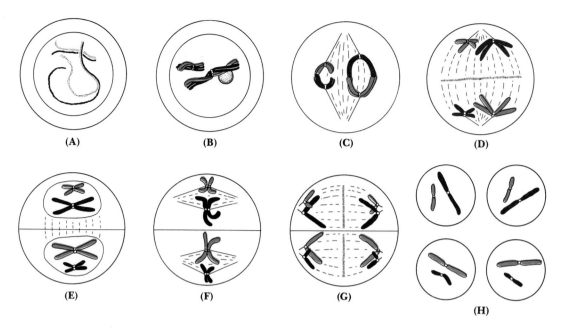

Fig. 16-2 Diagram of meiosis. (A–E): stage I. (A): early prophase, early pairing of homologous chromosomes. The chromosomes from one parent are in color, the other black. (B): late prophase; the pairing of homologous chromosomes is complete. (C): metaphase. (D): anaphase. (E): telophase. (F–H): stage II. (F): metaphase. (G): anaphase. (H): cell walls form, resulting in four microspores, each with the *n* chromosome number.

cell are reduced to two in the nucleus of each microspore (Fig. 16-2). Meiosis takes place in two stages (I, II), the second ordinarily following immediately after the first. The prophase of stage I is long, and is marked by several important events. Early in the prophase the chromosomes are greatly elongated. The elongated chromosomes undergo pairing (SYNAPSIS), in which two chromosomes become intimately associated, side by side (Fig. 16-2A). Pairing is not indiscriminate, but always occurs between a chromosome derived from one parent and the corresponding chromosome derived from the other parent. The two chromosomes thus associated are said to be HOMOLOGOUS, for they are similar in appearance and bear corresponding genes affecting the same characters. By the time they have paired, each chromosome is already

double, and consists of two sister chromatids, as in prophase of mitosis. Throughout the prophase, but especially during the latter part, the chromosomes gradually thicken and shorten so that, at the end of prophase, it can readily be seen that each pair of chromosomes consists of four chromatids (Fig. 16-2B).

At metaphase of stage I, the chromosomes move to the equatorial region of the spindle. Their centromeres do not lie along the equator, however, as in mitosis. Rather, the structurally double centromeres of the chromosomes appear to repel each other, with the result that the centromeres of the pairs lie toward the poles, with their corresponding arms at the equator (Fig. 16-2C). At anaphase I the paired chromosomes separate and move toward opposite poles of the spindle. The sister chromatids of each chromosome do not separate, as in mitosis. The arms of the two chromatids of each chromosome do, however, diverge, as though they were repelling each other (FIG. 16-2D). The telophase of stage I (FIG. 16-2E) is unlike that of mitosis in that nucleoli are not ordinarily formed, although a nuclear membrane develops. The chromosome number of the daughter nuclei has now been reduced to half that of the parent number (from four to two) and two nuclei are now present within the wall of the microspore mother cell. In some angiosperms a cell wall is laid down between the two nuclei, but in many others the wall formation is delayed until after four nuclei have been formed, at the end of stage II.

The second stage generally follows quickly, for the chromosomes already consist of two chromatids each, and are thus ready to enter metaphase. Prophase II is thus usually very short, and the nuclear membranes soon disappear. The chromosomes again become arranged in the equatorial region of a spindle, this constituting metaphase II (FIG. 16-2F). At anaphase II the centromeres of each pair of chromatids separate and the sister chromatids move toward opposite poles (FIG. 16-2G). The sister chromatids, once separated, are now called chromosomes. Cell walls are then laid down, and the nuclei pass through telophase and enter interphase. Four microspores thus result, each with the haploid chromosome number. The justification of a common definition of meiosis is now apparent: Meiosis consists essentially of two nuclear divisions, following each other in rapid sequence, in the course of which the chromosomes duplicate but once.

The pairing of homologous chromosomes in meiosis is an important step in heredity. The chromosomes contain genes in a linear arrangement and a definite order. For each gene on a chromosome there is an identical, or a related, gene in the corresponding position on the homologous chromosome. Typically, the gene at any position on the chromosome is one of two variants, but it may be one of several. The related genes that may occupy a specific position (LOCUS)

are called ALLELES, meaning "reciprocal" or "corresponding." The pairing of the homologous chromosomes in meiosis and then the separation of the pair, one to each of the new haploid nuclei, ensures the presence of one allele of each gene in each spore and gamete.

Returning to the cross between the tall and short peas, we may assume that the tall peas contained a pair of homologous chromosomes, each bearing the gene for tallness (*TT*, FIG. 16-3). In this plant the genes for tallness are identical, and such a plant is said to be HOMOZYGOUS for the genes at this particular locus. The plant is "pure"

Fig. 16-3 A monohybrid cross. The chart shows the behavior of the chromosomes carrying genes for tallness (T) and dwarfness (t). Since tallness is dominant over dwarfness, a phenotypic ratio of 3:1 results in the F₂ generation.

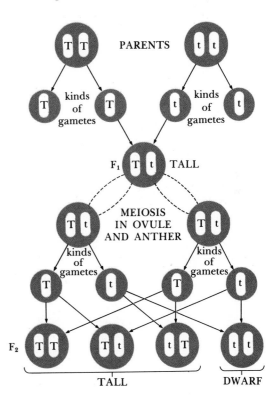

for that character and, if self-pollinated, will breed true, producing only tall offspring. Similarly, the dwarf plant bore identical genes for dwarfness at the same locus on the same pair of homologous chromosomes (*tt*, FIG. 16-3). It, too, is homozygous and will produce only dwarf plants when self-pollinated. Since the two genes that affect the length of the stem are identical, all the gametes of the tall plant will contain a chromosome bearing a gene for tallness. Similarly, all the gametes of the dwarf plant will contain a chromosome bearing a gene for short stature. The fusion of a gamete from a tall plant with a gamete from a short plant will produce a tall plant in the F_1 generation, because the gene for tallness (*T*) is dominant over that for dwarfness (*t*). The F_1 generation now possesses, among other chromosomes, a pair of homologous chromosomes carrying, at a specific locus, one allele for tallness and one for dwarfness, and the F_1 plant therefore is said to be HETEROZYGOUS (*Tt*) with respect to the genes that determine the length of the stem.

When the tall, heterozygous plant flowers, gametes of two kinds will be formed following meiosis in the ovule and anther. In the first meiotic division of each spore mother cell, the homologous chromosomes separate. Eventually, two kinds of eggs and two kinds of pollen grains with respect to the allele for height are produced. With self-pollination and fertilization, the two kinds of eggs will form all possible combinations with the two kinds of sperms to produce the F_2 generation.

As a result of these chance combinations, an approximate ratio of three tall plants to one dwarf plant may be expected. The actual result approaches closer to the expected result as the number of plants studied is increased. Mendel found, for example, that of 1064 plants in an F_2 generation, 787 were tall and 277 short—a ratio of 2.84:1. The theoretical expectation would be 798 tall plants to 266 dwarf plants. The difference between the observed and expected ratios was no greater than that which might result from chance fluctuations.

A cross such as that just described, in which only a single pair of alleles is considered, is called a MONOHYBRID CROSS.

Phenotype and Genotype The hybrid tall pea plants carried genes for both tallness and dwarfness. The plants all grew tall because the allele for tallness was dominant over the allele for dwarfness. If the F_1 plants are compared with the tall parents, it is impossible to tell by their appearance which plants are homozygous and which heterozygous for height. The characteristic of being tall is the PHENOTYPE of the F_1 generation. By phenotype is meant the expressed or apparent characters of an organism, as contrasted to the GENOTYPE, which is the genetic constitution of an organism, irrespective of its external appearance. For example, the phenotype tallness may result from either of the two genotypes, *TT* or *Tt*. In most cases these genotypes cannot be determined from the phenotypes; they must be identified by breeding experiments.

The 3:1 ratio observed in the F_2 generation is, of course, a phenotype ratio. Two-thirds of the tall F_2 plants indicated in FIG. 16-3 are heterozygous for height and, if self-pollinated, will not breed true. The genotype ratio of the F_2 plants, therefore, is 1:2:1. This follows from the fact that on the average, 25 percent of the plants of this generation are homozygous for the genes determining tallness, 50 percent are heterozygous, and 25 percent are homozygous for the genes determining dwarfness. This 1:2:1 ratio is not only a result of the chance combination of the gametes at the time of fertilization. It is also a result of the manner of segregation of the chromosomes (and genes) into different gametes at the time of meiosis.

Segregation of Genes A gamete can contain only one member of each pair of alternative genes, or alleles. The hybrid produced by crossing the tall and dwarf plants contained genes for both characters, although, as a result of dominance, only one of the two alternative genes was expressed. In each gamete of the hybrid, however, only one of the two alternative genes is present; at meiosis, one half of these gametes will receive the gene for tallness, and one half will receive the gene for dwarfness (Fig. 16-3). This separation of the genes for tallness and dwarfness is termed SEGREGATION. The concept of segregation of genes governing unit characters is known as Mendel's first law.

Dihybrids Let us consider another of Mendel's crosses, this time following two pairs of characters instead of one. This is known as a DIHYBRID CROSS, a dihybrid being an organism that is heterozygous for two pairs of alleles. Mendel crossed pea plants having round, yellow seeds with plants having wrinkled, green seeds (Fig. 16-4). Round and yellow turned out to be dominant over wrinkled and green. Since the genes influencing these characters are located on different chromosomes, they are inherited independently. The genotype of the parent with round and yellow seeds may be indicated by *RRYY* and that of the parent with wrinkled and green seeds as *rryy*.

The parents bear the genes for round and yellow, or their alleles, on two pairs of homologous chromosomes. In meiosis the homologous chromosomes separate, each passing to a different pole, and eventually their daughter chromatids enter different gametes. But since the parents are homozygous for the genes under consideration, the gametes of each parent are alike; one produces only *RY* gametes and the other only *ry*. The genotype of an F$_1$ generation resulting from a cross between these parents is *RrYy*, and the phenotype, because of dominance, is round yellow seeds.

Independent Assortment The behavior of homologous chromosomes during the first meiotic division is governed by chance. It is entirely a matter of chance to which pole a given

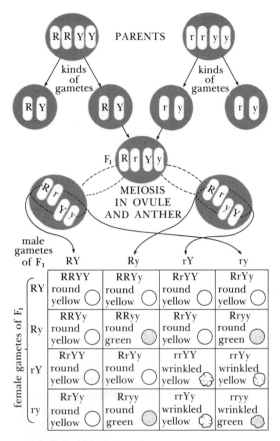

Fig. 16-4 A dihybrid cross between a pea plant with round, yellow seeds and one with wrinkled, green seeds. Chromosome assortment to male gametes is shown by arrows; identical assortment in female gametes is arranged vertically at the left of the checkerboard. All possible combinations of the four kinds of male gametes with the four kinds of female gametes result in a 9:3:3:1 phenotypic ratio in the F$_2$ generation.

chromosome of a homologous pair moves. Thus, in our dihybrid, the chromosome bearing gene *R* may move to the same pole as the chromosome from another pair bearing gene *Y* or, with equal likelihood, to the same pole as the chromosome bearing gene *y*. Similarly, the chromosome bearing gene *r* may move to the same pole as the chromosome bearing gene *Y* or to the pole toward which the chromosome bearing gene y is moving.

However, in normal meiosis the two members of a given pair of chromosomes always move to opposite poles; hence, their daughter chromatids are eventually included in different gametes. The chance combinations of the chromosomes of the different homologous pairs result in the formation of four types of gametes following meiosis in both the microspore and megaspore mother cells. An examination of the genetic constitution of these gametes will show that if the gene for round seeds alone is considered, half of the gametes produced by the F_1 generation carry the gene R and half the gene r. But every gamete carries not only the gene for seed shape but other genes also. Half the gametes must therefore carry the gene Y and half the gene y. But in any given gamete, it is entirely a matter of chance whether gene R is associated with gene Y or with y. And it is likewise entirely a matter of chance whether gene r is associated with Y or y. As a consequence, four types of gametes result: RY, Ry, rY, and ry. These four types are expected to be produced in approximately equal numbers. This chance combination of genes in the gametes of the hybrid is known as the law of independent assortment, or Mendel's second law.

Dihybrid Phenotype Ratio In order to see how gametes produced by the F_1 generation may combine, it is convenient to employ the checkerboard, a graphic device (FIG. 16-4) long in use to ensure that no possible combinations have been omitted. If the gametes of one sex are placed on the vertical side and those of the other are placed above and horizontally, the genotypes of the F_2 may be filled in by combining a gamete of one sex successively with each of the gametes of the other sex. An analysis of the F_2 plants obtained in this way shows that four phenotypes are now represented. Nine sixteenths have round and yellow seeds, three sixteenths have round and green seeds, three sixteenths have wrinkled and yellow seeds, and one sixteenth has wrinkled and green seeds. This constitutes the $9:3:3:1$ phenotypic ratio of the F_2 generation of a dihybrid cross. Another example of a dihybrid phenotype ratio is shown in FIG. 16-5, left. One of the original parents of the plant on which this ear was borne had blue kernels filled with starch, resulting in a smooth kernel. The other parent had white, wrinkled kernels. When blue smooth was crossed with white wrinkled, the F_1 produced ears with smooth, blue kernels. The F_1, self-pollinated, yielded F_2 (FIG. 16-5) ears bearing kernels in a ratio approximating 9 blue smooth, 3 blue wrinkled, 3 white smooth, and 1 white wrinkled.

Recombination An examination of the genotypes shown in FIG. 16-4 will reveal that some gene combinations are the same as those of the parents, but that most of them are new, differing from those of the parents. The formation of such new gene combinations is called RECOMBINATION. In the case of a dihybrid such as is shown in FIG. 16-4, the number of recombinations is seven. The total number of recombinant genotypes increases very rapidly with the number of pairs of genes. For example, with five pairs of genes, the number of recombinants is 241, and with ten pairs the number is more than 58,000. Such gene recombination is immensely important in plant and animal breeding and in evolutionary change under natural conditions. It will be repeatedly referred to in later pages.

The Test Cross It is sometimes desirable to test whether a plant showing dominant characters is homozygous or heterozygous for the genes determining these characters. If, for example, the F_1 plants of a cross between a plant with red flowers (R) and one with white flowers (r) bear red flowers, and in the F_2 the phenotype ratio is three red to one white, which of the F_2 red plants is homozygous and which heterozygous? This could be determined, in some cases at least, by self-pollination, but this is not always possible or

Fig. 16-5 Segregation of genes in corn. (Left): the F_2 generation resulting from a cross between blue smooth and white wrinkled. Four phenotypes result in a ratio approximately 9:3:3:1. (Right): the result of a cross between a corn heterozygous for the genes determining kernel color and a white homozygous recessive. The blue and white kernels are produced in approximately equal numbers.

practicable. Instead, the plants showing the dominant traits are crossed with the homozygous recessive type, a procedure known as TEST CROSSING. If such a cross results in all red flowers, the dominant plant was, of course, homozygous. The cross of a heterozygote with the recessive type, on the other hand, will produce a 1:1 ratio of red- and white-flowered plants.

Test crosses such as those described above are of general application in genetic experiments.

They also illustrate again the principle of segregation, and such crosses were used by Mendel as further proof of this principle. If a red-flowered plant is heterozygous, one half of the gametes (following segregation) should contain R and one half r. Only one kind of gamete, r, is formed by the homozygous recessive parent. Following cross-pollination, the zygotes formed should be Rr and rr in approximately equal numbers, and therefore half should produce red flowers and half should produce white.

Good examples of actual test crosses are found in experiments with certain varieties of corn. A plant bearing blue (B) kernels may be test crossed by pollinating with pollen from a strain bearing white (b) kernels. The progeny, when reared, may produce ears bearing blue and white kernels in an approximately 1:1 ratio (FIG. 16-5, right), indicating that the blue parent was heterozygous for the genes determining blue grains.

Test crossing should be distinguished from BACKCROSSING, to be discussed later. In backcrossing the hybrid is crossed with one of its parents, or the equivalent. This is done in plant breeding to increase the genetic contribution of one parent to the progeny, and is unrelated to the dominance or recessiveness of any gene.

GENETICS SINCE MENDEL

In the years following 1900, the rediscovery of Mendel's results served as a stimulus to thousands of scientific workers, and within a relatively short time the new science of genetics was firmly established. It advanced rapidly, and important discoveries were announced. Early students of genetics confirmed Mendel's laws by work not only on peas but also on many other plants and on animals. Such characteristics as color, branching, hairiness, leaf form, and doubleness in flowers were investigated in plants as varied as sunflower, cotton, wheat, beans, wallflower, snapdragon, and orchids.

These workers found that Mendelian principles held true not only for the more obvious external features of plants but also for more economically valuable characters. Investigations on the fruit fly (*Drosophila*), beginning about 1906, have yielded a wealth of information. Intensive and fruitful studies on the corn plant were initiated somewhat later. Much of what we know today about inheritance in plants and animals has come from breeding experiments on *Drosophila* and corn. Within recent years this knowledge has been increased by extensive studies on bacteria and viruses.

Experimental genetics soon showed, however, that Mendelian concepts are too simple to account for some of the results obtained. Numerous exceptions to the usual Mendelian ratios were encountered. Although the laws of segregation and independent assortment are fundamentally correct, the view that every character is the result of the action of a single gene has undergone modification. It has been found that although genes are inherited independently, they do not necessarily manifest themselves independently. A given character may be the result of the interaction or cooperation of several genes. Commonly, when several genes acting together determine a character, that character will not appear if any one of these genes is absent. Other characters—especially quantitative characters, such as those involving growth and size—show intermediate conditions. These are usually influenced by a number of genes, each of which makes a specific contribution but all of which interact to produce a cumulative effect upon the phenotype. Inheritance of this type is known as MULTIPLE-GENE or QUANTITATIVE inheritance.

Incomplete Dominance The original concept of dominance has undergone modifications also. It was once viewed as the complete covering up of the effect of one gene by another. Dominance of this type was later shown not to be an essential aspect of Mendelian inheritance. Dominance may exist in all degrees and variations. Many characters show partial, or *incomplete*, dominance,

and intermediates between two parents often result. A classical example of this was disclosed in early work on the breeding of four-o'clocks. In crossing a crimson-flowered plant with a white-flowered one, an F_1 generation with pink flowers was produced. These F_1 plants, when self-pollinated, produced crimson, pink, and white in the proportions $1:2:1$. When self-pollinated, the crimson and white forms bred true, whereas the pink again segregated in a $1:2:1$ ratio. Other examples of incomplete dominance are known, but most cases in which the offspring are intermediate between the parents are due to multiple genes. In practice, it may be difficult to distinguish between incomplete dominance and multiple-gene inheritance, and detailed investigations are often necessary to identify the underlying principle.

Linkage Mendelian ratios depend upon the independent assortment of the genes in meiosis. Such an assortment will occur only when the genes are located on different chromosomes. The chromosomes of plants and animals contain numerous genes, probably many thousands. Since the chromosomes are limited in number, each chromosome must carry a large number of genes. If two genes are located on the same chromosome, these genes will tend to be transmitted together to the offspring, so that independent assortment does not occur. Genes that are located on the same chromosome and tend to be inherited together are said to be LINKED.

To illustrate the concept of linkage, let us examine the crossing of two plants, one having the genotype *AABB* and the other *aabb*. The F_1 genotype will be *AaBb*. If independent assortment takes place, the gametes formed will be *AB*, *Ab*, *aB*, and *ab*, in equal numbers. If, however, both dominant genes are located on one chromosome, and the recessives on its homolog, only two kinds of gametes will be formed: *AB* and *ab*. Instead of the expected dihybrid $9:3:3:1$ ratio in the F_2 generation, a $3:1$ ratio will result, since the two genes are inherited as though they

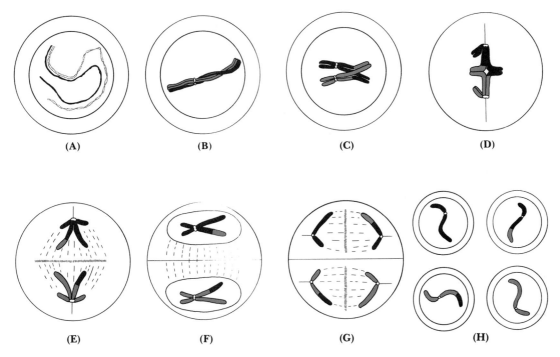

Fig. 16-6 Diagram showing crossing-over. (A–F): stage I. (A,B,C): prophase. (A): beginning of pairing (synapsis). (B): the paired chromosomes are twisted around one another. (C): two nonsister chromatids have "crossed over," that is, exchanged parts. (D): metaphase. (E): anaphase. (F): telophase. Stage II. (G): anaphase. (H): four microspores, two of them containing a crossover chromatid.

were a single gene. A number of genes that tend to be inherited together form a LINKAGE GROUP. In corn, about 360 different sets of alleles are known. These fall into ten linkage groups, representing the ten pairs of chromosomes.

Crossing Over Genes originally linked on the same chromosome do not always remain linked. Complete linkage, in which the gametes contain genes from only one chromosome of a particular pair, is, in fact, uncommon. Genes located within one chromosome may be exchanged for corresponding genes of the homolog by a process of CROSSING OVER, which occurs early in the first stage of meiosis.

For the sake of simplicity, FIG. 16-6 shows meiosis in a spore mother cell with only one pair of homologous chromosomes. The chromosomes pair (FIG. 16-6A, B), each chromosome consisting of two chromatids. During the pairing process the chromosomes tend to twist about one another. At this time two paired, nonsister chromatids of homologous chromosomes may break apart at corresponding points. Broken ends then become joined in new sequences, and two rearranged, or crossover, chromatids are formed (FIG. 16-6C). It should again be noted that this exchange of parts occurs between chromatids originating from different members of the same pair—that is, nonsister chromatids. In the late prophase, the two members of the chromosome pair begin to separate from one another, starting in the region of the centromeres. However, as a result of crossing over, the homologous chromosomes cannot completely separate, for adhesions between the chromatids bind the two chromosomes into a unit. At metaphase (FIG. 16-6D), therefore, the centromeres have repelled one another, but the homologous chromosomes remain united by their crossed-over long arms. At anaphase (FIG. 16-6E), they separate and thereafter form two telophase nuclei (FIG. 16-6F).

In the second stage of meiosis, the members of each pair of sister chromatids separate and pass to opposite poles of the daughter nucleus (FIG. 16-6G). This results in four nuclei, each containing one chromosome (FIG. 16-6H). The genetic consequence of crossing over, then, is the exchange of hereditary material between chromatids of homologous chromosomes. Genes formerly associated in one member of a pair of chromosomes may thus be exchanged to the other member of that pair. It may be noted that two genes located at different positions (LOCI) on a chromosome are more likely to be separated by crossing over as the distance between them increases.

We may consider a hypothetical example in which one member of a pair of homologous chromosomes bears gene A, which determines resistance to disease. It also bears gene B at another position, causing late maturity. On the homologous chromosome, at corresponding positions, are alleles a, determining susceptibility to disease, and b, causing early maturity. If the genes A and B, and a and b remain together on their respective chromosomes, a disease-resistant plant will always mature late, and a susceptible plant will always mature early. But if crossing over occurs between these positions, the part of a chromatid bearing gene A becomes attached to the part of a chromatid that bears gene b. The genes Ab are now located on one chromatid and aB on another. It is then possible to obtain plants that are not only resistant to disease but also early maturing.

Cytoplasmic Inheritance In addition to the Mendelian inheritance of genes contained in specific chromosomes, a small but growing number of instances are known in which a given character appears to be inherited only from the mother, and to remain largely independent of the germ plasm contributed by the sperm. In certain varieties of the common four-o'clock, for example, one branch may bear normal green leaves, another may be light green or nearly white, and a third show leaves with variegations of the two colors. Seeds from a flower on a green or pale branch will produce only green or pale leaves, without regard to the pollen used. Seeds from a variegated branch may produce green, pale, or variegated leaves, and will continue to show the same character in all seeds produced by the plants in succeeding generations.

This type of cytoplasmic inheritance is considered to be due to the transfer of proplastids (Chapter 3) in the cytoplasm of the egg cell. These proplastids divide and produce plastids as the plant develops. They carry within themselves the determinants of chlorophyll development and hence of leaf color. The proplastids of the cells may be of the normal green type, the pale type, or both may occur together. The proplastids thus constitute a part of the inheritance of the plant, its cytoplasmic inheritance.

An important type of cytoplasmic inheritance is responsible in part for MALE STERILITY, that is, the failure to produce pollen, in a number of plants—for example, corn, onions, sugar beets, grain sorghum, sugar cane, flax, and petunias. The use of male sterility in producing hybrid seed corn is discussed later in this chapter. The factor is transmitted in the cytoplasm of the egg cell, and all kernels produced by a male sterile plant will carry the male-sterile factor and will commonly produce only male-sterile progeny without regard to the pollen parent.

Cytoplasmic inheritance is assumed to be due to DNA molecules, analogous to those carried in the chromosomes. This genetic material is also present in the chloroplasts and in the mitochondria. A mitochondrion of the field

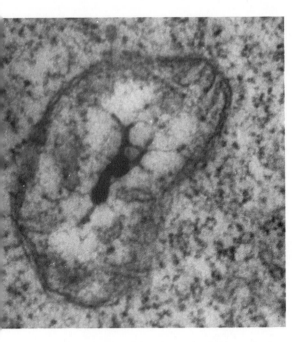

Fig. 16-7 Mitochondrion from the field mushroom (*Agaricus campestris*). The dark central mass is DNA.

mushroom, stained to show the presence of DNA, is shown in Fig. 16-7. The DNA of the mitochondria may possibly be responsible for the cytoplasmic inheritance of male sterility. Although cytoplasmic inheritance is interesting and at times valuable, it is a very minor factor in the total inheritance of plants.

GENES AND ENVIRONMENT

Among the numerous characters of a plant are size, form and arrangement of parts, internal structure, and rate of vegetative growth. The characters are determined by genes, or, collectively, the genotype, which provides instructions for phenotypic growth and behavior. The genes, however, never act alone, but always in an internal environment, the nature of which is often af-

fected by the external environment. The genotype (except for mutations) remains the same, but may express itself differently in different environments, giving rise to modifications of the phenotype. External factors of the environment which influence gene expression include the intensity, quality, and duration of light, temperature, supply of essential minerals, and atmospheric and soil moisture. Some characters remain unchanged under a wide range of environments; among these are the construction of the flower and internal features such as the nature of the pits on the side walls of the vessel elements. Other characters, such as the height of the plant, time of flowering, and size of fruit, may vary considerably with differences in the environment. From the genetic point of view, characters result from the expression of the genotype in a given environment; it is the genotype that is inherited, not the phenotype. The nature of every organism thus depends both upon the genes that it inherits and upon the environment in which it develops. The mechanisms by which genes are regulated was considered in Chapter 14 under "Differentiation and gene activity."

The following examples illustrate the effect of the environment upon gene expression. In a few varieties of sweet and pop corn, a red, gene-determined anthocyanin pigment develops, but only in those parts of the plant exposed to sunlight. The parts of the stalk covered by the leaf bases and the inner husks of the ear are without this red color, called "sun red." If sunlight is excluded by covering the developing ear with a black hood, the outer husks remain green, although these tissues still contain the genes for sun red. A number of examples of temperature-induced modifications of the phenotype are known. In certain corn mutants, a high temperature is necessary for the expression of the mutant phenotype. The seedlings and older plants are pale green when grown at 37°C (98.6°F). When grown at 20°C (68°F) the color of the mutant plant approaches the normal green.

It may again be emphasized that the variations

brought about by environmental modifications of the phenotype are not inherited. Only inherited variations are effective in evolutionary processes or in breeding.

PLANT IMPROVEMENT

Selection Modern man has always attempted to improve domesticated plants and animals. The best-known method of plant improvement, and one that has been practiced for centuries, is SELECTION. Because plants are variable, men have been able to select those plants that best serve their needs and to propagate these by seeds or by vegetative means. Seeds from the best plants were saved each year for planting in the next season. The next year the process was repeated and the best plants were again selected for further propagation.

The process of selection has numerous limitations. Because one can work only with the phenotype, the basis for selection is not always clear. Early plant breeders could not distinguish between the genetic and environmental factors that produced a good plant; hence, many errors were bound to occur. But even so, selection has worked and, continued for generation after generation, it has yielded valuable varieties of cereals, sugar beets, flax, cotton, and other cultivated plants. This method of plant improvement is still practiced, but nowadays it is commonly supplemented by hybridization.

Hybridization Through hybridization one can bring together the best genes of two or more individuals, varieties, or even species. Crosses are made between selected parents that differ in one or more genes (FIG. 16-8), and the progeny are grown for several generations. During this time, new gene combinations are brought about by chromosome assortment and crossing over. Eventually, forms with the desired phenotype and genotype are selected for propagation on a commercial scale.

Hybridization and selection create no new genes, but may make available the best combinations of genes already present in the parent stock. After a time, the limit of effective selection is reached, and further progress can be attained only by introducing new and different genes, frequently from closely related plants found in other parts of the world. To improve our wheat and other crops, botanists of the U.S. Department of Agriculture have searched the world for varieties that could be adapted to this country or that could be crossed with our crops to improve them in specific ways. The work of Mark Carleton, who in 1900 introduced several valuable varieties and strains of hard red winter wheat from Russia into the United States, is but one example of the continuing, painstaking endeavor to improve our food plants. Many of our present-day varieties of this type of wheat were developed from these original introductions and are now widely cultivated in the Great Plains area and elsewhere.

It is frequently desirable to incorporate one or a few new characters into a variety that is already highly improved. Such an introduction of new genes is accomplished by backcrossing, which is the mating of a hybrid with one of its parents or with an organism genetically equivalent. Backcrossing has been especially useful in the development of disease resistance in cultivated plants, an important part of the work of the plant breeder. Early work in this field consisted of the selection and propagation of resistant forms that occurred naturally. Disease-resistant plants can now be developed by crossing a resistant plant of poor qualities with a susceptible but otherwise valuable plant. Repeated backcrossing to the susceptible parent type, accompanied by selection for disease resistance, yields new plants containing the more desirable genes of both original parents.

Hybridization in Corn Indian corn, or maize, is one crop in which efforts to improve quality and productivity have met with marked success.

Fig. 16-8 Hybridization. The pollen from the stamen of one flower is placed upon the stigma of another flower. This is the first step in producing a hybrid.

The earliest method, and one that was undoubtedly used as soon as corn was domesticated, was mass selection. Seed ears were selected in the field from large, vigorous plants with well-developed ears, and the grains from these ears were mixed together and used for planting the following season. This procedure, continued, was fairly effective in improving corn, but after a time the yield became stabilized and very little further increase was obtained. The reasons for this lie largely in the nature of the corn plant. Corn is monoecious, and is more than 99 percent cross-pollinated. As a result of constant cross-pollination, corn varieties are heterozygous for most genes. The progeny of such mass-selected ears may therefore be expected to vary widely because of gene segregation and the combination of gametes carrying undesirable as well as desirable genes. A proportion of the progeny, in fact, will carry numerous genes in the homozygous condition, and the yield of such plants will be less than that of the heterozygous parents.

Dissatisfied with mass selection, corn breeders, beginning about 1896, adopted the ear-to-row method. Again, seed ears were selected from the field. A part of the grain from each of the ears was planted in a single row. Each row was harvested separately to determine the yielding ability of the different ears. The reserve grain from the ears that gave the highest yield in the test was planted the second year. This procedure, continued for a number of years, also yielded unsatisfactory results, and no varieties produced by this method were significantly more productive

Fig. 16-9 Reduced vigor of corn with successive generations of self-pollination.

than the best of the mass-selected strains. Sometimes, in fact, the yield actually decreased. One important reason for the inadequacy of this method was that the ears selected tended to be all of one type, and with continuous selection of single ears, the plants became genetically more and more alike. Pollination among the selected plants thus became increasingly equivalent to self-pollination, or inbreeding, which in normally cross-pollinated plants is frequently deleterious to yield. The basis for this is explained later in this chapter.

The modern method of corn improvement involves controlled hybridization of selected *inbred* lines. Such inbred strains are produced by self-pollination over a period of five or more years. Self-pollinated corn is produced by collecting pollen from a plant and dusting it onto the silks of the same plant. The silks are covered before selfing to prevent cross-pollination, and pollinated ears are again covered until mature.

The inbred strains produced by self-pollination decline in size and vigor, especially in the early generations (Fig. 16-9). A large number of the progeny, in fact, are so weak or sterile that

they die out or are discarded. This deterioration involves the operation of the following two important principles.

1. Heterozygous organisms may carry a large reservoir of deleterious recessive genes that are concealed by their dominant alleles. If a harmful dominant gene should occur, it is quickly eliminated by natural selection (Chapter 17). A harmful recessive, however, can remain indefinitely because it is covered by heterozygosity. Therefore, most deleterious genes in a group of interbreeding individuals are recessive.

2. If any normally cross-pollinated plant (such as corn) is inbred for a number of generations, the number of pairs of genes in the homozygous condition increases and the number of heterozygotes decreases. Almost complete homozygosity may be achieved in eight to ten generations. This principle is illustrated in its simplest form in Fig. 16-10, which shows what happens following self-pollination in an organism heterozygous for a single pair of alleles. The F_1 represents a heterozygous corn plant; the F_2, the first generation following selfing. In the F_2, 50 percent of the progeny are heterozygous, but in

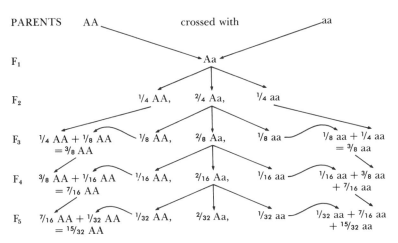

PARENTS   AA   crossed with   aa

F₁ appears in figure.

Fig. 16-10 Simplified scheme showing how continuous self-pollination results in homozygosity.

F_1 Aa

F_2 ¼ AA,   2/4 Aa,   ¼ aa

F_3 ¼ AA + ⅛ AA   ⅛ AA,   2/8 Aa,   ⅛ aa   ⅛ aa + ¼ aa
         = ⅜ AA   = ⅜ aa

F_4 ⅜ AA + 1/16 AA   1/16 AA,   2/16 Aa,   1/16 aa   1/16 aa + ⅜ aa
         = 7/16 AA   + 7/16 aa

F_5 7/16 AA + 1/32 AA   1/32 AA,   2/32 Aa,   1/32 aa   1/32 aa + 7/16 aa
         = 15/32 AA   + 15/32 aa

Fig. 16-11 Hybrid vigor. Hybrid corn is shown at center; inbred parent strains at left and right.

the next, or F_3 generation, only 25 percent are heterozygous, whereas 75 percent are homozygous. In the F_4 generation, 12.5 percent are heterozygous, and 87.5 percent are homozygous for one or the other of the two alleles. In the F_5 generation, 94 percent are homozygous, and, if carried to another generation, the F_6 will be 97 percent homozygous.

It is apparent, then, that inbreeding may decrease vigor by increasing the number of recessive, deleterious genes in the homozygous condition. The effects of many of these genes were formerly concealed by their dominant alleles, but since they are now able to express themselves, the result is a general decline in vigor. After about six generations of inbreeding, the corn becomes homozygous to the extent that no further deterioration is apparent (FIG. 16-9). Selection during the period of inbreeding eliminates the least desirable characteristics. Inbred strains that breed true and are remarkably uniform are finally produced, although the various strains may differ greatly from each other.

At the end of the period of inbreeding, two or more inbred lines are crossed and the resulting hybrid seed is sold to the farmer or gardener. Plants grown from such hybrid seed are taller and sturdier, and bear larger ears than their inbred parents (FIG. 16-11). Their crop yield is greater than that of the inbred lines and is commonly greater than the yield of the varieties from which the inbred lines were derived. Some hybrids yield 35 percent more than the naturally cross-pollinated plants from which the inbred lines were developed. Among the major advantages of the better hybrids is that they can be planted 2 to 3 or 4 times as thick as the old, open-pollinated strains without failing to produce grain. Instead of 7000 or 8000 plants per acre, growers now use 20,000 to 30,000 on better soils with heavy fertilization. In addition to yield, other desirable qualities, such as a strong root system and resistance to drought, disease, and insect and storm damage, are often combined in one strain. With respect to yield, it should be emphasized that hybrid corn does not necessarily produce bigger or better ears than the best open-pollinated plants. Open pollination is natural pollination, involving both selfing and crossing of more or less related plants. Some individuals of standard varieties produce large ears, but these are above average, whereas all the ears of hybrids tend to be large and uniformly shaped. Uniformity is especially important in commercial sweet corn, where it is important that all the ears ripen at the same time so that the entire field may be harvested in one picking.

The production of hybrid seed corn is a very considerable industry. Several methods are used. In making the SINGLE CROSS, the inbreds are planted together, commonly in the proportion of two rows of the seed parent to one row of the pollen parent. The tassels are pulled from the seed parents before the pollen is shed, or a male-sterile inbred is used, thus ensuring cross-pollination. Only the seed produced on the de-tasseled or male sterile rows is hybrid seed. The single cross has certain disadvantages, such as the low yield of pollen and seed, and thus high cost, but it is very uniform, and is widely employed in the production of hybrid sweet corn. Hybrid field corn is grown mainly from seed produced by the DOUBLE-CROSS method (FIG. 16-12).

The double cross involves four selected inbred strains, which may be designated as A, B, C, and D. Inbred B is pollinated by A, and C by D. The seed produced by the single crosses B × A and C × D is planted the next year in alternate strips; (frequently, six rows of the seed parent and two of the pollen parent). Formerly, the hybrid B × A was detasseled by hand and left to be pollinated by C × D. Small fields, particularly of the first cross, as of C × D, may still be detasseled by hand, but male sterility is now generally used to prevent pollen formation and self-pollination in the B × A plants that are to be pollinated by C × D for commercial seed production. The genes of the four parent lines are thus combined, and the yield and quality of the resulting seed are consequently high. It has been estimated that close to 70 million acres in the United States are planted with double-crossed corn each year.

The increase in growth of the F_1 plants that results from crossing existing varieties or inbreds is known as HYBRID VIGOR. It occurs in many other plants besides corn. Hybrid vigor usually appears in the heterozygous F_1 generation following the crossing of plants that are nearly homozygous as the result of natural or artificial self-pollination. The basis of hybrid vigor is not clearly understood, but one explanation is that it is the result of combining favorable dominant genes with unfavorable recessive alleles in the hybrid generation. Although each inbred line is weak because of the homozygous detrimental recessives, the same recessives are not likely to be present in the two or more inbred strains chosen to produce hybrid corn. Therefore, in the hybrid, the heterozygous condition is renewed, the dominants from one parent cover the recessives from the other, and hybrid vigor is the result. Hybrid vigor has been utilized in the production of tomato, summer squash, onion, sugar beet, tobacco, sorghum, and other useful plants.

The grower of hybrid corn is warned against saving seed from the resulting crop for sowing the succeeding year, for the use of such seed in-

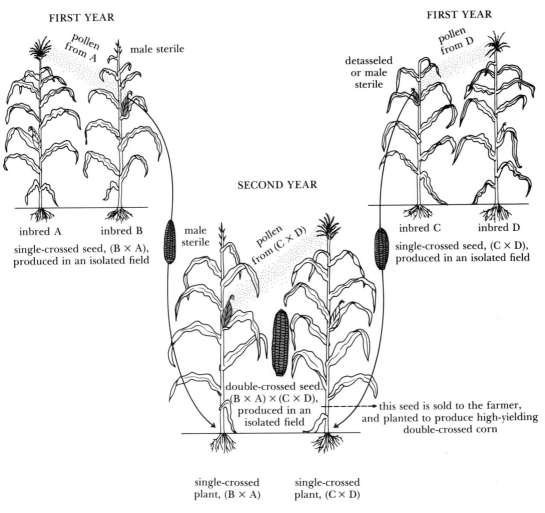

pollen from A

male sterile

inbred A   inbred B

single-crossed seed, (B × A), produced in an isolated field

male sterile

detasseled or male sterile

pollen from D

inbred C   inbred D

single-crossed seed, (C × D), produced in an isolated field

SECOND YEAR

pollen from (C × D)

double-crossed seed, (B × A) × (C × D), produced in an isolated field

→ this seed is sold to the farmer, and planted to produce high-yielding double-crossed corn

single-crossed plant, (B × A)

single-crossed plant, (C × D)

Fig. 16-12 Hybrid corn. Method of producing single and double crosses.

variably results in reduction in yield. This is illustrated in Fig. 16-9, which shows the reduction in vigor of corn with inbreeding, and in Fig. 16-10, which shows how self-pollination results in increased homozygosity. Corn from a commercial field planted with hybrid seed is produced by open pollination among closely related plants.

This process is equivalent to a large amount of self-pollination, or inbreeding, and thus yields are reduced sharply if this seed is replanted. Gametes bearing recessive genes will combine with genes of similar heredity, and recessive genes will no longer be masked by their dominant alleles. In order to obtain maximum yields,

Fig. 16-13 Commercial hybrid onion seed field in California. The light rows are the female parents, which are male-sterile. The dark rows are inbred onions producing normal pollen. This is carried by insects to the male-sterile plants, which produce the hybrid seed. Hybrid onions have advantages similar to those of hybrid corn.

hybrid corn seed must be obtained each season from the continued crossing of inbred lines. Many commercial growers devote their entire energies to this task.

Male Sterility The most laborious operation in the production of hybrid seed corn is the removal by hand of the tassels of the seed parents. This expensive procedure is now being largely replaced by the use of male-sterile seed parents. Male-sterile plants produce seed normally, but viable pollen is not formed; hence self-pollination is impossible. The male-sterile factor (Ms) is carried in the cytoplasm, and all of the direct progeny of a male-sterile plant will be male-sterile, regardless of the pollen used. The Ms factor is not expressed, however, in the presence of a nuclear gene called the restorer (R).

To use male sterility in hybrid seed corn production, the Ms factor must be introduced into a desirable inbred line, and the restorer (R) must be absent. This line will then carry the cytoplasmic factor $MsMs$ for male sterility and the nuclear genes rr, indicating the absence of the restorer gene (R) in the nucleus. To form a male-sterile inbred from a desirable, standard inbred, a male-sterile line is crossed and backcrossed for six or more generations with pollen from the desirable inbred, using male-sterile plants always as the seed parent. Selection during this period is used to ensure the absence of the restorer gene (R). The chromosomes of inbred B have now been transferred into male-sterile cytoplasm. The line is reproduced by using normal inbred B as a pollen parent and Ms-inbred B as the seed parent. Note that the inbred B pollen parent must not carry the restorer gene (R), and that the inbred A, used in the B $\times$ A cross, must also be recessive (rr) for the restorer. Otherwise the single cross B $\times$ A will produce pollen and hand detasseling will be required.

In the second year (FIG. 16-12) the inbreds B $\times$ A and C $\times$ D are planted in alternate strips with pollination by C $\times$ D. Beginning in the late 1960s, farmers mixed 90 or 95 percent of seed of the male-sterile hybrid B $\times$ A with 5 or 10 percent of pollen-producing C $\times$ D. This latter hybrid *must* carry the restorer gene RR, or the farmer who buys the seed will produce no corn. Ten percent of the plants will provide more than enough pollen for the field, and the 90 percent of male-sterile plants will produce more corn because they use no food in pollen production. With this plan, 90 percent of the plants produce only corn and 10 percent produce pollen as well as corn.

Male sterility has also been found in other plants of economic importance. Male-sterile onions are used in the commercial production of hybrid onion seed, where cross-pollination by hand would be impractical (FIG. 16-13).

Hybrid Vigor and Vegetative Propagation A large proportion of our food and ornamental plants are heterozygous for many desirable genes and possess hybrid vigor to a marked degree. This vigor would be lost, and undesirable genes would show their effects, if such plants were grown from seed. This is one reason why many plants are propagated by cuttings, budding, grafting, and other kinds of vegetative propagation (Chapter 11). Among the food and ornamental plants reproduced vegetatively are white and sweet potatoes, sugar cane, apple, pear, cherry, peach, apricot, strawberry, peony, dahlia, and many others. Valuable plants produced by hybridization may be preserved and propagated by asexual methods without loss of vigor or the appearance of undesirable qualities that might occur through segregation and recombination.

SUMMARY

1. The characters of an organism are determined by genes.
2. Genes are located on chromosomes and exist in pairs, called alleles. Chromosomes containing allelic genes are called homologous. One member of each pair of homologous chromosomes comes from each parent.

3. In the absence of mutation, each gene maintains its identity, generation after generation.

4. Meiosis is the process by which the chromosome number is reduced from the diploid to the haploid. It consists essentially of the pairing of homologous chromosomes and the separation of the members of each pair, followed by two nuclear divisions, so that each of the resulting four nuclei contains but one daughter chromosome from each original pair.

5. An organism in which the two genes at the same position on homologous chromosomes are identical is said to be homozygous for that gene. Thus an individual containing either *TT* or *tt* is homozygous for those genes.

6. An organism in which the two genes at the same position on homologous chromosomes are unlike is said to be heterozygous for that pair of genes. Thus the individual containing *Tt* is heterozygous.

7. As a result of the separation of the chromosomes at meiosis, a gamete contains but one gene of a pair, never both. The separation of the members of a pair of genes at meiosis results in segregation. This is Mendel's first law.

8. As a result of the behavior of the chromosomes at meiosis, the members of each pair of genes segregate independently of the members of other pairs, provided they are not on the same chromosome. This is known as Mendel's second law, the law of independent assortment.

9. A gene for a character is dominant when it shows itself in the hybrid to the exclusion of the contrasting character caused by the recessive gene.

10. The phenotype of an organism is the external appearance—that is, the sum total of the visible characters. The genotype is the aggregate of the genes, irrespective of whether these are expressed.

11. The result of a cross between two organisms differing in only a single gene is called a monohybrid. The F_2 phenotype ratio of such a cross is $3:1$ and the genotype ratio is $1:2:1$.

12. The result of a cross between two organisms differing in two pairs of genes is known as a dihybrid. The F_2 phenotype ratio in a dihybrid cross is $9:3:2:1$.

13. A cross between parents differing in one or more genes results in new gene combinations in the progeny. The formation of such new combinations is termed recombination.

14. Genes are inherited independently (unless they are located on the same chromosome), but do not necessarily manifest themselves independently. Genes that are located on the same chromosome tend to be inherited together, a condition that is known as linkage.

15. Genes may become unlinked in inheritance as the result of crossing over at meiosis. Crossing over is the exchange of parts of two chromatids between members of a pair of homologous chromosomes.

16. Genes are inherited, characters are not. A character is the expression of a gene or genes in a given environment, and the phenotype of an organism is the result of both the inherited genotype and the environment.

17. Selection, with or without hybridization, is the outstanding method of plant improvement.

18. The most important method of improvement in corn is brought about by hybridization between inbred lines. Such lines exhibit hybrid vigor. The seed from a hybrid corn crop should not be planted because the yield will decrease as a result of natural inbreeding.

19. Inbreeding of corn (or other normally cross-pollinated plants) causes a decrease in vigor and yield because it increases the number of recessive, deleterious genes in the homozygous state.

20. Male sterility in corn is inherited from the

mother (seed) plant, and the progeny will fail to produce pollen as long as a restorer gene is not introduced. The restorer *must* be introduced in the last hybrid cross; otherwise, the farmer's field will produce no pollen and no corn.

Mechanisms of Evolution

THE WORD EVOLUTION IM-
PLIES CHANGE. Changes have occurred through the ages, both in living
things and in the inorganic world. We live in a
universe of continual change, and our earth is no
exception. The "everlasting hills" are not everlasting but are gradually being eroded away. In
the future, as in the past, continents will rise and
sink, and the face of the earth will undergo vast
transformations. And just as our earth has
evolved, so have the plants and animals upon it—
a process we call organic evolution.

According to the concepts of organic evolution, all living plants and animals are descended
from different, and in most cases simpler, kinds
of life that lived in the past. The differences
between organisms of the past and those of the
present are the result of structural and physiological modifications acquired during innumerable successive generations. The theory of organic
evolution is one of the most important scientific
generalizations. It is supported by a great many
kinds of evidence, drawn from genetics, from the
comparative anatomy of both plants and animals,
from paleontology, from geographical distribution, from classification, from comparative physiology and biochemistry, and from selection
among domesticated plants and animals.

Although the idea of evolution can be traced
back to the ancient Greeks and followed as a thin
thread through the Middle Ages, it remained a
vague philosophical speculation until the early
nineteenth century. The idea flowered in the mid-
nineteenth century, and today is a fundamental
part of the biological sciences and our culture in
general.

This chapter does not attempt to trace the history of the idea of evolution or to amass the evidences for it. Our present concern is not
primarily with the facts of evolution but with the
processes involved. Before attempting to present
the modern point of view with respect to evolutionary mechanisms, we shall discuss two earlier
explanations of evolution, for our story would be
incomplete without a picture of the work of Jean
Lamarck and Charles Darwin.

THEORIES OF LAMARCK AND DARWIN

Inheritance of Acquired Characters In 1809, the French naturalist Jean Baptiste de Lamarck published a book, *Philosophie Zoologique,* in which he presented an explanation of evolution based upon the inheritance of acquired characters. By "acquired characters" Lamarck meant those modifications of structure that occur during the life of a plant or animal in response to environmental influences. Lamarck believed that all features and characteristics acquired during the development of an individual plant or animal were inherited by its progeny, and that such changes were cumulative from generation to generation. To explain the acquisition of new characters, at least with respect to animals, Lamarck stated that a change in the environment would cause a change in the needs or wants felt by an animal organism. Because of these needs, new organs would eventually develop as the result of efforts put forth by the animal.

The further development of these new organs or the loss of characters was explained on the basis of the theory of use and disuse. Lamarck believed that the frequent use of any organ would improve its functioning, or bring about the enlargement or further development of that organ; conversely, if an organ were no longer used, it would cease to function and finally disappear. He theorized that the results of changes occurring during the life of the individual—the acquisition of new parts and their further development through use or loss through disuse—are transmitted to the progeny of the individuals in which these changes have occurred. According to Lamarck, the environment is not a direct molder of animal species but influences the direction of evolution by changing the needs and requirements of the organism. These, in turn, stimulate the production of new parts, the greater use of parts already existing, or the disappearance of parts no longer needed.

Among the numerous examples cited by Lamarck to support his theory was the giraffe.

This animal browses upon the foliage of trees, and the forelegs and neck were assumed to have lengthened as a result of efforts to reach this food. The increase in length was passed on to its progeny, to become still more pronounced in successive generations. Lamarck used the cave salamander, which he believed had lost its eyes as a result of living in the dark, to illustrate the effect of disuse. Many other examples of use and disuse in animals were cited.

Since he considered that plants do not carry on activities comparable to those of animals, Lamarck believed that plants were changed by the direct effect of the environment. For example, the more favorable environment supplied to wild plants when brought into cultivation caused them to change gradually into the familiar plants of garden and field. Conversely, he cited the hypothetical example of the seed of a meadow grass that by chance is transported to a dry, barren, and stony area exposed to the wind. If the seed germinates and the small, struggling plant lives and reproduces, there will arise a race different from the meadow plants from which it originated.

The hypothesis that explains the origin of new forms of life on the basis of the inheritance of acquired characters has been abandoned by modern biologists. No one denies that organs or structures are affected by use or disuse, but experiments extending over more than a century have shown that such effects are not inherited, nor is it possible to fit such a theory into our modern knowledge of hereditary mechanisms and processes. This highly speculative and facile early theory has been discussed here because of its historical interest and because people today who are untrained in scientific methods tend to advance similar explanations for adaptations. The acceptance of this point of view usually involves a teleological approach that impedes further progress and investigations. It may be noted that the explanation of adaptation on the basis of inheritance of acquired characters was revived in recent years in the now discredited work of the Russian botanist T. Lysenko.

Natural Selection In 1859, Charles Darwin published his monumental work *On the Origin of Species by Means of Natural Selection, or the Preservation of Favoured Races in the Struggle for Life*. In this book, which had profound sociological as well as scientific effects, Darwin established the fact of organic evolution upon a firm foundation, and advanced the theory of natural selection as the explanation of the process. The essential aspects of Darwin's theory may be divided into three interrelated parts.

VARIATION Darwin noted, and spent a great deal of time recording, the variability that exists in plants and animals. No two individuals derived from the same parents are exactly alike, even when inhabiting the same locality. He observed that some character variations were transmitted and others were not, but, lacking the modern knowledge of hereditary mechanisms, he was unable to recognize the two fundamentally different types of variations, one inherited and the other caused by the environment and thus not handed down from one generation to the next.

THE STRUGGLE FOR EXISTENCE The normal rate of increase of a species is geometric. Calculations show that if all the progeny of any of a number of species should survive and again reproduce, this species would populate the earth in a few generations. Let us assume, for example, that an annual plant produces 100 seeds (many produce several thousand), that these seeds give rise to 100 plants in the next growing season, and that each of these produces 100 seeds also. At this rate of increase by the end of the eighth consecutive season of growth, one hundred thousand billion mature plants would be produced as the direct descendants of the first plant, and if they were corn planted at the usual rate, they would cover the cultivatable land of the earth.

Because the tremendous reproductive rates of plants and animals result in overproduction, only a certain proportion of the offspring will survive. This differential survival is the result of what Darwin termed the "struggle for existence." Survival of a species depends upon reproduction despite the disadvantageous effects of the total environment. which includes not only soil and climatic factors, such as extremes of temperature and drought, but also the action of other organisms, including predators and parasites. Animals tend to deplete their food supply; plant seedlings frequently grow under conditions in which only the most vigorous and fortunate will survive (FIG. 17-1). But if, despite all destructive forces, the number of offspring reaching maturity, and capable of reproducing, balances or exceeds the number in the previous generation, then species survival seems assured.

NATURAL SELECTION Since organisms are exposed to adverse forces under natural conditions, those variants within a population that are best adapted to the prevailing conditions will be the most likely to survive. Many of the variations within or between species are of no known im-

Fig. 17-1 Competition among seedlings of the weed hemp-nettle (*Galeopsis tetrahit*). These seedlings compete with one another for light, water, and mineral nutrients. Only a few can survive.

portance to the organism, but some are deleterious and others are advantageous—that is, of *survival value.* Plants with leaves deficient in chlorophyll will make less food, grow more slowly, and fail to survive. Individuals possessing even slightly unfavorable variations may be eliminated; an apparently inconsequential loss in efficiency may be fatal. On the other hand, variations that favor pollination, permit growth in poor locations, improve seed dispersal, or reduce winter injury usually have survival value, and the plants that possess them are favored over those that do not.

Darwin recognized, although he did not emphasize to the same extent as does modern evolutionary theory, that the significance of natural selection does not lie primarily in the survival of the individual, but rather in success in the production of offspring. In any population there will be some variant individuals better adapted than others to either existing or new environmental conditions. The progeny of these individuals possessing favorable variations will, to the extent that they inherit the genes responsible for these variations, tend to survive and reproduce in greater numbers than will the progeny of the less well adapted. Thus, as generation follows generation, the better adapted individuals will be increasingly represented. This is differential reproduction—reproduction that tends to increase the frequency of the better adapted progeny as compared with the less well adapted. Differential reproduction is the result of natural selection, which can be defined as the effect of the environment upon the survival of heritable characteristics, thus bringing about changes in a population between one generation and the next.

The terms "struggle for existence" and "survival of the fittest," commonly associated with natural selection, are given meanings never intended by Darwin when they are used to suggest that struggle involves actual combat between animals. This may occur with some animals, but for the most part the "struggle" is to obtain, within the total environment, food and territory. "Survival," also, refers more to the survival of the

genes present in a plant or animal, through its reproduction, than it does to the survival of the individual.

In summary, the environment tends to select those forms possessing beneficial variations and to eliminate those with deleterious variations. This is the process that Darwin termed NATURAL SELECTION. Natural selection was so designated in contrast to ARTIFICIAL SELECTION, as exercised by man in the improvement of domestic plants and animals. Artificial selection of plants and animals under domestication offers evidence that supports the theory of natural selection.

THE MODERN VIEW

During the first quarter of this century, Darwinism tended to be forgotten in the enthusiasm for the new science of genetics. The findings of genetics, however, have merely clarified many aspects of natural selection, and have not contradicted the basic deductions that Darwin made. In particular, explanations have been found for the origin of the variations acted upon by natural selection.

We saw in Chapter 16 that inheritance may be changed by the segregation and recombinations of genes in sexual reproduction. Improvement of domestic plants and animals has largely been due to such processes, and natural selection also tends to perpetuate the best gene combinations within any interbreeding population. A vast number of changes are possible with recombination alone. But the number of such changes would be limited where it not for the fact that new units of inheritance are continually being added to the population. These new units arise by a process called MUTATION. Because environments do not remain uniform, genetic change through mutation and recombination is necessary if species are to survive.

Mutation A mutation is usually discovered by the appearance in the progeny of an heritable character not present in the immediate ancestors. Such modifications are the result of alterations in the hereditary materials of the cell. Mutations may be classed as (1) *gene* mutations and (2) *chromosome* mutations. A gene mutation has been considered to be due to a change in the arrangement of the nitrogen bases within the segment of a DNA molecule that constitutes a gene. When this gene is duplicated, the change is preserved, resulting in a new gene. Chromosome mutations may be due to new arrangements of the old genes by the INVERSION of short chromosome segments within a single chromosome, or the TRANSLOCATION of a segment of one chromosome to another, nonhomologous chromosome. Translocation should not be confused with crossing over, where segments are exchanged between the homologous chromosomes of a pair. Other changes involve DELETION (loss) or DUPLICATION of chromosome segments. Although such changes introduce no new genetic material, they may cause phenotypic changes, and they frequently result in intersterility between the mutant plant and its parents. If such mutants can survive, they are already well on the way to becoming new species. Still another type of chromosome mutation, an increase in the number of chromosome sets, is discussed later in this chapter.

It is often difficult or impossible to distinguish between gene and chromosome mutations, and both are important in evolutionary processes. Nevertheless, it is generally held that gene mutations, in part because they give rise to new alleles, are the main bases for long-range evolution, and are responsible for the majority of structural and physiological differences that are involved in evolutionary change.

The ultimate causes of gene mutations are unknown, but repeated experiments have shown that the frequency with which they occur may be increased by x-rays, ultraviolet rays, radiations from radioactive substances, or by certain chemical treatments or high temperatures. Although all genes presumably have the potentiality of mutation, they tend to be remarkably stable, and the probability that any one specific gene will mutate is only on the average about one in a million. Even so, since even "simple" organisms such as *Drosophila* may contain from 5000 to 10,000 genes in each cell, there is still a high degree of probability that a gene mutation of some kind will occur every few generations in the life of most organisms.

Mutations appear to occur at random, and their effects cannot be predicted. Some genes mutate more frequently than others, and mutations are known to occur at a more rapid rate in some plants and animals than in others. Mutations may arise at any stage in the development of the organism and may take place either in body cells or in reproductive cells. Both dominant and recessive mutations are known to occur, but changes to the recessive state may not be immediately apparent in diploid organisms.

Rarely, mutations cause a large change in the phenotype; most produce only small changes. These changes are often so small that they can be detected only with difficulty. They may be neutral, beneficial, or deleterious. Harmful mutations may cause the organism to be less well adapted. There are many mutations that bring about the death of the cell or organism. A cell that lacks an essential respiratory enzyme dies quickly; a mutant plant that lacks the ability to produce chlorophyll may grow for a short time, but cannot survive the exhaustion of its food supply (FIG. 17-2). Most mutations that have been investigated have been found to be detrimental. The reason may be understood if one compares an organism to a complicated machine that is running well; if a minor adjustment is made at random, the machine is almost certain to function less efficiently. Nevertheless, beneficial mutations occur in sufficient number to permit evolutionary progress through natural selection. They can bring about changes with sufficient rapidity to account for the progress of organic evolution as recorded in the strata of the earth's surface. The most remarkable aspect of natural selection is that it is able to sift out the small number of favorable mutants and retain these

Fig. 17-2 (Left): a normal corn seedling and one bearing lethal genes that result in chlorophyll deficiency soon after germination. (Right): the same plants three weeks later. The seed reserves of the nonchlorophyll plant are exhausted and the plant is dying.

for future generations, despite the fact that the overwhelming majority of mutations are harmful.

The effect of gene mutations upon the phenotype is often complex. Many genes cause changes in several characters, rather than a single one. For example, a mutant recessive gene in lima beans causes not only male sterility (failure of normal pollen to develop) but also the development of abnormal seedlings. A single dominant mutant gene in field corn affects the plant in such a way that, in addition to the husk around the ear, smaller husklike bracts are distributed over the entire ear, each one covering from one to four grains (FIG. 17-3). This same mutant gene produces several other effects; the plant branches freely at the base (tillering), its leaf blades are abnormally long and slender, the tassel is unbranched, the outer husks enclosing the individual ears are greatly increased in number, and the stem nodes above the ear are

more numerous than in the normal type. A third example can be cited from research on the mold *Neurospora*, now used extensively in experimental genetics. A single gene mutation gave rise to a mutant unable to synthesize a growth substance (adenine) that the wild type can manufacture. Moreover, this mutant is purple in color, instead of pink, the usual color for *Neurospora* (Plate 7). Many other examples could be cited from both plants and animals.

When a mutant gene is known to produce several effects, it is possible that each of these may be caused by the direct action of the gene. Yet no such case has been demonstrated. It is believed, rather—and proved in some instances—that the mutant gene has but a single primary effect upon some metabolic step. This change, in turn, brings about a number of other reactions that may be more or less remote consequences of the original action.

Conversely, a single character is frequently affected by genes at many loci. For example, more than 30 genes are known to affect eye color in the tiny fruit fly *Drosophila*. Finally, the effect of combinations of genes upon the phenotype is often very different from what would be expected from a knowledge of their individual effects. Thus the characteristic expression of genes can be altered by the presence of other genes, a phenomenon known as GENE INTERACTION. For instance, a certain gene *A*, may be neutral in combination with another gene *B*, deleterious with a gene *C*, but useful with a gene *D*.

Somatic Mutations Mutations, either genic or chromosomal, may occur in somatic (body) cells as well as in sexually reproductive cells. Among the mutations that arise in this way are *bud mutations* or *bud sports*. A single branch may bear leaves, flowers, or fruits different from those on other parts of the plant. Such mutations are found in many kinds of flowering plants, both woody and herbaceous. Some of our most valuable fruits, such as varieties of apples, and seedless oranges and grapes,

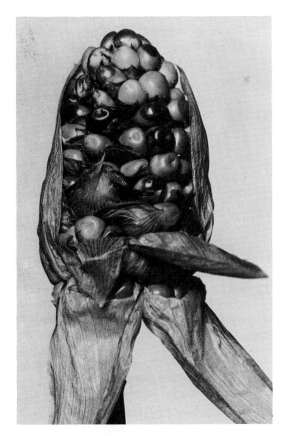

Fig. 17-3 Corn mutant called "teopod." Enlarged bracts on the ear cover 1–4 grains.

Fig. 17-4 Bud mutation in the Emperor grape. (Left): a normal seeded grape. (Right): a seedless mutation.

have originated in this way (FIG. 17-4). Bud mutations have also been found in ornamental plants such as chrysanthemums, ferns, roses, and dahlias. Sometimes these mutations can be reproduced by seed, but it is often necessary to propagate them vegetatively. The familiar African violet (*Saintpaulia*) is noted for the number of somatic mutations that have arisen in leaf cuttings. Such mutant forms differ in size, form, and color of leaves and flowers, and are maintained by vegetative propagation. Of interest is the finding that the shoot of each new plantlet arises from a single epidermal cell of the petiole or base of the blade. Exposure, before planting, of the base of the petiole to a certain dosage of x-rays brought about a 21 percent increase in the rate of mutations.

The Evolutionary Significance of Sex Reproduction by a sexual process occurs in nearly all (perhaps all) kinds of plants and animals, and is one of the major features common to both great groups of living things. Asexual reproduction, however, also occurs. It is absent in mammals and very rare in other vertebrates, but it is common in the lower animals. Among plants, vegetative reproduction is found in species that also reproduce sexually. It is common even in the flowering plants, the most highly evolved in the plant kingdom, and many species are known in which sexual reproduction is subordinate to vegetative propagation. The water hyacinth of the southern states and California is an example of such a condition. This is an obnoxious, yet attractive, weed that may form such great masses of floating plants as to clog streams (Plate 3). The plant is capable of forming seeds, but it reproduces chiefly by means of rhizomes. In more primitive plants such as the algae and fungi, asexual reproduction is an efficient means of increasing in number. It may be more effective than sexual reproduction. What, then, is the significance of sexual reproduction? What is its evolutionary value?

The importance of sexual reproduction in evolution becomes apparent when we consider the significance of variation in relation to natural selection. Sexual reproduction involves not only the fusion of two kinds of gametes but also the entire mechanism of independent assortment of homologous chromosomes and segregation and recombination of genes, all of which comprise the essentials of Mendelian heredity in its modern meaning. Sexual reproduction makes possible a variety of combinations of genes from different individuals and thereby increases the number of variations upon which selection may operate. The greater number of combinations that result from sexual reproduction directly increases the probability that some will be favorable. In organisms that reproduce only asexually, such a variety of combinations can be brought about only by the occurrence of a large number of mutations over a long period of time. Such organisms, therefore, are handicapped in competition with those that reproduce sexually.

A simple example will illustrate the selective advantage of sexual over asexual reproduction. Let us assume two different but related organisms, reproducing only asexually (FIG. 17-5). In one of these a mutation *A* occurs, and in the other a different mutation *B* takes place. Let us further assume that the possession of either of these particular genes alone is of no great value to the plant but that both, acting together, will cause the organism to be better adapted. In a species that reproduces only by asexual methods, a combination of the genes *A* and *B* could result only if the new mutation *B* arose in one of the *A* individuals or if an *A* mutation arose in one of the *B* individuals. Either of these events is highly improbable, since mutations are of relatively rare occurrence. If, on the other hand, these organisms could reproduce sexually, individuals pos-

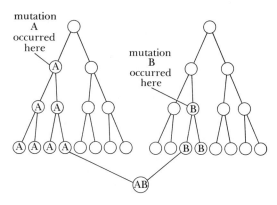

Fig. 17-5 Diagram illustrating how gene mutations, occurring in different organisms that normally reproduce asexually, could be combined in a single individual if sexual reproduction occurred.

sessing mutation *A* could hybridize with those containing *B*, thus bringing the two genes into one organism.

Sexual reproduction thus brings together independently occurring mutations into diverse combinations. In the absence of sexual reproduction, such combinations would be very much less likely to occur. The number of combinations of mutations made possible by sexual reproduction is striking. If 100 mutations occur in an asexual population, 101 types of individuals could result, the original type and the 100 mutants. On the other hand, 100 mutations in a sexually reproducing population could result in 100 pairs of genes, each pair consisting of a normal and a mutant gene. These could be combined sexually to form 3^{100} different kinds of genotypes. This is an astronomically large number, amounting to the figure 51,534,762 followed by 40 zeros. Only a fraction of these potential combinations might be realized, but the example illustrates the tremendous increase in variability and possible gene combinations that sexual reproduction may produce. With such an increase in the amount of variability, there must be a proportionate increase in the number of favorable variations, thus greatly increasing the efficiency of the evolutionary process.

This process of recombination is valuable not only in relation to new mutations but also in producing new combinations of genes already present in the population. This source of variability was unsuspected by Darwin and has come to light only as a result of genetic studies in the twentieth century. Such gene recombinations are analogous to dealing hands from a shuffled deck of cards. Nothing new is added to the deck in the process of shuffling, but new combinations appear when the hands are dealt. Just as some of these combinations are advantageous in a card game, so are some new combinations of old genes advantageous to the plant when subjected to natural selection.

Cross- and Self-Pollination Sexual reproduction in itself does not necessarily produce the genotypic variability upon which selection may act to bring about evolutionary change. Such variability, as pointed out previously, arises chiefly from the segregation and recombination of genes that have long existed or recently arisen by mutation, and it will occur in a population of a species that is normally or frequently cross-pollinated. The very large number of structural features and other factors that prevent self-pollination are thus useful adaptations because cross-pollination leads to heterozygosity and greater variation upon which selection can operate.

In self-pollinated plants, on the other hand, almost complete homozygosity may exist. As discussed in connection with self-pollination in corn (FIG. 16-9), self-pollination in a heterozygous population reduces the number of heterozygotes by half, and, if continued, will halve the remaining number in each successive generation. Deleterious genes in the homozygous state are eliminated more or less rapidly by natural selection, and the individuals of a population become highly uniform. Sexual reproduction by means of strict self-pollination is thus essentially the equivalent of vegetative reproduction as far as variation is concerned. Under such conditions, the only changes that would occur would be those resulting from new mutations.

Cross-pollination is probably the ancestral condition from which self-pollinated species have evolved independently in many different evolutionary lines. Self-pollination is not harmful to normally self-pollinated species; in many such plants, in fact, it may have survival value, since pollination is such plants is more likely to occur than in those dependent upon cross-pollination. It is especially useful in annuals. Strictly self-pollinated species can flourish for long periods of time, but it may be assumed that, lacking variability, they would die out if the environment should change and they could not spread into new environments, for they would not possess new gene combinations that would allow for adaptation.

Complete self-pollination, however, rarely, if ever, occurs. In self-pollinated cultivated plants, such as barley, oats, wheat, flax, soybeans, kidney beans, and tobacco, and in many wild plants, there is occasional cross-pollination; 1–3 percent may be considered typical. In the wild, permanent homozygosity through self-pollination would be a handicap to survival of the species, and it is probable that such forms have been eliminated. In each generation, or perhaps only at intervals, a few seeds will result from crossing between plants with different genotypes. Following segregation, new genotypes will be produced that may be better adapted than the parents to a changing environment or that may be able to occupy new environments.

EVOLUTION AND ADAPTATION

As a result of evolutionary processes, new forms of life became adapted to their environment. The term adaptation includes the acquisition of all characters—structural, physiological, and biochemical—that enable organisms to survive and reproduce in a particular environment.

Adaptation is sometimes visualized in terms of single individuals, but it should be viewed as the result of changes that occur in a POPULATION (Plate 8). A population may be defined as a relatively isolated group of sexually reproducing and interbreeding individuals of the same species, possessing a common "gene pool"—the total number of genes in the zygotes that developed into the individuals of the population. A population has continuity in space and time, whereas an individual lasts only for a single generation and contains only a small part of the gene pool of the entire population. The individual is therefore merely a temporary carrier of genes, which through the process of sexual reproduction may become available to other individuals of the population. A population may vary in size, from a few individuals to thousands or millions; it may constitute an entire community, or may be intermixed with other species.

Adaptation results from three fundamental processes: (1) mutation, (2) subsequent recombination, and (3) the selective action of the environment on the resulting hereditary variations. Although single genes play a role in adaptation, this role is accomplished in combination with other genes of the genotype. Thus, the value of a mutation within the population depends upon the genes with which it eventually becomes associated in the genotype. Selection acts upon the phenotype, which in turn is determined by the genotype, and the nature of the genotype is governed by the kinds of genes brought together by recombination of the various genes in the gene pool of the population. Certain of the resulting phenotypes survive and, following reproduction, contribute more of their genes to subsequent generations that do those of their competitors that are less well adapted.

We may now visualize a population long established in a particular environment. Although variable and with much heterozygosity, with no two genotypes exactly the same, the range of phenotypes is still confined by the limits of adaptive fitness. New mutations will continue to arise, additional gene combinations giving rise to new phenotypes will be formed, but these are not likely to be useful in that particular environment and will be eliminated by selection. It is otherwise, however, if the population is subjected to a slowly changing environment. Through successive generations the population now becomes adapted to the new environment. Mutation continues, and new and old mutations are incorporated into gene combinations that produce phenotypes favorable to survival and reproduction under the new conditions. But how do these favorable gene combinations arise? Most mutations, as we have seen, are deleterious, and most gene combinations are nonadaptive. Adaptation involves changes in a large number of characters, and since mutations and recombinations are a matter of chance, it is improbable that the proper favorable gene combinations can arise by chance alone. Even a continuous process of selection that eliminates the gene combinations present in the unfit—those that leave fewer progeny than the fit—will not produce the improbable combination of characters that mark the adapted organism. Elimination of unfavorable genes and gene combinations does not ensure the presence of favorable gene combinations.

But selection does more than eliminate the unfit; it is visualized as a positive factor in the evolution of adaptation as well as a negative one. This is because selection acts also to preserve genes and gene combinations that improve adaptation. The percentages, or relative frequencies, of such adaptive genes thus tend to increase in the population as one generation follows another. If such genes increase in relative frequency, there will also be an increased tendency for combinations of such genes to increase in the individuals of that population. Selection thus acts to promote the proportionate increase of individuals containing mutations and combinations of mutations that improve adaptability. The best of these combinations are then available for further recombination and selection. The creative role of natural selection is well expressed by R.A. Fisher, a prominent British geneticist, in his often quoted statement, "Natural selection is a mechanism for generating an exceedingly high degree of improbability."

Evolution, then, is not a random process, and adaptation, which is the achievement of evolution, is not accidental. It is now clear that mutations do not occur *in order to* produce a beneficial effect upon the organism. They do not occur in response to a present or anticipated need nor in response to the demands of a particular environment. Organisms do not purposefully adapt themselves to the environment; they become adapted if their favorable genes and gene combinations are preserved in the population by natural selection to the extent that they become more reproductively effective in the existing environment.

EVOLUTION AND THE ORIGIN OF SPECIES

The manner of origin of the numerous kinds (species) of living things that inhabit the earth has been a puzzle to thinking men for centuries. Many people recognize such groups of plants as maples, oaks, gentians, sumacs, blueberries, goldenrods, irises, mustards, milkweeds, willows, and poplars. But these groups, or genera, in turn comprise different kinds of milkweeds, mustards, and oaks, and these different kinds are examples of species. It is with the origin of these species that geneticists, cytologists, taxonomists, and other students of evolutionary problems have been concerned since the time of Darwin, and more especially during recent decades, in which the findings of Darwin have been modified and interpreted in the light of Mendelian principles of segregation and recombination.

Many difficulties are involved in the study of species and of their origin. Some species are sharply defined, but others grade into related species through intermediate forms. Some of these intermediate forms may represent incipient species, for species are arising today, as in the past. Most evolutionary changes take place so slowly, however, that they do not become evident within any easily recorded period of time, and so

the problem is attacked indirectly. By growing plants in experimental gardens, by studying their mutations and their chromosomes, by collecting them and studying their distribution, it is frequently possible to make reasonable inferences concerning their evolutionary history. Although there is no general agreement on the definition of a species, it may be defined ideally as a group of individuals that are morphologically distinguishable from related kinds, and that will not cross, or that cross with difficulty, with related species. All existing species have come from pre-existing species, but so complex are environmental factors and living organisms that species are believed to have arisen in different ways; there is no one solution to the problem of their origin.

The chief mechanisms involved in the formation of new plant species may, according to modern views, be grouped under two main heads: (1) reproductive isolation and (2) species hydridization.

Reproductive Isolation This concept of the origin of new species postulates that an ancestral, cross-pollinated, and widely distributed species has become broken up into two or more smaller groups. If these groups breed within themselves but are isolated from each other by reproductive barriers that restrict or prevent hybridization between them, they may become different species.

Reproductive isolation is commonly preceded by spatial isolation—that is, the physical separation of populations due to distance or to different habitats within the same area. Spatial isolation will prevent or reduce cross-pollination, with its continual exchange of genes, and thus make possible the development of populations genetically different from one another. These differences may manifest themselves externally and internally to the extent that such populations are called races. However, spatial isolation alone cannot produce new species. The differences that characterize species and separate one from another are the result of adaptation to different environments. These differences are maintained only in the absence of hybridization between the species. If reproductive mechanisms that restrict or prevent hybridization arise, these differences can be maintained even in the absence of spatial isolation. Such reproductive isolating mechanisms can be external or internal. Among external mechanisms are different seasons of blooming and structural modifications of the flower that prevent cross-pollination between populations. Internal factors include failure of foreign pollen to germinate upon the stigma or retardation of growth of foreign pollen tubes in the style, and failure of hybrid zygotes to grow into normal embryos. The accumulation of mutations that prevent or restrict interbreeding is normally a gradual process that extends over long periods of time.

Hybridization The term hybrid is applied in various ways. Geneticists define a hydrid as the result of a cross between any two individuals differing in one or more gene pairs. In this sense, cross-pollinated plants within a species are hybrids. From the standpoint of the origin of species, however, we may define a hybrid as a cross between individuals belonging to different species or to previously isolated races. Despite isolating mechanisms, hybrids do occur in nature, and the formation of such hybrids has been shown to be more common than was previously supposed. Numerous hybrids are encountered in ferns, hybrid conifers are well known, and an extensive list of hybrids in flowering plants has been compiled. Species hybridization is believed to be important in the production of new kinds of plant life, although the role that it plays is probably less important than that of reproductive isolation.

Although individuals that are not closely related may be sufficiently compatible to produce a hybrid, such hybrids are commonly sterile or only partly fertile, with much reduced seed yield or only occasional seed production. Such sterility is frequently due to differences between parental genes and chromosomes, which make normal meiosis in the hybrid difficult or impossible or which prevent the gametes from functioning normally. The occasional seed produced by such hybrids, however, may be significant from the standpoint of the origin of new kinds. Moreover, the F_1 hybrids may backcross with one or the other of the parental species, and the progeny of such a cross is more likely to be fertile than is the progeny of a cross between two F_1 hybrids. Such crosses will result, following segregation and recombination in later generation, in a great variety of gene combinations that will be acted upon by natural selection. Most of these combinations will be eliminated, but some may possess superior adaptations, or may be particularly well adapted to habitats formerly unavailable. If they also become reproductively isolated, they may be recognized as new species.

Polyploidy In contrast to the slow evolution of new species from hybrids as discussed above, new species may arise in a single step following interspecific hybridization if the resulting cross is POLYPLOID. Polyploidy is the condition in which the nuclei of the cells of the sporophyte generation contain more than two chromosome sets. In polyploid plants there may be three, four, or more sets of chromosomes, and such plants are called triploids ($3n$), tetraploids ($4n$), pentaploids ($5n$), hexaploids ($6n$), and so on, all the chromosome numbers usually constituting multiples of some basic haploid number. Soon after polyploids were discovered it was inferred that such forms must originally have been derived from diploids, and this inference has been confirmed repeatedly. The origin of new forms by polyploidy has been termed cataclysmic evolution.

Tetraploids are the most common type of polyploids, and these arise through a doubling of the chromosome number during some phase of the life cycle. This is known to occur spontaneously, but it may be brought about by artificial treatment. Probably the most common cause of doubling is the failure of normal meiosis in the spore mother cells of an F_1 hybrid, resulting in the formation of gametes with the diploid chromosome number. The chance fusion of two such gametes produces a tetraploid plant. A triploid usually arises by the fusion of a diploid and a haploid gamete. For example, whereas most commercial apple varieties are diploid, with 34 chromosomes, others, such as Baldwin and Gravenstein, have 51 chromosomes in their body cells. These are triploids, having probably arisen by the fusion of an unreduced or diploid egg carrying 34 chromosomes with a sperm bearing the usual set of 17 chromosomes. Doubling may also occur in the zygote or body cells, such as those of the proembryo or vegetative bud. In such cases the chromosomes become duplicated at mitosis, but a new cell wall does not form and the new nucleus contains the $4n$ chromosome number. Doubling of the chromosome number in the cells of a vegetative bud may result in a tetraploid branch, the rest of the plant remaining diploid.

The polyploid condition is widely distributed in the plant kingdom. It is found in some algae and a number of mosses, but is rare in the fungi. It occurs in species scattered throughout the vascular plants, in groups such as the horsetails, ferns, conifers, and flowering plants. One authority has estimated that, among the angiosperms, 30 to 35 percent of the species are polyploids, while other investigators believe the proportion to be even higher. Studies have shown that polyploidy is relatively infrequent in woody angiosperms, somewhat more common in annual herbs, and most frequent in herbaceous perennials. It is rare among animals. It has been found in the brine shrimp, certain salamanders, a sowbug, a few insects, and a fish, but in the animal kingdom it is probably of little evolutionary significance.

Polyploids are commonly classified into two groups: (1) AUTOPOLYPLOIDS, which arise by the multiplication of the chromosomes within a single species. They may be produced by individuals following self-pollination, or by cross-pollination between different individuals of a species. Autopolyploids are usually sterile,

owing to chromosome duplications. (2) ALLOPOLYPLOIDS, on the other hand, arise following hydridization between species. If the chromosome number of the hybrids becomes doubled, they commonly become fertile and relatively true-breeding. These are clearly defined new species, for they possess structural and physiological characteristics different from related forms. Further, they cannot be crossed with either of their original diploid parents or, if rarely they do cross, they form hybrids that are largely or entirely sterile. The most successful natural polyploids are allopolyploids.

A classical example of a natural allopolyploid is the cord or marsh grass (*Spartina townsendii*). This plant is a tall, coarse perennial with strong, creeping rhizomes. The species was first noticed in southern England about 1870. It spread rapidly and by 1906 covered thousands of acres along the south coast of England. It also crossed the English Channel and now grows extensively along the coast of France. There is sound evidence that this species arose suddenly by doubling of the chromosome number following hybridization between *Spartina maritima* ($2n = 60$), a European species of cord grass, and *Spartina alternifolia* ($2n = 62$), accidentally introduced into England from North America. The behavior of the chromosomes may be somewhat irregular, and a sterile F_1 hybrid has been found with $2n = 62$, the result of the fusion of $n = 30$ and $n = 32$ gametes. This $2n$ number, if doubled at meiosis, would give a $2n$ number of 124. The actual number of chromosomes in the body cells of *Spartina townsendii* has been found to be 120, 122, or 124. This is one of the best examples of an allopolyploid known to have arisen spontaneously in historic times.

Another example of a natural allotetraploid is found in *Iris versicolor*, the native blue flag of the northeastern United States and Canada, which extends westward into Minnesota and Wisconsin. Studies show that this species probably originated as a fertile, true-breeding hybrid between the southern blue flag (*Iris virginica*) and the arctic blue flag (*Iris setosa*). *Iris virginica* ($2n = 72$) is found from Virginia southward along the Atlantic coast, and also in the Great Lakes region south to Texas and Alabama. *Iris setosa* ($2n = 36$) and its varieties are primarily arctic and subarctic in distribution. *Iris versicolor* has the chromosome number to be expected if it were a tetraploid hybrid of these two species: the $2n$ number is 108. This species is believed to have originated during glacial times in the general region of the Great Lakes, when the two parent species grew in the same area. The hypothesis of speciation through polyploidy cannot be fully documented in this case. It is of interest, however, because it suggests that polyploidy has long been an important factor in speciation.

Evidence supporting the allopolyploid nature of a number of species has been provided by artificial synthesis from their presumed diploid ancestors. One of the earliest of these reconstructions was carried out by the Swedish geneticist Müntzing in the years 1924–1932. He worked with European species of plants known as hemp nettles, which are annual species of the mint family, belonging to the genus *Galeopsis*. From two of these species, each having a $2n$ chromosome number of 16, Müntzing produced, after several hybridizations, a third species. This artificial species had a chromosome number of 32, and was therefore a tetraploid. Moreover, it was in all essential particulars structurally identical with a naturally occurring member of the same genus, *Galeopsis tetrahit,* which grows in Europe and has been introduced into North America, where it is a widely distributed weed (FIG. 17-6). Müntzing demonstrated, therefore, as others have also, that species may arise by hybridization and chromosome doubling. A number of investigators have also crossed species in experimental gardens and produced new, fertile allopolyploids that have no counterpart in nature.

Numerous additional examples of polyploids could be cited. Many of our most important cultivated plants are of polyploid origin. Among these are banana, coffee, sweet potato, sugar cane, certain varieties of cotton and tobacco, the cultivated peanuts, the plum, sour cherries, some

Fig. 17-6 The hemp-nettle (*Galeopsis tetrahit*). This naturally occurring tetraploid has been artificially synthesized by hybridizations involving two other species.

kinds of apples and pears, the loganberry, strawberry, cultivated blackberries, all dahlias (both wild and cultivated), many varieties of roses, irises, lilies, tulips, and hyacinths, and a number of cereal and forage crops, such as the bread wheats and alfalfa. Autopolyploidy, for the most part, results only in variation within the species, but allopolyploidy may result in the immediate production of new and distinct species, and is therefore an important evolutionary process. Allopolyploids exhibit not only a combination of various characters of the parents, but frequently also favorable new combinations. Following natural selection, new forms emerge that are better able to compete or to flourish in environments different from those of their ancestors; in many cases they have a wider geographical distribution and can survive under greater extremes of temperature and water supply.

Induced polyploidy Induced or artificial polyploidy has now become an important tool, both in evolutionary studies and in plant breeding. Earlier methods of inducing chromosome doubling included use of chemical agents such as chloral hydrate, and also rapid and extreme changes in temperature. Then, in 1937, it was discovered that colchicine, an alkaloid, was more uniformly effective than any other agent. Colchicine is obtained from the corm and seeds of the autumn crocus (*Colchicum autumnale*), a member of the lily family. In high concentrations this drug is extremely toxic, but at low concentrations and with proper length of exposure, it has been shown to cause chromosome doubling in a number of plants. This occurs during mitosis. Prophase appears normal, but the spindle fibers of metaphase do not develop. The chromosomes therefore do not separate to opposite poles and a new cell wall does not form. The divided chromosomes return to the interphase within a nuclear membrane that now encloses twice the original number of chromosomes. Autopolyploids may be produced by direct chromosome doubling, and allopolyploids may be obtained by the same process following hybridization. Colchicine may be used to treat the seeds, seedlings, or the shoot apex of an older plant. It may be placed on the bud in the axil of a leaf to produce a tetraploid branch.

Autopolploidy may increase the size of the various organs, produce larger flowers (FIG. 17-7) or fruits, increase hardiness, change season of maturity, or in other ways modify developmental characteristics. It frequently increases the size of the individual cells, especially the guard cells and pollen grains, and the size of these in contrast to those of related diploids is often used as an indication of autopolyploidy. Allotetraploids, also, are often larger and more vigorous than the diploid hybrids from which they arose, but this is by no means always true. The most important advantage of artificial tetraploids is that by doubling the chromosome number and thus restoring fertility, it makes available for breeding purposes those hybrids that would otherwise be sterile. New polyploids are usually of no great economic value; they must be tested and selected just as ordinary hybrids. Nevertheless, it is confidently believed that induced polyploidy is a tool that will eventually give rise to many new plants of great value to man.

The preceding discussion does not pretend to be complete or to tell the entire story of the complexities involved in the formation of new species. It does, however, outline the most plausible explanations of the mechanism of evolution

Fig. 17-7 **Polyploidy in snapdragon (*Antirrhinum majus*). The flowers at the left were borne on a diploid plant. The larger flowers at the right were produced on a tetraploid plant obtained by treatment with colchicine.**

of the innumerable forms of plant life that have enriched the flora of our earth.

ORIGIN AND EVOLUTION OF CULTIVATED PLANTS

The origin of man's domesticated plants is of great interest, not only to botanists, but also to archeologists, anthropologists, and others concerned with the history of man and of human civilization. Because most of our basic food plants originated in prehistoric times, the clarification of their possible origin is an extremely difficult task. Nevertheless, while the origin of many such plants remains obscure, considerable light has been shed upon this subject in recent times by the efforts of plant breeders, cytologists, plant explorers, and other scientists. Information on the origin of domesticated plants is of more than academic importance—it may be, and often is, of practical importance in plant breeding. For example, the genes and gene complexes existing in the ancestral forms and nearest relatives of a cultivated plant sometimes may be utilized in hybridization for plant improvement.

Cultivated plants, as did wild plants, undoubtedly arose by the action of evolutionary mechanisms such as mutation, recombination, hybridization followed by polyploidy, and selection. Evolution under domestication, however, must have proceeded much more rapidly than it did under natural conditions. Variations that would have little chance to survive if acted upon by natural selection are preserved under artificial selection because they are useful or valuable to man. They are protected by cultivation, under which competition with other species is largely eliminated. The effect of artificial selection is shown by the fact that many domesticated crop plants, such as corn and wheat, can survive only when planted and harvested by man; they are unknown in nature. Under cultivation, also, favorable mutations have been observed and propagated. Hybridization, a significant evolutionary mechanism in wild plants, has probably been even more impor-

tant in the evolution of domesticated plants. Primitive man in his migrations accidentally brought together plants that previously had not hybridized because of geographical barriers. As plant breeding became systematic, new varieties were produced by combining, through hybridization, genes from plants obtained from widely separated parts of the world. Hybridization between species, accompanied by an increase in chromosome number, has undoubtedly been responsible, either in nature or under domestication, for the origin of many new species useful to man. The cultivated wheats, cottons, and tobacco are examples of such hybrid forms. For example, the origin of the bread wheats, which are hexaploids with the $2n$ number of 42, has been worked out with reasonable certainty. The story is a complex one, but it appears that such wheats are the final result of a series of natural hybridizations, involving a primitive wild wheat of the Near East and two weedy grasses still found in that general region. These chance hybridizations, all of which apparently occurred within the past 10,000 years, are responsible for the millions of acres of wheat that annually help to feed mankind. Some information is available on the origin of other economic plants, such as banana, cabbage, cotton, sunflower, beet, plum, strawberry, and maize. Corn, or maize, is native to the Americas, and archeological discoveries in Mexico have made possible a reconstruction of the reproductive parts of an ancestral type of wild corn that existed some 7000 years ago (Fig. 17-8). These parts consisted of a slender, spikelike axis bearing pistillate flowers below and staminate flowers above. The lower part, the cob, about an inch in length, bore an average of 55 kernels, each partly enclosed by floral bracts (glumes). The cob in turn was surrounded by two husks, which opened at maturity, allowing seed dispersal. This bisexual structure was probably borne terminally on the parent plant. Under domestication, the sexes became separated—the staminate flowers, in a tassel, remaining terminal, and the cob becoming lateral and borne high on the stalk. The husks increased in number, but did not open at maturity, the grains became adherent to the cob, and seed dispersal was no longer possible. Mutations at many loci, and hybridizations among various races and with wild relatives, led to the desirable changes that have been selected by man over more than 5000 years.

EVOLUTION AND RETENTION OF ANCESTRAL CHARACTERS

The conditions under which we assume that life arose no longer exist. Organisms, one-celled or multicellular, now arise only from pre-existing organisms. Since life began, the continuity between individuals has been maintained by genes and gene aggregations, and the changes that arise result from the modifications of these genes and gene combinations.

Evolution has by no means proceeded at the same rate in all groups of plants. Some groups have changed rapidly, whereas others have apparently remained essentially unchanged over long periods of time. The latter situation is especially true of organisms existing in stable environments such as the waters of the oceans. The study of those species that appear to have changed but little is of great value in appraising the changes that have taken place in other species or in larger groups.

As plants evolve, there is an accretion of small, discontinuous changes brought about by mutations. Of the thousands of genes in the chromosomes of the plant, relatively few may be involved in mutations that have produced new and perhaps better-adapted species. Mutations occurring at many loci are eliminated so that the genes then remain in effect the same throughout repeated generations; even in the most variable species there persist characters that resemble those of ancestral types. Many genes of the past are carried over into the present, and the racial history of an individual is as much a part of the organism as is the development from zygote to

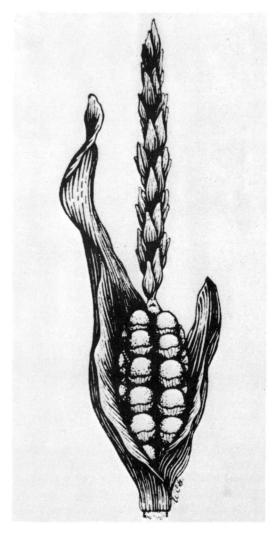

Fig. 17-8 Artist's reconstruction of wild corn, based upon archeological remains (twice natural size).

adult form. A detailed study of advanced types of plants, as compared with other less specialized forms, shows how deep the roots of evolution go. There is always much of the old in what seems to be the latest and new.

Since ancestral genes are preserved, it is to be expected that they will be responsible for the appearance of ancestral features. In plants, such

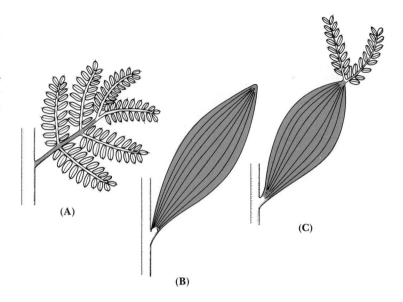

Fig. 17-9 Acacia. (A): bipinnately compound leaf of *Acacia discolor*. (B): phyllode from *Acacia cunninghamii*, the only leaflike organ on mature parts. (C): portion of seedling of *Acacia cunninghamii*, showing phyllode terminated by a portion of a compound leaf.

(A)

(B)

(C)

ancestral structures, sometimes in a vestigial or highly reduced form, may appear at any stage of development, or ontogeny—in the early reproductive stages preceding embryo formation, in the embryo itself, in the adult plant body, or in other stages of the life cycle. These features may be referred to as RETENTIONS of ancestral characters. Such retentions frequently offer clues to the ancestral history, or phylogeny, of the organisms concerned.

Since several examples of the persistence of ancestral traits are considered later at appropriate points, only one is cited here. In the large woody genus *Acacia*, of the legume family, most species have a twice-pinnately compound leaf (FIG. 17-9A). The remaining species bear at maturity what are apparently simple leaves (FIG. 17-9B). These bladelike structures, however, consist of a flattened petiole (called a PHYLLODE), together with a considerable portion of the main axis of the compound leaf. Of considerable interest is the production, by the seedlings and young plants of these species, of normally twice-pinnate leaves, phyllodes, and structures intermediate between these. Among the intermediate

structures are phyllodes bearing a portion of a compound leaf at the apex (FIG. 17-9C). Since phyllodes show evidences of modification, we may assume that the ancestors of the phyllode-bearing acacias bore compound leaves and that the leaflets have disappeared as a result of evolutionary change. The occurrence of compound leaves and intermediate forms (such as FIG. 17-9 C) in young plants of species in which the adult plants bear only phyllodes may therefore be interpreted as a retention of an ancestral feature.

SUMMARY

1. A distinction is made between the concept of evolution, which is old, and the scientific study of the mechanisms of evolution, which is comparatively new.

2. According to Lamarck, characters acquired during the development of an organism are inherited. In plants, such changes were considered to have resulted from the direct effects of the environment. In animals, it was thought that a change in the environment

brought about a change in the needs of the animal, new structures being produced to satisfy these needs. The further development of these structures or their disappearance in future generations was explained on the basis of use and disuse.

3. The main features of Darwinism are: (1) variability, (2) the struggle for existence, resulting from the high rate of reproduction of most organisms, and (3) natural selection, the action of the environment upon inherited variations, which results in the preservation of individuals with favorable variations and the elimination of those with unfavorable variations.

4. Modern views on the significance of natural selection place emphasis not on mere survival of the adapted individual, but on reproduction, which spreads the favorable genes of the adapted form throughout the population over successive generations. The best-adapted organisms not only tend to survive in greater numbers than the less well adapted, but also are able to leave more progeny.

5. A single gene may affect a single character, or it may affect several characters. Conversely, a single character may be affected by number of genes. The effect of a gene may vary greatly in different combinations.

6. Variation among individuals is brought about by mutation and recombination. Variation is necessary if evolution is to occur, but is not itself responsible for evolutionary change.

7. Sexual reproduction in plants that are normally or occasionally cross-pollinated is im-mensely important from an evolutionary standpoint because, through Mendelian mechanisms, it brings about more new combinations than does asexual reproduction or self-pollination.

8. Strictly self-pollinated plants are homozygous for most gene pairs and possess little more variability than those that reproduce only asexually. The formation of favorable gene combinations is difficult in such plants and in those reproducing only by vegetative methods, for such combinations are very unlikely to occur in the same line of descent. In most self-pollinated plants, however, occasional cross-pollination occurs.

9. Adaptation results from the action of natural selection upon hereditary variations. Through this action, favorable genes, and especially favorable gene combinations, are preserved and increased in frequency in a population.

10. New species may arise slowly following isolation or hybridization, or suddenly by hybridization followed by polyploidy. All of these methods involve adaptation to different environments and the operation of natural selection.

11. From an evolutionary standpoint, allopolyploids are considered more important than autopolyploids. Induced polyploidy by the use of colchicine is an important tool for plant improvement.

12. Certain characteristics of ancestors may be retained in the ontogeny of their descendants and may be useful in interpreting evolutionary relationships.

CAROLI LINNÆI

S:æ R:giæ M:tis Sveciæ Archiatri; Medic. & Botan.
Profess. Upsal; Equitis aur. de Stella Polari;
nec non Acad. Imper. Monspel. Berol. Tolos.
Upsal. Stockh. Soc. & Paris. Coresp.

SPECIES
PLANTARUM,

EXHIBENTES

PLANTAS RITE COGNITAS,

AD

GENERA RELATAS,

CUM

DIFFERENTIIS SPECIFICIS,
NOMINIBUS TRIVIALIBUS,
SYNONYMIS SELECTIS,
LOCIS NATALIBUS,
SECUNDUM
SYSTEMA SEXUALE
DIGESTAS.

Tomus I.

Cum Privilegio S. R. M:tis Sueciæ & S. R. M:tis Polonicæ ac Electoris Saxon.

HOLMIÆ,
IMPENSIS LAURENTII SALVII.
1753.

The Names and Kinds of Plants

T HE NAMING AND CLASSIFICA-TION of plants undoubtedly began in the earliest stages of civilization. Our own observations show that plants are of many kinds, and we immediately seek for a name to apply to any plant of interest. The naming of plants, however, was more than an intellectual pastime on the part of early man. It was, rather, a necessity, for he was dependent upon plants for a considerable portion of those materials that enabled him to survive, and it was impossible to communicate with others concerning these necessary objects without having names for them. Many primitive peoples of today, as well as in the past, apply common names to those plants that are conspicuous or that affect their welfare in any way. Plant names first appeared not in books but in the minds of men.

With increasing civilization, especially as knowledge grew concerning the uses of plants in medicine and as food, the necessity for plant names became even greater. And finally, as the numbers of known plants increased and as botanical travelers brought plants to centers of learning from far corners of the earth, it became desirable to group plants into large categories according to rational principles. The gathering, naming, and classification of plants in modern times is carried on primarily with the objective of showing the origins and relationship of plants, and of providing positive identification for the hundreds of thousands of different kinds of plants.

NAMES OF PLANTS

The Scientific Names of Plants The scientific name of every plant consists of two parts. The first part designates the *genus* and the second is the *specific epithet.* Collectively, the two parts constitute the name of the *species.*[1] A scientific name is not complete unless it includes both parts. Generic, or group, names are always nouns, and may have been taken from Latin or Greek. Specific names are commonly descriptive adjectives, but may be nouns. *Populus tremuloides* is the scientific name for the quaking aspen or poplar. *Populus* is the genus or generic name, a name that is probably derived from *arbor populi,* or tree of the people, since the Romans are supposed to have planted poplars in public places. There are many species of *Populus* but only one is *Populus tremuloides.* This name is given to a kind of poplar widely distributed in North America. The name itself refers to the trembling or fluttering of the leaves when the slightest breeze strikes the flattened petioles. All species of *Populus* have characteristics in common, but the species *Populus tremuloides* has characteristics of leaf, bark, habit, bud, and flower that distinguish it from other poplars.

The species is regarded as a group of closely related individuals. Species that are different but related by descent are grouped together in a genus. Thus, all of the different kinds of oaks are classed together in the genus *Quercus,* the ashes into the genus *Fraxinus,* and the peaches, cherries, and related plants into the genus *Prunus.*

VARIETIES are distinguished in a number of species. A variety is a unit of classification below a species and varying from it in certain inherited characters but not sufficiently to be considered a distinct species. The term variety is often misused in popular writing when a species is meant. Generic names are always capitalized, but both specific and varietal epithets are usually written with a small initial letter.

The nomenclature of cultivated plants follows in general the same principles as that of wild plants. The common garden plants that we know as cabbage, kale, Brussels sprouts, kohlrabi, cauliflower, and broccoli have all developed from a wild species, *Brassica oleracea,* and are considered varieties of this species. Plants under cultivation have usually been selected and hybridized. Bud sports and other mutations have arisen, and numerous forms now reproduce only vegetatively and not by seed. The problem of naming cultivated plants is so complex that authorities disagree on many points.

Scientific and Common Names When our present system of naming plants was adopted, in the mid-eighteenth century, Latin was the written language of scholars. Although Latin has diminished in importance, its use in scientific names is retained. Since it is not a spoken language, it does not change, and it is intelligible to scientific workers of all nationalities.

In addition to the scientific name, many plants, especially the more common, useful, or interesting ones, have common, or vernacular, names. Great numbers of plants, however, even among the larger and conspicuous forms, possess no common names at all. More than 25,000 species of orchids have been described, but probably only a few hundred of them have common names. Among the lower and microscopic forms of life, there are thousands of plants that can be referred to only by their scientific names.

An important asset of the scientific name is its relative stability. Once a plant is named, the name remains, or if it is changed, the change is made according to established botanical rules. The scientific name is the same wherever the plant is found. But common names often vary with each locality, country, or other geographic subdivision. They differ, of course, from one language to another. The plant we call corn (*Zea mays*) is known as maize in Britain. Corn in England refers to wheat or other small grains such as barley. Seven species of willow (*Salix*) are commonly called black willow. *Gleditsia tria-*

---

[1] It should be noted that the word "species" is both singular and plural. The use of the term "specie" in connection with a plant name is incorrect.

canthos, often known as honey locust, has 13 other common names. *Thuja occidentalis* is known as arbor vitae or white cedar, depending upon the locality.

Of all these confusions, the use of the same common name for different plants may well be the worst. *Carpinus caroliniana* and *Ostrya virginiana* are small trees native in North America. Both are called hornbeam, and both are called ironwood. Mayflower refers to several distinct and unrelated spring flowers. At least five trees and shrubs, all members of the genus *Prunus,* are called wild cherry. Common names frequently disguise relationships that are obvious from looking at the plant. All maples belong to the genus *Acer,* yet species of maples have such common names as box elder, moosewood, buckwood, mountain alder, black ash, and stinking ash—hardly clues that all are closely related. One soon discovers that Spanish moss, reindeer moss, and haircap moss have very little in common and that only one of the three is actually a moss.

Scientific names are being used more and more. Recent years have seen the introduction of a great number of plants, especially ornamentals, into cultivation. These have come from many countries and for the most part have no common names. Many people, both professional and amateur plant growers, are now using scientific names as a matter of course. The notion that such names are unpronounceable, difficult to remember, and unnecessary is fast disappearing.

CLASSIFICATION OF PLANTS

Categories of Classification To be complete, classification must place a given species in a definite position in the plant kingdom. Related species are grouped into *genera,* related genera are grouped into *families,* and these in turn into *orders.* Orders, in turn, are grouped into *classes,* and these into *phyla,* or *divisions.* Subclassifications of all these categories may be recognized as

subkingdom, subdivision, subfamily, subgenus, subspecies, and variety. With certain exceptions, the family name ends in -aceae; the lilies, for example, are grouped in the family Liliaceae and the pore fungi in the family Polyporaceae. The termination of the name used for the orders is -ales. The families containing the rush, lily, and amaryllis are grouped together with certain other families in the order Liliales. The complete classification of a cultivated, weeping variety of the soft, or silver, maple would read as follows:

Kingdom—Plant
  Subkingdom—Embryophyta
   Division—Tracheophyta
    Subdivision—Spermopsida
     Class—Angiospermae
      Subclass—Dicotyledoneae
       Order—Sapindales
        Family—Aceraceae
         Genus—*Acer*
          Species—*Acer saccharinum*
           Variety—*Acer saccharinum*
             var. *laciniatum*

It is often convenient to refer to any group of organisms, irrespective of size or rank in the foregoing hierarchy, as a *taxon* (pl. *taxa*). Thus, the class Filicineae is a taxon and so is the genus *Pinus.*

The Development of Classification The current system of naming and classifying plants began in ancient Greece. The earliest works reflected the practical need for names of medicinal, poisonous, and food plants, but it was not until the fourth century B.C. that a major step was taken to develop a body of organized botanical knowledge.

Theophrastus of Eresus (370–285 B.C.**)** Many men have contributed to the development of modern methods of naming and classifying plants, and among these Theophrastus was outstanding. He was born in the city of Eresus, on an island in the Aegean Sea. A pupil of Plato, and later a pupil and assistant of Aristotle, he embodied to the full extent the culture and learning of ancient Greece. For most of his life he lived in the midst of a botanical garden established by Aristotle at Athens, and there he taught and wrote books representing many fields of knowledge. We are chiefly interested in his work known as *Historia Plantarum* (*Enquiry into Plants*). This is the oldest known botanical work, and so great were its contributions that Theophrastus was later given by Linnaeus the title "Father of Botany."

The *Historia* is considered to mark the beginning of scientific botany, and its contributions to classification were numerous. Theophrastus classified plants under four groups—trees, shrubs, half-shrubs, and herbs—a grouping still in use. He recognized and described families among flowering plants, such as the carrot family, known today as the Umbelliferae. He perceived relationships among the conifers, the cereals, and the thistles and their relatives, and among the poplars, birches, and alders. He recognized genera, in the sense of a group of species, and applied to them the Greek names then in use. Many of these names have come down to us today as the scientific names of such genera as *Anemone, Asparagus,* and *Cydonia.*

The contributions of Theophrastus were more than taxonomic. He studied seed germination and the growth of the seedling, pointing out how seeds of different kinds germinate. He showed that the root is the first structure to appear from all seeds at the time of germination. He classified leaves and studied the arrangement of leaves on the stem. Although he worked centuries before the invention of lenses and the microscope, he studied the internal structure of the plant. He showed that not all roots are subterranean and that not all subterranean parts of plants are roots—that they may be bulbs, corms, or other modifications of stems.

Theophrastus made many observations on the distribution of plants and their relations to the environment, thus becoming the founder of plant geography and ecology. He observed, for example, that certain plants lived in marshes, others on shallow lake shores, and others on mountainsides. He noted the effect that wind, light, and shade, and the crowding of plants had upon growth. Of Theophrastus it may be said that he laid a firm foundation for a science upon which others have built.

The herbalists From the time of Theophrastus until the sixteenth century there was no real progress in the naming and classification of plants. Then began, however, the work of the *herbalists*—principally German, but also English and Italian—whose activities extended from about 1470–1670. The herbalists (FIG. 18-1) were concerned chiefly with the practical use of plants, primarily from a medicinal standpoint. Their numerous printed works, called HERBALS, contained descriptions of native and foreign plants. Their careful descriptions led inevitably to some kind of classification, since they were dealing with numerous species. In the latter part of this period, therefore, the herbals contain a certain dim foreshadowing of modern methods of classification along the lines of natural relationships. Some authorities consider that modern botany takes its origin from the attempts of the herbalists to discover and describe medicinal plants.

Carl Linnaeus (1707–1778) It is to a Swedish naturalist, Carl Linnaeus (ennobled von Linné), that we owe the modern methods of naming plants. He was the son of a clergyman and amateur horticulturist who lived in a small Swedish village. Following his struggles as a poverty-stricken student at the University of Uppsala, he became a physician and later a professor of botany and medicine at the university, where he did most of his work. He rejuvenated the botanic garden at Uppsala and studied plants sent to him from all parts of the world.

Before the time of Linnaeus it was the general custom to name plants with a single name

Fig. 18-1 Leonhard Fuchs (1501–1566), an outstanding German herbalist and physician. His principal work, *De Historia Stirpium*, appeared in 1542. It contained descriptions of about 500 native and foreign plants. The latter included Indian corn and pumpkin, described for the first time in a European work. *Fuchsia*, a genus of decorative flowering plants, was named after Fuchs.

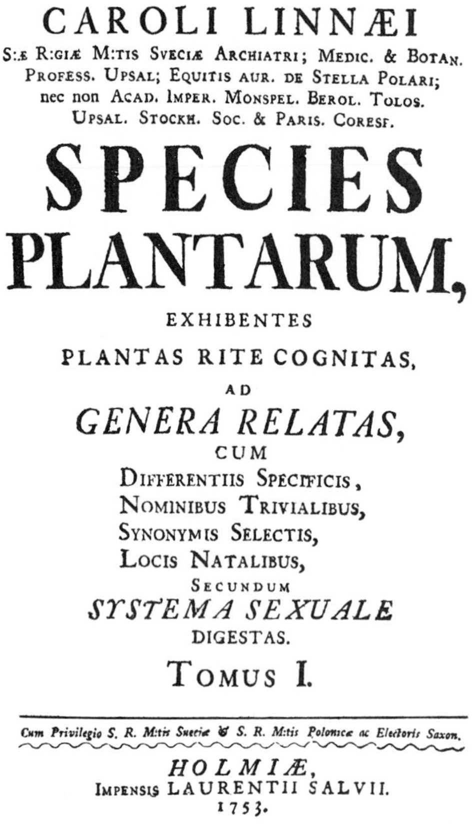

CAROLI LINNÆI

S:Æ R:GIÆ M:TIS SVECIÆ ARCHIATRI; MEDIC. & BOTAN. PROFESS. UPSAL; EQUITIS AUR. DE STELLA POLARI; nec non ACAD. IMPER. MONSPEL. BEROL. TOLOS. UPSAL. STOCKH. SOC. & PARIS. CORESP.

SPECIES PLANTARUM,

EXHIBENTES

PLANTAS RITE COGNITAS,

AD

GENERA RELATAS,

CUM

DIFFERENTIIS SPECIFICIS,
NOMINIBUS TRIVIALIBUS,
SYNONYMIS SELECTIS,
LOCIS NATALIBUS,

SECUNDUM

SYSTEMA SEXUALE

DIGESTAS.

TOMUS I.

Cum Privilegio S. R. M:tis Sueciæ & S. R. M:tis Poloniæ ac Electoris Saxon.

HOLMIÆ,

IMPENSIS LAURENTII SALVII.

1753.

Fig. 18-2 Title page of *Species Plantarum,* 1753.

followed by a set of descriptive nouns and adjectives. Linnaeus established what has come to be known as the BINOMIAL SYSTEM of nomenclature, in which each species is given a two-word name. This was an accomplishment of outstanding importance, for with the great increase in exploration and discovery, the number of known plants was increasing rapidly. The difficulties in dealing with these new plants were intensified by the long and involved terminology of new descriptions. This reform in nomenclature, previously suggested by others but never consistently applied, was set forth by Linnaeus in *Species Plantarum,* published in 1753 (FIG. 18-2). According to rules accepted by botanists, the nomenclature of many of the great groups of plants begins with this book.

In addition to establishing the practice of binomial nomenclature, Linnaeus also set up a system of classification that was more comprehensive than any previously devised. This system is usually designated as the "sexual" system because Linnaeus concentrated on the number of stamens and the relation of these to one another and to other floral parts. On this basis, Linnaeus set up 23 groups or classes of flowering plants. Thirteen of the classes are based on the number of stamens, from 1 to 20 or more. Two are based on the relative lengths of the stamens; four on connected stamens; one on the fusion of stamens and pistils; and three on the possession of imperfect flowers. A twenty-fourth group has no flowers and includes the ferns, mosses, fungi, and algae.

Kinds of Classifications According to the principles employed, three kinds of classifications are recognized: (1) artificial, (2) natural, and (3) phylogenetic. In practice, these may overlap.

The earlier pre-Darwinian systems of classification were largely artificial, that is, they were based upon a few convenient morphological characters. Such was the system of Linnaeus, and it resulted in such oddities as the distribution of the species of the mint family (Labiatae) between two of his classes, those species with two stamens into one class, and those with four stamens into another. As a matter of convenience, artificial systems of classification are still employed, for example, in some popular books on wild flowers. When plants are classified arbitrarily on the basis of flower color, time of blooming, habitat, form or arrangement of leaves, and whether they are woody or herbaceous, the classification is an artificial one.

In later pre-Darwinian times, natural systems came into use. These were based upon over-all resemblances in external morphology and, unlike artificial systems, involved as many characters as possible. The basic unit was the species; similar species were grouped into genera, genera with numerous characters in common were placed together in a family, and so on. At that period it was assumed that the discontinuities between species, between genera, and between families were the result of special creation. The evidence now indicates that such gaps are due to natural processes such as reproductive isolation, polyploidy, or extinction of intermediate forms. Natural systems of classification, then, involved only degrees of similarity, and the arrangement, or sequence, of the constituent groups implied no evolutionary relationships.

In the years following Darwin's *Origin of Species* (1859), the theory of evolution gradually replaced the concept of special creation. It was found that species are not fixed and unchanging, but have evolved from pre-existing species during geological time. It is now considered that, in general, similarities in structure are evidences of evolutionary relationships. Thus have arisen modern phylogenetic systems of classification based upon relationship by descent. Such systems utilize previously determined natural groups, and the categories—genera, families, orders—of the natural systems are arranged in a scheme that presumably reflects evolutionary relationships. If the species in any category, as a genus or family, are properly established, they are considered to be related by descent from a common ancestry. Those groups are presumed to be most closely related that have the greatest number of characters in common, and those most distantly related that possess the fewest characters in common.

Efforts to discover such phylogenetic relationships, to arrange groups of living organisms in a series that will show their near and distant relatives, form a thread that interweaves through much of biological investigation. These efforts involve difficulties, and evidence from many sources must be sought. The number of living plants is very great; new species and other groups have appeared and older ones have become extinct, and many species are genetically variable and unstable. The fossil record, so necessary in determining the evolutionary origin of any group, is usually fragmentary. Moreover, botanists exhibit a great variety of opinion on the relative importance to be assigned to various characters and groups of characters. These opinions, in turn, change with the discovery of new plants and new facts about plants. Thus, any phylogenetic scheme of plant classification is subject to change as our knowledge increases.

Phylogenetic Trends in Classification In arranging plants in a phylogenetic system it is assumed that the primary trend of evolution in plants (and in animals) has been from organisms relatively simple in structure to those of greater complexity. Ample evidence from both fossil and living plants indicates this sequence. The divergence that has resulted from the action of evolutionary mechanisms enables us to group organisms into species, genera, families, orders, classes, and divisions, and to arrange these in a sequence based upon the presence of primitive or advanced characters.

It is realized, however, that while the general evolutionary trend is progressive, it may, in many groups and for many characters, retrogress toward *reduction* or greater simplicity. The difficulties of phylogenetic studies are increased by this situation, which frequently involves the problem of whether a given condition or structure is simple because it is actually primitive, or whether it is simple by reduction. This difficulty is omnipresent, for evidences of reduction are found even in simply organized forms of life such as algae and fungi, as well as in the more advanced vascular plants. In many kinds of flowers, for example, certain parts, such as stamens, petals, or the entire perianth, may be absent. Externally, there may be no evidence that such parts were originally present, or they may be represented by glands, scales, or nectaries. Internally, vestigial vascular bundles may be present in positions that correspond to those in flowers in which such parts are present and well developed.

Another difficulty in the interpretation of

evolutionary relationships is CONVERGENT EVOLUTION. This term is applied to the independent development of similar structures in organisms that are unrelated or only distantly related. Such forms, although not descended from the same evolutionary stock, may show similarities that are mistaken for evidences of relationship. One example of convergent evolution is the independent acquisition of vessels in six different categories of vascular plants. In addition to the monocotyledons and dicotyledons, vessels occur in the Selaginellales, Equisetales, Filicales, and Gnetales (see classification of the plant kingdom in this chapter). Another example of convergence is found in the resemblance of unrelated organisms growing under similar environmental conditions. This is well illustrated by the resemblance between plants of the spurge and cactus families (FIGS. 15-2, 15-3). Numerous other examples of convergence could be cited. Efforts to distinguish between direct and convergent relationships are important aspects of evolutionary studies.

In spite of reduction, convergence, and other difficulties, studies on evolution in the plant kingdom will continue, the goal being a classification established upon relationship by descent. Such a classification may never be perfected, but the efforts put forth in the attempt will lead to a more accurate classification based upon genetic relationships, and to a better understanding of the evolutionary processes that have shaped the world of living things.

Characters Used in Classification Classification is usually based upon morphology, or form, external and internal. Physiological features may be important, but using them as criteria is frequently inconvenient and time consuming. There are, it is true, certain groups, such as the bacteria, in which morphological features do not suffice and must be supplemented by a study of physiological reactions.

The morphological features used in classification include both vegetative structures and those associated with sexual reproduction. The internal construction and external symmetry of the plant body, the presence or absence of vascular tissues, and the nature and arrangement of these when found, are vegetative features. Among flowering plants, the nature of the underground parts, whether bulb, rhizome, or corm, is significant in classification. Leaf features may be used, including the structure of the epidermis and the nature of epidermal hairs. In general, reproductive features are more widely employed than vegetative ones. Reproductive characteristics are less frequently correlated with physical features of the environment and therefore tend to be more stable over large groups of plants. An example of the use of reproductive features in the classification of flowering plants may be seen by comparing the bamboos, which frequently become treelike in size, with the grasses of our lawns. Both are members of the grass family and are classified on the basis of their flowers, which are fundamentally alike. On the same basis, trees such as the locust and mesquite are placed in the same family as the herbaceous alfalfa and clovers, and the tiny tuliplike flowers of asparagus relate it to the lilies. Flower structures show that the strawberry, apple, peach, and rose have much in common. They have all been placed in the rose family. Not only flowers, but also fruits and seeds are used in classification, especially in distinguishing one species from another.

THE PLANT KINGDOM

No classification of the plant kingdom can be final as long as our knowledge of plants is increasing. Thus the one presented here is only one of the latest of a series extending back to pre-Linnaean times. It is largely a natural system—that is, it places together groups that have many features in common—and it is phylogenetic only in that structurally less advanced groups are listed as preceding more advanced groups. The largest categories, or subkingdoms, are frequently regarded as artificial, and are employed here for purposes of convenience. The next largest categories, according to the rules of botanical nomenclature, are called divisions. The phylum, used by zoologists, is equivalent to the division, and the subdivision is equivalent to the subphylum. Twelve divisions are recognized. The plants of the last division, the Tracheophyta, are set apart by the possession of specialized conducting and supporting tissues. They are distributed among five subdivisions. In the first three of these, the leaves, if present, are simple and usually small, and leaf gaps are absent. In the Pteropsida (ferns) and Spermopsida (seed plants), leaf gaps are present and the leaves are commonly large.

The following classification of the plant kingdom is not complete, but it does provide a survey of the groups of plants that will be considered later. A few of the groups mentioned are given little or no further consideration. These are marked with an asterisk. Several groups of extinct plants and a few living ones are omitted from the list. The scope of the listing is generally limited to the classes and subclasses, but sometimes includes orders and families. The seed plants are treated in more detail in Chapters 27 and 28, in which some of the more important and common families are described briefly.

A CLASSIFICATION OF THE PLANT KINGDOM

SUBKINGDOM 1 *Thallophyta* (Thallus Plants)
Plant body not differentiated into root, stem, or leaf; vascular tissues absent; gametes enclosed only by a cell wall. Zygote not developing into an embryo while still enclosed within the female sex organ.

THALLOPHYTES WITH CHLOROPHYLL

DIVISION 1   *Chlorophyta* **(Green Algae)**   Approximately 7000 species. Chlorophyll, carotenes, and xanthophylls essentially as in higher plants. Food usually stored as true starch. Both sexual and asexual reproduction occur. Plant body unicellular, colonial, or multicellular, motile or nonmotile. Freshwater and marine species.

DIVISION 2   *Euglenophyta* **(Euglenoids)**   About 450 species. Green or colorless flagellates, usually with two flagella per cell (in some genera only one functional). Cells naked (true cell wall absent). Outer layer, the pellicle, composed of plasma membrane overlying interlocking pellicular strips that pass spirally along the cell. Food stored as a polysaccharide, paramylon. Vegetative reproduction by cell division in a longitudinal plane. Sexual reproduction unknown. Mostly freshwater, some marine.

DIVISION 3   *Cyanophyta* **(Blue-Green Algae)**   About 1500 species. Procaryotes: photosynthetic lamellae not organized into plastids; nuclear membrane absent. Sexual reproduction unknown, no organs of motility known. Plant body unicellular, colonial, or filamentous. Color due chiefly to the pigment phycocyanin, but in a few species a red pigment (phycoerythrin) predominant. Food stored as a carbohydrate (cyanophycean starch). Common on damp soil and rocks, and in fresh and salt water.

DIVISION 4   *Chrysophyta*   A diverse, perhaps artificial division, the classes held together by various combinations of characters. Color due to excess of various carotenoids; food stored primarily as the polysaccharide leucosin and oils; cellulose largely absent from the cell wall, pectin sometimes infiltrated by calcium carbonate or silica; flagella, when present, apical in position.

    Class I   *Xanthophyceae* **(Yellow-Green Algae)**   About 360 species. Unicellular or filamentous forms, color due chiefly to excess of beta-carotene.

    Class II   *Chrysophyceae* **(Golden-Brown Algae)**   About 650 species. Mostly unicellular or colonial flagellates, color due to excess of beta-carotene and the xanthophyll fucoxanthin.

    Class III   *Bacillariophyceae* **(Diatoms)**   About 10,000 species. Pigments as in Class II. Cell wall infiltrated with silica; in two parts, or valves, one fitting over the other like the parts of a pillbox.

DIVISION 5   *Pyrrhophyta* **(Dinoflagellates)**   The chief class of this division is the dinoflagellates, a group of about 1000 species. Plastids yellow-green or golden-brown; food stored as starch or as oil. Unicellular flagellates, usually with a heavy cellulose cell wall divided into plates and provided with furrows, one transverse and one longitudinal, each furrow accompanied by a long flagellum. Sexual reproduction rare or lacking. More common in salt than in fresh waters.

DIVISION 6   *Phaeophyta* **(Brown Algae)**   About 1500 species. Plastids brown, due to the xanthophyll fucoxanthin. Food stored chiefly as the complex polysaccharide laminarin or oil. Asexual reproduction, when present, by motile zoospores. Sexual reproduction isogamous or oogamous. Chiefly marine; plant body nonmotile, commonly large and complex.

DIVISION 7   *Rhodophyta* **(Red Algae)**   About 4000 species. Color due chiefly to a red pigment, phycoerythrin. Food stored as a carbohydrate (floridean starch) closely related to starch of higher plants. Sexual reproduction oogamous, by nonmotile gametes. Nonmotile, usually conspicuous, plants, found chiefly in marine environments.

THALLOPHYTES WITHOUT CHLOROPHYLL

DIVISION 8   *Schizophyta* **(Fission Plants)**   Unicellular procaryotic thallophytes, usually reproducing by cell division. About 1600 species.

           ORDER 1   *Eubacteriales* (True Bacteria)   A few autotrophic but most heterotrophic. About 1300 species.

           ORDER 2   *Actinomycetales* (*Actinomycetes*)   Plants usually composed of elongate branched threads. Reproducing by cell division or conidia. More than 200 species. Some 7–8 additional orders have been placed in this division.

Although a separate group, the *Virales* **(Viruses)** are conveniently placed near the Schizophyta.

DIVISION 9   *Myxomycota* **(*Myxomycetes* or True Slime Molds)**   450 species. Plant body a naked mass of protoplasm. Spores commonly produced in sporangia, but not forming hyphae upon germination.

DIVISION 10   *Eumycota* **(True Fungi)**   Plant body usually composed of hyphae.

      Class 1   *Phycomycetes* **(Algal Fungi)**   1400 species. Vegetative hyphae usually without cross walls. Sexual reproduction by zygospores or oospores.

      Class 2   *Ascomycetes* **(Sac Fungi)**   15,500 species. Hyphae with cross walls. Ascospores formed following sexual fusions.

      Class 3   *Basidiomycetes* **(Basidium Fungi, Club Fungi)**   15,000 species. Basidiospores formed following sexual fusions.

      Class 4   *Deuteromycetes* **(Fungi Imperfecti)**   15,000 species. Sexual reproduction unknown.

      The **Lichens** (*Class Lichenes*) are usually placed adjacent to the true fungi. Dual organisms, plants composed of an alga and a fungus. 17,000 species.

SUBKINGDOM 2   *Embryophyta*

Zygote developing into a multicellular embryo while still enclosed within the female sex organ (archegonium) or within an embryo sac. Gametes enclosed by a layer of sterile cells.

DIVISION 11   *Bryophyta*   Conspicuous plant body belongs to the gametophyte generation. Sporophyte unbranched, attached to the gametophyte. Plants without vascular tissue or roots. Sex organs (archegonia and antheridia) multicellular, with an outer layer of sterile cells.

      Class 1   *Hepaticae* **(Liverworts)**   About 6000 species. Plant body in most species with dorsiventral symmetry, thalluslike or leafy, usually prostrate. Sporophyte ephemeral, shriveling after maturity of spores. Capsule opening irregularly or by four valves.

      Class 2   *Anthocerotae* **(Horned Liverworts or Hornworts)**   About 320 species. Gametophyte with dorsiventral symmetry, thalluslike. Sporophyte provided with a basal meristem, elongated, not ephemeral, opening by two valves.

      Class 3   *Musci* **(Mosses)**   More than 14,000 species. Plant body leafy, with radial symmetry, erect or prostrate. Spores released following the falling of a lid from a capsule. Sporophyte not ephemeral, gametophyte long lived.

DIVISION 12   *Tracheophyta* **(Vascular Plants)**   Plant body belongs to the sporophyte generation. Plants with specialized conducting cells (xylem and phloem). Archegonia usually present (except in angiosperms).

Subdivision A   *Psilopsida*   Leaves and roots absent. Plant body composed of a branched aerial axis extending downward into the soil. Sporangia terminal or lateral. Mostly extinct.

> *ORDER RHYNIALES*   Ancient and simple vascular plants. All extinct. *Rhynia, Hicklingia, Hedeia, Cooksonia,* and others.
> *ORDER ZOSTEROPHYLLALES*   *Zosterophyllum* and other extinct genera.
> Other extinct orders
> *ORDER PSILOTALES*   Two living genera, *Psilotum, Tmesipteris.* (Compare with Psilotales, Chapter 25.)

Subdivision B   *Lycopsida*   Leaves simple, frequently small, irregularly or spirally arranged in vertical rows, never whorled. Sporangia borne singly in the axils of fertile leaves (sporophylls). Sporophylls commonly reduced and grouped in terminal cones (strobili).

> *ORDER 1*   *Lepidodendrales* (Giant Club Mosses) Known only as fossils. *Lepidodendron,* 300 species. *Sigillaria,* 400 species.
> *ORDER 2*   *Lycopodiales* (Club Mosses)   *Lycopodium,* 450 living species. *Phylloglossum,* 1 living species. *Lycopodites,* extinct. (*Asteroxylon* and *Baragwanathia* may be included among the fossil forms in a closely related order.)
> *ORDER 3*   *Selaginellales* (Small Club Mosses)   *Selaginella,* 700 living species. *Selaginellites,* extinct.
> **ORDER 4*   *Isoetales* (Quillworts)   *Isoetes,* 64 species. *Stylites,* 2 species.

Subdivision C   *Sphenopsida*   Leaves small and simple, arranged in whorls. Stems jointed. Several sporangia on modified sporophylls, these arranged in strobili. Known mostly as fossils.

> **ORDER 1*   *Sphenophyllales*   *Sphenophyllum.* Extinct, 100 species.
> *ORDER 2*   *Equisetales*   Horsetails (*Equisetum*). 25 living species. *Equisetites,* extinct.
> *ORDER 3*   *Calamitales*   *Calamites.* Extinct, 300 species.

Subdivision D   *Pteropsida*   Leaves usually large and complex. Sporangia on sporophylls.

Class 1   *Filicineae* **(The Ferns)**   Sporangia numerous, usually on the lower surface of the sporophylls, but may be terminal or marginal. Primary growth only. Fertilization by swimming sperms.

> *ORDER 1*   *Ophioglossales* (Adder's Tongue and Grape Ferns)   90 species.
> **ORDER 2*   *Marattiales* (Marattiaceous Ferns)   190 species.
> *ORDER 3*   *Filicales* (Typical Ferns)   About 15,000 species.

Subdivision E   *Spermopsida* **(The Seed Plants)**

Class 1   *Gymnospermae* **(The Gymnosperms)**   Seeds naked, pollen deposited on or near the ovules.

> *ORDER 1*   *Pteridospermales* (Seed Ferns)   Plants bearing seeds but with the habit of tree ferns. Seeds borne on the leaves, never in cones. Extinct.
> *ORDER 2*   *Bennettitales*   Mostly large and coarse plants, frequently with thick, unbranched stems bearing a crown of palmlike leaves. Male sporophylls usually leaflike, surrounding numerous stalked megasporophylls, each bearing a terminal ovule. Extinct.
> *ORDER 3*   *Cycadales* (Cycads)   Plants resembling palms. Male and female reproductive structures usually in cones. Sperms motile. 90 living species.
> *ORDER 4*   *Cordaitales*   The early conifers. Tall, slender trees, with elongated, simple leaves. Reproductive structures in catkinlike clusters. Extinct.

ORDER 5   *Ginkgoales*   One living species, *Ginkgo biloba,* the maidenhair tree. Sperms motile. Several extinct species.

ORDER 6   *Coniferales* (The Conifers) 550 species. "Softwoods" or "evergreens." Leaves mostly needlelike. Male and female reproductive structures usually in cones. Seeds usually on the surface of scales. Sperms nonmotile, conveyed to the egg by a pollen tube.

*ORDER 7   *Gnetales*   *Gnetum, Welwitschia, Ephedra.* 70 species.

Class 2   *Angiospermae* (**Angiosperms**)   The flowering plants. Seeds enclosed by carpels. Leaves typically broad. Pollen deposited on the stigma. Nonmotile sperms transferred to the ovule by a pollen tube.

Subclass A   *Dicotyledoneae* (**Dicotyledons**)   200,000 species. Embryo with two cotyledons; flower parts mostly in 4s or 5s; vascular tissue of the stem in a cylinder; cambium usually present; netted-veined leaves. Herbs and woody plants. The following are some of the larger families, together with the approximate number of species in each: crowfoot or buttercup (Ranunculaceae), 1500; mallow (Malvaceae), 1000; mustard (Cruciferae), 3000; spurge (Euphorbiaceae), 8000; rose (Rosaceae), 3000; legume (Leguminosae), 14,000; carrot (Umbelliferae), 3000; mint (Labiatae) 4500; potato (Solanaceae), 2500; and composites (Compositae), 20,000.

Subclass B   *Monocotyledoneae* (**Monocotyledons**)   55,000 species. Embryo with one cotyledon; flower parts usually in 3s; vascular tissue of the stem usually in scattered vascular bundles; cambium usually absent; leaves with parallel veins. Mostly herbs, some trees. Among the larger families with the number of species in each are the following: lily (Liliaceae), 4200; sedge (Cyperaceae), 4100; palms (Palmaceae), 2600; pineapple (Bromeliaceae), 1300; grass (Gramineae), 7500; and orchid (Orchidaceae), about 25,000.

SUMMARY

1. The binomial system consists of the genus and species name. In addition, some plants are distinguished by a variety name. Scientific names show relationships that may not be apparent in the common names.

2. In plant classification, the species is placed in a genus; the genus in a family; the family in an order; an order in a class; a class in a division, which forms one of the major groups of the plant kingdom. Intermediate groupings, such as subdivisions and subclasses, are also recognized.

3. The naming of plants goes back to antiquity. Systematic naming and classification began with the work of Theophrastus, who, among other contributions, recognized and described plant families and genera.

4. The herbalists of the sixteenth century stimulated plant classification, although their chief concern was with the medicinal uses of plants.

5. Linnaeus established the binomial system—a double name for each species—and developed a widely used artificial system of classification.

6. Systems of classification may be artificial, natural, or phylogenetic. Artificial systems are systems of convenience based upon a few features. They may have important but limited uses. Natural systems involved as many characters as possible and were in use before the acceptance of the theory of organic evolution. Phylogenetic systems use similarities in structure and other characteristics to arrange organisms in a sequence that indicates evolutionary relationships.

The Algae

T O MOST PEOPLE the word "plant" implies an organism with roots, stems, and leaves. Many kinds of plants do, of course, possess these organs, but a large part of the plant kingdom is made up of forms of life that have a much simpler plan. Among them are the pond scums, seaweeds, and the like, known collectively as the algae. These plants usually receive little popular recognition or attention. Many of them, indeed, are small, nondescript, and unattractive. Nevertheless, the algae are highly important components of the plant population of the world and are significant to man in many ways.

GENERAL CHARACTERISTICS

The algae have relatively undifferentiated tissues that never form true roots, stems, or leaves. Most algae contain chlorophyll and are photoauto-trophic. The plant body of the alga is known as a THALLUS. This term is used even if the plant is unicellular. The algae are classified, together with the bacteria and fungi, in the great subkingdom of nonvascular plants known as the Thallophytes.

The algae and other thallophytes are regarded as "low" forms of plant life—that is, closely related to the first and most primitive organisms on the earth. The adjective "low" refers to a lack of structural complexity rather than to any lack of adaptation to the environment. Thallophytes may be as well adapted to their mode of life as the "higher" plants such as ferns, conifers, and flowering plants. Although the algae as a whole have a relatively simple structural organization, many groups of algae are far more complex than others, not only in structural differentiation but also in reproductive methods. Algae exhibit a number of evolutionary trends, the study of which helps us to understand the involved rela-

453

tionships of the higher plants and the rich and varied plant life that now covers the earth and lives in its waters.

Although algae are easily overlooked, they are very common. They are predominantly aquatic, living either in fresh or in salt water and even occurring in desert water holes or in temporary lakes. But many species are terrestrial and are found growing in or on moist soil, on rocks, on snow, or on other plants or animals. The green "moss" that often stains the bark of a tree, especially on the north side, is usually an alga. Some algae grow actively on snow, giving the snow a decidedly pinkish color. Some species thrive in hot springs. Algal growths have been found in caves and mines hundreds of feet below the surface of the ground. These growths occur close to electric lights, which provide the radiant energy that makes photosynthesis possible. Generally, the adaptations of algae to varied environments are comparable to, if not greater than, those of any other plant group.

In all, there are about 30,000 species of algae, most of which are marine. Fresh-water species seem to have developed independently of salt-water forms. Comparatively few families of algae include both fresh- and salt-water species.

The various species exhibit a great variety in color, ranging from green, yellow-green, and blue-green to red, yellow, orange, olive, and brown. In form they include balls, threads, sheets, ribbons, and a variety of branching shapes. Some are massive, some delicate. Algae also vary tremendously in size. Many species are unicellular, swimming or floating plants of microscopic size. The largest algae, the kelps of the colder Pacific waters, may be well over 100 feet long. Most algae are small plants, but many of these make up for their lack of size by their abundance and the rapidity with which they reproduce. Many common species reproduce so rapidly that ponds and sluggish streams are choked and even parts of the ocean are colored by them.

Algae grow on the hairs of the three-toed sloth, and may be abundant enough to give a greenish tinge to the gray hair of the animal. And, in an entirely different environment, large colonies of algae (diatoms) cover the undersides of the blue whale, the largest living animal, and seem to thrive on this unusual habitat. These yellowish, microscopic algae are so abundant that the blue whale is sometimes called the sulfur-bottomed whale because of their color.

A number of green algae also live in a mutualistic relationship with fungi (Chapter 21) and with several kinds of animals. Some unicellular green algae live within protozoa. A large green paramecium may contain several dozen living algal cells. Algae are found in protozoa, foraminifera, rotifers, *Hydra* and other coelenterates, fresh-water sponges, the reef-building corals, and snails. Perhaps the most unusual relationship occurs between a unicellular, motile, green alga and a flatworm (*Convoluta*). The worm is only a few millimeters long, and occurs in large patches on sandy beaches of the northern coast of France. Four stages have been recognized in this relationship. In the first, the newly hatched colorless animal feeds actively for some days on minute organisms in the environment. During this period it usually also ingests cells of the alga —the worm is "infected." If, under experimental conditions, infection is prevented, the animal soon ceases to feed and eventually dies. During the second stage, the algae multiply rapidly in the body of the animal and eventually become arranged in rows beneath the surface, coloring it dark green. The worm continues to feed, but the food taken in is now supplemented by the products of the photosynthetic cells—sugars and perhaps lipids and proteins. The algae at this stage are thought to perform an important role in utilizing the nitrogenous wastes from the animal, which possesses no excretory system. In the third stage, relatively brief, the animal ceases to feed, and subsists entirely on the food materials produced by the green plant. Finally, in the fourth stage, as the eggs are produced, the worm digests its green cells and loses its green color. This relationship between plant and animal has been termed mutualism, but it is actually obligatory parasitism, for the worm is unable to live without the alga, which it eventually destroys.

Photosynthesis and Algal Pigments Like other phototrophic plants, algae are dependent for normal growth and reproduction upon such environmental factors as favorable temperatures and light, and adequate supplies of oxygen, carbon dioxide, and essential elements. Light is an important factor, limiting the depth at which algae can live—in marine algae the practical lower limit may be about 300 feet, but some species may grow at considerably lower depths. At increasing depths, light is selectively absorbed by sea water, beginning with the longer red, orange, and yellow wavelengths, leaving eventually only blue and green wavelengths. Thus, the light received by the photosynthetic lamellae may be lacking in both intensity and quality.

Algal pigments, other than the green, may contribute to photosynthesis. The pigments of algae are classified as (1) the chlorophylls, (2) the carotenoids, and (3) the *phycobilins*, of which there are two, *phycocyanin* and *phycoerythrin*. All divisions of the algae have chlorophyll *a*, and chlorophylls *b*, *c*, *d*, or *e* may be present. In all divisions except the green algae, the chlorophyll is largely masked by nonchlorophyllous accessory pigments, carotenoids or phycobilins. For example, the color of the brown algae is due chiefly to one of the xanthophylls; of the red algae, to phycoerythrin; and of the blue-green algae, to phycocyanin. These nonchlorophyllous accessory pigments enable algae to make more efficient use of available light, for they, as well as chlorophyll *a*, absorb light. The energy absorbed by the accessory pigments, however, does not function directly in photosynthesis, but is transferred to chlorophyll *a* and utilized by this pigment.

The Plant Body The plant body of the algae is highly varied. Some of the unicellular forms are motile, some nonmotile. They vary greatly in shape and most are microscopic in size. In many species the cells are grouped into aggregations known as COLONIES. These may be filamentous (threadlike) or assume other forms. The filamentous types of algae are usually multicellular, with the cells placed end to end forming branched or unbranched threads large enough to be seen with the naked eye. The basal cell of a filament is commonly specialized. It is known as a HOLDFAST and functions as an anchor that holds the filament to a rock or stick or, more rarely, to the sand and mud of the bottom.

The nonfilamentous algal colony is usually a spherical or platelike aggregation of cells with no basal holdfast. The cells within such a colony may be independent of one another or may be connected by strands of cytoplasm. Frequently, the entire colony is surrounded by a common gelatinous envelope. The green, blue-green, and yellow-green algae, and the diatoms, are groups that include many unicellular forms as well as filamentous and nonfilamentous colonies. In the red and brown algae, unicellular and nonfilamentous colonial forms are rare and simple filaments are uncommon. The thallus of these algae is usually composed of numerous, closely compacted filaments that in cross section look much like the parenchyma tissues of the higher plants. Many of the red and brown algae exhibit external and internal differentiation and attain considerable size.

Plastids are commonly the most conspicuous feature of algal cells as seen under the microscope. They vary in number from one to many and exhibit great diversity in form. In many algae a dense, specialized proteinaceous region known as a PYRENOID, is found in the plastid. In the green algae the pyrenoid is usually surrounded by a sheath of starch grains, lying within the chloroplast. In other algae, starchlike bodies may be found outside the plastid, in the cytoplasm.

Reproduction The methods of reproduction in the algae are almost as diverse as the nature of the plant body. The specialist is concerned with the detailed variations in the process because of their importance in classifying algae and because of their evolutionary significance. However, for the beginner, algal reproduction is best considered in general terms.

Asexual and vegetative reproduction In lower forms of plant life a distinction is sometimes made between asexual and vegetative reproduction. Asexual reproduction is brought about by the formation of ZOOSPORES or other spores not derived as a result of the sexual process. Reproduction by zoospores is common. As their name implies, zoospores are animal-like. They are unicellular and naked; that is, they have a cell membrane but not a true cell wall. They are motile, free-swimming cells that move by means of one or more FLAGELLA—whiplike threads that are attached terminally or sometimes laterally on the zoospore. A vegetative cell may form one or a number of zoospores, which are released with the rupture of the walls of the parent cell. Each zoospore may give rise to a new plant.

Vegetative reproduction results from cell division or fragmentation. Many unicellular algae reproduce by division of the cell, forming two daughter cells, each becoming a new individual. The daughter cells may separate immediately or they may remain together for a time, developing a superficial resemblance to true colonies. This may be the sole method of reproduction in such species, or it may be accompanied by other methods. Reproduction by fragmentation occurs when mature, nonfilamentous colonies split into two or more portions or when filaments or more complex bodies break apart. Vegetative reproduction is extremely common, not only in the smaller pond scums but also in the larger brown and red algae. In both groups, portions of the plant body may become detached, forming new individuals.

Sexual reproduction Sexual reproduction in the algae, as in higher plants, involves the production of gametes. A gamete is a cell that grows into a new individual only after fusion with another gamete. Two chief types of sexual reproduction are found in the algae. In the first, ISOGAMY, the gametes are approximately equal in size and frequently motile. If, as is commonly the case, the resulting zygote is thick-walled and serves as a resting stage, it is termed a ZYGOSPORE. In the second type, OOGAMY, the gametes are unequal in size and a large, nonmotile, female gamete (the egg) is fertilized by a small, motile, male gamete (the sperm). If the zygote does not germinate immediately but enters a period of dormancy, it is termed an OOSPORE. In oogamy the gametes are borne within specialized cells that are differentiated from the vegetative cells of the thallus. It is known that many algae are heterothallic (see Glossary); in these forms two compatible mating types are necessary for sexual reproduction.

In several groups of algae, especially the green, yellow-green, and golden-brown, the zygote is usually the only cell of the diploid generation. The division of the nucleus of the zygote, at germination, is by meiosis, and leads to the formation of four motile (rarely nonmotile) spores, each of which can produce a new plant. Such motile spores are also referred to as zoospores, but this leads to confusion with the zoospores that are produced asexually by the haploid plant. Therefore, spores associated with meiosis as a part of the sexual process have been termed MEIOSPORES. This term could be extended, where desirable, to other groups of plants as well as the algae.

CLASSIFICATION OF THE ALGAE

For many years the algae were classified primarily upon the basis of the most conspicuous pigments, and were therefore divided into four groups (classes); the blue-greens, greens, browns, and reds. Modern classification still takes into account the occurrence of certain pigments, but it also emphasizes other features such as cytology, life histories, the chemical nature of the reserve products and the cell wall, and the number, structure, and arrangement of flagella. The original four classes have thus been rearranged into a number of divisions, the number varying with the authority consulted. In this text, seven divisions are recognized (Chapter 18).

The Flagellates Before considering the divisions of the algae, it is desirable to deal briefly with the flagellates, a heterogeneous group of organisms identified in part by the presence of an elongated, whiplike appendage, the flagellum, by which motility is brought about. Some cells have but one of these; others, two; and a few, as many as six. Most flagellates are unicellular, but some are colonial. Some have a rigid cell wall, and others have only an outer limiting layer that may be rigid or flexible, allowing changes in body shape. Because the flagellates differ in their structure (FIG. 19-1), in their pigments, and in their storage products, they are divided among various divisions of the algae— the euglenoids, the greens, the yellow-greens, and the dinoflagellates. They are abundant, both in fresh and oceanic waters.

One of the most interesting features of the flagellates is the great diversity of their nutrition. Most are chlorophyll-bearing phototrophs, but in a considerable number of these the phototrophism is incomplete, since they are dependent upon an external supply of certain growth factors—one or more of the vitamins B_{12}, thiamine, and biotin. Such vitamins are produced, at least in large part, by bacteria in the environment. Not only flagellates, but also other small algae—nonmotile unicellular, colonial, and even multicellular—have been found to be deficient in the ability to synthesize one or more of these vitamins. Some photosynthetic flagellates can grow only in the light; others can also grow in the absence of light, heterotrophically, obtaining food in solution through the cell membrane. Some are always colorless heterotrophs. These may obtain their food in soluble form or, animal-like, take food in particulate form through a mouth part

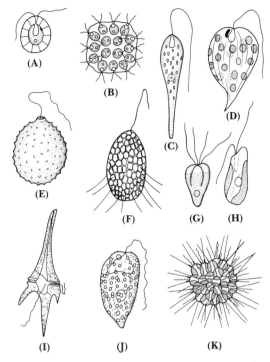

Fig. 19-1 **Flagellates belonging to different divisions of the algae. All are phototrophic except** *Astasia*, **which is a colorless saprophyte. (A):** *Carteria.* **(B):** *Gonium.* **(C):** *Astasia* **(D):** *Phacus.* **(E):** *Trachelomonas.* **(F):** *Mallomonas.* **(G):** *Pyramimonas* **(H):** *Cryptomonas.* **(I):** *Ceratium.* **(J):** *Gonyostomum.* **(K):** *Chrysosphaerella.*

and gullet (PHAGOTROPHISM). Finally, there are species that are not only chlorophyllous but also phagotrophic.

Reproduction is usually vegetative, commonly by longitudinal division of the cell into two equal halves. Sexual reproduction is found in some flagellates.

The plant-animal features of the flagellates have long raised questions of classification, and have made these organisms a subject of evolutionary speculation. One point of view is that, fundamentally, flagellates were photosynthetic plants and later acquired animal features, such as phagotrophism. It is held that many if not all of the colorless heterotrophs, whether they obtain food in soluble or particulate form, are related by descent to pigmented forms through similar morphology and storage products. On the basis of this concept the flagellates are regarded as plants with animal tendencies.

Another point of view is that flagellates are neither plants nor animals, but should be placed in a third kingdom, PROTISTA, coordinate with kingdoms PLANTS and ANIMALS. Here also are classified all other algae, the bacteria, slime molds, fungi, and protozoa. The protista are characterized by their simple structure—that is, they are unicellular or coenocytic (see Glossary), or if multicellular are not differentiated into specialized tissue systems. The use of the term protista to include microscopic and other simply constructed forms of life is convenient, for in classifying many of these organisms it is difficult to draw sharp distinctions between animals and plants.

Many living flagellates are probably specialized, as compared to their remote ancestors. Nevertheless, whether we call them plants or protista, the group as a whole enables us to visualize, even if dimly, ancient ancestral forms that on the one hand may have given rise to various groups of colonial and multicellular algae, and on the other, to protozoa and some multicellular animals.

DIVISION 1. THE GREEN ALGAE

The green algae include about 7000 species, both aquatic and terrestrial. A number of green algae live in the sea, but the group as a whole is more characteristic of fresh water. The green algae exhibit no high degree of differentiation. The individual plants are usually either single cells or filamentous or nonfilamentous colonies (FIG. 19-2). In a few genera such as the leaflike sea lettuce (*Ulva*) and the branched stoneworts (*Nitella, Chara*) (FIG. 19-3), the plant body is more complex, but in size it never approaches the larger red and brown algae. Green algae are either free-floating or attached, but even those that produce holdfasts may float in masses on the surface of the water.

The cell walls of green algae are composed chiefly of cellulose, and the cells usually contain a large central vacuole surrounded by a layer of cytoplasm. In the cytoplasm are embedded one or more chloroplasts. Each of these contains a pyrenoid, usually surrounded by an envelope of starch grains.

The green algae are of particular interest from an evolutionary standpoint, since the higher and more specialized land plants probably arose from ancient green algae. Certain present-day forms of green algae represent stages in this evolution, for it is likely that many living algae have changed but little over long periods of geological time and have remained relatively primitive. Such species have not given rise to more advanced types and may represent only low side branches on the evolutionary tree. But because they represent early roles of plants in the pageant of life, they are worthy of study. In discussing representative green algae, then, we may consider what light they shed upon the evolution of other and more advanced types of algae, and what clues they provide to the nature of the ancestors of the many kinds of plants that now cover the land.

Some Motile Green Algae As examples of motile species, we may consider the flagellates *Chlamydomonas* and *Volvox*.

Chlamydomonas The species of this genus are motile and unicellular. They are found chiefly in fresh water, especially in stagnant pools. The plant is spherical or ellipsoidal, with a cellulose cell wall (FIG. 19-4). Two flagella arise from the anterior end of the cell, and a single cup-shaped chloroplast occupies much of the space within it. The cell contains one or sometimes several pyrenoids. An EYESPOT, bearing a reddish pigment, is found near the flagellated end of the

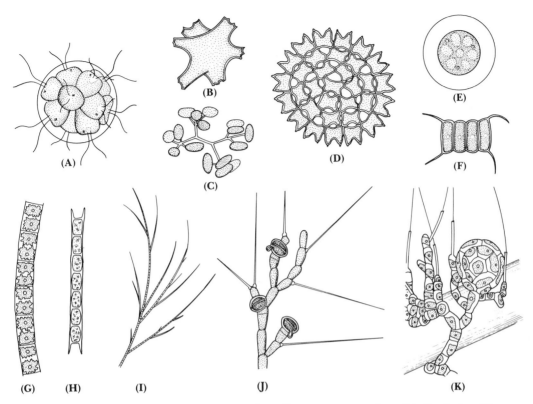

Fig. 19-2 Green algae. (A–F): unicellular and nonfilamentous green algae. (A): *Pandorina*. (B): *Tetraedron* (C): *Dictyosphaerium*. (D): *Pediastrum*. (E): *Asterococcus*. (F): *Scenedesmus*. (G–K): filamentous green algae. (G): *Zygnema*. (H): *Microspora*. (I): *Stigeoclonium*. (J): *Bulbochaete*. (K): *Coleochaete*.

cell. This eyespot is sensitive to light and determines the direction in which the organism will move in response to light. The cell also has two contractile vacuoles that lie near the point at which the flagella originate.

Asexual reproduction in *Chlamydomonas* is more specialized than in most flagellates. It involves the division of the cell into two to eight daughter protoplasts. This is preceded by the loss of the flagella of the parent cell. The daughter cells form cell walls and new flagella, and are then liberated by the gelatinization of the wall of the parent cell. Isogamous sexual reproduction also occurs in the genus.

Volvox The ancient flagellates probably evolved in several directions, giving rise to varied kinds of plant bodies. We may assume that one of these was the nonfilamentous colony that originated by the failure of the daughter cells of unicellular flagellates to separate after cell division, thus giving rise to a motile, multicellular plant body. A motile colony may have become nonmotile by the disappearance of flagella. In many nonfilamentous, colonial forms the cells are all alike. In some, however, certain cells have become differentiated as asexually or sexually reproductive, illustrating an early stage in the division of labor found in greater degree in the bodies of more structurally complex plants.

A plant known as *Volvox*, found in ponds and lakes, is a good illustration of a nonfilamentous motile colony. The plant is spherical (FIG. 19-5) and large enough so that a single one is visible to the naked eye. The cells in a *Volvox* colony vary in number from 500 to as many as 20,000. They are closely arranged in a single layer that constitutes the skin of a hollow sphere. The interior is filled with a watery mucilage, and the whole colony is surrounded by a gelatinous envelope. Each cell, in turn, is enclosed within a gelatinous sheath and is connected to adjoining cells by cytoplasmic strands. Except for the gelatinous wall, an individual cell is very like *Chlamydomonas;* it contains a single chloroplast, a pyrenoid, vacuoles, and an eyespot, and bears two flagella that point outward and contribute to the motility of the colony.

Reproduction of the *Volvox* colony is both sexual and vegetative; the latter type is brought about by the formation of several small colonies within the parent sphere. Certain vegetative cells enlarge and push into the central cavity. They begin a series of divisions that eventually result in the production of new colonies, which float free within the parent colony. These are released by the disintegration of the old colony. Sexual reproduction in *Volvox* (FIG. 19-6) is oogamous; two types of gametes form. Plants that reproduce sexually are usually without daughter colonies. Occasional cells of the parent colony enlarge without dividing and form nonmotile eggs. Other cells divide many times and produce

Fig. 19-3 Green algae with complex plant bodies. (Left): *Ulva.* **(Right):** *Nitella.*

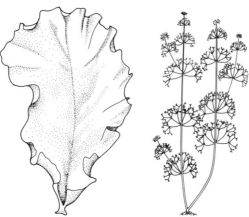

Fig. 19-4 *Chlamydomonas.* **(A): single cell. (B): asexual reproduction; daughter cells nearly mature (C): liberation of daughter cells.**

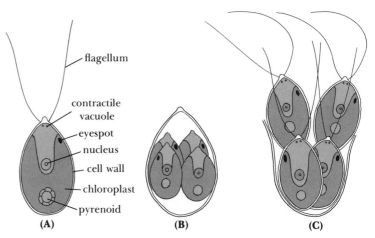

flagellum

contractile vacuole

eyespot

nucleus

cell wall

chloroplast

pyrenoid

(A) (B) (C)

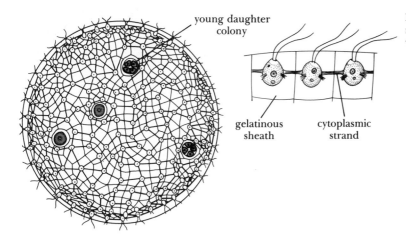

young daughter colony

gelatinous sheath

cytoplasmic strand

Fig. 19-5 *Volvox*. (Left): mature colony (Right): individual cells of a colony.

numerous motile sperms in flat or spherical, colonylike masses. After fertilization, the zygote enlarges and develops a thick wall. During this maturation the zygote becomes conspicuous by the development of an orange-red pigment. The zygote is released by the disintegration of the mother colony, and germinates after a dormant period. Following meiosis, four nuclei are formed. In at least one species, only one of these nuclei survives and becomes the nucleus of a single motile spore (meiospore). This emerges from the zygote and gives rise directly to a new colony.

Some Nonmotile Green Algae In this group we will survey briefly a few forms worthy of note.

Chlorella The species of this genus are aquatic, very small, spherical or ellipsoidal in shape, and consist essentially of a cell wall, a cup-shaped chloroplast, and a nucleus (Fig. 19-7A). No colonies are formed and sexual reproduction is unknown. Asexual reproduction is by the formation of up to 16 nonmotile spores within the parent cell. These are released by the disintegration of the surrounding cell wall and then enlarge to parent size. Plants such as *Chlorella* could have arisen from a green flagellate by loss of flagella,

Fig. 19-6 Portion of *Volvox* colony, showing sexual reproduction.

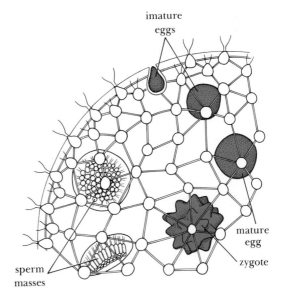

imature eggs

mature egg

zygote

sperm masses

and the nonmotile spores could be regarded as modified zoospores.

Chlorella is best known from laboratory studies on cells isolated and grown in cultures free from

Fig. 19-7 (A): *Chlorella*, two species, differing in size and shape. Formation of asexual spores is shown. (B): *Protococcus*, single cell and vegetative reproduction.

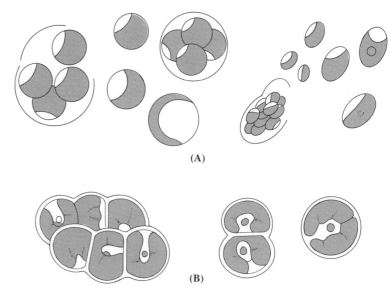

(A)

(B)

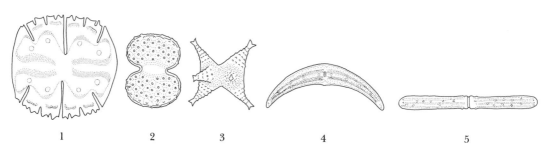

1   2   3   4   5

Fig. 19-8 Green algae, desmids. (1): *Micrasterias*. (2): *Cosmarium*. (3): *Straurastrum*. (4): *Closterium*. (5): *Pleurotaenium*.

Fig. 19-9 *Spirogyra*, a single cell of filament and portion of adjoining cell.

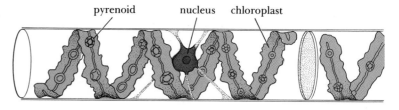

pyrenoid   nucleus   chloroplast

other organisms. It has been widely employed in studies of metabolic processes, especially photosynthesis. Particularly interesting, however, is the potential use of mass cultures of *Chlorella,* an efficient producer of protein, and similar unicellular algae as a source of food for an expanding world population.

Desmids   Desmids (FIG. 19-8) are aquatic, free-floating green algae, mostly unicellular, although sometimes the cells are united end to end forming a filamentlike colony. Most desmids are characterized by a median constriction that divides the cell into two equal halves, each containing one or two large chloroplasts (FIG. 19-8). Numerous species of desmids are known. They are common and widely distributed in waters of ponds and lakes.

Protococcus   This is one of the most common of the small green algae. It is terrestrial, growing as a thin green film on damp stones, walls, fence posts, and tree bark. The cell is spherical (FIG. 19-7B) and contains a single large, lobed chloroplast just within the cell wall. The only known method of reproduction is cell division, which may occur in any one of three planes. The resulting cells may separate or remain together for a time in clusters of two, four, eight, or even more cells. *Protococcus* is considered to be a reduced member of a family of filamentous species and therefore is not significant in the evolution of more specialized plants.

Filamentous Green Algae   In nonfilamentous colonial forms, cell division may occur in more than one plane. In the filamentous algae, however, new cell walls are usually laid down only at right angles to the long axis of the plant body. The result is the formation of an elongated thread, or filament, composed of cylindrical cells. The filamentous algae may have developed directly from flagellates in which loss of motility

was accompanied by the establishment of unlimited cell division in a single plane.

***Acetabularia* (*Mermaid's Wineglass*)**   The species of this genus are widely distributed in shallow waters of tropical and subtropical seas. The mature plant is a single, large cell. It is differentiated into a basal, lobed rhizoid or holdfast—a stalk up to 6 centimeters in length—and a saucerlike disk, about 1 centimeter in diameter, at the tip of the stalk (Plate 9). The single, large nucleus is located in one of the lobes of the holdfast. The plants grow erect, and usually occur in clusters. *Acetabularia* can be cultivated in the laboratory, and has been the subject of numerous experimental investigations, notably on nuclear function and the relations between nucleus and cytoplasm.

Spirogyra   The study of filamentous algae is most conveniently begun with a consideration of some species of the large, readily identifiable, and widely distributed plants known as *Spirogyra.* These plants, which form bright green masses on the surfaces of quiet ponds and streams, are often termed pond scum. The filaments are unbranched. Each cell contains one or, in some species, several large chloroplasts embedded in the cytoplasm just within the cell wall (FIG. 19-9). The plastids (chloroplasts) are ribbonlike bodies with uneven margins, spiraling from one end of the cell to the other. Pyrenoids, surrounded by starch grains, are embedded in the plastids at regular intervals and are conspicuous features of the cell. The cytoplasm surrounds a large, central vacuole. The nucleus, surrounded by a sheath of cytoplasm, lies in the center of the cell, where it is suspended by cytoplasmic strands extending through the vacuole to the peripheral cytoplasmic layer.

Vegetative reproduction in *Spirogyra* appears to be simple, since any cell or group of cells will continue to grow and form a filament. The action of water currents or the feeding of fish or small water animals produces the fragmentation that makes possible the formation of new plants. This action is enhanced by the seasonal dissolution of the intercellular substance between adjoining cells, which permits the filaments to disintegrate. Fragments of *Spirogyra* filaments overwinter in an inactive state and renew their growth in the spring.

Sexual reproduction (FIG. 19-10) involves the fusion of two nonmotile gametes, usually derived from different filaments, forming an ovoid or spherical zygospore. When two filaments lie side by side, a mucilagelike substance causes them to adhere. From each of two opposite vegetative cells of the filaments, budlike PAPILLAE now grow. These papillae are in contact from the beginning and the filaments are gradually pushed apart as growth proceeds. Finally the wall between the papillae dissolves, forming an open CONJUGATION TUBE between the two cells. The entire content (SUPPLYING GAMETE) of one cell then moves through the conjugation tube and fuses with the protoplast (RECEIVING GAMETE) of the other cell.

The process of conjugation is facilitated by the contraction of the two gametes, which allows the fused protoplasts to occupy the space within the cell previously filled by one gamete alone. This contraction is brought about by the loss of water from the cytoplasm, primarily from vacuoles.

A given filament usually produces only one kind of gamete—either supplying or receiving. Therefore, at the end of the period of conjugation, only cell walls remain in one filament, whereas all the cells of the adjoining filament contain zygospores. Gametic differentiation is so slight that sexual reproduction is classified as isogamous. After fusion, the chloroplasts and pyrenoids of the gametes may be identified for a time in the young zygospore, but a thick, opaque wall forms shortly, and the zygospore enters a dormant stage. After the disintegration of the walls of the parent cell, the zygospore sinks to the bottom and at the beginning of the next growing season germinates, forming a new filament (FIG.

19-10B). In some species, meiosis occurs during the maturation of the zygospore, but in others it is delayed until just previous to germination. Following meiosis, three of the four resulting haploid nuclei disintegrate, leaving a single functional nucleus. At germination, this haploid nucleus becomes the nucleus of the emerging first cell of the new filament.

Ulothrix The filament in *Ulothrix* is also unbranched. It consists of a row of short, cylindrical cells attached end to end (FIG. 19-11A). The basal cell of the filament is often modified into a holdfast. *Ulothrix* plants are found attached to rocks and other objects in streams and lakes, but they also occur, like *Spirogyra*, as free-floating masses on the surface of the water. Each cell contains but a single girdlelike chloroplast, open at both ends. The chloroplast may take the form of a complete cylinder or it may extend only partially around the circumference of the cell. It contains one or several pyrenoids.

Nonsexual reproduction in *Ulothrix* (FIG. 19-11B-D) occurs by fragmentation and by the production of zoospores. The latter may be produced in any cell of the filament with the exception of the holdfast. Zoospores are formed by the division of the entire protoplasm of a single cell into smaller units, commonly four to eight in number. The zoospores lack a definite cell wall, but each possesses a single chloroplast, an eyespot, and four flagella. When mature, the zoospores are liberated through a pore in a lateral wall of the cell. They are usually retained for a brief period in a thin, baglike extension of the parent cell, but this shortly disappears and the free zoospores swim about vigorously for several hours. They finally settle down, form holdfasts, and produce new filaments by the process of cell division.

The formation of a large number (8–64) of small, motile gametes in a single cell initiates sexual reproduction (FIG. 19-12). These gametes are very similar to zoospores, but possess two flagella

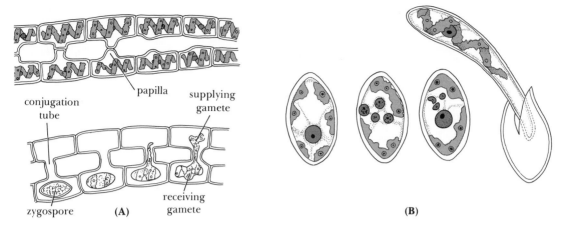

conjugation tube

papilla

supplying gamete

receiving gamete

zygospore

(A)

(B)

Fig. 19-10 *Spirogyra*, sexual reproduction. (A): stages in the formation of a zygospore. (B): stages in germination of zygospore. The nucleus divides twice, and in the course of these divisions meiosis occurs. Only one of the four nuclei survives.

Fig. 19-11 *Ulothrix*. (A): vegetative cells of a filament. (B): development of zoospores. (C): escape of zoospores. (D): zoospore germination.

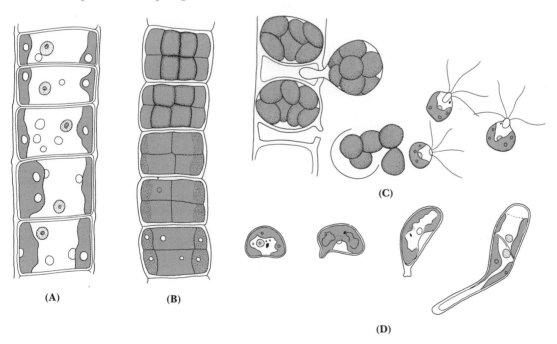

(A)

(B)

(C)

(D)

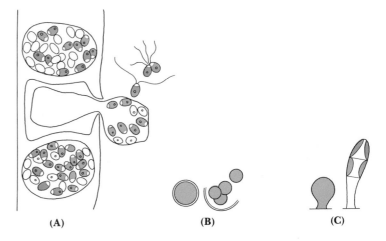

Fig. 19-12 *Ulothrix,* **sexual reproduction. (A): formation and fusion of gametes. (B): zygospore germination. (C): spore germination.**

(A)   (B)   (C)

instead of four. After escaping from the filament, pairs of gametes fuse to form a zygospore. The fusing gametes are derived from different filaments, but are alike in size and form; hence, sexual reproduction in *Ulothrix* is isogamous. After a period of rest, the zygospores germinate. The nucleus divides by meiosis, and motile or nonmotile spores (meiospores), usually four in number, are produced. Each of these grows into a new filament.

The production of flagellatelike zoospores by *Ulothrix* may be cited as an example of the retention of an ancestral character. The zoospores of *Ulothrix* may thus represent a stage during which plants were unicellular and similar to living flagellates.

Oedogonium *Oedogonium* is a common, widely distributed alga growing as an unbranched filament attached by a holdfast when young (FIG. 19-13A), but usually floating in masses when mature. The single chloroplast is cylindrical and netlike, with numerous pyrenoids (FIG. 19-13B). It reproduces nonsexually by fragmentation and by means of large, green, spherical or ovoid zoospores (FIG. 19-14). These are produced singly in vegetative cells and bear numerous flagella arranged in a whorl at one end. When the zoospore matures, the cell in which it has formed ruptures at the upper end and the zoospore

escapes. After swimming about for perhaps an hour, the zoospore settles down and produces a holdfast and a new filament.

Sexual reproduction in *Oedogonium* (FIG. 19-13B-D) is somewhat complicated, but the main features give an adequate picture of the process. Since the gametes are both large eggs and small sperms, the sexual process is oogamous. The eggs are produced singly in enlarged, specialized cells of the filament known as OOGONIA. The specialized cells containing the sperm are termed ANTHERIDIA. These short, disklike cells are found singly or in a row of 2 to 40 or more along the length of the filament. Each antheridium contains two sperms lying side by side. Except for size, these are very much like the asexual zoospores. In some species, oogonia and antheridia are produced in the same filament; in others they are segregated in different filaments.

The sperms escape from the antheridia and swim about freely. A single sperm enters the oogonium through a slit or pore in the oogonium wall and fuses with the egg. The fertilized egg develops a thick wall. As the filament decays, the zygote (oospore) is released and continues a period of dormancy that may last for several months. The oospore then germinates; the nucleus undergoes meiosis, and four zoospores (meiospores) are produced, each capable of growing into a new filament.

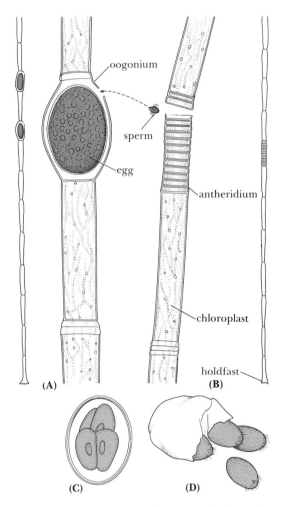

oogonium

sperm

egg

antheridium

chloroplast

holdfast

(A)   **(B)**

(C)   **(D)**

Fig. 19-13 *Oedogonium*, sexual reproduction. (A): female plant, and portion much enlarged, showing oogonium at time of fertilization. (B): male plant, and portion much enlarged, showing antheridia and escaping sperm. (C): zygote, division into four meiospores. (D): escape of meiospores from zygote.

Sex and Sex Differentiation Sexual reproduction is one of the most conspicuous features of land plants. Certainly this aspect of plants is significant to man, for the fruit and the seed, upon which man is largely dependent for food, are produced only as a result of the sexual process. It is therefore interesting to try to identify the

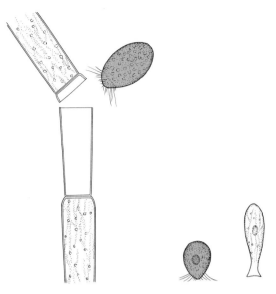

Fig. 19-14 *Oedogonium*, asexual reproduction. (Left): escape of zoospore from cell of filament. (Right): germination of zoospore.

stages that led to the highly specialized methods of sexual reproduction now characteristic of the higher plants.

We may seek among the algae for forms that represent the stages through which plants passed in the evolution of the more advanced methods of sexual reproduction. In this connection, both *Ulothrix* and *Oedogonium* are significant. *Ulothrix* represents the most primitive method of sexual reproduction, in which externally similar, motile gametes escape from the undifferentiated cells in which they are formed and fuse in the water. *Oedogonium*, on the other hand, illustrates the evolution of sexual differentiation (oogamy) in the production of unlike gametes, a large nonmotile egg, and a smaller, motile sperm. Moreover, it possesses sex organs—the oogonium and antheridium—which are differentiated from the vegetative cells of the plant body. Gametic union no longer takes place in the water after discharge of both gametes. The nonmotile egg is retained by the parent plant, and the sperm must swim to the egg if fertilization is to occur. One should not assume that *Ulothrix* and *Oedogonium* are themselves the ancestors of higher plants, but they do possess features in their life cycles that may represent stages in the evolution of our familiar plants.

In the higher plants—the flowering plants, conifers, and others—oogamy is a constant feature. The most primitive living land plants, vascular and nonvascular, are also oogamous. The earliest land plants may, however, have been isogamous, and the occurrence of oogamy in *Oedogonium* may only indicate that the potentialities for sexual differentiation were inherent in the green algae.

Alternation of Generations   Not only the origin of sexual reproduction but also the origin of the alternation of generations, closely associated with the sexual process, may be traced to the algae. In the life cycle of a flowering plant, the most conspicuous and dominant phase—the plant itself—belongs to the sporophyte, or diploid, generation. This is also true of all other vascular plants. The gametophyte (haploid) generation that follows is an inconspicuous and reduced phase of the life cycle.

The plant body is not always the diploid phase, however. In the algae there is considerable diversity with respect to the nature of the two generations. The plant body of most colonial—filamentous and nonfilamentous—green algae belongs to the haploid, or gametophyte, generation. The plant gives rise to haploid gametes, which fuse, forming a zygote. This zygote constitutes the sporophyte generation, for meiosis occurs at the time the zygote germinates. In *Oedogonium*, for example, the fertilized egg is the only diploid cell; all other structures of the plant, including the filament, asexual zoospores, gametes, and meiospores, belong to the gametophyte generation.

In *Spirogyra*, the nucleus of the zygospore gives rise, at the time of germination, to four nuclei, each with the *n* (haploid) chromosome number.

Three of these nuclei abort, but the fourth becomes the nucleus of the first cell of the new filament (FIG. 19-10B). The origin of the diploid body of the higher plants cannot be sought among such species as these, in which the plant body belongs to the gametophyte generation.

Among other types of life cycles found in the algae is that in which the diploid generation is the conspicuous phase in the life cycle, the haploid generation being small and probably much reduced. Such a life cycle, approaching that of the seed plants, is found chiefly among the brown algae. A third type is that in which the two generations are independent and much alike as to size. This life cycle is found in certain green algae (sea lettuce, for example), some brown, and most red algae. The red and brown algae, however, are not believed to have given rise to any higher forms of plant life, and they have become highly specialized in the sea.

The search for forms illustrating the steps in evolution taken by plants prior to their migration to the land is generally restricted to the green algae. Since most present-day green algae have continued to live in an aquatic environment, it is clear that they were not the direct progenitors of early land plants. Other ancient forms became adapted to a terrestrial mode of life and developed plant bodies of large size and complex structure. The ancestors of the present vascular plants probably arose among isogamous green algae in which the two generations were similar. The sporophyte generation gave rise, through evolutionary processes, to the modern plant body. The gametophyte generation, on the other hand, gradually became an inconspicuous part of the life cycle. The evolution of the early vascular plants of the land is discussed further in Chapter 25.

DIVISION 2. EUGLENOIDS

Like the flagellates as a whole, the euglenoids include green and colorless forms, some of the

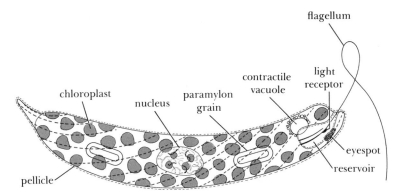

flagellum

Fig. 19-15 *Euglena,* diagram showing structural features.

chloroplast

nucleus

paramylon grain

contractile vacuole

light receptor

eyespot

reservoir

pellicle

latter obtaining food in soluble form and others engulfing solid particles. No photosynthetic euglenoid is known to be phagotrophic. The vitamins B_{12} and thiamine are required for growth by many of the euglenoids.

Few organisms have been more extensively studied than the various species (about 50 in number) of the genus *Euglena*. These organisms are relatively large, readily grown in bacteria-free cultures, and highly suitable for laboratory studies on various metabolic processes. *Euglena* is primarily a fresh-water genus, and "blooms" of *Euglena* often occur in acid or nitrogen-rich waters, forming a green scum. Most species are approximately spindle-shaped, with one end rounded and the other pointed (FIG. 19-15). Reproduction is vegetative, the cells dividing longitudinally. Sexual reproduction is unknown.

A true cell wall is absent, and the protoplast is separated from the environment by a *pellicle.* This consists of two parts. The outermost is the thin plasma membrane. Just within this is a system of flat, spirally arranged, elongated pellicular strips, 30–40 in number in the middle of the cell and fewer at the ends. The strips are flexible and elastic and mainly protein in composition. They interlock with one another by lateral ridges fitting into grooves, thus constituting a single unit that readily permits changes in shape in response to forces acting from within. The spiral striations that can be seen with the phase contrast micro-scope are the lines of overlap between the adjacent strips.

The cell organelles include a relatively large nucleus, chloroplasts, a contractile vacuole, and polysaccharide storage products called *paramylon* that may be associated with pyrenoids in the chloroplasts or free in the cytoplasm as rings or solid rods. The chloroplasts vary in number and shape, depending on the species; they may be disklike, ribbonlike, or assume other forms. At the rounded end of the cell is an opening connected by a canal to an enlarged area, the *reservoir.* On one side of the reservoir lies the *contractile vacuole,* an organelle apparently concerned with the water relations of the cell. It discharges into the reservoir. In some euglenoids, two flagella are attached to the back wall of the reservoir and extrude from the canal. In others, including *Euglena,* only one flagellum is evident on the outside of the cell; the second flagellum is vestigial, short, nonfunctional, and contained within the reservoir. A swelling near the base of the functional flagellum is probably the *light receptor,* which perceives the presence and intensity of light. Opposite the light receptor, and just within the reservoir wall, is the orange-red *eyespot.* This is believed to act as a light-absorbing screen that intermittently shades the light receptor as the cell rotates during swimming. *Euglena* moves toward a region of medium light intensity but away from high light intensity.

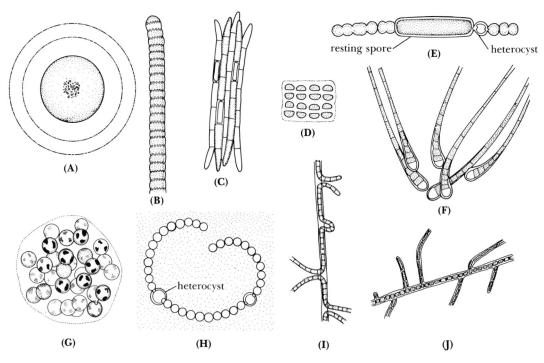

Fig. 19-16 Blue-green algae. (A): *Chroococcus*. (B): *Oscillatoria*. (C): *Aphanizomenon*. (D): *Merismopedia*. (F): *Anabaena*. (F): *Rivularia*. (G): *Microcystis*. (H): *Nostoc*. (I): *Scytonema*. (J): *Hapalosiphon*.

DIVISION 3. THE BLUE-GREEN ALGAE

The approximately 1500 species in this group are usually characterized by a bluish-green color, caused by an accessory pigment in addition to chlorophyll and the carotenoids. A red pigment is sometimes also present, and variations in the relative proportions of these pigments produce a considerable range of color in the plants of this class. The Red Sea owes its name to the occasional abundance of certain blue-green algae in which the red pigment is predominant.

Blue-green algae display little structural differentiation. Unicellular forms occur, but most blue-greens either are nonfilamentous colonies or grow in branched or unbranched filaments (Fig. 19-16). Cells and nonfilamentous colonies are surrounded by a gelatinous sheath, which may be very conspicuous. Sexual reproduction is unknown in blue-green algae. A few experiments indicate, however, that some recombination, possibly following fusion of vegetative filaments, may occur. Recombination mechanisms in the absence of sexual reproduction have been clearly demonstrated in bacteria (Chapter 20). Motile spores are not produced; the only known method of reproduction is vegetative, chiefly by cell division in unicellular forms and by cell division and fragmentation in colonies. Some filaments may also develop resting spores. These are thick-walled spores, which are resistant to heat and drying and which contain accumulated food materials. A new cell grows out of the resting spore at the time of germination. Some filaments also produce enlarged and seemingly empty cells known as heterocysts, whose function is not well understood.

The blue-green algae possess no organs of locomotion, but some of them are nevertheless capable of moving over surfaces by a gliding movement. The mechanism of this movement is unknown.

The cellular organization of the blue-green algae (Fig. 19-17) differs in a number of ways from that of the higher plants, but it closely resembles that of the bacteria (Chapter 20). Cell division takes place by a ringlike extension of the cell wall, like a diaphragm with decreasing aperture, and no cell plate is formed. The DNA material is concentrated in a three-dimensional network extending throughout the cell or concentrated in the central part. This network, which we may refer to as the nuclear body, is the genetic equivalent of the nucleus of higher organisms, but varies in shape from cell to cell, is not surrounded by a nuclear membrane, and has no mitotic figures. As in the higher plants, the chlorophyll is bound to lamellae (Chapter 5), but these are not aggregated into grana. The photosynthetic lamellae form a complex network, sometimes extending throughout the cytoplasm of the cell, but usually arranged in parallel layers in the periphery of the cell. Ribosomes are abundant, but mitochondria and endoplasmic reticulum are absent.

It is generally recognized that the cells of blue-green algae and bacteria, with their lack of a nuclear membrane, absence of mitosis, and lack of mitochondria and membrane-limited plastids (some bacteria are photosynthetic), are so different from all other kinds of organisms that there are two kinds of cells among living things. The presumably primitive cells of blue-greens and bacteria are termed PROCARYOTIC, and the cells of other algae, the fungi, and of all other organisms are called EUCARYOTIC (see Glossary). The bacteria and blue-greens are frequently referred to as the *lower* protista, whereas all other algae, the fungi, slime molds, and protozoa are designated as *higher* protista.

The blue-green algae are widely distributed and occur in varied habitats. Many are aquatic; others are found on damp soil, wet rocks, or attached to plants or animals. They are likely to be abundant in warm ponds rich in organic matter, for they require large amounts of nitrogen. A number of species thrive in polluted waters and are frequently an indication of organic pollution. Blue-green algae occur in hot springs in various parts of the world. It has been found that blue-greens, together with associated bacteria, are able

Fig. 19-17 **Electron micrograph of a single cell of a filamentous blue-green alga (*Tolypothrix*), showing cellular structures.**

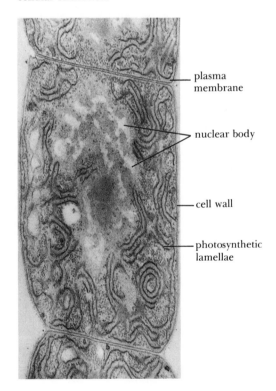

plasma membrane

nuclear body

cell wall

photosynthetic lamellae

to grow in such an environment up to 73°C (163°F), but not beyond. This is approximately the upper temperature limit at which life can exist in an active state—that is, nondormant and growing.

DIVISION 4. CHRYSOPHYTA

The plants of this division are usually divided into three classes; the yellow-greens, the golden-browns, and the diatoms. The last of these is considered separately and in some detail.

The yellow-green algae have yellowish-green plastids, the color due to an excess of beta-carotene. Some indication of the range of structure in this group can be seen in FIG. 19-18. The common and conspicuous *Vaucheria* (formerly classed with the green algae) is frequently studied. The various species may be either aquatic or terrestrial. They may be found growing in dense feltlike masses on the margins of ponds or moist streambanks, or may occur as thin films on garden soil and greenhouse flowerpots.

The plants are composed of tubular, occasionally branched threads or filaments. Terrestrial plants may be attached by colorless, rootlike branches, called rhizoids. The filament is multinucleate, but the nuclei are not separated by cross walls except where reproductive bodies occur. Such a multinucleate body is termed a COENOCYTE. The coenocytic condition is not confined to *Vaucheria,* but is also found in other algae, in fungi, and even in the tissues of higher plants. The cytoplasm lies just within the cell wall and surrounds a large, continuous central vacuole. In the cytoplasm are found nuclei, many disk-shaped plastids devoid of pyrenoids, and numerous oil droplets.

Reproduction (FIG. 19-18) is both asexual and by an oogamous type of sexual reproduction. Asexual reproduction is usually by large zoospores, formed singly in swollen, club-shaped sporangia that are cut off at the ends of branches. The zoospore is multinucleate, and the surface is covered by numerous flagella. These flagella are borne in pairs, and the zoospore is regarded as a compound structure representing a large number of biflagellate smaller zoospores that have failed to separate. The zoospore escapes from the sporangium through an apical pore, swims about for a short time, then settles down, withdraws the flagella, germinates, and produces a new plant.

When reproducing sexually, sex organs—oogonia and antheridia—are usually produced on the same filament, on the same lateral branch, or on adjoining branches. The oogonia are terminal or lateral outgrowths cut off by a wall from the main filament or fertile branch, and enclose a single, large, uninucleate egg, which contains plastids and conspicuous oil droplets. An antheridium is composed of the terminal portion of a lateral branch; it is commonly curved and contains a large number of minute biflagellate sperms. The sperms escape through one or more pores in the antheridium and enter the oogonium through a single pore. One of the sperms fuses with the nucleus of the egg. After fertilization, the zygote forms a thick wall and enters a dormant period. Following germination, the zygote grows directly into a new filament.

The golden-brown algae (FIG. 19-18), like the yellow-greens, are very diverse in form although most are motile unicellular or nonfilamentous colonial forms. The cells bear one or several large plastids containing, in addition to chlorophyll, an excess of certain carotenoid pigments. To this class belong the silicoflagellates and coccolithophores, important producers in the food web of the sea.

Diatoms The diatoms are unicellular or colonial algae (FIG. 19-19), widely distributed in both fresh and salt water. Most species are free-floating, but some live attached to plants or other objects. The cell wall is composed of two halves, or valves, one overlapping the other like the top and bottom of a pillbox (FIG. 19-19). Based on the shape of the cell, two major groups may be distinguished; the pennate diatoms, with bilateral sym-

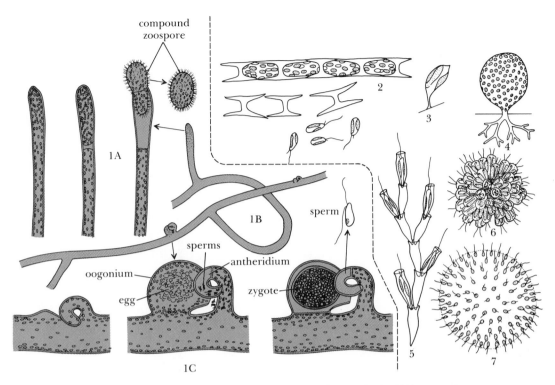

Fig. 19-18 Chrysophyta. (1–4): yellow-green algae. (1A–C): *Vaucheria.* **(1A): stages in zoospore formation and mature zoospore. (1B): habit of plant. (1C): stages in sexual reproduction. (2):** *Tribonema,* **vegetative cells and zoospore formation. (3):** *Characiopsis.* **(4):** *Botrydium,* **growing on soil. (5–7): golden-brown algae. (5):** *Dinobryon.* **(6):** *Synura.* **(7)** *Uroglenopsis.* **The last three often contaminate water supplies.**

metry (Fig. 19-19, A,B,C), and the centric diatoms, with radial symmetry (Fig. 19-20).

The cell wall is composed of an inner layer of pectin and an outer layer of silica (SiO_2), one of the world's most widespread minerals and a common constituent of glass. Silicon must be obtained in soluble form from the environment. When the pectin and the organic contents of the cell are destroyed, the remaining silica shell is transparent. The valves are covered with a great variety of striations, beads, ribs, pores, and other markings that make the diatoms objects of great beauty under the microscope (Fig. 19-20). Pores in the otherwise impervious shell allow contact

between the enclosed protoplast and the aqueous environment. Embedded in the cytoplasm are one to several plastids, bearing a golden-brown pigment that masks the chlorophyll. Food is chiefly stored in the form of oil, and this may often be seen in the cells in the form of large rounded drops.

Diatoms may propagate by a sexual process, but the chief method of reproduction is cell division (Fig. 19-19). The nucleus, protoplasm, and plastids divide and form two protoplasts, each lying within one of the valves. A new wall, constituting an inner valve, then grows over each of the exposed protoplasts. The daughter cells

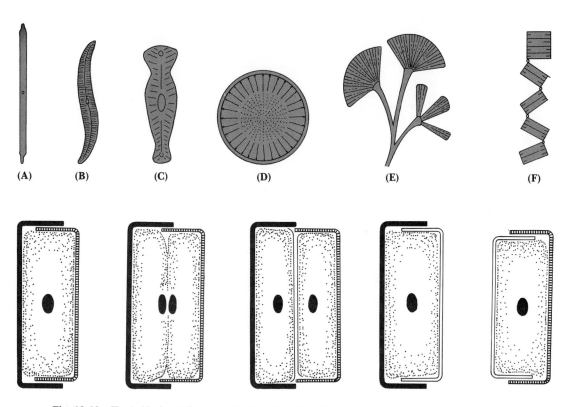

Fig. 19-19 (Top): kinds of diatoms. (A): *Synedra*. (B): *Pleurosigma*. (C): *Gomphonema*. (D): *Cyclotella*. (E): *Licmophora*. (F): *Tabellaria*. (Bottom): Vegetative reproduction in diatoms.

may separate or remain together in a colony, the cells of which are held together by mucilage.

The number of diatom species is large (about 10,000). The number of individuals now living or known to have lived in the geologic past is almost beyond reckoning. Most species are marine, and when these minute plants die, they fall to the sea bottom, and owing to their siliceous nature the cell walls are preserved indefinitely. Great deposits of this material, known as DIATOMACEOUS EARTH, are found in many parts of the world. The largest beds in the United States, some 1400 feet thick, are in California. The beds are sedimentary deposits originally laid down on the floor of the ocean and later raised above the level of the sea.

Because diatomaceous earth is inert chemically and has unusual physical properties, it has become an important and valuable material in industry. It makes an excellent filtering agent, which is widely used to remove coloring matter from products as diverse as gasoline and sugar. Being a poor conductor of heat, it is used to insulate boilers and steam pipes. Because it absorbs sound, it is used in soundproofing. It is also used in the manufacture of paints and varnishes. Because of its hardness, it is used as an abrasive in scouring and polishing powders.

Fig. 19-20 Photomicrograph of a diatom (*Arach-noidiscus ehrenbergii* var. *californica*), showing markings on cell wall.

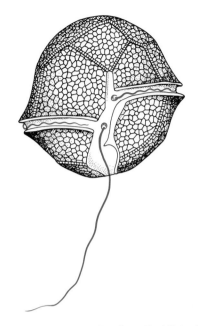

Fig. 19-21 A dinoflagellate (*Peridinium*).

DIVISION 5. PYRRHOPHYTA

Most members of the division Pyrrhophyta are dinoflagellates. These include a great variety of unicellular, motile species, some naked but the majority with cell walls. Their chief characteristic is the presence of two external grooves, or furrows, each containing a single flagellum (FIG. 19-21). One furrow is transverse and completely encircles the cell; the other is longitudinal and extends along one side only. The cell wall, when present, is frequently divided into polygonal plates of cellulose, closely joined together. Several plastids, bearing chlorophyll and a yellowish-brown pigment, are contained within the cell. The usual method of reproduction is cell division. The dinoflagellates are chiefly marine, although some species occur in fresh waters, sometimes in great abundance. A number of marine dinoflagellates, together with certain minute sea animals, are phosphorescent and emit so much light that they are very conspicuous at night, especially if the sea is disturbed. The dinoflagellates, together with the diatoms, play a fundamental role in the economy of the sea.

LIFE IN THE WATER—THE PLANKTON

Both fresh and salt waters contain a huge variety of living organisms. Some live on the bottom, some are attached, and some swim freely. Many float on or near the surface, are nonmotile or feebly swimming, and are subject to the action of tides and currents. These free-floating species are termed PLANKTON and are divided into two groups, the plant plankton (phytoplankton) and the animal plankton (zooplankton). Some constituents of the plankton, such as certain algae and jellyfishes, are large, but most are small, microscopic, or near-microscopic. In common usage the term plankton applies to the more minute rather than to the larger forms of floating life.

The phytoplankton is chiefly unicellular or colonial, and it contains species representing most of the classes of algae, together with some bacteria and fungi. The components of the zooplankton are unicellular or multicellular and include a great variety of small invertebrate animals, together with the larval stages of larger forms. Plankton is of the greatest significance to other forms of life, both aquatic and terrestrial. Its importance in the food chains of the water can hardly be overestimated.

Food chains of the sea We have already emphasized the fact that green plants support all animal life, including man. They constitute the fundamental or primary link of many diverse food chains. In the water, as upon the land, life is dependent upon the activities of autotrophic organisms, but here it is algae that serve as the base of all food chains. The oceans constitute about 71 percent of the earth's surface, with an area of approximately 140 million square miles. The volume of water in the seas is about 324 million cubic miles—about 11 times that of all land above sea level.

The oceans contain plants and animals of great diversity and complexity. So far as plant life is concerned, many essential mineral elements, such as calcium, sulfur, potassium, and magnesium, are usually present in quantity, as are the essential gases, oxygen and carbon dioxide. The plankton population varies from one region to another—in some areas it is sparse; in others, dense. The quantities of nitrogen and, especially, of phosphorus compounds are generally the factors that limit the development of populations.

The phytoplankton, supplied with sunlight and essential elements, synthesize organic food stuffs, just as do the plants of the land. As the animals on land are dependent upon the activities of the green leaf, so the fish and other aquatic forms of animal life are dependent, directly or indirectly, upon algae, and fish in turn are an important item in the diet of larger sea animals and man.

The most important components of the phytoplankton that constitute the ultimate elements of this web of life are diatoms and dinoflagellates. The diatoms are the more numerous, yet both are extraordinarily abundant in the seas, varying in numbers from time to time and from place to place. Diatoms, for example, have been found to occur on the east coast of the United States in numbers up to one to two million per gallon of sea water.

The dinoflagellates are less numerous and have somewhat less significance than the diatoms. Nevertheless, they occur in great numbers, sometimes in such abundance that the water of the ocean is discolored over great areas. Such outbursts occur in the Gulf of Mexico from time to time. The water assumes a yellowish- to reddish-brown color; hence the name "red tide." The number of dinoflagellates in one affected area was estimated to have reached more than 600 million cells per gallon. As a result of a toxin released into the water by the dinoflagellates, many millions of fish are killed and cast up on the beaches of southern Florida. This phenomenon has been reported in this area since 1844. Similar discoloration, or "blooming of the sea," due to dinoflagellates and other algae, has been reported from many other parts of the world.

Food relationships of the water, like those of the land, are numerous and intricate. As on the land, organisms may be classed as producers and consumers. Some food chains are simple, such as the one in which oysters consume diatoms and in turn are preyed upon by starfish or eaten by man. One such food chain constitutes a public health problem. A species of dinoflagellate found chiefly along the Pacific Coast of North America produces a powerful poison so toxic that one millionth of a gram will kill a mouse. Marine mussels consume large quantities of this dinoflagellate and concentrate the poisonous principle in the digestive glands and liver. The poison is not harmful to the shellfish, but the consumption of these mollusks by man during the summer months has resulted in severe illness and a number of deaths.

More commonly, many kinds of organisms intervene between the producers and the ultimate consumers. The phytoplankton is the food for the zooplankton. Outstanding members of the latter group are the copepods (FIG. 19-22), a subclass of the Crustacea. The latter also includes the lobster, sow bug, and barnacle. Copepods are small, free-swimming animals, varying in length from 1 to 10 millimeters, and very abundant. One estimate of the numbers in the Gulf of Maine (FIG. 19-23), for example, reports 6000 to 50,000 per square yard of surface.

Diatoms, dinoflagellates (FIG. 19-23), and other phytoplankton are consumed by copepods, and these are eaten by small fish, which are consumed by larger fish, such as bluefish, swordfish, and tuna, which in turn may serve as food for man.

The zooplankton devoured by larger animals include not only copepods but also larval stages of other crustaceans, minute worms, tiny mollusks, and other animal organisms. Many fish feed upon both phytoplankton and zooplankton at different stages of their life cycles, and large fish may consume both zooplankton and smaller fish. But the yield of animal life in the sea depends to a large degree upon diatoms. As flesh is grass, so fish are diatoms.

Fig. 19-22 A copepod, an important component of the zooplankton.

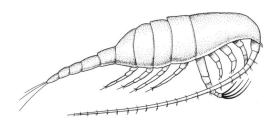

Fig. 19-23 Plankton from the Gulf of Maine. (Left): diatoms (chiefly *Thallassiosira*). (Center): a dinoflagellate (*Ceratium*). (Right): a copepod, *Calanus finmarchicus*, 2–5 mm long.

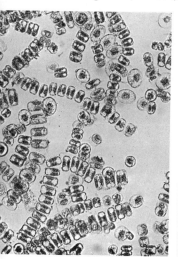
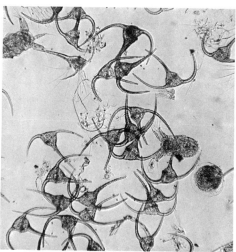

Food from the sea One of the most common of the layman's answers to the problem of food supply in the face of an expanding world population is greater exploitation of the oceans. Present estimates are that the oceans provide 15 percent of the world's consumption of animal protein. Cannot this be increased—are not the resources of the seas inexhaustible? More intense harvesting could take two forms: (1) harvesting at the producer level, that is, the phytoplankton, and thus avoiding the energy losses in the food chains; or (2) increasing the catch of fish. Harvesting the algae and other plankton would, of course, decrease the production of fish, and it has been pointed out that the energy expended in harvesting this crop would exceed that recovered by human consumption. Fishing activities are being expanded, but some species of fish are already overharvested, and some areas of the oceans show signs of overexploitation. The resources of the seas are not limitless, for it must be kept in mind that there are limiting factors in production, including the mineral nutrients required by the phytoplankton. These, chiefly nitrogen and phosphorus, are derived primarily from drainage waters from the great land masses, and must be present in the lighted zones where the phytoplankton live. They are lost to these areas by fishing, and by descent to the bottom of degraded plant and animal bodies, animal excretions, and other wastes. There are, it is true, areas of upwelling from great depths, which through energy fixation and transfer provide rich fishing grounds, but most such areas are already heavily fished. Even if the harvest of fish is doubled by new methods, the oceans, already in part overexploited and polluted, offer no panacea to the food demands of a rapidly increasing world population.

Food chains in fresh water The food relationships in fresh water are similar to those of the ocean. Diatoms are abundant, but the dinoflagellates are much less significant than they are in the sea. In addition to diatoms, unicellular and colonial forms of blue-green, green, golden-brown,

and other fresh-water algae are consumed by the zooplankton. Important components of fresh-water zooplankton, in addition to copepods, are water fleas, rotifers, and the larvae of many kinds of insects. Small fish, such as minnows, suckers, sunfish, bream, and gizzard shad, feed upon the zooplankton and are in turn eaten by game fish such as trout, pickerel, pike, perch, and bass. The larger fish also eat the larger zooplankton, together with larval and adult insects.

As a means of increasing food production, and for recreational purposes, streams and lakes are frequently stocked with fish. Before this is done, the waters concerned should be surveyed thoroughly to investigate such factors as temperature, degree of pollution, and especially the relative abundance of the natural foods of the fish. Stocking has sometimes been undertaken without a full knowledge of the ecology of the stream or lake concerned. It is a wasteful procedure unless the natural foods are present in sufficient amounts to support the growth of fish.

In some parts of the world, including North America, the yield of fish from farm ponds has been greatly increased by the addition of mineral fertilizers to the pond waters. These, under proper conditions, stimulate the growth of algae, and this additional phytoplankton starts a wave of growth moving up the entire food chain and ending with such food fish as bass. This method, in its many variations, has been an impetus to establishing farm ponds as an integral part of a sound, long-range, land-use program.

Algae and Water Supplies While algae are essential as producer organisms in both ocean and fresh-water food chains, they can also be highly deleterious. In the spring and autumn, not infrequently also in the summer months, the phytoplankton in ponds, streams, lakes, and reservoirs may become so abundant as to be extremely conspicuous. The water becomes cloudy and may have a brown, red, yellow, or green tinge. A floating mat or scum may develop. These manifestations of algal growth are termed "waterblooms." Such concentrations are extremely objectionable, not only in public water supplies but also in waters used for bathing, fishing, and other recreational purposes. The algae responsible fall in four groups: (1) blue-greens; (2) diatoms; (3) filamentous and nonmotile nonfilamentous greens; and (4) flagellates, including motile greens, euglenoids, dinoflagellates, and yellow-greens. Species belonging to more than 135 genera of algae have been recorded as important or common in waterblooms of North American fresh waters. Both living and decaying algae impart fishy, musty, or pig-pen odors and tastes to the water. Toxic decomposition products are formed by certain blue-green algae, and these are known to have caused the death of domestic animals that have drunk heavily infested water. The presence of algae in great numbers also causes difficulty in the filtration processes of water-treatment plants.

Various methods have been used to control such algal growth in lakes and reservoirs, including the use of chemical compounds. The most widely used of these is copper sulfate, commonly in concentrations of less than one part per million. This algicide has a number of disadvantages and its effect may be only temporary. For example, the chemical may destroy the major offenders, but more resistant species may survive and produce uncontrollable blooms as a result of the sudden release of nutrients accompanying bacterial decay of the dead algae.

Algal blooms may occur in relatively unpolluted lakes or reservoirs—that is, those with a low mineral nutrient content—but much more important are the problems that arise in lakes or streams that receive an unusual supply of nutrients, resulting in a bloom of large magnitude and long duration. This may come about when waters are polluted by raw or partly treated sewage (including household detergents), drainage waters from farm lands, or waste from industrial operations. The events that follow are a part of the general story of the deterioration of man's environment as the direct result of human activities. In these events, algae play a prominent role.

We may briefly consider, for example, the deterioration of the innumerable fresh-water lakes throughout the world. Under natural conditions, lakes age and eventually become extinct, that is, they fill in and degrade to dry land. Dissolved and particulate matter of inorganic and organic origin is carried into the lake through inlets. The resulting gradual enrichment of the water by nutrients, permitting the growth of algae and fish and other animals, is known as *natural eutrophication*. The organisms die and, together with mineral particles, add to the lake bottom. Under undisturbed conditions the extinction of a lake may take thousands of years. When polluted by sewage and other organic materials, however, the lake will usually undergo a rapid, artificial, or *cultural eutrophication*, which hastens the process of aging to the extent that many deleterious changes may be observable within a few decades. Sewage is rich in inorganic phosphates and nitrogen, and these stimulate a population explosion among the phytoplankton, which tend to change in composition, blue-green algae often predominating. The concentration of dissolved oxygen drops markedly, in large part through the activities of decay bacteria. The bottom layer of water is especially affected, often becoming so anaerobic that decay is inhibited, and this affects the bottom-dwelling animal life of the lake. The organic residues, sinking downward, build up the lake bottom, increasing the depth of the sediment. The increased nutrient supply may also stimulate the growth of rooted aquatic plants and attached filamentous algae in shallow waters along the lake margins, so that the lake shore encroaches upon the water as the bottom builds up.

An increased rate of eutrophication may also result from the deforestation of the drainage area of a lake. This may lead to increased erosion and transport of silt into the lake. The water may also be enriched by a higher rate of flow of nutrients from the land. Eventually a lake might be reduced to a pond, a pond to a marsh, and a marsh to a dry meadow.

All of these processes may occur even in the

world's largest lakes, especially those in heavily populated areas. Evidence of cultural eutrophication is found in many of the lakes of North America, notably in certain of the Great Lakes, which contain more than 20 percent of the world's fresh water. Accelerated eutrophication is most evident in Lake Erie. In certain areas of this lake, in 1930, the bottom-feeding may-fly nymphs (*Hexagenia*) occurred at a density of 400–500 per square meter; in 1961 there were few or none. These organisms require dissolved oxygen, and are important as food for fish. The catch of commercially important fish has been greatly reduced. For example, the catch of blue pike, with an average annual production of about 15 million pounds, was reduced to 1.4 million pounds in 1958, to 79,000 pounds in 1959, and to 1000 pounds in 1962.

Cultural eutrophication has also proceeded in the rivers of the United States, to the extent that relatively few have any similarity to natural waters. Rivers can die, as can lakes; they become choked with wastes whose decomposition uses the oxygen upon which aquatic life depends. The accelerated processes of cultural eutrophication can, however, be checked, and even to some extent reversed, if man will refrain from practices that hasten the natural aging of waters.

DIVISION 6. THE BROWN ALGAE

The brown algae are distinguished from other algae by their brown or olive-green color and also by the structure of the plant body and the reproductive organs. They are almost entirely marine, being widely distributed on ocean coasts, especially in the colder seas. About 1000 species have been described. Brown algae grow on rocks in water from 5 to 75 feet deep and extend shoreward to areas just covered by high tide. All species are multicellular; some are slender, simple forms, whereas others develop considerable size and external differentiation. Among the more common of the brown algae are the

Fig. 19-24 *Fucus*, portion of thallus. The larger swellings (receptacles) contain the sexually reproductive structures. The smaller swellings are air bladders.

rockweeds and bladderwracks, belonging to the genera *Fucus* and *Ascophyllum.* The rockweed *Fucus* is world wide in its distribution, chiefly in the north temperate zone. It ranges from 1 to 3 feet in length, and is attached in great masses to rocks exposed at low tide (FIG. 1-7). Air bladders along the sides of the thallus lift the forked branches toward the surface. The tips of some of the branches are enlarged and contain the sex organs (FIG. 19-24).

Another well-known genus of brown algae is *Sargassum.* Most species grow attached along the rocky shores of the temperate or tropical zones, but one, *Sargassum natans,* is the principal component of masses of floating seaweed in the North Atlantic. These masses vary in size from those composed only of a few plants to others several hundred yards across. The area in which they occur is known as the Sargasso Sea. This "sea" is restricted by ocean currents to an oval area of more than 2 million square miles, extending from the Bahamas to the Azores. The plants multiply indefinitely in the open sea by vegetative reproduction, their only known method of reproduction. Most species, however, are known to reproduce sexually. *Sargassum filipendula* (FIG. 19-25) illustrates the general appearance of plants of the genus. It is a foot or more long, much branched, and bears small, stalked, air bladders.

The largest brown algae, the KELPS, are among the giants of the plant kingdom. They reach their greatest size along the western coasts of North America (FIG. 19-26), where some attain a length of well over 100 feet. Kelp is usually composed of one or several expanded leaflike blades connected by a stalk to a rootlike holdfast that has grown firmly to some rock at the bottom (FIG. 19-27). Air bladders, rounded and hollow expansions of the stalk, are found on the thallus of many species. The internal tissues are frequently differentiated—so much so that they possess cells that structurally approach the sieve elements of the vascular plants.

Kelps, together with other algae, extract a number of chemical elements from sea water, notably potassium and iodine, and accumulate them in their tissues. Potassium chloride may constitute as much as 32 percent of the dry weight of kelps. The concentration of iodine in kelp may be as much as 20,000 times that of sea water. In the United States the commercial exploitation of kelp for these chemicals is unprofitable because of their availability from other sources. Many species of kelp are used as food for man, especially in the Orient. In northern Europe and other parts of the world they may also serve as feed for farm animals. Kelp has long been used locally as a fertilizer; it is high in nitrogen and potassium, but low in phosphorus.

Extracts of kelp are important in a number of industrial processes. Probably 50 percent of commercial ice cream is stabilized with alginates, colloidlike materials obtained from the cell walls of brown algae. The use of this substance imparts a smooth consistency to the frozen product and prevents the formation of large ice crystals during storage. Alginates are also used in the manufacture of pharmaceutical products, such as pills, ointments, and dentifrices, and in cosmetics such as shaving creams and lotions. Many other commercial uses of alginates could be listed.

Fig. 19-25 *Sargassum filipendula*, showing branching nature of plant body and stalked air bladders.

Reproduction in the Brown Algae Alternation of generations in the brown algae is generally well defined, the large and conspicuous plant body usually constituting the 2n generation. In a common pattern of life cycle, the 2n thallus, in localized areas, liberates motile, pigmented meiospores after meiosis. In the larger forms, such as the kelps, the meiospores are produced in such numbers that they may contribute importantly to the phytoplankton. For example, in a dense bed of *Nereocystis*, a large annual kelp widely distributed on the west coast of North America, the number of meiospores produced through the period June to September was estimated at 3

Fig. 19-26 Kelp on rocks at low tide, Puget Sound, Washington. *Egregia, Laminaria,* and *Fucus.*

Fig. 19-27 *Pelagophycus porra,* a large kelp of the Pacific Coast. Holdfast at lower right; air bladder at top.

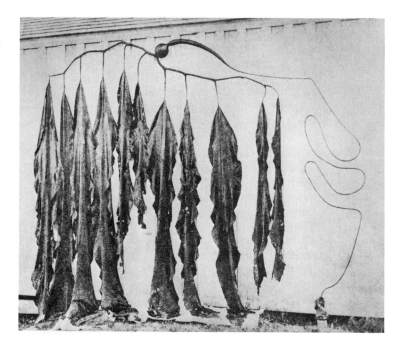

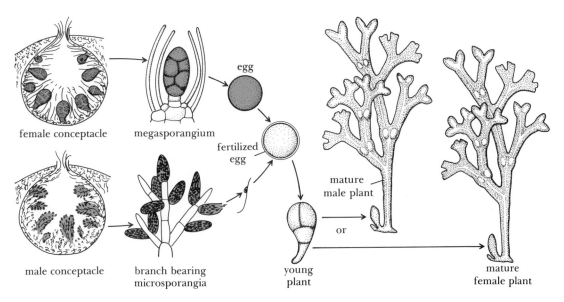

female conceptacle megasporangium

egg

fertilized egg

mature male plant

or

young plant

mature female plant

male conceptacle branch bearing microsporangia

Fig. 19-28 Sexual reproduction in *Fucus*. Gametophyte generation is shown in color.

million meiospores per liter per day. The meiospores of brown algae became attached and grow into multicellular gametophytic plants, sometimes approaching the sporophyte in size, but usually smaller and inconspicuous. The *n* generation bears the gametes, either isogamous and both flagellated or a nonmotile egg and a motile sperm. Following fertilization, the zygote develops into a new diploid sporophyte. The brown algae, like the greens, probably arose from flagellate ancestors.

The life cycle in one order of the brown algae, the Fucales, differs from all the others in two important respects: (1) the gametophyte generation is greatly reduced and (2) zoospores are absent. This order is represented by the genus *Fucus*, the species of which are widely distributed, common, and frequently the object of laboratory investigation.

Aside from fragmentation, *Fucus* reproduces only by a sexual process. The plant is oogamous. The eggs and sperm are produced on swollen branches of the thallus, known as *receptacles* (FIG. 19-24). The surfaces of the receptacles are marked by pores, visible to the naked eye. These open into spherical cavities, called *conceptacles*, that contain the gametes. In some species the eggs and sperms are produced in the same conceptacle, but more commonly they are found in separate cavities on different plants.

In one species commonly studied (*Fucus vesiculosis*), the sexes are segregated on different thalli (FIG. 19-28). Within the female conceptacles are numerous stalked megasporangia. The young megasporangia are unicellular, with a single $2n$ nucleus. Following meiosis, four megaspores are produced, and each of these divides again (by mitosis), eight cells resulting. These constitute the highly reduced female gametophyte, for no further divisions occur and each of the eight cells functions directly as a nonmotile female gamete, or egg. The male conceptacle bears numerous oval microsporangia arranged in clusters on short, branched filaments that arise from the bottom and sides of the conceptacle. Each microsporangium gives rise to four microspores following meiosis, and each microspore divides four times by mitosis, forming a reduced 16-cell male gametophyte. Each of these cells functions directly as a motile male gamete; a total of 64 gametes is thus produced within each microsporangium. Numerous colorless and unbranched hairs are found in both kinds of conceptacles.

When the gametes are mature, they are ejected into the water. Both eggs and sperms are at first surrounded by a thin membrane, but this shortly ruptures and the gametes are liberated. Free nonmotile eggs are found in two orders of the brown algae, but this condition is unusual because, in general, nonmotile eggs are not liberated prior to fertilization. Once freed, the sperms swarm around the egg until one sperm penetrates and fertilization occurs. There is some evidence that the egg produces a chemical attractant, but its nature is unknown. After fertilization the zygote develops a cell wall, becomes attached, and grows directly into a new plant.

The life cycle of *Fucus* is basically similar to that found in the vascular plants, especially the seed plants. Here, again, we find a dominant $2n$ sporophyte and the results of an evolutionary trend toward a reduced gametophyte, completely dependent upon the sporophyte. This reduction of the gametophyte generation has apparently occurred independently in many of the larger plant groups, and is not evidence of relationships among them.

DIVISION 7. THE RED ALGAE

The red algae are characterized by their distinctive red color, although certain species may show shades of green or brown, and by the nature of their reproductive structures. The spores and gametes are not flagellated, and sexual reproduction usually involves a highly specialized kind of oogamy. Although the red algae may have originated from the flagellates, actually there is

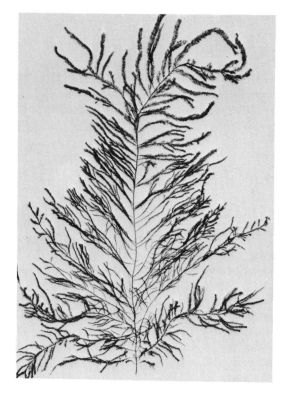

Fig. 19-29 The red alga *Dasya*.

no evidence among living forms to support such an hypothesis. They are probably most closely related to the blue-greens.

Like the browns, the red algae are almost exclusively marine, fewer than 3 percent of the 2500 species living in fresh water. They are widely distributed in the seas, chiefly in the warmer waters. Red algae grow on rocks in tidal zones and also in the water as deep as light can penetrate—300 feet or more below the surface. The plant body is relatively large, although never approaching the larger brown algae in size. Individual plants seldom exceed 3 feet in length. Some forms are filamentous, and a very few are unicellular, but most are complex in structure and may be branched and feathery (FIG. 19-29), or flat and expanded.

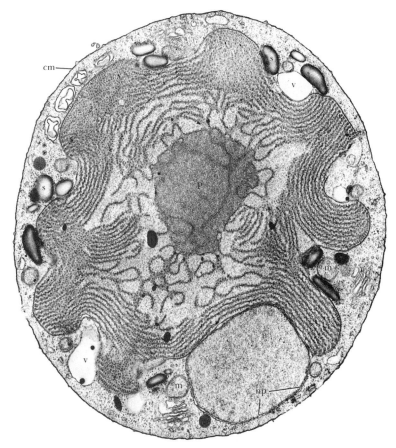

Fig. 19-30 Electron micrograph of a section through a one-celled red alga, *Porphyridium cruentum*. The single, large chloroplast is composed of mostly parallel lamellae. The granules associated with the lamellae contain phycoerythrin and phycocyanin. n: nucleus; np: nuclear pore; er: endoplasmic reticulum; cm: cell membrane; ch: chloroplast; p: pyrenoid; v: vacuole; s: starchlike bodies; m: mitochondrion. (14,850×)

A few genera of red algae are unicellular. A species of one of these, *Porphyridium cruentum*, is terrestrial, growing as a thin, gelatinous red film on soil and damp walls. It has been grown in pure culture and used extensively in experimental studies. Its ultrastructure is shown in Fig. 19-30.

Algae and Limestone Formation Many species of algae withdraw calcium from both fresh and salt water, and deposit it, in the form of calcium carbonate, in their cell walls or in gelatinous sheaths. The most significant forms in this category are the blue-greens and reds, but certain green algae and chrysophyte flagellates are also concerned. The blue-greens are chiefly important in fresh waters; they are responsible, for example, for the formation of extensive limestone deposits around hot springs and geysers in Iceland and in Yellowstone National Park. The red algae are the most important calcareous algae of the seas; in particular, they play a significant role in the development of coral reefs and islands. Although true coral results from the activities of minute sedentary animals, it is recognized that lime-secreting red algae may be as important, sometimes more important, in the formation of coral reefs as the coral organisms themselves. The encrusting red (coralline) algae bind together the skeletal material of the coral animals, and it is the integrated development of the animals and the algae that builds up the coral reef. In some reefs, atolls, and islands, the coral animals have been most important; in others, which should perhaps be called algal reefs or algal islands, the red algae have predominated. The calcareous red algae are best developed in the warm seas, but certain species also flourish in temperate and polar regions, where they form extensive banks of limestone in coastal areas. The algae are not only important in the present age in the formation of calcareous deposits, but also they have played a significant part in the production of ancient beds of limestone rocks, which may be 1000 feet thick.

Uses of Red Algae The red algae are used even more widely than the brown. Some of the red algae are among the most important food plants of Asiatic countries, where they are gathered in quiet bays and lagoons. Red algae are also eaten extensively by inhabitants of the rocky coasts of Europe and North America. "Laver," "dulse," and "Irish moss," or "carrageen," (Plate 10) are among the common names applied to species of red algae used for food or other purposes. The industrial uses of red algae are similar to those of the brown algae. They are used as a substitute for animal gelatin in manufactured puddings; as a sizing agent; in pie fillings, preserves, and candies; in toothpaste; as a filler and stabilizer in ice cream; and in many other ways.

An important use of the red algae is in the production of AGAR. This is a dried and bleached gelatinous extract obtained from several species of red algae. One of the most important uses of agar is as an essential ingredient in the preparation of media for the growth of bacteria and fungi. As such, it is indispensable in bacteriological laboratories because no adequate substitute for agar is known.

SUMMARY

1. The algae, in general, are phototrophic plants and the plant body is a thallus. They are usually found in aquatic habitats. With the bacteria and fungi, they are classified in the subkingdom Thallophyta.

2. Reproduction in the algae is commonly both sexual and nonsexual. Nonsexual reproduction occurs by cell division, fragmentation, and zoospore formation. Sexual reproduction may be isogamous or oogamous. Oogamy probably arose from isogamy.

3. The algae may be classified in seven divisions. Among the features that characterize the divisions are the kinds of pigments, the nature of the stored food, and the number, kind, and insertion of flagella.

4. Several divisions of the algae include flagellates. Most of these are phototrophic but some take in food in solution through the cell membrane or in particulate form (phagotrophism). They may require vitamins from external sources.

5. Attention is called to a classification of organisms into plants, animals, and protista, the latter including flagellates and all other algae, bacteria, slime molds, and fungi.

6. The green algae are the most significant from the point of view of evolution of higher forms of plant life. Unicellular flagellates may represent the ancestors of filamentous and nonfilamentous colonies. The filamentous *Ulothrix* and *Oedogonium* may illustrate stages in the evolution of sexual reproduction and sex differentiation.

7. In most green algae the zygote is the only cell of the diploid phase of the life cycle. If the zygote, at germination, produces motile spores, these may be called meiospores to distinguish them from asexually produced zoospores.

8. It is probable that alternation of genera-

tions first arose in the algae. Land plants may have arisen from green algae in which the gametophyte and sporophyte generations were independent and similar in size.

9. The euglenoids are flagellates in which true cell walls are absent. They include both green and colorless forms, some of the latter phagotrophic.

10. The blue-green algae are unicellular or colonial. Sexual reproduction is unknown. Plastids are absent, and the pigments are attached to lamellae that extend throughout the cell. The DNA material occurs in three-dimensional bodies, not bounded by a membrane. Mitosis does not occur, and mitochondria are absent.

11. The coencytic *Vaucheria* is an example of an organism in which nuclear division is not followed by formation of cell walls.

12. Algae lie at the base of all aquatic food chains. In the sea, these algae are chiefly diatoms and dinoflagellates. Copepods are important links between the phytoplankton and higher forms of animal life. Food chains, similar in general outline to those of the sea, occur also in fresh waters.

13. Algae are important in the deterioration of water supplies, especially when these are polluted by organic and inorganic wastes. The enrichment of waters by nutrients is known as eutrophication.

14. Brown algae include the rockweeds, bladderwracks, *Sargassum*, and the kelps. *Fucus*, a rockweed, is oogamous, both eggs and sperm being discharged into the water prior to fertilization. The plant body of *Fucus* belongs to the sporophyte generation.

15. The red and brown algae have a number of economic uses. Important products are agar, and fillers and stabilizers.

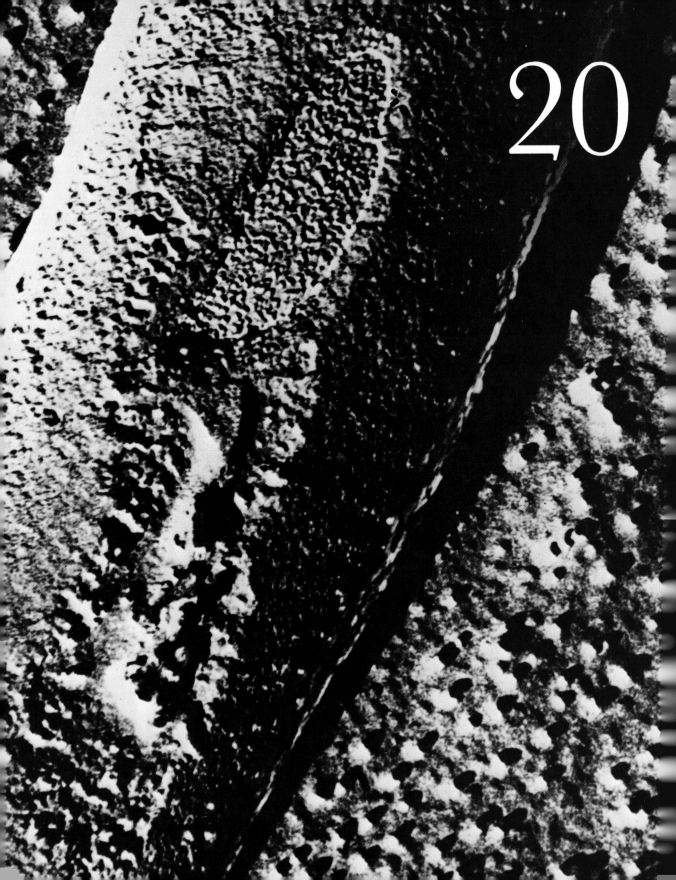

20

Bacteria and Viruses and Some Relations to Man

ORGANISMS, FOR PURPOSES OF CONVENIENCE, are sometimes classified as macroscopic or microscopic (microorganisms), the latter invisible to the unaided eye. These include a vast array of unrelated kinds of life, differing greatly from one another in form, life cycles, and ways of living. Many algae and some fungi are microorganisms, as are the protozoa, bacteria, and viruses, although, as we shall see, the viruses are probably not, strictly speaking, organisms. While there are great differences among microorganisms, the various techniques employed in studying them are very similar, and for this and other reasons they have come to constitute the subject matter of a single field of knowledge, MICROBIOLOGY. Of these microorganisms, we have already considered examples of those classified among the algae; the fungi are dealt with in later chapters. The topics of this chapter are the bacteria and the viruses.

THE BACTERIA

Bacteria affect man in innumerable ways. As causes of infectious diseases of man, their death toll has exceeded the devastation of all wars. Yet the number of pathogenic (disease-producing) bacteria is greatly exceeded by the number of nonpathogenic forms that play a beneficial role in the economy of nature. If, for example, the dead bodies of plants and animals were not reduced to simple compounds by the activities of bacteria and other microorganisms, the elements essential to plant and animal life would soon be tied up in organic compounds, leaving none for future generations.

Discoveries of the nature and activities of bacteria rank high among the scientific accomplishments of modern times. Pasteur, and other investigators of his day, identified bacteria as the causes of fermentation and disease. This knowledge laid the foundations for aseptic surgery, for

Photo at left shows external view of cell wall of *Bacillus subtilus* under electron microscope. The oval structure at top center is the surface of the plasma membrane exposed by removal of a portion of the cell wall.

the control of epidemics, for improved quarantine procedures, for modern methods of sewage disposal, for the purification of water supplies, and for measures to increase the purity of milk and other foods and to preserve them for longer periods. Studies on bacteria have also made possible their utilization in industry and in agriculture. Bacteria play an important role in our everyday lives—in more ways than most of us suspect—but to understand these various bacterial activities some knowledge of the structure and physiology of the bacteria themselves is essential.

The Classification of Bacteria   The discovery of bacteria by Anton van Leeuwenhoek in 1676 was followed by numerous attempts to classify them. Until the middle of the nineteenth century they were regarded by most workers as animals; later they came to be viewed as more closely related to plants than to animals. Even today their evolutionary relationships to other organisms are not clear. This situation has led to differing points of view among those who deal with these minute forms of life. By some they are placed in the protista, a group of structurally simple organisms that cannot readily be classified as either plants or animals (Chapter 19). By others they are classified in the plant kingdom, since it is considered that in most of their structural features they resemble plants more than they do animals. Most true bacteria possess rigid cell walls, unlike the flexible membranes of animal cells. They take in food only in soluble form through the cell walls, whereas animals characteristically ingest solid food, which is then digested within their bodies. Some authorities view bacteria as related to the fungi because they usually lack chlorophyll and because of the existence of a number of kinds of organisms interpreted as transitional between bacteria and fungi. In this text, bacteria are classified as plants and as the principal order of the division Schizophyta, coordinate with the fungi under the Thallophyta.

Size and Form   Bacteria are among the smallest of living things—so small, in fact, that the small-est unit employed in light microscopy is used to measure them. This is the micron (μ), 0.001 millimeter, or approximately $1/25,000$ of an inch. The spherical bacteria are about 1 micron in diameter; the rod-shaped forms range in width from 0.2 to 2 microns and in length from 0.3 to as much as 10 microns. It would take more than a thousand of one of the largest bacteria, and from 10,000 to 50,000 of a species of average size, placed end to end, to equal an inch. A single drop of liquid can contain up to 50 million bacteria. Most bacteria are barely visible under the high power of the ordinary microscope (about 400 power), and greater magnifications, especially with the electron microscope, are necessary to reveal details of their structure.

True bacteria are recognized as belonging to one of three types, depending upon their shape (Fig. 20-1). The majority are rod-shaped; these are the BACILLI (sing. BACILLUS). The second largest group is the COCCI (sing. COCCUS), spherical in form. Last, and smallest in number (probably fewer than 100 species) are the SPIRILLA (sing. SPIRILLUM), spiral or corkscrew-shaped bacterial.

Cell Structure   Bacteria are one-celled organisms. The cell consists of a rigid multilayered cell wall and cytoplasm containing hereditary material. As in the blue-green algae, ribosomes are abundant, but mitochondria are absent. Respiration is probably located in the plasma membrane, which is thus functionally comparable to the internal membranes of mitochondria. Reserve foods and pigments may be found within the cell. The cell wall is chemically complex, containing amino acids, sugars, and other substances that in some species include lipids. Cellulose is absent from the bacterial cell wall. The presence of a cell wall can be demonstrated by staining procedures, by electron microscopy, and by plasmolysis experiments. In some species the cell wall can be digested by the use of enzymes.

Many bacteria secrete an extracellular viscous or gelatinous material called SLIME, usually composed of polysaccharides with sometimes other

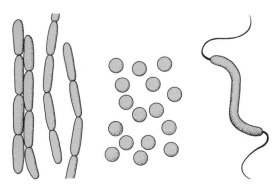

Fig. 20-1 The fundamental shapes of bacteria. (Left): bacilli. (Center): cocci. (Right): spirillum.

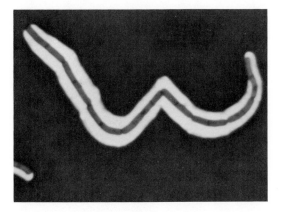

Fig. 20-2 Capsule of *Bacillus megaterium* (950×).

substances. The slime may be loosely attached, diffuse, and tending to spread into the environment, in which case it is termed LOOSE slime. But frequently the slime accumulates in a conspicuous layer around the cell wall and is attached to it. The cell is then said to possess a CAPSULE (FIG. 20-2). The slime, whether loose or as a capsule, presumably protects the cell in some degree against unfavorable environmental factors. Many bacteria develop capsules, and in some pathogenic bacteria there is a relationship between the presence of a capsule and virulence. Encapsulated bacteria may escape the destructive

action of the white blood cells that commonly ingest and destroy bacteria. Strains of the cocci (*Diplococcus pneumoniae*) that cause pneumonia are encapsulated. These are pathogenic, whereas mutant strains of pneumococci lacking capsules are nonpathogenic.

The nature of the bacterial genetic material has been extensively investigated, especially with the electron microscope. As in the blue-green algae, most or all of the DNA of the cell is concentrated in a three-dimensional "nuclear body" extending throughout the cell (FIG. 20-3). This is the genetic equivalent of the nucleus in other organisms. However, as in the blue-green algae, the nuclear body is not separated from the cytoplasm by a membrane, and mitosis does not occur. A bacterial cell may contain one or two, or even more, of these irregularly shaped bodies, for at certain stages of growth they divide at such a rapid rate that there is no time for cross walls to form between them, and the cell becomes multinucleate. It will be recalled that a cell of the type just described, in which mitosis and mitochondria are absent and the organelles are not bounded by membranes, is termed procaryotic (Chapter 19). A combination of genetic and physical studies has revealed that the nuclear body is composed of DNA in the form of a tightly packed, double-stranded, elongated thread about 1000 microns long. The strand is circular, that is, closed at both ends. The DNA macromolecule will, for convenience, be referred to as a chromosome, but it is not combined with protein as is characteristic of a true chromosome. The DNA molecule replicates, the products of the duplication becoming incorporated into different cells.

Motility Many species of bacteria are motile, as may readily be observed under the microscope. Motility is brought about by hairlike flagella (FIG. 20-4) composed entirely of protein that differs in composition from cytoplasmic protein. They originate in the plasma membrane, not in the cell wall. Flagella are observable under the light microscope only with high magnifications and

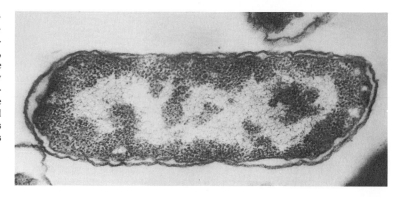

Fig. 20-3 Electron micrograph of section through a bacterial cell (*Hemophilus influenzae*). Note the cell wall, the plasma membrane, and the centrally located, irregularly shaped, nuclear body composed of fibrils of DNA. The darker granular material outside the nuclear body is composed of ribosomes (35,300×).

special staining methods. They differ in number and distribution in different species. Some bacteria have a single terminal flagellum, while others have one or a cluster of flagella at one or both ends of the cell. Still others are completely surrounded, terminally and laterally, as in Fig. 20-4. Most motile bacteria are either bacilli or spirilla; cocci are usually nonmotile.

Reproduction The usual method of reproduction in true bacteria is vegetative, brought about by transverse fission; thus, the origin of the name Schizophyta (fission plants). The cell becomes divided into two equal parts by the inward growth of a ring, as in the blue-green algae. The daughter cells may separate immediately, or they may remain adherent, forming chains or masses of cells.

Each daughter cell quickly enlarges and then divides again. Under favorable conditions, some bacteria divide as often as once every 20 minutes. At this rate of division a single cell could give rise to nearly 70 billion cells in 12 hours. In 24 hours a single cholera bacterium could theoretically produce descendants weighing 2000 tons. But such a rate of division could be maintained for only a short time. The exhaustion of food materials, water, and oxygen (for aerobes), the accumulation of toxic waste products, and changes in the pH of the environment drastically limit the bacterial population.

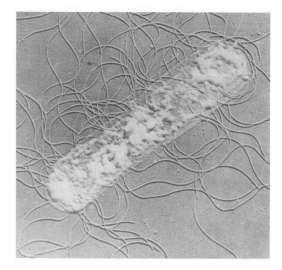

Fig. 20-4 Flagella of *Proteus vulgaris*, a saprophyte found chiefly in putrifying materials. (Electron micrograph; ~3400×.)

When bacteria adhere together after division, their arrangement depends upon the plane in which the cells divide. In bacilli, the plane of division is always at right angles to the long axis of the cell, and the daughter cells, if they remain attached end to end, form a chain. The cocci, on the other hand, divide in one or more planes, depending on the species; hence, various cell arrangements result (Fig. 20-5). STREPTOCOCCI divide in only one plane, with the planes of division parallel, and they remain together in chains. STAPHYLOCOCCI (and very closely related micrococci) divide in two or more planes in such a way that flat plates or loose masses like a bunch of grapes are formed. DIPLOCOCCI form groups of two or multiples of two. SARCINAE divide in three planes, producing cubes or cubical packets of eight or multiples of eight. Spirilla usually grow as single cells, although they sometimes occur in short chains.

Recombination in Bacteria Recombination is usually brought about by sexual reproduction, which, as previously discussed, includes both (1) the fusion of two haploid nuclei, forming a diploid nucleus containing all the genes of both parents, and (2) meiosis. Recombination as a result of the sexual process should be termed sexual genetic recombination. But recombination, which in a broad sense means any rearrangement of genetic material, can also be brought about by other mechanisms. Those found in bacteria are classed as follows: (1) transformation, (2) conjugation, and (3) transduction. The first two involve only bacteria, and are dis-

Fig. 20-5 Cell aggregations among cocci. (A): diplococci. (B): streptococci. (C): sarcinae. (D): staphylococci.

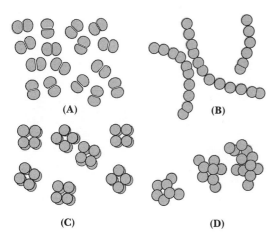

(A) (B)

(C) (D)

cussed below. The third, which involves in addition certain kinds of viruses, is considered later in this chapter. In all of these mechanisms, genetic material is transferred from a *donor* cell to a *recipient* cell, but usually only a portion of the genetic material is transferred. The recipient cell is thus incompletely diploid. For this reason, and because meiosis is unknown, bacteria cannot be said to reproduce sexually.

Transformation The discovery of this process by an English bacteriologist in 1928 was important, not only because it showed that hereditary factors could be transferred from one bacterium to another, but also because it laid the foundation for the later identification of DNA as the hereditary material. When mice were inoculated, either with a culture of living, noncapsulated (avirulent) pneumococci or with a heat-killed suspension of capsulated (virulent) pneumococci, no infection resulted. But if a mixture of living avirulent and heat-killed virulent organisms was injected, the mice died of pneumonia, and living, capsulated pneumococci were recovered at autopsy. It appeared that some transforming mechanism, derived from the heat-killed pneumococci, had converted the living, avirulent organisms, which had lost the capacity to produce a capsule, into forms able to synthesize such a structure and thus become virulent. In 1944, workers at the Rockefeller Institute (now University) identified this transforming substance as DNA. This was a landmark in the history of genetics—the first evidence that DNA possesses and transmits hereditary information.

Investigations since 1945 have confirmed this finding, and have demonstrated that genetic transformation in bacteria is complex, involving several stages. It is clear that in transformation the donor and recipient cells need not be in contact. Considerable evidence exists for the following simplified concept. Double-stranded fragments of DNA, each representing 1/100–2/100 of the bacterial chromosome, are released as a result of the breakdown of a donor cell (FIG. 20-6). These fragments, which contain one or

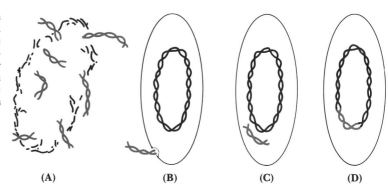

Fig. 20-6 Transformation in bacteria. (A): release of DNA (color) from donor cell. (B): uptake of DNA by the recipient cell. (C): pairing of donor DNA with similar region on recipient DNA molecule. (D): replacement of recipient genes by donor genes.

(A)   (B)   (C)   (D)

several genes, enter the recipient cells at several special entry sites. Following entry, the fragments of donor DNA replace similar regions of the recipient DNA, thus forming recombinants. Thus, a recipient, avirulent cell, which has lost by mutation the ability to form a capsule, now reacquires the gene by transformation. This mechanism has now been found in about 20 other species of bacteria, representing about 10 genera and including some plant pathogens. A number of different genes are known to be transferred.

Conjugation In the colon bacillus, mutant strains have been found that differ in their ability to synthesize particular growth factors. Two of these strains were used in a crucial experiment. Strain A could not synthesize the amino acids threonine and leucine, nor the vitamin thiamine. Strain B lacked the ability to make the amino acids phenylanaline and cystine and the vitamin biotin. But when large numbers of these two strains were grown together in a minimum medium (containing none of the six growth factors mentioned), occasional cells were able to grow. These were obviously recombinants, possessing characters of both parents. Further studies showed that physical contact is necessary between the two parent strains, and that genetic material is transferred from a donor cell to a recipient cell across a cytoplasmic bridge, possibly a tube, connecting the two cells (FIG. 20-7). This process is known as *conjugation*.

Fig. 20-7 Electron micrograph of colon bacilli in the process of conjugation. The recipient cell (right) is sensitive to the attached bacterial viruses (phages) to which the donor cell (left) is insensitive. Genetic material is believed to pass through a tube located at the point of attachment of the two cells (22,000×).

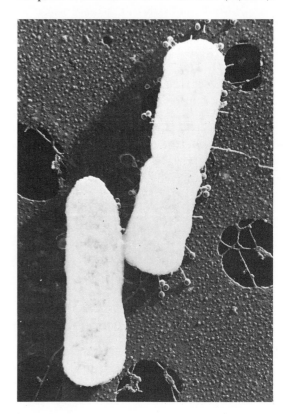

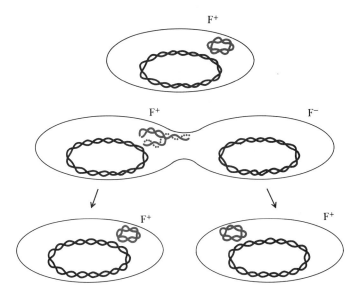

Fig. 20-8 Conjugation: only the episome transferred. Conjugation between a cell carrying the F factor (F⁺) as an episome (in color) and an F⁻ cell. Further explanation in text.

Further aspects of conjugation are complex, and can be considered here only in general terms. A striking feature of conjugation is the presence, in addition to the normal bacterial "chromosome," of a particle of DNA known as an *episome*. Episomes, which may be visualized as very short threads of DNA, may exist in either of two alternative states: (1) as episomes free in the cytoplasm and replicating independently of the DNA chromosome, or (2) as units integrated with, and constituting a part of, the normal chromosome. Episomes are not present in all bacterial cells, and when present are not injurious to the cell.

The episome, as does the chromosome, carries a group of genes, some of which govern the conjugation process. For this reason the episome is called the F, or fertility, factor. Cells that possess the F factor are donor cells (F^+). The recipient cells, which lack the episome, are F^- (FIG. 20-8). During conjugation, one of two events may occur:

1. If the F factor is free in the cytoplasm, it alone duplicates and one of the daughter episome particles passes from the donor to the recipient cell, which is then converted into a donor cell, F^+ (FIG. 20-8). In addition to the fertility factor, some episomes carry genes governing resistance to certain antibiotics and sulfa drugs. During conjugation these genes are transferred with the episome to recipient cells, converting recipient cells to donor, drug-resistant cells. Thus, through several cycles of such conjugations, an entire population of bacteria could be converted from drug sensitivity to drug resistance. The injudicious use of antibiotics, by favoring cells carrying episomes bearing genes for drug resistance, may produce a situation where certain antibiotics are no longer effective in disease control.

2. If the episome is integrated into the bacterial chromosome, it induces the replication of the bacterial DNA. A part (rarely all) of one of the daughter molecules is then transferred to the recipient cell (FIG. 20-9). The other daughter chromosome remains in the donor cell, which, unaltered, may divide in the usual way. The genes transferred to the recipient cell pair with similar genes, and, as in the case of transformation, recombination by means of integration occurs.

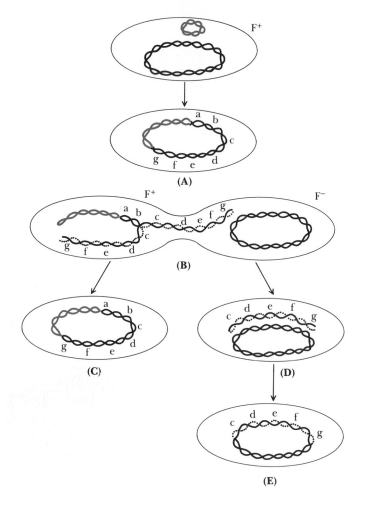

Fig. 20-9 Conjugation: a portion of a daughter chromosome is transferred. Conjugation between a cell carrying the F factor integrated into the bacterial chromosome and a recipient cell F⁻. (A): integration of episome into bacterial chromosome. (B): replication and transfer of donor DNA. (C): former donor cell. (D): integration of donor DNA into recipient chromosome.

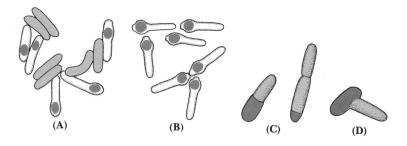

Fig. 20-10 Bacterial spores and spore germination. (A): botulinus cells and spores. (B): tetanus spores. (C, D): spore germination.

Spore Formation Some kinds of bacteria, under certain environmental conditions, form a resting spore (also called endospore). Usually only one spore forms in a cell, near the middle or toward one end. In most species the diameter of the spore is less than that of the parent cell (FIG. 20-10A), but in a few species the spore is greater in diameter and causes the cell wall to bulge (FIG. 20-10B). The spore becomes surrounded by a thick spore coat and remains within the parent cell until released by the disintegration of the cell wall. Under suitable environmental conditions the spore takes up water, the spore coat ruptures, and a growing cell emerges (FIG. 20-10C). The spore has a low water content, very low metabolic activity, and is resistant to heat, disinfectants, desiccation, and other environmental factors. Spore formation occurs almost exclusively in the bacilli. The ability of bacteria to form spores is important not only to these organisms but also to man, as will be seen in later paragraphs.

Spore formation in bacteria is not a method of reproduction, as is sometimes stated, but may be regarded as a modification of the vegetative cell capable of surviving under unfavorable conditions. Bacterial spores thus differ from the truly reproductive spores of the algae, fungi, and other spore-bearing plants.

Occurrence of Bacteria Bacteria may be found in any environment in which life can exist. They are found in the soil, in numbers from 10 million to 5000 million per gram, depending upon the soil organic matter, moisture, and temperature. They are found in both fresh and salt waters, and in the air, either free or attached to dust particles. Airborne bacteria and the spores of yeasts and molds are carried over the earth by winds, and are present in the air of the Arctic and Antarctic—although the numbers are small—as well as in tropical and temperate regions.

Bacteria are always associated with the human body. They are found on the surface of the body, but are notably numerous in the respiratory passages, the throat, and the alimentary canal. It is estimated that from one-fourth to one-third of the dry matter of human feces consists of bacteria. Most of these are harmless, but sewage—even from healthy persons—usually contains bacteria and viruses that are pathogenic in environments other than the intestinal tract. In addition, sewage may contain the causal organisms of bacillary dysentery and other intestinal infections. Sewage should therefore be treated to destroy or remove pathogenic bacteria that could pollute supplies of water used for drinking, swimming, or laundry purposes. Because such treatments may not remove all pathogenic microorganisms, a final step, unless the water supply is completely protected against pollution, is boiling.

The bacteria found in foods, particularly milk, are of special significance to man. Milk as drawn from a healthy cow usually contains less than 1000 bacteria per milliliter. As soon as the milk is drawn, however, it is exposed to contamination from many sources—the animal, the milking machine or the hands of the milker, the air, and the milk containers. By the time it reaches the consumer, even a good grade of raw milk may contain 50,000 bacteria per milliliter, and this increases to 20 million or about 1 million bacteria per drop by the time the milk begins to sour. Usually, none of these bacteria is pathogenic. Milk is such an effective carrier for pathogenic bacteria, however, that thorough sanitation and rapid chilling of the milk is necessary to reduce bacterial numbers and the probability of milk-borne infections. But sanitation processes may not be sufficient, and low bacterial counts are not necessarily safe, for at least two important diseases, tuberculosis and undulant fever, are transmitted directly through milk from infected animals. Pasteurization is the best safeguard in preventing all milk-borne diseases, and most states have laws prohibiting the sale of raw milk.

Colonies and Pure Cultures To aid in the study of bacteria, techniques have been developed to cultivate them in the laboratory on a nutrient medium, which contains food and other materials essential for bacterial growth. The bacteria are grown either in test tubes or in Petri dishes—shallow glass dishes with a loosely fitting cover. The nutrient medium may be liquid or solid. A common liquid medium is a nutrient broth containing meat extract to which has been added partially digested proteins known as peptones. In the growing of certain kinds of bacteria, sugars, such as glucose, lactose, or sucrose, are also added to the broth. A solid medium usually consists of a mixture of food materials incorporated into a gel of agar, which is obtained from red algae (Chapter 19). The agar itself serves merely as a solidifying agent and is not acted upon by most bacteria. The medium and containers must, of course, be sterilized to ensure that no organisms are present except those selected for study.

A growth of microorganisms in a nutrient medium is known as a CULTURE, and when all of the bacteria present belong to the same species, the culture is known as a PURE CULTURE (FIG. 20-11A). It is almost always necessary to isolate pure cultures in order to study structure and processes in microorganisms. For example, in studying the bacteria of milk, a drop of milk, containing a number of kinds of bacteria, is added to a tube of sterile, melted agar and thoroughly mixed to disperse the bacteria throughout the fluid. When this is poured into a Petri dish, it forms a thin layer of large area, and the dispersed bacteria are separated and distributed evenly in the agar, which soon becomes solid. The culture is then incubated. The individual bacteria are unable to move in the solid medium and each, by repeated division, forms a roundish mass, termed a COLONY. Since colonies, each containing millions of cells, may be formed by different species, they frequently differ in size, margin, color, and consistency (FIG. 20-11B). If the tip of a sterile needle is brought into contact with one of the colonies, some of the bacteria may be transferred to a tube of sterile nutrient agar. Because a colony is usually derived from a single cell or cells of a single species, the resulting growth is ordinarily a pure culture.

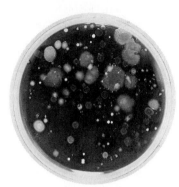

Fig. 20-11A (Left): a pure culture of a bacterium (*Serratia marcescens*) on agar.

Fig. 20-11B (Above): colonies; a Petri dish completely filled with solid medium was inverted and brought into contact with a recently scrubbed floor. The resulting colonies represent a number of species.

Colonies can also be obtained by exposing a Petri dish of nutrient medium to the air for a short time. After a few days of incubation the surface of the agar will show many colonies, each usually the result of the growth of a single cell that fell upon the medium. Pure cultures may be obtained from these colonies by the technique described above.

Bacterial Metabolism On the basis of their primary source of energy, bacteria may be grouped as autotrophic or heterotrophic. Autotrophs can grow in the absence of carbon from organic compounds. They are either photoautotrophic (phototrophs) or chemoautotrophic (chemotrophs). The phototrophs (photosynthetic bacteria) are relatively few in kind and of little ecological importance today, but studies on them have furthered our knowledge of photosynthesis. They are green or purple, containing characteristic chlorophylls. While their photosynthetic mechanisms differ in several ways from those of higher plants, they have in common with these the use of carbon dioxide as the source of carbon for cell syntheses, and the transformation of energy from light into chemical energy. The chemoautotrophs also obtain their carbon from carbon dioxide, but they obtain energy for synthetic processes by the oxidation of inorganic compounds rather than from light. Prominent among the chemoautotrophs are the nitrifying bacteria (Chapter 10), which oxidize ammonia to nitrite and then to nitrate, and the colorless, sulfur bacteria that obtain energy from the oxidation of sulfur or sulfur compounds to sulfuric acid. The latter forms occur in water, mud, and soil, in peat bogs, and in sulfur deposits. Some are important in the sulfur cycle in nature. It should be noted that the energy obtained by the chemoautotrophic bacteria, just as the energy obtained by phototrophs, is used in the synthesis of organic compounds. Some of these, in turn, are metabolized, yielding usable energy to the cell; the remainder are used in producing new cell materials.

Most bacteria are heterotrophic; that is, they obtain their carbon from organic compounds synthesized by other organisms, and so are either parasites or saprophytes. Insoluble food materials are first acted upon by extracellular enzymes such as cellulase, amylase, lipase, pectinase, and the proteases, which are secreted into the immediate environment. The soluble materials then move into the cell by diffusion or by active transport. Bacteria are able to produce enzymes that act upon a tremendous variety of organic compounds. It is probable, in fact, that every naturally occurring organic compound can be utilized as food by some kind of bacteria. Among these compounds are flesh, wood, leather, horn, the chitin of insects, organic acids, and other materials containing animal or plant protein, lipids or carbohydrates. Some bacteria decompose a wide variety of substances, while others are more restricted in their food relationships.

Foods are utilized by bacteria as a source of energy and as building compounds. Glucose, for example, used as an energy source, may also be employed in the construction of more complex carbon compounds. Similarly, amino acids, although used chiefly in the synthesis of proteins, may provide a source of energy. While all heterotrophs obtain their carbon from carbon compounds, nitrogen—necessary for the synthesis of protein and nucleic acids—is obtained from various sources, depending on the species. A few, such as the mutualistic and free-living nitrogen-fixing bacteria, use atmospheric nitrogen. Most heterotrophs, as do green plants, utilize the inorganic nitrogen of nitrates and ammonia, chiefly the latter. Many such heterotrophs may be cultivated in the laboratory on a medium containing only a source of reduced carbon (such as simple sugar), a salt of ammonia, and essential mineral elements. With these nutrients only, the organism is self-sufficient, able to synthesize amino acids, vitamins, components of nucleic acids, and other compounds essential to growth.

An excellent example in nature of the utilization of ammonia (ammonium ions) by bacteria is found in the digestive process of ruminants such as the cow, sheep, deer, and goat. In these animals the food passes from the esophagus into the rumen, or first compartment of the stomach, where its organic compounds are acted upon by enormous numbers of bacteria and other microorganisms and broken down into simpler compounds. The cellulose and other carbohydrates of the food are digested to soluble forms; the proteins, also, are digested by bacterial action and the amino acids in part fermented, with ammonia as one of the products. The rumen bacteria have the ability to synthesize all essential amino acids, either from other amino acids or from carbohydrates and ammonia. These compounds are then used by the bacteria in the synthesis of their own proteins. As the food moves along through the alimentary tract, most of the bacteria present are killed and digested by the cow's enzymes, yielding large quantities of essential amino acids to the animal. It is thus possible for ruminants to obtain from bacterial proteins the amino acids that were not present in the feed and which they would not be able to synthesize themselves. Because ammonia is utilized by bacteria, nonprotein, nitrogen-containing compounds are frequently included in the diet of cattle and sheep, such compounds in part replacing expensive feed proteins. The chief compound utilized is urea, which can be synthesized inexpensively. The urea as such is of little or no value to the animal, and is toxic if added in excess, but it is hydrolyzed by the rumen bacteria into ammonia and carbon dioxide, and the ammonia is used in the synthesis of amino acids that may later be used for protein synthesis by the animal.

Oxygen Relationships Bacteria differ greatly in their behavior in the presence or absence of atmospheric oxygen. Some are OBLIGATE AEROBES; they grow only in the presence of molecular oxygen. Among these are the nitrifying and nitrogen-fixing bacteria and the causal organisms of diphtheria and tuberculosis. Others, the OBLIGATE ANAEROBES, do not grow when exposed to air. In this group are the organisms that cause tetanus and gas gangrene (these multiply readily in deep wounds), blackleg of cattle, and botulism. A third (and the largest) group is composed of the FACULTATIVE ANAEROBES, which can grow in either the presence or the absence of air. Many of these are fermentative and can carry on respiration or fermentation, depending upon the availability of oxygen.

Fermentations As pointed out in Chapter 6, in fermentations the hydrogen obtained in glycolysis is combined with some organic compound. Carbohydrates, primarily sugars, are the chief compounds acted upon by fermentative bacteria. The fermentative processes differ with different organisms, and the products include various alcohols and organic acids, many of which are important in industry. Ethyl alcohol fermentation is carried on by both bacteria and yeasts, and will be discussed in Chapter 21.

One of the major fermentation reactions is lactic acid fermentation, characteristic of the normal souring of milk. If lactose (milk sugar) is present it is first hydrolyzed to two simpler sugars by the enzyme β-galactosidase. These sugars are then fermented to lactic acid:

$$C_6H_{12}O_6 \longrightarrow 2C_3H_6O_3 + \text{energy}$$
simple sugar   lactic acid

Lactic acid, formed during the fermentation of corn sugar or molasses, has many uses in pharmaceuticals and in the chemical industry. Lactic acid fermentation is involved in the production of many kinds of cheese, sauerkraut, silage, and fermented milk products.

Amino acids may be utilized in growth or respired, aerobically or by fermentative processes. The bacterial cell is impermeable to large protein molecules, but may be penetrated by amino acids resulting from the digestion of proteins by extracellular enzymes. The fermentation of amino acids results in the formation of such products as organic acids, ammonia, hydrogen, and carbon dioxide. In addition, compounds having a disagreeable odor, such as hydrogen sulfide (odor of rotten eggs), may be formed. Hydrogen sulfide is produced during the fermentation of the sulfur-containing amino acids cystine and methionine, which are particularly abundant in eggs. The fermentation of amino acids is especially important in sewage disposal.

Anaerobic respiration As previously mentioned, true anaerobic respiration is carried on by only a relatively small group of bacteria. These are able to combine the hydrogen from glycolysis with the oxygen of an inorganic oxygen-containing compound, energy becoming available in the process. Among such forms are the facultative anaerobic, denitrifying bacteria that convert nitrate to free nitrogen, and the obligate anaerobic methane bacteria that reduce carbon dioxide to methane as indicated by the equation

$$4H_2 + CO_2 \longrightarrow CH_4 + 2H_2O + \text{energy}$$
hydrogen   carbon   methane water
         dioxide

Methane, the chief component of natural gas, is also known as marsh gas, for it is the most important constituent of the bubbles that arise from the mud of swamps and ponds rich in organic matter. Methane is also formed in the anaerobic treatment of sewage, and in some sewage disposal plants it is collected and burned to provide light or power.

Bacteria and the Environment Bacteria require for growth not only energy and building materials but also moisture, essential elements, oxygen (for aerobes), favorable temperatures, suitable osmotic concentrations, and favorable pH. Some species also require growth factors, such as certain amino acids and vitamins, that they are unable to synthesize for themselves and must obtain from the environment. Bacteria vary considerably in their ability to live and grow in solutions of high osmotic concentrations. Many species are inhibited at concentrations at which others grow actively. The effect of osmotic concentration on the growth of spoilage bacteria is shown by the preservation of many kinds of foods by sugar or salt. Most species of bacteria exhibit optimum growth in a medium that is neutral or close to neutrality in reaction. A few bacteria, however, are relatively tolerant of acid conditions. Among these are the lactic acid bacteria, which flourish at pH 4 or lower. Even these forms, however, eventually cease growth because of the acid conditions they themselves have produced. The pH of the environment affects the activity of enzymes, spore formation, and other aspects of bacterial metabolism. It also affects the rate of bacterial destruction, whether by heat or by chemical compounds.

Temperature is an important factor in bacterial growth. The growth rate increases with temperature within a range that varies with different species. For each species there is a minimum and a maximum temperature below and above which no growth occurs, and an optimum at which multiplication takes place most rapidly. But even at optimum temperatures the growth rate may vary with the pH, the presence of inhibiting substances, or the nature of the food supply. Bacteria have been placed in three groups based upon their temperature ranges, but there is no sharp line between these, for maximum and minimum temperatures may overlap.

1. Thermophilic (heat loving) bacteria. These have a temperature range for growth of about 40–75°C (104–167°F). They are widely distributed in nature in hot springs, in soil, in compost heaps, manure piles, and other decaying organic matter. Both vegetative and spore forms of many thermophiles are unusually heat resistant, and these are deleterious in food processing, dairying, and other industries.

2. Psychrophilic (cold loving) bacteria. These will grow within the range 0–30°C (32–86°F), with an optimum for growth around 20°C (68°F). Psychrophiles sometimes cause spoilage in refrigerated and even frozen foods.

3. Mesophilic bacteria. These, intermediate in their temperature requirements, include the majority of bacteria, among them the common saprophytes of soil and water. The growth range is approximately 20–46°C (68–115°F). Some mesophiles, pathogenic for man, flourish at the temperature of the human body, 37°C (98°F). Because the optimum temperature for the growth of most bacteria is above room temperature (about 20°C), bacterial cultures are placed in an *incubator*, which is an enclosed cabinet in which a constant, suitable temperature can be maintained.

Identification of Bacteria In general, organisms are distinguishable one from another by their external form, or morphology. To some extent this is also true of bacteria. Preliminary studies on identification involve (1) gross morphology, including the size, texture, color, and shape of colonies, and (2) microscopic morphology of cells, both when living and when killed and stained. Microscopic studies reveal such morphological features as motility, presence of spores, and size, shape, and arrangement of cells. Structures invisible in living, unstained cells can be seen following staining, which may also aid in distinguishing between species of diverse chemical composition. But so numerous are the species of bacteria and so close is their resemblance to one another that identification by morphological criteria must be supplemented by physiological features. Physiological reactions are studied to determine whether an organism is aerobic or anaerobic, which sugars it ferments and which acids or gases are produced, the temperatures at which it grows best, whether it will liquefy a gelatin medium, and to observe other results of bacterial action. The identification of a given kind of bacterium involves far more than a quick inspection under the microscope; many complicated laboratory procedures are required. So numerous are the differences among bacteria that computers are frequently used in identification.

The Control of Bacteria The control of bacteria that cause disease and decay is a matter of great importance to health, industry, agriculture, and the processing and storage of food. Procedures aimed at such control—to inhibit the growth of bacteria or to destroy them—are in daily use, and involve the use of iodine, chlorine, some detergents, compounds containing mercury or silver, phenol, cresol, sulfonamides, formaldehyde, certain basic dyes such as crystal violet, the antibiotics, organic acids, ultraviolet light, drying, and heating. The canning of food, the pasteurization of milk, and the sterilization of surgical instruments are all procedures designed to control bacteria. Certain treatments will kill bacteria in the vegetative but not the spore stage. When all forms of life, including spores, are destroyed, the process is known as STERILIZATION. Heat is one of the most common agents used in the destruction of microorganisms.

Boiling is one of the common methods of applying heat. This process kills bacteria in the vegetative stage, for it coagulates the proteins of the cell, resulting in the death of the organism. Boiling is not a method of sterilization, for some spores are not killed. Fortunately, the bacteria responsible for most infectious diseases do not form spores and may therefore be killed by boiling. Food, dishes, cooking utensils, and clothing may be adequately treated by this method. Contaminated water may be made safe to drink by boiling.

A limited number of pathogens do form spores, however. This list includes the organisms causing anthrax, botulism, gas gangrene, and tetanus, together with certain species pathogenic to domestic animals. The boiling of water to make it safe for drinking is nevertheless effective, for these spore formers are either absent from water or do not bring about infection when introduced into the alimentary tract.

Since the spores of bacteria, pathogenic and nonpathogenic, may survive boiling, a sterilization procedure must be employed in the laboratory when it is essential that all organisms be killed. Sterilization by heat requires temperatures higher than boiling, and such temperatures may be attained in various ways. Of the two methods involving heat in general use, *dry heat* is the simplest. Spores are killed on materials placed in a hot-air sterilizer for 1½ hours at 120°C (248°F). Glassware, some kinds of instruments, and other materials are sterilized in this way. The use of dry heat is necessarily restricted to the sterilization of materials unaffected by drying or by the temperature used.

The more widely used method of heat sterilization involves the use of steam under pressure. Under atmospheric pressure, water boils at 100°C (212°F). As the pressure is increased, the boiling temperature rises. At approximately twice atmospheric pressure, the boiling point is 121°C, sufficient for sterilization when maintained for about 15 minutes. The necessary temperature is attained by heating water in a container, commonly known as an autoclave (FIG. 20-12), or feeding steam under pressure into it, until the gage reads 15 pounds. Articles to be sterilized are placed so that they are enveloped by superheated steam. The use of steam under pressure produces moist heat that kills spores at temperatures lower than those used in dry-heat sterilization. Hospitals routinely use the autoclave to sterilize surgical instruments, dressings, and other materials. Modifications of the autoclave are the process (pressure) cookers or retorts of canning factories that cook and sterilize thousands of cans of food at one time, and the smaller pressure cookers used in the home for home cooking and canning.

Another method of sterilization, and one that has come into wide use in recent years, is the employment of gas mixtures containing ethylene oxide. This method is useful for the sterilization of medical and surgical materials that are moisture sensitive and affected by heat, and which

Fig. 20-12 An autoclave, electric heat.

cannot be sterilized by either steam or dry heat. It is carried on in special sterilizers.

Bacteria in Relationship to Food The important agents in the spoilage of food are bacteria, yeasts, and molds. To preserve food, the growth of bacteria and other decay organisms must be retarded or inhibited, or better still, all such organisms upon or within the food should be destroyed. Among the more important means of food preservation are low temperatures, desiccation, the use of preservatives, and canning.

Low temperatures involve either refrigeration at temperatures commonly from 0–10°C (32–50°F), as in the home refrigerator, or freezing storage [−18°C (0°F) or lower], as in packaged frozen foods. At refrigeration temperatures, bacteria are not killed, and bacterial growth may still proceed, although slowly. Freezing is more effective than refrigeration because bacterial growth is prevented. When frozen food is brought to room temperature, it will spoil unless consumed promptly.

Since bacteria, in common with all organisms, require water for growth, dehydration below a critical point is a method of food preservation. Many kinds of foods, such as eggs, fish, meat, milk, fruits, and vegetables, may be preserved by dehydration. Bacterial growth begins as soon as water is added to dried foods, and the product will then be apt to spoil. An advance over older methods, which retains maximum flavor and appearance, is called freeze-drying. The food is frozen and then placed in a vacuum cabinet. The ice crystals in the food are transformed directly to water vapor, and about 98 percent of the water in the food is removed.

Sugar, salt, acids, and a number of other substances are added to preserve food. The preservative action of sugar and salt in large part results from plasmolysis of the bacterial cell. Certain molds are more tolerant to high concentrations of sugar than are bacteria, and are not uncommon on the surfaces of jams, jellies, and other preserved foods where bacteria will not grow.

Acids, chiefly acetic and lactic, have long been employed in food preservation. Foods may be preserved by the addition of vinegar or by allowing the acidity to increase during a lactic acid fermentation process. In an acid environment, bacteria do not multiply and may even be killed. Lactic acid, developed during a fermentation process, is involved in the preservation of sauerkraut, pickles, green olives, silage, fermented milks, and many kinds of cheese.

The processing of food with ionizing radiations (electrons, gamma rays) has been extensively studied, and bears promise for the future.

Canning All canning processes aim at the destruction by heat of the vegetative or spore forms of bacteria. In killing bacteria by heat, many factors must be considered. Some of the more important are temperature, length of exposure of foods to a specific temperature, pH, and the nature of the foods to be canned. Foods may be processed either at the temperature of boiling water or by steam under pressure.

Processing food by boiling is sometimes practiced. This is a safe and effective method only for strongly acid foods, such as apples, pears, pineapple, rhubarb, most varieties of cherries, peaches, plums, sauerkraut, and tomatoes. Most fruits fall within the range pH 3–4. Vinegar is sometimes added to bring such weakly acid foods as beets to the pH 4.5 that will permit safe canning by boiling. With boiling and in a strongly acid environment, many spores as well as all vegetative cells are killed. Any spores that survive will not germinate under the acid conditions (pH 4.5 or below) that prevail in many foods. Slightly acid foods, however (pH above 4.5), such as asparagus, beans, corn, peas, and meats, should be processed by steam under pressure. Slightly acid foods are processed commercially at temperatures of 110–120°C (230–248°F) for periods of 12 minutes to 1½ hours. If such foods are canned, as they sometimes are in the home, by boiling only, spores may survive, leading to later spoilage. Such spores may include those of the bacillus causing botulism, a deadly food-poisoning disease.

Pasteurization Pasteurization was first developed to prevent the spoilage of wines, and is still used in the production of wine, beer, cheese, and unfermented fruit juices. It is chiefly associated with the treatment of milk to eliminate pathogenic bacteria such as the organisms causing tuberculosis, undulant fever, septic sore throat, and scarlet fever. Two methods are used to pasteurize milk. In the older of these, the HOLDING process, the milk is heated to 62°C (143°F) for 30 minutes. The SHORT-TIME HIGH-TEMPERATURE (FLASH) METHOD has come into use increasingly in recent years. The milk, in thin layers, is exposed to a temperature of 72°C (161°F) for a minimum of 15 seconds. These time and temperature relationships were worked out for the thermal destruction of the organism causing tuberculosis, but they affect other pathogens as well. Following either treatment the milk is cooled rapidly to a temperature that retards bacterial growth.

Pasteurization is not a process of sterilization. The spores of bacteria are not killed and, since the temperature falls far short of boiling, neither are all the vegetative cells. This is notably true of some lactic acid bacteria, although the number of these is greatly reduced. Pasteurized milk will therefore normally sour, as will raw milk, although more slowly.

BACTERIA AND DISEASE

The fact that bacteria are the cause of many infectious diseases is now so commonly accepted that it is difficult to visualize the scientific thought of the times when this relationship was unknown. It was not until 1876 that a specific bacterium was shown to be the cause of a specific disease, anthrax, in animals. This was demonstrated by Robert Koch, a German physician, who also established, in 1882, that tuberculosis in man is caused by a particular organism, the tubercle bacillus. The habitats in which bacteria live, then, include not only soil, water, and decaying organic materials but also the living tissues of plants and animals. The activities of pathogenic bacteria in the body of man and the responses of the human body to these activities are of particular interest to us, since we are the potential victims. In this unresolved conflict between man and parasitic organisms, it is only in recent times that man has gained the upper hand, at least for many diseases.

Pathogenic Bacteria and Toxins The symptoms of bacterial diseases in man and animals are frequently due to toxins, poisonous substances produced by the invading bacteria. The toxins of such pathogens are of two types, EXOTOXINS and ENDOTOXINS. Relatively few pathogenic bacteria produce exotoxins. An important distinction between these two kinds of toxins lies in their chemical composition—exotoxins are proteins, whereas endotoxins are made up of a lipid-polysaccharide complex.

Exotoxins are also termed soluble toxins, for (with certain exceptions) they diffuse from the living bacterial cell into the surrounding tissues of the host, and often may be obtained, after filtration, from a liquid medium in which the bacteria have been growing. Such filtrates, injected into an experimental animal, will produce the symptoms of the specific disease. Certain exotoxins are among the most powerful poisons known, and are far more potent than aconitine, strychnine, or cobra venom. Among the pathogens producing exotoxins are certain staphylococci and streptococci, and the organisms responsible for diphtheria and gas gangrene. Exotoxins are also produced by some bacteria pathogenic to plants.

In contrast to the exotoxins, which are released from the living cell, endotoxins are bound to the cell wall, and are released only upon the death and disintegration of the bacterial cell. They are much less toxic than the exotoxins. Among the organisms producing endotoxins are those causing typhoid fever, whooping cough, Asiatic cholera, and epidemic meningitis.

Some pathogenic bacteria are highly specialized, affecting only one species or genus of plants or animals. Other kinds can infect several kinds of hosts, frequently unrelated. Undulant fever (brucellosis), anthrax, and tuberculosis occur in man and in certain warm-blooded animals. Less specialized bacteria sometimes have intermediate hosts. Plague, for example, is primarily a disease of rats and other rodents. The organisms may be carried from rat to rat or rat to man by fleas, or from man to man by the same agents or by droplet infection. With such complications, it is easy to see why the transmission of infectious diseases baffled even the best authorities before the concept of pathogenic organisms became established.

Food Poisoning and Food Infection Illness resulting from the consumption of food may be due to a number of causes. Poisonous products of plants and animals can cause illness. Some mushrooms and shellfish, for example, are poisonous when eaten. Foods may be contaminated with pesticides or a wide range of inorganic and organic compounds. But bacteria or their products are responsible for most cases of food poisoning. It is to the illness caused by these agents that the popular but wholly inaccurate term "ptomaine poisoning" is sometimes applied. Ptomaines are decomposition products resulting from the action of bacteria upon proteins. They appear only when foods have reached an advanced stage of decay, and there is no evidence that such substances are connected with attacks of food poisoning. Ptomaines taken by mouth are, in fact, considered harmless. The bacteria responsible for food-borne diseases are of several kinds, and the chief types of illness that they cause are discussed below.

Staphylococcus food poisoning This is the most common kind, and is usually involved when large numbers of people become ill at the same time after a banquet or a picnic. Many kinds of food have been the source of staphylococcus food poisoning. Prominent among them are precooked meats, custard-filled bakery goods, some salad dressings, and gravy—all excellent media for the growth of bacteria when allowed to stand without proper refrigeration.

Several strains of staphylococci are involved, and an exotoxin is produced by them in the food before it is eaten. It is this toxin that is responsible for the illness, and not the bacteria themselves. The toxin is soluble and resistant to heat. It is therefore impractical to destroy the toxin of infected foods by cooking.

Contaminated food frequently cannot be identified by either taste or smell. A short interval, usually 2–4 hours, intervenes between the consumption of the food and the appearance of the symptoms. The chief method of control is sanitation and careful refrigeration. Staphylococcus food poisoning is not fatal to otherwise healthy people, and recovery takes place within a few days. It is estimated that more than $1/2$ million cases occur yearly in the United States.

Salmonella food infection   A second and less common food-borne disease is caused by bacteria of the genus *Salmonella*. Other species of this genus cause paratyphoid fever, and the group is closely related to the bacilli that cause typhoid fever. Unlike those of staphylococcus poisoning, the symptoms are due to the bacteria themselves. No exotoxin is produced. This is a true infection, and the symptoms do not appear until the bacteria have had time to multiply in the body. A period of 8–24 hours therefore elapses between the eating of the infected food and the advent of illness.

Although infection has been traced to several kinds of foods, including milk, meat is usually involved. *Salmonella* also infects domestic animals, and if it is not detected in slaughtering, infections in man may result from the consumption of such meat when it is insufficiently cooked. The methods of control involve inspection, sanitation, thorough cooking, refrigeration and, for milk, pasteurization. The mortality from this type of illness is low.

Botulism food poisoning   The organism causing botulism is a large, motile, anaerobic spore-form-ing bacillus belonging to the same genus (*Clostridium*) as that causing tetanus. It is a soil saprophyte, widely distributed in many parts of the world. The disease is caused by the consumption of food containing a preformed exotoxin, and although not common, is extremely serious, with a mortality of about 70 percent. In many outbreaks, all who eat the food die. The toxin is the most potent known. It is estimated that 1 ounce of it would be enough to kill 400 million people. Most cases of botulism have been traced to the consumption of home-canned food processed by boiling only, particularly meats, fish, and low-acid vegetables such as string beans, peas, asparagus, spinach, beets, and corn. In the United States only one outbreak of botulism (in 1963), with two fatalities, is known to have resulted from commercially canned food since 1925. The food in this instance was improperly canned and the bacilli gained entrance through defective seams after sterilization was completed.

The vegetative forms of the botulism organism are killed by canning at boiling temperature, but the spores may survive and germinate under the anaerobic conditions created. After germination, the organism multiplies and liberates the exotoxin into the food. Spoilage of such food may or may not be evident. Taste or smell is not a reliable indicator. As pointed out previously, slightly acid foods should be canned only by steam under pressure, which destroys the spores of all bacteria. Any home-canned food that has not been processed in this manner should be removed from the container and boiled vigorously for 15 minutes before it is eaten. The exotoxin is a protein and is destroyed by boiling.

Bacterial Diseases of Plants   As do the parasitic fungi, bacteria frequently attack cultivated crops and ornamental plants. They cause reduction in yield and sometimes total failure of the crop. In the United States and its possessions more than 200 species or varieties of bacteria are currently recognized as plant pathogens. This number far exceeds the number of species pathogenic to man and other animals. The bacterial plant pathogens are more closely related to the soil bacteria than to animal or human pathogens. All are rod-shaped and nonspore formers. It is curious that apparently none attack cellulose and relatively few hydrolyze starch, two of the major constituents of plants. Young plant tissues, especially in the seedling stage, are much more susceptible to infection than older tissues. Some bacterial plant pathogens attack but a single host; others are restricted to a genus or a family of plants. Several of the wound parasites (crown gall, bacterial soft rots), on the other hand, attack hundreds of species over many plant families. Only one bacterium pathogenic to man and animals is known to infect plants also. In 1955 it was shown that *Pseudomonas aeruginosa*, isolated from a human abscess, was pathogenic to the potato tuber and produced soft rot.

There are few economic plants that are not attacked by at least one bacterial disease. Among these diseases are fire blight of apple and pear; bacterial soft rots of vegetables, fruits, iris, and calla; bacterial blights of bean, peas, rice, cotton, Persian walnut, and delphiniums; crown gall on numerous woody plants; citrus canker; bacterial wilts of alfalfa, geranium, tomato, corn, and vine crops; angular leaf spot of cucurbits; black rot of crucifers; bacterial canker of tomato (Fig. 20-13); and wildfire of tobacco. The effect of bacterial invasion upon the host varies greatly. In the vascular diseases, such as wilt of tomato, potato, and tobacco, the bacteria invade the vessels of the xylem and the plant suddenly wilts. The wilting results from the plugging of the vessels by the loose slime produced by great numbers of bacteria. In other diseases, the parenchyma cells of the leaf, stem, or root are disintegrated by a toxin or by pectin-digesting enzymes. The organism causing crown gall stimulates an overgrowth of tissues that resembles cancer. The most common route of invasion of plants by bacteria is through wounds, but entrance into the host may also take place through stomata, hydathodes, lenticels, and the flower. Bacteria may be transmitted by higher animals, insects, and mites; by splashing

Fig. 20-13 **Bacterial canker of tomato caused by** *Corynebacterium michiganense.*

rain, flowing water, and wind-blown dust; by man through cultivation, pruning, and transporting diseased plant material; and through infected seeds, bulbs, tubers, or other plant parts.

Antibiotics When microorganisms grow together in the soil, in water, or in artificial culture media, their relation to one another may assume various forms. Sometimes the association is beneficial to one, sometimes to both, of the associated organisms. On the other hand, one organism may be antagonistic to another, either because it creates conditions unfavorable for growth or because it produces one or more toxic organic compounds. These substances, at low concentrations, inhibit the growth of other microorganisms and even destroy them, and are known as ANTIBIOTICS. In common usage the term antibiotic is applied to substances of biological origin that interfere in any way with the metabolism of other organisms (Fig. 20-14).

Beginning about 1943, antibiotics have been produced commercially and applied to the treat-

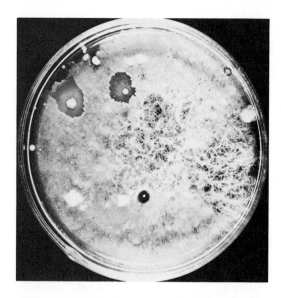

Fig. 20-14 Antagonism among microorganisms. The agar was seeded with 1 ml of a 1:10,000,000 dilution of soil and incubated at 28°C for five days. The clear areas around three of the colonies indicate that these colonies have produced an antibiotic that has diffused into the surrounding medium and inhibited the growth of other organisms.

ment of disease in man and other animals. They also have nonmedical uses, as in animal nutrition and plant disease control. The most important antibiotic-yielding organisms have been obtained from the soil. A tremendous number of cultures of soil microorganisms have now been grown and tested for antibiotic properties. Many have been found to produce antibiotics, but relatively few of them have been useful in therapy. Prominent among the reasons for this is that antibiotics vary greatly in their toxicity to their host, man, animal, or plant. Some are extremely toxic; others can be used externally but not internally. Antibiotics are of two basic types; antibacterial and antifungal. The same organism may produce both types. *Streptomyces griseus*, for example, produces both the antibacterial streptomycin and the antifungal cycloheximide (Actidione). The antibiotics vary in the nature of

their inhibiting action. For example, penicillin blocks cell wall formation in bacteria, and erythromycin is believed to interfere with the incorporation of amino acids into proteins. The inhibitory mechanisms of many antibiotics are still obscure.

The most important antibiotics are obtained from (1) the actinomycetes, (2) the bacteria, and (3) the molds (*Penicillium*) in the true fungi. Still others have been reported from other groups of fungi, including yeasts, and from algae, lichens, and seed plants.

Actinomycetes   Most of the important antibiotics have been obtained from the actinomycetes. These are sometimes termed the "mold" bacteria, for in some respects they are intermediate between the true bacteria and the fungi. The organisms are usually composed of branching threads about 1 micron in diameter, thus approaching the size of bacteria. In many species the threads break apart into cells that resemble cocci or bacilli, and these segments continue to divide. In others, reproduction is funguslike; special reproductive cells, called CONIDIA, are formed at the ends of serial threads, singly or in chains (FIG. 20-15). The cells of the actinomycetes are procaryotic, however, and in spite of some resemblances to fungi are in most systems of classification placed either in the true bacteria or in a separate but closely related order. They occur chiefly in the soil in numbers varying from 1 million to 500 million per gram of surface soil. The characteristic odor of moist, newly turned soil is caused by volatile compounds produced by certain actinomycetes. Most are saprophytes, but a few are pathogenic for plants or animals.

Among the early antibiotics obtained from actinomycetes are streptomycin, found in 1944; Chloromycetin (now prepared synthetically) in 1947; Aureomycin, in 1948; neomycin, in 1949; and Terramycin, in 1950. A considerable number have since been found, and new ones are constantly being discovered. Actinomycetes

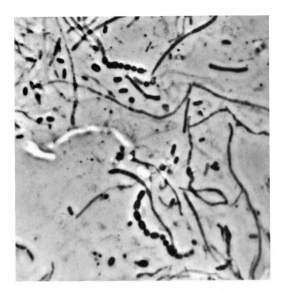

Fig. 20-15 *Streptomyces griseus*, the actinomycete from which the antibacterial streptomycin and antifungal cycloheximide (actidione) are obtained. This same organism also produces other antibiotics. Note the spores (conidia) in chains.

produce not only antibiotics against bacteria, but also antifungal antibiotics, such as nystatin.

The bacteria Although antibiotics from bacteria are greatly outnumbered by those from actinomycetes, a number have been found and the usefulness of several established. Among these are bacitracin, isolated from a contaminated wound in 1945, and polymyxin, isolated in 1947. The latter is produced by a spore-forming soil bacterium.

Molds Penicillin, a valuable antibiotic against bacterial infections, for medical purposes is produced by certain species of *Penicillium* (chiefly *Penicillium chrysogenium*), a genus of greenish molds that commonly grows on bread, cheese, citrus fruits, leather goods, and decaying organic matter. While useful against many bacterial in-fections, penicillin is ineffective against others. Another species of *Penicillium* produces an antifungal antibiotic, griseofulvin.

Production of antibiotics The details of the industrial methods by which antibiotics are produced vary widely, but the same general principles apply for both actinomycetes and molds. A pure culture of the organism is inoculated into a sterile nutrient solution in large tanks containing from 1000 to 20,000 gallons. These tanks are referred to as fermentation tanks, but the organisms are aerobic, and sterile air must be pumped through the medium for normal growth. After a growth period of hours or days, the medium is filtered, and the crude antibiotic is extracted from the filtrate and purified.

VIRUSES

It has long been known that certain ultramicroscopic agents cause disease in animals and plants. These are the viruses, originally called filterable viruses, because they pass through filters, the pores of which are small enough to hold back even small bacteria. Hundreds of disease-producing viruses have been discovered. They have conveniently been classed in three groups:

1. Animal viruses, causing disease in man and other animals.
2. Plant viruses, which for years were known only in flowering plants. In 1962, however, a disease of the cultivated mushroom was found to be caused by a virus, and in 1963 a virus was discovered that attacks certain blue-green algae.
3. Viruses infecting bacteria and actinomycetes.

Among the virus diseases of man are influenza, measles, shingles, mumps, chicken pox, the common cold, warts, virus pneumonia, dengue, poliomyelitis, infectious hepatitis, rabies, yellow fever, and smallpox. Animal diseases include foot-and-mouth disease of cattle, dog distemper, swine influenza, and infectious myxomytosis of rabbits, together with various infections of birds and insects. Some malignant tumors in animals (and possibly also in man) are caused by viruses. Among economically important plants affected are potato, tomato, sugar cane, corn, cacao, sugar beet, wheat, peach, bean, rice, cucumber, strawberry, apple, raspberry, tobacco, and many ornamentals. Viral diseases are generally most serious in plants that are vegetatively propagated by man. The virus infection may kill localized areas, the entire plant, or only reduce vigor and thus yield.

All attempts to multiply viruses on nonliving media have failed. They are known to be reproduced only within the living cells of a susceptible host. For experimental purposes, viruses are maintained and multiplied by repeated inoculation of host plants or animals, or by cultivation on cultures of living tissues or on the membrane surrounding the developing chick embryo.

The extremely small size of nearly all virus particles makes them invisible under the highest powers of the light microscope, and they are usually studied with the electron microscope. While bacteria may be measured in microns, virus particles are so small that they are measured in millimicrons. One millimicron equals one-thousandth of a micron, or one-millionth of a millimeter. Viruses range in size (FIG. 20-16) from the pox viruses, just visible under the light microscope, to those approximately the size of a large protein molecule. The shapes also vary considerably. Some viruses are oval; others are spherical or near spherical, although most, if not all, such forms are polyhedral, with numerous facets (FIG. 20-17). Still others are rigid rods or flexible threads (Fig. 20-18). Most bacterial viruses are tadpolelike in form, with a head and tail (FIG. 20-16).

Because viruses cannot be multiplied on nonliving media, they have been isolated and purified by a combination of chemical and physical methods, such as precipitation by alcohols or salts followed by differential centrifugation. Fol-

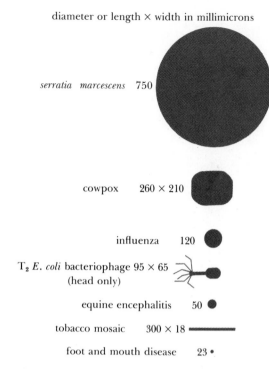

Fig. 20-16 Comparative sizes of bacteria and viruses. *Serratia* is a short bacillus.

lowing purification, the chemical composition of viruses has been determined. Most viruses consist only of a sheath of protein surrounding a core of nucleic acid. The protein sheath, in turn, is composed of regularly arranged protein subunits (capsomeres), varying in number in different viruses (FIG. 20-19). The nucleic acid is either DNA or RNA, but never both in a given virus. The nucleic acids of nearly all bacterial viruses (phages) are DNA. In addition, some animal viruses, as, for instance, the pox viruses and most insect viruses, contain DNA, but most animal viruses, together with the viruses of higher plants, contain only RNA. But whether it is DNA or RNA, the nucleic acid is the substance responsible for the specific characteristics of the virus—the material that conveys the hereditary information.

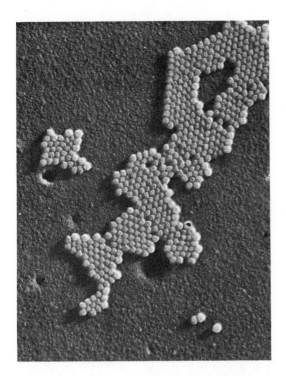

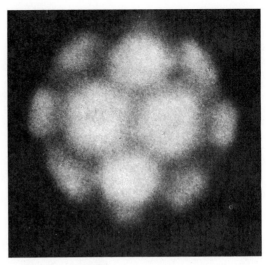

Fig. 20-19 Virus of turnip yellows mosaic (~2,000,000×). This is among the smallest of the viruses, 20–30 mμ in diameter. The protein shell is composed of 32 protein subunits.

Fig. 20-17 Virus of poliomyelitis, Type 2 (~49,000×). The virus is about 28 mμ in diameter.

Fig. 20-18 Cymbidium mosaic virus, causing a leaf disease of orchids. The sinuous rods are about 480 mμ long (21,000×).

Reconstitution of Viruses In one of the spectacular advances in knowledge of the nature of viruses (1956) it was discovered that the rod-shaped tobacco-mosaic virus could be separated into two constituents and that these could be recombined to form virus again. This much-studied virus has been found to consist of a thick-walled cylinder of protein (about 94 percent of the virus) surrounding a core of ribonucleic acid (FIG. 20-20) arranged in a spiral throughout the length of the protein cylinder. By chemical treatment the virus protein was separated from the core of RNA. These two components, under suitable conditions and in the absence of host cells, recombined to form virus particles that looked like the originals, and like them, caused infection of the tobacco leaf. The protein cylinder alone was noninfectious. The pure virus RNA, however, was infectious, although not to the same degree as the original virus or the reconstituted particle. The RNA alone is partly degraded by an

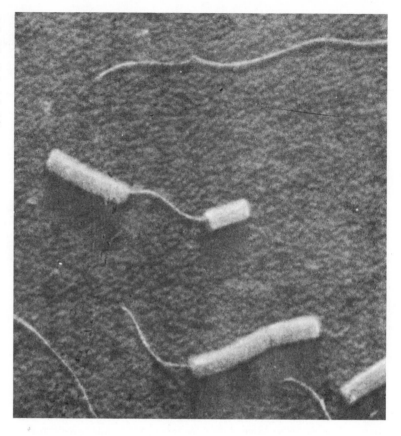

Fig. 20-20 Partly degraded tobacco mosaic virus (217,500×). The protein sheath was removed or partly removed by treatment with phenol. The thread at the top is RNA, from which the sheath was entirely removed. The considerable length of this thread is due to uncoiling of the RNA.

enzyme in the host cell from which it is normally protected by the protein coat.

A later experiment not only afforded additional proof that the infectivity of the newly formed particles was due to the nucleic acid but also demonstrated that RNA is the genetic determinant, and thus has the same role as DNA in some other viruses and in cells. The RNA and protein of two different strains, which we call A and B, were separated. The RNA of strain A was then combined with the protein of strain B, a "hybrid" strain resulting. This strain was inoculated into a tobacco leaf, where it caused infection. When the virus was recovered from the leaf, it was found that the RNA of the recovered virus belonged to strain A and that the protein of

the virus also belonged to strain A. This indicated that the RNA in the progeny of the hybrid virus had brought about the synthesis of its own type of protein rather than the kind with which it was originally supplied.

Viruses—Some Miscellaneous Aspects Virus diseases in man and animals resemble those caused by bacteria in that recovery from many virus diseases is followed by immunity or resistance to further infection, and such immunity frequently prevails for life. Measles, smallpox, yellow fever, and canine distemper are examples of virus diseases from which recovery usually protects against further infection throughout life. Another important characteristic of viruses is that they mutate under both natural and experimental conditions. Specific viruses, introduced into a new host or new tissue, or grown under abnormal conditions, produce strains that differ in their virulence, in their host range, in the symptoms that they produce, in the amounts, kinds, and arrangements of the amino acids of which their proteins are composed, and in other ways. Among plant viruses, more than 100 strains, or mutants, of tobacco-mosaic virus have been recognized, and similar but less numerous variations are known in other viruses.

Viruses and viral diseases were first studied in plants (1892, 1897). Plant viruses (more than 400 are known) are named and classified in various ways: (1) according to the host plants they infect and the typical symptoms of the disease produced (examples are peach yellows and rosette, curlytop of beets, dahlia ringspot, pea streak, potato leaf roll, cranberry false blossom, and phloem necrosis of elm); (2) their morphology as revealed by the electron microscope; (3) their thermal death point, ultracentrifuge sedimentations, diffusion gradients in various media, and chromatographic analyses, as determined by biochemical-physiological tests.

In an important group of virus diseases the virus infection results in MOSAIC DISEASES, so termed because the leaves of the infected plants commonly present a patchwork appearance, the result of the presence of irregular light green or yellow areas in the darker green of the leaf caused by the destruction of the chlorophyll by the virus. The foliage may also be mottled or wrinkled (FIG. 20-21). Among the important cultivated plants subject to mosaic diseases are tobacco, tomato, potato, cucumber, bean, sugar beets, and sugar cane. Some mosaic viruses cause a striking variegation, or "breaking," in the flowers of certain ornamentals, such as gladioli, pansies, and wallflowers. The response is best known in tulips. The diseased flowers exhibit streaks or lines of contrasting colors in the perianth (FIG. 20-22). Some varieties of broken tulips have long been in cultivation, and show no cumulative

Fig. 20-21 Potato mosaic. (Left): normal leaf. (Right): diseased leaf. The leaflets are dwarfed and mottled and the surfaces are deeply wrinkled.

Fig. 20-22 Rembrandt tulip, showing "broken" color, the result of virus infection.

weakening effect of the infection. In other varieties the infected plants are less vigorous each year, and die in 3–5 years. This condition was once considered normal, and varieties of tulips were named by growers and sold at high prices, according to the various color patterns. The tulip break viruses are easily spread by cutting infected flower stems and then healthy ones.

Most virus diseases are highly infectious and are readily transmitted from diseased to healthy hosts. The methods of transmission have been studied intensively, for these bear an obvious relation to disease prevention. Mosquitoes are important among the agents that carry animal virus directly from one infected human being to another or from some animal that serves as a reservoir of infection. Yellow fever, dengue, and equine encephalitis are spread in this way. Rabies is transmitted only through a wound, usually by the bite of a rabid animal. Many viruses are spread by contact or by droplets in the air.

The methods by which plant viruses pass from plant to plant are even more numerous. Under experimental conditions, viruses can be transmitted by grafting diseased parts on healthy plants, by rubbing sap of diseased plants into the leaves of healthy plants, or by transferring insects to healthy plants after they have fed on a diseased plant. Under natural conditions the most common means of spread are insects, usually those with sucking mouth parts, such as aphids, leafhoppers, mealybugs, whiteflies, and thrips. Mechanical transmission, as by rubbing of leaves against one another, is also believed to occur. The tobacco-mosaic virus, like many other viruses, is very stable, and is not destroyed during the curing process. The fingers of smokers of infected tobacco may transmit the infection to growing crops of tobacco or tomatoes (which the virus also infects) when the plants are handled. Virus particles remain infective in dried tobacco leaves for over 50 years, or in clothing for 2 weeks, and will even stand boiling for 10 minutes. A few viruses are seed-borne—many, fortunately, are not; all are readily distributed by vegetative propagation, as in the potato or in fruit trees.

The Bacterial Viruses Of outstanding interest among the viruses are those that attack bacteria. These were discovered independently in 1915 and in 1917 by an English and a French scientist, and were named by the latter BACTERIOPHAGE, now usually shortened to PHAGE. Beginning about 1940 great advances have been made in knowledge of phages and their relation to the bacterial host. Most of these studies have been made on certain phage strains that attack a single strain of the colon bacillus. This knowledge has contributed to a better understanding of the mechanisms of inheritance at the molecular level, not only in phages and other viruses but also in organisms.

The phages have most of the characteristics of other viruses, but differ in external form. They are typically tadpolelike in shape, with a large polyhedral head and a cylindrical tail to which six tail fibers are attached (FIGS. 20-16, 20-23). The tail may be conspicuous or very short but does not provide the virus with motility. Isotope experiments have shown that the DNA is localized in the head portion, and surrounded by a protein sheath. The tail, also, is protein. The DNA is believed to be composed of a single,

Fig. 20-23 Electron micrograph of phage of colon bacillus, strain T_2 (21,000×).

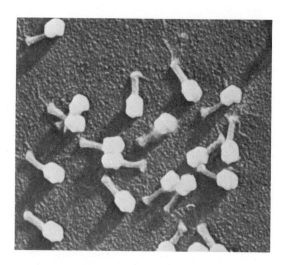

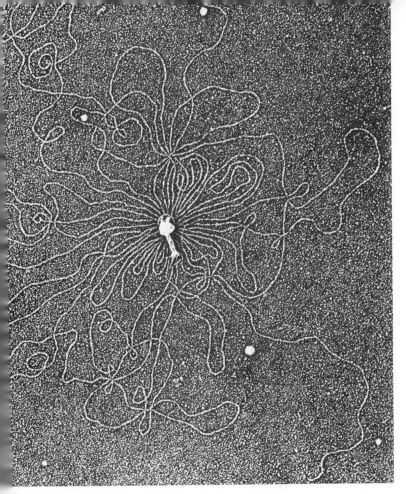

Fig. 20-24 The DNA macromolecule of bacteriophage T₂, liberated from the virus head by osmotic shock. Two ends of the DNA strand are seen at the top center and bottom right. These free ends may be the result of the breaking of the closed DNA strand during preparation (64,000×).

closed, double-stranded macromolecule, about 50 microns long, packed tightly within the head (Fig. 20-24).

When virulent phages are mixed with a suspension of susceptible bacteria, the virus particles come in contact with the bacterial cells. The first stage of infection is the attachment, tail first, of the phage to a particular site on the cell wall (Fig. 20-25). This is achieved by the adsorption of the tail fibers to the wall. An enzyme, presumably carried by the tail, then digests the wall at the point of contact, and the DNA thread is expelled from the head through the cylindrical cavity in the tail into the host cell. The empty protein head and the tail are left on the outside, still attached to the cell wall. After an average interval of about

Fig. 20-25 Electron micrograph of a strain (T₅) of bacteriophage attached to a cell of its host, the colon bacillus (*Escherichia coli*), 10 minutes after mixing at 37°C (36,720×).

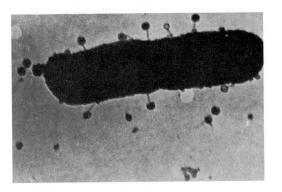

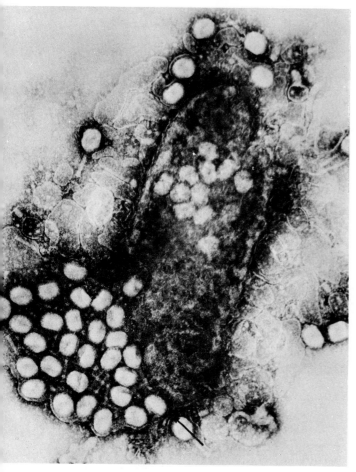

Fig. 20-26 Incomplete lysis of a colon bacillus following infection by phage. Some phages are still within the wall of the host cell; others are outside the original cell but still held within the cell membrane. The free phages are probably from other lysed cells. Note arrow (lower right) pointing to cell membrane (40,000×).

considered here only in general terms. If cells are broken open during periods up to about 12 minutes after infection, no phage particles are found. After this, if cells are broken at intervals, increasing numbers of mature particles are found until the host cell lyses. Within the host cell at least two important activities must proceed: (1) the production of more phage DNA, thus eventually providing a DNA molecule for each new particle; and (2) the synthesis of new phage protein that will provide the heads and tails of the new phage particles. All this is accomplished by the introduced DNA, which takes control of the metabolic machinery of the cell, "borrowing" the host enzymes and materials. In addition, certain new enzymes are synthesized by the bacterial cell under the direction of the information supplied by the viral DNA. New phage DNA instead of bacterial DNA is synthesized, and amino acids are assembled into phage protein instead of bacterial protein. These are complicated biochemical processes. The energy to drive these reactions is also supplied by the host cell. Near the end of the latent period, all phage components are assembled. The DNA molecule is condensed and then incorporated within the head protein, the tail becomes attached, and the host cell lyses, releasing mature phage particles.

The effects of phage infection of bacteria can be seen with the naked eye by a simple but highly important procedure in the study of bacterial viruses, called the PLAQUE ASSAY. A Petri dish containing a solid layer of nutrient agar is first prepared. To this is added a thin layer of melted agar containing large numbers of susceptible host bacteria and a known volume of a highly diluted suspension of bacteriophage. The thin top layer of agar is allowed to solidify, and the dish is incubated. The bacteria grow vigorously, their colonies quickly becoming confluent and covering the entire surface, except for a number of clear spots, the plaques (FIG. 20-27). Phage particles in the suspension infect neighboring bacteria, which lyse, yielding a new crop of phages. These in turn infect neighboring bacteria and so on, the plaques gradually enlarging to several millimeters in diameter. They may vary in size

½ hour (called the LATENT period), the bacterial cell suddenly undergoes disintegration (LYSIS), and a number (from 30 to 100 or more) of new, mature phage particles is released (FIG. 20-26).

The mechanism of phage multiplication during the latent period is a matter of great interest and has been extensively investigated, but can be and appearance, depending upon the phage and the host. Evidence from several sources indicates that each plaque is initiated by a single phage particle, much as bacterial colonies arise from a single cell. The number of plaques and the volume and dilution of the suspension may now be used to calculate the number of virus particles in the original suspension.

There are many kinds of phages, differing in particle size, in host range, and in other ways. They range in size from 10 millimicrons to 200 millimicrons. No one phage can attack all kinds of bacteria; a phage active against dysentery bacilli, for example, does not attack staphylococci, and vice versa. Phages occur in nature wherever bacteria are numerous, as in the soil, polluted waters, and decaying materials of many kinds. It is improbable, however, that massive bacterial destruction such as that which occurs in a test tube takes place in most natural environments. Following the discovery of bacteriophage, and particularly in the 1920s, much effort was expended to show that phages could be used in medical therapy, but all such efforts have been fruitless. The reasons for this have never been fully determined.

Phages are of considerable practical importance in industries that use bacteria. If the bacterial cultures employed in fermentations or in the dairy industry become contaminated with phage, they must be discarded and thorough sanitation undertaken. Phages may also attack actinomycetes used in the production of antibiotics.

Temperate phages We have so far considered only those phages that lyse their host cells shortly after infection, releasing infective particles. Such phages are termed *virulent*, or *intemperate*. It has been found, however, that in some strains of bacteria, and for some phages, a phage may be carried within the bacterium for a number of cell generations without causing lysis (FIG. 20-28). Such phages are called *temperate*. The strains of bacteria involved are termed *lysogenic*, and the heritable, noninfectious form in which the phage exists in such lysogenized cells is known as *prophage*. The prophage does not multiply in-

dependently, as do virulent phages; rather, it is inserted into and constitutes a part of the DNA macromolecule of the host cell and is replicated with it (FIG. 20-28A, B, C).

Occasionally, however, some individuals of a population of lysogenic bacteria are somehow induced to undergo lysis, releasing infective particles and resulting in the death of the cells (FIG. 20-28D, E). Induction of virulence is brought about by the detachment of the prophage from the host chromosome. The phage genes, suppressed while in the prophage stage, now begin functioning, and bring about the formation of phage DNA and phage protein. The prophage-phage cycle is thus seen to be similar to the situation in conjugation, in which episomes are found to exist free or integrated into the bacterial chromosome. The number of bacterial cells induced to undergo lysis can be increased by such special treatments as short exposures to ultraviolet light.

Fig. 20-27 Phage plaques. Explanation in text.

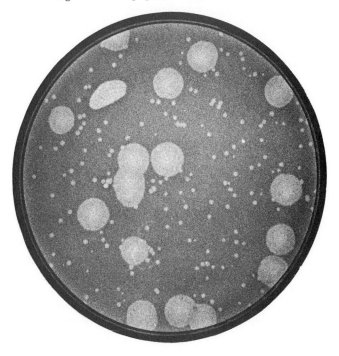

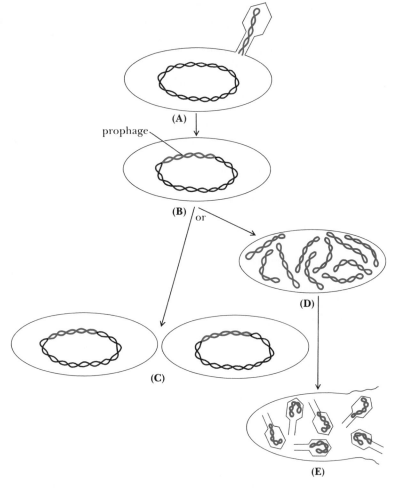

Fig. 20-28 Lysogeny. (A): infection of sensitive cell by temperate phage. (B): insertion of phage DNA into bacterial chromosome; the host cell has now been lysogenized. (C): lysogenized cell now divides, reproducing both bacterial and prophage DNA. (D): induction of virulence and reproduction of phage DNA. (E): incorporation of phage DNA into mature phage particles and lysis of cell.

prophage

(A)

(B) or

(D)

(C)

(E)

The lysogenized cells of the culture that produced the virulent phage are immune to the attacks of the infective particles, which infect other sensitive strains of bacteria. Some individuals of such a strain are lysed within a relatively short time, but others are lysogenized; that is, they continue to harbor the phage in a noninfectious form for a number of cell generations. This behavior of the temperate phages has been compared to the condition commonly found when two organisms live together in a host-parasite relationship. The best-adapted parasite, from the standpoint of survival and reproduction of the species, is one that lives with the host in a balanced condition, utilizing sufficient material from the host to allow growth and multiplication of the parasite, but not killing the host.

Transduction We have now laid the background for a consideration of transduction (FIG. 20-29), mentioned earlier in this chapter as one of the methods by which recombination can

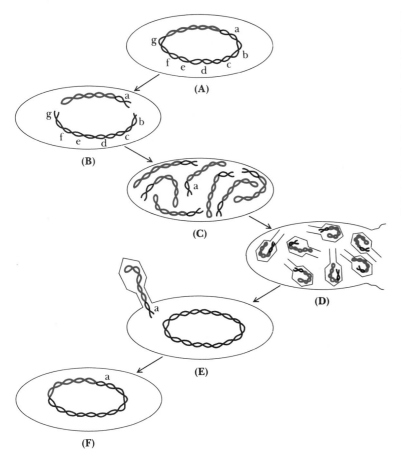

Fig. 20-29 Transduction. (A): strain B, lysogenic cells, bearing phage DNA in the prophage state (color). (B): detachment of prophage, together with part of host chromosome *a*. (C): replication of phage DNA and attached fragment of host chromosome. (D): incorporation of DNA into phage head and lysis of donor cells. (E): infection of strain A (recipient cell) carrying donor DNA fragment. (F): insertion of donor DNA into recipient chromosome, resulting in recombination.

occur in bacteria in the absence of sexual reproduction. In transduction, genetic transfer from donor to recipient cells is conveyed by phage particles. This mechanism was discovered in 1952 by using experimental procedures similar to those employed in the discovery of conjugation. Two mutant strains of *Salmonella* were used; one, which we shall call strain *A*, lacked the ability to synthesize three essential amino acids; the other, strain *B*, was unable to synthesize two other amino acids. These strains, after growth in

media containing the lacking amino acids, were placed together in a medium containing only mineral salts and carbohydrate. Recombinants that could grow in the medium then arose; that is, they were able to synthesize all their amino acids.

That contact between cells, and therefore conjugation, was not necessary for this genetic exchange was shown by the following experiment. The two strains were placed together in the same culture chamber, separated by a filter that was freely permeable to the culture medium but impermeable to the *Salmonella* cells. After some hours it was found that recombinants had arisen, but only on the side of the filter on which strain *A* had been growing. It appeared, then, that some sort of filterable agent was responsible for the transmission of genetic factors from the strain *B* donor to the strain *A* recipient cells. Several experiments showed that this agent was not free DNA, as in transformation in the pneumococcus. One of these experiments consisted of treating the filterable agent with the DNA-specific enzyme deoxyribonuclease, without destroying its activity.

Further studies showed that strain *B* was lysogenic, carrying a prophage (FIG. 20-29A), and that strain *A* was sensitive to infection by phage from strain *B*. It is visualized that during the lysis of strain *B* (FIG. 20-29B, C, D), a very small fragment of the host chromosome becomes lodged in the head of the infective phage particle as this is organized from the pool of virus DNA in the host cell. This "contaminated" phage particle then infects the recipient lysogenic cell (strain *A*), and the DNA fragment of the donor cell is incorporated into the DNA macromolecule of the recipient cell, thus bringing about recombination (FIG. 20-29E, F).

Although questions have been raised concerning the evolutionary significance of the recombination mechanisms (transformation, conjugation, and transduction), it seems reasonable to assume that they play the same role in bacteria as sexual genetic recombination plays in other organisms. That is, they bring about variations upon which natural selection may operate.

Viruses versus Organisms From the time of the first recognition of the existence of viruses, investigators have raised the question of whether viruses are organisms. On the one hand, they are constructed of organized organic compounds of a nature and quantity specific for each kind of virus. As in other biological objects, hereditary characteristics are transmitted by nucleic acids. They mutate and multiply. On the other hand,

virus particles isolated from the host in purified form are entirely inert and exhibit none of the characteristics usually attributed to the living state. They behave like microorganisms only when associated with the complex mechanisms of living cells. They are, then, obligate parasites. But such parasites, among cellular organisms, can carry on respiration and synthesize a part, at least, of their requirements for growth, whereas viruses, even in host cells, are dependent for energy and all growth materials upon the host. Even within the host they produce no respiratory enzymes nor any enzymes catalyzing synthetic reactions. Even multiplication, which occurs only within the host, differs from reproduction in organisms. The virus particles do not grow and divide, particles do not arise from pre-existing particles, in comparison to the cells of organisms, which always arise from pre-existing cells. At present it seems best to define a virus as a noncellular ultramicroscopic particle of organic matter, containing only one kind of nucleic acid, which can multiply only in living cells. Whether viruses are living or nonliving is really not important from a practical standpoint; in many respects they behave like microorganisms and are dealt with as such.

SUMMARY

1. Bacteria are classed by some as protista, by others as plants and placed in the thallophytes adjacent to the fungi.
2. The bacterial cell is procaryotic. The organelles are not bounded by membranes, mitochondria are absent, and there is no mitotic mechanism. The nuclear body is composed of a closed, double-stranded, DNA macromolecule, the genetic equivalent of the nucleus of other organisms.
3. Vegetative reproduction of bacteria is by transverse fission. Spore formation is not a method of reproduction, but provides resis-

tance to high temperatures, desiccation, and other unfavorable environmental factors.

4. Sexual reproduction is unknown in bacteria, but recombination can occur by transformation, conjugation, or transduction.

5. Microorganisms are studied in pure culture, growing on a medium that may be liquid or solid.

6. Bacteria are either autotrophic or heterotrophic. Autotrophs, like higher plants, use the carbon from carbon dioxide in the synthesis of cell materials. Some, the phototrophs, obtain energy from light; others, the chemoautotrophs, from the oxidation of inorganic compounds. Heterotrophs obtain their carbon and energy from the oxidation of organic compounds.

7. Some bacteria are able to synthesize all organic compounds necessary for growth when cultivated on simple media containing sugars and essential elements. Others are deficient in the ability to synthesize certain growth factors, such as amino acids and vitamins, and must obtain these from the environment.

8. Bacteria carry on many kinds of fermentations. Lactic acid is a common product of the fermentation of sugars. Amino acids may also be fermented, with the formation of organic acids, ammonia, hydrogen, and carbon dioxide.

9. Some bacteria are obligate aerobes, others obligate anaerobes, but most are facultative anaerobes. True anaerobic respiration is carried on by a small group of bacteria.

10. The growth requirements of bacteria include not only organic compounds as a source of energy and building materials but also moisture, essential elements, oxygen (for aerobes), and suitable temperatures, osmotic concentrations, and pH.

11. Bacteria are identified by a combination of morphological features and physiological behavior in pure culture.

12. The use of heat is an important method of bacterial control. Spores are not killed by boiling, but steam under pressure, as in the autoclave, can destroy spores as well as vegetative cells of bacteria.

13. To preserve food, microorganisms must be destroyed or their growth inhibited. This may be accomplished, among other procedures, by low temperatures, drying, heat, as in canning, and by the use of preservatives, such as sugars, acids, and antibiotics.

14. Foods may be canned at boiling temperatures or by steam under pressure. Neutral or slightly acid foods should be canned only by steam under pressure. The spores of botulism may survive in slightly acid foods canned by boiling only.

15. Pasteurization can destroy pathogenic bacteria, since these, in general, do not form spores.

16. Bacteria pathogenic for man and animals produce two kinds of toxins, exotoxins and endotoxins. These differ in their chemical composition and solubility.

17. Most food-borne diseases are caused by bacteria—staphylococcus and botulism food poisoning and salmonella food infection. The exotoxin of botulism organisms can be destroyed by boiling.

18. The most useful antibiotics against bacteria are obtained from bacteria, actinomycetes, and molds.

19. Viruses are noncellular ultramicroscopic organic particles that multiply only within living cells. They contain only one kind of nucleic acid, either DNA or RNA. Most viruses consist only of a protein sheath surrounding a core of nucleic acid. They produce no respiratory or synthetic enzymes and are dependent on the host cell for energy and all growth materials. Their method of multiplication is unlike that of cells. Some bacterial viruses, the intemperate phages, disintegrate their hosts shortly after infection. Others, the temperate phages, are carried within the host cells for a number of cell generations in a noninfectious form. This is the condition of lysogeny.

The Fungi: Phycomycetes, Ascomycetes, Lichens

THE FUNGI ARE A LARGE GROUP OF PLANTS. Many thousands of species have been described, and the list of species is extended each year. The large number of species is matched by the great importance of fungi to man. Together with bacteria, the fungi are the principal agents of decay and so, through the decomposition of organic matter, they play an essential part in the nutrition of green plants. Fungi are used for food, in medicine, and in industrial processes. They cause diseases of man, other animals, and plants.

It is in their role as causes of plant diseases that fungi have their greatest economic importance. Hundreds of millions of dollars are spent annually in the United States for sprays and dusts used to combat fungal enemies of crop plants. Millions more are spent in breeding plants resistant to fungus attack. An army of scientists in agricultural experiment stations and at university and industrial laboratories is constantly at work on problems directly or indirectly related to fungi. These problems involve the treatment of seeds and soils; treatment of wood, fabrics, and other materials; methods of crop production and other agricultural practices; and the effectiveness of inspections, quarantines, and other precautionary measures. Fungi are more dangerous to plants than bacteria are to us.

The city dweller may believe that fungi are important only to the farmer or the forester. Yet, as a consumer, he must pay for the ravages of plant disease and for the efforts expended to control such disease. Plant diseases mean increased costs for food, clothing, and shelter for all of us. For example, our nations's forests lose more than 20 billion board-feet to decay and other types of diseases each year—enough lumber to construct a wooden sidewalk a mile wide from New York to San Francisco. And in numerous ways in daily life the layman is even more directly concerned with fungi. He paints his house not simply for appearance but also because paint protects the wood from the action of decay fungi. If he discovers dry rot (caused by fungi) in its beams, the costs of repair may be heavy. Fungi may spoil his stored fruits and vegetables, and he may find mildew on his roses and other fungous diseases on his shade trees and in his vegetable garden. He and his pets may suffer from fungal diseases. About 15 percent of all human allergies are caused by fungi. The count of *Alternaria* and other mold spores is frequently publicized, together with that of ragweed pollen, in mid- and late summer.

The Nature of Fungi   Fungi, like algae and bacteria, are thallophytes. They are characterized by two important features: (1) they are nongreen plants, devoid of chlorophyll; and (2) the plant body is characteristically composed of branching, filamentous (threadlike) structures termed HYPHAE. These may occur in an interlacing tangle collectively known as MYCELIUM. The hyphae contain nuclei and cytoplasm and are either continuous (nonseptate) or divided by cross walls (septate). In the latter case, each segment of a hypha may contain one, two, or many nuclei. In most fungi the chief component of the cell wall is CHITIN, a polysaccharide that makes up the exoskeleton of insects and crustaceans. Cellulose is found in the cell walls of some of the lower fungi, but is absent in most fungi. Callose, a polysaccharide, and other organic materials have also been reported in many fungi.

Fungi reproduce by vegetative and by asexual and sexual reproduction. Examples of vegetative reproduction are fragmentation, in which the mycelium breaks into pieces, each fragment capable of growing into a new individual, and budding, as in the yeasts. Asexual reproduction is by spores not associated with the sexual process. Sexual reproduction is highly varied, although always involving the fusion of two haploid nuclei and meiosis. Spores (meiospores) are also associated with the sexual process. In some fungi, such as the mushrooms and puffballs, the spores are borne within or upon more or less massive "fruiting" bodies, formed by com-

pact aggregations of numerous hyphae. In other fungi, such conspicuous structures are absent, and the spores are closely associated with vegetative (somatic) hyphae.

Most fungi reproduce both sexually and asexually. Asexual reproduction is generally more important for propagation of the species. Large numbers of individuals are produced and the asexual cycle is often repeated several or more times during a season, whereas the sexual stage of most fungi is produced only once a year.

Since fungi lack chlorophyll, they must be either saprophytic or parasitic. Some species are restricted to one or the other of these modes of nutrition—they are either obligate parasites, such as the rusts and powdery mildews, or obligate saprophytes, such as many of the large and fleshy fungi. Between these lies a great variety of nutritional relationships. Many fungi have both parasitic and saprophytic phases in the same life cycle; others may be either parasitic or saprophytic.

Fungi flourish in any environment that provides proper conditions for growth. They thrive in the soil, in the living and dead remains of plants and animals, and on other organic materials such as foods and leather. In the wet tropics, shoes and clothing may become covered with mold almost overnight. Food is obtained from these sources through the action of a wide range of enzymes, including proteases, cellulase, amylase, and pectinase. Like bacteria, most fungi can be grown in pure culture on nutrient media. As with other organisms, growth is affected by temperature, moisture, pH, and oxygen supply. Free water, or high humidity, is nearly always essential for the growth of fungi. Most fungi grow best at a pH of about 6. Some, however (like *Rhizoctonia*), will grow within a wider pH range (2–12) than any other kind of plant or animal life. Fungi vary in their food requirements. Many can grow in the laboratory on simple media containing only a carbon source, such as glucose or sucrose, and mineral salts, including a source of nitrogen. From these they can synthesize their own proteins and other growth compounds, including vitamins. Others, such as the pink soil mold *Neurospora,* an important organism in genetic investigations, require in addition the vitamin biotin. Still others require not only biotin but also thiamine and other vitamins.

The fungi are aerobes, although their oxygen requirements may be low. The commercial production of organic acids and antibiotics by fungi takes place in the presence of oxygen, and is associated with aerobic respiration. A considerable number of fungi, including many yeasts, are able to carry on fermentative processes, but these proceed with available cell material. Growth does not occur unless some free oxygen is available.

Fungi and Plant Disease The annual loss in the United States from diseases of farm and garden plants, turf, nursery and florist crops, and of forest trees is estimated at more than $4 billion. About 1000 parasitic plant diseases are caused by bacteria, viruses, and nematodes, but many more thousands are the result of infection by fungi. This is a parasitic relationship in which infection of the host is preceded by the lodging of spores or other reproductive structures, such as fragments of mycelium, upon some plant part, commonly a leaf. With suitable moisture and temperature the spores germinate and form short threads, the GERM TUBES. These tubes penetrate into the host through natural openings (stomata, lenticels, hydathodes), wounds made by nematodes, insects, man, or other agents, or through the unbroken cuticle.

Within the host, the hyphae of certain pathogenic fungi grow between the cells, sometimes thrusting out special food-absorbing hyphae (HAUSTORIA) into adjoining cells. Haustoria may be knoblike, elongated, or branched in a rootlike manner (FIG. 21-1). The hyphae of other fungi penetrate directly into the cells of the host. The symptoms of the disease soon appear, and these are commonly so characteristic as to be recognized readily by the plant pathologist or the experienced layman. The names of many of the symptoms are descriptive—such as canker and wilt, rot, blight, mildew, leaf spot, rust, scab, and

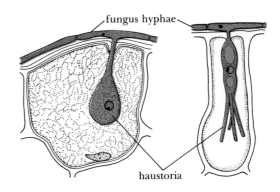

Fig. 21-1 **Haustoria of powdery mildews in leaf epidermal cells of host.**

smut. Among other symptoms associated with specific diseases are rolling of leaves, deformation of fruits, dwarfing of the entire plant or certain organs, and dropping of blossoms or fruits. Chlorophyll may be reduced in amount, or active leaf tissue destroyed. Parasites may kill root cells, thus indirectly affecting the aerial parts. Cells may be stimulated to divide at an abnormal rate, producing tumorlike tissues. The components of the host cell wall may be digested by enzymes that dissolve pectic substances, cellulose, and even lignin. The damage caused by plant disease is commonly serious and may lead to destruction of the crop.

Diseases in plants have been recognized for centuries. Aristotle, Theophrastus, and Pliny recorded them, and religious writings such as the Old Testament indicate that rusts, mildews, and blights have long been important problems. Until the middle of the nineteenth century, however, disease in plants was explained on the basis of bad weather, poor soil, or supernatural intervention. In ancient Rome a feast was celebrated and animals sacrificed to Robigus, the god of wheat, that the grain might be saved from destructive rusts. Although records of plant disease during the Middle Ages are scanty, it is known that there were epidemics (epiphytotics) of wheat rust through these centuries, as in ancient times. Outbreaks of the ergot fungus, causing a disease of cereals and poisoning of those who consume the diseased grains, have been known since the sixth century.

Crop pestilence and failure in modern times are more familiar to us. The middle of the nineteenth century, for example, was marked by two destructive plagues, both introduced into Europe from the Western Hemisphere. Powdery mildew of grapes was first observed in England in 1845. It spread rapidly, and by 1854 had reached every grape-growing country in Europe. The damage was enormous, and it increased every year. In 1854 it reduced the grape crop of France by 75 percent, with a total destruction of the vines in some areas. Only the belated discovery that mildew could be controlled by sulfur checked the disease. Even more serious was the potato blight epiphytotic of 1845–1847, which swept over Europe and caused the greatest distress in Ireland, where more than a million people died of famine. Another notable epiphytotic was the rust of the coffee leaf, which, beginning in 1869, swept over the island of Ceylon. No method of control was successful in halting the devastation; planters were ruined, banks failed, and the cultivation of the crop was abandoned. The disease spread throughout the East, and the center of coffee production shifted to the Western Hemisphere.

One of the most serious epiphytotics known in the United States was that of chestnut blight. This disease, widely distributed over the world, was first discovered in this country in 1904. The rapidity of its spread was phenomenal, and by 1914 the fungus had become distributed throughout the range of the native chestnut, destroying millions of trees. The native chestnut, one of the most important trees from the standpoint of lumbering and other industries, is doomed. A similar fate threatens the American elm, one of our important shade and ornamental trees. The threat here is the Dutch elm disease, caused by a fungus introduced into the United States in 1930.

Aquatic as well as land plants are subject to disease. In the late spring of 1931, an epiphytotic, the "wasting" disease, swept through the great beds of eelgrass growing in the Atlantic coastal waters of North America. Eelgrass (not a grass) is a herbaceous monocotyledon with narrow, ribbonlike leaves up to 5 feet in length. By the summer of 1932 this once abundant plant had largely disappeared. The causal agent was later traced to an amoebalike organism related to the slime molds—not a true fungus—that had suddenly increased in virulence. The epiphytotic occurred also in European waters, from the Mediterranean to Sweden.

In many places the eelgrass has again become abundant, but the effects of the epiphytotic were serious, for the plant is important as a producer organism in many food chains. The seeds, rhizomes, and other parts are fed upon by several kinds of waterfowl. These died off or sought other feeding grounds when the epiphytotic struck. The eelgrass also served as a source of food for a variety of small marine animals. As the eelgrass died out, these decreased in numbers, thus affecting the harvest of shellfish and the catch of cod, flounders, and other commercially valuable fish. Other periods of scarcity of the eelgrass have occurred, but none comparable to the catastrophe of 1931–1932.

THE SLIME MOLDS

Before taking up the fungi proper, consideration should be given to the slime molds, a unique group of organisms whose origin and relationships have not been agreed upon. In schemes of plant classification they are associated with the fungi, but while they possess fungal characteristics, they are usually excluded from the true fungi because of certain features of their structure and life histories.

In nature the slime molds are to be found in moist woods, living on rotting logs, damp soil, leaf mold, sawdust, decaying fleshy fungi, or other organic matter. Following periods of rainy weather they may occur on the leaves of grasses or other plants on lawns. The vegetative phase

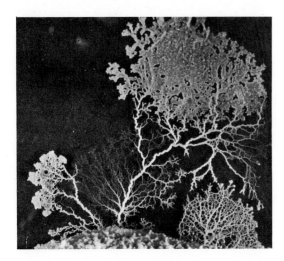

Fig. 21-2 Plasmodium of a slime mold, *Physarum polycephalum*.

and contain numerous, walled resting spores that are dispersed by wind. In additon to the spores, the sporangia often contain an intricate network of fine threads. The color of the spores, the shape and size of the sporangium, and the pat-

Fig. 21-3 Slime molds. (A, B): *Stemonitis*. (A): sporangium after discharge of spores, showing network in which spores were embedded. (B): habit of plants. (C, D): *Hemitrichia*. (C): cluster of sporangia. (D): single sporangium enlarged, after spore discharge. (*Stemonitis* is about 15 mm high; *Hemitrichia*, about 2 mm.)

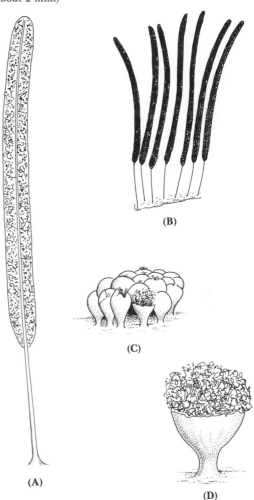

consists of a naked (without cell walls), multinucleate mass of protoplasm called a PLAS-MODIUM. This commonly assumes the form of a fan-shaped network with veinlike branches continuous along the outer margin, the whole often brightly colored (FIG. 21-2). The plasmodium flows over the surface in an amoeboid fashion, obtaining food by surrounding and engulfing particles of organic material. This food is digested within food vacuoles, and waste products or undigested particles are left behind as the plasmodium moves ahead. In the laboratory, plasmodia of some species have been maintained by feeding them bacteria, yeast cells, pulverized rolled oats, or oatmeal agar. The absence of cell walls in the plasmodium, amoeboid movement, and the method of obtaining food are features characteristics of animals rather than of plants.

Although the vegetative stage of a slime mold is animal-like, reproduction is plantlike. Under certain environmental conditions the plasmodium gives rise to groups of thin-walled SPORANGIA, or spore cases, very similar to those of many fungi. The sporangia (FIG. 21-3) are small (only a few millimeters high), sessile or stalked,

Fig. 21-4 *Lycogala epidendrum,* a common slime mold. Usually found on rotten logs. Each fruiting body is about ¼ inch in diameter and is composed of numerous fused sporangia. The plants are white when young (as in photograph), becoming pink and finally brown when mature.

tern of the inner network are characteristic of each species. In a few kinds of slime molds, the sporangia are fused to the extent that they lose their identity in a mass, thus forming a relatively large and conspicuous fruiting body (FIG. 21-4).

A sexual stage intervenes between the resting spore and the plasmodium. Upon germination, the spores produce from one to four naked gametes, each with two flagella. The gametes may feed and divide by fission for a time, but eventually fuse in pairs to form a zygote. The plasmodium is produced directly from the zygote by increase in size and numerous mitotic divisions of the diploid nuclei. Young plasmodia may combine with zygotes or other plasmodia of the same species and thus increase in size. Meiosis occurs in the sporangium prior to spore formation, and the spores and gametes are thus haploid.

The slime molds are of little or no economic value, but a few species have been the subject of intensive laboratory studies. These provide a large amount of protoplasm free from cell walls, and so can be used in studies of protoplasm. They have been widely used in experimental

studies of physiology and development in the lower organisms. Particular attention has been given to protoplasmic streaming in *Physarum polycephalum*. When one of the larger veins of a plasmodium is examined under the microscope, the protoplasm is seen to be streaming rapidly, the velocity being the greatest recorded for any organism. After streaming in one direction for a short period—commonly less than a minute—the flow is reversed, and this rhythmic backward and forward movement continues for the duration of active growth.

The combination of plant and animal characteristics have caused the slime molds to be classified variously as plants, as animals, and as protista. Almost certainly they have long existed on the earth, but have given rise to no known higher forms of life. Slime molds, although essentially naked masses of protoplasm, are extremely resistant to x-rays and other forms of radiation. This may indicate a very early form of life on earth when radiation levels were much higher than at present. Because of their peculiarities, slime molds well illustrate the infinite variety of living things.

THE TRUE FUNGI

Classification of the Fungi As with other groups of organisms, the classification of the fungi presents difficulties. Traditionally, however, the fungi have been grouped into four classes (Chapter 18). In the first three of these—the Phycomycetes, Ascomycetes, and Basidiomycetes —sexual reproduction is present. The phase of the life cycle that includes the union of gametes and the spores associated with sexual reproduction is termed the PERFECT STAGE. The phase of the life cycle in which no sexual fusion occurs and in which only asexual spores are found is known as the IMPERFECT STAGE. The nature of the spore or spores of the perfect stage and the methods of their formation are the most important features characterizing each of these three groups and are essential in determining the class into which a given fungus is placed.

In great numbers of fungi, however, the perfect, or sexual stage is unknown, and such species are placed in a fourth (artificial) group, the Fungi Imperfecti. In the course of study, the perfect stages of members of this group have occasionally been found, whereupon that species is transferred to the class to which it belongs. The nuclear cycle in many fungi with sexual reproduction is complex, and will not be described completely for all forms discussed in the following pages.

PHYCOMYCETES (Algal Fungi)

The Phycomycetes are commonly referred to as the algal fungi because members of the group, in structure and methods of reproduction, resemble certain of the green algae. This is a heterogeneous group that has been subdivided by various workers into a number of classes, as many as six. The group is also referred to as the lower fungi, in distinction to the higher fungi, the Ascomycetes and Basidiomycetes. Some, presumably primitive, are one-celled and form no hyphae, but most produce vegetative hyphae with no cross walls (nonseptate). Conspicuous fruiting bodies are not formed. Some are aquatic, a habitat considered more primitive than the terrestrial. The asexual spores (SPORANGIOSPORES) may be either motile or nonmotile, and are typically produced within a thin-walled sac, the SPORANGIUM. The motile spores, which may bear either one or two flagella, are called zoospores, as in the algae. This type of spore is produced in the true fungi only in the Phycomycetes. The spore of the perfect stage is commonly either a zygospore, the product of the fusion of two equal sex organs (gametangia), or an oospore, resulting from the fusion of a large egg with a smaller male gamete. That sex hormones are involved in sexual reproduction in the Phycomycetes and other fungi is well established by numerous experimental investigations. Most Phycomycetes are haploid, except for the single diploid-zygote nucleus, which undergoes meiosis shortly after gametic fusion. This simple type may represent a generalized kind of life cycle from which more complicated kinds have evolved.

Certain of the aquatic Phycomycetes are termed water molds, and these may be very destructive in aquaria and fish hatcheries. Among the terrestrial species, the bread mold fungus and certain other common and economically important forms provide us with a perspective of the class as a whole.

Bread Mold The bread mold fungus (*Rhizopus stolonifer*), a member of a group termed the black molds, is one of the most common and widely distributed of all the fungi. It can be seen in the home as a cottony, moldlike growth on bread and other foodstuffs (FIG. 21-5). Such a growth, produced after the germination of the asexual spores, consists of masses of tangled and branched hyphae, from which stalked, globular sporangia grow (FIG. 21-6). These contain the asexual spores. The sporangia are usually borne in clusters, each on a specialized hypha, the SPORANGIOPHORE. A cluster of sporangia is supported by a group of short, rootlike hyphae, the RHIZOIDS, which absorb food and water. New clusters of sporangia and rhizoids arise from still other specialized hyphae, the STOLONS, which extend outward from a parent clump like the runners of a strawberry plant.

As the mold matures, the swollen tip of a sporangiophore enlarges, forming a sporangium (FIG. 21-7). Nuclei, cytoplasm, and food materials flow into this enlarged apex. Most of the nuclei and the greater part of the cytoplasm accumulate in an outer peripheral zone, leaving an inner area with few nuclei and little cytoplasm. A dome-shaped layer of vacuoles now appears between the inner and outer zones. The vacuoles flatten and fuse; then a wall forms, separating the sporangium into two parts: an outer fertile zone, and an inner sterile area, known as the COLUMELLA. The protoplasm of the fertile zone

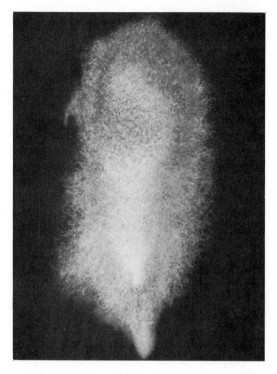

Fig. 21-5 The bread-mold fungus (*Rhizopus stolonifer*) growing on a carrot.

now becomes divided into a great number of sporangiospores (as many as 70,000) by a process of furrowing. Bits of cytoplasm, each containing several nuclei, become delimited from one another and from the surrounding wall, forming oval spores with cell walls ornamented by minute striations.

The sporangial wall is extremely fragile and disintegrates at maturity, releasing the spores. The columella persists after the rupturing of the sporangium. The spores, carried away by air currents, germinate under suitable conditions of warmth and moisture. Each spore forms a germ tube, which grows rapidly, branches, and soon gives rise to a mass of hyphae.

Sexual reproduction The spore of the perfect stage in bread mold and other black molds is a

zygospore. Sexual reproduction is by conjugation (FIG. 21-8) and is similar in several respects to that of *Spirogyra*. Where two hyphae from genetically different mycelia come in contact, specialized side branches, the PROGAMETANGIA, may form. These are adherent at the tips from the beginning, and as they increase in size, they gradually push apart the hyphae from which they arose. Soon a cross wall, enclosing numerous gamete nuclei, is laid down behind the apex of each progametangium. The two contiguous portions thus isolated are the GAMETANGIA. The remaining portions of the specialized branches, behind the gametangia, are termed the SUSPENSORS. The wall between the pair of gametangia now breaks down, the contents mingle, and the fused gametangia develop into the zygospore. The young zygospore enlarges, develops an extremely thick, warted wall, and enters a period of dormancy. In microscopic preparations of the mature zygospores, portions of the suspensors may be observed, still attached to either side of the zygospore.

Fig. 21-6 The bread-mold fungus. (A): structures associated with asexual reproduction by spores. (B): portion of hypha. (C): spore germination.

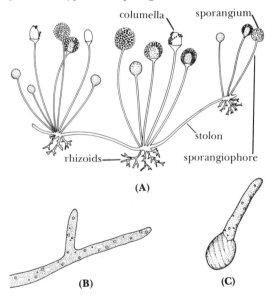

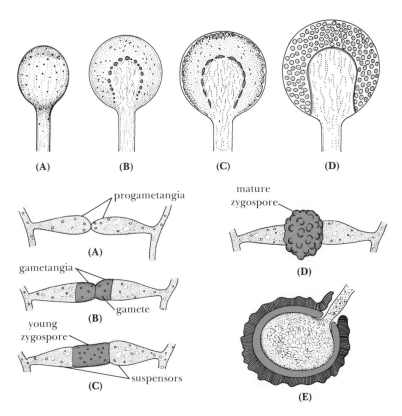

Fig. 21-7 Bread-mold fungus. Stages in development of the sporangium.

Fig. 21-8 Sexual reproduction in the bread-mold fungus. (A–D): stages in the formation of the zygospore. (E): germination of the zygospore.

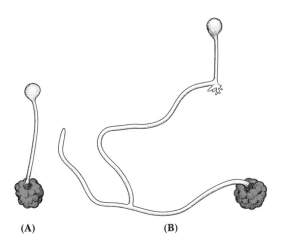

Fig. 21-9 Methods of germination of zygospores of *Rhizopus*. (A): the zygospore gives rise directly to a sporangium. (B): sporangiophore produced after sparse growth of vegetative hyphae.

Dormancy in the zygospore lasts for one to several months, and germination has been reported by only a few investigators. The zygospore germinates by either of two methods: (1) commonly it gives rise directly to a sporangiophore, which grows erect and produces a sporangium at the apex (FIG. 21-9A); (2) less commonly, perhaps, it produces a single hypha, or germ tube, which grows horizontally for a short distance, perhaps branching sparsely. The germ tube sends down rhizoids at the tip, and at this point a single sporangiophore arises, forming a sporangium (FIG. 21-9B). The zygospores of related black molds behave similarly. It is generally agreed that the subsequent events in the life cycle are as follows (FIG. 21-10):

Shortly after the zygospore is formed, some of the gamete nuclei fuse in pairs, each member of a pair being derived from a different gam-

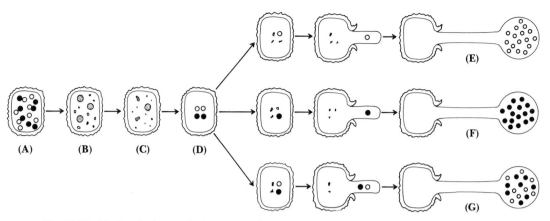

Fig. 21-10 Nuclear fusion, meiosis, and distribution of meiospore nuclei to sporangia in *Rhizopus stolonifer*. (A, B): fusion of gamete nuclei in zygospore. (C, D): meiosis. (E, F): three meiospore nuclei degenerate and a nucleus of one mating type enters the sporangium. (G): only two nuclei degenerate, sporangium contains nuclei of both mating types.

etangium. The remaining haploid nuclei probably degenerate. Since the gametes, consisting chiefly of nuclei, are alike as to size, sexual reproduction in this fungus is isogamous. During zygospore dormancy, all but one of the diploid nuclei disappear, and meiosis occurs in the single surviving diploid nucleus at the time of zygospore germination. Degeneration continues, for commonly only one of the tetrad of nuclei resulting from meiosis survives and produces, by successive mitoses, the haploid nuclei of all the spores in the sporangium. The single sporangium produced at germination may be one of three types: (1) some contain only plus spores (see next paragraph); (2) some contain only minus spores; and (3) some, fewer in number, contain spores of both mating types. The last condition may be explained by the survival in some zygospores of two nuclei of the tetrad rather than only one.

Homothallism and heterothallism The black molds can be divided into two distinct groups. In one group, zygospores are readily obtained, either in nature or from pure cultures. In the second group, which includes *Rhizopus stolonifer* and certain other species, zygospores rarely occur. This condition was a baffling one until it was demonstrated that the zygospores of the second group are formed only when hyphae of two different strains, produced from different spores, are brought together. One of these is termed the *plus* strain; the other, the *minus* strain. A species with two such unisexual strains, or MATING TYPES, is termed HETEROTHALLIC. A bisexual species, in which zygospores develop readily on hyphae produced from a single spore, is HOMOTHALLIC.

In the light of this discovery, it is now easy to obtain the zygospores of species in which zygospores were formerly difficult to obtain. The plus and minus strains are carried in pure culture, and when planted together in a nutrient medium, they will form zygospores. If the two strains are inoculated together at opposite sides of a culture dish, the hyphae will grow until they meet and form a line of zygospores across the middle of the dish (FIG. 21-11). The inoculation of two plus or two minus strains in the same dish will fail to produce zygospores. After the discovery of heterothallism in the black molds, the condition was found to be widely distributed in many other groups of the fungi.

Nutrition of bread mold   The bread mold fungus is one of many fungi able to obtain food from either nonliving materials or living tissues. In the laboratory (and often in the home), it grows upon bread or some other readily available food substance. From this, through enzymatic action, it obtains soluble food that makes possible its rapid growth. The bread mold fungus may also live parasitically. It is one of the most common and destructive fungi encountered in the storage and shipment of fresh fruits and vegetables. Sweet potatoes, grapes, tomatoes, plums, peaches, cherries, muskmelons, and strawberries are attacked by the mold. *Rhizopus* also causes an important boll rot of cotton. Much of the loss from Rhizopus rot in food products can be avoided by proper cooling and ventilation, and by handling so as to avoid bruises and wounds that permit the invasion of the tissues by the hyphae of the fungus.

Water Molds   Water molds are primarily aquatic, although certain species also live in the upper layers of the soil. Members of a closely related group, chiefly the genera *Aphanomyces* and *Pythium,* cause the well-known and destructive "damping off" of seedlings in greenhouse and garden soils. Among the common species of water molds are members of the genera *Saprolegnia* and *Achlya.* These live saprophytically in water on dead animal tissues, such as those of insects and fish. Several species of *Saprolegnia* are also parasitic on wild fish and in aquaria and fish hatcheries. Entrance into the host takes place chiefly through abraded areas or wounds. The molds probably do not invade uninjured tissues. Cultures of water molds can commonly be obtained by placing a dead housefly or other animal material in water from a pond or stream. Within a few days the material will be covered by a cottonlike mass of hyphae, visible to the naked eye.

Other hyphae extend into the tissues of the decaying animal.

A microscopic examination of the external hyphae will show that some of these are club-shaped and have been transformed into sporangia by the formation of a transverse wall at the base of the enlarged portion (FIG. 21-12, left). After the wall is laid down, the protoplasmic content of the sporangium becomes divided into numerous small zoospores, each with two flagella. These are discharged through a pore in the tip of the sporangium.

The sex organs of *Saprolegnia,* oogonia and antheridia (FIG. 21-12, right), are also formed by the enlargement of the tips of specialized hyphae and the formation of cross walls. The oogonium is large, spherical or oval, and contains several eggs. The antheridial branches are pressed against the oogonium and send out one or more

Fig. 21-11   A mold (***Mucor hiemalis***), showing plus and minus strains. A line of zygospores has formed where hyphae of the two strains came together.

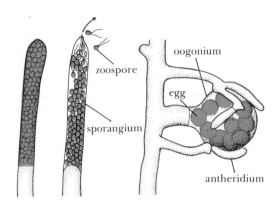

Fig. 21-12 *Saprolegnia.* (Left): reproduction by zoospore formation. (Right): sexual reproduction.

fertilization tubes that penetrate the eggs. Male nuclei pass through these tubes, bringing about fertilization by fusion with the nuclei of the eggs. Following fertilization, the zygotes (oospores) form a thick wall and pass into a resting stage. Germination of the zygote takes place either by the formation of a hypha or by the production of zoospores. Sexual reproduction in *Saprolegnia* is thus oogamous, in contrast to the isogamy found in the bread mold fungus.

Late Blight of Potatoes The fungus (*Phytophthora infestans*) that causes late blight of potatoes is one of the most destructive that attack cultivated plants. The disease is of considerable historical interest. A famine resulted in 1845–1847 from an epidemic of potato blight that raged over Europe. It was especially destructive in Ireland, where the potato was a mainstay in the diet of the peasantry. The resulting famine led to widespread starvation and death. As a direct result, thousands of families emigrated to the United States and the movement westward continued for a generation or more.

The distress created by the epiphytotic stimulated investigations of the cause and nature of the disease and of plant diseases in general. At that time the theory of spontaneous generation was still held, and although it was recognized that a fungus was associated with the disease, the prevailing belief was that such a fungus was the result, not the cause, of the disease. As a consequence of the many investigations that followed, proof of the pathogenic nature of many fungi was obtained. On the basis of the knowledge of the life histories of fungi acquired in these studies, effective methods of control were eventually devised.

The fungus that produces potato late blight grows in the leaves, petioles, and stems, where it produces dead areas (FIG. 21-13), and may finally kill the entire top of the plant. The hyphae grow between the host cells, sending long, curled

Fig. 21-13 Early symptoms of late blight (*Phytophthora infestans*) on lower surface of potato leaf. The dark areas have a moldy appearance, due to the reproductive bodies of the fungus. The infected areas spread rapidly and the leaves and frequently the crop are destroyed.

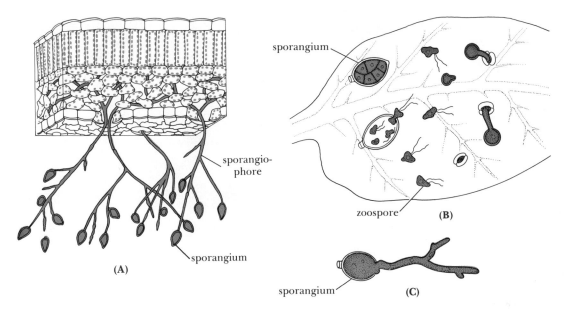

Fig. 21-14 The late blight fungus. (A): production of sporangia on lower side of potato leaf. (B): indirect germination. (C): direct germination.

haustoria into neighboring cells. As the disease progresses, numerous branched sporangiophores grow out through stomata on the lower side of the leaf (Fig. 21-14A). The sporangiophores give rise to large numbers of lemon-shaped sporangia. Wind and splashing rain spread the sporangia to new potato plants.

When a sporangium lands upon the leaf of a host plant, its behavior is most unusual. It functions either as a sporangium or as a spore. At an optimum temperature of about 12°C (54°F), through a process of *indirect* germination (Fig. 14B), the protoplasmic content of a sporangium becomes divided into numerous zoospores, each with two flagella. The zoospores escape through a pore at the apex of the sporangium, swim about in rain water or dew, and finally settle down and form germ tubes that penetrate the tissues of the leaf through a stoma.

At higher temperatures [optimum about 24°C (75°F)] a less common, *direct* germination occurs (Fig. 21-14C). The sporangium behaves like a spore, producing a germ tube that penetrates the

cuticle and grows into the leaf. The method of indirect germination may be interpreted as an example of the retention of an ancestral character. Although the late blight fungus is a terrestrial plant, it exhibits a primitive feature in its life cycle when it reproduces by motile spores. Such a method was probably used by the ancient, aquatic ancestors of this and other fungi.

The sporangia are not only carried to the leaves but also are washed into the soil, where they germinate and penetrate the potato tuber. As a result of such infections, or of infections during harvest, the tubers may be rotted by harvest time or develop rot during storage (Fig. 21-15). If blight-infected tubers are dumped in cull piles or are used for seed pieces the following spring, the hyphae grow up within the new tissues sprouting from the potatoes. The fungus soon sporulates on the young leaves and stems. The first infections of the season occur when the resulting sporangia are transported by splashing rain or are blown by the wind and infect nearby potato plants.

Fig. 21-15 Cut surface of potato tuber partly rotted by late blight fungus.

Fig. 21-16 Downy mildew (*Plasmopara viticola*) on fruits of grape.

Sexual reproduction in the late blight fungus is oogamous, but oospores had seldom been observed, until 1955. It is now known that the fungus is heterothallic, that there are two mating types, and that oospores are formed in abundance only when cells from one mating type come into contact with cells from the other type. Interestingly enough, both mating types, so far as is now known, are largely confined to Mexico, where they occur in a 1:1 ratio. Since the blight fungi found in North America and most of the rest of the world usually belong to only one mating type, oospores are only rarely produced. It may be expected that these discoveries in sexual reproduction in the blight fungus will lead to a better understanding of its genetic mechanisms and pathogenicity.

Even today, with an arsenal of effective fungicides, late blight of potatoes is an important disease. In one county in Maine an average loss of about 16 percent of the potatoes in storage was reported in one year, although growers had spent more than a million dollars spraying the crop to control the disease.

The potato blight fungus also causes late blight of tomatoes. The symptoms on tomato leaves are similar to those of the potato, and the fruits are readily infected and subject to rot at all stages of development. Late blight of tomatoes is common in certain southern states; elsewhere it is not of great importance but occasionally appears in epiphytotic form. Such an epiphytotic swept through the eastern half of the United States in 1946, reducing the crop by nearly 50 percent, and causing a loss estimated at 50 million dollars.

The Downy Mildews These fungi are closely related and have a life cycle similar to that of the fungus causing late blight. They are so named because of the moldlike appearance of the surfaces of the leaves and other affected organs of the host plants. The more common and destructive downy mildews are those of cabbage, hops, lettuce, onion, cucumber, tobacco, spinach, lima bean, pea, and grape (FIG. 21-16). Downy mildew (*Sclerospora*) of corn is one of the most destructive diseases of this crop plant in many parts of Asia, particularly in the lowland tropics.

As in the late blight fungus, deciduous sporangia may form germ tubes in some species. In others, zoospores form, and in still others, germination may occur by either of these two methods. All of these species are oogamous, producing antheridia and oogonia within the tissues of the host. In the downy mildew of grapes, a white, moldy growth that kills the tissues is produced on the leaves. The young shoots, tendrils, and fruit may also be attacked. It was in connection with this disease, epiphytotic in the vineyards of France, that Bordeaux mixture was discovered in 1882. This well-known combination of copper sulfate, lime, and water, used as a spray, was long one of the most important weapons in the attack on plant disease. Bordeaux mixture has now largely been replaced by more effective, organic fungicides, less toxic to plants.

ASCOMYCETES (Sac Fungi)

The Ascomycetes, or sac fungi, constitute one of the largest groups of the fungi. They include the powdery mildews, yeasts, many of the common black and blue-green molds, the cup fungi, morels, truffles, and many others. Ascomycetes also cause such important diseases as chestnut blight, Dutch elm disease, brown rot of stone fruits, peach leaf curl, apple scab, drop and watery soft rot of vegetables, black rot of sweet potato, common leaf spot of alfalfa, scab and foot rot of cereals, black knot of plum, stalk and ear rot of corn, and blast of rice. Many species are saprophytic, occurring in rich soil, on decaying wood, or wherever there is available food and moisture.

The hyphae of the Ascomycetes are septate, that is, divided by cross walls. The septum is not complete, but is perforated in the center. Cytoplasm and nuclei may move from one hyphal cell to another through these pores, so that the hyphae are essentially coenocytic.

The most common type of asexual spore, the CONIDIUM, is formed singly or in chains, generally at the apex of a specialized hypha, the CONIDIOPHORE (FIGS. 21-23, 21-24, 21-26). Such spores are nonmotile and germinate by the formation of a germ tube. The same mycelium that produces the conidia may later give rise to the spores of the perfect stage, the ASCOSPORES. These develop within a saclike, enlarged end of a hypha, known as an ASCUS (FIG. 21-17B). In most Ascomycetes the asci, usually associated with sterile hyphae, are borne upon or within definite "fruiting" bodies, of which the most important kinds are as follows (FIG. 21-17):

1. The asci are not enclosed but form a palisadelike *fertile layer* (FIG. 21-17A) on the surface of the fruiting body, which is commonly fleshy, as in the cup fungi (FIG. 21-19) and morels (FIG. 21-21).

2. The asci are enclosed within a minute (less than 1 millimeter in diameter), flask-shaped structure, the PERITHECIUM, (FIG. 21-17 C), which has an opening at the top through which the ascospores escape.

3. The asci are inclosed within a globose body, the CLEISTOTHECIUM, (FIG. 21-17D), similar to the perithecium, but a pore is absent and the ascospores are liberated following the breakdown of the cleistothecial wall. Release of the ascospores from the ascus is accomplished in various ways. In some species the ascus disintegrates, but commonly the spores are ejected through a terminal pore or slit.

The nature of the sexual process associated with ascospore formation is common to many Ascomycetes and is shown diagrammatically in FIG. 21-18. Sex organs (gametangia) that differ in size—male (antheridia) and female (ASCOGONIA) —develop within a young fruiting body (for example, at the base of a cup or perithecium). In homothallic species both kinds of gametangia arise from the same hyphal system; in heterothallic forms they arise from hyphae of different mating types. The antheridium comes in contact with the elongated receptive neck of the ascogonium, a pore develops, and the nuclei from the antheridium move into the ascogonium. Nuclear fusion does not occur at this time, however. The ascogonium now gives rise to a number of budlike processes, each of which grows into an ASCOGENOUS HYPHA. Each of these receives, early in growth, one or more sets of paired nuclei, one member of each pair originally derived from the antheridium and one

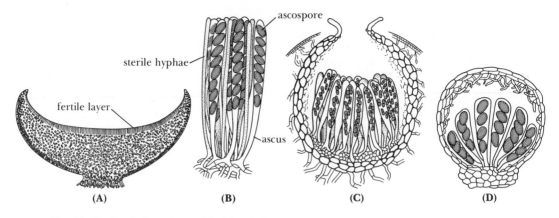

Fig. 21-17 Vertical sections of fruiting bodies, asci and ascospores in Ascomycetes. (A): cup, showing fertile layer composed of asci in palisadelike arrangement. (B): asci, ascospores, and sterile hyphae. (C): perithecium of apple scab fungus (*Venturia inequalis*) buried in dead tissues of leaf. (D): cleistothecium of a powdery mildew (*Microsphaera*).

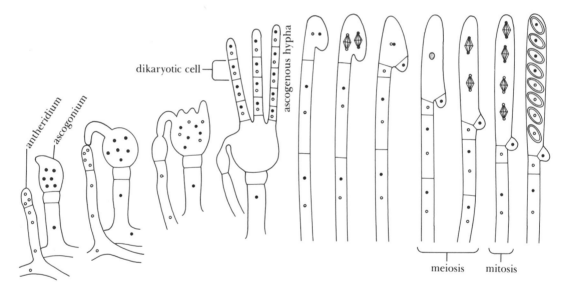

Fig. 21-18 Diagram showing nuclear behavior in the perfect stage of an Ascomycete.

from the ascogonium. These pairs now undergo a series of simultaneous mitoses, which provide a pair of nuclei for each of the subsequently formed cells of the hypha.

The terminal cell of an ascogenous hypha next bends back upon itself, forming a hook. The paired nuclei of the cell divide simultaneously, and cross walls are laid down in such a way that three cells result, a new terminal cell with two paired nuclei, and two cells with one nucleus each. The paired nuclei now fuse, and the cell containing this diploid nucleus—typically the only diploid nucleus in the life cycle—constitutes the ascus mother cell. The diploid nucleus undergoes meiosis, and four haploid nuclei are formed. Each of these nuclei again divides by mitosis, and cell walls form around each of the resulting eight nuclei and their associated cytoplasm, forming the ascospores.

The life cycle of an ascomycete conforms in a general way to the condition in other organisms that have an alternation of an extensive haploid phase with a single-celled diploid phase. Although the vegetative hyphae are frequently composed of multinucleate cells, the nuclei are haploid. There are, however, complications, and one of them is the delay in time between the fusion of the gametangia and nuclear fusion, the latter equivalent to fertilization. This intervening period is marked by the presence of paired nuclei, known as DIKARYONS, in the ascogenous hyphae. Numerous ascogenous hyphae are formed from each ascogonium, and all hyphae may branch. These features are useful, among other ways, in that they increase the number of paired nuclei available for the formation of diploid nuclei, and thus increase the number of asci and ascospores. This dikaryotic phase is relatively short in the Ascomycetes, but, as we shall see, is characteristic of much of the life cycle in a typical Basidiomycete.

Among the heterothallic Ascomycetes is the pink mold *Neurospora*, widely employed in genetic investigations. Both conidia and ascospores are produced in the life cycle. Four of the ascospores in a single ascus belong to one mating type, A, and four to the other, a. Starting at one end of the ascus, four pairs of spores may be counted, the members of each pair genetically alike but differing from the members of the other pairs. Single ascospores may be isolated, germinated, and cultured separately in the order in which they occur in the ascus, thus making possible an analysis of each of the four products of a single meiosis. In *Neurospora crassa*, experiments (including hybridizations) have identified some 368 genes and located them specifically on the seven chromosomes that comprise the haploid number.

A general understanding of the varied nature of the Ascomycetes and of their significance to

Fig. 21-19 Fruiting bodies of a cup fungus (*Peziza*).

man is best attained by a consideration of a number of representatives of the class.

The Cup Fungi and Morels The cup fungi, which are classified in several orders and families, are characterized by a leathery, fleshy, or gelatinous cuplike, disk-shaped, or vase-shaped fruiting body. The cups (FIGS. 21-19, 21-20) range in size from a fraction of an inch to as much as 5 inches in diameter, and may be brightly colored. Some species are saprophytes, living on dead wood or humus, and their fruiting bodies may be found in woodlands from spring to autumn. In most of these a conidial stage in unknown. Others are important parasites on cultivated plants, and produce conidia as well as ascospores in the life cycle. The discharge of ascospores in the cup fungi is accompanied by a phenomenon known as puffing, most easily observed in the larger cup fungi. When the spores are ripe, turgor in the ascus is high; a shock, as when the fruiting body is touched or blown upon, causes all the asci to discharge simultaneously, throwing into the air a visible cloud of spores (FIG. 21-20).

The morels, with about 15 species, are found in many parts of the world, chiefly in the temper-

Fig. 21-20 Spore discharge in a cup fungus (*Sclerotinia*). Spores discharging in a cloud above.

Fig. 21-22 Powdery mildew on willow (*Salix*) leaf.

Fig. 21-21 A morel (*Morchella esculenta*).

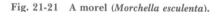

ate zone, and are sometimes called the sponge or honeycomb "mushrooms." They are edible, and are prepared for the table in the same manner as mushrooms. They grow in rich soil in early spring, and are usually identified without difficulty. The fruiting portion (Fig. 21-21) consists of a thick, whitish, hollow stalk terminated by a conical yellowish-brown cap that darkens with age. Some species are little more than an inch in height, but others attain heights of 4–5 inches. The surface of the cap is folded and convoluted, and the asci are borne in palisadelike layers lining the depressions.

The Powdery Mildews In the powdery mildews the asci and ascospores are produced within cleistothecia rather than in open cups. The powdery mildews take their name from the white, powdery or mealy appearance of the fungus on the leaves, petioles, young stems, buds, and even flowers of infected plants. The kinds of plants attacked include cereals, forage and lawn grasses, clovers, grapes, apple, hops, willows (Fig. 21-22), oaks, lilacs, roses, zinnias,

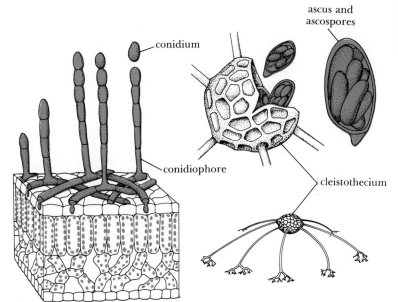

conidium

conidiophore

ascus and
ascospores

cleistothecium

Fig. 21-23 Stages in the life
cycle of a powdery mildew.
(Left): conidial stage. (Right):
cleistothecial stage.

phlox, cherries, cucurbits, peas, and beans. Some powdery mildew fungi are capable of attacking only one species of host plant, while others can infect over 350 kinds of plants in widely scattered plant families. Many powdery mildew fungi are further subdivided into a number of physiological races, each attacking a limited number of host plants.

The fungus is superficial in its growth. Early in the growing season it consists of masses of tangled hyphae that obtain their food by sending haustoria into the living epidermal cells. At this time the fungus forms enormous numbers of conidia (FIG. 21-23, left) in chains at the ends of erect conidiophores. The conidia are carried by wind to other parts of the same plant or to neighboring plants where they germinate.

The cleistothecia (FIG. 21-23, right) are formed toward the end of the growing season and are just visible to the naked eye as black dots in the mildew patches on the leaf surface. They bear appendages that are commonly curled or branched at the tip, in a manner characteristic of each genus. The cleistothecia, containing one to several asci, survive the winter, and in the spring

both cleistothecia and asci absorb water and swell. The cleistothecium breaks open, and the asci discharge the spores by bursting. The number of ascospores varies from two to eight, depending on the species. In warm climates, many species never form cleistothecia, reproducing solely by conidia. Various fungicides, including sulfur and cycloheximide, are used to control powdery mildews. Plant breeders have also been able to develop mildew-resistant varieties of numerous plants.

Aspergillus and Penicillium These genera are commonly called the black, blue, or green molds. They occur in patches or thick films that are black, blue-green, brown, or yellow in color, depending on the species and on the material on which the fungus is growing. The species of these genera are exceedingly common and widely distributed. They grow as saprophytes on all kinds of organic materials such as bread, cured meats, woolen and cotton fabrics, leather goods, grains, and wood. The housewife finds them as a greenish or black mat on the surface of jelly or jam. The fungus is not poisonous, and if the mold layer is removed, the food is still edible. These and other molds are able to thrive under acid conditions and in high concentrations of sugar or salt that inhibit the growth of bacteria. They grow as parasites as well as saprophytes and cause decay in fruits and vegetables in transit and storage.

Several species of *Aspergillus* cause serious diseases in poultry, other domesticated and wild birds, as well as in cattle, sheep, horses, and, more rarely, in man. The group of diseases, known collectively as *aspergilloses* (sing. *aspergillosis*), is an infection of the respiratory tract that produces symptoms resembling those of tuberculosis. Infection occurs from inhalation of spores produced on moldy grain, litter, or other organic matter. *Aspergillus* and other fungi may also be involved in infections of the ear in man. *Aspergillus fumigatus* is the species most often associated with aspergilloses. This and other species of *Aspergillus*, together with species of *Penicillium*, are also important in the heating and spoilage of moist hay and grain.

Asexual reproduction of *Aspergillus* and *Penicillium* is by conidia, borne in chains at the ends of erect conidophores. The unbranched conidiophore of *Aspergillus* (FIG. 21-24B) bears numerous short stalks (*sterigmata*) from which arise the chains of conidia. The *Penicillium* conidiophore (FIG. 21-24C) is usually branched, and each branch bears a cluster of elongated sterigmata that produce the chains of conidia. The conidia are produced in enormous numbers and it is these that are responsible for the characteristic color of the various species of *Penicillium*. They are readily distributed by air currents. Some species, when grown in pure culture on specialized media, also develop cleistothecia.

Certain species of these genera are of positive economic importance and are utilized in the preparation of food and other materials useful to man. Species of *Aspergillus*, growing on sugar, yield citric and other organic acids. A powerful amylase, known commercially as Taka-Diastase, is produced by *Aspergillus oryzae*. It is obtained by growing the fungus on wheat bran or cooked rice starch. The enzyme, which has a limited use in medicine, is capable, under suitable conditions, of hydrolyzing 3000 times its own weight of starch in 10 minutes. A useful feature of *Aspergillus fumigatus* is the production of an antibiotic effective against the organisms causing amoebic dysentery.

Camembert, Roquefort, Gorgonzola, and Stilton cheeses owe their characteristic properties and flavors to species of *Penicillium*. The outstanding use of *Penicillium* is in the production of penicillin, a powerful antibiotic agent in disease control (Chapter 20).

Fig. 21-24 *Aspergillus* and *Penicillium*. (A): *Aspergillus*, entire plant. (B): *Aspergillus*, unbranched conidiophore bearing chains of conidia. (C): *Penicillium*, branched conidiophore and conidia.

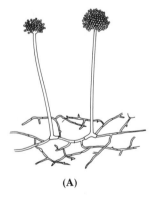

(A)

(B)

(C)

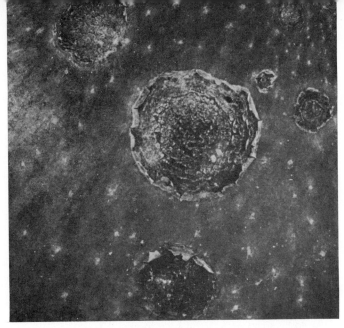

Fig. 21-25   Symptoms of apple scab. (Left): symptoms of the disease on apple leaf. (Right): apple scab (enlarged) on surface of the fruit.

Apple Scab   Apple scab is an excellent example of a plant disease in which a knowledge of the life cycle of the disease organism is necessary in planning effective control measures. Apple scab is practically universal wherever apples and crabapples are grown. Among its symptoms are raised and discolored areas on the leaves and scabby spots on the fruits (FIG. 21-25). The entire fruit is often dwarfed and distorted. The fungus (*Venturia inequalis*) that causes the disease is parasitic during one stage of its development and grows saprophytically during its other phase. The saprophytic stage is found on dead apple leaves on the ground. These contain hyphae that grow and produce perithecia within the leaf tissues (FIG. 21-17C). In the spring, the ascospores are expelled during wet periods and carried by the wind to the flower buds and immature leaves, where they initiate primary infections. The periods of ascospore discharge may extend over several weeks. The hyphae formed following infection in flowers, young leaves, or fruits grow parasitically just beneath the cuticle.

Numerous short conidiophores are soon produced, which push up and rupture the overlying cuticle. Each conidiophore then cuts off successively a number of conidia at its tip (FIG. 21-26).

The conidia from the primary infections are carried by rain to other leaves and to fruits, where they bring about secondary infections. Conidia from both primary and secondary infections may produce new infections throughout the season under proper conditions of temperature and moisture. The fungus returns to the saprophytic stage when the leaves, containing hyphae, fall to the ground in the autumn.

The application of protective sprays or dusts to the trees is the principal means of controlling apple scab. Such treatments must begin early in the spring and cover the entire period of ascospore discharge. They should also be continued throughout much of the growing season, to prevent secondary infection by conidia that may have developed in spite of earlier spraying. Early spraying, before the bloom period, is also effective in eradicating or killing the fungus in the leaves within several days after the start of infection. A fungicide may therefore be both eradicative and protective.

Ergot Ergot is a disease of cultivated cereals, such as wheat, barley, and rye, and also of more than 200 wild and cultivated grasses. It is caused by the Ascomycete *Claviceps purpurea* and other species and races of fungi of the genus *Claviceps*. The disease becomes apparent with the formation of elongated, dark bodies, the ERGOTS, which protude from the head of the grain (FIG. 21-27). This stage is the result of the growth of hyphae within the ovary from infection occurring at the time of flowering. As growth proceeds, the ovary tissues are destroyed and replaced by a cottony mass of mycelium covered with layers of short conidiophores bearing oval conidia at their tips. The conidia become mixed with a nectarlike secretion, called honeydew, which is attractive to flies and other insects. While feeding on the honeydew, they become contaminated with conidia and transport the spores to healthy flowers, where new infections occur. As infection proceeds, conidial production gradually ceases, the mycelial mat hardens, and the large, horny, brown to purplish-black ergots, composed of compacted masses of thick-walled hyphae, develop in place of kernels. The ergot is a resting body and falls to the ground, where it persists during the winter, or it may be harvested with the grain. When sown with the seed or left on the soil surface, the ergots germinate in the spring by the formation of globose perithecial heads, each seated upon a stalk about ½ inch long (FIG. 21-28). Buried in the tissues of each head are numerous perithecia from which ascospores are discharged. These soon infect the flowering grains and other grasses.

In Europe, during the Middle Ages, peasants who made their black bread from ergot-infected rye or wheat often became very ill with a disease known as ERGOTISM. Among the symptoms were convulsions and gangrene of the hands and feet. Cattle that fed on ergot grain developed

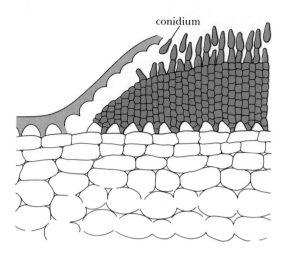

Fig. 21-26 Apple scab fungus, conidial stage on fruit.

Fig. 21-27 Ergots on heads of rye.

gangrene of the ears, tails, and hoofs. Pregnant cows aborted, and mortality was high.

Modern methods of milling grain eliminate ergots during the processing, and the disease now occurs chiefly in domestic animals feeding upon wild grasses. In 1951, however, an epidemic of ergotism in man occurred in a small village in France, resulting in severe illness and several deaths. Among other effects, ergotism causes contraction of the smooth muscles and the walls of the blood vessels. This contraction, since it affects small blood vessels, results in a decreased blood supply and brings about gangrene. It has been found that the poisonous effects of the ergot fungus are due to several powerful alkaloids. Ergot is one of several kinds of fungi used in modern medicine. The refined alkaloids are utilized medicinally in the treatment of migraine and in the practice of obstetrics. The alkaloids are no longer obtained from wild ergot, however, but from high-yielding strains of *Claviceps* grown in submerged culture with an adequate oxygen supply. Lysergic acid, also produced both by the wild and cultured fungus, may be used in a chemical semisynthetic process to obtain lysergic acid diethylamide (LSD), a psychedelic drug.

Yeasts The yeasts are Ascomycetes that form no fruiting bodies. They differ from other fungi in that the plant body is not composed of hyphae but consists of a single spherical or oval cell enclosed by a cell wall. In many yeasts, such as the brewers' or bakers' yeast (FIG. 21-29), a vacuole occupies much of the space within the cell. Granules of various kinds of reserve food may be observed within the cytoplasm, and special staining techniques reveal the presence of a nucleus.

Yeasts reproduce both vegetatively and by the formation of ascospores. Vegetative reproduction may be by transverse division (fission), or by budding, or both. Most yeasts reproduce by budding. In budding, the cell wall protrudes outward in a small localized area. The swelling thus formed receives cytoplasm from the parent cell, increases in size, and becomes constricted at the base. The nucleus of the parent cell divides, and one of the daughter nuclei moves into the bud. The bud either becomes detached or, together with the parent cell, continues to bud, thus forming a colony of cells (FIG. 21-29). A single yeast cell, under optimum conditions for growth, may give rise to more than 20 buds before it ceases to reproduce. These buds arise from different areas of the surface of the parent cell; buds do not arise from the same site from which one has previously grown.

Asci and ascospores are produced when yeasts are grown under conditions unsuitable for continued vegetative activities. Ascospores are somewhat more resistant to unfavorable environ-

Fig. 21-28 The ergot fungus. (A): perithecial heads growing from ergot. (B): section through perithecial head, showing embedded perithecia. (C): ascus with eight threadlike ascospores, greatly enlarged.

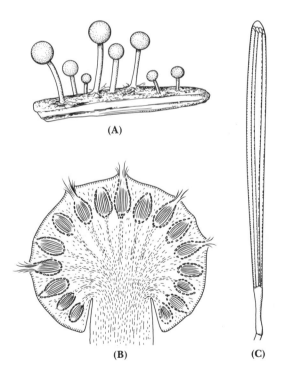

(A)

(B)   (C)

mental conditions than are vegetative cells and may enable yeasts to survive extremes of temperature or prolonged drying. The formation of ascospores involves a sexual process, as in other Ascomycetes. Several life cycle patterns are found in yeasts, but only that found in the heterothallic brewers' yeast (*Saccharomyes cerevisae*) and its various strains will be described. Here are found two clearly defined generations,

a diploid and a haploid (FIG. 21-30). The large, ellipsoidal, vegetative cells ordinarily seen under the microscope and used in industrial processes belong to the diploid generation. These multiply extensively by budding, producing large numbers of diploid cells. At the time of spore formation, the diploid nucleus undergoes meiosis. The nucleus divides, and the daughter nuclei divide again. Cell walls then develop around each nucleus and associated cytoplasm, forming four haploid ascospores. The cell wall of the original diploid cell thus functions directly as an ascus.

The ascospores are of two mating types, two of them belonging to a plus strain and two to a minus strain. The ascospores swell and press against the ascus, finally rupturing it and releasing the spores. These now begin budding, and reproduce vegetatively for a time. These haploid cells are round or spherical, and smaller than the more elongated, oval diploid cells. The haploid cells are potential gametes, for two of them, belonging to different mating types, may fuse to form a large diploid cell.

The apparent simplicity of the yeast cell and life cycle does not imply that yeasts are primitive fungi. On the contrary, yeasts are probably derived from some group of the Ascomycetes with a more complex somatic organization and life cycle.

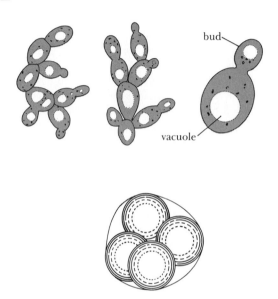

Fig. 21-29 Brewers' yeast (*Saccharomyces cereviciae*). (Above): vegetative reproduction. (Below): ascus and ascospores.

Fig. 21-30 Life cycle of brewers' yeast (*Saccharomyces cereviciae*). Diploid and haploid stages of the life cycle; the haploid stage is indicated in color.

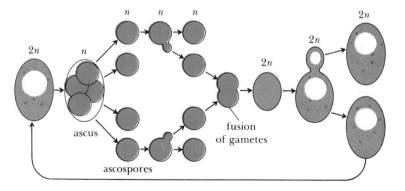

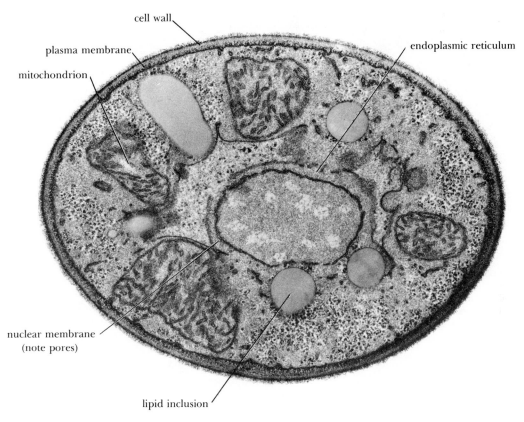

cell wall

plasma membrane

mitochondrion

endoplasmic reticulum

nuclear membrane
(note pores)

lipid inclusion

Fig. 21-31 The torula yeast (*Candida edulis*). Note the various cell organelles. (Electron microscope; 44,410×.)

Yeasts may be hybridized in the laboratory. To accomplish this, individual spores are planted in a nutrient medium, where they give rise to colonies of haploid cells. After selection for suitable qualities, haploid cells from different strains are brought together to produce hybrid diploid cells. These hybrids may differ from other strains in their ability to synthesize vitamins, to ferment sugars, or to produce better yeast from the standpoint of nutrition or keeping qualities.

The yeasts are of direct and outstanding importance to man. They are utilized in medicine, in baking and brewing, as a source of enzymes, in the production of alcohol, and in many other ways. The yeasts used in baking, brewing, and in the production of wines and other alcoholic beverages are usually varieties or strains of the bakers' yeast. Many strains, developed for particular uses, are maintained as pure cultures in industrial laboratories. Considerable

Fig. 21-32 Open vat of wort being fermented by yeast. The thick foam produced by escaping carbon dioxide excludes air from the fermenting liquor beneath.

and carbon dioxide, as shown by the over-all equation

$$C_6H_{12}O_6 \longrightarrow 2C_2H_5OH + 2CO_2 + \text{energy}$$

simple sugar   alcohol   carbon
(glucose or fructose)   dioxide

If only complex sugars such as maltose or sucrose are supplied, these are digested to simple sugars by the yeast enzymes. The sugar is taken into the yeast cell, where it goes through the steps of fermentation, outlined in Chapter 6. Under strictly anaerobic conditions the energy obtained is sufficient to maintain the cells, but most strains do not produce new cells unless some free oxygen is supplied. With high sugar and low oxygen concentrations, fermentation will be rapid. The alcohol and carbon dioxide are waste products as far as the yeast plant is concerned. The accumulation of the alcohol inhibits the action of the yeast, but various strains differ in their tolerance, from a low concentration in brewers' yeast (3–5 percent) up to about 14 percent in wine yeasts. Wines with a higher concentration are fortified by the addition of alcohol. Beverages with a still higher alcoholic content, such as brandy or whiskey, are produced by fermentation followed by distillation.

In the production of bread, the carbon dioxide formed during the fermentation of sugar present in the dough is responsible for the rising of the dough and for the porous texture of the resulting loaf. Most of the small amount of alcohol produced is driven off during the baking process.

Although yeasts produce many enzymes, such as proteases, maltase, and sucrase, they do not produce the extracellular enzyme amylase that will digest starch. The lack of this enzyme is not significant in the production of compressed yeasts, wines, and some industrial alcohol, for the materials upon which the yeast is grown consist of molasses, fruit juices, or other substances rich in sugar. In brewing, however, and in the production of some kinds of beverages such as whiskey, starchy materials such as corn, rice, rye, or barley are used. The starch is usually con-

attention has been given to the cultivation of yeasts as a source of protein, deficient in human diets in many parts of the world. For this purpose, yeasts other than *Saccharomyces* are advocated, notably the torula yeast (*Candida edulis*). The plant (Fig. 21-31) is cultivated on raw materials containing sugar, supplemented with ammonium salts. It is a poor fermenter, but grows rapidly and produces high yields of protein. After removal from the culture medium, the yeast is dried and used directly. Currently, this yeast has greater use as a feed supplement for farm animals, especially poultry, than as human food.

Yeast plants may produce many enzymes capable of acting upon various organic materials. As do many other organisms, yeasts ferment various sugars. One of the most common fermentations results in the formation of alcohol verted to sugar by the use of malt, which is ground, partly germinated barley (Chapter 6). During germination a high percentage of amylase is produced.

In the production of beer, malt is suspended in water in large mixing tanks. Adjuncts to the malt, in the form of cooked starch or other carbohydrate material, are commonly added. The resulting *mash* is kept warm to allow further conversion of the starch to sugars by the malt enzymes. The liquid portion, or *wort,* is then separated from the grain residues and boiled with hops. The hops not only contribute flavor but also inhibit the growth of spoilage bacteria while allowing the yeast cells to grow freely. The hops are removed by filtration, and the cooled wort, which contains maltose and other sugars, proteins, and mineral salts, is fermented by the addition of special strains of yeast (FIG. 21-32). The complete fermentation process requires from one to two weeks.

In the manufacture of compressed and dried yeast, selected strains are inoculated into tanks of sugar-cane or sugar-beet molasses. Since the manufacturer is interested in the yield of yeast, not alcohol, the liquid containing the yeast is aerated rapidly by the introduction of air near the bottom of the tank. Respiration of the yeast is therefore mainly aerobic, and vegetative growth is stimulated. Aeration is used also in producing brewers' yeast and usually in the early stages of fermentation.

THE LICHENS

The lichens are a very large group of plants (about 17,000 species) widely distributed on every continent and in varied habitats. Lichens are found growing on the bark of trees, on decaying wood, on rocks, and on the soil. They are adapted to survival under great extremes of heat, cold, and drought. Some lichens superficially resemble mosses, but are not at all related to these plants. According to the form of the plant body, lichens are classified into three

Fig. 21-33 Rock tripe (*Umbilicaria*), a foliose lichen. The rock tripes are composed of leathery sheets, circular, or the margins may be torn and ragged. They are attached to the rock at a central point. The thallus is large, from a few inches to as much as a foot in diameter.

groups: (1) crustlike (crustose), (2) leaflike (foliose), and (3) shrublike (fruticose).

The many species range in size from minute types to large and conspicuous forms, such as the rock tripes (FIG. 21-33) and the hair, or beard, lichens (FIG. 21-34), the latter attaining a length of several feet. Lichens, or their reproductive organs, are often brightly colored (Plate 11). Masses of lichens, especially of the genus *Cladonia*, which includes the reindeer lichens or reindeer "moss" (FIG. 21-35), cover great areas in northern arctic and subarctic regions. These plants serve as pasturage for musk ox, caribou, and other wild browsers and for domesticated reindeer. Thus, as do other phototrophic organisms, lichens can serve as the basis of food chains.

A lichen is a composite plant; that is, it is composed of two organisms, an alga and a fungus.

Fig. 21-34 A beard lichen (*Usnea*), a fruticose lichen with a threadlike thallus.

The fungus is usually an Ascomycete; the alga may be a green or a blue-green, unicellular or filamentous. Although dual in nature, a lichen behaves as a single functional unit. The external and internal structures, and the kinds of organic compounds produced, are relatively constant from one generation to another, and thus lichens may be classified into species, genera, and families like other organisms. The bulk of the plant body is composed of more or less loosely interwoven hyphae of the fungus (FIG. 21-36). The algal cells, which in the foliose forms usually occur in the upper part of the thallus, are in close contact with hyphae that send haustoria into the photosynthetic cells. In many species, specialized hyphae, the rhizoids, aggregated into strands, extend downward and attach the plant to the substratum. The thallus is spongelike and rapidly imbibes water from rain, fog, or dew. Growth of the lichen thallus is slow, in general only a few millimeters in diameter per year.

The nutritional and other relationships among the component organisms of a lichen are usually cited as a classical example of mutualism. The

Fig. 21-35 Reindeer lichens. (A): clumps growing on forest floor. *Cladonia alpestris* (left) and *Cladonia rangiferina* (right); the pen shows comparative size. (B): individual branches of *Cladonia rangiferina*, from clumps shown in (A), illustrating the erect, fruticose habit.

(A)

(B)

alga provides the fungus with energy-yielding organic compounds, and probably also with vitamins such as biotin and thiamine. The fungus provides the alga with water, essential elements, and protection from high light intensities. The essential elements are obtained primarily from solutes in the water with which the thallus comes in contact, and absorption can occur over the entire surface of the plant. Many species of lichen algae are known to occur free in nature, but with a few exceptions the fungi of lichens are found only as components of a lichen thallus. A delicate balance exists in the relationship between fungus and alga. The fungus occasionally parasitizes the algal cells, and the "mutualism" is not always beneficial to both.

The ascospores of the fungus are produced in perithecia or, more commonly, in small cuplike or disk-shaped fruiting bodies on the surface of the thallus (Figs. 21-36B, 21-37) or on specialized stalks or branches. The spores are discharged, fall upon some substratum, and produce hyphae upon germination. If these hyphae by chance encounter a suitable species of

alga, a new thallus will arise. Such chance encounters are probably of infrequent occurrence, and lichen reproduction is chiefly vegetative, by fragmentation or specialized dispersal units. In fragmentation, portions of the thallus are broken off and carried away by wind or water. Perhaps the most common of the specialized dispersal units are the SOREDIA (Fig. 21-36C), which are composed of one to several algal cells closely enveloped by fungus hyphae. They originate within the algal layer of the thallus and are extruded upward through cracks, appearing on the surface as powdery clumps or patches. The soredia are dispersed by wind or rain and each soredium may produce a new plant.

As pointed out in Chapter 15, lichens may be important as pioneers on bare rock (Fig. 21-37), where they gradually break up the surface, thus initiating soil formation. This is largely a mechanical process, for the thallus, when wet, is gelatinous and tightly adherent to the rock surface. As it dries, the thallus contracts, detaching minute fragments from the underlying surface. These fragments may become

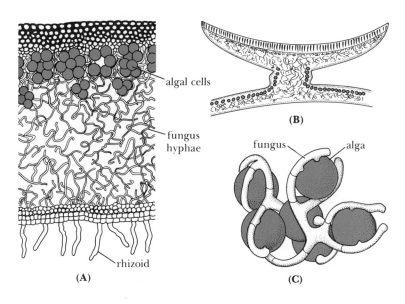

Fig. 21-36 Lichen thallus and reproduction. (A): vertical section of lichen thallus. Algal cells are intermingled with fungus hyphae in the upper part of thallus. (B): vertical section of a cup containing asci. (C): soredium.

algal cells

fungus hyphae

rhizoid

(A)

(B)

fungus   alga

(C)

Fig. 21-37 A foliose lichen (*Parmelia*) growing on rock. Note cups on surface of thallus.

incorporated into the thallus, where they are slowly dissolved by lichen acids.

Lichens have several economic uses. In addition to serving as food for animals, they are used in dyeing processes, in the perfume industry, and as a source of the litmus used in the chemical laboratory.

SUMMARY

1. The fungi are parasitic or saprophytic thallophytes important to man in two chief ways: (1) as agents of decay, and (2) as major causes of destructive plant diseases.

2. A fungus is a nongreen plant whose body is usually composed of branching, threadlike structures called hyphae. The reproductive spores arise from specialized hyphae. In many fungi, hyphae are aggregated into more or less massive fruiting bodies.

3. The slime molds, or slime fungi, are not true fungi. They form no hyphae and possess both plant and animal characteristics.

4. The true fungi are commonly grouped into four classes: the Phycomycetes, Ascomycetes, Basidiomycetes, and Fungi Imper-

fecti. The first three of these reproduce sexually (have a perfect stage). Fungi Imperfecti are those for which no perfect stage is known.

5. The Phycomycetes are characterized by nonsepate hyphae not aggregated into definite fruiting bodies. Reproduction is both asexual and sexual. The asexual spores may be motile or nonmotile, and are usually produced within a sporangium. The Phycomycetes are the only true fungi that produce motile spores (zoospores). The spore of the perfect stage is either a zygospore or an oospore. Examples of Phycomycetes are the bread mold fungus, water molds, late blight fungus, and the downy mildews.

6. The bread mold fungus reproduces chiefly by asexual spores produced in sporangia, but sexual reproduction also occurs. This is isogamous and heterotallic. Heterothallism is found in many groups of the fungi. Two mating types, or plus and minus strains, are involved.

7. The late blight fungus produces small sporangia on branched sporangiophores. The sporangia germinate directly or indirectly, depending upon the temperature. The motile spores produced by indirect germination may be cited as an example of the retention of an ancestral character.

8. The Ascomycetes are characterized by definite fruiting bodies, septate hyphae, and the formation of ascospores in asci. The fruiting bodies, which contain the asci, are usually cups, perithecia, or cleistothecia. The ascospores are the spores of the perfect stage. The spores of the imperfect stage are usually conidia.

9. The sex organs are formed at the base of a fruiting body. A dikaryotic phase intervenes between gametangial fusion and the formation of the diploid nucleus.

10. The morels are common edible Ascomycetes. The asci are produced in depressions on the surface of the stalked cap.

11. The powdery mildews grow superficially on the surface of the host. They reproduce by conidia and ascospores.

12. *Penicillium* and *Aspergillus* are the chief genera of the black and blue-green molds. They may be distinguished by the nature of the conidiophore—branched in *Penicillium* and unbranched in *Aspergillus*.

13. Apple scab is a good example of a plant disease in which a knowledge of the life cycle of the causal organism is necessary to control the disease.

14. Ergot is a disease of grasses in which the developing grain is replaced by a compact mass of hyphae, forming the ergot. Reproduction is by both conidia and ascospores.

15. The plant body of the yeasts is a single cell. Reproduction is by fission or budding and by ascospores. In some yeasts the life cycle includes clearly defined diploid and haploid generations.

16. In alcoholic fermentation enzymes convert a simple sugar into alcohol and carbon dioxide, with the release of a limited amount of energy. Yeasts do not produce the enzyme amylase.

17. A lichen is composed of a fungus and an alga associated in a mutualistic relationship. Ascospores are formed, but reproduction is chiefly vegetative, by fragmentation or specialized dispersal units such as soredia.

22

The Fungi: Basidiomycetes and Fungi Imperfecti

THE BASIDIOMYCETES INCLUDE most of the larger, conspicuous species so often noticed in the fields and woodlands. The fungi commonly known as mushrooms, toadstools, puffballs, and bracket fungi are Basidiomycetes. Here also are classified the rusts and smuts, responsible for diseases of economically important plants, together with many other common and important groups.

As in the Ascomycetes, the hyphae of the Basidiomycetes are septate, and the cross walls contain a pore that permits nuclear movement between cells. The nonmotile spores of the perfect stage are called BASIDIOSPORES, and, like ascospores, are produced immediately following sexual fusion and meiosis. The basidiospores are borne externally, however, on a club-shaped, specialized hypha, the BASIDIUM, whereas ascospores develop within a closed sac. No specialized sex organs are produced by the Basidiomycetes.

In addition to basidiospores (meiospores), other kinds of spores may occur, and these will be discussed when the fungi that produce them are studied. First, in terms of general interest, are Basidiomycetes with conspicuous organized "fruiting bodies"—gelatinous, leathery, fleshy, spongy, corky, or woody structures that bear the basidia.

BASIDIOMYCETES WITH FRUITING BODIES

The Gill Fungi (Agaricaceae) The gill fungi are mostly saprophytic, fleshy fungi. The hyphae spread through the soil or dead wood and produce the conspicuous fruiting body that consists of a STALK supporting a CAP. In most of the gill fungi the stalk is centrally attached to the cap, and the entire fruiting body is umbrella- or

---

Photo at left shows the sulfur polypore (*Polyporus sulfureus*), growing on a fallen tree. Fungus edible when young.

Fig. 22-1   Field mushroom (*Agaricus campestris*). (Left): view of underside of cap previous to rupturing of the partial veil. (Right): mature fruiting body showing gills and annulus.

mushroom-shaped (FIG. 22-1). But in some species the stalk is attached to one side, and it may be short or even absent. The size, shape, texture, and color of both stalk and cap are characteristics upon which identification of these Basidiomycetes is based. The basidiospores are produced on the surfaces of thin, sheetlike structures, the *gills,* which are borne on the underside of the cap, where they radiate from the stalk to the margin.

The Meadow, or Field, Mushroom   The meadow, or field, mushroom (*Agaricus campestris*) grows wild in lawns, pastures, and gardens, and a variety of it is the chief mushroom cultivated and grown commercially. It serves admirably to illustrate many of the features of the gill fungi. The vegetative portion of the plant—the hyphae, frequently combined into whitish strands—branches through the soil just beneath the surface (FIG. 22-2). Hyphae are always present but usually escape notice.

The reproductive phase is initiated by the formation of small enlargements on these strands.

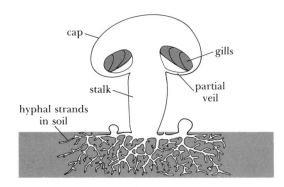

Fig. 22-2   Longitudinal section through button stage of mushroom.

These swellings increase in size and break through the surface as small balls, which constitute the BUTTON stage. A longitudinal section through a button (FIG. 22-2) shows the immature stages of the stalk, cap, and gills, together with a thin tissue, the PARTIAL or INNER VEIL, which connects the margins of the cap to the upper part of the stalk. As growth proceeds and the cap expands, the partial veil is ruptured, revealing the gills. In this species the gills are at first pink, but change to brown or purple as the spores ripen. Remnants of the veil usually persist on the upper part of the stalk as a ragged ring, the ANNULUS (FIG. 22-1). The presence or absence of an annulus is another of the numerous features used to identify the various species of gill fungi.

A cross section of a gill (FIG. 22-3A) shows that it is wedge-shaped, and that the bulk of the tissue, like that of the rest of the mushroom, is composed of compacted hyphae. On both sides of the gill are palisadelike layers of specialized hyphae, slightly enlarged or swollen at the tips. The most conspicuous elements of this layer are the mature basidia, each of which bears four basidiospores on short stalks (STERIGMATA) (FIG. 22-3B). Immature basidia are also found, and interspersed among them are hyphae that resemble young basidia but are sterile and never produce spores. A variety of the field mushroom bears but two basidiospores, and this two-spored form is the one grown commercially for food.

The formation of basidiospores in the field mushroom (FIG. 22-3C) is typical of the gill fungi. The young basidium, together with the hyphae from which it arises, is typically binucleate. The nuclei in the basidium fuse, and such a fusion is regarded as a true sexual union. Meiosis now occurs, producing four daughter nuclei. Following this step, a single daughter nucleus migrates through a sterigma into each of

Fig. 22-3 Reproduction in the field mushroom (*Agaricus campestris*). (A): gills of mushroom in vertical section. Arrows indicate path of falling spores. (B): portion of gill bearing basidia, greatly enlarged. (C): (1–5), stages in development of basidium and basidiospores; (6), basidium after discharge of spores.

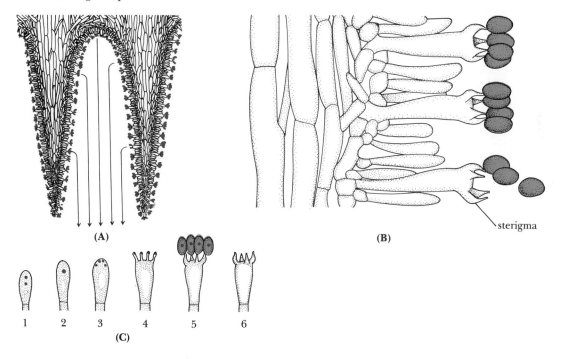

(A)

(B)

sterigma

1   2   3   4   5   6

(C)

the four basidiospores. (In the cultivated two-spored form, two nuclei migrate into each basidiospore.) When the spores are mature, they are discharged successively, over short intervals, into the space between two gills. The force of ejection is only enough to carry the spore a horizontal distance of about 0.1 millimeter across this narrow space; then it falls downward. Only one set of basidiospores is borne on a basidium, but the maturation of successive crops of basidia and basidiospores extends the period of spore discharge over several days. It has been estimated that some 10 billion spores may be produced by a large field mushroom. The value of the stalk is now apparent. It raises the cap and the gills above the soil surface, and the resulting space underneath allows air currents to carry away the falling spores.

In a suitable environment the basidiospores germinate and give rise to hyphae. The field mushroom, unlike most Basidiomycetes, is homothallic, and the mushroom can arise from hyphae produced by a single spore. Since this fungus is homothallic, the two nuclei that fuse in the young basidia are alike sexually. In heterothallic gill fungi, on the other hand, the spores are of different mating types, and a fruiting body cannot arise from a single spore. The spores of heterothallic Basidiomycetes give rise, on germination, to septate hyphae containing haploid nuclei of only one mating type. If the hyphae of opposite mating types come in contact in the soil, they fuse, and nuclei of both mating types migrate into the hyphal system of their opposites. The invading nuclei move through the septal pores to the growing tips of the hyphae, and there become associated in pairs with the established nuclei. Each such pair is a dikaryon, as noted in the description of the sexual process in the Ascomycetes. As the hyphae elongate, the paired nuclei divide synchronously, and new cell walls are laid down, separating the pairs of daughter nuclei. Although the two genetically different nuclei maintain their identities, many characters of the organism are under their joint control, and a dikaryon is thus in many respects

functionally equivalent to the diploid state. The dikaryotic mycelium spreads extensively in the soil, and after a period of growth and accumulation of reserve foods, fruiting bodies are initiated. The dikaryotic phase persists throughout the formation of the fruiting body and comes to an end only with the fusion of the paired nuclei in the young basidium. Thus, while in the Ascomycetes the dikaryotic phase, consisting primarily of the ascogenous hyphae, is of limited extent, in most Basidiomycetes it comprises the greater part of the life cycle. The dikaryotic mycelium may even persist from year to year and give rise to successive crops of fruiting bodies.

The fruiting bodies of the gill and other fleshy fungi, growing in lawns or pastures, are sometimes produced singly, sometimes in a circular colony called a "fairy ring" (FIG. 22-4). The perennial mycelium of the fungus begins growth in the center of the ring and gradually spreads outward in all directions. The older mycelium dies when the organic material of the soil is exhausted, but the younger mycelium produces annually a succession of fruiting bodies. The diameter of such rings varies with age and rate of growth, from a few to many feet. Studies on fairy rings in the grassland of Colorado cite a species of *Agaricus* that formed a ring 180 feet in diameter with an estimated age of 250 years.

The Spore Print The characteristics used to distinguish one species of mushroom from another include not only external structure but also the color of the spores. The spore color is not necessarily correlated with the color of the gills, which often changes with age, darkening as the cap matures. Individual spores are much too small to be seen with the naked eye. Ordinarily, spore color is determined by bringing the fungus indoors and making a *spore print*. The stalk of the mushroom is removed close to the cap and the cap is placed, gill side down, on a sheet of white paper. Since some spores are white or very light colored, a duplicate cap is often placed upon black paper as well. The cap is covered with a bowl or tumbler to prevent the spores from being disturbed by air currents. After several hours the cap is carefully lifted, and the spores that have fallen are seen in a radiating pattern that duplicates the arrangement of the gills (Fig. 22-5). The color may be observed easily.

The spore colors of the various genera of gill fungi are white, yellowish, brown, pink to salmon-colored, dark brown to purplish brown, and black. The spores of the common field mushroom are purplish brown.

Edible Mushrooms Many gill fungi are edible, but the eating of others is followed by distressing symptoms or even death. The edible species are popularly known as mushrooms, the poisonous ones as toadstools. Most of the edible fungi are gill fungi, but some have teeth, tubes, or cavities instead of gills. All the true puffballs (Fig. 22-15) are edible. Some edible fungi, such as the truffles and morels, are Ascomycetes, but most are Basidiomycetes.

Edible and poisonous species are very similar and are frequently found in the same genus. It is popularly believed that tests are available by which the inexperienced may distinguish edible from poisonous fungi. To this there is but one answer: there is no test, method, or rule of thumb for distinguishing edible from poisonous species; one must be able to recognize the species by its botanical characteristics. It is commonly believed, for example, that a silver spoon or coin will tarnish if cooked with poisonous fungi. Nothing could be further from the truth. It is sometimes asserted that the cap of an edible form will peel readily, whereas that of a poisonous one will not. This is not true. Neither are fungi upon which insects or other animals are observed to feed necessarily safe to eat. No test may be depended upon, and if the fungus cannot be positively identified or if any doubt exists, it should not be eaten.

Among the more conspicuous and readily recognized edible fungi are the black-spored, inky-cap mushrooms of the genus *Coprinus*. Sev-

Fig. 22-4   A fairy ring.

Fig. 22-5   Spore print of destroying angel (*Amanita verna*).

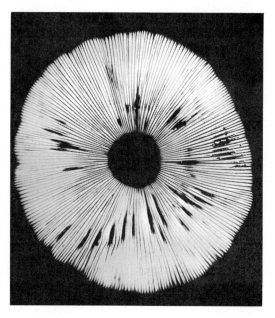

Fig. 22-6 Shaggy-mane mushroom (*Coprinus comatus*). (Left): edible stage. (Right): the same fungus with caps partly digested, allowing escape of spores.

eral species are common in lawns, gardens, and around rotting tree stumps. One of the most striking is the shaggy mane (*Coprinus comatus*) with a barrel-shaped, scaly cap (FIG. 22-6). With a few exceptions, the gills of the species of this genus constitute a type of spore-liberating body different from that of the field mushroom, and this may be illustrated by the shaggy mane. The gills, instead of being horizontal, are vertically arranged so that the gill margins are adjacent to the stalk. The gills are very thin, with their sides almost parallel, in contrast to the wedge shape of the gills of the field mushroom, and they develop very close together so that insufficient space is available for the liberation of the spores into the spaces between the gills. The escape of the spores is made possible by a curious phenomenon called autodigestion, that is, self-digestion by the organism's own enzymes.

When the fruiting body has attained full size, the cap begins to expand from a barrel shape to a bell shape. The tensions resulting from the expansion of the cap bring about a separation of the ends of the gills so that the spaces between them are similar to those between the gills of the field mushroom. The spores of *Coprinus* mature successively from the base to the top of each gill,

Fig. 22-7 *Amanita*, button stage. The universal veil is ruptured by the growth of the cap and stalk.

and the spores around the margin of the cap are thus the first to be discharged. At this stage the process of autodigestion sets in at the margins of the cap, and the products tend to accumulate as black, inky drops. The autodigestion process, moving upward, destroys the spore-free portions of the gills and thus allows the cap to open out still further, from a bell shape to an umbrella shape (FIG. 22-6), with a consequent continued production of spaces between the lower ends of the gills. As the spores are discharged, the spore-free area is digested; the spores lying higher on the gills are released next, continuing until the entire cap has disappeared.

Fig. 22-8   The destroying angel (*Amanita verna*).

Poisonous Gill Fungi   More than 70 species of gill fungi are known to cause symptoms of poisoning in man. This number is not large when compared to the several thousand species that are nonpoisonous. The most poisonous forms are found in the genus *Amanita*. Several relatively common species are so rich in toxic substances as to be fatally poisonous. The genus contains edible species as well, but only an expert should try to distinguish them.

In the amanitas and a few other gill fungi, in addition to the partial veil that extends from the edge of the cap to the stalk, a second veil, the UNIVERSAL VEIL, is present in the early stages. This extends over the entire button and is ruptured as growth proceeds (FIG. 22-7). When the fruiting body is mature, the remnants of the universal veil may still be evident as fragments on the surface (FIG. 22-9) or along the edges of the cap. A basal portion, the VOLVA, may also be found enclosing the bottom of the stem. The volva is commonly cuplike (FIGS. 22-7, 22-8), but it may be scaly or shelflike. It is frequently buried or partly buried in the soil. The annulus (remains of the partial veil) is fragile and may disappear at maturity. The presence of both an annulus and a volva identifies the amanitas, but it may not be inferred that all gill fungi without a volva are safe to eat. Certain other poisonous species do not possess this structure.

It is convenient to classify the poisonous amanitas into two groups. The first, composed of about six species, includes the most poisonous of the amanitas. In the eastern and central parts of North America, one of the most common of these is *Amanita verna*, the destroying angel (FIG. 22-8). This species is pure white, and the annulus and cuplike volva are usually prominent. The symptoms of poisoning appear 6–15 hours after consumption. By this time the toxic substances have been well absorbed, and fatalities may amount to more than 50 percent of those eating the fungus. The species of this group, in fact, are responsible for most (perhaps 90 percent) of the fatal mushroom poisoning in North America. The poisoning results from the effects of a number of toxic peptides classified in two groups, the phallotoxins and the amatoxins.

The toxicity of the species of the second group is much less than that of the first. Symptoms may become evident in 15 minutes to 3 hours after eating, and fatalities, at least in North America, are few. The best-known member of the group is the cosmopolitan *Amanita muscaria*, or fly amanita (FIG. 22-9). The cap is yellow to orange and is covered by prominent scales, the remnants of the universal veil. The volva is usually scaly and forms a series of rings at the base of the stalk. Another species of this group, more toxic than *Amanita muscaria*, is *Amanita pantherina* (panther fungus), said to be the most poisonous species of the Pacific northwest. The cap is gray-brown to dingy yellow, and bears the wartlike remains of the universal veil. The toxicity of these species was formerly believed to be due to an alkaloid called muscarine, but it is now known that three relatively simple chemical compounds, not alkaloids, are responsible.

The Pore Fungi (Polyporaceae) The most conspicuous plants in this group are the large, leathery or woody bracket, or shelf fungi frequently observed growing on living and dead trees. The visible portion of the fungus is the fruiting body produced by vegetative hyphae located in the wood. Removal of such a fruiting body from a tree in which the hyphae are living parasitically does not destroy the fungus or constitute a method of control. The hyphae will continue to feed upon the wood and will form new fruiting bodies. Infection within the tree may extend 4–6 feet above and below the fruiting body.

The pore fungi, together with certain gilled species, are largely responsible for decay in living trees (FIG. 22-10), telephone poles, fence posts, and structural timbers. Each year in North America tree diseases ruin as much timber as is cut. The loss is greater than that caused by fire and insects combined. The fungus spores are deposited on the surfaces of wounds caused by fire, lumbering, pruning, insects, and storms. After spore germination, the hyphae spread into adjoining tissues, which they digest by means of cellulase and other enzymes. An important method in preventing wood and wood products from rotting is to poison the food supply of the fungus by treating the wood with some preservative, such as coal tar creosote or compounds containing zinc or copper. Wounds in shade and fruit trees are frequently covered with a protective paint or other wound dressing to prevent fungus invasion. Most pore fungi growing on living trees will continue to grow after the tree is dead.

The fruiting body of the pore fungi is shelflike or stalked, and is either annual or perennial (FIG.

Fig. 22-9   The fly amanita (*Amanita muscaria*).

Fig. 22-10   Species of pore fungi parasitic on the wood and bark of a beech tree.

Fig. 22-11 Fruiting body of a shelf fungus (*Fomes applanatus*). (Top): lower (pore-bearing) surface. (Bottom): section view, showing the annual zones of growth.

22-11). In most bracket fungi the bulk of the fruiting body is composed of vertically arranged cylindrical or angled tubes, the ends of which appear as pores on the lower surface. The basidia and basidiospores line the sides of these tubes (FIG. 22-12), and when discharged, the spores fall downward through the pores and are carried away by currents of air.

In the perennial species (FIG. 22-11) the tubes grow downward each year. In some species these new additions form a distinct layer, and it is possible to count them and so determine the age of the fruiting body. It is not rare to find perennial fruiting bodies with 15 such layers, and as many as 80 have been recorded. The spores in such perennial forms are discharged continuously, day and night, over a period of several months, and in one common species (*Fomes applanatus*) it has been determined that spore fall continues for 6 months. Calculations have shown that, for each square foot of pore surface, about 30 billion spores are liberated in 24 hours. The total number of spores released from such a surface over the entire period of discharge would therefore amount to more than 5000 billion. In favorable circumstances, the discharged spores may be visible as a fine white cloud beneath the lower surface of the fruiting body.

Most of the pore fungi are too tough to be edible, but several annual species are tender when young and may be used as food.

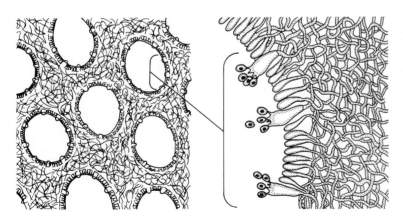

Fig. 22-12 Reproduction in a pore fungus. (Left): cross section, small portion of fruiting body; the openings are cavities of the tubes. (Right): portion of one tube, greatly enlarged, showing the basidia.

Fig. 22-13   A coral fungus (*Clavaria stricta*) growing on decaying stump.

Fig. 22-14   A puffball (*Lycoperdon*).

The Coral Fungi (Clavariaceae)   The coral fungi, or clavarias (FIG. 22-13), are fleshy fungi common in many localities. They usually grow in woods, in rich humus soil, or on decaying stumps and fallen trees. The common name is derived from the resemblance of many species to some of the branching colonies of marine corals. They vary in height from less than an inch to 8 or 10 inches, and exhibit nearly every color and shade, from white, gray, and all shades of brown to yellow, pink, rose, violet, and purple. The basidia and basidiospores are borne in a continuous layer over the surfaces of the erect branches. The clavarias are not widely used for food, although their fleshy structure when young renders them suitable for such purposes. One species is reported as doubtfully poisonous.

Fig. 22-15   A giant puffball, *Calvatia gigantea*. It has been estimated that a fungus of this size will produce enough spores to cover the state of Pennsylvania with 20 spores per square foot.

The Puffballs (Lycoperdaceae) and Earthstars (Geastraceae) The puffballs are common and widely distributed. Species grow in lawns, meadows, cultivated soil, on tree stumps, and decaying logs. The fruiting body is round or pear-shaped (FIG. 22-14), lacking any conspicuous stalk. It consists of an outer sheath that encloses the spore-bearing tissue. The latter, when young, is divided by walls into minute cavities lined with basidia and basidiospores. At this stage the fruiting body is white or cream-colored within. Later, as the spores mature, the walls break down and the interior of the fruiting body becomes a mass of yellow or brown, dry, powdery spores intermingled with branched hyphal threads. In some species, the sheath scales away, thus allowing the spores to escape; in others (FIG. 22-14), an apical pore forms as the fruiting body matures. All of the true puffballs are edible when young and are highly prized as food. The giant puffball (FIG. 22-15) is the largest edible fungus. Specimens sometimes exceed a diameter of 3 feet and contain many billions of spores.

In the closely related earthstars (*Geastrum*) the sheath is composed of two layers (FIG. 22-16). The outermost of these layers splits into lobes that curve outward, revealing an inner layer surrounding the spore-bearing tissue. The spores escape through a single pore. The earthstars, like the puffballs, are edible when young.

Fig. 22-17 A stinkhorn (*Dictyophora duplicata*). In this species, the outer layer of the stem becomes separated and hangs down as a veil. Note the carrion flies eating the spores.

The Stinkhorn Fungi (Phallaceae) Stinkhorns (FIG. 22-17) grow where there is ample organic matter such as humus, rotted tree stumps, or decaying sawdust. The mature plant consists of a hollow stalk with swollen tip, on the surface of which the spores are borne. These are embedded in a gelatinous material having the strong fetid odor of decaying meat, whence the common name. Certain flies, beetles, and other carrion-eating insects, attracted by this odor, carry the spores away on their bodies. During the early stages of development, the fruiting body of the fungus is subterranean, with the stalk enclosed in a tough membrane the size and shape of a hen's egg. After the spores are mature, the stalk ruptures the enclosing membrane and rapidly ex-

Fig. 22-16 An earthstar (*Geastrum*).

pands upward, carrying the viscid mass of spores into the air.

THE SMUTS AND RUSTS

The smuts and rusts are Basidiomycetes that lack a conspicuous fruiting body. The fungus consists of vegetative hyphae, together with spores of various kinds. The rusts and smuts are parasites and cause serious plant diseases that result in great economic loss and necessitate expensive control measures.

The Smuts The name "smut" refers to the black and dusty masses of spores formed within the tissues of the host plants. The spores are of two general types, TELIOSPORES and basidiospores. The first of these are produced within the cells of hyphae. The hyphal walls gelatinize, and the protoplasm within each cell becomes converted into a spore with a thick, dark, spore wall. The gelatinized walls disappear, setting the spores free. As the tissues of the host rupture, the spores are released and disseminated, chiefly by wind. Smut fungi invade and destroy various tissues of their hosts, usually the reproductive organs, such as the flower and seed (FIG. 22-18), but vegetative tissues may be attacked as well. In some smut diseases, the host tissue becomes greatly swollen at the time of teliospore formation; in others, the parts attacked are merely consumed by the fungus without the formation of swellings or galls.

The most important smut diseases are found in the cereal grains—wheat, barley, rye, oats, sorghum, millet, and corn. Other plants attacked include onion, violet, carnation, lawn and forage grasses, the knotweeds and the sedges. The life cycles of the various smut fungi differ in certain details, and these differences often determine the methods of control. More than 1100 species of smut fungi are recognized, causing losses to American farmers estimated at hundreds of millions of dollars each year.

Common, or boil, smut of corn, a well-known and readily observable disease of all varieties of corn, occurs in nearly all regions where the host is grown. In North America the losses from smut are variable and difficult to measure, ranging from a trace up to 6 percent or more in localized areas. The symptoms of the disease may appear on any part of the plant, including the tassels, husks, ears, leaves, and stalks. The first indication of the smut is the appearance of swellings of various sizes. These are at first covered with a glistening, greenish- to silvery-white membrane of host tissue. This is later ruptured. exposing a black, greasy to powdery mass containing many millions of heavily spined teliospores (FIG. 22-19). The spores are blown about and fall upon the soil or on plant parts, where they either germinate or, more commonly, enter a period of dormancy that lasts until the next growing season. When animals eat diseased tissues, the teliospores may pass unharmed through the animal and be carried in the manure.

At the time of its germination in the soil or manure, the teliospore normally produces an elongate, transversely septate tube, the basidium (FIG. 22-20). This is typically four-celled, although numerous deviations occur. Each cell of the basidium produces basidiospores, and these may form secondary basidiospores by budding in a yeastlike fashion. Successive crops of basidiospores are possible, for the basidium is able to live saprophytically in rich soil or plant litter for some time. The basidiospores are carried to the corn by wind and rain, and there they infect meristematic tissues in the plant. These infections are localized in the immediate zone surrounding the infection. The smut hyphae stimulate the parenchyma cells in this region to further division, and the swelling that results contains not only the hyphae and spores of the fungus but also tissues of the host.

Experiments have shown that the corn smut fungus, like many other fungi, is heterothallic. When the corn plant is infected following germination of a basidiospore of one strain only, the

Fig. 22-19 Common smut (*Ustilago maydis*) of corn.

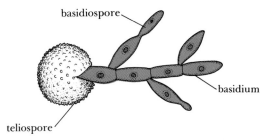

Fig. 22-20 Corn smut: teliospore, basidium, and basidiospore.

short, slender infection hypha is but weakly parasitic, and no teliospores are formed. Hyphae from basidiospores of opposite strains may fuse within the host tissue, however, forming a thicker hypha containing numerous pairs of nuclei. The fungus spreads in the host tissue, and eventually cell walls are laid down between each pair of nuclei. The nuclei in these cells now fuse, and the cells are converted into teliospores. The nucleus in each teliospore is then diploid, the only diploid nucleus in the life cycle. Meiosis occurs when the teliospores germinate, and the haploid basidio-spores subsequently formed on the basidium belong to two different strains.

The control of corn smut is difficult. In other cereal smuts, the fungus is carried either as hyphal threads within the seed or as spores on the exterior of the seed. Seed treatment is therefore a widely practiced method of control. In corn smut, on the contrary, the fungus does not overwinter in or on the grain, and seed treat-ment is ineffective. The most feasible means of control lies in the development of smut-resistant varieties of corn.

The Rusts The name "rust" comes from the reddish, yellow, orange, chocolate-brown, or black powdery appearance of the spores produced on the host. More than 2000 species of rusts are known, all parasitic on seed plants or ferns. No other group of fungi is so injurious to plants useful to man. Many of our most impor-tant crop, forest, and ornamental plants are parasitized by rusts.

The life cycle of many rusts is extremely complicated, involving as many as five kinds of spores that follow one another in a definite sequence. In some rusts, all five spore forms are present; in others, certain of them are absent from the life cycle. In some species, as few as two kinds are developed. All rusts produce telio-spores.

The rusts may be divided, on the basis of their host relationships, into two groups. One, the HET-EROECIOUS rusts, includes species that require two unrelated hosts for the completion of the normal life cycle. Certain of the spores are produced on one host and other spores on a sec-ond, or alternate, host. In the other group, the AUTOECIOUS rusts, the life cycle is completed on the same host species, frequently on the same plant. Among the important heteroecious rusts are white pine blister rust, which has currant and gooseberry as alternate hosts; cedar-apple rust on apple, crabapple and certain species of juniper (*Juniperus*); crown rust on oats, nu-merous grasses, and several species of buckthorn (*Rhamnus*); and black stem rust on cereals, many grasses, and certain barberries. Among the more common autoecious species are the rusts of bean, hollyhock, snapdragon, chrysanthemum, rose, carnation, blackberry, and flax. The evolutionary history of heteroecism and autoecism in the fungi is as yet unknown.

The condition of having alternate hosts is not entirely restricted to the rusts, but is most clearly exhibited in this group. Somewhat similar rela-tionships exist in the animal kingdom. The liver fluke, for example, lives within the body of the sheep, cow, pig, or even man, and also passes part of its life cycle within the tissues of certain fresh-water snails. Similarly, certain protozoans are dependent upon two hosts, such as the orga-nism responsible for malaria, which lives in man and in female *Anopheles* mosquitoes.

Black stem rust of cereals and grasses The life history of the black stem rust fungus (*Puccinia graminis*) is a classic story illustrating the complexities of the life cycle of many of the rusts and of the protection of our food supply. The disease has ravaged grain fields for centuries and is found in nearly every region where cereals are important crops. It has appeared in epiphytotic proportion on numerous occasions in the United States, especially in the great spring wheat areas of the upper Mississippi Valley. Black stem rust also causes heavy losses in winter wheat, oats, and barley as well as in certain forage and lawn grasses. In the United States, stem rust has reduced grain yields by over 300 million bushels in a single year. The black stem rust fungus produces five kinds of spores in its complex life cycle.

THE CLUSTER CUP STAGE The chief host of the CLUSTER CUP stage is the common barberry (*Ber-beris vulgaris*), a shrub introduced from Europe during colonial times and now widely distrib-uted. (The Japanese barberry, *Berberis thunbergii*, widely cultivated as a hedge plant, is not infected

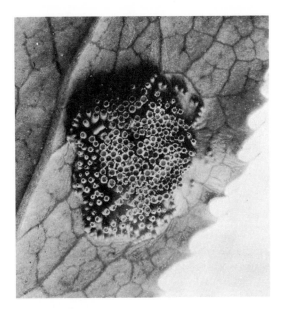

Fig. 22-21 Portion of leaf of barberry (*Berberis vulgaris*) bearing cluster cups of *Puccinia graminis.*

by the wheat rust fungus.) In the spring, the orange cluster cup stage appears on the lower surfaces of the barberry leaves as small swollen areas bearing groups of AECIA, or cluster cups (FIG. 22-21). The young cups are closed and buried in the tissue of the leaf, but as growth proceeds, they push through the leaf surface and open, exposing the spores within.

The cup itself is composed of a single layer of whitish hyphae, and the uppermost cells, forming the rim of the cup, readily slough away, leaving a ragged margin. Within the cup are the binucleate AECIOSPORES, yellow to orange in color, extending in chains from specialized stalk cells at the base of the cup (FIG. 22-22). When mature, the aeciospores fall from the cup are are carried away by currents of air.

THE RED AND BLACK RUST STAGES The aeciospores formed on barberry are incapable of infecting this plant, but they cause rust on wheat,

small grains, and grasses. If they fall upon a susceptible host plant in favorable weather—high humidity, frequent fogs, heavy dews or light rains, and temperatures of 20°C (68°F) or above—they germinate by sending germ tubes through stomata into the tissues of stem and leaf. Spores produced on barberries may infect cereals and grasses hundreds of miles away. It has been calculated that a single infected barberry bush may produce 70 billion aeciospores. This helps to explain how infection, usually amounting to an epiphytotic, can occur in grain fields near infected barberries.

The hyphae of the rust fungus ramify through the small grain or grass tissues and produce great numbers of unicellular UREDOSPORES, which burst through the host epidermis, forming elongated, reddish-brown pustules. These constitute the red rust, or early summer, stage. The uredospores are oval, covered with short spines, and are borne singly on the ends of specialized hyphae (FIG. 22-22, right). From the original infections on small grains or grasses that appear in late spring or early summer, uredospores are carried by air currents to other cereal and grass plants, where new infections occur until the grain is ripe. A single red rust pustule may produce as many as 350,000 uredospores, which are capable, under favorable conditions, of immediate germination. Since the time between germination and the appearance of new pustules is only about ten days in warm, moist weather, the rapid spread of rust through a field of grain is assured. Moreover, the red spores may be carried upward by air currents and blown more than 1500 miles without losing their viability.

As the grain begins to ripen, another type of spore, the two-celled teliospore, appears. These elongated spores, with thick, brownish walls, appear black in masses and cause what is known as the black rust stage. Teliospores may develop in a pustule that has been producing uredospores, or in new pustules, appearing after late infection by uredospores. The pustules of the red and black rust stages are alike except for the kinds of spores produced in them (FIG. 22-23).

Fig. 22-22 *Puccinia graminis.* (Left): longitudinal section through a cluster cup, showing chains of aeciospores. (Right): portion of pustule of red rust stage. The growth of the uredospores has pushed up the epidermis.

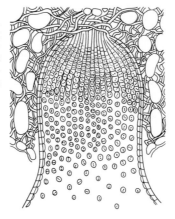
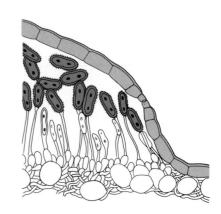

The teliospores are resting spores that survive the winter on straw or stubble and germinate in the early spring. At this time, each of the two cells of the teliospore gives rise to an elongated four-celled basidium, and each basidium produces four basidiospores (Fig. 22-24). These basidiospores are carried in the air to the young leaves of the barberry. Here they produce a germ tube that penetrates the epidermis of the barberry leaf, and the aecial stage subsequently develops. Of the four types of spores thus far discussed, only one, the basidiospore, infects the barberry; two, the aeciospore and uredospore, infect wheat or other grasses; one spore, the teliospore, which produces the basidium, is unable to infect any host directly (Fig. 22-25).

The reduction in yield from rusted plants is due in large part to the shriveled and shrunken grains formed in place of normal kernels. The hyphae of the fungus destroy chlorenchyma tissue and utilize large quantities of food that should be available for storage in the seed. In addition, the destruction of the epidermis at the site of pustule formation permits increased transpiration. Rapid use of food by the fungus weakens the stem and roots of the plant as well as reducing the filling of the grain. Rusted plants lodge (fall down) readily, and their weakened root systems cannot supply the plants with enough moisture in dry weather.

Fig. 22-23 *Puccinia graminis.* Pustules of black rust stage on wheat stem. A pustule of the red rust stage appears the same except for the reddish-brown color.

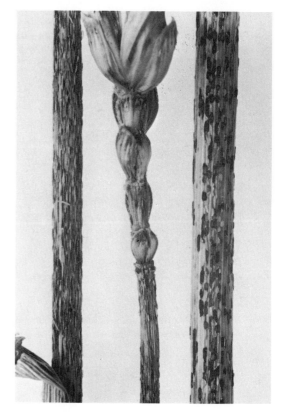

SPERMATIA AND HETEROTHALLISM The last of the five kinds of spores of the stem rust fungus is commonly involved in events leading to sexual reproduction. The aecia that occur on the lower surface of the barberry leaf have structures known as SPERMAGONIA associated with them, usually on the upper surface. These are small, globular, peritheciumlike cavities, lined with hyphae and opening by a pore. The tips of the hyphae that line the chambers pinch off minute colorless spores, the SPERMATIA (FIG. 22-26).

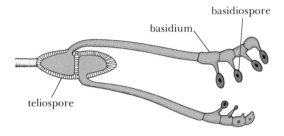

Fig. 22-24 *Puccinia graminis*. teliospore, basidium, and basidiospores.

Fig. 22-25 Diagram of life history of *Puccinia graminis*. The sexual stage is omitted.

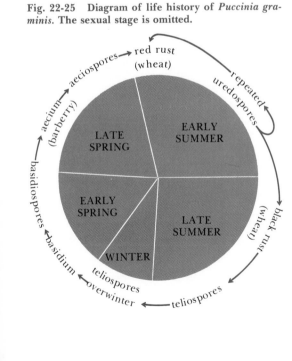

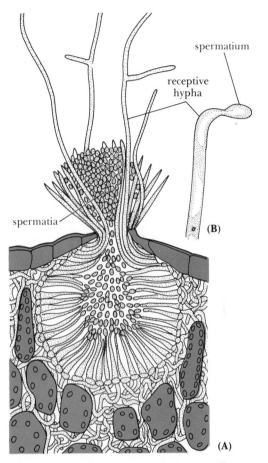

Fig. 22-26 *Puccinia graminis*. (A): spermagonia and spermatia. (B): fusion of spermatium with receptive hypha.

Other receptive hyphae grow from the side walls and project outward through the pore.

The stem rust fungus is heterothallic, and the basidiospores produced in the spring are of two mating types, plus and minus. When the barberry leaf is infected by a basidiospore of either mating type, haploid hyphae develop in the tissues. The hyphae branch and grow vigorously, giving rise to a cluster of spermagonia on the upper side of the leaf and to a group of immature aecia just beneath the lower surface. Each

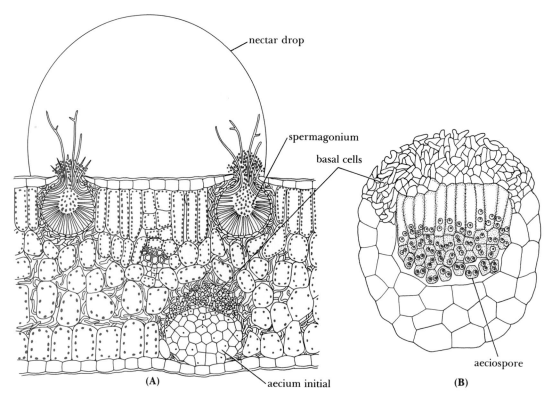

Fig. 22-27 *Puccinia graminis*. (A): cross section of barberry leaf, showing two spermagonia on upper surface covered by a drop of nectar, and an aecium initial near the lower surface. The spermagonia and aecium initial arose from a single basidiospore of one mating type. The cells of the aecium initial are all uninucleate. (B): young aecium, enlarged. The aeciospores are now binucleate.

Fig. 22-28 Diagram of complete life cycle of *Puccinia graminis*.

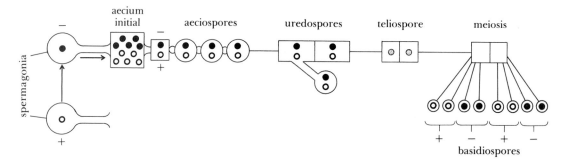

aecium initial at this stage consists only of an upper, closely packed weft of hyphae, the basal cells, and a lower group of larger, parenchyma-like cells (Fig. 22-27A). Unless the basal cells become dikaryotic, the aecium develops no further, forms no spores, and is functionless. The further progress of the aecium to maturity follows only after the transfer of spermatia of opposite mating type to the spermagonia with which the aecium initials are associated.

This transfer is brought about by insects. The cell walls of the spermatia are gelatinous, and the spores swell and extrude through the pore into a drop of nectar (Fig. 22-27A) attractive to insects, especially flies. These visit the spermagonia and carry spermatia from one cluster to another. A spermatium of one mating type, when brought into contact with a receptive hypha protruding from the spermagonium of another mating type, fuses with it (Fig. 22-26). A pore is formed, and the nucleus of the spermatium passes into the receptive hypha and downward into the underlying hyphae in the leaf. This migration is accompanied by nuclear division, the nuclei passing through the pores in the hyphal cross walls. The invasion of the basal cells of the aecium initial brings about the dikaryotic state, for pairing occurs between the local and the invading nuclei, one member of each pair belonging to a different mating type. It is now evident that the function of the spermatium is the initiation of dikaryotization in the life cycle. Heterothallism, and the function of the spermatia, were first demonstrated in 1927 by the Canadian botanist J. H. Craigie.

Chains of binucleate aeciospores are now produced (Fig. 22-27B). The aeciospores displace the parenchymalike cells of the aecium initial, become surrounded by a thick-walled cup, and eventually burst through the lower epidermis of the leaf and are discharged into the air (Fig. 22-22, left). Following germination on a susceptible cereal or grass plant, the binucleate aeciospores produce hyphae in the host tissue. All the cells of these hyphae are binucleate. This condition persists until, as the teliospore matures

at the end of the growing season, the two nuclei in each cell fuse. When the teliospore germinates the following spring, the diploid nucleus in each cell undergoes meiosis, and four haploid nuclei are formed. Each nucleus migrates into a basidiospore. As a result of segregation of sex factors during meiosis, two of these basidiospores belong to the plus strain and two to the minus strain. The complete life cycle of the black stem rust fungus is shown diagrammatically in Fig. 22-28. Sexual processes of the same general type have also been demonstrated in more than 20 other rust fungi.

Control of black stem rust It is apparent that a typical life cycle, as outlined above, is possible only if barberries are present. The most obvious method of control of this disease is to eradicate these alternate hosts wherever found. This principle has wide application in all rust diseases where two hosts are essential in the life cycle. The destruction of the common barberry as a method of control of wheat rust was, in some degree, practiced both in Europe and in North America, even before botanists had conclusive proof that the disease on the barberry and the rust on the wheat were caused by the same fungus. It was not until 1865, in fact, that a German botanist, Heinrich Anton De Bary, demonstrated the connection between the aecial stage on the barberry and the uredial and telial stages on wheat. But only in modern times have unified and widespread efforts been made to eradicate barberry over all wheat-growing areas. The first comprehensive efforts began in the United States in 1918 as a result of the great losses in cereals during World War I. Federal, state, and local agencies in 19 northern states, comprising more than a million square miles, now cooperate in barberry eradication, and more than 500 million barberry bushes have been destroyed since 1918.

Although epiphytotics of black stem rust have been more or less steadily reduced both in number and severity by the campaign against the barberry, they still occur. Cereals and grasses can be infected from only two possible sources—

aeciospores from the barberry, and uredospores. The number of aeciospores may be greatly reduced or virtually eliminated by barberry eradication. In the northern United States and in Canada, uredospores do not live through the winter. In more southerly states, and in Mexico, however, they may survive the winter on cultivated cereals and wild grasses. In the spring, these uredospores are carried upward by air currents, borne northward, and dropped upon fields of cereals during rains. Fortunately, the warm moist weather conditions favorable for the development of severe epiphytotics do not often occur. In the wheat-growing areas of countries with mild climates, such as Australia and Argentina, the uredospores may also live from one growing season to another, and the barberry is not essential to the life cycle.

Physiologic races The stem rust fungus affects not only wheat but also other grasses. The species *Puccinia graminis* has therefore been divided into seven varieties, named according to their chief hosts. One variety usually attacks wheat; another, oats; another, principally rye; and still others parasitize wild and cultivated grasses. Within each of these varieties, there are in turn numerous PHYSIOLOGIC RACES, or FORMS, that differ from one another principally in the extent to which they infect specific varieties of the host plant. These races are so numerous that they are numbered rather than named. *Puccinia graminis* var. *tritici*, which chiefly attacks wheat, for example, is subdivided into more than 300 races. Fortunately, only a dozen or so of these races cause much damage in the United States each year. There are more than 30 known physiologic races of *Puccinia graminis* var. *avenae*, which principally attack oats, and about 14 races of the variety *secalis*, which infect chiefly rye. Physiologic races are also found in many other species, perhaps the majority of pathogenic fungi.

The existence of physiologic races greatly complicates the production of resistant varieties of wheat, oats, other cereals, and grasses. Some races are predominant in one region and some in another; thus, a variety of wheat resistant to rust in one region may not be resistant in another. Stress in plant breeding is therefore placed upon the production of cereal varieties resistant to the most important races of rusts in the region in which varieties are to be grown. Further complications may appear as races previously unimportant become prevalent and new races of rust arise. An example of the former is a race called 15*B*, which has been termed the most virulent race of stem rust ever to appear in North America. It was first recognized in 1939, but did not become abundant enough to cause damage in wheat fields until 1950. In that year it spread widely in the United States, occurring in 16 states. It attacked nearly all the commercial varieties of wheat grown in North America. The spread of this race has made necessary the breeding of new rust-resistant varieties of wheat.

Physiologic races of the stem rust fungus, for the most part, are believed to originate by hybridization during growth upon the alternate host, the barberry. The significance of heterothallism in the stem rust fungus is now apparent. This form of sexual reproduction results in hybridization, and hence the reassortment of the genes of the two parents. The resulting new combinations are the source of the origin of new races that may attack previously rust-resistant varieties of wheat. Eradication of the barberry is important because it not only eliminates this host but also largely prevents the formation of new races of the rust fungus. Hybridization on the barberry is not, however, the only source of new and virulent hybrid strains of the rust fungus. In recent years it has been demonstrated that somatic hybridization can occur on the grain or grass host during the repeating uredostage. Binucleate hyphae, originating from uredospores of different races of the rust fungus, may fuse and exchange nuclei. The actual mechanism of somatic hybridization requires further research, and the extent to which this gives rise to virulent strains is not known.

Fig. 22-29   Aecia of apple rust on leaf of apple.

Fig. 22-30   Apple rust on cedar. The gall-like structure has exuded gelatinous horns containing teliospores.

Other rusts with alternate hosts   Certain aspects of the life cycles of two other rust fungi should be noted. One is the cedar-apple rust fungus (*Gymnosporangium juniperi-virginianae*), which produces spermagonia and aecia on the leaves of the apple (FIG. 22-29) and related hosts. The fruit is also affected. Aeciospores from the apple infect certain junipers, chiefly the red cedar (*Juniperus virginiana*), upon which teliospores are formed. No uredostage occurs in this species. Prior to the appearance of the teliospores, the cedar is stimulated and forms globular brown galls, varying in diameter from a fraction of an inch to as much as 2 inches. After spring rains, these galls extrude gelatinous orange-colored horns (FIG. 22-30) composed of two-celled, stalked teliospores. The teliospores germinate in a few days and form basidia and basidiospores while still attached to the spore horns. The basidiospores are then carried by wind and infect the apple and crabapple. With no repeating stage, the life cycle can easily be broken by eradication of one of the susceptible alternate hosts from the vicinity (1–3 miles) of the other. The disease is also readily controlled by spraying apple trees with a rust-protecting fungicide. Rust-resistant apples, crabapples, and junipers are also available from nurseries.

The white pine blister rust (FIG. 22-31) is one of the more serious of the rust diseases from an economic standpoint, for it affects the five-needle soft pines of the northeast, Great Lakes states, and northwest, which are among the most valuable of our timber trees. The alternate host is wild and cultivated species of currants and gooseberries (*Ribes*). Quarantine regulations prevent planting of susceptible currants and gooseberries in areas where white pines are important timber trees. The wind-borne aeciospores are carried from pine to currants and gooseberries, and basidiospores are carried from these plants to pine. The aeciospores formed on

Fig. 22-31 Blisters (aecial stage) of white pine blister rust (*Cronartium ribicola*) on white pine (*Pinus strobus*).

TABLE 22-1
Some Heteroecious Rusts

Common name and principal host	Alternate host
Corn rust	*Oxalis*
Poplar rusts	Larch, Douglas fir, and other hosts
Rust of stone fruits	Anemone, hepatica, buttercup, and meadowrue (*Thalictrum*)
Witches' broom of fir and spruce	Chickweeds
Ash rust	Marsh and cord grasses
Southern fusiform rust on hard pines	Oaks

pine may be carried several hundred miles to the alternate host, but spores from susceptible wild or cultivated currants infect pines only within 1000 feet. Eradication of currants and gooseberries, an important method of controlling the disease, is therefore necessary only within a short distance from stands of pine. Recently, widespread aerial and ground spraying of timber stands, using a systemic fungicide, has eradicated blister rust cankers from large areas in the Pacific Northwest.

A few additional examples of the numerous kinds of heteroecious rusts are listed in Table 22-1.

FUNGI IMPERFECTI

The perfect stage is absent or unknown in many fungi, and such plants are grouped in the Fungi Imperfecti. With some exceptions, however, the Fungi Imperfecti are Ascomycetes; they have septate hyphae and their imperfect stages are similar to those of known Ascomycetes. Most are probably reduced forms that have lost the power to reproduce sexually, but the sexual stages of some may not yet have been found. The group contains over 10,000 named species, many of which are of direct economic importance. Others are saprophytic on decaying organic matter and in the soil. They usually reproduce by fragmentation or by one or several kinds of conidia. The conidia are borne in various ways, and may be one-celled, two-celled, or many-celled. Among the numerous plant diseases caused by members of this group are early blight of tomato and potato, leaf spot of beets, wilts of numerous plants, apple pink rot and blotch, stalk rot of corn, anthracnose of bean, cucurbits, pecan and walnuts, tomato leaf mold, peach scab, and white rot of onion.

Fungi That Attack Man Of direct personal interest are the fungi that cause disease in man. These diseases may be classified as: (1) superficial infections and (2) deep, or systemic, infections. Many species of Fungi Imperfecti are involved in these diseases. In some cases the microscopic appearance of the fungus is sufficiently distinctive for identification, but usually it must be grown in pure culture for accurate diagnosis of the disease.

Superficial infections These infections constitute the majority of the fungous infections of man. Most result from the parasitic growth of hyphae of a group of closely related fungi known as the DERMATOPHYTES, or RINGWORM fungi. These may cause ringworm of the scalp (found chiefly in children), of the smooth skin, of the nails, and of the hands and feet. Ringworm of the feet is known popularly as athlete's foot. The dermatophytes may produce conidia of two kinds: small, one-celled, and large, multicellular spores (FIG. 22-32). They are primarily parasites on man, other animals, or both, but may also grow saprophytically on nonliving organic matter. Among domestic animals in the United States, ringworm infections are observed most frequently in calves, but they also occur in horses, cats, and dogs, and on the combs of chickens and turkeys.

Another superficial infection, called candidiasis, may occur on the skin, but the fungus usually involved (*Candida albicans*) is primarily a parasite of the mucous membranes. It is especially well known as the cause of thrush, an infection of the membranes of the mouth, most commonly found in newborn infants. Thrush was formerly an extremely common disease, sometimes reaching epidemic proportions, but it has declined greatly in frequency and severity.

Systemic infections Unlike the superficial infections, which are contagious, the systemic infections usually are not transmitted from one person to another. The organisms are widespread in natural habitats, living saprophytically on vegetation and in the soil, and invasion occurs commonly by inhalation or through small wounds made by splinters or thorns. Although deep in-

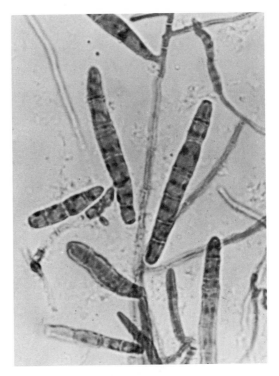

Fig. 22-32   Multicellular conidia (macroconidia) of a dermatophyte (*Trichophyton mentagrophytes*).

fections are not common, they are usually chronic, serious, and sometimes fatal. The infection may be generalized, or specific parts of the body may be involved, depending upon the disease. Among the affected organs and tissues are the brain and spinal cord, the lymphatic tissues, the kidneys, lungs, liver, and bone marrow.

Some systemic diseases, such as sporotrichosis, cryptococcosis, and histoplasmosis, are widely distributed. Others, such as North American blastomycosis and valley fever (coccidioidomycosis), are more limited in occurrence. Histoplasmosis is one of the common diseases in the United States and is also worldwide in distribution. The causal organism (*Histoplasma capsulatum*) may live either as a parasite on man or animals, or as a saprophyte in the soil. Like many other pathogenic fungi, it occurs in two distinct forms. In the parasitic phase, in living tissues, it forms yeastlike cells that reproduce by budding. In pure culture, and in the soil, it is moldlike, with hyphae that bear two kinds of conidia. The disease is not transmitted from host to host, but infection is acquired by breathing spore-contaminated dust derived from soil high in organic matter. It is usually mild, but may terminate fatally.

PLANT DISEASE CONTROL

The control of plant disease is fundamental to the successful growing of most kinds of plants. In order to devise effective methods for the control of a given disease, it is first necessary to make a correct diagnosis. Then, if the disease is caused by a parasite, the life cycle of the parasite and the conditions under which it thrives must be established. With this knowledge, plant pathologists usually find some phase of the life cycle vulnerable to attack. The time and mode of infection, the methods of transmission of the fungus, bacterium, or virus, the method of overwintering, the weather conditions involved —these and other factors are important in plant disease control.

An illustration of the need to know the life history of a pathogen before control measures can be applied is found in the history of the use of copper in plant disease control. In 1845, during the European potato late blight epiphytotic, Charles Morren, professor of agriculture at the University of Liége, wrote a letter to a Brussels newspaper advocating "lime and marine salt, mixed with a small quantity of the sulphate of copper" for the control of the blight. Morren, not knowing the life history of the fungus, recommended that farmers "powder with these mixed substances the soil which has been planted with sick potatoes." If he had advocated placing the mixture on the foliage of healthy plants, the epiphytotic might have been controlled and many thousands of human lives saved.

Plant diseases vary in severity from year to year and from one locality to another, depending on (1) the environment (chiefly the amount and frequency of rains or heavy dews, relative humidity, and temperature); (2) the relative resistance or susceptibility of the plant host; and (3) the parasite. For disease to develop, all three factors must be present and in balance. For example, if the environment is favorable for a disease and the causal organism is present, but the host plant is highly resistant, little or no disease will develop. Similarly, if the parasite is present and the host is susceptible, but the environment is unfavorable, disease usually does not appear.

These relationships form the basis for three general principles of plant disease control: (1) the resistance of a susceptible plant is increased; (2) the environment is made less favorable for the parasite and more favorable for the host plant; and (3) the disease organism is killed or prevented from reaching the plant, penetrating it, and producing disease. Some applications of these principles are considered in the following paragraphs.

Disease Resistance The development of disease-resistant varieties and hybrids is an outstanding method of disease control. Once a highly resistant or immune variety or hybrid is obtained, other control methods are usually unnecessary. Further, the initial expense of developing such a variety is small in comparison with the cost of other methods of control. A number of diseases can be controlled only by growing resistant varieties. Among the results of breeding for disease resistance are varieties of asparagus, snapdragon, and cereals resistant to rust; sweet corn, cucumber, squash, and carnation to powdery mildew; beans, cabbage, sugar cane, corn, and spinach to mosaic; beets, muskmelon, lima bean, grape, and tobacco to downy mildew (FIG. 22-33); and many other resistant varieties of field, garden, and ornamental plants. The increased value of farm products

Fig. 22-33 Lima beans (left) resistant to downy mildew (*Phytophthora phaseoli*). Those at right were killed by the fungus.

resulting from the use of disease-resistant varieties and hybrids amounts to hundreds of millions of dollars annually.

Unfortunately, as pointed out previously, pathogenic fungi, bacteria, and viruses may form new races that can attack formerly resistant varieties. Clinton is an example of an excellent oat variety that had to be discarded because of the appearance of new races of crown rust to which it was very susceptible. Even though the appearance of such races has complicated the problem of developing resistant varieties, there has been constant over-all progress in reducing losses from plant diseases by this method.

Cultural Practices Plants can frequently be made more disease-resistant by adjusting soil fertility. Many diseases are less serious where an adequate level of essential elements is present in the root zone, ensuring uniform, steady growth. Plants are frequently most susceptible to disease when nitrogen is high in relation to other factors, especially in hot, wet weather. Excess nitrogen leads to a lush, tender growth that is more susceptible to many fungal, bacterial, and viral parasites.

The environment may be made less favorable for the parasite and more favorable for the host plant by improved air circulation. Spacing of plants by pruning, thinning, and using a correct rate of seeding or planting aids in keeping the foilage dry, reduces shading, and increases plant vigor. Fungi, with the exception of some powdery mildews and certain others, require free moisture on plant surfaces for 3–12 hours for germination and infection.

The control of weeds also can reduce disease. Weeds harbor diseases and insects that may transmit bacterial and viral diseases; they reduce air circulation; and they compete for water, sunlight, and essential elements.

Crop rotation Many pathogenic fungi and bacteria are soil-borne, and may persist saprophytically in the soil for one to many years. The continuous growing of the same or related crops commonly produces an increase in numbers of these pathogens in the soil. Such conditions can often be controlled by a 3- or 4-year rotation that "starves out" many organisms that cause diseases of the shoot or underground parts. Rotation, however, does not always control soil-borne pathogens, for they may be normal inhabitants of the soil or may be able to persist (as does clubroot of crucifers) for ten years or longer, so that rotation is not effective. Weeds, as well as crop plants, may be susceptible to crop disease organisms and so carry the diseases through crop rotations, even though these rotations involve nonsusceptible crops.

Sanitation The prompt collection and burning, burying, or composting of diseased plant material is one of the home gardener's best means of plant disease control. Sanitation is also widely practiced by the successful florist, nurseryman, orchardist, vegetable grower, and forester. The severity of many diseases can be at least partly reduced by carrying out simple sanitation procedures.

Eradication of alternate hosts The elimination of alternate hosts as a means of control for various rust diseases was considered earlier in the chapter.

Quarantines The majority of our most destructive insect pests and plant diseases, such as the cotton boll weevil, Mediterranean fruit fly, gypsy moth, Japanese beetle, chestnut blight, white pine blister rust, citrus canker (later eradicated), and Dutch elm disease have come to us from other countries. If allowed unrestricted entry, new invasions of plant enemies from foreign countries could cost Americans immense sums of money in reduced production and increased expenditures for pest control. The Federal Plant Quarantine Act of 1912 was enacted to protect American agriculture from such enemies. The Act and its amendments restrict the entrance into the United States of soil, plants, plant products, or plant parts such as nursery stock, cuttings, seeds, or fruits that might harbor harmful bacteria, fungi, viruses, nematodes, insects, or other pests. An increasing volume of plant pests is being intercepted by plant quarantine inspectors each year. Interstate quarantines are also in effect in parts of the United States, such as California, Arizona, and Florida.

Chemical controls Fungicides and bactericides are chemical and physical agents that kill or inhibit the growth of fungi and bacteria. Some are inorganic, such as sulfur and compounds containing mercury, copper, or zinc. In recent years, however, chemists have synthesized a host of organic compounds that have been found more effective than older fungicides in plant disease control. These are complex compounds known to the layman by trade names not important here. The mechanism of action varies, but in a number of cases the fungicide or bactericide penetrates into the cytoplasm of the bacterial cell, fungous spore, or germ tube, and interferes with the action of specific enzyme systems. Fungicides and bactericides are used to protect plants by stopping fungi and bacteria before they become established in the plant; to eradicate disease organisms that have not penetrated too deeply into the plant; or to kill the organisms throughout the plant, a practice known as chemotherapy.

SPRAYS AND DUSTS Most fungicides are protectant in nature. They are usually applied as sprays or dusts (FIG. 22-34) at 3- to 14-day intervals prior to the arrival of fungous spores. The spore, or germ tube growing from it, is destroyed by the action of the fungicide, and infection is prevented. If the parasite has already invaded the host, spraying or dusting usually has little or no effect. There are a few exceptions, to be mentioned later, but it is generally true for the diseases encountered by the homeowner. This aspect of plant disease is usually misunderstood by the layman, who, noticing the symptoms of disease on house or garden plants, inquires, "What can I do about it?" The disease has usually progressed so far that nothing can be done for the diseased plant. But applying a protective fungicide can often prevent the infection from spreading to other parts of the same plant or to nearby healthy plants.

SEED TREATMENT As pointed out earlier (Chapter 13), the most vulnerable stage in the life cycle of a seed plant is from germination to establishment of the seedling as an independent plant. During this period the tissues of the seedling are growing rapidly and are readily attacked by fungal enemies. When planted, the seeds may decay and the seedlings are subject to root rots and to damping-off, in which the seedlings are killed just before or after they have emerged from the ground. The sources of infection are the spores and mycelium on the seed and soil-borne pathogens. The spores and mycelium may be destroyed, or the seed and seedling protected against soil-borne fungi by coating the seed, before planting, with a fungicide (FIG. 22-35). Many kinds of vegetable and field crop seeds are treated by seedsmen before they are sold to the grower.

Fig. 22-34 Spraying apples (*Pyrus malus*) to control fungous and insect pests.

Fig. 22-35 Seed treatment of sorghum (*Sorghum vulgare*). (Left): seed treated with a fungicide (Captan). (Right): seed untreated; the seeds are covered with a fungous growth.

SOIL TREATMENT Plants may also be protected from soil pathogens by treating the soil, before sowing, with fungicidal compounds that are dispersed in the form of a gas, liquid, or solid. These may be injected into the soil, mixed with soil, or spread on the surface and watered in. Small amounts of soil may be disinfested by heating in an oven at 160°F for an hour. Soil treatment using steam, electric, or dry heat, or a soil fumigant is commonly applied to potting and greenhouse soils and to outdoor seedbeds in which seedlings are grown for later transplanting.

ERADICANT FUNGICIDES Eradicants are applied as foliage sprays, seed treatments, or soil drenches to kill or check fungal parasites after they have penetrated into the host and become established. For example, certain fungicides are used by apple growers to eradicate the apple-scab fungus in lesions up to five days after the start of infection. Certain methods of seed treatment may be eradicative by destroying organisms inside the seed coat. Eradication of pathogens is also possible in dormant bulbs, corms, tubers, and other propagative plant parts. Powdery mildews grow on the surface of the host, sending haustoria into the cells beneath. Since the fungus is largely external, a protective fungicide, applied to the leaf surface, also has an eradicative effect. The removal of diseased tissue from decayed or cankered areas in trees may be classed as eradication. In theory, the removal of such infected material should be continued until sound healthy wood is reached. In practice, unless discoloration or other sign of disease is present, it is difficult to determine how deeply the disease agent has penetrated into the wood. After the diseased wood is removed, a disinfectant, followed by a protective wound dressing, is applied to the exposed surface to prevent further infection. It will be seen from the preceding discussion that it is not always possible to distinguish clearly between protection and eradication.

CHEMOTHERAPY This control method involves the treatment of the diseased plant by the use of a systemic fungicide (or bactericide) that can be translocated throughout the host. Such a compound may prevent multiplication of the pathogen, inhibit its growth, or neutralize toxins. It may act against both invading and established pathogens. Many compounds have been tested, including the antibiotics streptomycin, cycloheximide, and griseofulvin. These are applied as dusts or sprays, or to the soil. The toxicity of the systemic fungicide to the host is often a problem, and the commercial control of plant disease by the use of such fungicides is effective for only a few diseases.

While plant pathologists have accomplished much in plant disease control, many problems still remain. Finding the solution to these problems is the challenge to the next generation of plant pathologists and plant breeders.

SUMMARY

1. The Basidiomycetes include the largest fungi and those that receive most popular attention. They have septate hyphae and nonmotile spores; the spore of the perfect stage is the basidiospore. Some have large and fleshy fruiting bodies; others, the rusts and smuts, do not. Examples of Basidiomycetes with conspicuous fruiting bodies are the gill fungi, pore fungi, coral fungi, puffballs, and stinkhorn fungi.

2. In the gill fungi, the basidia and basidiospores are borne on the surfaces of gills radiating under a fleshy cap. These include most of the edible and poisonous fungi. No common tests can distinguish between edible and poisonous species.

3. A common pattern of life cycle includes the following: (a) Basidiospores of different mating types germinate and produce hyphae; (b) the hyphae fuse and dikaryons

are established throughout most of the life cycle, including the fruiting bodies; (c) fusion of dikaryons in the basidium, resulting in the only diploid nucleus in the life cycle; (d) meiosis in the basidium, with the production of four basidiospores (meiospores).

3. The pore fungi are parasitic or saprophytic on living or dead trees. They are annual or perennial. The spores are borne on the sides of tubes that open to the air by pores. The pore fungi are the most important agents of wood decay.

4. The coral fungi are saprophytic, growing on rich soil or decaying wood. Puffballs are edible fungi with a single or double sheath surrounding the spores. Stinkhorn fungi attract carrion-eating insects, which disseminate the spores.

5. The conspicuous spore of the smut fungi is the teliospore. This spore germinates and forms an elongated basidium, which produces the basidiospores. The host plant is stimulated to produce tumorlike outgrowths composed of teliospores and host tissue.

6. A rust fungus is autoecious if it requires but one host, and is heteroecious if it requires two hosts belonging to different species.

7. The black stem rust fungus produces five kinds of spores. Two of these, the aeciospore and spermatium, are produced on the barberry. Two, the uredospore and teliospore, are produced on wheat or other grasses. The basidiospore grows from the basidium and is not produced on any host.

8. The stem rust fungus is heterothallic. The basidiospores are of two strains, plus and minus. The infection of the barberry leaf by either kind results in the formation of spermagonia on the upper surface of the leaf and of immature cluster cups in the tissues near the lower surface. If spermatia are carried from a spermagonium of one strain to a spermagonium of another, the associated cluster cups mature and produce binucleate aeciospores.

9. The aeciospores are carried to a cereal or grass plant, where they cause the red rust stage. The uredospores of this stage are followed by the two-celled teliospores of the black rust stage. As the teliospores mature, the two nuclei in each cell fuse. When the teliospore germinates, the fusion nucleus undergoes meiosis and each cell produces basidia and basidiospores.

10. Eradication of the barberry is effective as a control measure against stem rust, not only because it eliminates the aeciospores but also because it largely prevents hybridization and the origin of new physiologic races. Stem rust can, however, occur even in the absence of the barberry if uredospores that have overwintered in southerly regions are blown northward.

11. Other important heteroecious rusts are those of apple, the alternate host being chiefly red cedar, and white pine blister rust, which has currants or gooseberries as the alternate host.

12. The Fungi Imperfecti usually reproduce by means of conidia, a perfect stage being absent or unknown. Included in this group are many forms causing plant diseases, together with most of the fungi responsible for skin and systemic diseases of man.

13. The principles of plant disease control include: (1) development of disease resistance, either by breeding or by cultural practices that improve soil fertility; (2) making the environment less favorable for the parasite and more favorable for the host, such as improvement in air circulation and weed control; (3) killing of the parasite or preventing it from reaching the host and causing disease. This category includes crop rotation, sanitation, eradication of alternate hosts, quarantines, and the use of fungicides and bactericides as protectant sprays and dusts, and as seed and soil treatments.

23

The Liverworts and the Mosses

P EOPLE ARE PRONE TO DESIGNATE as "mosses" all small plants growing on the soil, on rocks, on trees, or even in the water. Such growths are recognized by botanists, however, as algae, liverworts, mosses, lichens, or flowering plants, depending on their structure. Even plants bearing the common name of "moss" may not be mosses at all. Sea moss belongs to the red algae, Iceland moss and reindeer moss are both lichens, and Spanish moss is an angiosperm. The club mosses, also, are not mosses, nor are they related to mosses. The liverworts and true mosses are large, diverse, and interesting groups. They live on soil, rocks, and wood, and sometimes in the water.

Many kinds of evidence support the belief that terrestrial forms of plant life are descended from algalike plants that once flourished in ancient seas. The nature of these algal ancestors is not known. They were probably green algae, perhaps similar to certain species now living in fresh waters and along ocean coasts. The descendants of these plants that migrated from the water to the land have evolved along two different lines. One group of primitive land plants developed a system of conducting and supporting cells—that is, a vascular system. In the second line of evolution, no such specialized supporting and conducting cells ever evolved. Plants in this group have remained small and inconspicuous. With no tissues that translocate food, water, and essential mineral elements over any distance, and without specialized supporting cells, they do not attain any considerable size.

The plants of today that have come from this second line of evolutionary development are the liverworts, hornworts, and mosses, known collectively as the bryophytes. In all probability, these plants have undergone considerable change since plants invaded the barren land. They are not, however, so far as we know, the ancestors of complex modern plants. They therefore constitute a blind alley in evolution.

Individual liverworts and mosses are usually inconspicuous. Yet they are noticeable and even attractive when growing in masses. In general, they are poorly adapted to conditions of terrestrial life and, for the most part, they are plants of moist, shady environments. Certain bryophytes, however, especially mosses, are able to survive periods of drought. Growth is renewed when moisture again becomes available.

The life cycle of the bryophytes exhibits the same pattern that is found in the algae and continues up through the flowering plants. A gametophyte generation alternates with a sporophyte generation. But unlike the condition in the flowering plants, the plant body belongs to the gametophyte generation, is independent, and is the most prominent feature of the life cycle. The sporophyte generation is commonly less conspicuous, is always attached to the gametophyte, and never leads an independent life, although it may manufacture some food.

In addition to certain vegetative features showing greater specialization of internal structure and body form, the bryophytes differ from their more primitive algal progenitors in two important ways: (1) The sex organs, ARCHEGONIA and ANTHERIDIA, which contain the eggs and sperms, are multicellular. (2) After fertilization, the developing embryo is retained within the archegonium, and the young sporophyte, for a time at least, is nourished by the tissues of the parent plant.

Many bryophytes are homothallic, with both kinds of sex organs borne on the same gametophyte. Others are heterothallic, with the two kinds of sex organs on different gametophytes. In heterothallism the segregation of sex factors takes place at meiosis; two spores of each tetrad produce male gametophytes and two give rise to female gametophytes. In some species a pair of sex chromosomes has been identified.

THE LIVERWORTS (Hepaticae)

Some 8000 species of liverworts have been described. Most are plants of moist environments. They are inconspicuous, except in masses, and so escape casual notice, or, if noticed, are usually confused with mosses, as most liverworts are mosslike in their external appearance. All grow prostrate, or nearly so, on the ground, on the bark of trees, or on rotted wood. They are often seen on moist rocks and soil along woodland streams. Most liverworts are land plants. The relatively few aquatic species are secondarily aquatic; that is, they are land plants that have become readapted to the aquatic environment of their ancestors.

The liverworts are commonly classified into two groups: the THALLOSE liverworts (FIG. 23-1) and the LEAFY liverworts. In both groups the plant body is dorsiventral in form; that is, it is differentiated into an upper, or *dorsal*, surface and a lower, or *ventral*, surface. The sex organs are borne on the dorsal surface. The body, growing close to the soil, is attached to the soil by numerous threadlike rhizoids, comparable in function to root hairs. The plant body of a liverwort exhibits certain features that evolved in early nonvascular land plants and which were not possessed by their aquatic ancestors. Among these are the rhizoids just mentioned and other adaptations to a land habitat, such as a cuticle covering an epidermis, and thick-walled spores adapted to dissemination by air.

Thallose Liverworts The thallose liverworts are conspicuously lobed. Each time the thallus divides, it forms two more-or-less equal branches. Growth occurs through the activities of one or more apical cells located in the notch of each fork of the thallus. The fancied resemblance of the lobed thallus to the human liver is responsible for the names "Hepaticae" and "liverworts" applied by early botanists.

Ricciocarpus natans, an example of a thallose liverwort, is an aquatic, floating plant (FIG. 23-2),

Fig. 23-1   Plant body (gametophyte) of a thallose liverwort (*Conocephalum conicum*) growing on soil.

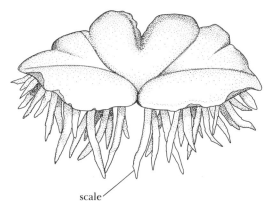

Fig. 23-2   *Ricciocarpus natans* plant body; an aquatic form. The ribbonlike scales absorb water.

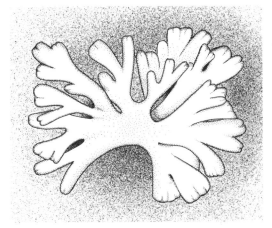

Fig. 23-3   *Riccia frostii* growing on damp soil.

which may also grow, although less commonly, on moist soil. Closely related to it are various species of *Riccia*, which are more branched and which commonly form rosettes when growing on damp ground (FIG. 23-3). *Ricciocarpus* and the related *Riccia* are simple liverworts, although this simplicity is generally considered to be due to reduction from ancestral forms of greater

complexity and not to inherently primitive characteristics.

The germination of the spore of *Ricciocarpus* is followed by the development of the lobed, somewhat heart-shaped plant, about ½ inch wide. Large masses of these thalli may be found floating on the surfaces of ponds and quiet streams. Attached to the ventral surface are a few scat-

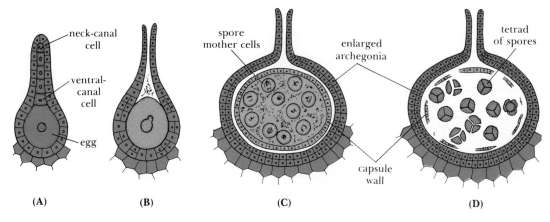

Fig. 23-4 *Ricciocarpus.* (A): mature archegonium. (B): fertilized egg. (C): mature sporophyte (capsule wall and spore mother cells). (D): spore tetrads. The capsule wall has disintegrated, and the spores are contained within the enlarged base of the archegonium. Gametophyte generation shown in color.

tered rhizoids and numerous brownish-purple, ribbonlike scales (FIG. 23-2). Both rhizoids and scales absorb water. When the plants grow upon the soil, the rhizoids become more numerous and the scales are greatly reduced. The upper surface of the thallus is divided into small areas. In the center of each is an inconspicuous pore that opens into a large internal air space. The upper part of the gametophyte is composed of photosynthetic tissue (chlorenchyma) and large air spaces. Beneath is solid parenchyma tissue, largely lacking in chloroplasts, in the cells of which starch may accumulate (FIG. 23-5).

The plants reproduce by fragmentation of the thallus through the middle and by spores formed following a sexual process. Most species of *Riccia* are homothallic. The sex organs (FIGS. 23-4, 23-5) of *Ricciocarpus* are located at the bottoms of furrows on the upper side of the thallus. Their features resemble those of a whole series of higher terrestrial plants. The flask-shaped archegonium, the female sex organ, contains the egg in its enlarged base. Just above the egg, like a plug holding it in place, is the VENTRAL CANAL CELL. The neck of the archegonium is filled with

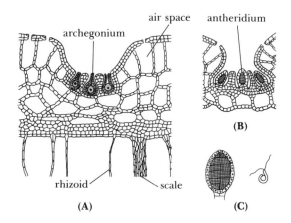

Fig. 23-5 *Ricciocarpus.* (A): diagram of cross section of thallus, showing archegonia borne at the bottom of a furrow. (B): cross section of thallus, region of antheridia. (C): single antheridium and sperm.

a linear series of NECK CANAL CELLS. The antheridium (FIG. 23-5B, C) is an oval structure with a wall that is one layer of cells in thickness. This wall encloses a mass of minute cells that develop into the sperms, or ANTHEROZOIDS. At maturity, the antheridium bursts and the sperms escape. While this is happening, the neck canal cells and the ventral canal cell in the archegonium degenerate to form a protoplasmic mass that extrudes from the archegonial tip. The sperms swim, and reach the archegonium through water held in the furrows. They become entangled in the mucilaginous extrusion and make their way down the neck. One sperm finally penetrates the egg (FIG. 23-4B).

The fertilized egg enlarges and becomes surrounded by a thin cellulose membrane. The zygote then undergoes a series of cell divisions and forms a globular mass of cells, the EMBRYO. The cells of the embryo are at first all alike, but eventually an outer layer of cells becomes distinct from the cells that it encloses. This outer layer is the CAPSULE WALL (FIG. 23-4C). As the embryo matures, the cells within the capsule separate, become globular, and function as spore mother cells. With the enlargement of the embryo, the base of the archegonium increases in size and becomes two cells thick.

Each spore mother cell formed in the capsule now undergoes meiosis, forming four spores. In *Ricciocarpus*, the number of chromosomes in the fertilized egg, the spore mother cells, and the cells of the capsule is 8. Therefore the *n* number in each spore, after meiosis, is 4. In other liverworts, an *n* number of 9 and a *2n* number of 18 is common.

The spores formed from each spore mother cell tend to remain together in globular groups of four, called tetrads (FIG. 23-4D), like the tetrad of microspores in an angiosperm. As the spores ripen, the capsule breaks down, leaving them free in the cavity of the enlarged base of the archegonium. The spores are finally released by the decay of the thallus, and they germinate the following season.

The life cycle of *Ricciocarpus* (FIG. 23-6) illustrates, in its general features, the life cycle of all liverworts and mosses. The spore, the plant body, the sex organs, and the gametes all belong to the gametophyte, or *n*, generation. After fertilization and doubling of the chromosome

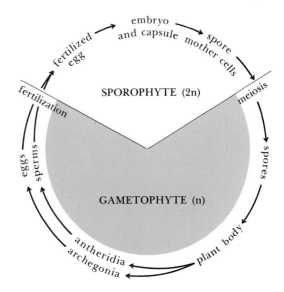

Fig. 23-6 Life cycle of *Ricciocarpus*.

number, a zygote is formed that grows into a capsule containing spore mother cells. These are the structures composing the sporophyte, or *2n*, generation. The gametophyte generation of liverworts and mosses is phototrophic. The embryo, capsule, and spore mother cells of *Ricciocarpus* are reported to contain droplets of a green pigment, but it is likely that little or no food is synthesized. During development, the sporophyte is dependent upon the gametophyte for water and for most, if not all, food materials. In other liverworts and in mosses the sporophyte develops chloroplasts, produces starch, and is probably not so dependent upon the gametophyte.

Marchantia Most of the thallose liverworts are more complicated in structure than *Ricciocarpus*. The thallus (gametophyte) is more specialized, and the female sex organs in many species are raised above the thallus on stalked structures termed RECEPTACLES. In *Marchantia*, an attractive and well-known genus of liverworts, the antheridia also are located on stalked receptacles (FIG. 23-7).

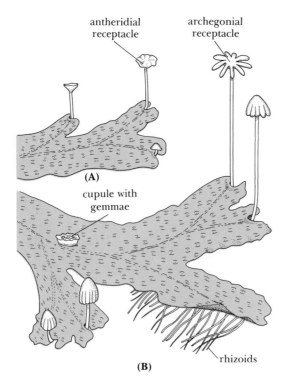

antheridial receptacle

archegonial receptacle

(A)

cupule with gemmae

rhizoids

(B)

Fig. 23-7 *Marchantia polymorpha*. (A): male plant. (B): female plant.

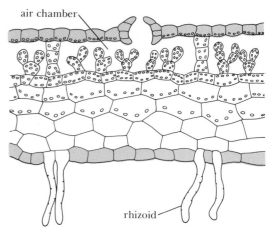

air chamber

rhizoid

Fig. 23-8 *Marchantia*, cross section of thallus.

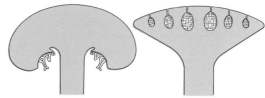

Fig. 23-9 *Marchantia*. Diagram showing position of archegonia (left) and antheridia (right).

Marchantia polymorpha is a widely distributed plant of damp ravines and other moist, shady situations. A number of observations indicate that this plant is likely to invade areas in which other vegetation has recently been destroyed by fire, provided sufficient moisture is present. Under such conditions the plants may flourish as a dense matted growth for several years, gradually being replaced by mosses, grasses, and the seedlings of woody plants. The thallus of *Marchantia* and certain closely related liverworts forms a broad, branching ribbon, which spreads over the ground, anchored by numerous rhizoids. The surface of the thallus is marked by diamond-shaped plates, which indicate the position of internal air chambers. A section through the thallus shows these air chambers occupying

the upper portion of the thallus and covered by an epidermis (FIG. 23-8). Each chamber opens to the outer air by a chimneylike pore, analogous to a stoma. From the bottoms of the air chambers arise chains of cells, each cell containing numerous chloroplasts. The basal portion of the thallus is composed of colorless, closely packed cells that frequently contain starch grains.

Marchantia is heterothallic, and sexual reproduction involves two kinds of plants—male plants, which bear the antheridial receptacles, and female, which bear the archegonial receptacles (FIG. 23-7, Plate 12). The stalks bearing the receptacles are vertical branches of the thallus. The archegonial receptacle is somewhat expanded and umbrellalike, with fingerlike lobes, usually nine in number, around the

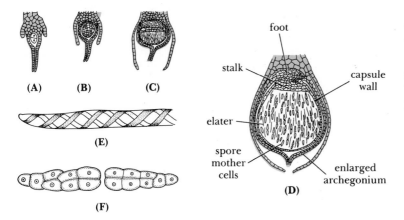

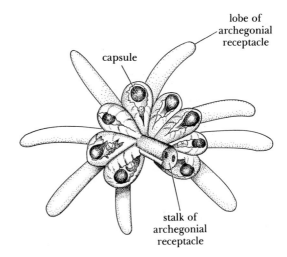

Fig. 23-10 Stages in development of sporophyte of *Marchantia.* (A): first division of zygote. (B): many-celled embryo. (C): older embryo; the foot has begun to penetrate the tissues of the gametophyte, and stalk cells are differentiating. (D): mature sporophyte. (E): portion of elater. (F): group of spore mother cells, enlarged. Gametophyte generation in color.

margin. The archegonia are borne in rows between these lobes, with their necks pointing downward (FIG. 23-9, left). The antheridial receptacles are disklike, with scalloped edges (FIG. 23-9, right). Fertilization takes place prior to the elongation of the receptacular stalks. The antheridia burst at the apex and the sperms are splashed by rain water to the vicinity of the archegonia on neighboring plants. Fertilization takes place as in *Ricciocarpus.*

The sporophyte generation begins with the fertilized egg. After fertilization, the stalks of the receptacles lengthen. The zygote divides repeatedly, forming a multicellular embryo within the base of the archegonium, which enlarges with the growth of the embryo (FIG. 23-10A–C). During the development of the embryo, a cylindrical sheath grows from the base of each archegonium and surrounds the embryo. A sheet of tissue also grows downward on either side of a row of archegonia.

The embryo is at first spherical, but soon a basal portion, the FOOT, grows into the tissues of the receptacle and functions as an absorbing organ. The bulk of the embryo forms a capsule, which is separated from the foot by a zone of undifferentiated cells, the STALK. The capsule contents differentiate into spore mother cells, grouped into vertically arranged rows or clusters, and into ELATERS, slender elongated threads with

spirally thickened inner walls (FIG. 23-10D–F). Following meiosis and the formation of spore tetrads, the stalk elongates, the enlarged archegonium is ruptured, and the capsule is pushed downward (FIG. 23-11). The capsule then dries and opens, setting free a loose, cottony mass of spores, which are disseminated by wind. The escape of the spores is aided by the elaters,

Fig. 23-11 *Marchantia,* view of archegonial receptacle from below. Note the mature capsules.

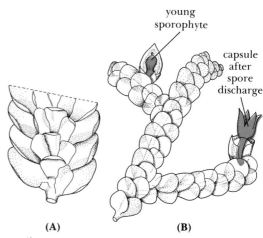

Fig. 23-13   A leafy liverwort (*Porella*). (A): lower surface of axis, bearing leaves. (B): plant with attached sporophyte.

Fig. 23-12   Mosses and liverworts. The compact masses on the branches and, to some extent, on the main trunk of the tree in the foreground consist of mosses and liverworts. The principal genus of liverworts is a leafy liverwort, *Porella*. The long, stringy, pendulous masses are composed of a small club moss, *Selaginella oregana* (Chapter 26). Olympic National Park, Washington.

which are hygroscopic. They coil and twist as they dry, undergoing jerking movements that loosen the spore mass, thus exposing the spores to air currents. All of the sporophytes do not mature, yet careful estimates have shown that about 7 million spores are formed on each archegonial receptacle.

In its earlier stages of development, the sporophyte generation of *Marchantia* is entirely dependent, nutritionally, upon the tissues of the gametophyte. Later, however, the stalk, capsule wall, elaters, and even the foot of the sporophyte become green. The cells of these tissues are abundantly supplied with chloroplasts and possess a considerable capacity for photosynthesis. Food manufactured by the sporophyte thus supplements the supply received from the gametophyte, and so the sporophyte is not entirely dependent upon the gametophyte for food, as it is for water and minerals. At maturity the capsule wall splits and the spores are set free.

In addition to sexual reproduction, many species of liverworts reproduce vegetatively. In *Marchantia* and other thallose liverworts, the older parts gradually die away as the plant grows over the soil. After the death of the thallus at the base of a fork, the two branches become separate plants. Several liverworts, including *Marchantia*, also produce specialized structures, known as GEMMAE, which bring about vegetative reproduction. The gemmae are borne in cuplike structures (CUPULES) on the upper surface of the thallus (FIG. 23-7). In *Marchantia* the cupules are bowl-shaped and the gemmae are minute, lens-shaped bodies attached by a short stalk to the bottom of the cupule. Gemmae are washed free of the thallus by rain water and may be carried some distance from the parent plant. When the gemmae lie flat on the soil, rhizoids develop from the lower surface, and a new thallus is produced.

The life cycle of *Marchantia*, although apparently complex, follows a familiar and relatively simple pattern. The gametophyte generation includes the spore, the thallus, the two types of receptacles, sex organs, and gametes. The sporophyte generation, initiated by the fertilized egg, includes the foot, stalk, capsule, elaters, and spore mother cells. The sporophyte generation, although capable of photosynthesis, receives a portion of its food supply from the gametophyte.

Leafy Liverworts The leafy liverworts, sometimes known as the scale mosses, constitute the largest group of the liverworts. They grow profusely as mats or carpets on rotten logs, on damp soil, and as epiphytes on the trunks and branches of trees (FIG. 23-12). The plant body is typically dorsiventral. In some it is thalluslike, composed of a narrow midrib with winglike extensions on either side. In most, however, the dorsiventral plant body is composed of an axis bearing leaflike expansions (FIG. 23-13A). There is little or no internal differentiation of tissues, and stomata are absent. The "leaves" are borne in three ranks. Two of these are lateral, extending outward on either side of the axis. The lateral leaves are sometimes divided into two parts. The third rank of leaves, absent in some genera, arises from the lower surface. The leafy liverworts are sometimes confused with the true mosses, but they may be distinguished by anyone who will observe their vegetative structures carefully. The mosses are RADIALLY SYMMETRICAL; that is, the leaves are attached all around the stem, in contrast to the liverwort pattern just described. In addition, the moss leaf (with some exceptions, such as the peat mosses and certain others) possesses a midrib, whereas such a structure is absent in the leafy lobes of the liverworts.

The sex organs of the leafy liverworts are borne upon the green or gametophyte plant. The antheridia are borne in the axils of the leaves, whereas the archegonia are terminal in position, at the apex of the main shoot or of its branches. The sporophyte possesses a foot, a stalk, and a capsule, which opens by four valves (FIG. 23-13B). The young sporophyte is usually enclosed within a protective sheath, from which it emerges as it nears maturity. Investigations have shown that the immature sporophytes of certain leafy liverworts are green and contain starch grains. Although they are not nutritionally independent of the gametophyte, they probably do produce a considerable portion of their own food supply.

HORNED LIVERWORTS
(Anthocerotae)

The horned liverworts, or hornworts, are a small group related to other bryophytes but sufficiently different to justify placing them in a separate class. *Anthoceros* is the best-known genus, and its species are relatively common on the edges of streams or lakes or frequently along damp ditches, paths, and roadsides. The dark green thalluslike gametophyte is small, lobed, and approximately circular in outline. There is little internal differentiation of tissues. The cells of the gametophyte generation usually contain only a single large chloroplast, which includes a body termed a pyrenoid, similar in structure and probably in function to the pyrenoids of certain green algae but more complicated in structure. Structures showing similarities to the stomata of the sporophyte (FIG. 23-14C) have been found on both the upper and lower surfaces of the gametophytes of several species of *Anthoceros* and other hornworts. These have been interpreted as modified stomata, no longer functional but homologous with those of the sporophyte. In most species archegonia and antheridia are borne on the same thallus, deeply embedded in

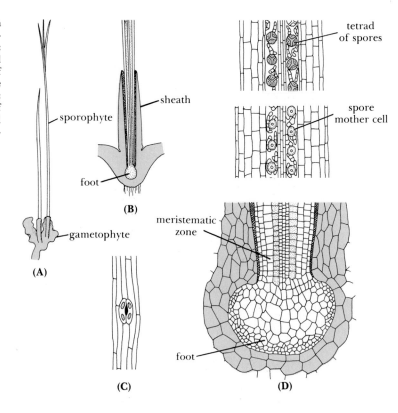

Fig. 23-14 *Anthoceros*, a horned liverwort. (A): gametophyte and sporophyte. (B): diagram of longitudinal section through base of capsule. (C): stoma of capsule as seen in surface view. (D): details of structures at base of capsule as seen in longitudinal section; gametophyte generation in color.

sporophyte

sheath

foot

(B)

gametophyte

(A)

tetrad of spores

spore mother cell

meristematic zone

foot

(C)

(D)

Fig. 23-15 *Anthoceros*, cross section of mature capsule; gametophyte generation in color.

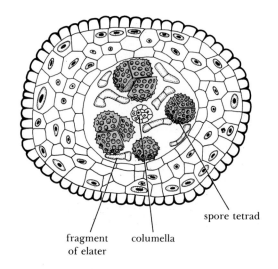

fragment of elater

columella

spore tetrad

its tissues. The sporophyte that develops from the fertilized egg is a cylindrical, spikelike capsule, slightly tapering toward the apex (FIG. 23-14A). It is usually about a centimeter in length, but in one California species it attains a length of 5–6 centimeters—over 2 inches. The base of the capsule is surrounded by a sheath of gametophyte tissue. The capsule base extends downward as a foot, an organ of attachment and absorption, deeply sunken in the tissue of the thallus (FIG. 23-14B).

The structure of the capsule of *Anthoceros* resembles, in certain aspects, the capsule of moss plants, a condition that may be viewed as an example of convergent evolution. A cross section of a mature capsule (FIG. 23-15) reveals a small group of sterile cells, the COLUMELLA, in the center. Surrounding the columella is a hollow cylinder containing elaters and tetrads of spores. The columella and cylinder of spore-producing tissue extend vertically through the length of the capsule. Lying outside these is a zone of sterile cells, and this zone in turn is covered by a well-developed and cutinized epidermis, interrupted by stomata similar to those found in vascular plants (Fig. 23-14C). Each of the cells of the sterile zone contains one or several chloroplasts, the number varying with the species. The presence of the chloroplasts makes the mature sporophyte largely independent of the gametophyte for food, although still dependent upon it for water and minerals. At maturity, the capsule wall splits and the spores are set free.

After an early period of growth the capsule elongates by the activity of a meristematic zone at its base—a developmental feature unique in the bryophytes. This zone gives rise to the cells of all kinds found in the mature capsule—sterile as well as spore-producing tissue (Fig. 23-14D). Thus, while spores ripen and are shed from the upper part of the capsule, new spores are continually produced below. In some species the capsule continues to grow and produce new spores as long as the gametophyte lives.

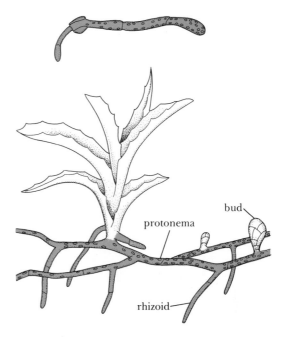

Fig. 23-16 Moss gametophyte. (Top): spore germinating and forming a protonema. (Bottom): protonema, bearing buds, rhizoids, and erect leafy plant.

MOSSES (Musci)

People notice the mosses more commonly than the liverworts because they grow in more accessible places and usually are more conspicuous. The mosses, in contrast to the liverworts, are radially symmetrical; that is, their leaves arise from all sides of a central axis. In many species the stems are erect. Others have prostrate stems that give rise to erect branches. In other species the plant is entirely prostrate and creeping. Such plants may appear to be flattened as a result of the downward bending of leaves and branches.

Although the lateral appendages on the moss stem are called leaves, they are not homologous with the leaves of vascular plants. They are more leaflike than the lobed thallus of the leafy liverworts, and in the true mosses they usually possess a definite midrib.

The mosses include some 14,000 species, most of which belong to one order, the true mosses, *Bryales*. These are the most common mosses. The sphagnum, bog, or peat mosses, *Sphagnales*, form a second order, and the rock mosses, *Andreaeales*, a third. The rock mosses are arctic and alpine plants, dark in color and small in size. The true mosses and the bog mosses will be described in more detail, although all three orders are closely related and the differences among them are not very great.

The True Mosses A consideration of the life cycle of a typical moss plant starts with the spore, the first structure of the gametophyte generation. In a moist environment, a moss spore ger-

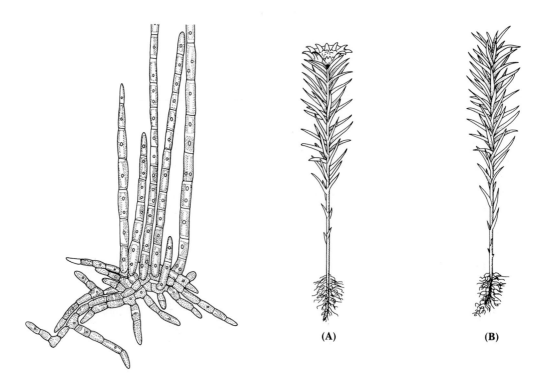

Fig. 23-17 *Stigeoclonium*, a common aquatic green alga, with plant body differentiated into a prostrate and an erect portion.

Fig. 23-18 Male and female plants of the pigeon-wheat moss (*Polytrichum juniperinum*). (A): antheridial plant. (B): archegonial plant.

minates and forms a PROTONEMA, a much-branched, threadlike structure that is divided into cells by a series of cross walls (FIG. 23-16). Branches of the protonema spread over the soil surface. The cells of these branches contain chloroplasts. Rhizoids growing from the protonema penetrate the soil; these are brown or colorless and devoid of chloroplasts. The threadlike branching character of the protonema is very suggestive of certain green algae. It has been suggested, in fact, that the protonema represents a portion of the plant body of an ancient algal ancestral type. Such a form possessed a prostrate system of branching filaments that gave rise to erect branches. The erect portion, according to this hypothesis, was ancestral to the leafy stem of the moss, whereas the prostrate

portion has been retained as the protonema. Green algae with such branching systems are living today (FIG. 23-17).

The leafy moss plants, the conspicuous structures of the gametophyte generation, originate on the protonema as pear-shaped buds. These buds, through the activity of a single apical cell, give rise to the leafy stems. The erect stems (FIG. 23-18), in turn, develop their own rhizoids and become independent of the protonema, which soon disappears. The leafy moss plants are usually only a few inches high, but they vary in size from minute forms just visible to the naked eye to aquatic and tropical species a foot or more in height. The leaves of the mosses are green and sometimes possess specialized photosynthetic tissues. But in most species they are only one cell in thickness, except along the midrib.

The path of water movement in the moss gametophyte has long been a subject of investigation. Physiologically and anatomically, mosses appear to fall into two groups. In one, rhizoids are conspicuous and a well-defined central strand, surrounded by a cortical-like tissue, is found in the axis of the leafy plant. In some species the cells of this strand are thick-walled, but usually they are thin-walled. The strand is not vascular tissue, but nevertheless it has been found, by the use of fluorescent dyes among other techniques, that water absorbed by the rhizoids is translocated upward through the central strand. But the strand in these plants is not the only path of water, for internal conduction is supplemented by water rising from the rhizoids over the external surface of the plant, moving from one leaf to another as thin capillary films. Mosses with a central conducting strand usually grow on soil or humus, not on rocks or bark. In the second group, rhizoids are few or absent and the central strand is absent or poorly developed. The leafy plant can absorb water directly from dew or rain through almost any part of the external surface, and there is no internal conduction.

The sex organs are produced at the tips of the leafy stems. In some mosses both archegonia and antheridia are borne on the same plants, and sometimes on the same apex (homothallic); in others, such as *Polytrichum,* the haircap or pigeon wheat mosses, they are borne on different plants (heterothallic). In the widely distributed cord moss, *Funaria hygrometrica,* a budlike branch, bearing archegonia, arises as a lateral structure some distance below the summit of the antheridial shoot (FIG. 23-19). After fertilization and during the development of the sporophyte the antheridial head withers and is easily overlooked.

The leaves surrounding the sex organs are frequently modified. This is particularly true of the terminal leaves on stems or branches bearing antheridia. Such leaves are commonly short and

may be pale pink or rose. They form a cluster or rosette, superficially resembling a miniature flower (FIGS. 23-18, 23-19). The antheridia, usually stalked and somewhat club-shaped, consist chiefly of a jacket of cells, with one or several cap cells at the apex, surrounding a mass of sperm-producing tissue (FIG. 23-20). The archegonia are also stalked and greatly elongated (FIG. 23-21). The wall at the base is several cells thick, and the neck contains a considerable

Fig. 23-19 The cord moss, *Funaria hygrometrica.* (A): antheridial and archegonial heads on the same plant but different branches. (B): the mature sporophyte develops after fertilization.

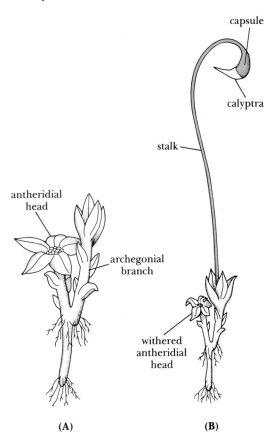

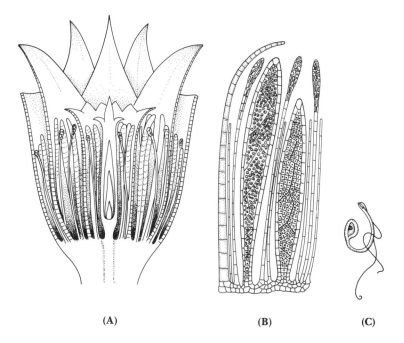

Fig. 23-20 The moss antheridium (from *Polytrichum*). (A): section through antheridial head; club-shaped hairs are scattered among the antheridia. (B): two antheridia, enlarged. (C): sperm.

(A)   (B)   (C)

number of neck canal cells, which disintegrate as the archegonium matures. Scattered among the sex organs are modified hairlike structures, which contain chloroplasts. These hairs may aid in the retention of water around the archegonia and antheridia.

At the time of fertilization, the antheridium absorbs water and swells. The cap cells are forced off, and the sperms emerge. Water is essential for fertilization. In the larger mosses, the sperms are splashed by raindrops from antheridial to archegonial heads. In smaller species, the sperms may spread from the antheridia to the archegonia through a film of water that often covers the tops of the plants. The sperms swim to the apex of the archegonium and down the narrow passage that leads to the egg. One sperm fuses with the egg cell, fertilizing it. Fertilization may occur in more than one archegonium, but usually only one sporophyte develops.

The sporophyte generation begins with the fertilized egg, which develops rapidly into an elongated spindle-shaped embryo. From the embryo grows an elongated stalk, or SETA, with a capsule at the upper end (FIG. 23-22A). At the lower end of the stalk is the foot (FIG. 23-22B), buried in the tissues of the leafy gametophyte plant. The production of the foot, stalk, and capsule is the result of the activities of two apical cells, one located at each end of the spindle-shaped embryo. The capsule, when mature, may be erect, pendulous, or bend at right angles to the stalk. Characteristics of the capsule are important in the identification of the various species of mosses. As the embryo increases in size, the lower part of the archegonium enlarges and elongates and is converted into a CALYPTRA, or cap. As the sporophyte lengthens, the calyptra is ruptured at the base and borne upward, remaining for a period of time (which varies with the species) as a thin brownish or greenish cap or hood covering the capsule (FIGS. 23-19B, 23-22).

The spores are produced within the capsule in a cylindrical zone that surrounds a column of sterile tissue, the columella (Fig. 23-23). In some species, the spore-producing cylinder extends the length of the capsule. In others, spore production is limited to the upper portion of the capsule. All the cells of the spore-bearing area develop into spore mother cells, which, after meiosis, give rise to spores. Internally, the stalk of the capsule develops differentiated tissues similar to those of the leafy gametophyte plant.

The embryo develops chloroplasts while very young and still enclosed within the archegonium. As the sporophyte ripens, chloroplasts become more abundant throughout the stalk and outer tissues of the capsule. Such chloroplasts usually contain starch grains. The manufacture of carbohydrates is facilitated, in many genera, by the presence of stomata in the epidermal tissues of the capsule. When the sporophyte is fully ripe, the chlorophyll fades and the capsule turns a pale or rich brown. The photosynthetic capacity of the sporophyte generation is undoubtedly sufficient to permit the stalk and capsule to be largely independent of the gametophyte generation for food materials, but the sporophyte is still dependent upon the leafy plant for some food and for water and essential mineral elements.

Fig. 23-21 Moss archegonia (from *Polytrichum*). (Left): section through archegonial head. (Right): single archegonium, enlarged.

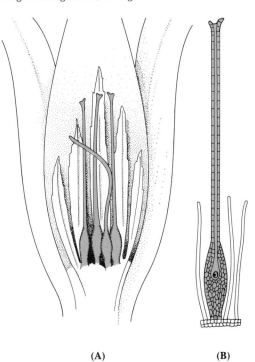

(A)   (B)

Fig. 23-22 Moss plant, sporophyte and gametophyte (*Polytrichum*). (A): sporophyte (foot, stalk, and capsule) attached to gametophyte. (B): lower part of stalk, showing the foot. (C): calyptra. Gametophyte generation in color.

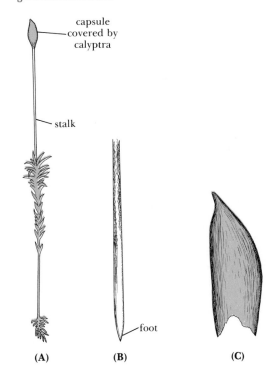

capsule covered by calyptra

stalk

foot

(A)   (B)   (C)

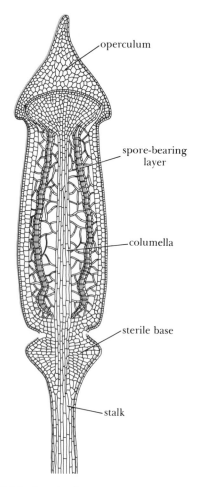

operculum

spore-bearing
layer

columella

sterile base

stalk

**Fig. 23-23 Longitudinal section of a moss capsule
(*Polytrichum*).**

As the capsule matures (Fig. 23-24) and the spores within it ripen, a lid, or OPERCULUM, becomes evident at its apex. Just within the mouth of the capsule, and under the operculum, one or several rows of teeth also develop. After the capsule dries, the calyptra and the operculum fall away, the spores are released and are carried away by currents of air. The release of the spores is facilitated by the circles of teeth around the mouth of the capsule. These teeth are hygroscopic and control the release of the spores. They bend outward when the air is dry, and the spores are released by wind shaking the capsule. When the air is humid, the teeth return to their former position, preventing the escape of the spores.

In summary, the gametophyte, or *n*, generation begins with the formation of spores. Each spore germinates, producing a filamentous protonema composed of two parts: threadlike branches that lie flat upon the soil and contain chloroplasts, and nongreen rhizoids that penetrate the soil. Specialized branches of the aerial portion of the protonema grow into leafy stems with radial symmetry. The sex organs are borne at the apex of the erect stems. The gametes, eggs and sperms, are the last cells of the gametophyte generation. From the fertilized egg—which is the first cell of the sporophyte, or 2*n*, generation—a foot, stalk, and capsule develop. The spore mother cells, produced within the capsule, are the last cells of the sporophyte generation. The calyptra, which covers the ripe capsule for a time, belongs to the gametophyte generation, since it has developed from the archegonium. The complete life cycle of a moss is shown in Fig. 23-25.

The Peat, or Bog, Mosses These mosses are represented by a single genus, *Sphagnum,* which contains a large number of species (about 350). Peat mosses are characteristically plants of wet, boggy areas with an acid reaction. In such places the plants grow in thick, soft masses, sometimes covering great areas. In both the true and the peat mosses, the stems branch extensively as they grow. The older parts die and partly decay, and the terminal parts grow on from year to year. Sphagnum is worldwide in distribution, but it is most abundant in the colder temperate zones where wet soils occur together with high relative humidity and comparatively low temperatures. Sphagnum is usually an important plant of muskeg, moor, bog, and tundra. Either alone or

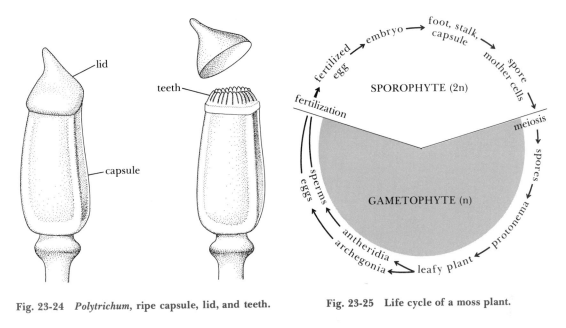

Fig. 23-24 *Polytrichum*, ripe capsule, lid, and teeth.

Fig. 23-25 Life cycle of a moss plant.

combined with the partly decayed remains of small vascular plants, sphagnum forms the organic material known as peat, which may accumulate in beds of considerable thickness.

The leaves of sphagnum (FIG. 23-26) are minute and crowded. They are one cell thick and lack a midrib. The cells of the leaf are of two types, large and small, alternating with each other, forming a netlike pattern. The smaller cells contain chloroplasts, and so do the larger cells when young. But as the leaf matures, the chlorophyll and protoplasm disappear from the larger cells. The cell walls develop spiral or ringlike thickenings, and circular openings appear on either the upper or lower surface. These large empty cells, because of their structure, are able to absorb large quantities of water by capillarity. They form a natural cellulose sponge.

Because of this property, sphagnum or peat moss has been utilized in many ways. When dried

and sterilized, it has been employed as a substitute for gauze in surgical dressings. It is added to heavy or sandy soils as a mulch to improve their texture. It is often used as a medium for starting seeds and rooting cuttings. Great quantities of peat moss are used by nurserymen as packing, in transplanting, and in other operations with plants.

The life cycle of sphagnum is like that of the true mosses, with some variations in detail. The spore, upon germination, produces a protonema that is broad and flat rather than threadlike. The leafy plants arise from this protonema and are without rhizoids. The sex organs are borne on side branches just below the apex of the main axis. From the fertilized egg a globular capsule (FIG. 23-27) develops, attached by a foot to the tissues of the gametophyte. When young, the capsule contains chloroplasts and is only partly dependent upon the gametophyte for food. As maturity approaches, the chlorophyll disappears

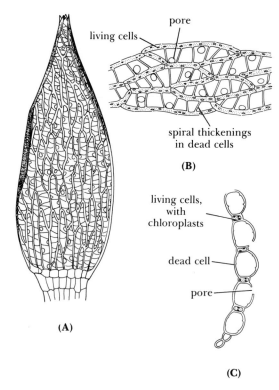

pore

living cells

spiral thickenings
in dead cells

(B)

living cells,
with
chloroplasts

dead cell

pore

(C)

(A)

Fig. 23-26 **Structure of leaf of** *Sphagnum*. **(A): entire
leaf, surface view. (B): portion of (A), much enlarged.
(C): cross section of portion of leaf.**

and the capsule turns from green to brown or
black. A stalk is present, but does not elongate.
When the capsule is mature, the upper portion
of the archegonial branch elongates and carries
the capsule upward. Internally, the capsule consists of a dome-shaped mass of spores that
overlies a massive columella.

Vegetative Reproduction Many mosses, perhaps all, are able to spread by vegetative
reproduction as well as by spores. By this method
individual plants are enabled to form dense

masses of plants extending over considerable
areas. There are two general methods of vegetative reproduction:

1. By branching of the leafy stems. These may
branch by the formation and development of
buds at the base of the stem. Ultimately these
branches may be detached by the decay of the
basal parts and become independent plants. The
branches may also take the form of stolons,
leafless or with only small leaves, which creep on
or beneath the soil and grow upward as leafy
stems.

2. By protonemata. The original protonema
may continue to grow for some time, producing
several to many leafy stems. More important is
the production of secondary protonemata,
which, like the original, give rise to leafy plants.
The secondary protonemata are produced most
abundantly as branches of the rhizoids at the

Fig. 23-27 *Sphagnum*. **(A): leafy plant bearing capsules. (B): longitudinal section of mature capsule.**

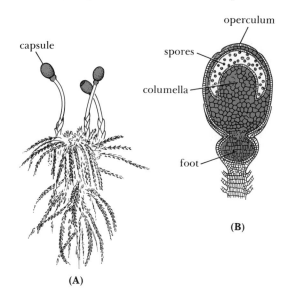

operculum

capsule

spores

columella

foot

(B)

(A)

base of the leafy stems, but they may also grow directly from the stems and even from leaves.

EXPERIMENTAL STUDIES ON BRYOPHYTES

Various bryophytes have long been the subject of cytological, genetic, and physiological investigations. Many bryophytes are easily grown on artificial media, and numerous investigations have been made on the roles of external factors (such as the intensity and quality of light) and internal factors (such as growth substances) on growth and differentiation at various stages of the life cycle. Prominent among the results of such studies is the finding that certain bryophytes exhibit clearly defined photoperiodism—in some the structures associated with sexual reproduction are initiated under long days and in others under short days. A number of thallose liverworts, such as *Marchantia polymorpha* and *Conocephalum conicum,* and certain leafy liverworts, fall in the long-day category. Thus, it is possible to obtain reproductive material of *Marchantia polymorpha,* widely used for instructional purposes, during the winter months by adding artificial light to increase the daylength. In contrast to *Marchantia* and *Conocephalum,* species of *Riccia, Anthoceros,* and *Sphagnum* have been found to be short-day plants.

EVOLUTION AND RELATIONSHIPS OF THE BRYOPHYTES

The oldest presumed fossil remains of bryophytes are found in the Upper Devonian, and consist of scanty remains of thallose liverworts much like those living today. Although the bryophytes are thus very ancient, the fossil record sheds no light on their evolutionary history, and it is therefore to living forms that we must look in any attempt to understand their early evolution. The following broad outline presents a modern concept of some of the events in the evolution of the bryophytes.

The liverworts, hornworts, and mosses represent several and divergent evolutionary lines. Each of these may have evolved independently from different, although probably similar, aquatic green algal ancestors with no potentialities for the development of vascular tissues in either generation. In all these evolutionary lines, during the earlier stages of the transition from the water to the land, both the gametophyte and sporophyte generations were probably independent, green, and with erect shoots and radial symmetry, as in the ancestors of the vascular plants. The two generations were probably much alike except that one bore spores and the other, isogametes. At a later stage of the transition the sporophyte became attached to, and partly parasitic upon, the gametophyte generation, and both generations tended to become reduced. The chlorophyll and plastids in the sporophyte generation of mosses, liverworts, and hornworts, and the presence of stomata in the capsule of mosses and *Anthoceros,* are thus interpreted as ancestral features, retentions from the time when the sporophyte was an independent plant. The sporophyte of *Anthoceros,* because of its resemblance to a postulated ancient leafless axis, its complex internal structure—including a basal, persistent meristem—and its relatively long life, is regarded as the least reduced of any bryophyte. Stages in the reduction of the sporophyte may be found among the liverworts, from the condition in the leafy forms where a foot, stalk, and capsule are present to that in *Ricciocarpus* in which only a spherical capsule is found.

The plant body, the chief part of the gametophyte generation, did not attain large size. This could be attributed to several factors, among them the lack of a vascular system and the survival value of low stature and consequent greater likelihood of fertilization. In some evolutionary lines, the leafless axes (probably now much reduced in size) produced outgrowths, the "leaves." Those forms that retained a radial symmetry gave rise to the mosses of today. Other leafy axes, the forerunners of the leafy liverworts, developed dorsiventral symmetry. The ancestral gametophytes of *Anthoceros* and the thallose liverworts may never have developed leaves—the modern gametophyte could be regarded as a flattened and prostrate axis. As previously noted, modified, vestigial stomata have been reported on both surfaces of the gametophyte of species of *Anthoceros* and other hornworts. All living bryophytes, then, are thought to be reduced in one or more respects and to represent the culmination of various evolutionary lines that gave rise to no higher forms of land plants. They became stranded in a backwater of the main evolutionary stream.

The preceding outline is admittedly speculative, but it does conform to the concept, developed in Chapter 18, that evolutionary change is not always associated with increase in complexity but rather is frequently accompanied by reduction.

ECOLOGY OF THE BRYOPHYTES

The liverworts and mosses grow on soil, on damp sand, on rocks, and on the trunks and limbs of standing and prostrate trees. They also grow in water and among other plants of fields and meadows, bogs, and marshes. Some species flourish in a variety of habitats; others are restricted to specific, limited environments. Certain mosses, for example, are exacting with respect to the soil reaction and are confined either to acid or to alkaline soils.

In general, mosses are more tolerant of shade than are higher forms of plant life, and this in part accounts for their ability to invade lawns and replace grass in shady spots. The mosses include many xerophytic forms as well as mesophytes and hydrophytes. Bryophytes are likewise adapted to great extremes of temperature, for they range from the arctic zone to the tropics and grow in the vicinity of hot springs. They reach their greatest development in cool, moist forests, such as those of the west coast of North America and the mountains of the tropics.

The bryophytes are an important component of the flora of the earth and play a significant role in the economy of nature. This is partly the result of the great number of individual plants produced by vegetative propagation. Mosses are so prolific that they form great masses or carpets covering the soil. Another characteristic of ecological importance is the ability of mosses to hold water, which is trapped among the leaves and stems. Many woodland mosses share with the sphagnums the ability to absorb water through their leaves. By their structures and their mode of life, mosses contribute in many ways to the modification of their own environment.

The significance of the mosses as soil formers following lichens or other lower forms of plant life on bare rock surfaces has already been mentioned (Chapters 15 and 21). The retention of water by masses of leafy liverworts and mosses growing on fallen trees and other organic material hastens the processes of decomposition and hence the organic enrichment of the soil. Absorbing but little water from the substratum, they do not dry out the soil but protect it from desiccation. As a result of their ability to retain water, natural beds of mosses undoubtedly act as seed beds for herbaceous and woody flowering plants and for conifers.

One of the roles of the bryophytes is in the retardation of erosion. Carpets or feltlike masses of moss plants possess a greater water-retaining power than do layers of dead leaves. They therefore slow down the rapid runoff of rain water and melted snow. In addition to this, dense stands of moss collect and hold particles of soil. Insignificant as the individual plants of this group may appear, they play a part, together with other and more advanced forms of plant life, in making and changing man's environment.

SUMMARY

1. The bryophytes, one of the divisions of the plant kingdom, include the liverworts, hornworts, and mosses. These are probably very ancient groups, representing non-vascular branches of the early land plants that evolved along evolutionary bypaths.

2. Alternation of generations in the bryophytes is clearly defined. The gametophyte, or *n,* generation begins with the spore. This gives rise to a plant that ranges in complexity from the thalluslike body of *Ricciocarpus* to the protonema and leafy shoots of the mosses. The sporophyte, or *2n,* generation begins with the fertilized egg. The sporophyte may be simple, consisting only of a capsule and spore mother cells, as in *Ricciocarpus,* or relatively large and complex, as in the foot, stalk, and capsule of the mosses.

3. The plant body, the chief structure of the gametophyte generation, is phototrophic. The sporophyte generation is dependent upon the gametophyte generation for water and essential mineral elements. In most bryophytes, the sporophyte is able to manufacture a considerable part of its own food supply.

4. The bryophytes are advanced in their methods of reproduction as compared to the algae. These advances include the development of multicellular sex organs—archegonia and antheridia. After fertilization, the developing embryo within the archegonium is nourished by the tissues of the gametophyte.

5. The plant body of the thallose liverworts, of *Anthoceros,* and of most leafy liverworts, is dorsiventrally symmetrical. The gametophyte of the mosses is radially symmetrical. The sex organs of some bryophytes are produced on different plants.

6. The male gametes in all bryophytes are motile, and fertilization occurs only in the presence of water.

7. Among the factors responsible for the low stature of the gametophyte of bryophytes is a lack of a vascular system and the survival value of low height in relation to fertilization in the presence of free water.

8. The leafy liverworts and mosses, because they retain water, are significant in decay processes and in retarding erosion. The sphagnums are used in growing and propagating higher plants.

9. In addition to sexual reproduction, the bryophytes reproduce vegetatively by several methods. It is vegetative reproduction that accounts, in large degree, for the occurrence of liverworts and mosses in large masses.

The Ferns

AMONG THE EARLY LAND PLANTS, two lines of descent are evident. One of these gave rise to the nonvascular liverworts and mosses. The other lineage developed a vascular system and became the progenitor of the great assemblage of vascular plants (Tracheophyta), including the ferns, club mosses, horsetails, conifers, and flowering plants, either living or known only as fossils. In these groups is the majority of plants that we know today.

The living ferns include some 15,000 or more species. They have been classified, together with several groups of extinct ferns and fernlike plants, in the subdivision Pteropsida (Greek *pteris*, fern; *opsis*, appearance) of the vascular plants (Chapter 18). Other spore-bearing vascular plants are dealt with in Chapters 25 and 26. Many plants of these subdivisions (Psilopsida, Lycopsida, and Sphenopsida, Chapter 18) are extinct, and the living forms are unfamiliar to most people. These unfamiliar plants can be better understood through the study of commonly known plants with a similar life history. This chapter is devoted, therefore, to an account of the ferns, with particular reference to their reproductive structures. While the ferns exhibit great diversity, it is possible to gain an understanding of most characteristics of the group by studying species of the polypody family of the order Filicales (Chapter 18). The discussions in this chapter of form, structure, and life cycles apply in particular to that group.

SOME GENERAL FEATURES

Age, Distribution, and Habitat Ferns and fernlike plants are among the oldest groups of vascular plants. Fossils of early ferns are found in rocks of the Middle Devonian, about 390 million years old. The remains of fernlike plants are so abundant in the Carboniferous, some 345 million years old, that this period has been termed the "Age of Ferns." However, many of the fernlike plants of the Carboniferous were not true ferns but seed ferns (Chapter 27). Most Devonian and Carboniferous ferns differed from those of today, although in their methods of reproduction they were undoubtedly similar to living species.

Ferns are widely distributed, from the moist tropics to beyond the Arctic Circle. The greatest number of species is found in the moister parts of the tropics, especially at higher elevations. But ferns flourish also in temperate regions, where they are familiar in woodlands and meadows, along roadsides and stream banks. They range in size from tiny, floating water ferns to the tree ferns of the tropics, which attain heights of 20–60 feet and bear crowns of huge leaves often 6–12 feet long (FIG. 24-21). Ferns have assumed many of the same growth forms as flowering plants. They are generally found in moist and shaded habitats, but some are able to grow in strong sunlight and even in markedly xerophytic environments. Adaptations to dry environments may be evident in stiff leathery leaves, sometimes with a coating of wax or overlapping scales. Most ferns of the temperate zone are terrestrial, growing upon soil or rocks. But some ferns are climbers, and many—largely confined to the tropics—are epiphytes, living at various levels from just above the ground to the canopy of the forest. A remarkable adaptation to this mode of life is found in the bracket epiphytes (*Drynaria*) in the tropics and subtropics of the Old World (Africa, Asia to Australia). These are low-level epiphytes, growing on tree trunks. In addition to large green leaves (fronds), specialized, short, sessile bracket leaves develop on the surface of the trunk. These grow over the rhizome, collect humus, and protect the roots from desiccation (FIG. 24-1). The bracket leaves are persistent, and the strong veins that remain, even after the mesophyll decays, hold the humus as in a woven basket.

Because of the diversity of their habitats and

Fig. 24-1 Bracket epiphyte (*Drynaria sparsisora*), Australia. Note the broad expanded leaves (fronds) and the bracket leaves against the tree trunk.

great numbers of species and individuals, ferns should be viewed as having considerable ecological importance. Among other roles, they help hold and form soil, they may constitute stages in succession, and they participate in overall cycles of decay. Some ferns occur in such numbers as to exclude other kinds of vegetation. An example is the bracken fern (*Pteridium aquilinum*), which in many parts of the world takes over pastures and abandoned agricultural land in large areas.

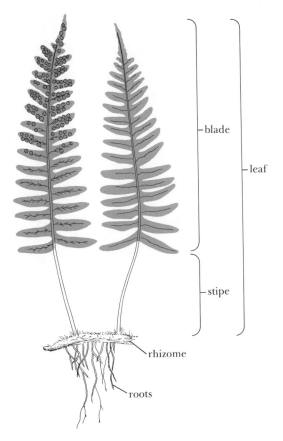

Fig. 24-2 Plant body of a fern (common polypody, *Polypodium virginianum*)**. (Left): lower surface of frond. (Right): upper surface.**

External Form Like other vascular plants, the ferns have roots, stems, and leaves. The stems live from year to year, and ferns are therefore perennial. In temperate climates, the leaves usually die in the autumn, although certain species are evergreen. While the stems of some ferns are upright, even becoming treelike in form, in most they are rhizomes, sometimes creeping over the surfaces of rocks or tree trunks, but usually buried in the soil. The rhizomes are slender and elongated, with scat-

tered leaves (Fig. 24-2), or short and compact, with clustered leaves. The parenchyma cells of the rhizomes are usually filled with starch, and the rhizome is thus an efficient storage organ. The root system consists of numerous wiry or hairlike adventitious roots arising from the stem.

Many ferns reproduce by vegetative means. The creeping rhizomes branch as they grow through the soil or leaf mold. If the rhizomes die at the base, the branches persist as independent plants. This is usually the way in which colonies, composed of numerous individuals, are formed. Some ferns increase in number so rapidly by vegetative propagation that they are classed as weeds. The manner of growth of several species of ferns, in which the rhizome is short and robust and the leaves are in compact clusters, has attracted considerable interest. The branching rhizomes gradually radiate outward from a common center and, as the older, central parts die and decay, a ringlike group of plants is formed. Similar rings, formed by gilled fungi, are termed "fairy rings," and the same term may be applied to the ferns. One of these rings, formed by the cinnamon fern, is recorded as having a diameter of more than 12 feet and an estimated age of more than 300 years.

Fern leaves, commonly called FRONDS, consist of two parts, the stalk, or STIPE, and the BLADE (Fig. 24-2). The blade may be simple (undivided), but more typically it is pinnately compound. Each of the larger divisions of the blade of such a compound leaf is termed a PINNA (Fig. 24-3), and each pinna, in turn, is often divided or dissected in various ways. The way in which the fern leaf grows is one of its most striking features. The leaf of a flowering plant begins its growth with a short initial phase, characterized by the activity of an apical meristem. This is followed by generalized growth throughout the leaf. In the fern leaf, all the tissues are formed, as are those of a stem, by long-continued apical growth. The tip of the growing fern frond is tightly coiled like a watch spring. As growth proceeds, the more rapid elongation of the cells of the inner side of the leaf causes the tip to

Fig. 24-3 Portion of a fern frond divided into pinnae, which in turn are dissected.

Fig. 24-4 Manner of growth of fern fronds. The "fiddleheads" of the interrupted fern (*Osmunda claytoniana*).

uncurl slowly. The coiled, immature fern frond is popularly known as a "fiddlehead" (FIG. 24-4). Fiddleheads of larger ferns are among the attractive spring sights in woodlands and swamps. This unfolding of the leaf is one of the features that distinguishes most ferns from other kinds of vascular plants.

Internal Structure   The structure of the blade of the fern leaf, as seen in cross section, is very similar to that of the leaf of a flowering plant. But since most ferns grow in the shade, the leaf is usually thin and contains large intercellular spaces. The distinction between palisade and spongy tissue, so typical of many flowering plants, is not always evident.

The vascular system of ferns varies widely in different species, but includes most of the cells and tissues found in the stems and roots of seed plants. In some ferns the vascular tissues of the stem are arranged in the form of a solid rod of xylem surrounded by phloem. In others, the conducting tissues are arranged in a circle of separate vascular bundles (FIG. 24-5, above). Less common, and not typical of a fern stem, is the condition found in the bracken fern, in which several vascular bundles traverse the pith (FIG. 24-5, below). A cambium is usually absent in the ferns and all of the tissues are primary in origin. Numerous thick-walled sclerenchyma cells, located just under the epidermis or associated more or less closely with the vascular tissues, are found in the stem of many ferns. These cells contribute to the rigidity of the stem and compensate in part for the lack of secondary tissues.

As in the seed plants, leaf traces of vascular tissues depart from the stem into the bases of the leaves. A leaf gap is present with the leaf traces.

Fig. 24-5 Cross sections of fern stems (rhizomes). (Top): common polypody (*Polypodium virginianum*). (Bottom): brake fern (*Pteridium aquilinum*).

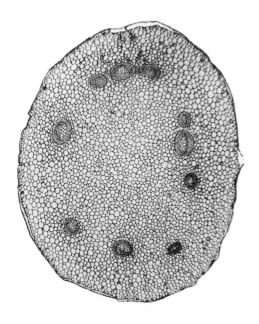

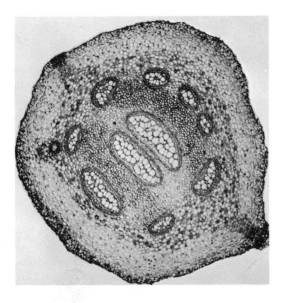

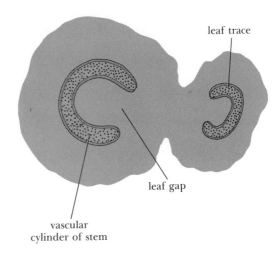

Fig. 24-6 Leaf trace and leaf gap in cross section of rhizome of maidenhair fern (*Adiantum pedatum*).

A simple form of leaf trace is seen in the maidenhair and hay-scented ferns and consists, as seen in cross section, of a horseshoe-shaped band of conducting tissue (FIG. 24-6). In other species, two or even many leaf traces connect the vascular tissues of the stem with those of the leaf.

THE LIFE CYCLE OF FERNS

The life cycle of a fern, like that of other sexually reproducing plants, comprises two alternating generations. The large and leafy fern plant, the conspicuous part of the sporophyte generation, bears spores, which fall upon the soil. The germination of a spore results in a plant unlike the sporophyte—a small, green, flat, heart-shaped body, the PROTHALLUS. This prothallus, bearing the sex organs and gametes, is the principal structure of the gametophyte generation. After fertilization, the egg gives rise to a new fern plant, which grows and reaches spore-bearing maturity.

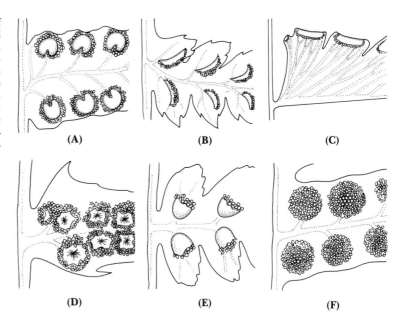

Fig. 24-7 Sori and indusia of ferns. (A): marginal shield fern (*Dryopteris marginalis*). (B): lady fern (*Athyrium filix-femina*). (C): maidenhair fern (*Adiantum pedatum*); false indusium. (D): Christmas fern (*Polystichum acrostichoides*). (E): bulblet fern (*Cystopteris bulbifera*). (F): common polypody (*Polypodium virginianum*); indusium absent.

(A) (B) (C)

(D) (E) (F)

The life history of the ferns was a mystery until it was explained by the German botanist Wilhelm Hofmeister in the middle of the nineteenth century. Even the spores were not discovered until 1669, and their role in the life cycle, after they were observed, was not understood. During the Middle Ages and later, ferns were the object of numerous superstitions. For example, it was widely believed that if the "seeds" (spores) of certain kinds of ferns were gathered on St. John's Eve (June 23) and placed in the shoes, they would cause the wearer to be invisible. This belief is reflected by Shakespeare in *King Henry IV:* "We have the receipt of fern-seed, we walk invisible."

Sori and Indusia If the sporophyte is observed closely, specialized spore-producing organs may be seen. The spores of a fern are always produced within multicellular containers, the SPORANGIA. The sporangia usually develop in clusters. In most ferns they are borne on the lower surface of the leaf. An examination of the back of the leaf with the naked eye or the hand lens will reveal small brown or black clusters composed of numerous sporangia. Each of these clusters is termed a SORUS (FIG. 24-7), or, more popularly, a "fruit dot." Many people believe that the sori are some form of insect life, harmful to the plant. They are, of course, a part of the normal life cycle. The sori occur in various arrangements on the leaf surface. They may be found near the margin, scattered over the surface, or in two rows, one on either side of the midrib of the pinna or, if the pinna is subdivided, on either side of each subdivision.

Associated with the sori of many ferns are thin, membranous structures, known as INDUSIA (FIG. 24-7), produced as outgrowths of the leaf surface. The form of the indusium varies with different species and genera and is a feature of great value in identification. It may be circular or kidney-shaped and attached to the leaf by a central stalk. It may be elongated and attached along one side, forming a sort of hood. The indusium commonly covers the entire sorus, at least when young, and is assumed to be protective in function. In some species a structure called a FALSE indusium is formed by the inrolling of the leaf margin.

Dimorphism In many ferns, leaves of two kinds may be recognized. One kind is the sporophyll, referred to as the FERTILE leaf, which bears sporangia. In contrast, the other kind is the STERILE leaf, which does not bear sporangia and which is entirely vegetative in function. This condition, in which two kinds of leaves, fertile and sterile, are borne on the same plant, is termed DIMORPHISM. In some ferns there is little difference in form between the fertile and the sterile fronds. Both types are photosynthetic; the fertile leaf, in addition, bears sporangia (FIG. 24-8). In other species, termed strongly dimorphic, the fertile leaf has wholly lost the capacity for carrying on photosynthesis and has become so greatly modified that it has lost much of its resemblance to a frond (FIG. 24-18). In still other species, as in the interrupted fern, only certain pinnae have become differentiated into spore production.

The Sporangium Individual sporangia are so minute that they must be studied under the microscope if we are to see the details of their structure. The sporangium (FIG. 24-9) consists of a hollow, lens-shaped case on a short stalk. Flattened cells with thin walls form the sides and enclose a number of spores, commonly 64. Connecting the sides and arching over about two thirds of the sporangium is a row of cells with peculiarly thickened walls. This row of cells, termed the ANNULUS, performs an essential function in the discharge of the spores. On one side the annulus is replaced by thinner-walled cells extending to the stalk. Two of these cells are termed LIP cells.

The mechanism involved in the discharge and dissemination of fern spores is one of the more remarkable in the plant kingdom. The sporangium dehisces (breaks open) and the spores are violently discharged, falling near the parent plant or carried away by air currents. Two stages are involved in the ejection of the spores. In the first (FIG. 24-9A,B), the sporangium opens slowly, carrying the majority of the spores with it. In the second stage, the annulus, acting like a spring, suddenly snaps forward again, and the spores are thrown into the air (FIG. 24-9C).

The opening of the sporangium is brought about by the annulus. The outer and side walls of the annulus cells are thin and flexible, but the inner tangential and radial walls are greatly thickened. When the sporangium is mature, each cell of the annulus is full of water. If the air is dry, this water tends to evaporate through the

Fig. 24-8 Dimorphism in the crested wood fern (*Dryopteris cristata*). (Left): fertile frond. (Right): sterile frond. Both kinds of fronds are photosynthetic, but only the fertile frond bears sporangia.

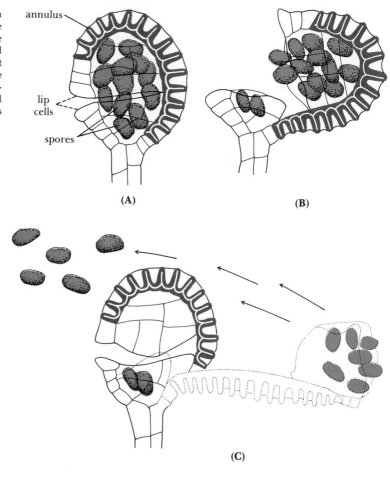

Fig. 24-9 The fern sporangium, stages in spore discharge. (A, B): the top of the sporangium is bent backward by the annulus, carrying most of the spores with it. (C): the top of the sporangium suddenly returns to its original position, throwing the spores into the air.

annulus

lip cells

spores

(A)

(B)

(C)

thin outer and side walls. This evaporation loss cannot be replaced, and the volume of water in the cells of the annulus decreases.

Two forces are operating within the cells of the annulus. These are the cohesion of the water molecules and the adhesion of the water to the cell walls. The cohesive forces prevent the rupture of the water and the consequent formation of vapor bubbles. The adhesion of the water to the walls causes them to be pulled inward. The thin outer wall of the annulus is drawn in, thus pulling the thick radial walls of the annulus backward. As a result of this stress, the

sporangium splits open between the lip cells. The lateral walls of the sporangium are torn, and the annulus slowly straightens out and bends backward. As it does so, it carries the bulk of the spores with it in a cup formed by the side walls of the sporangium.

As water continues to evaporate from the cells of the annulus, a point is finally reached at which the water within the cells no longer resists the increasing pull exerted by the thick, elastic radial walls. The water breaks in one cell, and the resulting shock causes a bubble of water vapor to form suddenly in each cell of the annulus. This

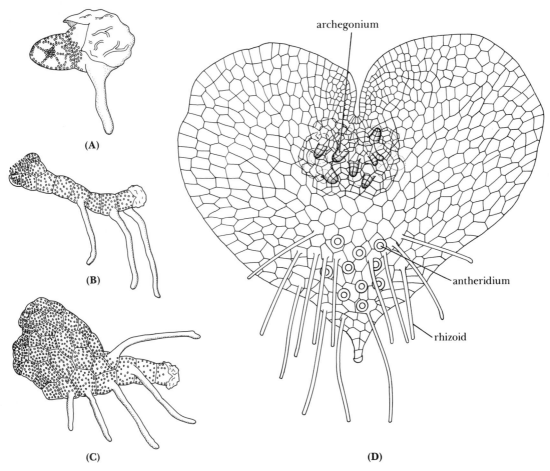

Fig. 24-10 The prothallus. (A): spore germination. (B, C): stages in the development of the prothallus. (D): mature prothallus, lower surface.

releases the tension on the elastic radial walls, which then spring back into their first position. As a result, the top of the sporangium recoils with a speed that the eye cannot follow, and the spores are catapulted an inch or two into the air, where they may be carried away by air currents.

The Prothallus Under suitable conditions the spore absorbs water. Its wall breaks and the first cell of a new plant body, the prothallus, appears. Cell divisions shortly result in a filament that becomes attached to the soil by the formation of several colorless rhizoids (FIG. 24-10A, B). With the exception of the rhizoids, the cells of the prothallus are green, and nutrition is photo-trophic. Further cell division tends to be localized at the apex of the filament (FIG. 24-10C). After some weeks of growth, a flat, green, commonly heart-shaped body, perhaps ¼ inch in diameter, is formed. This is the mature prothallus, or gametophyte (FIG. 24-10D), which grows closely appressed to·the soil (FIG. 24-11). For the most part, the prothallus is only one cell thick, but the central part, just below the SINUS, or indentation, is several cells thick, and forms a cushion. The prothallus bears the sex organs and rhizoids on the lower surface (FIG. 24-10D), adjacent to the soil. Prothalli are usually found in damp, shaded spots, for they are easily killed by drying.

The Sex Organs The sex organs, archegonia and antheridia, are multicellular (FIG. 24-12). Each archegonium contains a single egg. A number of sperms are formed in each antheridium. In most ferns, both antheridia and archegonia are produced on the same prothallus. If the prothalli grow too closely together, only antheridia may develop. Under any conditions the antheridia are usually the first to appear and are most numerous on the basal region of the prothallus, scattered among the rhizoids. They are dome-shaped and project from the surface.

The mature antheridium (FIG. 24-12A) consists of a jacket of cells surrounding a cavity that contains the sperms. The sperms are spirally coiled and bear numerous flagella (FIG. 24-12B). The antheridial jacket is only one cell thick and is composed of three cells. Two of these extend entirely around the antheridium; the third, the CAP cell, forms a lid.

The archegonia are located chiefly on the central thickening of the prothallus near the notch. The archegonium, as in the bryophytes, is flask-shaped and consists of an enlarged basal portion, buried in the tissue of the prothallus, with a neck projecting from the surface (FIG. 24-12C). The basal portion contains a large egg and another cell, the ventral canal cell, just above the egg at the base of the neck. The neck itself is composed of a cellular jacket that surrounds a row of two (rarely three or four) neck canal cells, between which cell walls fail to form.

Fertilization occurs when water is present between the lower surface of the prothallus and

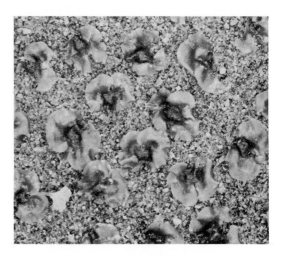

Fig. 24-11   Fern prothalli growing on soil.

the soil. The antheridium absorbs water and swells. The cap cell either tilts upward or becomes detached, thus permitting the sperms to escape. The mature archegonium likewise absorbs water. The ventral and neck canal cells separate from the neck of the archegonium and are extruded to the outside through the ruptured terminal cells of the neck. The sperms, attracted by a diffusible secretion, swim to the mouth of the archegonium and move through the canal toward the egg (FIG. 24-12D). Although several sperms may enter the archegonium, only one penetrates the egg, fuses with it, and forms a zygote. Occasionally more than one egg on the same prothallus is fertilized. Usually, however, only one develops into a plant capable of independent existence.

The production of motile sperms and the necessity for water in the fertilization of ferns are highly significant from an evolutionary point of view. Ferns as a group are relatively advanced kinds of plants, with a highly organized plant body and a long history of survival and development. But in spite of their terrestrial habitat, they are, in a sense, amphibious in their mode of life. On the one hand, they are dependent upon a terrestrial environment for vegetative existence and for the release and dispersal of their spores. On the other hand, like their algal progenitors, they depend upon water as a medium through which the sperms can reach the egg. The life cycle of the fern illustrates, once again, the retention of ancestral features, for these plants have retained, in their ontogeny, the motile sperms associated with the sexual processes of their aquatic ancestors.

Ferns are not alone among living vascular plants in the retention of such a primitive hereditary feature, for swimming sperms are also found in other spore plants and in several groups of seed plants of ancient lineage. This aquatic mode of fertilization may be regarded as a characteristic of the early land plants that has been retained by the ferns and certain other plants through the ages. A somewhat analogous situation is found in the breeding habits of some amphibians such as the frog, toad, and salamander. Such animals, although terrestrial or even arboreal as adults, deposit their eggs in the water, where the progeny pass through larval stages. The amphibians emerged from the water and began life on the land somewhat later in geological time than did the earliest land plants. They have, like the ferns, preserved in certain phases of their life cycles an ancestral manner of living.

The Embryo and the Young Sporophyte The fertilized egg remains in the base of the archegonium and develops into a spherical mass of cells (Fig. 24-13A). From this embryo, as growth proceeds, are developed the organs of the young sporophyte—the foot, first leaf, stem, and primary root (Fig. 24-13B). The foot is the first organ to be formed, by division of cells of the upper part of the embryo. It grows upward between the cells of the prothallus, from which it obtains food. From other regions of the embryo are developed in rapid succession the first leaf and the primary root, followed by the embryonic stem. The young sporophyte is at first dependent upon the tissues of the gametophyte for food and water. The first leaf and the root soon emerge, and the first leaf grows upward through the notch in the prothallus and becomes expanded and green. The root penetrates the soil, and at this stage the sporophyte is an independent plant (Fig. 24-13C). The primary leaf, which is not long-lived, is followed by a series of leaves, increasing in size and complexity. As the secondary leaves develop, the prothallus shrivels and disappears. The primary root is soon replaced by adventitious roots, arising from the stem. Most ferns grow for several years before they are sufficiently mature to produce sporangia and spores.

Fig. 24-12   Sex organs and fertilization. (A): antheridium. (B): sperm. (C): archegonium. (D): fertilization.

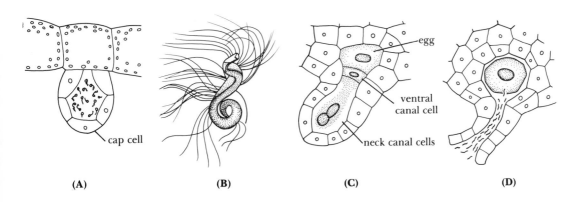

(A)   (B)   (C)   (D)

egg

ventral
canal cell

neck canal cells

cap cell

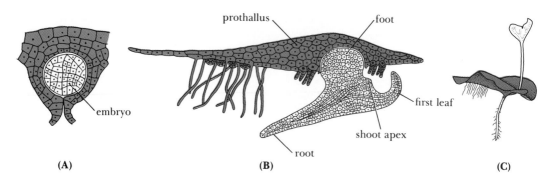

prothallus   foot

embryo

first leaf

shoot apex

root

(A)   (B)   (C)

Fig. 24-13 The embryo and young sporophyte. (A): enlarged archegonium surrounding the embryo. (B): section through prothallus and attached sporophyte; the sporophyte at this stage is still dependent upon the prothallus. (C): the sporophyte becomes independent of the prothallus.

Alternation of Generations Two contrasting phases of the life cycle have been described. The first of these, the sporophyte, bears the spores; the second, the gametophyte, bears the sex organs and gametes. The sporophyte is long-lived, is organized into root, stem, and leaf, and possesses a specialized system of conducting tissues. The gametophyte is relatively short-lived, is thalluslike, and lacks vascular tissues. The gametophyte is able to manufacture its own food. This is also true of the sporophyte, except for its earliest stages, when it is physiologically dependent upon the gametophyte.

A study of the life cycle of a fern with respect to the diploid and haploid phases may begin with the spore mother cells, produced within the young sporangium (FIG. 24-14A). The number of such cells varies with the species, but common numbers are 12 and 16. The spore mother cells possess the diploid ($2n$) chromosome number and belong to the sporophyte generation. Each spore mother cell, at maturity, undergoes the usual two meiotic divisions to form a tetrad of four spores (FIG. 24-14B). Hence, 16 spore mother cells will produce 64 spores. These spores contain the haploid (n) chromosome

Fig. 24-14 (A): young sporangium containing spore mother cells. (B): tetrad of spores and mature spores.

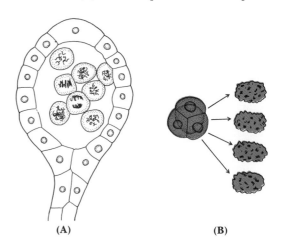

(A)   (B)

number and constitute the first cells of the gametophyte generation. A spore gives rise to a prothallus, all the cells of which, including the sex organs and gametes, are also haploid.

When a sperm fuses with an egg, the

chromosome number is doubled and the fertilized egg becomes the first cell of the new $2n$ sporophyte generation. All the cells of the fern plant, including those of the root, the stem, the leaves, the sporangia, and the spore mother cells, are diploid. The normal life cycle of the fern (Fig. 24-15) thus involves differences in the chromosome number of alternating generations. These differences are instituted at fertilization and at the time of meiosis within the sporangium.

Apogamy and Apospory   Since, in the usual life cycle, the sporophyte generation is $2n$ and the gametophyte generation is n, it might be assumed that the chromosome number is the cause of the great morphological differences between the two generations. Evidence that this assumption is not valid is shown by the following observations.

In a number of ferns it has been observed, both under natural conditions and in laboratory cultures, that the prothallus is able vegetatively to produce a sporophyte. A shoot and a root apex are organized within the tissues of the thickened archegonial region near the notch, and the resulting embryo grows and forms a leafy plant. This plant is, of course, haploid. Since gametes are not involved in the production of the sporophyte, this process is termed APOGAMY (without gametes). Similarly, in several species, the $2n$ plant is known to form, both naturally and in artificial culture, vegetative prothalli having the $2n$ chromosome number. Such prothalli develop from the margins or petioles of very young leaves or even from the margins of mature leaves. Superficial cells produce filamentous structures that grow into heart-shaped structures with sex organs. Since the prothalli do not develop from spores, this process is called *apospory*. Apogamy and apospory also occur in the mosses.

It is apparent that the chromosome number does not determine the growth pattern of either the prothallus or sporophyte. How this is established is not known, but it is speculated that the pattern is fixed early in development by the interaction of genes and the environment of the developing spore and zygote. Briefly, the spore, free from any surrounding tissue, develops the two-dimensional prothallus. The fertilized egg, on the other hand, is surrounded by a jacket of archegonial cells that may determine the planes of cell division as the zygote develops into the three-dimensional embryo and young plant.

Evolution and Diversity of Ferns   The ferns, since their origin in early Paleozoic times, have evolved into a network of evolutionary lines whose relationships have been much debated. Some living groups are considered to be related to similar forms of the Carboniferous. Others are more modern and apparently arose in the Mesozoic or even the Tertiary, and many are not known from the fossil record. A brief survey of representatives of a few of the families of living ferns will indicate, although inadequately, the present great diversity among ferns. Such a study

Fig. 24-15   Life cycle of the fern.

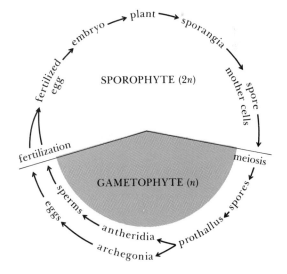

Fig. 24-16 A grape fern (*Botrychium dissectum* forma *obliquum*); about 8 inches high.

will also reveal that sporangia are not always found in sori; that sori are not always on the lower surface but may be marginal on specialized branches or pinnules of the frond; that indusia may not cover sori but arise from beneath as flaps or a cup; and that the structure of the sporangium is not always the same as that found in the polypody family.

Adder's Tongue and Grape Ferns (Ophioglossales) This order, with but one family and three genera, is believed to be very ancient, although fossils are unknown. Only the grape ferns (*Botrychium*), with some 23 species in the subarctic and temperate zones, are considered

here. They are mostly small plants, a few inches high, although one North American species may attain a height of more than 2 feet. The aerial portion of the shoot is usually a single frond composed of a sterile, expanded, leaflike portion and an erect fertile spike (FIG. 24-16). The sporangia (FIG. 24-17A) are borne in two rows, one on each side of the divisions of the fertile spike, and are very large, containing from 1500 to 2000 spores. When mature, they split open transversely and the spores fall away—an annulus is absent and there is no mechanism of spore discharge. The prothallus is a flattened, tuberlike body up to ¾ inch long, buried in the soil, and obtaining food through the activities of an associated fungus. This mode of life of the prothallus is found not only in the Ophioglossales but also in certain primitive groups of the Filicales.

Filicales This order, to which most living ferns belong, is divided into some 14 families, the number varying with the authority consulted. Only a few of these are discussed here.

Osmunda family (Osmundaceae) The osmundas, with 3 living genera and 20 species, are large coarse ferns with a geologic record extending back to the Permian. The common (in eastern North America) cinnamon fern (*Osmunda cinnamomea*) is strongly dimorphic (FIG. 24-18). The sporangia, borne on the margins of the much-reduced segments of the fertile frond, are not arranged in sori and an indusium is lacking. The sporangia are large, short-stalked, globose structures, each producing up to 500 spores. The annulus is rudimentary, and there is no mechanism of spore discharge—the spores fall away as the sporangium opens along a line that extends across the top and down the sides (FIG. 24-17B). The prothalli are heart-shaped, large, fleshy, and long-lived.

Climbing fern or curly grass family (Schizeaceae) This is another ancient family, with relatives in the Carboniferous. The family includes 4 genera and 118 species, chiefly in the

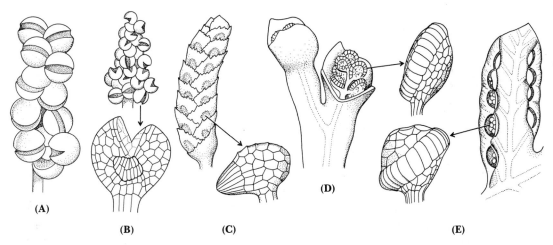

Fig. 24-17 Fern sporangia. (A): sporangia of grape fern (*Botrychium dissectum* forma *obliquum*) in two rows on a division of the fertile spike. (B): cinnamon fern (*Osmunda cinnamomea*), cluster of sporangia and single sporangium. (C): climbing fern (*Lygodium heterodoxum*), fertile segment of a frond and single sporangium. The sporangia are borne singly on the fertile segment and covered by a membrane. (D): filmy fern (*Hymenophyllum polyanthos*); marginal sorus and single sporangium. The sporangia are arranged along an axis and the annulus is oblique. (E): tree fern (*Cibotium*, Dicksoniaceae), marginal sori and single sporangium with oblique annulus.

tropics, but two species, a climbing fern (*Lygodium*) (Fig. 24-19) and a curly grass (*Schizaea*), occur in eastern North America. Either the leaves are divided into sterile and fertile parts, or some leaves are fertile and others sterile. In a tropical climbing fern the twining leaves, with apparently unlimited growth, may attain a length of 100 feet or more—the longest leaves of any plant. The large sporangia, characterized by a caplike annulus (Fig. 24-17C) are not clustered in sori. The sporangium dehisces longitudinally, but there is no mechanism of spore discharge. It is probable that the ancestors of these plants in the Carboniferous attained much greater size than present-day forms.

Filmy ferns (Hymenophyllaceae) The filmy ferns are so-termed because the leaves are extremely thin—usually but one layer of cells thick except for the veins. The 600–650 species are mostly restricted to permanently moist habitats in the tropics and subtropics, where they grow on the soil or as epiphytes. A few species extend northward into the eastern United States. Leaves may be upright (Fig. 24-20) or creeping, with fronds ranging from less than an inch to nearly 2 feet in length. The sporangia are short-stalked, with an oblique annulus (Fig. 24-17D). They are attached to an elongated axis and enclosed at the base by a cuplike or two-lobed indusium. The filmy character of the leaf indicates reduction and an adaptation to conditions of high humidity—these forms are, ecologically and in other ways, highly specialized.

Tree ferns These belong to two families, Cyathaceae (Fig. 24-21) and Dicksoniaceae, not considered to be closely related. Together they include over 800 species. In the Cyathaceae the sori are on the lower surface of the frond; in the Dicksoniaceae, they are marginal on the edges of the frond segments (Fig. 24-17E). The

Fig. 24-18 Cinnamon fern (*Osmunda cinnamomea*), strongly dimorphic. A cluster of spikelike fertile fronds is seen in the center, surrounded by the broad, green, sterile fronds.

Fig. 24-19 A climbing fern (*Lygodium palmatum*) from eastern North America. Portion of frond (about 30 inches long) with sterile and fertile leaflets.

sporangium, with its oblique annulus, is similar in both families and also resembles that of the filmy ferns. The tree ferns are characteristically found in tropical mountain forests and several species are cultivated in conservatories.

Polypody family (Polypodiaceae) These are the "modern" ferns, which probably developed to their present forms in the Tertiary but assumed their basic features in the Mesozoic. This is the largest family of ferns (some 7000 species), and includes most of the ferns around us. For convenience we shall consider all of these as one family, but it is so diverse that authorities have

divided it into a number of families of which the polypodies are only one. The life cycle is that described previously in this chapter, with a green, heart-shaped prothallus and a long-stalked sporangium with a vertical, interrupted annulus that opens by a transverse split (FIG. 24-9). The sporangia are usually in sori on the lower surface of the leaf, but may be marginal. The sorus is commonly associated with an indusium, but may be naked.

Among the various evolutionary developments in the ferns, none is more interesting than the sporangium with its unique annulus. The sporangium of the Polypodiaceae, with its spe-

Fig. 24-20 An epiphytic filmy fern (*Hymenophyllum polyanthos*), Central America; about 7 inches high.

cialized spore-discharge mechanism, is believed to be the most advanced type. But the sporangia of the filmy ferns, the tree ferns, and others with an oblique annulus approximating to this in structure are also able to expel their spores with some force. In other Filicales, as the Schizeaceae and Osmundaceae, no spore discharge mechanism is found. Such a mechanism would become of great biological value as the number of sporangia in a sorus increased, with resultant crowding and lack of room for lateral dehiscence. The oblique or vertical annulus would allow the sporangium to open upward and utilize the free space above.

SUMMARY

1. Ferns, as a race, have a long history, extending back at least to the Middle Devonian. They are mostly terrestrial plants that have become adapted to a number of different environments.

2. The fern stem, with some exceptions, is a rhizome. Vegetative reproduction by rhizomes is common. The leaves (fronds) of most ferns are coiled when young, and develop by apical growth. The vascular tissues in general resemble those of seed plants, but a cambium is lacking and all tissues are primary in origin.

3. Fern spores are produced within special structures, the sporangia. These are borne on the lower surfaces of the expanded fronds, on the margins, or on specialized fronds or parts of fronds. The sporangia may be borne in a cluster called a sorus. A sorus may be naked or may have an indusium associated with it. The annulus of the sporangium, if present, may be rudimentary, apical, oblique, or, as in the polypodies, vertical in position.

4. Fern fronds may be of two kinds, fertile and sterile, a condition termed dimorphism. The differences between fertile and sterile fronds may be slight or great.

5. In the Ophioglossales, Osmundas, and Schizeaceae there is no mechanism of spore discharge. In other families mentioned—the tree ferns, filmy ferns, and polypodies—the spores are forcibly ejected from the sporangium. The mechanism of this expulsion is unique to the ferns. The factors involved are the nature of the cell walls of the annulus, evaporation of water, the cohesion of water within the cells of the annulus, and the adhesion of water to the annulus cell walls.

6. The spore germinates and in most ferns develops a heart-shaped prothallus bearing rhizoids and multicellular sex organs

Fig. 24-21 A tree fern (*Cyathea arborea*), Puerto Rico.

(archegonia and antheridia) on the ventral side. The sperms are motile and swim to the archegonium, where the egg is fertilized. This aquatic method of fertilization may be regarded as an example of retention of an ancestral feature.

7. The fertilized egg develops into an embryo sporophyte composed of a foot, a primary leaf, a stem, and a primary root. The young sporophyte absorbs food and water through the foot and is, for a time, dependent upon the gametophyte.

8. The root penetrates the soil, the first leaf extends into the air, and the sporophyte soon becomes an independent plant. The prothallus shrivels and disappears. The primary root is replaced by adventitious roots and the primary leaf by secondary leaves.

9. A clearly defined alternation of generations is present in ferns. The prothallus is the chief structure of the gametophyte generation, which also includes the spores and the gametes, eggs and sperms.

10. The fertilized egg is the first cell of the sporophyte generation, which also includes the embryo and the mature plant with sporangia and spore mother cells. The n (haploid) chromosome number of the egg is doubled at the time of fertilization, and the $2n$ (diploid) chromosome number of the sporophyte generation is reduced to the n number when meiosis occurs in the spore mother cells within the sporangium. In the ferns, as in all other vascular plants, the sporophyte generation dominates the life cycle.

11. The morphological differences between the sporophyte and gametophyte generations are not caused by their chromosome numbers, respectively diploid and haploid. This is shown by the occurrence of diploid prothalli (apospory) and haploid fern plants (apogamy).

Early Land Plants and Their Evolution

HUNDREDS OF YEARS of botanical investigation have yielded a great wealth of information about plants. During the past century or so, this knowledge has been increasing at a geometric rate. The accumulation of these facts has scientific and practical value, but their organization into patterns that produce basic principles is of even greater value. Of all such principles in science, the concept of evolution has been one of the most fruitful. Not only has it brought existing facts together into a logical structure, but also it has stimulated investigations into neglected fields that, in turn, have enriched and amplified the principle of evolution.

RECORDS OF THE EARTH'S PAST

Fossils, the remains and imprints of living things found in the earth's crust, have for centuries been objects of curiosity and wonder. A number of Greek and Roman writers of the pre-Christian era recognized that fossils represented the remains of organisms that once existed on earth. Such concepts were forgotten during the Middle Ages, and instead, curious explanations were advanced to explain such phenomena. During these times it was not admitted—in fact, it was denied—that fossils were the remains of living things that had existed in a bygone age. They were variously considered to be freaks of nature, products of vital forces in the earth or results of a petrifying juice. The late seventeenth and early eighteenth centuries saw a great controversy on the "flood theory." According to this, all fossils were formed following the great deluge, the Flood of Noah. On the basis of this hypothesis all fossils were of the same geologic age, since they had all been formed at the same time.

A truer appreciation of fossils grew slowly, and during the eighteenth century there gradually developed an understanding that fossils are evidence of events in the history of the earth and that plants and animals had inhabited the earth long before the age of man. With the development of the doctrine of descent, during the nineteenth century, there came about a reawakened interest in fossils as tangible and direct evidence of how living plants and animals came to be. The theory of evolution suggested that gradual change over long periods of time could account for present species. Geologic studies showed that such periods of time had passed in the history of the earth. Paleontology, the study of past life by means of fossils, is an important science upon which both botanists and zoologists depend. In addition to its rich scientific contribution, this science has had direct practical applications in locating oil, coal, and other mineral deposits of economic importance.

Some extinct plants differed greatly from the plants of today; others were closely similar. The differences become greater in degree with the age of the rock in which the fossils are found. The older the rock, the more marked are the differences from modern plants. Through the correlation of time, complexity of structure, and similarity to living plants, a sequence in evolutionary development of land plants can be traced. A knowledge of the comparative morphology of living plants is thus essential to an understanding of the plants of the past, whose fossil record is usually fragmentary. The reverse is also true; a knowledge of extinct plants provides a basis for understanding the evolutionary steps that have led to the plants of today. The fossil record shows that the pattern of vegetation on the earth has undergone continued change throughout the ages. Great groups of plants have arisen and flourished, and some have become important or even dominant elements in present-day plant life. Others waned but left descendants that, as "living fossils," supply evidence of the nature of their ancestors. Many other groups have become extinct. The plants of today are, then, the living members of some groups that had their origins millions of years ago.

Plant Fossils Most plant fossils are found in sedimentary rocks. Such rocks originate as sand or mud, which is transported and deposited under water. Plant remains, buried by such deposits, sometimes become fossilized as the inorganic material is consolidated into hard rock. Plant fossils are also found in beds of clay and unconsolidated volcanic ash. Since the plant remains were for the most part transported from sites where the plants grew, to other areas where the deposits were laid down, the fossils are usually fragmentary. They may consist of leaves or leaf parts, portions of stems or roots, spores, seeds, or cones. Seldom are all the organs of a plant attached as when it was alive.

Plant fossils may be classified as *compressions, impressions, casts, molds,* and *petrifactions.* A compression (FIG. 25-1) is formed when a plant part, such as a leaf, is deposited upon a soft surface, such as mud, and then covered with a fine sediment. The plant material is compressed by the weight of the sediments, and internal structures are usually not preserved. A thin, carbonaceous film remains, however, which may show surface structure such as cuticle, epidermis, stomata, and hairs. Impressions are similar to compressions, but retain no organic matter; the plant material has entirely decayed, and only the outlines of the plant part may be seen. Casts and molds are also related. If embedded plant parts, such as stems, later decay, there may be left in the rock a cavity or mold that retains the shape and size of the part. A cast results from the filling of such a mold with inorganic materials (FIG. 25-2).

The most important type of plant fossil is the petrifaction, in which plant tissues, such as wood, roots, or reproductive organs, are embedded in an inorganic matrix. This matrix is the result of the infiltration into the tissues of such compounds as silicon dioxide, calcium or magnesium carbonate, and iron sulfide, which then precipitate. Petrified wood is a well-known example. By the use of special methods, such as cutting, grinding, polishing, and peeling, thin sections of petrifications may be prepared for microscopic study (FIG. 25-3). These reveal almost as many

Fig. 25-1 Compression of part of a frond of a seed fern (*Neuropteris gigantea*) from the Pennsylvanian.

Fig. 25-2 *Calamites*, pith cast. The pith disappeared and was replaced by sedimentary material that hardened into rock. The markings are impressions of the inner surface of the wood.

details of internal structure as similar preparations made from living plants—even, in some examples, including nuclei and chloroplasts. It was long believed that the original plant material in a petrifaction was replaced, molecule by molecule, by mineral substances such as silica, the replacement being so gradual that the mineral skeleton preserved was the exact replica of the original tissue. Modern investigations show, however, that the original carbon compounds, or their modified remnants, of the cell walls of a petrifaction are still present, although such compounds have usually undergone chemical alteration.

Although the structure of great numbers of extinct plants is partially known from their fossils, the record of the past is still so incomplete that the problem of tracing the genealogies of living plants is one of great difficulty. The reasons for this imperfection in the fossil record are many. Fossilization occurred only under specialized conditions, which may have existed only in local areas for comparatively limited periods of time. Many plants grew in high, rocky, or arid regions and decayed where they lived, since conditions were unfavorable for fossilization. Fossils in many parts of the earth's crust are inaccessible, and great parts of the records of the plant life of the past have been destroyed by erosion, earth movements, or exposure to volcanic heat. With

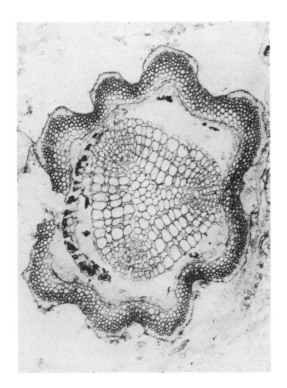

Fig. 25-3 Cross section of a petrifaction, the stem of *Sphenophyllum*, from the Pennsylvanian of Illinois.

all these limitations, it is remarkable that so many fossils have been found and that such minute details of structure can be seen.

The Earth's History Change has been an outstanding characteristic of the history of the earth. At certain periods this change has been accelerated. These shorter, accelerated periods were characterized by rock folding and mountain building, and by changes in the relative positions of great areas of land and sea. They have alternated with longer periods of relative stability and quiet, during which the predominant activity was extensive erosion and deposition. These longer intervals, terminated by major changes, are termed ERAS. Subdivisions of these, often terminated by minor geologic revolutions, are known as PERIODS and EPOCHS. A study of the layers (strata) of sedimentary rock laid down in each interval has enabled geologists to construct a geological time scale (Table 25-1) by which the various rock formations are classified according to their respective ages.

The age of the rocks of each of these periods has been determined by several methods. The most important is based upon the rate of disintegration of radioactive elements that may occur in the rocks. The eras and periods are identified partly by physical differences but chiefly by the kinds of plant and animal fossils found in each. The correlation between the various geologic periods and the occurrence of certain fossils constitute the most important kinds of evidence for organic evolution. The simplest and most impressive fact is that the oldest fossiliferous rocks contain the remains of only the most primitive plants and animals, the younger formations yielding successively more and more highly evolved organisms.

Early Records of Life The earliest indications of life are found in the rocks of the Precambrian, an era that includes some five-sixths of all geological time. The age of the earth is estimated at about 4.6 billion years, and the Precambrian extends to the beginning of the Cambrian, some 600 million years ago.

What little we know of the life of the Precambrian suggests that it was exclusively aquatic and largely confined to shallow seas. For many years the chief evidence of life in the Precambrian was the nature of certain kinds of limestones, called stromatolites, found in various parts of the world and occurring from the Precambrian to the Recent. Stromatolites are rounded, nodular, or irregular masses, showing banded structure, and are believed to have been produced by blue-green algae that precipitated calcium or calcium-magnesium carbonates from the surrounding waters. The algae have been so

TABLE 25-1

Divisions of Geologic Time and Some of the More Important Kinds of Plants and Animals in Each Division

Era	Period	Epoch	Important Plants and Animals	Million Years Ago to Present
CENOZOIC	Quaternary	Recent Pleistocene	Age of man and herbaceous flowering plants	1.5–2
CENOZOIC	Tertiary	Pliocene Miocene Oligocene Eocene Paleocene	Age of modern life: birds, mammals and flowering plants	13 25 36 58 63
MESOZOIC	Cretaceous		Transition from Jurassic ferns and gymnosperms to expansion of angiosperms; rise of modern conifers	135
MESOZOIC	Jurassic		Climax of Cycads, Bennettitales, Ginkgoales, forests of conifers, giant reptiles, first birds	181
MESOZOIC	Triassic		Decline of seed ferns, establishment of conifers, ferns, horsetails, cycads, Bennettitales, Ginkgoales. Early dinosaurs, primitive mammals (late Triassic)	230
PALEOZOIC	Permian		Decline of Carboniferous floras, rise of early conifers, rise of reptiles, first beetles	280
PALEOZOIC	Upper Carboniferous (Pennsylvanian) Lower Carboniferous (Mississippian)		Greatest development of Cordaitales and seed ferns; giant horsetails (*Calamites*) and club mosses (*Lepidodendron*); ancient conifers, herbaceous Lycopodiums, ancient ferns. Winged insects; spread of amphibia. First reptiles. Greatest beds of coal (Pennsylvanian)	310 345
PALEOZOIC	Devonian Upper Middle Lower		Middle and Upper: Increasing diversity of plant life; early ferns; lycopsids and sphenopsids; development of forest trees; transition of vertebrates to the land. Lower: Rhyniales and other simple forms of vascular land plants	405
PALEOZOIC	Silurian		First unquestionable record of vascular land plants; psilopsids	425
PALEOZOIC	Ordovician		Lime-secreting algae, many marine invertebrates, first vertebrates (jawless fishes)	500

(Table continues.)

	Cambrian		First abundant fossils; lime-secreting and other algae, marine invertebrates	600
PRECAMBRIAN	Upper		Appearance of eucaryotic green algae and one-celled animals (about 1 billion years ago)	
	Middle Lower		Bacteria and blue-green algae, in rocks approximately 3 billion years old	4600

modified by biological and geological processes, however, that cellular structures are usually not preserved. The rocks associated with certain Precambrian stromatolites in southern Rhodesia have been assigned a minimum age of 2.7 billion years.

In view of the absence of structurally preserved organisms in the Precambrian, reports of investigators, beginning in 1954, of the findings of fossilized cellular remains of simple thallophytes were of great interest. The oldest of these fossilized organisms were found in the Lower Precambrian of South Africa, in rocks dated at approximately 3.1 billion years. They consisted of rod-shaped bacteria and spheroidal bodies that are comparable to modern, unicellular, blue-green algae. Chemical analyses of the rock in which they were embedded revealed the presence of hydrocarbons, probably derived from chlorophyll, and of several amino acids, believed to have been synthesized by living organisms. If bacteria and simple algae existed at this remote period, we may certainly project the origin of life itself much further back in time.

Studies on younger rocks—Middle Precambrian—dated at 1.9 billion years, from Ontario near Lake Superior, have also yielded microfossils. The most abundant organisms were filamentous blue-green algae and threadlike bacteria not unlike some living species. Microfossils are also reported from the Upper Precambrian. Fossils of the first unicellular animals (foraminifera) have been found in Montana in rocks 1.1 billion years old. In Australia a rich flora was found in rocks about 1 billion years old. These were diverse in kind, including some 30 species of organisms—filamentous blue-green algae, colonial bacteria, and aquatic fungi. Most significant of all were unicellular and colonial green algae in which nuclei could be seen, indicating that eucaryotic organisms had evolved by the late Precambrian.

Beginning with the Cambrian, fossils are abundant, and the record of evolution in the plant kingdom can be read more easily. The plant life of the Cambrian, Ordovician, and Silurian was apparently largely aquatic, and consisted of algae descended from the simpler forms of the Precambrian. These algae included forms that resemble living blue-green, green, and red algae, together with brown algae (FIG. 25-4). It is probable, in fact, that most of the great divisions of the algae had arisen in the sea long before there were plants upon the land. The warm Paleozoic seas teemed with animal life, indicating a plentiful supply of plant food. Among the prevalent animal types were corals, mollusks, sponges, crustaceans, worms, and other invertebrates. The first vertebrates, primitive jawless fishes, are found in the Ordovician, early in the Paleozoic.

EARLY LAND PLANTS

All lines of evidence indicate that modern terrestrial plants are descended from early migrants from the seas to the land more than 400 million years ago. Great obscurity, naturally, surrounds this remote migration, which constituted one of the most important events in the history of the

Fig. 25-4 The Cambrian: plant and animal life of shallow seas. (A–L): algae. (B): calcareous blue-green algae; these forms produced great cliff masses. (E, F, G): other blue-greens. (C, L): green algae. (H, I): brown algae. (A, D, J, K): red algae. (1–7): invertebrate animals. (1) calcareous sponges. (2): brachiopods. (3): trilobites. (4): mollusks. (5): annelid worms. (6): onychophores. (7): jellyfish.

plant kingdom. The immediate progenitors of land plants were probably green algae. It may be assumed that the red and brown algae, before there was vegetation upon the land, had become about as specialized as they are today. It is hardly likely that land plants arose from such highly evolved forms. On the other hand, the green algae were, and still are, more primitive in body form and methods of reproduction. They exhibit greater plasticity in their adaptation to various environments, a factor important in plant evolution.

The transition from an aquatic to a terrestrial mode of life was undoubtedly gradual. In these changed habitats those plants that evolved structures permitting growth in a dry environment became the first land plants. Many radical changes were necessary before these new land plants could prosper—certainly plants could become erect and attain any considerable size only after a system of conducting and supporting cells, or vascular system, had developed. Other alterations were the development of a cuticle and epidermis that could resist desiccation, stomata, specialized absorption structures, and thick-walled spores that could survive dispersal by wind. Unfortunately, no transitional forms between ancient algae and fully established land plants have been found in the fossil record. Rather, throughout geological time from the Upper Silurian and into the Middle Devonian, fossils of simple vascular plants, adapted to a terrestrial life, are found. There is no evidence that vascular plants existed before the Silurian, and it was probably in that period (perhaps during the latter part of it) that the transition to the land occurred. The early land plants, as might be expected, were diverse in kind, since they were free to exploit new habitats with little or no competition. The best known of these early plants may be classified in the subdivisions Psilopsida (Greek *psil*, bare; *opsida*, appearance) or Lycopsida, of the division Tracheophyta (Chapter 18).

The Psilopsida The subdivision Psilopsida is characterized by leafless stems and absence of roots. The stems were sparsely and dichotomously branched; in more advanced forms, a central axis bore repeatedly dichotomized branches. Secondary tissues were absent. The sporangia were either (1) terminal, usually elongated, and opening by a longitudinal slit; or (2) lateral, more or less kidney-shaped, opening by a lateral slit, and aggregated into spikes or scattered along the axis.

Rhyniales This order includes those fossil Psilopsida in which the axis bore relatively few dichotomizing branches and the sporangia were terminal. The group is not large, consisting of about nine genera, but has long been of interest to students of plant evolution. The fossil remains of these genera have been found in many parts of the world, including North America. The genus *Rhynia*, the best known representative of the order, is discussed in the next section.

Rhynia This genus was named after the village of northern Scotland where the fossils of the plants were found in strata (considered as Middle Devonian, or possibly Lower Devonian), at least 370 million years old. Two species are known, one perhaps 9 inches, the other about 20 inches high. The plant consisted of an underground rhizome and erect leafless stems (FIG. 25-5). Roots were absent, absorption taking place through rhizoids, similar to root hairs, arising directly from the rhizome. The branching of the aerial parts was largely dichotomous, that is, each division was composed of approximately equal parts. Some of the aerial branches were vegetative (sterile), and others terminated in sporangia containing numerous spores with cutinized walls. The spores were usually in tetrads and, when found separated, exhibited the lines or ridges marking the planes of contact with other spores of the tetrad (FIG. 25-6).

The presence of tetrads is evidence that meiosis occurred in *Rhynia*. The plant body therefore belonged to the sporophyte generation, but nothing is known of the gametophyte generation, which may have resembled a portion of a rhizome. Cross sections of the stem show a thick-walled epidermis, stomata, and a cuticle. Inside the epidermis is a cortex, some cells of which contained chloroplasts, indicating that this tissue carried on photosynthesis. A pithless strand of conducting tissue, composed of a central core of tracheids surrounded by phloem (FIG. 25-7), ran up the center of the stem.

OTHER RHYNIALES Other of the early land plants classified in this order are shown in FIG. 25-5. Among these are *Hicklingia*, *Hedeia*, and *Cooksonia*. *Hicklingia*, from the Lower Devonian of England, was perhaps 8 inches in height and composed of sparsely dichotomized stems in a tufted or bushlike arrangement. Only the terminal part of *Hedeia* is known; it bore its terminal sporangia in a three-dimensional cluster, with all the tips at the same level. *Cooksonia* (FIGS. 25-5, 25-8), with several species, was probably the smallest and simplest of the group. It was probably less than 6 inches high and composed of a slender dichotomously divided stem with terminal sporangia containing cuticularized spores in tetrads. It has been found in the Upper Silurian of Bohemia and the Lower Devonian of Australia and England; it is, to date, the oldest known vascular land plant.

EVOLUTIONARY SIGNIFICANCE OF THE RHYNIALES Two views have been advanced with respect to the origin of the great subdivisions of the vascular plants (Chapter 18 and FIG. 25-18) such

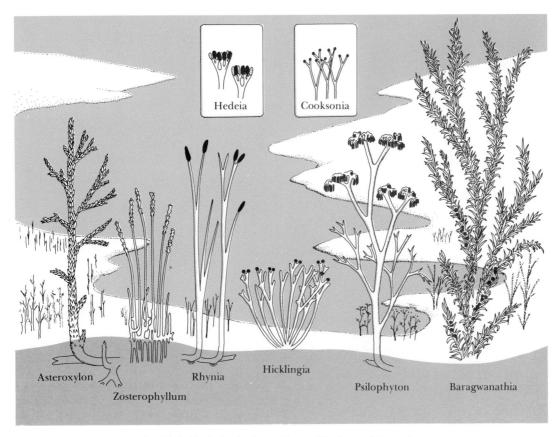

Fig. 25-5 Early land plants, Upper Silurian and Devonian.

Hedeia

Cooksonia

Asteroxylon

Zosterophyllum

Rhynia

Hicklingia

Psilophyton

Baragwanathia

Fig. 25-6 Aerial spore of *Rhynia*.

as the Psilopsida, Lycopsida, Sphenopsida, and Pteropsida. One is that these arose from different groups of the green algae, possibly at different times and places, and were distinct from the beginnings of vascular plant life upon the land. The other concept is the older, and was for a time out of favor. This is that the great subdivisions arose from ancestors that, at least in considerable degree, resembled the Rhyniales. Although there is still no general agreement, increasing knowledge of the fossil record now makes this second theory seem the more plausible. On this basis we may visualize the earli-

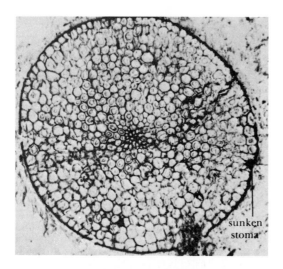

Fig. 25-7 Cross section, stem of *Rhynia*. A solid rod of xylem and phloem appears in the center.

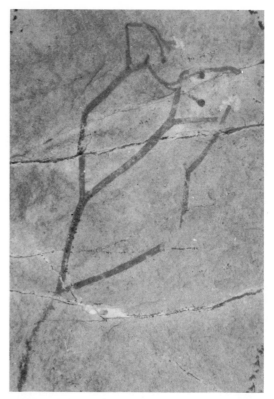

Fig. 25-8 *Cooksonia hemispherica*, from the Upper Silurian of central Bohemia. The natural size of this fragment was about 2¼ inches.

est land plants as small, leafless, and green, with a dichotomously branched axis bearing terminal sporangia at the tips of some of the branches. Roots were absent, but absorbing rhizoids arose from underground portions of the axis. Among other features, these plants would have been distinguished from algae by the possession of a simple strand of conducting tissue. The Rhyniales, then, in their external form, internal structure, and method of reproduction, enable us to visualize the nature of the ancestral vascular plant invaders of the land.

Evolution in the Devonian As suggested above, it is difficult to identify the first terrestrial ancestors of the higher plants on the basis of the fossil record. The period at which they lived was remote; many forms of life may have existed but have not been preserved; and the age of the strata in which such early plants are found may not represent their first appearance—they may have been survivors from a still earlier day. Even from Lower Devonian time, which extended over

some 15 million years, a number of plants are found that exhibit what may be interpreted as modifications of a more primitive type. For example, among the Psilopsida and possibly related to Rhyniales was an order represented by *Zosterophyllum* (Fig. 25-5), a widely distributed genus with several species. These simply constructed plants resembled the Rhyniales in that they were rootless and leafless, but the vascular strand was relatively more massive and the sporangia were lateral and somewhat kidney-shaped. This group has been regarded as derived from the Rhyniales and as representing the remote ancestors of the subdivision Lycopsida, the early fossils of which are considered in a later section.

We may visualize that, from forms similar to the Rhyniales of the Lower Devonian, there evolved larger and more complex psilopsid plant bodies, branching freely and producing numerous terminal clusters of sporangia. Such a stage could be represented by *Psilophyton princeps* (FIG. 25-5) from the early Middle Devonian of Canada. In height it may have ranged from 1 to 3 feet. This plant, and others, may have foreshadowed the still more complex bodies of the Sphenopsida, the extinct fernlike plants, and the many other kinds of spore-bearing plants that flourished in the Middle and Upper Devonian. Before the end of the Devonian (which extended over some 60 million years) all of the great subdivisions of the vascular plants had become well established—even primitive seeds have been reported from the Upper Devonian. In contrast to the origin of the phyla of the animal kingdom, all of which probably arose in the seas, all of the great groups of the vascular plants as well as the liverworts and mosses among nonvascular plants arose on the land.

The Psilotales   In many systems of classification, two small living genera, *Psilotum* (FIG. 25-9) and *Tmesipteris*, of limited distribution in the tropics and subtropics, comprise the family Psilotaceae, the single family of the order Psilotales in the Psilopsida. These plants, although anomalous in several ways, show some resemblances to the Rhyniales. They are slender, rootless, and terrestrial or epiphytic plants with erect or pendulous stems. *Tmesipteris* is about 8 inches long; *Psilotum* varies from a foot to as much as 3 feet in height. The plant body (sporophyte) consists of a rhizome bearing aerial branches and covered with rhizoids. *Psilotum* is said to be leafless; the aerial stem of *Tmesipteris* bears small, leaflike appendages.

A consideration of one species, *Psilotum nudum*, is sufficient to indicate many of the important features of the group. This species occurs in the southeastern United States, where it is known as

Fig. 25-9   *Psilotum nudum*, from Jamaica.

the "whiskfern." It is easily cultivated in greenhouses. The aerial stem is frequently and dichotomously branched (FIG. 25-9). Photosynthesis is carried on by the green branches, which are well equipped with stomata. The stem contains a slender strand of primary vascular tissue; in the rhizome this is devoid of a pith, but in the aerial stem the center of the strand is composed of sclerenchyma tissue.

The aerial stems bear two kinds of appendages: (1) *sterile*, which are small and scalelike and contain no vascular tissue (FIG. 25-10A); and (2) *fertile*, larger and more complicated structures (FIG. 25-10B, C, D). Each fertile appendage, in

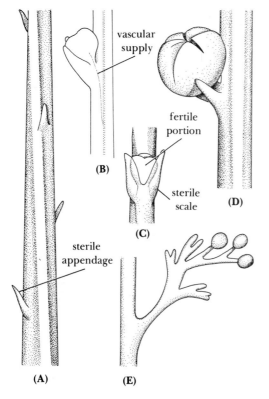

Fig. 25-10 *Psilotum nudum*. (A): stem bearing sterile appendage. (B, C, D): fertile appendage. (B): lateral view of young fertile appendage showing vascular supply to fertile portion. (C): face view, young fertile appendage, fertile portion and bilobed sterile scale. (D): mature fertile appendage. (E): hypothetical ancestral sporangial shoot from which the modern *Psilotum* fertile appendage may have been derived by condensation.

turn, is composed of two parts: (1) a fertile portion, which consists of a very short stalk, or branch, vascularized by a branch of the vascular system of the stem and bearing a single three-lobed and three-chambered sporangium (FIG. 25-10B). (2) A sterile outgrowth of the fertile branch, appearing as a forked scalelike structure just beneath the three-lobed sporangium. It contains no vascular tissue.

The sporangium dehisces through three clefts, one in each of the sporangial chambers, and the numerous spores are disseminated by wind. They fall upon the soil and give rise to prothalli. The mature prothallus (FIG. 25-11) is nongreen and subterranean, growing ½ inch or more beneath the surface. It is brownish, elongated, and cylindrical; may branch irregularly or dichotomously; and develops rhizoids that penetrate the surrounding soil. The prothalli are small, varying in length from 1 to 18 millimeters and in diameter from ½ to 2 millimeters. Numerous archegonia and antheridia (containing motile sperms) are scattered over the surface. A feature of interest is the occasional presence, in some of the larger prothalli, of a slender thread of vascular tissue extending back several millimeters from the apical region.

Shortly after germination the young prothallus is invaded by hyphae of soil fungi. The prothallus undoubtedly obtains food from the soil through these associated hyphae and the relation between prothallus and the fungi may be mutualistic. After fertilization, an embryo develops. This is composed only of an absorbing foot and an embryonic stem; neither embryonic leaf nor root is present.

The Psilotales, because of the apparently undifferentiated nature of the plant body, have long been regarded by many students of phylogeny as related to the Rhyniales. They are considered by these investigators as constituting a branch of the Rhyniales—perhaps at a somewhat higher level of structural organization—that has persisted from the Devonian to the present. This interpretation, however, is controversial, and other opinions have been held. The Psilotales have no fossil record. Moreover, there is ample evidence that these plants, while apparently simple in construction, may be simple in part because of reduction. For example, the single three-lobed and three-chambered sporangium of *Psilotum* is to be interpreted as the result of the fusion of three sporangia. Further, the fertile appendage (which has no counterpart in extinct plants of the Devonian) probably represents a condensed branch system, reduced from a larger and more complex system of sterile and sporangium-bearing appendages (FIG. 25-

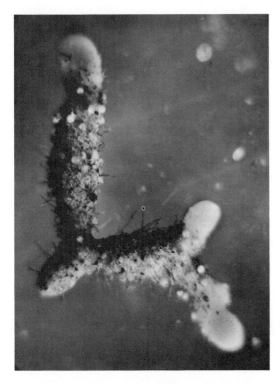

Fig. 25-11 *Psilotum nudum,* mature gametophyte bearing sex organs.

Fig. 25-12 *Baragwanathia.* (A): compression of portion of leafy shoot showing the long, simple leaves. (B): terminal portion of shoot, showing a fertile zone with large kidney-shaped sporangia. (C): portion of sporangium-bearing region of B, enlarged.

10E). Careful and detailed studies of certain primitive ferns, and comparison of various aspects of their structure and life cycles with those of the Psilotales, have in fact resulted in the transfer (in 1968) of the Psilotaceae to the Filicales (ferns). On the basis of these studies, *Psilotum* and *Tmesipteris* may still be regarded as among the most primitive of living vascular plants, but their ancestry is unknown.

The Lycopsida As examples of early land plants contemporaneous with the extinct Psilopsida, we may now consider a few of the fossil members of the Lycopsida. An important characteristic of this division is the presence of a single sporangium on the upper side of a fertile leaf (sporophyll), near the base, in the leaf axil, or attached to the stem just above the leaf base. Prominent among the living Lycopsida are the club mosses (*Lycopodium*), plants ranging in height from a few inches to more than 5 feet in one species (Chapter 26). The oldest of the extinct Lycopsida is the Lower Devonian *Baragwanathia* from Australia, named after W. Baragwanath, an Australian geologist. This was a dichotomously branching plant with stems varying up to 2 inches in diameter and from 3 to as much as 6 feet in length (FIG. 25-5). The stems, which contained a central column of pithless vascular tissue, bore numerous spirally arranged leaves, a millimeter wide and as much as 4 centimeters long, each with a single vein (FIG. 25-12A). At the base of the leaves were kidney-shaped sporangia containing spores (FIG. 25-12B, C).

Another extinct lycopsid is *Asteroxylon mackiei* (FIG. 25-5) from the Middle Devonian. The plant was probably less than 2 feet high, and the erect branching stems were covered with short, awl-shaped leaves about 4 millimeters long. The aerial stems arose from leafless rhizomes, small branches of which penetrated the peat in which the plants grew and probably served as absorbing organs. The vascular cylinder of the stem was pithless, composed only of primary tissues, and gave off minute vascular bundles, or "leaf traces," that, however, extended only to the bases of the appendages. The sporangia were large, somewhat kidney-shaped bodies attached by stalks to the axis, and interspersed among the leaves. *Baragwanathia* and *Asteroxylon* were early land plants whose ancestors may have been psilopsids with lateral sporangia. They show many resemblances to the club mosses of the modern world.

SOME STAGES IN PLANT EVOLUTION

It is impossible to trace the exact steps in the progression from primitive to more highly evolved kinds of plants. Nevertheless, important stages in the evolution of the plant kingdom are evident, and certain of them are discussed in subsequent sections.

Evolution of Life and the Cell   The concept of organic evolution embraces the view that life evolved spontaneously from nonliving matter. The possible steps by which this was brought about have been the subject of extensive discussions and experiments that can be considered only briefly here. It is believed that life evolved some time between the stabilization of the earth's mass, some 4.6 billion years ago, and the appearance of the first fossils, some 3.1 billion years ago. The first living cells may have appeared 3.4–3.3 billion years ago, before the first unaltered sedi-

mentary rocks were deposited, about 3.2 billion years ago. Thus, the early stages in the evolution of life are not preserved in the fossil record.

Current theory about the origin of life involves the concept of an essential continuity between nonliving matter and living organisms. Although this evolution was continuous, two general stages have been recognized: (1) a period of chemical evolution—that is, evolution at the molecular level—during which organic molecules were synthesized from nonliving sources; and (2) the organization of these molecules into living matter. In attempting to reconstruct the steps leading to the origin of life, we may begin with the materials available for the production of organic molecules, the energy sources, and the conditions prevailing on the primitive earth. The atmosphere was reducing, that is, oxygen was absent, and was composed primarily of ammonia (NH_3), methane (CH_4), hydrogen, and water vapor. Some of this ammonia and methane was dissolved in the waters of the ancient seas. Sulfur, phosphorus, and other elements were also available. Energy sources were abundant—short-wave ultraviolet radiation from the sun, radiation from which the earth is now protected by layers of ozone, perhaps being the most important; radioactivity, which was greater then than now, and thermal (heat) energy from volcanic lavas. Under these conditions, chemical reactions led to the formation of many of the kinds of organic molecules, including amino acids, that now compose the living cell. These organic molecules accumulated, over long periods of time, as a "soup" in the shallow waters of the primordial pools and seas. If such compounds were formed today—an unlikely event under prevailing environmental conditions—they would be oxidized, or consumed by heterotrophs.

The preceding theory is supported by numerous experiments carried on in an artificial, simulated prelife environment. In one of the earliest of these experiments (1953), a mixture of methane, ammonia, hydrogen, and water vapor in a sealed glass chamber was subjected to an electric discharge (supplying heat, an electron stream, and ultraviolet light). After a week's exposure, several of the presently known amino acids had formed, together with a number of other organic compounds. Since 1953, hundreds of experiments, using various mixtures, techniques, and energy sources, have been performed. Reactions resulting from the application of energy to mixtures of "primitive" gases or, stepwise, to the products resulting from these reactions have yielded not only amino acids but also many other complex organic molecules such as nitrogen bases, the simple sugars ribose and deoxyribose, and small peptides. Ultraviolet irradiation of solutions containing a nitrogen base (adenine), 5-carbon sugars, and phosphorus compounds have yielded adenosine phosphates, including ATP. The possible origin of proteins and perhaps of the protocell was shown by heating a mixture of 18 dry amino acids for 3–4 hours at 170°C. An amber-colored liquid resulted; this was treated with boiling water, and upon cooling, the great numbers of minute (about 1 millimeter in diameter) spheres, called proteinoids, that resulted had many of the properties of natural protein. It is postulated that such production of protein could have occurred on hot lava and that the products were washed into the sea by rainfall or tidal currents. On the basis of these and many other experiments it seems very probable that the building blocks of the living organism were abundant before life began.

There is, of course, a considerable gap between the complex organic molecules of the primordial soup and the first living cell. Living substance is not dispersed but is localized in individual cells separated from the environment by a membrane. The most prominent hypothesis accounting for this transition holds that, at an early stage in evolution, the large organic molecules tended to combine into molecular aggregates, held together by attractive forces. These aggregates became separated from the solution as droplets, with membranes, called *coacervates,* large enough to be visible under the microscope. Many kinds of coacervates could form, depending upon the materials enclosed, and some were more stable than others.

Coacervates then constituted a new level of structure and activity; chemical reactions could continue within them, materials could move in from the environment, and they could increase in volume and weight. As reactions became more and more numerous, new compounds were formed. The catalysts at first may have been metals such as iron or copper, perhaps in organic combination; it was only later that enzymes were formed. The components of the nucleic acid molecule, synthesized in the primordial soup, became organized within the protocell into nucleic acids, together with mechanisms for their replication and for the synthesis of enzyme proteins. Countless small evolutionary steps, governed by well-known chemical and physical principles, led to the living cell. This was indeed spontaneous generation, resulting from the capacity of carbon-containing molecules to react spontaneously in specific ways. It is generally believed to have occurred, however, during only one phase of the earth's history; once life evolved, all life arose from pre-existing life.

One difficulty with this hypothesis is the deleterious effect of ultraviolet light upon nitrogen bases, nucleic acids, and protein. However, the emerging organisms and their constituent molecules may have occurred at depths where only visible light could penetrate, or in the lower layers of the primordial soup, where they were shielded by the organic particles above them.

It is presumed that the earliest units of life were anaerobic procaryotes, subsisting upon the organic matter around them. But these organic compounds would eventually have been depleted and all life would have come to an end if photoautotrophs, releasing oxygen, had not evolved. Photosynthesis is probably a very ancient process—some of the earliest fossils, believed to be related to modern blue-green algae, were certainly photosynthetic, and other early life forms may have been also. As a result of the release of oxygen in photosynthesis the concentration of this gas slowly increased in the atmosphere. For more than a billion years, however, life remained anaerobic and procaryotic, until the concentration of oxygen allowed the formation of a layer of ozone that shielded the earth from the ultraviolet rays of the sun and permitted aerobic respiration. These new conditions made possible the evolution of the more advanced aerobic eucaryotes, unicellular and even multicellular plants and animals. These forms, which appeared between one and two billion years ago, with their organized nucleus, other organelles, and sexual reproduction, gave rise, through evolutionary change, to the great diversity of life in the modern world.

The Migration to the Land As pointed out previously, the fossil record supplies no evidence of the nature of the plants that migrated from the sea to the land. The earliest fossils of vascular plants so far discovered represent organisms well adjusted to a terrestrial life. Nevertheless, an hypothesis, called the homologous theory, based upon studies both of living algae and of living and extinct vascular plants, has long been widely accepted as providing a satisfactory picture, not only of evolutionary events during the invasion of the land but also of the nature of the aquatic ancestors of the vascular plants. The major features of this theory are described below.

The aquatic ancestors In the shallow waters of ancient seas there lived green algae with a prostrate system of filaments that gave rise to erect filaments, thus providing greater exposure to light and to dissolved gases. The cells of the erect filaments became divided by longitudinal walls in such a way that a multicellular, three-dimensional simple or branched shoot developed (FIG. 25-13). Clusters of cells grew out from the lower part of the erect axis, and together with the prostrate filaments, anchored the plant. Internally, elongated parenchyma cells and a system of apical cells constituted the beginnings of a vascular system and a scheme of unlimited growth.

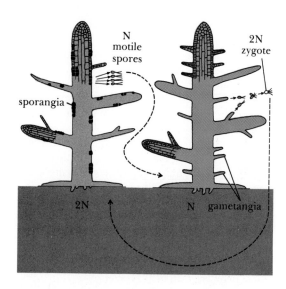

Fig. 25-13 Diagram showing hypothetical nature of the algal ancestors of vascular land plants. Haploid generation in color.

Important in tracing the origin of alternation of generations as manifested in living vascular plants is the concept that this ancestral form occurred in two phases (FIG. 25-13): (1) One, the $2n$ or sporophyte, bore superficial sporangia containing motile spores produced following meiosis. The spores, upon germination, produced (2) the n or gametophytic plant, identical with the sporophyte except that it produced gametes instead of spores. The gametes, probably isogametes, were formed within multicellular gametangia, the forerunners of archegonia and antheridia. The fusion of motile gametes in the water produced a zygote, which in turn gave rise directly to the sporophyte phase.

The transition to the land During the migration to land, plants may have lived in tidal zones or in brackish pools formed as the land slowly rose from the sea, and were thus gradually or periodically exposed to the drying effects of the sun and air. This went on for ages of time, a half-million to a million years, and during this time modifications arose that made possible plant survival and reproduction in the new environment. The spores lost motility and became cutinized, allowing dispersal by wind. A cutinized epidermis and stomata evolved. The cells of the central parenchyma strand became more elongated and then differentiated into vascular tissue. During this progressive adjustment to the land, a very significant trend developed with respect to the relative size and complexity of the two generations. The gametophyte generation became reduced, and the sporophyte underwent greater elaboration and increase in size (FIG. 25-14). The greater size of the sporophyte, of course, permitted more effective spore dispersal. Elaboration of the sporophyte involved the invasion of the substratum by the basal part of the vascularized erect system, thus providing not only additional support but also absorption of water and nutrients. Reduction of the gametophyte was useful because it brought the gametangia closer to the soil level, where moisture was more readily available. Oogamy may then have evolved, the gametangia becoming converted into archegonia and antheridia—structures that persist in many kinds of plants, vascular and nonvascular, to the present day. Following the evolution of oogamy, the female gamete was retained upon the gametophyte, and after fertilization gave rise to the plant body (FIG. 25-14). From this primitive type may have evolved a number of evolutionary lines, differing in their mode of branching, in the position of their sporangia, and in the nature of the gametophytic stage.

It is postulated that the gametophyte, although living for a shorter time than the sporophyte, yet was relatively long lived, remained photosynthetic, and retained for a time a limited amount of vascular tissue. It eventually became subterranean and mycorrhizic, assuming the form of a short, fleshy, cylindrical, sometimes branching axis that bore sex organs on all surfaces and grew slowly by an apical meristem.

Gametophytes of this general type are found in several primitive families of the Filicales, including the Psilotaceae. In other ferns (Ophioglossales) and in *Lycopodium* the subterranean gametophytes may be greatly elongated or short and tuberlike, and long-lived in the soil. These, as well as the subterranean prothalli of the Filicales, could be regarded as ancestral features evolved during the migration to the land in pre-Devonian times. The origin of the familiar green heart-shaped prothallus is obscure, but it may be postulated that it was derived, following several evolutionary steps, from the fleshy, cylindrical prothallus of primitive ferns. In the course of this evolution it became flattened and surface-living. Evidence that the gametophyte generation once had vascular tissue is also available—such as the occasional occurrence of tracheids in the gametophyte of *Psilotum*. Tracheids also appear in fern prothalli that produce apogamous embryos, and their formation may be induced in prothalli cultivated on artificial media containing a high concentration of sugar (FIG. 25-15).

Fig. 25-14 Hypothetical early land plant, reduction of gametophyte generation. Diploid generation shaded; haploid generation in color.

Fig. 25-15 Vascular tissue in fern prothallus cultured on agar-hardened mineral medium containing 3 percent sucrose.

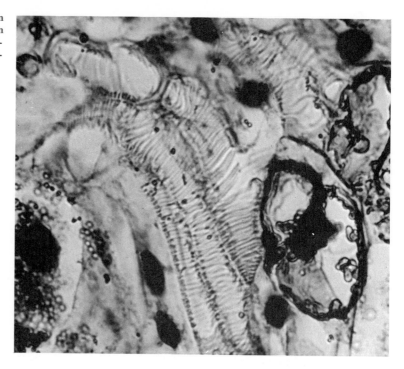

Fig. 25-16 (A–F): stages illustrating theory of evolutionary origin of large expanded leaf such as that of a fern. Explanation in text.

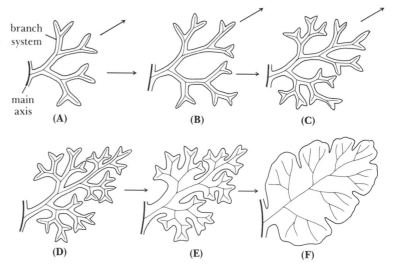

Plant Body Components In most vascular plants the embryo produces a shoot, composed of stem and leaves, and usually a primary root. The origin of the root is obscure. According to one theory, it originated from a subterranean branch of the leafless stem of plants similar to some of the Rhyniales. It acquired a root cap and retained the pithless vascular cylinder characteristic of the stem of many ancient plants.

The evolutionary origin of the leaf has long been a subject of inquiry. The leaf has probably originated in several ways. In the Lycopsida it is thought to have arisen from the stem as a superficial outgrowth that eventually became vascularized. In other vascular plants the leaf is considered to represent a flattened and expanded portion of a branching axis. This general concept, as applied to the origin of leaves of ferns and other lower vascular plants, is illustrated in Fig. 25-16. In Fig. 25-16A, a three-dimensional branched system became flattened into a two-dimensional dichotomous system. Several processes now became operative. First, the branches of each dichotomy developed unequally, one becoming stronger than the other, which became laterally displaced (Fig. 25-16B, C). This process, called overtopping, resulted in a central axis with lateral branches (Fig. 25-16D). Second, these branches became flattened and thus expanded (Fig. 25-16E). Finally, by a process called webbing, parenchyma tissue became continuous between the branches, and a blade was formed (Fig. 25-16F). This theory, as applied to the leaf of ferns and other lower vascular plants, has some support in the fossil record; that it applies to the leaf of seed plants has been questioned. The simple dorsiventral leaf of an angiosperm, for example, may represent only one of the ultimate divisions of a branch system.

Heterospory and the Seed Most of the lower vascular plants produce but one kind of spore, a condition known as HOMOSPORY. Most ferns are homosporous, as are certain other living plants: the Psilotales, the club mosses (*Lycopodium*), and the horsetails (*Equisetum*). The spores in these forms germinate on the soil, producing a gametophyte that is nutritionally independent of the sporophyte and which bears both kinds of sex organs on the same prothallus. In a number of lower forms, however, the condition of HETEROSPORY is found. Here there are two kinds of spores as, for example, in the little club mosses (*Selaginella*, Chapter 26). One, the megaspore, is much larger than the other, the microspore. These two types of spores are borne in different sporangia, designated as mega- and microsporangia. In heterospory, as in homosporous forms, each spore gives rise to a prothallus, but this is greatly reduced and is retained wholly or largely within the spore wall, developing at the expense of previously stored food. Although the term heterospory is commonly used to refer to spores that differ distinctively in size, a more important distinction is that they are each unisexual, for the megaspore produces only a female prothallus, and the microspore only a male prothallus. The sex organs and gametes are thus restricted to different gametophytes.

The modern seed plants—the conifers and other seed plants, including the flowering plants—are also classed as heterosporous. If based upon spore size, the term is a misnomer, for measurements have shown that the megaspore may be either larger or smaller than the microspore, or the two kinds of spores may be approximately the same size. These plants, then, are only functionally heterosporous. Their microspores and megaspores might just as appropriately be termed male and female spores, thus placing the emphasis upon the kinds of gametophytes and gametes each produce, rather than upon spore size. Whatever terminology is used, the term heterospory, in connection with seed plants, will probably be retained as a matter of custom and convenience.

It is generally postulated that the seed plants, living and extinct, arose from heterosporous ancestors, which, instead of discharging the megaspore, acquired the habit of retaining it within the sporangium. The megaspores may

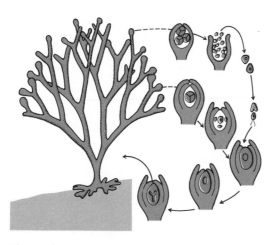

Fig. 25-17 Hypothetical early seed plant; the mega-spore produces a female prothallus upon the female plant where fertilization occurs. Diploid generation shaded; haploid generation in color.

originally have been larger than the microspores, but became reduced in size when they were permanently retained within the megasporangium. The megaspore germinated and formed a female prothallus, still enclosed within the spore wall (Fig. 25-17). In ancient heterosporous plants, the megaspore contained only minute amounts of food because its small size restricted the amount of storage. This is also true of modern plants. But in seed plants a further advance occurred. By means of the intimate contact of the wall of the megaspore with the parent plant, the female gametophyte received food for its development from the parent rather than from materials stored in the spore itself.

The microspores (male spores) also germinated and formed male gametophytes. These gametophytes, still surrounded by their spore walls, were then discharged from the sporangium, disseminated by wind or water, and fertilization occurred upon the parent plant. The male gametes were at first motile and reached the egg only in the presence of rainwater or dew. Later, a pollen tube developed, and the sperm

became nonmotile. In some living seed plants, the gametes within the pollen tubes still have organs of motility.

After fertilization, the embryo and female prothallus, together with an accessory structure, the seed coat, became the seed. The seeds of living plants—of oaks and pines, tomato and corn—have lost many of their ancestral characters, but have also retained structural features that give us, at least in broad outline, an understanding of their past history. We may view the seed as the culmination of a series of evolutionary steps that began with a simple prothallus in an ancient world.

In the modern conifers and certain other seed plants, the female gametophyte, prior to fertilization, becomes greatly enlarged. It produces archegonia and eggs and it accumulates food material in abundance. After fertilization, the zygote is nourished by this food as it develops into the embryo. The accumulation of a supply of food prior to fertilization may be compared to the condition in the reptiles and birds. The yolk of the egg of these animals corresponds to the accumulated food of the female gametophyte.

In the flowering plants a different condition prevails. Here the food supply necessary to the development and support of the embryo is received from the parent plant after fertilization has occurred, and no food is stored unless fertilization has taken place. This is somewhat analogous to the condition in most mammals, in which the embryo is nourished by the parent through the umbilical cord. This pattern in both mammals and flowering plants is considered an improvement in the reproductive process because no energy is expended in accumulating reserve food for an embryo until an embryo exists. In living seed plants, as a general rule, the seed is detached from the parent plant only when the embryo is fully formed and well supplied with food.

The megaspore, with its small supply of food material, is the resting stage in the lower heterosporous plants, but in modern seed plants the resting stage occurs later, after the embryo has developed. The success of the seed plants is due in no small measure to the reproductive advantage obtained when the spore is retained, protected, and supplied with food. This may be contrasted with the condition in which the small megaspore was set free and the prothallus developed independently. Seeds, as did heterospory, evolved at different times and in different evolutionary lines. Some of these lines have become extinct; others, chiefly the conifers and angiosperms, have become conspicuous among living plants.

The Flower The geological record has not yet yielded fossils of the earliest flowering plants, and the transitional stages in the evolution of the flower are unknown. The earliest fossils that are unquestionably angiospermous have been found in the Lower Cretaceous, perhaps 120 million years old. Fossilized wood and pollen grains from the Jurassic have been attributed to angiosperms, and a palmlike leaf has been reported from the Triassic, but convincing evidence that flowering plants existed previous to the Cretaceous is lacking. Nevertheless, the well-defined angiospermous characters exhibited by the earliest known Cretaceous fossils have led to the belief that angiosperms must have originated prior to the Cretaceous. Just how far back is quite uncertain, but some authors believe that angiosperms must have passed through a period of evolutionary development extending from the late Paleozoic (Permian).

Arising from some ancient seed plants, the flowering plants evolved a further structure, the carpel, which envelops and protects the seed. In all other seed plants, the pollen is conveyed to the immediate vicinity of the ovule, but in the flowering plants it is deposited upon a specialized receptive surface, the stigma. The flowering plants, in numbers of species and individuals and in adaptability to a wide range of environmental conditions, are even more successful than the gymnosperms and are regarded as the most highly evolved members of the plant kingdom.

Plant Relationships As pointed out earlier in this chapter, the plants of today are living members of groups that had their origins in a distant past. Evidences of evolution are observable in the changes in form and structure throughout geological time and in the variety of form and structure in living plants. On this basis some botanists have attempted a phylogenetic system—in everyday language, a "genealogical tree" of the plant kingdom—with the various categories arranged in uninterrupted evolutionary lines, showing the precise way in which the great assemblages are related by descent. Largely because of gaps in the fossil record, however, such a "tree" has little validity based on the present state of our knowledge. It is possible, however, to present an arrangement that illustrates evolutionary divergence, certain evolutionary trends, and expresses those natural relationships that serve as a basis for classification. Such a whole view of the plant kingdom is shown in FIG. 25-18.

SUMMARY

1. Evolution in the plant kingdom has in general proceeded from primitive and structurally simple forms to organisms of greater complexity.
2. Much of the evidence for organic evolution is derived from the study of fossils and from comparative studies on the vegetative and reproductive structures of living plants.
3. The various types of fossils include compressions, impressions, casts, molds, and petrifactions. Since conditions for the formation and preservation of fossils have frequently been lacking, the fossil record is fragmentary and incomplete.
4. Geologic time is divided into eras, periods, and epochs. The oldest unquestionable fossil remains, chiefly bacteria and blue-green algae, have been found in the lower Precambrian, about 3.1 billion years old.

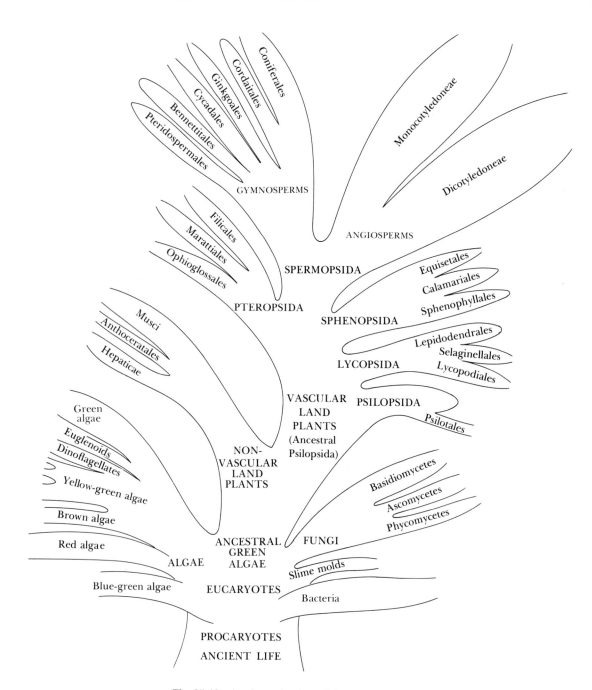

Fig. 25-18 A schematic view of the plant kingdom.

The record of the caryotic green algae begins about 1 billion years ago.

5. Fossils of early vascular land plants have been found in rocks of the Silurian and especially in the Devonian. The nature of the first land plants is unknown, but it is thought that they were similar to the members of the order Rhyniales, of the Psilopsida. Representatives of all other great divisions of the Tracheophyta, possibly derived from the Psilopsida, have been found in later Devonian times.

6. The two genera of the Psilotaceae (Psilotales) are among the most primitive of living vascular plants. They have long been classified in the subdivision Psilopsida and regarded as surviving members of this otherwise extinct group. New evidence, however, indicates that the Psilotaceae should be classified in the Filicales (ferns).

7. The first stage in plant evolution, like that of animals, was the evolution of the living cell. Among important later stages were the migration of plants from the sea to the land, the elaboration of the plant body into a root and a leaf-bearing shoot, and the evolution of heterospory, the seed, and the flower.

8. The first land plants probably arose from green algae in which the gametophyte and sporophyte generations were equal in size. During the migration and establishment of vascular plants upon the land, the gametophyte generation gradually became much reduced while the sporophyte generation became more elaborated and constituted the dominant plant body.

9. The success of the seed plants is due, in considerable degree, to the advantage of retaining the megaspore upon the parent plant and supplying it with food that can be used by the embryo. The food supply is transported to the female gametophyte before fertilization in all seed plants except the flowering plants. In these, the food supply of the embryo is accumulated after fertilization.

26

The Club Mosses and the Horsetails

T HE CLUB MOSSES (LYCOPSIDA) and the horsetails (Sphenopsida) are plants with a long history, extending from the Devonian to the present, about 400 million years. The living members of these two ancient groups are of no great significance to man, nor are they important components of the earth's present vegetative cover. But in the past, many of the species now extinct were very prominent. In some cases they were the dominating plants of the great swamp forests of the Carboniferous. A knowledge of the structure and of the methods of reproduction of the surviving members of these groups will provide many clues to an understanding of the extinct species, which can be studied only from their fossil remains. Because of their ancient lineage and former great abundance, the club mosses and horsetails must be included in any study that attempts to survey the plant life of the earth.

THE LYCOPSIDA

Most of the living species of the Lycopsida are grouped in two genera, *Lycopodium* and *Selaginella*. The 450 or so species of *Lycopodium* are known as the club mosses and are homosporous. *Selaginella* includes about 500 species, known as the small club mosses, or spike mosses. These are heterosporous. Once again, the common names are misleading, for these plants are not mosses or even related to mosses.

Lycopodium The club mosses are found in general in much the same habitats as ferns. Most species are tropical or subtropical, but some flourish in temperate climates, especially in the cool moist forests of the Northern Hemisphere. Some grow rooted in the soil, but many of the tropical species are epiphytes, with their roots in

Fig. 26-1 Nodding Clubmoss (*Lycopodium cernuum*) scrambling over a moss-covered clay bank, New Zealand. Only a portion of the plant is shown.

the crevices of the bark of the trunks or branches of trees. The Nodding Clubmoss (*Lycopodium cernuum*) is one of the most widely distributed of the soil-rooted species, occurring in the tropics and subtropics of both the Old and New Worlds. The plants grow 2–5 feet in height, resembling small trees, with numerous repeatedly forking branches from a main stem (FIG. 26-1) They sometimes form dense thickets. In some areas the plant may grow to a length up to 15 feet when supported by surrounding vegetation. At maturity many of the lateral branches are terminated by spore-bearing conelike structures 10–15 mm long (FIG. 26-1).

The club mosses native to the temperate parts of North America are green, terrestrial plants of varied habit and size (FIG. 26-2). The aerial stems may be clustered and rooted at the base, or trailing and creeping, with erect branches arising from horizontal stems that grow over or under the surface of the soil. The erect branches vary from 2 or 3 inches to 6 or 8 inches in height. Some species, in which the erect parts are bushy, are known as ground pine or ground cedar. The horizontal stems are elongated by terminal growth, often forming long runners. As the older parts die, the younger branches continue growth, a single plant thus giving rise to a number of separate plants. This vegetative method of reproduction frequently produces large colonies in woodlands and meadows.

Most temperate zone species of *Lycopodium* are evergreen perennials, and are sometimes—unfortunately, for reasons of conservation—sold as Christmas decorations. The leaves of the club mosses are numerous, small, and in some species so scalelike and inconspicuous that the branches may be mistaken for leaves. The branching is sometimes dichotomous, the paired branches being equal or one somewhat larger than the other. A primary root system is absent in adult club mosses. All roots are adventitious, arising from the bases of the stems or the underside of the prostrate parts of the shoot. A cross section of a club moss stem (FIG. 26-3) reveals an epidermis, a wide cortex, and a pithless vascular cylinder composed of xylem and phloem. The xylem is deeply and irregularly lobed, with the phloem lying between the lobes of the xylem.

The life cycle The life cycle of *Lycopodium* is generally very similar to that of the fern. The club mosses are homosporous. The numerous spores are produced in large, kidney-shaped sporangia, which are borne singly in the axils of the sporophylls. In one primitive group of species, the sporophylls are similar to the vegetative leaves in their size, shape, position, and capacity for photosynthesis (FIG. 26-4A). The sporangia are scattered along the leafy shoot, although they tend to form fertile zones that alternate with vegetative zones.

Fig. 26-2 Club mosses. (Left): shining club moss (*Lycopodium lucidulum*). (Right): common or running club moss (*Lycopodium clavatum*).

In a second, and more advanced, group of club mosses, the sporophylls are grouped into conelike structures called STROBILI. These are sessile at the ends of leafy branches or are raised on long stalks. The sporophylls are greatly reduced, and are scalelike (FIGS. 1-4, 26-4B), and yellowish. The compact appearance of the strobili, together with the small leaves and the general habit of the plants, suggests the common name of club moss.

The spores are formed from spore mother cells after meiosis. They are tetrahedral and usually have radiating lines on one side (FIG. 26-4A). The walls have thickened ridges laid down in various patterns. The mature spores escape through a slit in the wall of the sporangium and are disseminated by air. When a patch of club mosses in which the spores are ripe is disturbed, as by someone walking through it, a yellowish cloud arises, composed of millions of spores.

The prothallus

The spores of the club mosses germinate on or beneath the soil surface and eventually give rise to mature prothalli bearing the sex organs, archegonia and antheridia. The prothalli vary considerably in size and form in different species, but all are relatively small and tuberlike. They fall into various categories. In one type, found in many tropical and a few temperate zone species, the prothallus is an erect,

Fig. 26-3 *Lycopodium,* cross section of stem.

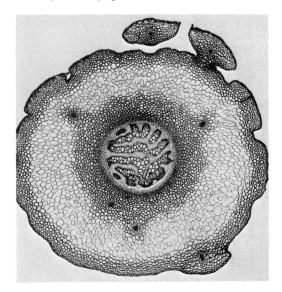

short, stout, ovoid or cylindrical structure, and is lobed at the top in some species. The basal portion is yellowish white, buried in the soil, and bears rhizoids. The upper part is green and photosynthetic. The prothalli are small (2–3 millimeters in diameter) and short-lived, living but a few months. The basal portion is early invaded by intracellular fungus hyphae, which

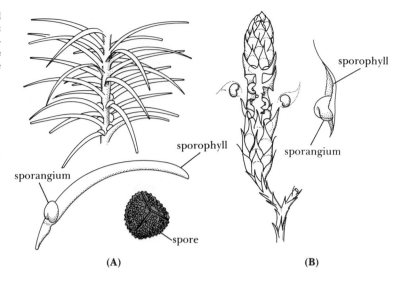

Fig. 26-4 Sporophylls and sporangia of *Lycopodium*. (A): *Lycopodium lucidulum;* a strobilus is absent. (B): *Lycopodium clavatum;* the sporophylls are grouped in a strobilus.

(A)

(B)

Fig. 26-5 Subterranean prothalli of *Lycopodium*, with attached sporelings. (Left): prothalli of *Lycopodium complanatum* var. *flabelliforme*. (Right): prothalli of *Lycopodium obscurum*.

appear to be necessary to the later growth of the prothallus.

Another, and evolutionarily more advanced, type of prothallus is that found in most species of club mosses of the temperate zone. This is without chlorophyll and develops entirely underground, buried at depths of 1–4 inches (Fig. 26-5). As with the aerial type, fungus hyphae are associated with these underground structures, but even more abundantly, occurring chiefly in a zone near the periphery of the prothallus. These hyphae make food available to the prothallus from the surrounding organic matter. The prothallus is commonly said to be saprophytic, but, as pointed out in Chapter 5, it is the fungus that is saprophytic, not the higher plant. The relationship between fungus and prothallus may be mutualistic, but if so, the advantage to the fungus in the association is not known.

The subterranean prothalli vary considerably in size and form, but are generally much larger (from 1–2 centimeters in length or width) than the aerial type. Depending upon the species, they are cylindrical, carrot-shaped, disklike, or lobed and convoluted. In both the green and nongreen types the sex organs are borne on the upper surface of the prothallus, sunken in the tissues. The biflagellate sperms reach the archegonia by swimming through a film of free water on the surface of the prothallus. After fertilization the $2n$ zygote produces an embryo that grows into a leafy plant. More than one egg may be fertilized, and several sporophytes may arise from the same prothallus. The subterranean type of prothallus may live several years following fertilization, continuing to supply food to the young sporophyte even after the latter has become green and grown upward some distance above the soil. Such buried prothalli are difficult to find in nature, and are usually discovered by finding the young sporophyte and tracing the stem downward to the attached prothallus.

It has long been known that the spores of the temperate zone species with deeply buried prothalli germinate and form mature prothalli only over long periods of time. A German investigator reported in 1885 that the spores of one species, placed on soil, required 3–5 years for germination and 6–8 years for the growth of the prothallus to the formation of sex organs. Two other species required 6–7 years to germinate and 12–15 years for the maturation of the prothallus. The buried position of the prothalli in the soil is explained by this delay in germination, for during this period the spores are carried downward by seepage water. In spite of numerous attempts, it was not until 1957 that the spores of two species of the underground type were induced to germinate and produce prothalli with sex organs in culture media containing minerals and sucrose. This was accomplished by treating the spores with concentrated sulfuric acid or other agents that acted upon the spore wall, allowing water to enter. The prothalli thus cultured were not colorless, tuberous, and fleshy as in nature, but green and much branched, with the sex organs borne on the branches. Even when cultured in this manner, six weeks were required for germination, and sex organs did not appear until four to five months after germination.

Selaginella The small club mosses are highly diverse in form (Fig. 26-6). Some are creeping and flat; others form tufts or mats; still others grow erect. The genus is much more conspicuous in the moist tropics than in the temperate zone. A number of tropical species branch freely, and the erect forms may attain a height of a foot or more. The delicacy and beauty of the foliage of some tropical species has resulted in their extensive cultivation in conservatories and by florists. One species, *Selaginella lepidophylla*, native in Texas and Mexico, forms a dense green rosette during the growing season. As the season becomes dry, the stems curl inward until the plant assumes the form of a compact ball. The balls break at the base and roll away in the wind. During the next rain, the plant absorbs moisture, the stems unroll, and growth may be resumed. This species of *Selaginella* is called the resurrection plant or bird's-nest moss.

The temperate zone species of *Selaginella* are

Fig. 26-6 Selaginellas. (Left): *Selaginella hamaetodes*, Central America. (Right): *Selaginella apoda*, North America.

mostly small and mosslike. They occur over a wide range of habitats, from moist, shady banks to dry, rocky slopes. The leaves are small and overlapping, and are arranged in spirals or in vertical rows along the slender stems. The vascular tissues may be arranged in a pithless cylinder, or as a group of two, three, or five or more irregularly distributed bundles. The xylem of some species contains true vessels.

The life cycle As in *Lycopodium*, the sporangia are borne in the axils of sporophylls, which are always clustered in a sessile strobilus, usually four-sided. The cones range in length, depending on the species, from ¼ inch to 2–3 inches. Unlike *Lycopodium*, *Selaginella* is heterosporous. It forms two kinds of spores: small and numerous (up to 600 in a sporangium) red, yellow, or brown microspores, borne within a microsporangium, and much larger thick-walled megaspores, commonly four in number, borne within a megasporangium (FIG. 26-7). The

megasporangium is usually much larger than the microsporangium, and is locally distended or lobed by the growth of the spores within. The four large megaspores are formed from a single spore mother cell. A number of spore mother cells are originally present, but all, with the exception of one, degenerate previous to meiosis, and the degenerating cells contribute to the nutrition of the megaspores.

The sporophylls are termed microsporophylls or megasporophylls, according to the kind of sporangium that each bears in its axil. Otherwise they are alike, green, and somewhat smaller than the vegetative leaves. Both kinds of sporophylls and their sporangia are usually borne on the same strobilus. The megasporangia may occur at the base and the microsporangia in the upper portion, or the two kinds may be intermingled.

Male and female prothalli The sporangia open and the spores are discharged with some force by a mechanism involving cells with walls thickened

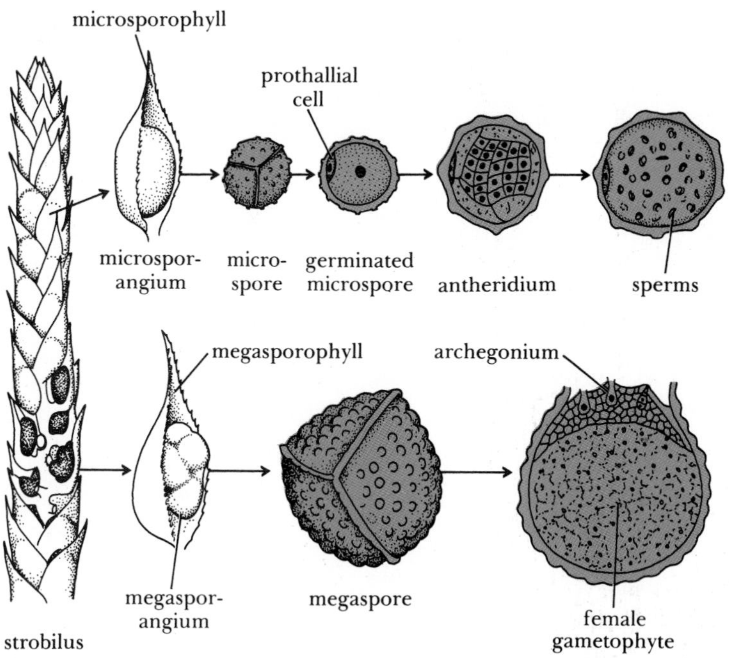

Fig. 26-7 *Selaginella*; sporophylls, sporangia, spores, male and female prothalli.

microsporophyll

prothallial cell

microsporangium · microspore · germinated microspore · antheridium · sperms

megasporophyll · archegonium

megasporangium · megaspore · female gametophyte

strobilus

in much the same manner as those of the annulus of the fern sporangium. The microspore produces a male prothallus; the megaspore, a female prothallus. In the homosporous *Lycopodium* and the ferns, germination results in a prothallus produced outside the spore wall. In *Selaginella* the prothalli develop mostly or entirely within the spore wall (FIG. 26-7), and germination may begin before the spores are shed from the parent plant. The male prothallus is formed entirely within the wall of the microspore. The first stage in germination consists in the formation of two cells. One of these is large and occupies most of the space within the spore. The other, the *prothallial* cell, is small and lens-shaped, and lies close to the spore wall. The prothallial cell is believed to be vestigial, the remains of vegetative tissue of a prothallus that was once multicellular. The larger cell divides many times. Some of the cells resulting from these divisions form a jacket of cells, the antheridium. Others form a group of centrally placed cells from which, following further divisions, many biflagellate sperms develop. The prothallial cell and antheridial jacket then degenerate, leaving the mature male gametophyte, which now consists only of numerous sperms contained within the microspore wall. Depending on the species, the microspores are discharged from the sporangium at various stages previous to sperm formation, and development of the male gametophyte continues on the soil. When the sperms are mature, the spore wall cracks open and the sperms are set free.

The megaspore germinates, forming a multicellular female gametophyte, or prothallus, which grows until it fills the space within the spore wall (FIG. 26-7). The lower part of the prothallus contains a considerable amount of accumulated food, later used in the growth of the young sporophyte. At a certain stage (again, varying with the species) the germinated megaspore is shed and prothallial development continues on the soil. Continued growth of the female prothallus finally causes a rupture of the spore wall and a cushion of tissue protrudes, bearing a number of embedded archegonia. In some species, rhizoids develop among the archegonia and extend into the soil. If microspores and megaspores fall in close proximity, the sperms swim to the egg in a film of water, fertilization occurs, and a young sporophyte develops. The young plant remains attached for some time to the female gametophyte, which is still held within the spore wall (Fig. 26-8). The female prothallus, even after exposure to light, probably never develops chlorophyll in nature. Certainly the female gametophytes of the two species of *Selaginella* that were grown on culture media never developed chlorophyll, even after some months of exposure to light. The complete life cycle of *Selaginella* is shown in Fig. 26-9.

The kind of heterospory found in *Selaginella*, with large, free megaspores and small microspores, occurs in relatively few other groups of living vascular plants, chiefly the water ferns and quillworts (*Isoetes*, Chapter 18). It was, however, present in a number of groups of extinct plants. *Selaginella* and *Isoetes* are not related to the water ferns, and heterospory in the latter group probably arose independently.

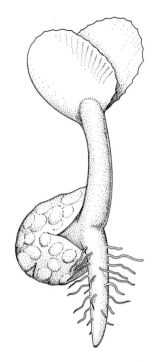

Fig. 26-8 Young sporophyte of *Selaginella apoda* still attached to the female prothallus. The prothallus, in turn, is enclosed within the megaspore wall.

Fossil Club Mosses Among the most conspicuous of the many unusual plants of the Carboniferous period were the giant club mosses. Two genera, *Lepidodendron* and *Sigillaria*, are well known. The lepidodendrons were large trees, frequently 100 feet or more high and 3 feet or more in diameter at the base (Fig. 26-10). The trunk branched dichotomously, and large cones, in some species more than a foot in length, terminated the branches. The lepidodendrons were heterosporous, with both kinds of spores produced in the same strobilus. The leaves were deciduous and narrow, and varied in length from a few inches to 2 feet or more. The trunk contained some secondary wood, but the greater part was composed of a thick layer of parenchymalike cork, produced by successive layers of cork cambium. The bark of the main trunk and branches was covered with spirally arranged LEAF CUSHIONS, diamond-shaped in outline. The leaf cushion was composed of the enlarged basal part of the leaf, which underwent abscission slightly above the base, leaving a leaf scar in the center or upper part of the leaf cushion (Fig. 26-11). As the tree aged, the increase in thickness of the cork layers led to the sloughing off of the leaf cushions at the base of the trunk, which became rough and ridged. Impressions of the bark showing leaf cushions are among the most common and characteristic fossils in Carboniferous strata.

The sigillarias were also large trees (Fig. 26-12), but they differed from the lepidodendrons

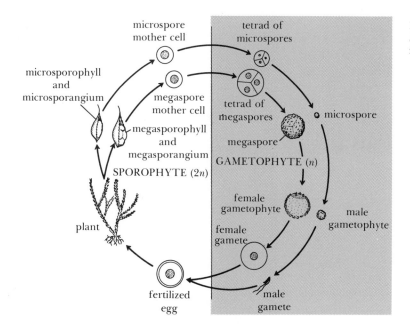

Fig. 26-9 Life cycle of *Selaginella*; haploid generation in color.

in several ways. The trunk was sparsely branched. It bore, at the top, a large cluster of elongated narrow leaves, sometimes 3 feet in length. The trunk of *Sigillaria* also bore leaf cushions, but less prominently raised, and most of the surface of the cushion was covered by a leaf scar (FIG. 26-13). The leaf cushions of *Sigillaria* were spirally arranged, as in *Lepidodendron*, but because of the presence of longitudinal ribs or ridges they appear to be aligned in vertical rows, a feature that distinguishes impressions of the bark from those of *Lepidodendron*. *Sigillaria* was also heterosporous, but the strobili were usually borne in whorls on the main axis, just below the apex of the trunk.

The relationships of the modern, living club mosses to *Lepidodendron* and *Sigillaria* are obscure. Students of these plant groups generally consider that the latter genera represent an evolutionary line separate from that which produced the living club mosses, *Lycopodium* and *Selaginella*. The giant club mosses flourished in the Carboniferous, waned, and became extinct,

leaving no descendants. The living forms are probably descended from ancient low-trailing plants, known from impressions in sedimentary rocks. One such genus, *Lycopodites*, found in the Upper Devonian and Carboniferous, resembles the living species of *Lycopodium* in which the sporophylls are not aggregated into strobili. A heterosporous, small club moss, *Selaginellites*, has been found from the Upper Carboniferous to the Triassic. The modern genera, *Lycopodium* and *Selaginella*, thus have their roots in a distant past, more than 375 million years ago.

THE SPHENOPSIDA

The living Sphenopsida are all classified in one genus, *Equisetum*, commonly known as horsetails. Several species are very common and cosmopolitan in distribution. Most horsetails grow in wet or swampy environments such as shallow ponds, marshes, or stream banks. But some thrive in the comparatively dry soil of meadows, roadsides,

Fig. 26-10 *Lepidodendron.*

Fig. 26-11 Surface of bark of *Lepidodendron*, showing leaf scars and diamond-shaped leaf cushions.

and even on barren railroad embankments. The horsetails vary in height from less than a foot to about 3 feet, but several species are taller. One tropical form (Fig. 26-14) may attain a height of more than 30 feet under favorable circumstances.

The Plant Body The plant is composed of perennial underground rhizomes from which

aerial branches arise. In some species the aerial stems are evergreen and perennial; in others they are annual, appearing early in the growing season. The stems (Fig. 26-15) are cylindrical, conspicuously jointed, and marked with longitudinal ridges. They may be branched or unbranched. The scalelike leaves, fused by their margins into a cylindrical sheath, are borne in whorls at the nodes. The branches, if present, also arise in whorls at the nodes. Their bushy symmetry suggests the common name of these plants. Photosynthesis is carried on chiefly by the stems and branches, for the leaves are minute and in many species devoid of chlorophyll. Strobili are borne terminally on the main stem or the branches. In some species, all the stems produce strobili (Fig. 26-15, left); other species, such as the widely distributed *Equisetum arvense*, are dimorphic, with two kinds of aerial stems, sterile and fertile (Fig. 26-15, right). In this

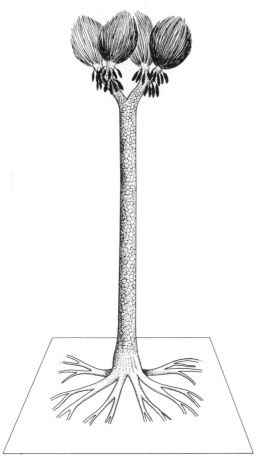

Fig. 26-12   *Sigillaria.*

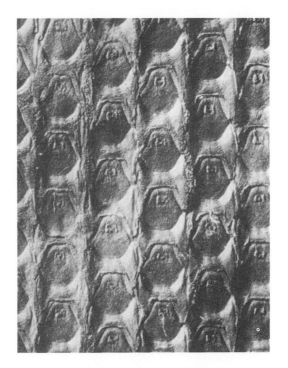

Fig. 26-13   Leaf scars of *Sigillaria.*

Fig. 26-14   (Right): *Equisetum giganteum,* Central America. The hat shows relative size.

Fig. 26-15 *Equisetum.* (Left): strobili of common scouring rush (*Equisetum hyemale* var. *affine*). (Right): strobili of common horsetail (*Equisetum arvense*). This species has both fertile and sterile aerial shoots.

Fig. 26-16 (Left): cross section, stem of *Equisetum*. (Right): portion of section, more highly magnified.

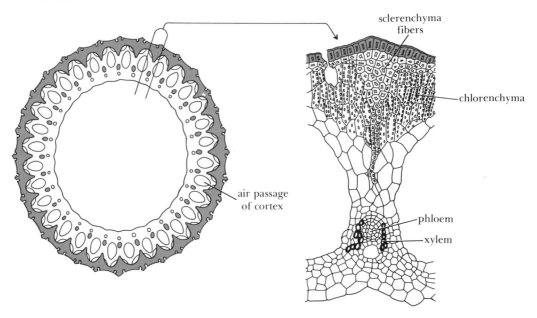

species the fertile shoot lacks chlorophyll and has but a brief existence, dying to the ground after the spores are shed. The sterile shoots continue to grow vigorously, becoming bushy plants that persist until autumn.

Internally (FIG. 26-16) the stem of horsetails is usually hollow, except at the nodes. The wide cortex contains air passages that extend from one node to the next. A cambium is absent. The vascular tissues are reduced to a ring of small vascular bundles, each bundle composed of a group of phloem cells and two small masses of xylem, one on either side of the phloem. True vessels have been observed in the xylem of the rhizomes of five species of *Equisetum*. Sclerenchyma fibers, lying in the outer cortex, are chiefly responsible for mechanical support of the stem. Both these fibers and the epidermal cells are strongly impregnated with silica, in one species to such an extent that it was formerly used for scouring pots and pans and received the common name "scouring rush" (FIG. 26-15).

The life cycle The life cycle of *Equisetum* is similar to that of *Lycopodium* and the ferns. The strobilus consists of a short, thick axis bearing clusters of whorled *sporangiophores,* which in turn bear the sporangia (FIG. 26-17). The sporangia are elongated, saclike structures, which are attached to the inner side of the stalked, shield-shaped sporangiophores. Each sporangium contains a large number of spores formed following meiosis in the spore mother cells. When the spores are ripe, the sporangium opens by a longitudinal slit on the inner side. The axis of the strobilus then elongates, separating the sporangiophores (FIG. 26-15) and leaving intervening spaces in which the spores may lodge before they are carried away by air currents and fall upon the ground. After spore dispersal, the strobilus shrivels, but the stem upon which it was borne may continue as a vegetative shoot.

The spores of the horsetails contain numerous chloroplasts and the mature spore wall consists of four layers. The outermost of these is deposited in the form of four spirally arranged strips, which later split away as four elongated appendages that have flattened tips and a common point of attachment. These spore appendages, or *elaters* (FIG. 26-17), are hygroscopic, coiling and uncoiling rapidly with changes in the humidity of the air. The resulting movements within the spore mass very likely contribute to the expulsion of the spores from the sporangium. In addition, the elaters probably play a role in dispersal. Since they are expanded when dry, they serve as "wings" that aid in spore dispersal by light breezes. It has also been suggested that when the spores are conveyed over a damp area where the air humidity is high, the elaters coil, thus reducing the buoyancy of the spores so that they settle upon the damp soil where conditions are suitable for germination.

The prothallus The spores of *Equisetum* germinate on the soil (FIG. 26-18A-D). They absorb

Fig. 26-17 *Equisetum,* **sporangiophore, sporangia, and spores. The spore elaters are shown in the uncoiled position (above) and in the coiled position (below).**

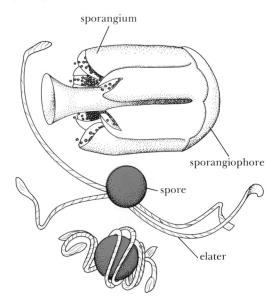

sporangium

sporangiophore

spore

elater

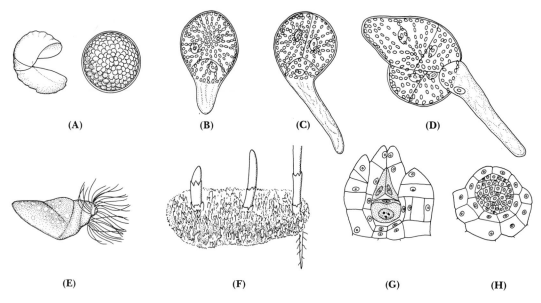

(A)   **(B)**   **(C)**   **(D)**

(E)   **(F)**   **(G)**   **(H)**

Fig. 26-18 *Equisetum.* (A–D): stages in spore germination. (E): sperm. (F): prothallus. (G): archegonium. (H): antheridium.

water, swell, and cast off the elaters and the second spore wall. A cell wall divides the spore into a larger and a smaller cell, the later shortly elongating into a colorless rhizoid. The larger cell continues to divide and eventually gives rise to a green prothallus (Fig. 26-18F), which lives an independent existence for some weeks. The prothallus is small, only a few millimeters in diameter, and consists of a basal cushion from which arise numerous, erect green lobes and which is attached to the soil by rhizoids. *Equisetum* is homosporous and bears both kinds of sex organs (Fig. 26-18G, H) on the upper surface of the cushion of the same prothallus, or the antheridia may be produced on the green lobes. In culture, however, some species are reported to produce two kinds of prothalli: unisexual, bearing only antheridia, and bisexual, bearing both kinds of sex organs. The sperms (Fig. 26-18E) are motile and swim to the archegonium in the presence of water. Several sporophytes may arise from the same prothallus if more than one egg is fertilized.

Fossil Horsetails Prominent among the extinct Sphenopsida are the giant horsetails (*Calamites*). Petrifactions, casts, and compressions of these plants show that they attained their maximum development in the Upper Carboniferous. The calamites strongly resembled the modern horsetails, but reached the dimensions of trees (Fig. 26-19). Some grew to a height of 40 feet or more. Stem fossils of *Calamites* showing nodes and longitudinal ridges strikingly similar to those of *Equisetum* are very common. Such fossils do not, however, represent the surface of the trunk, which was relatively smooth, but are casts of the pith cavity (Fig. 25-2). The grooves represent the vascular bundles that projected into the pith. The pith cavity, in some species, attained a diameter of more than a foot.

The leaves (Fig. 26-20) were narrow and sometimes needlelike, longer than those of *Equisetum*, and were the chief organs of photosynthesis. They were borne in whorls, the number in a whorl varying from 4 to 40, depending on the species. The leaves were not fused, or only

Fig. 26-19   A Carboniferous swamp forest. The trees in the center and at the left are *Calamites*. The large tree at the right is *Sigillaria*. The plant at lower left with whorled leaves is *Sphenophyllum*.

Fig. 26-20 Compression of whorls of leaves of *Calamites*.

slightly so, contrary to the condition in the modern horsetails. Many of the nodes of the main stem bore large whorls of branches. In addition to the treelike habit, another important difference between the calamites and modern horsetails was the presence of secondary wood and a thick bark. The sporangia were borne on various kinds of strobili, some similar to those of *Equisetum*, others more complex in structure. Most species were homosporous, but some were heterosporous, and the spores did not bear elaters. There is little doubt that *Equisetum* and *Calamites* are closely related, but it is highly unlikely that *Calamites* is a direct ancestor of *Equisetum. Equisetum* is also very ancient, and probable fossils (*Equisetites*) have been described from the early Mesozoic. They may even have lived contemporaneously with *Calamites* in the Carboniferous. *Calamites* and *Equisetum* may have arisen from a common ancestral stock, far back in the Paleozoic. The calamites became extinct, but the horsetails have continued to flourish to the present day.

FOSSIL FUEL—COAL

Coal is one of the most important of all the materials obtained from the earth's crust. The world's reserves of coal are estimated at 5000 billion tons, of which about 42 percent is in the United States. This abundant "mineral" fuel made possible the industrialization of the Western world and is of fundamental importance to our technology and industrial productivity.

Coal has been found in all geological strata from the Precambrian to the Pleistocene. Extensive beds have been found in the Permian, Triassic, Jurassic, and Upper Cretaceous. For the earth as a whole, however, the most extensive and most valuable deposits are of Carboniferous age, more than 300 million years old, sometimes called the "Age of Coal."

Coal is derived entirely from partly decomposed and consolidated plant materials. The plants of the Carboniferous concerned in coal formation were numerous in kind, but the most important groups were (1) the lepidodendrons, sigillarias, and their relatives; (2) the ferns and seed ferns; and (3) the primitive, coniferlike Cordaitales (Chapter 27). These plants, growing in low swampy forests, died and partly decayed where they lived. Their roots, stems, leaves, and spores formed beds of peat, similar to those being formed today in many parts of the world. Decay was only partial, for the stagnant waters of the swamps in which the plants grew excluded oxygen. Wood-destroying fungi did not live under such conditions, and anaerobic bacteria disintegrated wood but slowly. And so, year by year, the peat accumulated; wood and bark, leaves, twigs, spores, and pollen grains formed a mass of plant remains in all stages of decomposition. This accumulation of peat could proceed the year around, for the climate of the Carboniferous, in central Eurasia, western Europe, and North America, where the coal beds were laid down, was generally uniformly warm and temperate.

The peat was gradually transformed into coal. The vast swampy areas of peat and swamp forests—far more extensive than any in the world today—subsided and were covered by the sea. The peat deposits were buried by sedimentary clays; sand and silt washed in from the surrounding lands; and the sediment was converted into hard rock. The pressure of the overlying rock, and the heat resulting from this pressure, transformed the peat into LIGNITE, an immature, or low-grade, coal. The organic material was greatly compressed and underwent many physical and chemical changes, during which gases, especially hydrogen and oxygen, were driven off.

Further compression over long periods of time changed the lignite into bituminous, or "soft," coal, having an average carbon content of about 66 percent. Intermittent fluctuations in sea level resulted in successive layers of coal seams. As the seas receded and left vast inland swamps, swamp forests again became established and peat formed, to be covered in turn by sediments as the sea level rose. As many as 100 seams of coal have been found, one lying over the other and separated by layers of sedimentary rock. The rate of coal formation was very slow. It is estimated that 1 foot of solid peat may have formed in about 100 years, and that 3 feet of peat are compacted into 1 foot of bituminous coal. Favorable conditions over periods of millions of years made possible the accumulation of beds of coal over great areas in Pennsylvania and adjacent states. The well-known Pittsburgh seam averages about 7 feet in thickness over more than 2000 square miles. Isolated pockets of soft coal 50 feet thick or more have been found.

The final stage in coal formation—anthracite, or "hard," coal—was produced by even greater pressures and temperatures applied to the coal layers. These pressures resulted from the upward and lateral thrusts of rock strata in periods of mountain building. Where the folding was greatest, the coal beds were altered from bituminous to anthracite coal. Volatile matter was driven out, and the carbon content increased to more than 80 percent. Most of the anthracite mined in the United States is found in eastern Pennsylvania. It was formed when the folding of upthrusting rocks resulted in the building up of the Appalachian Mountains. Since anthracite is found only in folded strata, it is much less abundant and more difficult to mine than bituminous coal.

In spite of the increasing use of oil and gas, coal is still one of the important raw materials of our civilization. By its aid we make iron and steel, tools and machines. Converted into electricity, it lights our cities. From coal are obtained innumerable products used in the chemical industries—dyes, medicines, resins, plastics, synthetic rubber, solvents, and perfumes. Products obtained from coal are used to surface our roads and preserve our wood products from decay. And this significant resource of modern man resulted because green leaves trapped the energy of sunlight, hundreds of millions of years ago.

SUMMARY

1. The Lycopsida and Sphenopsida are groups with a long geological history. Most members of the groups are extinct. The chief living representatives of the Lycopsida are species of *Lycopodium* (club mosses) and *Selaginella,* known as small club mosses.

2. *Lycopodium* is homosporous. The sporangia are borne either in the axils of sporophylls similar to vegetative leaves or in the axils of modified and specialized sporophylls that are grouped in strobili.

3. The life cycle of *Lycopodium* resembles that of the fern. Several kinds of prothalli are produced. Chief among these are (1) prothallus short-lived and composed of a green aerial portion and a soil-buried basal portion that is usually invaded by fungus hyphae; (2) prothallus long-lived, non-green, entirely subterranean, and obtaining all food from the soil through the agency of an associated fungus. In the two types, both kinds of sex organs are borne on the upper portion of the same prothallus. A motile sperm fertilizes the egg, and the zygote grows into a leafy plant.

4. In the life cycle of *Lycopodium* the gametophyte generation includes the spore, prothallus, sex organs, and gametes. The structures of the sporophyte generation are the fertilized egg, the embryo, the leafy plant, sporangia, and spore mother cells.

5. *Selaginella* is heterosporous. The sporangia are borne in the axils of sporophylls, which are always grouped in strobili. The microspores and megaspores are produced within microsporangia and megasporangia.

6. In *Selaginella* the male and female prothalli are produced mostly or entirely within the spore wall. In many species the spores germinate within the sporangium before they are shed. After the microspores are shed, the sperms are produced. The megaspore germinates and produces a multicellular, female prothallus containing accumulated food. The megaspore with the female prothallus falls to the ground before the prothallus is mature. The spore wall splits, the prothallus protrudes, archegonia and antheridia are formed, and motile sperms bring about fertilization. The young sporophyte remains attached to the female prothallus until all accumulated food is exhausted.

7. In the life cycle of *Selaginella* the gametophyte generation includes the microspores and megaspores, male and female prothalli, the sex organs, and gametes. The sporophyte generation includes the fertilized egg, the embryo, the leafy plant, microsporangia and megasporangia, and microspore and megaspore mother cells.

8. The Lycopsida were far more important elements of the earth's vegetation in the past than in the present. The giant club mosses, *Lepidodendron* and *Sigillaria,* were large trees, conspicuous in Carboniferous forests. They became extinct early in the Mesozoic. The modern club mosses did not evolve from the giant club mosses and are not in the same evolutionary line.

9. The horsetails (*Equisetum*) are the only living representatives of the Sphenopsida. The stems are hollow and jointed, and the reduced and fused leaves are borne in whorls at the nodes. The branches, when present, are also borne in whorls. The vascular bundles are arranged in a ring, and cambium is absent. The stems contain a high concentration of silica.

10. *Equisetum* is homosporous. The sporangia are borne on shield-shaped sporangiophores that are aggregated into terminal strobili. The spores fall upon the soil and produce green prothalli with a number of erect, green lobes. Both kinds of sex organs

are usually produced on the same prothallus; motile sperms bring about fertilization.

11. The gametophyte generation of *Equisetum* includes the spores, prothalli, sex organs, and gametes. The sporophyte generation includes the fertilized egg, the embryo, the conspicuous plant, the sporangiophores, sporangia, and spore mother cells.

12. The calamites, related to the modern horsetails, were the most conspicuous representatives of the Sphenopsida in the Carboniferous.

13. Coal is composed of buried plant remains that have been subjected to heat and pressure. The principal plant groups that formed the coal of the Carboniferous were lepidodendrons, sigillarias and their relatives, the ferns and the seed ferns, and the Cordaitales.

27

The Seed Plants: Gymnosperms

THE VASCULAR PLANTS (Tracheophyta) deserve detailed attention because of their importance in the plant kingdom and to man. Four of the subdivisions of the Tracheophyta—the Psilopsida, Lycopsida, Sphenopsida, and Pteropsida—were considered in previous chapters. The fifth and last subdivision is the Spermopsida, or seed plants. Two classes comprise this subdivision. One of these is the gymnosperms (Gymnospermae), the subject of this chapter. The other is the angiosperms (Angiospermae), already considered in Chapters 12 and 13 and treated further in Chapter 28. Of all the tracheophytes the gymnosperms and angiosperms are of the greatest significance to man. The grouping of these classes into the same subdivision is a matter of convenience, the chief basis being their possession of seeds.

GENERAL FEATURES

The gymnosperms are of ancient lineage, some of them extending back into the Lower Carboniferous. Many are extinct and are known only as fossils. The name *gymnospermae* means "naked seeds." It was applied because the seeds are borne on the surface of an appendage and not enclosed, as in the angiosperms, within a pistil. The gymnosperms include very diverse groups, representing a number of evolutionary lines. It was once thought that these were not related, but evidence is accumulating that the group as a whole may have descended, directly or indirectly, from an assemblage of Upper Devonian plants called the progymnosperms.

The gymnosperms are heterosporous, producing microspores and megaspores. The mega-

671

spore is retained within the megasporangium, where it germinates and forms the female gametophyte. This megasporangium, together with the female gametophyte that it encloses and the integument that surrounds it, is called an ovule. After fertilization, an embryo is formed and the integument matures into a seed coat. The resulting seed may enter a resting stage, growth being resumed at the time of germination. The female gametophyte and the embryo are nourished by means of food obtained from the parent plant.

The microspores of gymnosperms germinate and form pollen grains. These are dispersed by wind and either come into direct contact with the ovule or are deposited in its immediate vicinity. Pollen tubes are produced by the pollen grains of living gymnosperms, but these structures have not been found in extinct orders and possibly never evolved in those groups. A notable feature of the gymnosperms is that probably all of the fossil orders and two of the living ones (the Cycadales and Ginkgoales) retained the swimming sperm of their aquatic ancestors. With certain exceptions, the reproductive structures of the gymnosperms are borne in strobili, which are usually conelike.

Fig. 27-1. Plants of a Carboniferous forest. (Left): a calamite. (Right): a seed fern (*Neuropteris*). (Center): a trunk of *Sigillaria*. The plant below the seed fern is *Cordaites*.

EXTINCT GYMNOSPERMS

The Seed Ferns (Pteridospermales) The Carboniferous strata are rich in fossils of fernlike plants, largely in the form of compressions of fronds. Some of these fossils were undoubtedly the fossils of true ferns, for they bear typical fern sori. Most, however, were for a long time found only in the sterile condition. It was suspected that these sterile fronds were not ferns but perhaps belonged to some other group. This suspicion was confirmed in the early years of the twentieth century by the discovery of fossil fronds bearing microsporangia and seeds. The plants bearing such structures were, then, seed plants and not ferns, and hence they were called pteridosperms, meaning "seed ferns."

The seed ferns varied greatly in structure and habit. Some were undoubtedly large plants (FIG. 27-1), resembling modern tree ferns; others were smaller, with slender stems that were probably supported by surrounding vegetation. The seed ferns differed from true ferns not only in the nature of their reproductive structures but also in having a cambium that, in many species, formed a considerable amount of secondary wood. The compounded fronds were large, often several feet long. The pollen sacs and seeds were borne not in cones but directly upon the fronds. The seeds were borne in various ways, including marginally on the divisions of the frond, replacing one of the ultimate divisions of the frond, on the upper or lower surfaces, or ter-

Plate 10 Red algae. Harvesting Irish moss (*Chondrus crispus*), Nova Scotia. (Terry L. Lyon, University of California, Davis.)

PLATE 11 Crustose lichens growing on sandstone in the Rocky Mountains of Alberta, Canada, elevation 5900 feet. *Lecanora muralis* (yellow); *Xanthoria elegans* (orange); *Rhizocarpon disporum* (dark gray); and *Lecidea tesselata* (ashy white). (Dr. C. D. Bird, University of Alberta.)

PLATE 12 *Marchantia.* (Left): portion of plant showing rhizoids. (Robert Fenn, Manchester Connecticut Community College.) (Right): female plant with archegonial receptacles and gemmae. (M. V. S. Raju, University of Saskatchewan.)

PLATE 13   A cycad (*Cycas media*), with seeds borne on leaflike structures arranged in loose clusters. The plants are commonly about ten feet high with leaves up to four feet long. This genus occurs from Japan to Australia, India, and Madagascar.

PLATE 14   Portion of drift log, probably of *Araucarioxylon arizonicum*, Petrified Forest National Park, Arizona. (Professor J. P. Amsden, Dartmouth College.)

PLATE 15 Cross section, needle of a pine (*Pinus palustris*), showing the two vascular bundles characteristic of the leaf of hard pines. (James Hannaford, Dartmouth College.)

PLATE 16 Flower of Mountain Lady's Slipper (*Cypripedium montanum*), western North America. The sterile stamen, which arches over the opening of the lip, is conspicuous. (Mrs. Katharine M. Kirkham, Norwich, Vermont.)

minally (FIG. 27-2). Fertilization was probably accomplished by swimming sperms. The resulting seeds varied in length from a fraction of an inch to 2 inches or more. The pteridosperms were most important and widely distributed in the Carboniferous, but fossils show that they lived from the Lower Carboniferous into the Jurassic. In speculations on the origin of the angiosperms it has frequently been suggested that these arose from the seed fern complex. Convincing evidence from the fossil record, however, is lacking.

The Bennettitales As the Paleozoic was characterized by ferns and seed ferns, and the Cenozoic by flowering plants, so the Mesozoic has been termed the Age of Gymnosperms. Conspicuous among these early gymnosperms were the Bennettitales (named after the British botanist J. J. Bennett, 1801–1876). These plants were undoubtedly a striking and common feature of the vegetation in many parts of the world during Upper Triassic, Jurassic, and early Cretaceous times (FIG. 27-3).

The reproductive bodies were borne in strobili. Each strobilus was composed primarily of a central, conical enlargement that bore the seeds, and a group of microsporophylls attached in a whorl around the base of the central column. In some groups the strobilus contained both male and female organs; in others, the male and female organs were in different strobili on the same or separate plants. In some of the Bennettitales the stem was relatively slender and branched (*Wielandiella, Williamsonia*, FIG. 27-3). The better-known and more specialized *Cycadeoidea* (FIG. 27-3) had a barrel-shaped, unbranched trunk less than 3 feet in height, covered by an armor of persistent leaf bases. It bore a terminal crown of palmlike leaves. Bisexual strobili were produced in the axils of the leaves along the sides of the trunk. It was long believed that the microsporangia in this genus were borne upon a whorl of pinnately compound microsporophylls, which later expanded in a flowerlike manner. Investigations indicate, however, that the microsporangia were fused in a mass around the central seed-bearing region and that the pollen-bearing structures did not expand and open as the restoration (FIG. 27-3) indicates.

The conditions that led to the extinction by the end of the Cretaceous of the group are not known, but it is possible that the low, slowly growing Bennettitales were eliminated in the competition for light and water by the taller, rapidly growing, broadleaved angiosperms that

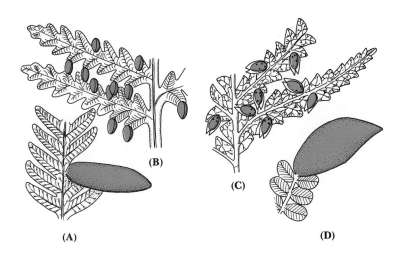

Fig. 27-2 Seeds of seed ferns. (A): *Alethopteris*, seed replacing a segment of the frond. (B): *Pecopteris*, seeds marginal. (C): *Emplectopteris*, seeds on upper surface of the frond. (D): *Neuropteris*, seed terminal on frond.

(B)

(C)

(A)

(D)

ARAUCARIOXYLON
TRIASSIC

WIELANDIELLA
TRIASSIC

WILLIAMSONIA
JURASSIC-CRETACEOUS

CYCADEOIDEA
JURASSIC-CRETACEOUS

MESOZOIC GYMNOSPERMS

Fig. 27-3 *Araucarioxylon* is an extinct member of a family of living conifers. *Wielandiella, William-sonia,* and *Cycadeoidea* belonged to the Bennettitales. The first two were slender and branched. *Cycadeoidea* had a short, thick, unbranched trunk.

spread over the earth during the Cretaceous period. Among such trees were poplars, laurels, breadfruit, magnolias, oaks, sycamore, and many others. These newer trees not only grew more rapidly but also produced greater numbers of readily disseminated seeds.

The Cordaitales These extinct forest trees were named after the Austrian botanist A. K. J. Corda (1809–1849). In size, the Cordaitales were comparable to modern conifers (FIG. 27-4). Their tall trunks usually branched only at the top. These high branches bore numerous strap-shaped leaves, which varied in size in the differ-

ent species, from a length of 6 inches to as much as 3 feet. The leaves were tough and leathery with internal xeromorphic modifications, such as abundant sclerenchyma tissue. Compressions of cordaitalean leaves are common fossils in the shales associated with bituminous coal. The structure of the wood was like that of some living conifers.

The pollen sacs and ovules of the Cordaitales were borne separately in catkinlike clusters on fertile side branches. Pollen tubes have not been found, and fertilization was probably brought about by swimming sperms. This group attained its greatest development in the Upper Carboniferous, but lived from the Lower Carboniferous into the Permian. Primitive cor-

daitalean forms may well have constituted the ancestral stock from which the modern conifers arose.

LIVING GYMNOSPERMS

The Cycads (Cycadales). The cycads are a small group of 10 living genera and about 100 species, found in the tropics and subtropics of both Eastern and Western hemispheres. Although fossils of cycads are not very common, the group was probably widely distributed from late Triassic to early Cretaceous times, and contemporaneous with the Bennettitales. Since the Cretaceous, the cycads have become restricted in numbers and distribution. Four genera are found in the Western Hemisphere, the largest, *Zamia*, occurring in Florida and from Mexico to Chile. The other six genera of the cycads are native to Australasia or South Africa.

The stem is typically unbranched, with a terminal crown of long, leathery, compound leaves. The plants are palmlike in appearance (FIG. 27-5). In some species the stem is very short and tuberlike; in others it is columnar. Although certain cycads have been known to attain a height of more than 30 feet, most are usually not more than 6 feet high. Several species are cultivated in conservatories as ornamental plants. Cultivated cycads are often confused with palms.

The cycads are entirely dioecious (FIG. 27-5). The pollen sacs are borne on scalelike micro-

Fig. 27-5 A cycad (*Zamia tuerckheimii*), Central America. (Left): plant bearing pollen cones. (Right): plant bearing seed cone.

sporophylls in compact cones. With the exception of one genus, the seed-bearing megasporophylls are also arranged in well-organized cones. In the genus *Cycas*, however, the seeds are borne on leaflike structures arranged in a loose cluster around the apex of the stem (Plate 13). The female cones are frequently large and out of proportion to the size of the plant. One species, from eastern Australia, is reported to have female cones 3 feet in length and weighing up to 85 pounds. Pollination takes place by wind. As the pollen grains germinate, pollen tubes containing very large, motile sperms are produced. These are the largest motile gametes in the plant kingdom. They may be as much as 180 μ in diameter, with 20,000 spirally arranged flagella, and are visible to the naked eye

The cycads are commonly held to have arisen from the seed ferns, probably during the late Paleozonic. The Bennettitales, also, are thought to be derived from the seed fern complex, but independently of the Cycadales and along entirely different evolutionary pathways.

The Ginkgoes (Ginkgoales) This order is represented by only a single living species, the ginkgo, or maidenhair, tree, *Ginkgo biloba* (FIG. 27-6). This most unusual and even mysterious tree was probably native in China, but whether it still exists in that country in a wild state is not known. It has long been grown there under cultivation. The ginkgo is a large tree, known to reach a height of more than 100 feet. It is strongly excurrent in growth, at least for many years. The leaves are broadly fan-shaped, 2–4 inches across, and usually divided by a deep depression into two symmetrical halves (FIG. 27-7). The dichotomous (division into two more or less equal parts) venation and general shape of the leaves resemble that of the leaflets of the maidenhair fern, whence the common name of the tree. Unlike most living gymnosperms, ginkgo is deciduous. The tree is widely cultivated for shade and ornamental purposes in North America and other parts of the world.

The ginkgo is dioecious; that is, the pollen and seeds are borne on different trees. The male strobilus is a short, catkinlike aggregation of paired microsporangia. The ovules are borne not in strobili but singly or in pairs on the end of a long stalk. After pollination by wind, pollen tubes with large, motile sperms are produced and fertilization takes place. The ovules develop into

Fig. 27-6 Ginkgo trees, showing habit of growth.

Fig. 27-7 *Ginkgo biloba,* leaves and seeds.

yellowish drupelike seeds (Fig. 27-7), about the size of a large cherry, but with a distinctive fetid odor that people will cross the street to avoid.

The Ginkgoales are of very ancient lineage, and the order contains numerous extinct genera and species. The group originated in the Permian, and by Jurassic times they had become worldwide in distribution. The order began to decline in the Cretaceous, and by the Tertiary was represented by the single genus *Ginkgo.* Species of this genus, however, continued to be an important forest tree, especially in the Northern Hemisphere. Fossils of Tertiary ginkgoes are abundant in the strata of America, Europe, and Asia. *Ginkgo* persisted during the Pleistocene but only in eastern Asia. It became restricted to a single species, the sole survivor of a group that for millions of years occupied a prominent place in the world's vegetation.

The Conifers (Coniferales) Several orders of gymnosperms are extinct. In others, the living descendants of ancient groups are greatly reduced in numbers, kinds, and distribution. But the conifers are the one exception, for this order contains numerous living species, many forming extensive forests in both the Northern and Southern hemispheres. Although a flourishing group today, the conifers, as are other gymnosperms, are very ancient, extending back into Upper Carboniferous times. Many of the petrified logs of the Petrified Forest National Park in Arizona are remains of an extinct conifer (*Araucarioxylon arizonicum*) of Upper Triassic Age (Fig. 27-3, Plate 14). These extinct conifers have several living relatives, none of which occur in the Northern Hemisphere.

The conifers are mostly evergreen and include such well-known trees as pines, spruce, fir, cedar,

Fig. 27-8 Pollen cones of eastern white pine (*Pinus strobus*). About natural size. Note the dark, pointed bract at the base of each cone.

Douglas fir, hemlock, and the sequoias. Two genera found in North America, larch and bald cypress, are deciduous. Some species of juniper and yew are shrubs, but neither the Coniferales nor any other order of gymnosperms includes herbaceous plants.

The Coniferales (meaning cone-bearing plants) are characterized by the production of conelike strobili. These are of two kinds, SEED CONES and POLLEN CONES. The seed cones, when young, and previous to fertilization, are known as OVULATE CONES. The pollen strobili are conelike, relatively small, and borne singly or in clusters (FIG. 27-8). They are frequently a bright shade of red or yellow. The pollen cones are ephemeral, lasting but a few days, and they often go unnoticed. They shrivel and drop from the tree soon after the pollen is shed.

The seed cones are woody (FIG. 1-3). They differ greatly in size, depending upon the species, ranging in length from less than $\frac{1}{2}$ inch to nearly 2 feet—a length attained by the cones of the sugar pine (*Pinus lambertiana*) of the northwestern United States. In certain conifers, such as the junipers and yews, the ovulate strobilus is highly modified and is not conelike at all but resembles a berry. Although some species of conifers are dioecious, most are monoecious. The two kinds of cones are usually borne on separate branches of the same tree.

LIFE HISTORY OF PINE

Pines (genus *Pinus*) are the best known and in many regions the most common of the conifers. They are plants of the Northern Hemisphere; only one species extends below the equator. Of about 105 species in the world, 32 species of pine are listed for North America. Pines do not range as far north as some other conifers, but are abundant in diverse habitats and climatic zones. Because they are attractive and of great value as timber trees, they have been studied in detail. The life cycle of the pine is perhaps the best known of any of the conifers. In its main outlines it illustrates the characteristic reproductive processes of all other conifers.

The Pollen Cones and Pollen   The pollen cones of pine are produced in clusters (FIG. 27-8). A pollen cone (FIG. 27-9A) is composed of a central axis to which are attached numerous, closely aggregated, and spirally arranged microsporophylls. That each cone is a short lateral branch is shown by its position in the axil of a modified leaf (bract), which is conspicuous before the pollen is shed (FIG. 27-8). The much reduced, scalelike microsporophylls bear two elongated microsporangia on the lower surface (FIG. 27-9B). The pollen cones appear in the spring, and at this time the contents of the microsporangium become differentiated into many microspore mother cells. These undergo

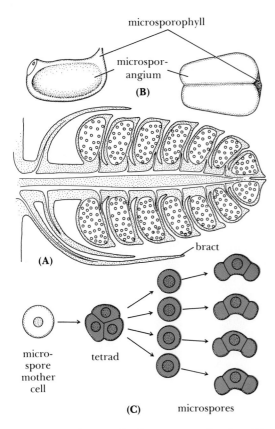

(A)

(B)

(C)

microsporophyll

microspor-
angium

bract

micro-
spore
mother
cell

tetrad

microspores

Fig. 27-9 Male reproductive structures in pine. (A): longitudinal section of pollen cone. (B): microsporophyll and microsporangia, view from side and below. (C): development of microspores.

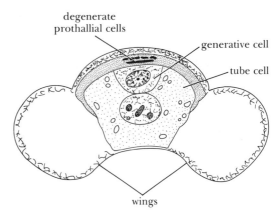

degenerate
prothallial cells

generative cell

tube cell

wings

Fig. 27-10 Mature pollen grain of pine.

meiosis, each forming a tetrad of microspores (FIG. 27-9C). The nucleus of each spore now contains the reduced, or *n*, chromosome number, and the microspore therefore constitutes the first cell of the male gametophyte generation. Each microspore develops, over most of its surface, a thick wall composed of an inner and an outer layer. The microspores shortly separate and enlarge. The outer layer of the spore wall expands on either side, forming a wing, or bladder, filled with air (FIG. 27-10).

The microspore now germinates while still within the sporangium. The nucleus divides, and after further nuclear divisions a mature pollen grain is formed (FIG. 27-10). This consists of a much reduced, partly developed male gametophyte enclosed within the microspore wall. Two of the four cells that now compose the pollen grain are termed PROTHALLIAL CELLS. These cells soon degenerate. The prothallial cells are vestigial and constitute the only remaining vegetative cells of the male gametophyte generation. The two remaining cells are a small GENERATIVE cell and a larger TUBE cell.

When the pollen grains are mature, the axis of the pollen cone elongates somewhat, separating the microsporophylls. The microsporangia (now called the POLLEN SACS) open by a longitudinal slit and the pollen grains are liberated and carried away by wind. The pollen is produced so abundantly that it may form a cloud of sulfur-colored dust in the vicinity of the tree.

Winged pollen grains similar to those of pines are also found in the spruces, firs, and other conifers. On the other hand, in larch, hemlock, Douglas fir, and many others the pollen grains are wingless. It is commonly stated that the wings aid in dissemination by wind. This may be so, although wingless pollen grains are generally larger than those with wings. More plausible is an explanation based upon the structure of winged grains. In wingless grains the pollen grain wall is uniformly thickened. In winged grains the wall on the lower side, between the wings, is thin and forms a shallow furrow (FIG. 27-10), the site of emergence of the pollen tube. As the pollen grain dries, the wings close over this furrow, thus providing protection against drying.

Fig. 27-11 Ovules and ovulate scale of *Pinus strobus* at time of pollination. (Left): as seen from above. (Right): as seen from below.

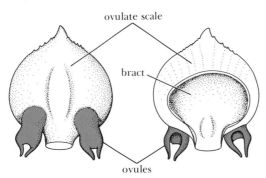

ovulate scale

bract

ovules

The ovulate cone and the ovule   The ovulate cone is composed of a central axis that bears the ovulate scales (FIG. 27-11). Each scale is borne in the axis of a scalelike bract. Both the bract and the scale enlarge as the cone grows; the scale, however, eventually becomes many times larger than the bract, which in the mature cone is found at the base of the scale and partly fused with it. Two inverted ovules develop on the upper side of the ovulate cone scale. Their micropyles point downward toward the axis of the cone. The ovulate cone scale was long believed to be a megasporophyll. Modern investigations have shown, however, that the scale is not a simple sporophyll but a complex and reduced structure, composed of a number of fused parts. The position of the scale in the axil of a bract indicates that it is a much modified, short lateral branch and is therefore homologous to a single pollen cone. Each ovule (FIG. 27-12) consists of an integument surrounding and united with a megasporangium. The integument is prolonged as a short tube beyond the megasporangium. The opening of this integument forms the micropyle. Further prolongations of the integument result in the formation of an arm on either side of the micropyle. The megasporangium con-

Fig. 27-12   Pine. (Left): the ovule at the time of pollination. (Right): shortly after pollination. The micropylar arms have become swollen, almost closing the micropyle.

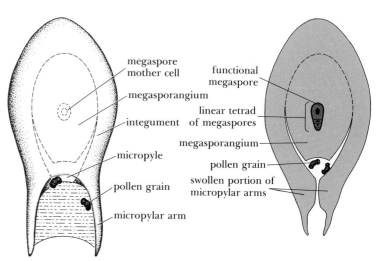

megaspore mother cell

megasporangium

integument

micropyle

pollen grain

micropylar arm

functional megaspore

linear tetrad of megaspores

megasporangium

pollen grain

swollen portion of micropylar arms

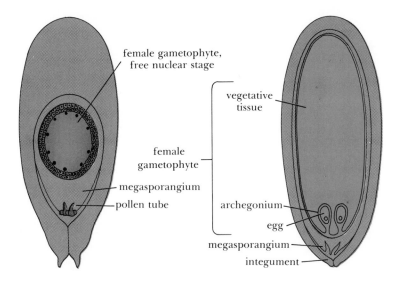

Fig. 27-13 Later stages in the growth of the pine ovule. (Left): development of the female gametophyte, end of the first season of growth; the micropyle is completely closed. (Right): mature female gametophyte; the eggs are ready for fertilization.

tains a single, large megaspore mother cell; this cell, following meiosis, gives rise to four megaspores, arranged in a row. Three of the megaspores disintegrate, but the fourth, that farthest from the micropyle, becomes the functional megaspore.

The female gametophyte The germination of the megaspore consists of enlargement and division of the nucleus. Nuclear divisions continue until a large number of free nuclei are formed. Division eventually ceases, and cell walls form between the nuclei (FIG. 27-13, left). The resulting multicellular body is the female gametophyte (FIG. 27-13, right). During the later stages of its development, two to five archegonia become differentiated at the micropylar end of the gametophyte. A mature archegonium in pine is much reduced in comparison to the archegonium of a fern. It consists chiefly of a small and variable number of neck cells, located just above a large egg (FIG. 27-14, left). A sheath of specialized cells, resembling the enlarged base of an archegonium, surrounds the egg. This sheath is not, however, a part of the archegonium proper, but is composed of modified cells of the female gametophyte.

Fig. 27-14 The archegonium and pollen tube. (Left): archegonium and egg. (Right): pollen tube and mature male gametophyte. Diagrammatic.

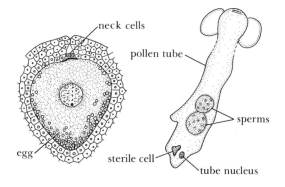

The female gametophyte of the pine is comparable in function to the endosperm in those angiosperm seeds that accumulate food in this tissue. Both the female gametophyte and endosperm supply food during seedling development. The female gametophyte differs from the endosperm, however, in several particulars.

It is formed before fertilization, its nuclei possess the *n* chromosome number, and it produces archegonia. The endosperm of the angiosperm, on the other hand, is formed after fertilization, its nuclei possess the 3*n* chromosome number, and it produces no archegonia.

Pollination and fertilization The pollen is discharged during a period of only a few days. This period may occur at any time during the spring or early summer, depending upon the latitude and upon the species concerned. At the time of pollen discharge, the small ovulate cones are growing erect at the ends of branches and the scales are spread apart (FIG. 27-15). The pollen grains drift downward between the scales, lodging upon the rim of the micropyle and upon the micropylar arms on either side (FIG. 27-12, left). A fluid secretion extends downward from the megasporangium, filling the micropylar canal. This fluid, after coming in contact with pollen grains, is shortly withdrawn, and the pollen grains are carried upward with it into a saucer-shaped depression at the top of the megasporangium. The basal regions of the micropylar arms then become swollen, thus closing the micropyle and providing protection for the pollen grains on the megasporangium (FIGS. 27-12, right, and 27-13, left). The arms wither, and the ovulate scales grow until they become closely pressed together (FIG. 27-13, left). In many pines the cone gradually bends downward as a result of the curvature of the stalk upon which it is borne.

The pollen grain germinates, producing a pollen tube that penetrates the tissue of the megasporangium (FIG. 27-13, left). The tube nucleus advances with the end of the pollen tube. The generative cell divides into a STERILE CELL and a SPERMATOGENOUS CELL, the spermatogenous cell again dividing and forming two nonmotile male gametes, or sperms, one commonly larger than the other. A fully developed male gametophyte in pine thus consists of two degenerated prothallial cells, a tube nucleus, a sterile cell, and two sperms (FIG. 27-14, right). The significance

Fig. 27-15   Ovulate cones of white pine at time of pollination.

Fig. 27-16 Cones of white pine. (Left): ovulate cone at the end of the first season of growth. (Right): seed cone, end of the second season of growth; the cone has opened, allowing the seeds to fall away.

of the sterile cell is unknown; it is possibly a vestigial structure. The pollen tube next penetrates between the neck cells of the archegonium and discharges its contents into the egg. The larger of the sperms fuses with the egg nucleus, and the remaining male gametophyte nuclei disintegrate.

The Time Element in Pollination, Fertilization, and Seed Maturity   In the pines, a considerable time interval elapses between the first appearance of the ovulate cones and the maturity of the seed. In all pines whose life cycles have been investigated, the cones and seeds do not mature in the same growing season in which pollination occurs. The interval between pollination and fertilization is a very long one in contrast to the condition in the angiosperms, where fertilization usually occurs within a few hours or days after pollination.

In the eastern white pine (*Pinus strobus*), for example, the ovule has advanced only to the spore mother cell stage at the time of pollination.

Growth continues slowly during the first season, and by fall the ovule and cone (FIG. 27-16, left) have enlarged several times. The megaspore germinates and forms a few (perhaps 32) free nuclei. The pollen tubes, which have penetrated the megasporangium, cease to grow with the approach of winter. Growth is renewed in the spring and proceeds rapidly. By late spring or early summer of the second year, the female gametophyte is fully formed and the eggs are ready for fertilization (FIG. 27-13, right). This occurs about the middle of June, approximately a year after pollination. The ovule at the time of fertilization is approximately the size of the mature seed. The cone continues to enlarge, and matures by late summer of the second year, after which the scales spread apart and the seeds are dispersed (FIG. 27-16, right).

This condition, in which many months elapse between pollination and fertilization, is found in other conifers as well as in pines. But most of the north temperate conifers, such as spruce, fir, larch, and hemlock, have an interval of only a few weeks between pollination and fertilization.

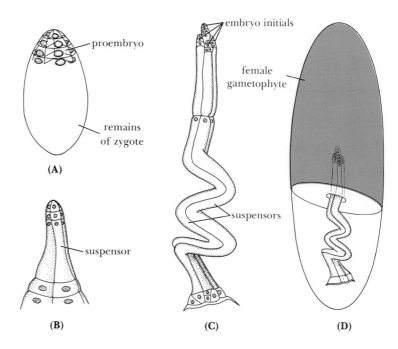

Fig. 27-17 Stages in the development of the embryo of pine.

(A) proembryo — remains of zygote

(B) suspensor

(C) embryo initials — female gametophyte — suspensors

(D)

The seeds of these conifers ripen in the same season in which pollination takes place.

The embryo, seed, and seedling The fertilized egg gives rise, after nuclear divisions and cell-wall formation, to a PROEMBRYO (FIG. 27-17A) composed of 16 cells, arranged in 4 tiers of 4 cells each. The proembryo is located at the end of the egg opposite the neck of the archegonium. The cells of the proembryo just below the apex now elongate to many times their former size, forming the SUSPENSORS (FIG. 27-17B, C). The apical cells of the proembryo divide, and *each* of the four cells of this tier gives rise to an embryo. The elongation of the suspensors, which become extremely sinuous and twisted, pushes the immature embryos into the tissue of the female gametophyte, where a cavity is developed by enzymatic action (FIG. 27-17D). During the further development of the embryos, only one survives, the remainder being eliminated as a result of competition within the female gametophyte.

The mature embryo of pine (FIG. 27-18) is composed of a plumule, hypocotyl, radicle, and a variable number of cotyledons (about eight). It is embedded in the female gametophyte, which in turn is surrounded by a hard seed coat derived from the integument of the ovule. In a number of pines a portion of the upper surface of the scale becomes fused to the ovule and splits off as a wing attached to the mature seed. Such winged seeds are dispersed by wind and seeds may be carried many feet from the parent tree.

The seeds of pine, and to some extent other conifers, form an important source of food for seed-eating birds, rodents, and other animals. Pines, together with oaks, are among the most important woody plants as a souce of food for North American wildlife. For example, pine seeds make up approximately 70 percent of the diet of three birds—the redcrossbill (*Loxia curvirostra*), Clark's nutcracker (*Nucifraga columbiana*), and the white-headed woodpecker (*Dendrocopos albolarvatus*). Dozens of other birds and mammals use pine seeds as food. This consumption of pine and other coniferous seed by wildlife is known to diminish reproduction of the trees, especially if the crop is small. Conversely, a low seed yield may affect the size of the population of a number of wildlife species, especially rodents. The seeds of one conifer, the pinon pine of the southwestern United States, were long used as a staple in the diet of the Indians of that region.

Germination of pine seed (FIG. 27-18) is similar to that of many angiosperms. The radicle emerges from the seed coat and penetrates the soil. The hypocotyl becomes arched as it grows, and appears above the soil. It then straightens, carrying the cotyledons and seed coat upward. The cotyledons absorb food from the seed until this food is exhausted; the seed coat then drops away. The green cotyledons manufacture food and persist for a considerable time—one or two growing seasons. The plumule, deep within the whorl of cotyledons, develops the stem and leaves of the mature sporophyte, which, once started, may grow for several hundred years. The complete life cycle of pine is shown diagrammatically in FIG. 27-19.

SOME MODERN CONIFERS

A study only of the reproductive cycle of a pine is inadequate to an understanding of the conifers as a whole, especially in relation to their diversity. The conifers are too large, too common, and too

Fig. 27-18 Seed and seedling of pine. (Above): the seed. (Below): stages in germination.

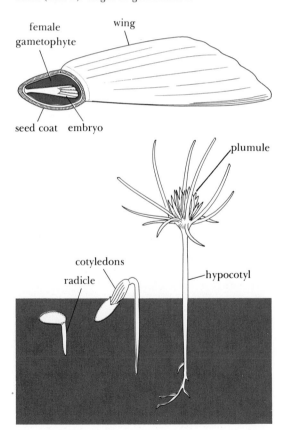

Fig. 27-19 Life cycle of pine.

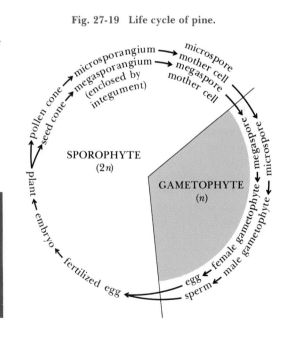

important a group to dismiss after a consideration only of methods of reproduction, important as this may be. The conifers (Coniferales) include about 550 species, classified in seven families, of which the pine family is the best known. Only five of these families are discussed here. The conifers provide us with a very large proportion of our lumber and of raw materials for the pulp and paper industry. Their wood is adaptable to a great variety of uses and is easily seasoned and manufactured. They include many species famous for their beauty as well as for their value as timber.

Fig. 27-20   Eastern yew (*Taxus canadensis*).

Yew Family (Taxaceae)   The small family of yews includes both trees and shrubs. The female reproductive structure is not conelike but consists of a single seed surrounded by a fleshy covering. In the yews (*Taxus*) this covering is cuplike, about the size of a large pea (FIG. 27-20), and scarlet at maturity. The yews are widely employed in ornamental plantings. They are shapely and slow growing, and have attractive, glossy, short, flat leaves. Various species of the family are native in North America, Europe, and Asia. The English yew, a tree native to Europe, western Asia, and North Africa, is long-lived, and numerous specimens of great size are recorded. The wood of this tree was once prized for making the bows that brought fame to English archers.

Fig. 27-21   A hard pine, western yellow pine (*Pinus ponderosa*), with three needles in a cluster. Two 1-year (immature) cones and a mature cone that has opened and shed its seeds.

Pine Family (Pinaceae)   The pines make up only part of the pine family. They are easily recognized by their long, needlelike leaves, which usually occur in clusters of two, three, or five. The pines are conveniently classified into two groups: the soft, or white, pines, with five needles in a cluster, and the hard pines, with two or three needles in a cluster (FIG. 27-21). Another feature distinguishing these groups can be seen in a cross section through the middle of a needle. The soft pines contain but one vascular bundle running the length of the leaf, whereas in the hard pines there are two such bundles (Plate 15). The wood of the hard pines is usually

heavier and harder than that of soft pines, and the distinction between the zones of the growth ring is greater so that the surface of a board exhibits a more pronounced pattern.

One soft pine, the eastern white pine (*Pinus strobus*), was one of the most important timber trees in North America. It was so prized for lumber that the most accessible stands were exploited more than a century ago, and only remnants now remain of the great forests of this species that once covered much of southern Canada and the northeastern and north central sections of the United States. The wood is soft, easily worked, moderately durable, and is used for many kinds of construction. A number of white pines are native in the Rocky Mountain and Pacific Coast states. Of these, the sugar pine and the western white pine are the most important commercially.

Among the more important eastern two- and three-needled pines are the red pine of the northeastern and lake states and species of southern yellow pines. The latter grow very rapidly and flourish on cut-over and burned-over areas in poor, sandy soils. The yellow pines constitute one of the most important crops of the South and are constantly used in reforestation. Prominent among the southern hard pines is the long-leaf pine, *Pinus australis* (FIG. 1-3), having needles up to 18 inches long. This species and the slash pine (*Pinus elliottii*) are the source of most of our rosin and turpentine. Yellow pines, notably the ponderosa pine, are also important in the western part of North America. The two-needled Austrian and Scotch pines, native to Europe, are commonly used for ornamental planting in the East and Midwest. The pines have been a conspicuous feature of the earth's forests since the late Mesozoic.

In 1957 it was discovered (by counting growth rings) that individual trees of the bristlecone pine (*Pinus aristata*), growing in an arid mountain region of east-central California, surpass the longevity record of the giant sequoias, previously believed to include the most ancient trees. Several bristlecone trees were found to be 4000 years old or more, and one more than 4600 years. In 1965 an even older tree—about 4900 years old—was found in eastern Nevada at an altitude of 10,700 feet (FIG. 27-22). The tree was 21 feet in circumference at a height of 18 inches above ground. Most of the tree was dead, but it bore a living shoot 11 feet high.

The Pinaceae include a number of important genera in addition to the pines. The larches (*Larix*), also known as tamaracks, are deciduous conifers of northern forests, attaining their largest size in western mountains. A species introduced from Europe is widely cultivated as an ornamental tree. Most of the short needles of larch are borne in close clusters on short side branches (FIG. 27-23). The hemlocks (*Tsuga*) are widely distributed in cool, moist forest soils. Four species are found in North America. The needles are short, flat, and spirally arranged on the branchlets, but in three of the species the leaves, by a twist of their bases, seem to grow from opposite sides in two-ranked, flat sprays (FIG. 27-24). About 45 species of fir (*Abies*) are known from the subarctic and temperate regions of the Northern Hemisphere. Twelve of these are native to North America. The firs are popularly known as balsam or fir-balsam. The leaves are short, blunt, and flat, resembling those of hemlock, but leave a round, flat scar when they fall. The cones, unlike those of other North American conifers, grow erect (FIG. 1-3), and do not fall complete; instead, the scales fall away singly, leaving intact a pointed central axis that usually persists for several years. Seven species of spruce (*Picea*) are found in North America, four in the west and three in the east. These are among the most commercially valuable of our trees. The leaves (needles) are short and flattened or four-sided, often sharply pointed. When the needles fall, they leave short, peglike projections on the branchlets (FIG. 27-25). One of the most important trees in western North America is the Douglas fir (*Pseudotsuga*). It attains a height second only to redwood and sequoia. The leaves are flat, resembling those of fir. The cones are readily recognized, for the bracts pro-

Fig. 27-22   A bristlecone pine (*Pinus aristata*), 4900 years old, in eastern Nevada. The chief living portion is the treelike shoot at the left.

Fig. 27-23   Branchlet of American larch (*Larix lari-cina*).

Fig. 27-24   Leaves and cones of hemlock (*Tsuga cana-densis*).

Fig. 27-25   White spruce (*Picea glauca*).

Fig. 27-26   Douglas fir (*Pseudotsuga menziesii*).

Fig. 27-27 (Left): redwood (*Sequoia sempervirens*). (Right): big tree, or giant sequoia (*Sequoiadendron giganteum*).

trude prominently from beneath each scale (Fig. 27-26). Fir, spruce, and Douglas fir are widely used as Christmas trees.

Redwood Family (Taxodiaceae) This interesting family is limited in its range and number of species. The most prominent North American species are the big tree, or giant sequoia (*Sequoiadendron giganteum*), the closely related redwood (*Sequoia sempervirens*), of the West Coast, and the bald cypress (*Taxodium distichum*), found in the southeastern United States and extending up the Mississippi Valley to Indiana and Illinois. In ancient times, however, the family was widely distributed over the Northern Hemisphere. The ancestors of the present-day sequoia and redwood go back to the Jurassic, and batlike flying reptiles may have rested in their branches. The bald cypress is almost as old.

One of the most exciting scientific discoveries of modern decades was made in 1946 when a new living conifer, the "dawn redwood" (*Metasequoia*), was found growing in central China. This tree, a close relative of redwood and sequoia, was a prominent forest tree during the Tertiary. It disappeared from North America by the end of the Miocene, but in Asia it continued to the present day in a limited area, where it escaped scientific notice.

Included in the family are the tallest and greatest in bulk of living things. The world's tallest tree is a redwood, 372 feet high. The greatest in bulk is the General Sherman tree, a giant sequoia growing in Sequoia National Park. This tree is 273 feet high, with a circumference of 84 feet at the base. It is difficult to determine the age of a living sequoia, but the General Sherman is conservatively estimated to be at least 3500 years old, and may be 4000 years.

The leaves, or "needles," of redwood form flat sprays. Those of the giant sequoia are scalelike, and similar to eastern red cedar. The cones of both trees are small (Fig. 27-27), and both trees have reddish-brown fibrous bark, a foot or more thick on large trees, and highly resistant to fire.

The bald cypress (Figs. 9-14, 27-28) also attains considerable size and age. The leaves appear in two ranks and, still attached to lateral branchlets, fall away at the end of each growing season.

Fig. 27-28 Bald cypress (*Taxodium distichum*).

Cypress Family (Cupressaceae) This family of some 18 genera is worldwide in distribution. A number of genera and species are found in North America, but only a few will be mentioned. The members of the family are mostly trees, with a few shrubs, and the adult leaves are usually small, scalelike, in pairs or threes, and closely pressed against the stem. Among important western trees are western red cedar (*Thuja plicata*) and Port Orford cedar (*Chamaecyparis nootkatensis*). In the east, red cedar (*Juniperus virginiana*) (FIG. 27-29), and white cedar or arborvitae (*Thuja occidentalis*), (FIG. 27-30) are widely distributed.

Many species of cypress (*Cupressus*) and other members of the family are cultivated as

Fig. 27-29 Red cedar (*Juniperus virginiana*).

Fig. 27-30 Arborvitae, or white cedar (*Thuja occidentalis*).

ornamentals, and many horticultural varieties have been developed. All members of the family except the genus *Juniperus* bear small, dry cones. In *Juniperus* the cone scales are fleshy and grow together, forming a berrylike structure about the size of a small pea and blue at maturity (FIG. 27-29).

The use of the word "cedar" in the popular names of certain members of this family is misleading. The only true cedars are members of the genus *Cedrus,* of the Pinaceae. They are native in the Himalayas, Asia Minor, and Lebanon. The Cedar of Lebanon, the famed cedar of the Bible, belongs to *Cedrus.* Cedar is used in North America as a common name, applied to various kinds of trees in different parts of the region.

Araucaria Family (Araucariaceae) No species of the araucaria family are native in North America today. Representatives of the two genera are found in the Southern Hemisphere— in the Philippines, New Zealand, the East Indies, and Australia. The group is very ancient, however, and was abundant in North America during the Mesozoic (FIG. 27-3, Plate 14). A species of *Araucaria,* called Paraná pine, is one of the most valuable timber trees in South America, in certain areas taking the place of the pines of the north. Several araucarias, such as the monkey puzzle from Chile, the bunya-bunya from Australia, and the Norfolk Island pine, are cultivated in greenhouses and grown out-of-doors in the warmer parts of the United States, especially along the West Coast.

SUMMARY

1. The gymnosperms, together with the angiosperms, are classified in the subdivision Spermopsida. The gymnosperms include both living and extinct forms. All are heterosporus, with naked seeds.

2. The extinct orders are the seed ferns, the Bennettiales, and the Cordaitales. The Bennettitales bore reproductive structures superficially resembling flowers. The Cordiatales were coniferlike trees that bore their reproductive organs in catkinlike structures. In the seed ferns, the pollen sacs and ovules were borne directly on the fronds.

3. The living groups of the gymnosperms are the Cycadales, the Ginkgoales, the Coniferales, and the Gnetales (not considered in this text). The cycads, a small group of palmlike plants now restricted to the tropics and subtropics, were contemporaneous with the Bennettitales during the Mesozoic. The pollen sacs and ovules are usually borne in cones. The ancient Gingkoales now survive as a single species, *Ginkgo biloba.* This tree is deciduous and broad-leaved, with drupelike seeds.

4. Pollen tubes are produced in all living gymnosperms, and swimming sperms in the Cycadales and Ginkgoales. Motile sperms were probably formed also in the extinct orders of gymnosperms.

5. The Coniferales are the most important living order of gymnosperms. All are woody plants, and most are evergreen. Many are of great economic importance as a source of lumber, paper pulp, and in other ways.

6. The conifers produce two kinds of strobili, pollen cones and seed cones. Most conifers are monoecious. The pollen cones are short-lived; the seed cones are usually woody. The ovulate (seed) cone scale is a highly reduced, complex structure, a much modified short lateral branch located in the axil of a bract. It is homologous with a single pollen cone.

Important Features of the Pine Life Cycle

7. Each microspore mother cell following meiosis, gives rise to four microspores. These microspores germinate within the

microsporangium and form pollen grains, each of which contains a partly developed male gametophyte, consisting of two degenerate male prothallial cells, a generative cell, and a tube cell.

8. The pollen grains are carried by wind to the ovulate cone, where they germinate on the surface of of the megasporangium. The pollen tube penetrates the tissue of the megasporangium and ceases growth. Growth is resumed the following spring and fertilization takes place about a year after pollination.

9. The megaspore mother cell gives rise to a tetrad of four megaspores arranged in an axial row within the megasporangium, which becomes invested by an integument. Three of the megaspores degenerate; the fourth germinates and produces the female gametophyte bearing archegonia and eggs.

10. At the time of fertilization the mature male gametophyte consists of two male prothallial cells, a tube nucleus, a sterile cell, and two sperms. The pollen tube penetrates the neck of the archegonium and discharges the contents of the gametophyte, except the prothallial cells, into the egg. One sperm fuses with the egg; the other nuclei degenerate.

11. The zygote divides and forms the proembryo. The four apical cells of this proembryo initiate four embryos, three of which are later eliminated.

12. Two growing seasons intervene between pollination and the maturity of the seed. In the first season, the pollen germinates, the female gametophyte begins development, and the ovulate cone increases in size. In the second season, the female gametophyte matures, fertilization takes place, the seed cone develops to full size, and the seed ripens.

28

The Seed Plants: Angiosperms

T HE DOMINANT PLANT LIFE of the geological era in which we live are the angiosperms. They are the products of a long line of evolutionary development that has culminated in the highly specialized organ of reproduction that we know as the flower. Flowering plants are the most conspicuous elements of the present-day landscape except in those forest areas where conifers predominate. The fact that angiosperms are dominant does not mean that their evolution has come to an end. There are numerous lines of evidence that this group is continuing to change and is likely to remain a significant one. A number of angiosperm families have expanded greatly in recent times, and will probably become even more significant components of the earth's flora as the centuries pass.

The two great subdivisions of the angiosperms—the dicotyledons and the monocotyledons—include about 300 families and more than 250,000 species. The most successful and important of these families is, in all probability, that of the grasses, with 7500 species. Grasses have colonized great areas of the earth's surface where, because of soil or climatic conditions, most kinds of woody plants do not grow. Directly or indirectly, the grasses constitute the main source of man's food supply.

Many factors are responsible for the success of the angiosperms. Among these are (1) their ability to survive and reproduce in almost all kinds of environments; and (2) the production of flowers, fruits, and seeds. Although the conifers and other groups of vascular plants are widely distributed, none flourish in so many different habitats as do the angiosperms. They live in all kinds of soils, under great extremes of temperature and rainfall, in deserts, in arctic regions, in prairies, meadows, and marshes, in the water and in the air, in mutualistic relationships with fungi and as parasites. They include trees, shrubs, vines, and herbs, annuals and perennials, succulents, and forms with underground storage organs. In most respects the angiosperms exhibit greater variability and plasticity than do the gymnosperms or any other vascular plants, and seem better fitted to survive in new or changing environments.

The flower, fruit, and seed are structures that have undoubtedly been important in the success of the flowering plants. The flower, a unique angiosperm structure, produces pollen, and ovules that develop into seeds. Many flowering plants, such as grasses, are either self- or wind-pollinated. Thousands of other species are pollinated through the agency of insects. The initiation of the endosperm at the time of fertilization ensures a supply of food for the embryo and for the resulting seedling during germination. Many species of angiosperms mature quickly and possess a high reproductive capacity, producing great numbers of seeds. Although the number of seeds produced by an angiosperm may be only a small fraction of the number of spores produced by a fungus or a fern, their structure and food supply make them much more efficient. Under many conditions a dozen seeds may produce more new plants than a million spores. Numerous modifications of seeds and fruits permit rapid and effective dispersal over considerable distances and by many natural agencies.

SIGNIFICANCE OF ANGIOSPERMS

Although man obtains from the conifers such materials as lumber, paper, and turpentine, he finds among the angiosperms a host of economically useful plants that yield food, fibers, oils, and medicines. All important food plants are angiosperms. Although woody and herbaceous perennials are widely used, it is the herbaceous annuals that are the basis of modern agriculture. Our most useful annuals have been cultivated for thousands of years. They include such cereals as corn, wheat, rye, sorghum, oats, barley, millet, and rice (FIG. 28-1), and such legumes as soybean, lentil, pea, and kidney bean.

Man, in the hunting period of his history, was for the most part only indirectly dependent upon flowering plants, through the intermediary of animals that consumed leaves. But ancient man did not live on meat alone. Even as a hunter he consumed wild fruits. Later he learned to gather the seeds of wild annuals and, still later, to plant and harvest them. In some parts of the world he abandoned his migratory mode of life and came to depend upon the products of the harvest. Thus began (some 10,000 years ago) not only agriculture but also modern society. Most of the annuals that man has used for food were first cultivated in prehistoric times, and their origin is difficult to trace. The seeds of these herbaceous annuals furnish a great store of food for man and his domestic animals. The important civilizations have had their basis in the cultivation of annual crop plants. The evolution of the angiosperm seed has made possible the rise of great nations and the concentration of large populations.

The flowering plants not only played a determining role in the development of human society, but also made possible the evolution of modern types of birds, insects, and warm-blooded animals. Although the angiosperms first became prominent in the Lower Cretaceous, 130 million years ago, it was in the Upper Cretaceous and Tertiary that they spread over the earth and became the dominant plant life. Early mammals lived in the Triassic, some 200 million years ago, but these were small and inconspicuous, and probably restricted to specialized environments and to feeding upon insects, worms, reptile eggs, and small animal life. During the Jurassic and Cretaceous, the mammals slowly became diversified. It was during the Tertiary, 10–60 million years ago, however, that the mammals evolved and spread more rapidly, and this was preceded by the development of the angiosperms and was in a number of ways dependent upon these plants. The evolution of the herbivores, in particular, was correlated with the world-wide spread of grasses, legumes, and other herbs in mid-Tertiary times. Although birds are known to

Fig. 28-1 Angiosperms. Rice and tea in Japan.

have existed as far back as the Jurassic, the evolution of modern kinds was coincident with the coming of the angiosperms. Most modern birds feed either upon the seeds, fruits, or buds of flowering plants, or upon insects that in turn feed upon angiosperms. The rapid spread of the angiosperms also favored the evolution of several orders of insects. The interdependence between flowers and insects (and some birds), which is one of the marvels of the living world, became established. It is difficult to visualize the kinds of plant and animal life that would inhabit our globe today had there not occurred that series of fortuitous events that gave rise to the angiosperm flower, fruit, and seed.

DISTINGUISHING FEATURES OF ANGIOSPERMS

Several earlier chapters have been concerned, in whole or in part, with the structure and methods of reproduction of the angiosperms. This group of plants is so important that it is useful here to attempt to summarize these aspects, and at the same time to compare briefly the angiosperms with that other great group of living seed plants, the Coniferales. Although the conifers occupy a subordinate position in the life and activities of mankind as compared with the flowering plants, it is these two groups, of all the vascular plants, that are the most useful to man. The characteristics of the angiosperms can best be summarized by a consideration of two aspects: the nature of the plant body, and the structures and mechanisms concerned in reproduction.

The evolution of the conifers proceeded in a direction different from that of the flowering plants. The conifers are all woody perennials. So also are thousands of species of angiosperms, but many thousands of others are perennial or annual herbs. The great group of monocotyledons is largely herbaceous.

Underground, the angiosperms develop not only tap roots, a common root type in the conifers, but also many root modifications that are useful in the accumulation of food and in vegetative reproduction. Many angiosperms also have underground stems in which food is accumulated, or which serve as a means of vegetative propagation. These and many other modifications of the plant body have enabled the angiosperms to occupy innumerable kinds of habitats to which the conifers are not adapted. As we have indicated, it was the herbaceous annuals that early became important to man as a direct and important source of his food supply. Such shortlived plants produce seed within a few months after planting, and this in large part explains why annuals were first chosen for cultivation.

The leaves of conifers are usually longer lived than those of angiosperms, remaining attached to the plant for two or more years, thus giving rise to the popular name of evergreens. This evergreen habit is associated with reduction in the size and surface of the leaves, which are mostly small and needlelike or scalelike. The leaves of the flowering plants, on the other hand, like the stems and roots, exhibit a high degree of variation and adaptation to various environmental conditions. They are mostly thin and expanded, although in general form they vary widely from the long, narrow leaves of the grasses to leaves that are oval or round.

Great variations in the size of leaves are also found. In some plants of specialized habitats, the leaves are so small as to be recognized only with difficulty. At the other end of the scale are those of the cultivated banana, the blade of which alone may be more than 12 feet long and 3 feet wide. The traveler's-tree, a relative of the banana, has leaf blades 18–20 feet or more in length. The circular leaves of the royal water lily of Guyana and the Amazon may attain a diameter of more than 6 feet. Even these dimensions may be exceeded by certain perennial herbs of the tropics.

Internally, the wood of angiosperms is more specialized and complex than that of conifers. Conifer wood is composed mainly of thick-walled tracheids and ray cells. The wood of an angiosperm, such as oak, maple, or birch, contains a greater variety of cells, including the important vessel elements, not found in conifers.

It is the flower, however, that distinguishes the angiosperms from all other groups of plants. The flowers may be borne in clusters or singly; they vary greatly in the number and complexity of their parts; but they are primarily involved in reproduction, and their activities result in the production of fruits and seeds.

The enclosure of the ovules within a protective covering is a unique characteristic of the angiosperms. The ovule of the conifers is not enclosed but is borne naked on the surface of a cone scale. The upper portion of the angiosperm pistil is modified, forming a stigma, which constitutes a receptive surface for the pollen grains and a site for their germination. The ovary, sometimes with other parts of the flower, matures into the fruit, an organ found only in the flowering plants.

The angiosperms, although highly specialized, have retained, among other ancestral features, microspores and megaspores. The microspore produces the pollen grain, and the megaspore gives rise to the embryo sac. The embryo sac contains the highly reduced female gametophyte— more highly reduced than in any other group of vascular plants. After a process of double fertilization, also unique to the angiosperms, an embryo and endosperm are formed. The female gametophyte of the conifers, on the other hand, is a prominent tissue, rich in accumulated food that is used by the seedling during germination. The gametophyte generation of the conifers still retains the ancestral archegonia, but these structures have disappeared in the ovule of the angiosperms.

Such comparisons as the preceding one are useful in view of the fact that the two groups compared are the most important of the living seed plants. It should not be inferred, however, that the angiosperms are derived from the conifers—certainly not from any resembling those living today or known from the fossil record.

CLASSIFICATION OF ANGIOSPERMS

The angiosperms are divided into two subclasses: the dicotyledons, familiarly known as the dicots; and the monocotyledons, or monocots. The dicots are the larger group, with approximately 200,000 species grouped into more than 250 families. Many botanists believe that the dicots are the older of the two groups, and that the moncots branched off from the dicots at an early stage in evolution, when only the most primitive dicots—probably only woody forms—had evolved.

The classification of the angiosperms into the subclasses dicots and monocots is based upon a number of differences, both vegetative and in the structure of the flower and seed. The characteristics of the two groups are enumerated in Chapter 18 and need not be repeated here. Each of the subclasses is made up of orders composed of one or more families. The family, in turn, includes genera that have in common many external features, especially those of the flower. These genera, at least in most families, are presumed to be derived from the same ancestral stock, and therefore closely related. Among the characters used in classification into families are the position of the ovary, superior or inferior, the number and arrangement of the floral parts,

the type of placentation, the fusion or lack of fusion between members of the same whorl or between different whorls, the manner in which the flowers are clustered together on a flower-bearing branch or inflorescence, and the structure of the seed and fruit. Families differ greatly in size; a few are composed of a single genus, while others contain more than 100 genera. The family is utilized more commonly than the order when dealing with the classification of the angiosperms. The grouping of the angiosperms into families was greatly aided by the work of the French botanist A. L. de Jussieu, who in 1789 named and described about 100 families, most of which are still recognized. The groupings were expanded by later taxonomists, and approximately 300 families are now generally recognized.

In flowering plants, as well as in other organisms, the evidences of relationship as revealed in many similarities in structure have led to attempts to arrange the orders and families in a phylogenetic classification somewhat resembling a genealogical family "tree." A number of such systems has been proposed, but all are unsatisfactory in at least some respects and no one of them has been generally accepted. Even so, it is possible to characterize families and other groups as relatively primitive or relatively advanced, mostly on the basis of the structure of the flower, which, perhaps more than any other part of the plant body, exhibits evidence of great evolutionary change. The nature of such changes is best revealed by comparison with a primitive flower. No flower more primitive than any now living, however, has been found in the fossil record; therefore we are restricted to a comparison with the flowers of the most primitive living angiosperms. These are believed to be found in the order Ranales, which includes a number of families distributed chiefly in the tropics and subtropics of the Northern and Southern hemispheres, but with some temperate zone representatives. Most members of the order are trees or shrubs, but some are herbaceous. Examples of woody families are the magnolia and custard-apple; of herbaceous families, the crowfoot, including the buttercups, and water lilies. There is general agreement that the families of the Ranales, especially the woody ones, should be placed at the base of any phylogenetic system, for they have retained, in one or more organs or parts, features that are considered to indicate the ancestral, primitive conditions in flowering plants.

Studies on floral evolution have led to the conclusion that contemporary plants with primitive flowers have the following characteristics: they possess a perianth; they are bisexual and regular, with numerous carpels and stamens; and the parts of the flower are distinct from one another—that is, they are not fused. Further, the carpels and stamens, sometimes the perianth parts also, are spirally arranged rather than in whorls. Even though the magnolias, buttercups, and related families possess a greater number of these primitive characteristics than any other group of families, it does not follow necessarily that they are the most primitive angiosperms that ever lived; they may be regarded, rather, as surviving remnants that have retained many of the primitive features of the earliest flowering plants. No plants living today, even though they have primitive characteristics, are considered to be the ancestors of other flowering plants; they have merely retained certain ancestral features that have furnished us with a series of clues, in the absence of geologic evidence, to the evolutionary history of the angiosperms.

The flowers of various families of angiosperms provide abundant evidence of modification from a primitive type. It is generally recognized that evolution has proceeded from a relatively simple condition to one of relative complexity, and this is illustrated by various modifications associated with cross-pollination by insects. As pointed out earlier (Chapter 18), however, evolution has also proceeded by reduction, leading to simplicity rather than complexity. This is especially well shown by such modifications in flower structure as the disappearance of stamens or pistils in unisexual flowers, the absence of petals, reduc-

tion in the number of stamens or carpels, or a decrease in the number of ovules. Other evolutionary trends exhibited are fusion and change in symmetry. Some of the more outstanding directions in which floral evolution has proceeded are summarized below:

1. From a spiral to a whorled arrangement of floral parts. The whorled arrangement has arisen from the spiral by a shortening of the floral axis, bringing the parts closer together.

2. From numerous to few floral parts. The possession of many stamens is a more primitive condition than few stamens, and a flower with two whorls of stamens is more primitive than one with a single whorl. Many carpels (simple pistils) preceded few carpels.

3. From distinct, or separate, to united parts. Separate sepals or petals are more primitive than united perianth parts, and separate carpels preceded carpels that are fused into a compound pistil. The stamens in some families are united by the fusion of their filaments or by their anthers.

4. From the superior to the inferior ovary.

5. From bisexual to unisexual flowers.

6. From regular to irregular.

7. From both perianth whorls present to petals absent, or the entire perianth absent or present in highly reduced form.

The phylogenetic classification of flowering plants is based upon relationship by descent, and utilizes the largest possible number of characters. It is now recognized that not only floral characters but also other features—such as pollen structure, chromosome number and form, organic compounds, and internal anatomy—may be important in revealing genetic relationships. Those families, of course, are presumed to be most closely related that have the greatest number of features in common, and those most distantly related that possess the fewest features in common.

As pointed out in Chapter 18, the possibility of convergent evolution must be kept in mind in any interpretation of evolutionary relationships. For example, the inferior ovary, the fusion of petals or other floral organs, zygomorphy, and reduction in number or disappearance of stamens or carpels have evolved in many lines of descent. The corolla in orchids and snapdragons is zygomorphic, but these families are not closely related. Similarly, the sunflower and parsnip are not related even though both have inferior ovaries, nor are the primroses, with petals fused in a corolla tube, related to the honeysuckles. In spite of convergence and other complicating factors, it is believed that many families of flowering plants are closely related and that these families must have had a common origin from some more primitive ancestral stock.

SOME FAMILIES OF DICOTYLEDONS

A number of dicotyledonous families are surveyed briefly in the subsequent pages. This limited survey attempts to focus attention upon certain groups of botanical or economic importance. Family names, both common and scientific, are frequently derived from the most common or most important genus in the family. Numerous manuals and reference works supply details of these and other families of dicots.

Fig. 28-2 Flower of the tulip tree (*Liriodendron tulipifera*). The three sepals beneath the petals cannot be seen.

Magnolia Family (Magnoliaceae) As mentioned before, the magnolia family is relatively primitive. All the hundred or so species in ten genera are either trees or shrubs, often with large, attractive flowers. A number are widely planted as ornamentals. The stamens and carpels, simple pistils, are usually numerous and spirally arranged on an elongated axis (receptacle). Several species of *Magnolia*, together with the tulip tree (*Liriodendron*), are native to the eastern and southeastern United States. The family as a whole is largely restricted to eastern North America and eastern Asia. The only existing relative of the tulip tree (FIG. 28-2), one of the largest and most valuable of our eastern trees, is a Chinese species. Fossil remains of

Magnolia and *Liriodendron* indicate, however, that these genera were widely distributed throughout the Northern Hemisphere in Upper Cretaceous times, and that the family is probably among the most ancient of the living dicotyledons. Like the sequoias and ginkgo, they are survivors of the past and have become extinct over great areas of their former range in the Northern Hemisphere.

Crowfoot Family (Ranunculaceae) Among the better-known members of this family are the buttercups (FIG. 28-3) and crowfoots (*Ranunculus*), marsh marigold, the hepaticas, anemones, columbine, larkspurs, meadowrues, and clematis. A number of species contain substances causing

Fig. 28-3 Ranunculaceae. Flower of common buttercup (*Ranunculus acris*).

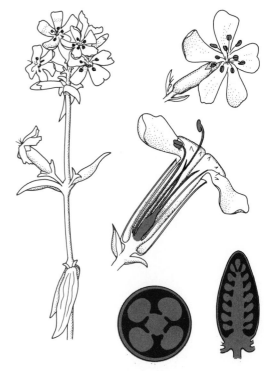

Fig. 28-4 Caryophyllaceae. Flower of soapwort (*Saponaria officinalis*). The manner of attachment of the ovules is shown in the cross and longitudinal sections of the ovary at the lower right.

forage poisoning in sheep, cattle, or horses. The root of the cultivated aconite, or wolfsbane, is extremely poisonous and is the source of the drug aconitine. The leaves of this plant are also poisonous when eaten. Most of the 1500 species of the family are perennial herbs, widely distributed in the temperate and arctic regions of the Northern Hemisphere. A number of crowfoots are aquatic or semiaquatic, growing in ponds or wet places. Many species are cultivated as ornamentals.

The flowers of the crowfoot family are typically regular and bisexual, with the stamens and carpels—in some species the perianth also—spirally arranged. The stamens are commonly numerous, and the carpels usually free and numerous. Some members exhibit advanced characters, such as irregular flowers in *Aconitum* and *Delphinium*, but the family as a whole is generally considered the most primitive of herbaceous dicotyledons.

Pink Family (Caryophyllaceae) The pink family includes about 1400 species, mostly annual or perennial herbs widely distributed in north temperate regions. Economically it is chiefly important for the large number of weeds and ornamental plants that it contains. The pinks (including sweet william and carnation) and baby's-breath are perhaps the best known, but many others are cultivated in flower and rock gardens. Among the weeds are corncockle, chickweeds, and other forms found in lawns, meadows, and grain fields. A bitter substance called *saponin* occurs in some species of soapworts of this family. This is contained in a mucilaginous juice that may be expressed from the roots. When mixed with boiling water, it makes a froth and may be used as a substitute for soap.

The stems are often swollen at the nodes and the leaves are opposite. The flowers (FIG. 28-4) are regular and usually bisexual. A calyx and corolla, each composed of four or five separate parts, are usually present, and the stamens are eight to ten in number. The compound ovary is superior. A distinctive feature of the family is that the ovules are attached to a central column arising from the base of the ovary. This condition is also found in the primroses, in which, however, the petals are not separate but are fused in a corolla tube.

Poppy Family (Papaveraceae) The poppy family includes about 250 species, mostly annual, biennial, or perennial herbs. All are native to the Northern Hemisphere. A notable, although not unique, feature of the family is the production of a colorless or milky juice (latex). Opium is obtained from the latex of the unripe fruit of the opium poppy, widely cultivated in the East. The two to three sepals fall away as the flowers open, and the petals, usually four to six, are commonly crumpled in the bud. The stamens are generally numerous, in several whorls. The carpels are several to many, usually fused into a compound ovary with parietal placentation. The fruit is usually a capsule, dehiscing by pores or flaplike valves. In the poppies (*Papaver*) themselves (FIG. 28-5), the stigma is expanded into a conspicuous flat disk. In the poppies, also, much enlarged placentae are formed as outgrowths from the fused margins of adjacent carpels. These placentae extend as flangelike structures toward the center of the ovary, and bear numerous ovules. The family contains many valued ornamentals.

Mustard Family (Cruciferae) The mustard family is worldwide in distribution, with about 3000 species of annual, biennial, or perennial herbs. The flower, with some exceptions, is uniform in structure and so distinctive that it readily identifies the family (FIG. 28-6). There are four sepals, four petals, and six stamens, four of which are longer than the other two. A special kind of small, podlike fruit develops from a compound pistil. The fruit usually splits by the formation of two valves. Prominent among the garden flowers in the mustard family are candytuft, sweet alyssum, stocks, and wallflower. Among garden vegetables are cabbage, cauliflower, kohlrabi, broccoli, kale, Brussels sprouts, horseradish, watercress, radish, rutabaga, and turnip. Many common and pernicious weeds are species of mustards. Several species of mustard (*Brassica*) are cultivated for their seeds, which are the source of medicinal and table mustard. A cruciferous herb about 2 feet tall, known as

Fig. 28-5 Flower and young fruit of oriental poppy (*Papaver orientalis*). (A): flower bud; the sepals fall away as the flower opens. (B): open flower. (C): cross section of young fruit, showing carpels and placentae of compound pistil. (D): longitudinal section of flower, placentae and stigma.

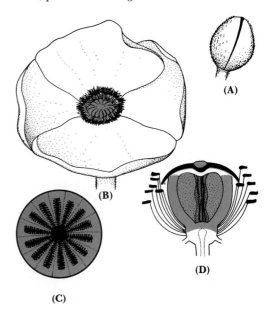

(A)

(B)

(D)

(C)

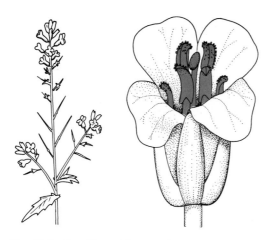

Fig. 28-6 Cruciferae. Flower of winter cress (*Barbarea vulgaris*). Flowers in racemes.

dyer's wood (*Isatis tinctoria*), native in Europe, was for centuries cultivated for the blue dye yielded by the leaves. Caesar reported that the ancient Britons used woad for staining their bodies blue.

Rose Family (Rosaceae) The rose family, of about 3000 species, ranks high among those whose members contribute to the welfare and pleasure of mankind. It includes a great variety of plants useful to man, either as ornamentals or as food plants. Among the ornamentals are the spiraeas, ninebark, the hawthorns, flowering quince, mountain ash, the roses, flowering almond, and the flowering cherries. Innumerable other herbs, shrubs, or trees are cultivated in gardens or greenhouses for the beauty of their form, foliage, or flowers. Important plants of the rose family grown for their fruits are apples and pears, blackberry, raspberry, and strawberry, and the stone fruits—plum, cherry, peach, prune, and apricot. The sweet almond is cultivated for its edible seeds.

The rose family is heterogeneous, and the flower varies greatly in structure. The sepals and petals are five in number, and the stamens are usually numerous (FIG. 28-7). The sepals, petals, and stamens are attached to the rim of a floral tube. The carpels vary in number from one to many, free, forming one or more simple pistils, or fused, forming a compound pistil. The ovary is superior or inferior, as in the apple. The fruit may be dry or fleshy, and includes achenes, pomes, drupes, and a variety of other types.

Pea Family (Leguminosae) The pea family is one of the largest families of flowering plants, containing about 500 genera and more than 15,000 species. The name hardly indicates the diversity of the group, which includes herbs, shrubs, vines, and trees growing under a great variety of environmental conditions, from semi-arid plains and grasslands to forests of the temperate and moist tropical regions. In importance to man, the pea family is exceeded only by the grass family. Even a list of some of the more important plants fails to convey completely the economic value of the family. Cultivated as food

Fig. 28-7 Rosaceae. Flower of blackberry (*Rubus allegheniensis*). Flowers in racemes.

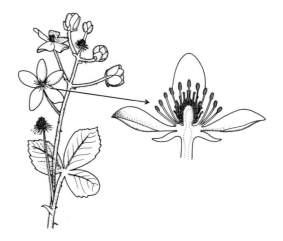

for domestic animals are the clovers, field peas, alfalfa, vetch, lupines, and lespedeza. Among the legumes edible by man are the many varieties of peas and beans, soybeans, peanuts, and lima beans. Ornamental plants include wisteria, Scotch broom, sweet peas, lupines, the redbuds, acacias, and many other woody and herbaceous plants.

From legumes we also obtain oils, proteins, gums, resins, drugs, and honey. The black locust is the chief North American leguminous tree used for lumber, but many tropical species, such as the rosewoods and cocobolo, have long been utilized for cabinetwork. A tropical American tree, logwood, is the source of the dye hematoxylin, used to some extent in industry and widely employed in the staining of plant and animal tissues for microscopic study. Other North American leguminous trees or shrubs include honey locust, Kentucky coffee tree, mesquite (*Prosopis*), and paloverde (*Cercidium*). It will be recalled that the legumes occupy a unique place in agriculture because of the part they play in the improvement of soil fertility. On the negative side, a number of legumes are weeds. Several native perennial legumes of the West and Southwest, known as loco weeds, cause poisoning when eaten by cattle, horses, or sheep.

The leaves of plants in the pea family are usually alternate and compound. A distinctive feature of the family as a whole is the fruit, which is podlike and developed from a simple pistil. When ripe, it splits typically along two sides. The structure of the flower varies to the extent that three subfamilies are recognized. The largest of these is distinguished by a type of flower similar to that of the garden pea or sweet pea (FIG. 12-11). The stamens may be separated or united by their filaments in various ways.

The flower of alfalfa (FIG. 28-8) is structurally well adapted to cross-pollination by bees. The stamens and stigma are held under tension, enclosed within the keel. When the keel is pushed downward by a bee, the stigma and stamens are released and snap upward (the flower is "tripped"). The exposed stigma receives

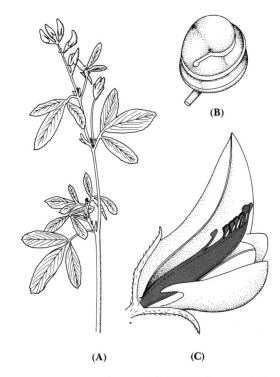

Fig. 28-8 Leguminosae. Alfalfa (*Medicago sativa*). (A): portion of flower-bearing shoot. (B): spirally coiled fruit. (C): flower; nine stamens are fused into a tube by their filaments; one is more or less separate.

pollen carried by the bee from flowers on other plants, and pollen is carried away from the tripped flower on the hairy body of the bee. Alfalfa is usually self-incompatible.

Mallow Family (Malvaceae) The mallows include about 1000 species and are widely distributed in both tropical and subtropical climates. The flowers are commonly large and showy, with five sepals and five petals. The filaments of the numerous stamens are fused into a tube that surrounds the superior ovary (Fig. 28-9). The family contains a number of attractive wild flowers, together with some troublesome weeds. Among the forms cultivated as ornamentals are hollyhock, flowering maple, and rose of Sharon. The capsules of okra, or gumbo, are used in soups and stews. Several members of the family yield stem fibers of commercial importance. The cottons, of which there are many species, are the most valuable plants of the family (Fig. 28-10). The hairs on the seed form the cotton of commerce. The two types most commonly cultivated in the United States are Upland, the most important, and American Egyptian (also called Pima).

Cactus Family (Cactaceae) The cacti, with more than 1500 species, are primarily New World plants. Although some cacti are plants of hot deserts, many other species live where temperatures far below freezing occur. They grow in the prairies of central North America and at high altitudes in the Andes of South America. They are found as far north as British Columbia in the west, and in the east a few have spread into Nova Scotia. Most cacti have thick, fleshy stems that bear spines, considered to be reduced leaves.

Fig. 28-9 Malvaceae. Flower of musk mallow (*Malva moschata*).

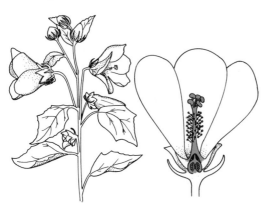

These stems, often bizarre in shape, accumulate water and carry on photosynthesis.

The cacti are classified into three groups, or tribes. The species of the most ancient and most primitive of these are strictly tropical and bear broad, expanded foliage leaves (Fig. 28-11, left) similar to those of other broad-leaved plants. These plants have woody rather than fleshy stems and bear spines as well as true leaves. The plants of the second tribe, represented by the prickly-pear cacti, usually bear, in addition to spines, small, reduced, awl-shaped leaves (Fig. 28-11, center) that soon fall from the plant. In the third and largest tribe, represented by the barrel cactus, true foliage leaves have disappeared, and only spines are to be seen (Fig. 28-11, right). Microscopic studies on some forms, however, reveal rudimentary, ephemeral scales in the position of leaves, without vascular tissues and only about 0.1 millimeter high. These three tribes of the family thus reveal an evolutionary series in the reduction of the leaf from a structure obviously a leaf to one recognizable as such only on a comparative basis.

The peyote cactus (*Lophophora williamsii*), native to the deserts of northern and central Mexico, is well known as a hallucinogen. The plant is spineless, and is composed of a small domelike head and an elongated carrotlike root. The crown, sliced off and dried, is a hard brownish disk known as a mescal button. The root remains in the soil and regenerates new crowns. The plant contains a number of alkaloids, of which mescaline is the most important. Other cacti have also been used as hallucinogens, and certain of these contain alkaloids similar to or identical with those of *Lophophora*.

The attractive flowers of cacti are sometimes very large, up to 6 inches in diameter. In some species the blooming period is very short, the flowers being open for but a single day or, as in the several species of night-blooming cereus, opening at night and closing in the morning. The flowers have numerous sepals and petals, with the sepals colored like the corolla. The ovary is inferior, and the stamens are numerous (Fig. 28-

Fig. 28-10 Malvaceae. Cotton plant; branch, flower, and the mature, green, and unopened boll (capsule).

12). Some cacti are small, others shrublike in size, whereas the giant saguaro (FIG. 2-20) grows to 40 feet or more in height.

Parsley Family (Umbelliferae) The members of the parsley family, comprising nearly 3000 species, are almost always herbaceous. The family includes a number of aromatic herbs used for seasoning—anise, caraway, dill, fennel, parsley, and coriander. Other members of the family are used as vegetables—carrot, parsnip, and celery. A final group are weeds, some of which are poisonous. Among the latter is the poison hemlock (*Conium maculatum*), introduced from Europe and common in many parts of the United States. The toxic principles of the hemlock were known to the ancients, and this plant is believed to have been the source of the poison administered to Socrates by the Athenians in 402 B.C. All parts of the plant, including the roots, stems, leaves, and seeds, are toxic to man and to farm animals. The several species of the genus *Cicuta* (water hemlock), native to North America, are also very poisonous when eaten.

The scientific name of the parsley family is derived from the arrangement of the flowers, which are grouped in umbrellalike clusters (UMBELS). The individual flowers (FIG. 28-13) are

Fig. 28-11 Cactaceae. Cacti showing transitions from broad, expanded leaves to the condition in which only spines (modified leaves) are found. (Left): Barbados gooseberry (*Pereskia aculeata*), a cactus with broad leaves. (Center): prickly pear (*Opuntia*); spines occur together with minute fleshy leaves that are short-lived. (Right): barrel cactus (*Ferocactus*); true leaves have disappeared and only spines occur.

Fig. 28-12 Cactaceae. Flowers of giant cactus or saguaro (*Carnegiea gigantea*), Arizona.

Fig. 28-13 Umbelliferae. Flowers of golden alexanders (*Zizea aurea*). Note arrangement of the flowers in umbels.

usually small. The sepals are minute or absent, and the five petals are commonly white in color. The five stamens are attached to the top of the inferior ovary. The fruits are dry and divide into two parts at maturity.

Heath Family (Ericaceae) The members of the heath family are almost always woody and include about 2000 species, mostly shrubs. Here are found a large number of attractive ornamentals such as the heaths, heather, mountain laurel, the rhododendrons, and the azaleas. Many species, wild or cultivated, have edible fruits, such as blueberry, cranberry, huckleberry,

and mountain cranberry. Other wild species include trailing arbutus, bearberry, wintergreen, and Labrador tea. Salal, madroño, and various species of manzanita are common western shrubs or small trees. Many of the heaths have evergreen leaves. Huckleberries, blueberries, and cranberries are sometimes placed in this family, and sometimes in a closely related family, the Vacciniaceae.

The sepals and petals are four or five in number, and there are twice as many stamens as there are petals. The petals are usually fused into an urn-shaped tube, and the ovary is superior or, in some groups such as blueberries and cranberries, inferior (FIG. 28-14).

Mint Family (Labiatae) The mint family includes about 4000 species, mostly annual or perennial herbs, widely distributed over the world. The family name is well known, for many mints develop aromatic oils in the stems and leaves. Some, such as spearmint and peppermint, thyme, sage, rosemary, lavender, and pennyroyal, are of economic importance for this reason. Many are cultivated as garden ornamentals. The stems of mints are usually square, with opposite or whorled leaves. The five petals are fused into a two-lipped (bilabiate) tube (FIG. 28-15), whence the scientific name of the family. The fruit is divided into four one-seeded nutlets.

Nightshade Family (Solanaceae) The nightshade family is large, consisting of more than 2000 species of herbs, shrubs, or small trees in the tropics and temperate regions. The potato, native to the temperate Andes, has become a staple food in many parts of the world. Other plants with edible parts are tomato, eggplant, red pepper, and ground cherry. Tobacco is another member of the family from tropical America. Petunias, *Salpiglossis*, and many others are cultivated for their flowers. Jimson weed, henbane, belladonna, and some of the nightshades contain poisonous alkaloids. Belladonna is the source of the drug atropine. The corolla of the nightshades is typically composed of five fused petals and is wheel-shaped or tubular. The stamens, usually five but sometimes fewer, are adnate to the corolla, and the ovary is superior (FIG. 28-16.)

Figwort or Foxglove Family (Scrophulariaceae) The members of this large family, about 4000 species, are widely distributed over the world. Most species are herbs, but some are small shrubs, trees, or vines. The family is chiefly important for the large number of house plants and garden ornamentals, including *Calceolaria*, snapdragon, the speedwells, beard tongues, Kenilworth ivy, and many others. The common foxglove (*Digitalis*) is a frequently cultivated bien-

nial, and preparations made from the leaves have important uses in treatment of diseases of the heart. The common mullein (*Verbascum thapsus*), a conspicuous biennial weed of this family introduced from Europe, is general throughout temperate North America in pastures and waste places. Its flowers are pale yellow and the stem and leaves are covered with a heavy mat of hairs, giving them a woolly appearance. The stems are tall, usually unbranched, and may attain a height of 8 feet.

The family has perfect flowers, and the perianth is usually irregular, the five fused petals forming a corolla tube (FIG. 28-17). The compound ovary, containing numerous ovules, is superior, and the carpels have been reduced to two. The stamens are typically four, in some species two, rarely five, in number, a fifth stamen occasionally being represented by the filament only. The stamens are usually in two pairs, one with longer and one with shorter filaments. The flowers are usually cross-pollinated by insects, commonly bees. These flowers often resemble the bilabiate flowers of the mints, but their

Fig. 28-14 Ericaceae. Flower of low sweet blueberry (*Vaccinium angustifolium*).

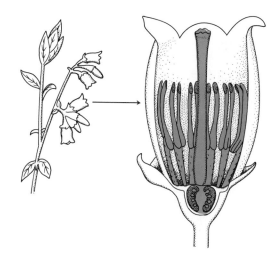

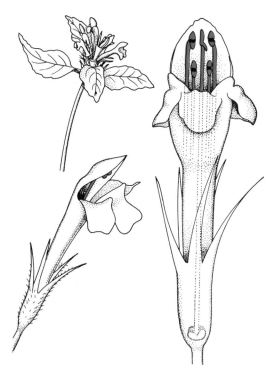

Fig. 28-15 Labiatae. Flower of hemp-nettle (*Galeopsis tetrahit*).

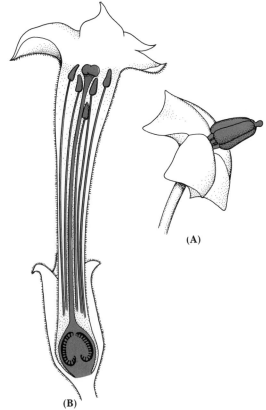

Fig. 28-16 Solanaceae. (A): flower of potato (*Solanum tuberosum*). (B): flower of tobacco (*Nicotiana tabacum*).

ovaries ripen into many-seeded capsules, in contrast to the four-lobed and four-seeded fruit of the Labiatae.

The Composites (Compositae) The composites are the second largest family of flowering plants, with about 20,000 species, mostly annual or perennial herbs. The flowers are very small and are closely grouped into compact heads, commonly mistaken for a single flower. The individual flowers are of two kinds. In one, the DISK flowers, the corolla is tubular. In the other, the RAY flowers, the corolla is strap-shaped and petal-like on one side. The head of composite flowers is commonly composed of disk flowers, making up the main body of the head, and ray flowers, along the margins (FIG. 28-18). The ray flowers in such heads are frequently sterile and produce no seeds. In some species only disk flowers or only ray flowers are present (FIG. 28-19).

The stamens are fused by their anthers and form a cylinder around the style; the ovary is inferior. The fruit, an achene, is popularly called a seed. A calyx is either absent or modified into plumes, hairs, spines, or hooks, often useful in dissemination of the fruit. An example is the downy parachute of the dandelion.

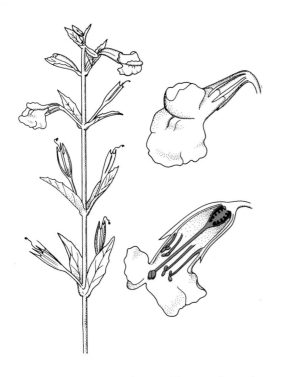

Fig. 28-17 Scrophulariaceae. Flower of monkey flower (*Mimulus ringens*).

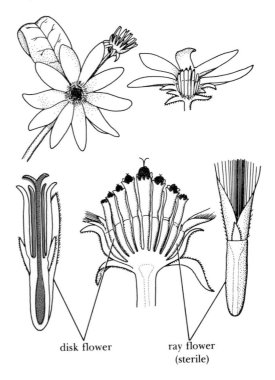

disk flower   ray flower
                     (sterile)

Fig. 28-18 Compositae. Head of flowers, bearing both ray and disk flowers, of thin-leaf sunflower (*Helianthus decapetalus*).

To list the more common composites is a task in itself. Among the kinds used for food are lettuce, artichoke, salsify, dandelion, Jerusalem artichoke, and endive. The leaves of tarragon are used for seasoning. The common sunflower is cultivated in many parts of the world for the flattened oval fruits, "seeds," which are rich in oil and are used as food for man and animals. A large number of composites, such as chrysanthemum, cosmos, zinnias, dahlias, calendulas, and marigold, are garden ornamentals. Burdock, boneset, grindelia, camomile, and arnica have been used medicinally. The white snakeroot (*Eupatorium rugosum*) when eaten by cattle, horses, and sheep, causes trembles, or milk sickness. The disease is transmitted to people who drink milk from affected cows or use butter made from such milk. The flowers of pyrethrum,

a species of *Chrysanthemum,* are the source of a well-known insecticide. The sagebrushes, ragweeds, and their relatives are important causes of hay fever. Finally, a number of composites are weeds, troublesome to farmers and gardeners—the thistles, dandelion, hawkweeds, cocklebur, and many others.

The composites are a highly evolved group of plants, probably extending back to the Upper Cretaceous. Few groups are more familiar and widespread, from the asters and goldenrods of eastern autumn roadsides to the sagebrush of western plains and the golden-yellow composites of the Pacific states.

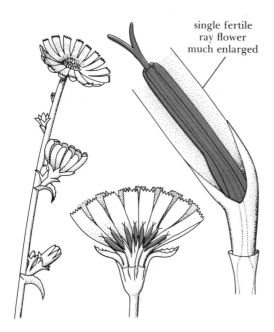

single fertile
ray flower
much enlarged

Fig. 28-19 Compositae. Chicory (*Cichorium intybus*).

SOME FAMILIES OF MONOCOTYLEDONS

The monocots include about 55,000 species, classed in more than 40 families. Since their beginning, they have developed many specialized structures that are frequently similar to, but not identical with, those of dicots. Some monocots, such as the palms, are treelike, but most are herbaceous annuals or perennials. Widely distributed in many kinds of environments, monocots are especially prominent in most tropical and subtropical climates. The aerial stems branch sparsely or not at all, and are frequently sheathed by the bases of the leaves. Most monocot leaves are simple, but some are compound, as in the palms. In a number of families the flowers are irregular, as in the orchids, or much reduced and inconspicuous, as in the grasses. Only a few of the more important of the numerous families are included in the subsequent paragraphs.

Lily Family (Liliaceae) Although named for a common, conspicuous genus (*Lilium*), most of the 4000 species of the lily family are not lilies at all. Asparagus is an important food plant. Onions, leeks, chives, and garlic are usually classed with the lilies, but some botanists place them in the closely related amaryllis family. Many lilies are prized as garden or house plants. These include the easter lily, regal lily, tiger lily, and other colorful species. The day lily, lily-of-the-valley, and the Mariposa lily are not included in the genus *Lilium*, but belong in the family. Other ornamentals are tulips, hyacinths, and autumn crocus (*Colchicum*). The alkaloid colchicine from this plant is widely employed in the production of polyploids. Plants poisonous to man or livestock include red squill (*Urginea*), used as a rat poison; false hellebore (*Veratrum*); and species of death camas (*Zigadenus*). Conspicuous among the attractive wild plants in this family are trilliums, bellworts, adder's-tongues, and clintonia.

The Liliaceae are mostly perennial herbs with a rhizome, bulb, or corm. The flower is regular, with six petal-like perianth parts in two whorls. The ovary is usually superior, and the placentation is axile (FIG. 28-20).

There has been much disagreement concerning the placement of genera in the families of the Liliales, the order to which the Liliaceae belong. For example, *Yucca*, once held to be a lily, and *Agave*, formerly placed in the amaryllis family, are now classified in the family Agavaceae of the Liliales. This family is composed of coarse, shrubby, or treelike plants with long-lived leaves (FIG. 28-21).

Palm Family (Palmae) Palms of various kinds supply the primary needs of millions of people in tropical regions (FIG. 2-3). Several thousand species of palms are known. Many are small, but some grow more than 100 feet high, rivaling the larger deciduous trees in size. Most palms are tropical, but a dozen or so are native to the subtropical parts of this country, and a number of others are cultivated. In the Upper Cretaceous

Fig. 28-20 Liliaceae. Flower of clintonia (*Clintonia borealis*).

Fig. 28-21 Agavaceae. Joshua trees (*Yucca brevifolia*).

and Lower Tertiary, palms were widely distributed in the Northern Hemisphere, extending into Canada. Modern palms are a source of food (coconuts and dates), lumber, fibers for clothing, thatch for native houses, vegetable ivory, oils and waxes, sago starch, and many other products.

Grass Family (Gramineae) The grasses, one of the most important to man of all the plant families, include about 7000 species, many of worldwide distribution. They grow under a wide range of climatic conditions, from the warm tropics to within the Arctic Circle. Many bamboos, which are grasses, are woody plants, known to attain a height of 120 feet, but most grasses are annual or perennial herbs. The leaves are usually narrow and long, and the base ensheathes the stem, which is frequently hollow

between the nodes. Perennial grasses spread by stolons or rhizomes and grow in tufts or bunches, or in dense masses forming a heavy sod. Bunch and sod grasses have colonized vast areas of the earth's surface.

The grasses, directly or indirectly, are the chief source of the food supply of man and of many kinds of animals. The shoots of young bamboo are used as food, and the stems have endless uses in house construction, in furniture, in paper, and in baskets and utensils for peoples living in tropical climates. The grains, or cereals, are outstanding. Wheat has been cultivated for more than 7000 years. Rice is a staple food for more of the world's peoples than any other plant. The stems of sugar cane (FIG. 5-1) have long been used as a source of sugar.

Important among the forage grasses in pastures and open range are brome grasses, buffalo grass, rye grass, meadow fescue, timothy, Kentucky blue grass, Bermuda grass, and many others. Some grasses are persistent weeds. Others are used for lawns, golf courses, and other ornamental planting. Grasses are valuable as food for wildlife; many birds and small mammals depend for their diet largely upon the grains of wild and cultivated grasses.

The flowers of the grasses are small, reduced, and borne in dense spikes or open, branching clusters (PANICLES). Such spikes or clusters are composed of units called SPIKELETS. Each spikelet, in turn, consists of an axis, usually bearing several flowers (FIG. 28-22A) together with a number of specialized bracts. In some plants, such as timothy (FIG. 28-22B), but one flower is borne on a spikelet. At the base of the spikelet are found two empty bracts (*glumes*), and each flower is tightly enclosed within two other bracts (*lemma, palea*). An individual flower is composed of a pistil, stamens (typically three in number) and usually two scales (LODICULES) below the ovary. These lodicules represent the highly reduced perianth. Some grasses are self-pollinated, but wind pollination is more common. The stamens emerge from the bracts at flowering, and shed their pollen. The two feathery stigmas protrude and catch the wind-borne pollen.

Orchid Family (Orchidaceae) The orchids comprise the largest family of flowering plants, with about 25,000 species. Like the composites, the orchids are a very diverse group. All are herbaceous perennials. Contrary to popular opinion, they are not all tropical, nor do they all grow upon other plants. Many are subtropical, and many others thrive in temperate regions. About 140 species are native to North America. Many orchids grow as epiphytes, but many others, including most temperate zone species, are terrestrial, growing in the soil. A few are colorless (Chapter 5).

The orchids have little value except as ornamentals. Vanilla flavoring is extracted from the pods of a tropical orchid, but a synthetic flavoring is commonly used. A very large number of tropical species, especially of the genus *Cattleya,* are cultivated for their odd and striking flowers, and many hundreds of horticultural

Fig. 28-22 Gramineae. (A): many-flowered spikelet of love grass (*Eragrostis*) forming seeds; the two basal bracts (glumes) are always empty. (B): one-flowered spikelet of timothy (*Phleum pratense*). (1): complete spikelet. (2): the glumes removed, showing the lemma and palea associated with each flower. (3): the naked flower, consisting of three stamens, an ovary bearing two feathery stigmas, and two lodicules at the base of the ovary, which constitute the reduced perianth.

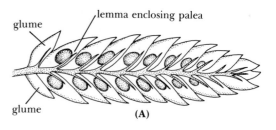

(A)

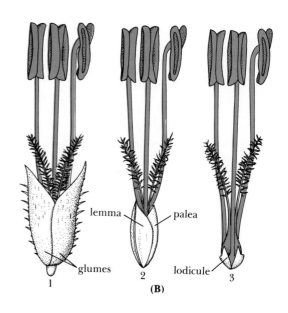

(B)

Fig. 28-23 Orchidaceae (*Cattleya gigas*). The "lip" on the lower side of the flower is a single modified petal.

varieties have been produced by hybridization. Among our native orchids, some of which are cultivated out-of-doors, are the lady's-slipper, showy orchids, pogonia, calypso, and lady's tresses. The picking of native orchids is forbidden by conservation laws in many states.

The orchid flower is highly modified and adapted to cross-pollination by insects. It is constructed upon the usual monocotyledonous plan, but this is obscured by the fusion of various parts. The ovary is inferior. The three sepals are usually distinct, but in some species two sepals are fused into one organ. The three petals are modified so that two form "wings" and the central one forms a conspicuous cuplike lip on one side of the flower (FIG. 28-23). Usually only one of the stamens is functional, but in some species two are.

An important feature of the orchid flower is the COLUMN formed by the fusion of the stamens with the style. In the lady's-slipper the central lip is much enlarged and slipper shaped, and encloses a large cavity, open at the top (Plate 16). The column lies at the back of the flower. One of the three stamens is sterile and arches over the stigma. The two fertile stamens are located on

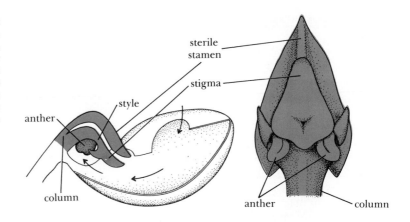

Fig. 28-24 Pollination in lady's-slipper (*Cypripedium*). (Left): flower from the side, with portion of the slipper removed, the arrows show the route of a small bee as it enters and leaves the flower. (Right): the column, which terminates in the stigma and bears the stamens.

either side of the stigma (FIG. 28-24). Small bees, entering the oval opening at the top of the lady's-slipper, feed upon juicy hairs at the bottom. Because escape by the path by which they entered is prevented by the overarching wall of the slipper, the insects can leave only through two small openings at the back of the flower, one on either side of the column. They creep under the stigma and squeeze through one of these openings, but in doing so, they brush against an anther, picking up some of the sticky pollen. When they visit another flower, the bees escape in the same way, this time depositing pollen on the stigma as they creep under it. In the absence of insect visits, pollination fails and no seeds form. The orchids, grasses, and composites each illustrate, in different ways, various culminations in the evolution of the modern flower.

SUMMARY

1. The angiosperms are the dominant plants of the modern world. They are adapted to a great variety of environments and include a greater number of species than any other group of plants.

2. The importance of the flowering plants to man is due, in large part, to their secondary tissues, to their ease of propagation by vegetative methods, and to their habit of accumulating food in seeds and fruits, and, to a lesser extent, in roots, tubers, and other underground parts.

3. Although the roots, stems, and leaves of angiosperms exhibit a wide range of specialization, the flower, fruit, and seed are probably chiefly responsible for the success of the angiosperms.

4. Primitive man could not have advanced beyond the hunting and fishing stage of culture without flowering plants, especially annuals, which are the foundations of modern agriculture. The evolution of birds, mammals, and insects depended upon the evolution of the angiosperms.

5. The angiosperms are divided into the monoctyledons and dicotyledons. The latter group is probably the older of the two. The two groups are distinguished by the number of cotyledons, the venation of the leaves, the arrangement of the vascular tissue in the stem, the presence or absence of a cambium, and the number of parts in each whorl of the flower. In both dicots and monocots the flowers may be irregularly symmetrical or reduced.

6. Of all the plant families, the orchids, grasses, and composites best illustrate evolutionary culmination along divergent lines.

Some Comparisons between Conifers and Angiosperms

7. The conifers are all woody trees or shrubs. The angiosperms include trees, shrubs, and great numbers of herbs—annual, biennial, and perennial.

8. Although the conifers are widely distributed, they have not become adapted to so great a range of environmental conditions as have the angiosperms.

9. The chief type of conducting cell in the wood of angiosperms is the vessel element. The most important kind of conducting cell in conifer wood is the tracheid.

10. The structures associated with reproduction in the conifers are borne in conelike strobili, with the ovules and pollen in different strobili. Angiosperms are characterized by the possession of flowers; ovules and pollen are usually borne in the same flower.

11. In the conifers, the seeds are borne on the surface of a scale. In the angiosperms, the seeds are enclosed in an ovary, the basal part of a pistil, composed of one or more carpels. From the ovary, or in some cases from this structure and adjacent parts, is produced the fruit, an organ found only in angiosperms.

12. In conifers, the pollen grains germinate on or near the ovule. In angiosperms, the pollen grains germinate on the stigma of the pistil.

13. Both conifers and angiosperms are termed heterosporous and both retain the megaspore within the megasporangium upon the parent plant. The female gametophyte of conifers is a relatively large multicellular tissue that is produced before fertilization and whose nuclei contain the n chromosome number. The female gametophyte of angiosperms is highly reduced, consisting of only a few cells and nuclei. In angiosperms an endosperm develops after fertilization and its nuclei contain the $3n$ chromosome number. The endosperm develops after double fertilization, a process found only in angiosperms.

14. Both angiosperms and conifers have lost the swimming sperms of their aquatic ancestors, and the sperm is conveyed to the egg through a pollen tube. The conifers, however, still form archegonia, structures probably characteristic of the earliest land plants and found in the bryophytes and all other vascular plants except angiosperms.

15. Many conifers still retain vestiges of the vegetative cells of the male gametophyte generation. In the pine, such tissues are represented by the degenerate prothallial cells within the pollen grain. No such vegetative cells have been retained by the male gametophyte generation of the angiosperms.

Glossary

Abscisic (ăb-sĭs-ic) **acid:** a plant hormone having a growth-inhibiting action. It promotes abscission and is associated with the onset and maintenance of dormancy.

Abscission (ăb-sĭzh′ŭn): the dropping of leaves, flowers, fruits, or other plant parts.

Absorption (ab-sorp′shŭn): sucking up or imbibing, as a sponge absorbs water.

Accessory: term applied to buds located above or on either side of the main axillary bud.

Accessory fruit: fruit containing tissues developed from other flower parts in addition to the ovary. Most accessory fruits develop from a flower with an inferior ovary, the floral tube constituting the accessory part. In some fruits the receptacle may be the accessory part.

Achene (ă-kēn′): a small, dry, one-seeded indehiscent fruit in which the seed coat is not adherent to the pericarp. (Compare with **Grain.**)

Actinomorphic (ăk-tĭ-nō-môr′fĭk): *See* **Regular.**

Active transport: movement across a cellular membrane; dependent upon respiratory energy and upon reactions within the membrane. It may result in accumulation against a diffusion gradient.

Active water absorption: the absorption of water by roots, brought about by osmotic and other forces developed in the roots.

Adaptation: changes in the composition of a group of organisms or in the physiology or structure of a single organism, resulting in the individual or group being better fitted to the environment.

Adenosine triphosphate (ă-dĕn′ō-sēn trĭ-fŏs′fāt): called ATP, the major source of usable chemical energy in metabolism. On hydrolysis ATP loses one phosphate and becomes adenosine diphosphate (ADP) plus usable energy, and on further hydrolysis it is changed to adenosine monophosphate (AMP) with further release of energy.

Adnate (ăd′nāt): fusion of unlike parts, as stamens and petals. (*See* **Connate.**)

ADP: *See* **Adenosine triphosphate.**

Adsorption (ăd-sôrp′shŭn): the adhesion of liquid, gaseous, or dissolved substances to the surface of a solid body, resulting in a concentration of the adsorbed substance.

Adventitious (ăd′věn-tĭsh′ŭs): plant organs, such as shoots or roots, produced in an unusual position or at an unusual time of development.

Aeciospores (ē′sĭ-ō-sporz): spores formed within an aecium (cup); one of the types of spores produced in the life cycle of certain rust fungi.

Aecium (ē′shĭ-ŭm): in the rust fungi, a cup-shaped structure within which aeciospores are produced.

Aerobic (ā′ĕr-ō′bĭk): growing or proceeding only in the presence of oxygen.

Aerobic respiration: *See* **Respiration.**

After-ripening: term applied to the metabolic changes that must occur in some dormant seeds before germination can occur.

Agar (ä′gär): a gelatinous substance obtained from red algae. Used as a solidifying agent in preparation of nutrient media for growing microorganisms and for other purposes.

Aggregate fruit: fruit developed from a flower having a number of pistils, all of which ripen together and are more or less coherent at maturity.

Alkaloids (ăl′kȧ-loidz): organic compounds with alkaline properties, produced by plants. These substances which constitute the active principles of many drugs and poisons of plant origin, have a bitter taste and are sometimes poisonous. All alkaloids contain carbon, hydrogen, and nitrogen; most contain oxygen also. Examples: nicotine, morphine, quinine, caffeine, strychnine.

Allele (ă-lēl′): one of the two or more alternative, contrasting genes that may exist at a particular locus on a chromosome. An organism may be homozygous for some alleles and heterozygous for others.

Allopolyploid (ăl′ō-pŏl′ĭ-ploid): an organism with more than two sets of chromosomes, derived from different species, in its body cells.

Alternate: term applied to the arrangement of the leaves or buds that occur singly at a node.

Alternation of generations: the alternation of gametophytic and sporophytic generations in the life cycle. The sporophyte develops from the zygote and produces spores. The gametophyte develops from the spore and produces gametes. The cells of the sporophyte generation contain twice as many chromosomes as those of the gametophyte.

Amino (ă-mē′nō) **acids:** nitrogen-containing organic acids, the "building stones" or units from which protein molecules are built.

Ammonification (ă-mōn′ĭ-fĭ-kā′shŭn): the formation of ammonia by decay organisms of the soil acting upon proteinaceous compounds.

AMP: *See* **Adenosine triphosphate.**

Amylase (ăm′ĭ-lās): an enzyme that hydrolyzes starch.

Amyloplast (ăm′i-lō-plăst): a colorless plastid containing one or more starch grains.

Anabolism (a-năb′ō-liz′m): constructive, or synthetic metabolism, such as photosynthesis, assimilation, and the synthesis of proteins.

Anaerobic (ăn-ā′ēr-ō′bĭk): proceeding in the absence of or not requiring oxygen.

Anaerobic respiration: a type of respiration, found only in bacteria, in which the hydrogen released in glycolysis is combined with the bound oxygen of inorganic compounds, such as carbon dioxide.

Anaphase (ăn′a-fāz): a stage in mitosis in which the chromatids of each chromosome separate and move to opposite poles.

Angiosperm (ăn′jĭ-ō-spûrm): a plant with seeds enclosed in a developed ovary (fruit).

Angstrom (ang′strŭm): one hundred-millionth of a centimeter; one ten-thousandth of a micron.

Annual: a plant in which the entire life cycle is completed in a single growing season.

Annual ring: an annual growth layer as seen in cross section. The term applies primarily to wood grown in the temperate zones. (*See* **Growth ring.**)

Annulus (ăn′ŭ-lŭs): (1) in gill fungi, a ring on the stalk, the remnant of the partial or inner veil; (2) in ferns, a specialized ring of cells around the sporangium.

Anther (ăn′thēr): the upper portion of a stamen, containing the pollen grains.

Antheridium (ăn′thēr-ĭd′ĭ-ŭm): a male sex organ containing sperms; multicellular in Bryophytes and Tracheophytes.

Anthocyanins (ăn′thŏ-sī′a-nĭnz): natural pigments, usually of shades of red or blue, occurring in solution in the cell sap.

Antibiotics: substances of biological origin that interfere with the metabolism of microorganisms.

Antipodals (ăn′tĭp′ŏ-dăl′z): three (sometimes more) cells of the mature embryo sac located at the end opposite the micropyle. They are usually functionless and are believed to be vestigial.

Apical dominance: influence exerted by a terminal bud in suppressing the growth of lateral buds.

Apical meristem: meristematic cells at the tip of the stem or root; region of initiation of the primary tissues. (*See also* **Shoot apex.**)

Apogamy (a-pŏg′a-mĭ): formation of a sporophyte from vegetative cells of the gametophyte.

Apospory (a-pŏs′pŏ-rĭ): formation of a gametophyte from vegetative cells of the sporophyte.

Archegonium (är′kĕ-gŏ′nĭ-ŭm): female sex organ containing an egg; it is unicellular in the Thallophytes and multicellular in the Bryophytes and Tracheophytes.

Ascomycete (ăs′kŏ-mī-sēt′): a fungus whose spores are borne in an ascus (sac).

Ascospore (ăs′kŏ-spōr): a fungus spore produced within an ascus following sexual fusion and meiosis.

Ascus (ăs′kŭs): the saclike structure of the perfect stage of an ascomycete, containing ascospores.

Asexual (a-sĕk′shoo-ăl) **reproduction:** (1) any reproductive process that does not involve the union of gametes; (2) in lower forms, reproduction by spores not associated with the sexual process.

Assimilation: the conversion of food materials into protoplasm and cell walls.

Atom (at′ŭm): the smallest unit of matter that can take part in the formation of molecules.

ATP: *See* **Adenosine triphosphate.**

Autoecious (ô-tē′shŭs): term referring chiefly to rust fungi that complete the life cycle on one species of host plant.

Autopolyploid: an organism with more than two sets of chromosomes in its body cells, all derived from a single species.

Autotrophic (ô′tŏ-trŏf′ĭk): organisms able to manufacture all of their own food, as most green plants and some bacteria.

Auxin (ôk′sĭn): a natural growth hormone or a synthetic growth regulator that promotes cell enlargement.

Axil (ăk′sĭl): angle on upper side between leaf and stem.

Axile (ăk′sĭl; sīl): belonging to or located in the axis.

Axillary: term applied to buds or branches occurring in the axil of a leaf.

Bacillus (ba-sil′ŭs): a rod-shaped bacterium.

Back-crossing: the crossing of a hybrid with one of its parents or with a genetically equivalent organism.

Bacteriophage (băk-tēr′ĭ-ŏ-fāj): a virus that attacks bacteria.

Bark: an inclusive term for all tissues outside the cambium.

Basidiomycete (ba-sĭd′ĭ-ŏ-mīsēt′): a fungus whose spores are borne on basidia (clubs).

Basidiospore (ba-sĭd′ĭ-ŏ-spōr′): a spore of Basidiomycetes produced following a sexual process; a spore of the perfect stage.

Basidium (ba-sĭd′ĭ-ŭm): the structure of Basidiomycetes that bears basidiospores.

Berry: a simple fleshy fruit such as a grape or tomato.

Biennial: a plant that normally requires two growing seasons to complete the life cycle. Only vegetative growth occurs the first year; flowering and fruiting occur in the second year.

Bilabiate (bī-lā′bĭ-ăt): two-lipped, usually applied to the corolla. Each lip in turn may be lobed or toothed. Bilabiate corollas are common among the mints and figworts.

Bilaterally symmetrical: term applied to flowers capable of division into two equal halves along a single longitudinal plane. Such flowers are also called *irregular,* as opposed to radially symmetrical.

Binomial system: a system in which the scientific name for any plant is composed of two parts.

Biosphere (bī-ŏ-sfēr′): the sphere of living organisms including the land, water, and air.

Bisexual: (1) a flower having both functional stamens and pistil; a perianth may be present or absent. (2) any organism having both sexes.

Blade: the broad, expanded part of a leaf.

Bordered pit: *See* **Pit.**

Bract: a modified, usually reduced leaflike structure.

Bromelain: a protease found in the fresh juice of the pineapple.

Bryophyte (brī′o-fīt): any liverwort or moss plant.

Bud: (1) an embryonic shoot; (2) vegetative outgrowth from a yeast cell.

Bud mutation: a genetic change in a bud that causes it to give rise to a branch, flower, or fruit that differs from other parts of the plant.

Bud scales: specialized organs, which cover the shoot apex and embryonic leaves of a winter bud.

Bud-scale scars: scars encircling the twig, left by the fall of bud scales of terminal bud of the previous year.

Budding: (1) the method of vegetative reproduction in yeasts; (2) a form of grafting.

Bulb: an underground storage organ, composed chiefly of enlarged and fleshy leaf bases.

Bundle scar: marks on a leaf scar formed by the breaking of the vascular bundles (leaf traces) that pass from the twig into the petiole.

Bundle sheath: a special sheath of parenchyma cells surrounding the vascular bundles of the leaf.

Bundle-sheath extensions: somewhat thick-walled cells extending from the bundle sheath of larger veins to the upper or lower epidermis of the leaf or to both upper and lower epidermis.

Button: an immature mushroom before the expansion of the cap.

Callus (kăl′ŭs): (1) wound tissue; parenchyma tissue formed on or below a wounded surface; (2) undifferentiated tissue; a term used in tissue culture.

Calorie: the amount of heat necessary to raise the temperature of 1 gram of water 1 degree on the centigrade scale.

Calyptra (ká-lĭp′trá): the hood or cap of a moss plant. partially or entirely covering the capsule, and formed from the archegonial wall.

Calyx (kā′lĭks): the first (beginning from below) of the series of floral parts, composed of sepals. The calyx is usually green and somewhat leaflike, but may be colored like the petals.

Cambial zone: a region of thin-walled cells between the xylem and phloem. Composed of cambium and its recent derivatives that have not yet differentiated into mature cells.

Cambial initials: the cells of the cambium. They are of two kinds: ray and fusiform initials.

Cambium (kăm′bĭ-ŭm): *See* **Cork cambium; Vascular cambium.**

Cap: the top or head of a fleshy, stalked fungus.

Capillarity: the movement of a liquid due to surface forces; especially observable in capillary tubes.

Capillary soil water: water held in the soil against the force of gravity; the chief source of the water absorbed by roots.

Capsule (kăp′sūl): (1) a dry fruit that develops from a compound pistil and opens in various ways, allowing the seeds to escape; (2) a slimy layer around the cells of certain bacteria; (3) the spore case in liverworts and mosses.

Carbohydrates: a large group of organic compounds composed of carbon, hydrogen, and oxygen, the ratio of hydrogen to oxygen being 2:1, as in water.

Carnivorous: feeding upon animals; opposed to herbivorous. The term refers also to plants that are able to utilize proteins obtained from trapped animals, chiefly insects.

Carotenes (kăr′ŏ-tēn′z): orange or yellow, unsaturated hydrocarbons, such as β-carotene ($C_{40}H_{56}$).

Carotenoids (ká-rŏt′ĕ-noid′z): yellow or orange (occasionally red) pigments found in the plastids of plants. (*See* **Carotenes** and **Xanthophylls.**)

Carpel (kär′pĕl): a floral organ bearing ovules, usually along the margins; the unit of structure of a compound pistil.

Casparian (kăs-pâr′ĭ-ăn) **strip:** strips that seal the radial and transverse walls of the endodermis and prevent diffusion in the walls between adjoining protoplasts.

Cast: a type of fossil that results from the filling of a cavity formed by the decay of plant tissues.

Catabolism (kȧ-tăb′ṓ-lĭz′m): the degradative aspects of metabolism, such as respiration and digestion.

Catalyst (kăt′ȧ-lĭst): a substance that affects the rate of a reaction but is chemically unchanged at the end of the reaction.

Catkin (kăt′kĭn): a pendulous spikelike inflorescence of unisexual flowers. Found only in woody plants.

Cell: the structural unit composing the bodies of plants and animals; an organized unit of protoplasm, in plants usually surrounded by a cell wall.

Cell division: the formation of a cell plate between two nuclei, resulting in two daughter cells.

Cell plate: a membranelike structure that forms at the equator of the spindle during early telophase; the predecessor of the intercellular layer.

Cell sap: collective term for the fluid content of the vacuole.

Cellulase: an enzyme that hydrolyzes cellulose.

Cellulose: a carbohydrate, the chief component of the cell wall in most plants.

Centromere (sĕn′trō-mēr): that portion of the chromosome to which the spindle fiber is attached.

Character: the phenotypic expression of a gene.

Chemoautotrophic (kĕm′ō-ô′tō-trŏf′ĭk): organisms (bacteria) able to manufacture their own basic foods with chemical energy. (*See* **Autotrophic.**)

Chlorenchyma (klȯ-rĕng′kĭ-mȧ): general term applied to parenchyma cells that contain chloroplasts.

Chlorophylls (klō′rȯ-fĭl′z): green pigments located in plastids; necessary to the process of photosynthesis.

Chloroplast: a specialized cytoplasmic body containing chlorophyll. Other pigments also present.

Chlorosis (klȯ-rō′sĭs): reduced development or loss of chlorophyll.

Chromatid (krō′mȧ-tĭd): term applied to the parallel threads resulting from chromosome duplication before they separate from each other.

Chromatin (krō′mȧ-tĭn): nuclear materials, primarily nucleoproteins, that stain deeply with basic dyes.

Chromoplast: a specialized cytoplasmic body containing carotenoids or other pigments, except chlorophyll.

Chromosome (krō′mȯ-sōm): structural units in the nucleus that preserve their individuality from one cell generation to another; the site of the genes.

Class: a group of plants ranking above an order and below a division (phylum).

Cleistothecium (klis′tȯ-thē′shĭ-ŭm): a completely enclosed envelope surrounding asci and ascospores.

Climax: the terminal community of a succession, which maintains itself relatively unchanged unless the environment changes.

Clone (klōn): a group of plants, often many thousand in number, that have had a common origin and that have been produced only by vegetative means such as grafting, cutting, or division rather than from seed. The members of a clone may be regarded as the extension of a single plant.

Cluster cups: term applied to the aecia of the rust fungi.

Coccus (kŏk′ŭs; pl., **Cocci,** koksī): a spherical bacterium.

Coenocytic (sē′nȯ-sĭt′ĭk): multinucleate, the nuclei not separated by cross walls.

Cohesion theory: a theory used to explain the rise of water in plants.

Coleoptile (kō′lḙ-ŏp′tĭl): a sheathlike, pointed structure covering the shoot of grass seedlings; interpreted as a portion of the cotyledon.

Collenchyma (kŏ-lĕng′kĭ-mȧ): elongated, living cells with variously thickened primary cell walls. A flexible, supporting tissue.

Colloid (kŏl′oid): a dispersion of small particles or very large molecules of one substance in another, such as glue in water.

Colony: (1) In algae, an aggregation of closely associated cells, the units of which function independently of each other but do not usually occur separately. In some colonies a certain degree of division of labor may be apparent, but in most the cells are alike. (2) Of bacteria, a mass of individuals, usually derived from cells of a single species. (3) An ecological term referring to a group of plants becoming established in a new situation.

Columella (kŏl′ŭ-mĕl′ȧ): the sterile central part within a mature sporangium or capsule.

Columnar: an unbranched tree stem, as in palms.

Community: an assemblage of organisms living together and interacting with each other in a characteristic natural habitat.

Companion cell: a small, specialized parenchyma cell associated with the sieve-tube elements of flowering plants.

Compensation point: the light intensity at which the rate of photosynthesis and the rate of respiration in the leaf are equal.

Competition: the effect of a common demand by two or more organisms on a limited supply of food, water, light, minerals, etc.

Complete flower: one with four sets of floral parts: pistil, stamens, petals, sepals. (*See* **Incomplete flower.**)

Compound leaf: a leaf divided into two or more parts, or leaflets.

Compound pistil: a pistil composed of two or more fused carpels.

Compression: a kind of plant fossil resulting from the weight of accumulated sediments upon plant organs, such as leaves.

Conceptacle (kŏn-sĕp′tȧ-k′l): a hollow structure containing sex organs.

Conidiophore (kŏ-nĭd′ĭ-ȯ-for′): a simple or branched hypha on which conidia (spores) are produced.

Conidium (kŏ-nĭd′ĭ-ŭm; pl., **Conidia**): an asexual fungus spore not contained within a sporangium. It may be produced singly or in chains.

Conjugation (kŏn′jŏo-gā′shŭn): (1) in eucaryotes, the union of two gametes that are not visibly differentiated; (2) the recombination mechanism in bacteria that most closely resembles sexual reproduction in other organisms.

Connate (kŏn′āt): the union or fusion of similar parts, as petals fused in a corolla tube. (*See* **Adnate.**)

Controlling gene: a gene that controls the expression of a structural gene. Controlling genes are of two types: operator genes, which directly control structural genes, and regulator genes, which control operator genes. (*See* **Structural gene.**)

Convergent evolution: the independent development of similar structures in forms of life that are unrelated or only distantly related.

Cork: a secondary tissue produced by a cork cambium; polygonal, cells, nonliving at maturity, with walls infiltrated with a waxy or fatty material resistant to the passage of gases and water vapor.

Cork cambium: a lateral meristem producing cork in woody and some herbaceous plants.

Corm: a short, thickened, underground stem, upright in position, in which food is accumulated.

Corolla (kȯ-rŏl′ȧ): the second (beginning from below) of the series of floral organs; composed of petals.

Corolla tube: a tubelike structure resulting from the fusion of the petals along their edges.

Cortex: the outer primary tissues of the stem or root, extending from the primary phloem (or endodermis, if present) to the epidermis; composed chiefly of parenchyma cells.

Cotyledons (kŏt′ĭ-lē′dun′z): the leaves (seed leaves) of the embryo, one or more in number.

Covalent bond: a chemical bond in which electrons are shared by two atoms, thus binding them together.

Critical light period: the dividing line between day length favorable to vegetative growth and that which induces flowering.

Crop rotation: the practice of growing different crops in regular succession to aid in the control of insects and diseases, to increase soil fertility, and to decrease erosion.

Cross-pollination: the transfer of pollen from the anther of one plant to the stigma of a flower of another plant.

Crossing over: an interchange of parts between the chromatids of two homologous chromosomes at meiosis.

Cuticle (kū′tĭ-k′l): a varnishlike layer covering the epidermis.

Cutting: a severed portion of a plant used for propagation; usually refers to a stem cutting.

Cytochromes: a series of pigmented iron-containing protein molecules. Some function in electron transfer in respiration; one, cytochrome *f*, in photosynthesis.

Cytokinins: hormones associated primarily with cell division.

Cytology (sī-tŏl′ȯ-jĭ): the branch of the biological sciences that deals with the structure and processes of protoplasm and the cell.

Cytoplasm (sī′tȯ-plaz′m): the protoplasm of the cell exclusive of the nucleus.

Day-neutral plant: a plant whose flowering is insensitive to length of day.

Deciduous (dė-sĭd′ū-ŭs): (1) falling of parts at the end of the growing period, such as leaves in autumn, or fruits or flower parts at maturity; (2) broad-leaved trees or shrubs that drop their leaves at the end of each growing season, as contrasted with plants that retain their leaves for more than one year.

Dehiscence (dė-hĭs′ĕns): the opening of an anther, fruit, or other structure, permitting the escape of reproductive bodies contained within.

Dehydrogenation (dė-hī′drȯ-jĕn-ā′shŭn): the removal of hydrogen from a molecule, as in cellular oxidation.

Deliquescent (del′ĭ-kwĕs′ĕnt): (1) mode of branching of a tree in which the trunk divides into many branches, leaving no central axis; (2) liquefying, or melting away, as of gills in the genus *Coprinus*.

Denitrification: the process by which nitrogen is released from the soil by the action of denitrifying bacteria.

Deoxyribonucleic (dė-ŏk-sĭ-rī′bȯ-nṳ-klē′ĭk) **acid (DNA):** one of the two main kinds of nucleic acids; the genic material of organisms and some viruses.

Deoxyribose (dė-ŏk′sĭ-rī′bōs): a 5-carbon sugar, ribose, reduced by the removal of an atom of oxygen.

Derepression: *See* **Repression.**

Dermatophytes (dûr′mȧ-tȯfit′z′): fungi that attack the skin, hair, and nails.

Detached meristem: a pocket of meristematic tissue derived from the shoot apex and giving rise to an axillary bud. It is located in the axil of a developing leaf.

Determination: the process in which a group of meristematic cells (such as an organ primordium) is fixed in a particular developmental pathway.

Development: the combination of growth and differentiation that gives rise to the organized plant body.

Dicaryon (dī-kăr-ē-on): in the higher fungi, a pair of associated but unfused nuclei, brought together by the fusion of cells from different parents.

Dichogamy (dī-kŏg′ȧ-mĭ): the condition in a perfect flower in which the anther and stigma mature at different times, thus preventing self-pollination.

Dichotomy (dī-kŏt′ō-mĭ): the division or forking of an axis into two more or less equal branches.

Dicotyledon (dī-kŏt′ĭ-lē′dŭn): also, dicot; a flowering plant with two seed leaves or cotyledons.

Dictyosome (dĭk-tēa-sōme): an aggregation of flat or curved and hollow disklike structures containing secreted products. The disks are often dilated along the margins. Also called Golgi body or Golgi apparatus.

Differentially permeable: term applied to membranes that allow some substances to pass through more readily than others.

Differentiation: the physiological and morphological changes that occur in a cell, tissue, or organ during development.

Diffuse-porous: hardwoods in which the vessels are approximately the same size throughout the growth ring.

Diffusion: the net movement of a substance as a result of the independent motion of its individual molecules, ions, or colloidal particles, from a region of higher diffusion pressure to one of lower diffusion pressure of that substance.

Diffusion pressure: the activity of a specific kind of molecule (particle) as the result of the combined effects of concentration, temperature, and pressure.

Digestion: the conversion of complex, usually insoluble foods into simple, usually soluble and diffusible forms by means of enzymatic action.

Dihybrid (dī-hī′brĭd): a cross between parents differing in two pairs of genes.

Dimorphism (dī-môr′fĭz′m): the condition of having two forms, such as sterile and fertile leaves in ferns, or sterile and fertile shoots in horsetails.

Dioecious (dī-ē′shŭs): bearing staminate and pistillate flowers or pollen and seed cones of conifers on different individuals of the same species.

Diploid (dĭp′loid): having two sets of chromosomes; the *2n* number characteristic of the sporophyte generation.

Disk flowers: the tubular flowers that compose the central part of a head of flowers in most Compositae;

contrasted with the flattened, ray-shaped (ray flowers) on the margins of the head.

Dispersal unit: any detached part of a plant involved in dispersal. May be a seed, a fruit, or a portion of vegetative plant body.

Division: the largest category of classification of plants according to rules of nomenclature; an aggregation of classes; synonymous with phylum as used by zoologists.

DNA: *See* **Deoxyribonucleic acid.**

Dominant: (1) A gene that expresses itself to the exclusion of the expression of its allele. (2) A character possessed by one parent of a hybrid that appears in the hybrid to the exclusion of the contrasting character (recessive) from the other parent. (3) An ecological term; a species that to a considerable extent controls the conditions for existence of its associates within an ecosystem. The dominance may be due to numbers or to size of the individual dominants.

Dormancy: a period of growth inactivity in bulbs, buds, seeds, and other plant organs.

Dorsal (dôr′săl): in liverworts, the upper surface of the plant body; the upper surface of a fern prothallus.

Dorsiventral (dôr′sĭ-věn′trăl) **symmetry:** having distinct upper and lower surfaces.

Double cross: in corn, a cross involving four inbred strains and two seasons.

Double fertilization: in angiosperms, the fusion of the egg and sperm (resulting in a *2n* fertilized egg) and the fusion of the second male gamete with the polar nuclei (resulting in a *3n* primary endosperm nucleus).

Drupe (drōōp): a simple fleshy fruit in which the inner part of the ovary wall develops into a hard, stony or woody endocarp, as in the peach.

Dry weight: moisture-free weight, obtained by drying at high temperatures for a sufficient period of time.

Early wood: that portion of a growth ring produced early in the growing season. (*See* **Spring wood.**)

Ecology (ē-kŏl′ō-ji): the science of the relationships between organisms and their environment.

Ecosystem: an interacting system of one to many living organisms and their nonliving environment.

Elaters (ĕl′ȧ-tēr′z): (1) elongated cells with spirally thickened walls, associated with the spores in the capsules of liverworts and mosses; (2) clubbed hygroscopic bands attached to the spores of the horsetails.

Electron: a particle of negative electricity that revolves about the nucleus of an atom and balances the positive charge of a proton.

Embryo (ĕm′brĭ-ō): the rudimentary plant formed in a seed or within the archegonium of lower plants.

Embryo sac: the female gametophyte of angiosperms, consisting typically at maturity of the egg and two synergids, two polar nuclei, and three antipodal cells.

Endocarp (ĕn′dŏ-kärp): the inner, leathery, woody or stony part of the wall of a fruit, as in a drupe or pome.

Endodermis (ĕn′dŏ-dûr′mĭs): a one-celled layer of specialized cells, frequently absent in stems but usually present in young roots, which separates the pericycle from the cortex.

Endoplasmic reticulum (ĕn′dŏ-plăz′mĭk rĕ-tĭk′ŭ-lŭm): a cell organelle; a continuous three-dimensional irregular network of membrane-bounded tubules or hollow plates that extends throughout the cytoplasm.

Endosperm (ĕn′dŏ-spûrm): a triploid, nutrient tissue of the angiosperm ovule that persists in some seeds and is used by the embryo and seedling at germination.

Endotoxins: toxins retained within a bacterial cell until freed by the disruption of the cell.

Energy: the capacity for doing work. Among the various forms of energy are radiant, heat, electrical, chemical, and kinetic.

Enzyme (ĕn′zīm): a proteinaceous catalytic agent that increases the rate of a particular transformation of materials in animals and plants.

Epidermis (ĕp′ĭ-dûr′mĭs): the outermost layer of cells of the leaf and of young stems and roots.

Epiphyte (ĕp′ĭ-fīt): a plant that grows upon another plant but is not parasitic upon it.

Epiphytotic (ĕp′ĭ-fī-tŏt′ĭk): common and widespread occurrence of a plant disease.

Episome: in bacteria, a unit of genetic material that can exist independently of the bacterial chromosome.

Erosion: the removal of soil and other materials by natural agencies, primarily water and wind.

Essential elements: elements required by plants for normal growth and development.

Ethylene: a gaseous hormone that has an inhibitory effect upon growth. It inhibits elongation and promotes ripening of fruits.

Etiolation (ē′tĭ-ŏ-lā′shŭn): the condition characterizing plants grown in the dark or in light of very low intensity.

Eucaryotic (ŭ-kăr′ĭ-ŏt′ĭk): organisms in which mitosis occurs; with membrane-bounded nucleus and other organelles.

Eutrophication: the enrichment of waters by organic or inorganic nutrients, ultimately leading to a state of increased productivity of aquatic life.

Evapotranspiration: total water loss from an area, by evaporation from the soil and by transpiration.

Evolution: the history of the development of a race, species, or larger group of organisms that, following modifications in successive generations, has acquired characteristics that distinguish it from other groups.

Excurrent (ĕks-kûr′ĕnt): the manner of tree growth in which the main stem continues to the top of the tree, giving rise to smaller, lateral branches.

Exocarp (ĕk′sŏ-kärp): the outermost layer of the pericarp or fruit wall; often the mere skin of the fruit.

Exotoxins (ĕk′sŏ-tŏk′sĭnz): soluble toxins that diffuse out of a living cell into the environment.

Eyespot: a small, pigmented structure in algae that may be sensitive to light.

F_1: the first filial generation following a cross. F_2 and F_3 are the second and third generations.

Fallowing: keeping land free of crops and weeds so that it can accumulate and store rainfall for later crops.

False berry: an accessory fruit in which both the ovary wall and the floral tube are fleshy at maturity, as in a cranberry.

False growth ring: the result of a stimulated late growth that gives the appearance of two growth rings produced in one year.

Family: a category of classification above a genus and below an order; composed of one or (usually) a number of genera. The suffix of the family name is usually -aceae.

Fats: organic compounds containing carbon, hydrogen, and oxygen, as in carbohydrates. The proportion of oxygen to carbon is considerably less in fats than it is in carbohydrates. One of the three kinds of plant foods. Fats in the liquid state are called oils. (*See* **Lipid**.)

Fatty acids: organic acids that may form a part of fat molecules.

Fermentation: a respiratory process in which the hydrogen released in glycolysis is recombined with pyruvic acid to form alcohol, lactic acid, or other products. No additional energy is made available during these steps.

Ferredoxin: an iron-containing protein functioning in electron transfer in photosynthesis.

Fertilization: the union of two gametes to form a zygote.

Fertilizers: materials added to the soil to provide elements essential to plant growth or to bring about a balance in the ratio of nutrients in the soil.

Fibers: greatly elongated, thick-walled, supporting cells, tapering at both ends. The two principal types are wood and sclerenchyma fibers.

Fibril (fī′brĭl): submicroscopic threads, composed of

cellulose molecules, that constitute the form in which cellulose occurs in the cell wall.

Fibrous roots: a root system in which the roots are finely divided.

Field capacity: *See* **Field percentage.**

Field percentage: the normal upper limit of the available capillary water.

Filament (fĭl′ȧ-mĕnt): (1) the stalk of the stamen, supporting the anther; (2) a term that describes the threadlike bodies of certain algae.

Fission (fĭsh′ŭn): the division of a unicellular organism into two equal daughter cells.

Flagellum (flȧ-jĕl′ŭm): a whiplike organ produced by some cells (spores, sperms, bacterial cells), which makes motility possible.

Floral tube: a cup or tube formed by the fusion of the basal parts of the sepals, petals, and stamens. This tube may be free from the ovary (ovary superior) or fused with the ovary wall (ovary inferior).

Florigen: the postulated flowering hormone.

Flower: the reproductive structure of the angiosperms; a group of sporophylls accompanied by a perianth. The perianth may be greatly reduced or lacking.

Flower bud: a bud that contains only one or more embryonic flowers.

Fluorescence (floó-ō-rĕs′ens): reemission of absorbed radiation at a longer wavelength than the original.

Follicle (fŏl′ĭ-k′l): a dry fruit derived from a simple pistil and opening along only one side.

Food: an organic compound that can be respired to yield energy and that can be used in assimilation.

Food chain: a group of plants and animals linked together by their food relationships.

Food web: the interconnecting food chains of a community.

Foot: in liverworts, mosses, and many vascular plants, the part of the embryo that remains in contact with gametophytic tissue, absorbing food from it and serving as an organ of attachment.

Fossil (fŏs′ĭl; -′l): a natural object, preserved in the earth's crust, that supplies information about a plant or animal of past geologic ages.

Frond (frŏnd): the leaf of a fern.

Fruit: a ripened ovary (or group of ovaries) containing the seeds, together with any adjacent parts that may be fused with it at maturity.

Fungicide (fŭn′jĭ-sīd): a toxic substance causing destruction or inhibition of growth of fungi.

β-Galactosidase (gā-lăc′tō-sĭd-ase): an enzyme that acts upon milk sugar (lactose).

Gametangium (găm′ė̇-tăn′jĭ-ŭm): general term applied to any cell or organ in which gametes are formed; usually restricted to reproductive bodies of lower plants such as algae and fungi.

Gamete (găm′ēt): a protoplasmic body capable of fusion with another gamete.

Gametophyte (gȧ-mē′tȯ-fīt) **generation:** the haploid (*n*) phase of the life cycle.

Gemmae (jĕm′ē): budlike structures in the liverworts that become detached and may develop into new plants.

Gene (jēn): a segment of a DNA molecule, composed of several hundred nucleotides, that specifies the number, kind, and arrangement of the amino acids in a polypeptide chain.

Gene interaction: the condition in which the usual expression of one gene is modified by the presence of other genes.

Gene mutation: a change in the nucleotide sequence of DNA as a result of errors in duplication.

Gene pool: the number and kinds of genes available to an interbreeding population of plants or animals.

Gene recombination: the formation of new gene combinations as a result of the sexual process. In a broad sense, any rearrangement of genetic material brought about by gene transfer.

Generative cell: (1) One of the two cells (the other is the tube cell) formed immediately following germination of the microspore of flowering plants. It divides to form two male gametes. (2) One of the cells of the mature pollen grain of pine.

Genetic code: the sequence of bases in the DNA molecule which determines the arrangement and kinds of the amino acids in a polypeptide chain.

Genetics (jė̇-nĕt′ĭks): the science of heredity.

Genotype (jĕn′ȯ-tīp): the genetic constitution, latent or expressed, of an organism, as contrasted with the phenotype; the sum total of all the genes present in an individual.

Genus (jē′nŭs): a group of closely related species that can be distinguished from other groups.

Geotropism (jė̇-ŏt′rȯ-piz′m): a growth movement in response to the influence of gravity.

Germ tube: the short, slender, tubular outgrowth first produced by a germinating spore.

Germination: resumption of growth of an embryo or spore.

Gibberellins: a group of growth hormones that promote both cell division and cell elongation.

Gill: a plate on the under side of the cap in the gill fungi (Basidiomycetes).

Girdling: the removal from a woody stem of a ring of bark extending inward to the wood; also called ringing.

Glucose: grape sugar, or dextrose, a 6-carbon sugar.

Glumes (gloomz): bracts, two in number, at the base of a spikelet of grasses.

Glycolysis (glī-kŏl′ĭ-sĭs): a respiratory process in which sugar is changed anaerobically to pyruvic acid with the liberation of a small amount of useful energy.

Golgi (gôl′jē) **body, Golgi apparatus:** *See* **Dictyosome.**

Grafting: a union of different individuals in which a portion, the scion, is inserted on a root or stem, the stock.

Grain: the fruit of the grass family; a small, dry, one-seeded fruit that does not open at maturity and in which the seed coat is fused with the pericarp.

Gram atomic weight: the quantity of an element which has a weight in grams equal numerically to the atomic weight of the element.

Gram-molecular weight (mole): the weight in grams numerically equal to the sum of the atomic weights of the atoms in the substance.

Grana (grăn′à): minute, waferlike bodies contained within the chloroplasts. The grana contain the chlorophylls and carotenoids and are the site of the photosynthetic process.

Gravitational water: water that the soil is unable to retain against the force of gravity.

Green manure: a fresh green crop that is plowed under to increase the organic matter and nitrogen content of the soil.

Ground meristem: the meristematic forerunner of the ground tissue system (primarily pith and cortex).

Ground tissues: in general, all plant tissues exclusive of the dermal and vascular.

Growth: irreversible increase in number and size of cells due to division and enlargement; usually accompanied by cell differentiation.

Growth hormone (hôr′mōn): a hormone (produced within the plant) that regulates growth.

Growth ring: a layer of growth as seen in the cross (transverse) section of a woody stem. (*See* **Annual ring.**)

Growth regulator: a synthetic chemical that affects growth.

Guard cells: specialized crescent-shaped epidermal cells surrounding a stoma.

Guttation (gŭ-tā′shŭn): the exudation of liquid water from plant leaves.

Gymnosperm (jĭm′nŏ-spûrm): a plant, as pines, with seeds not enclosed in an ovary.

Habit: characteristic form or bodily appearance of an organism.

Habitat (hăb′ĭ-tăt): the natural environment of a plant; the place where it is usually found.

Half-life: the amount of time required for half of the radioactive molecules in a substance to decompose. Because decomposition varies with concentration, the half-life is a relatively short part of the total decomposition period.

Haploid (hăp′loid): the reduced, or *n*, chromosome number, characteristic of the gametophyte generation.

Hardwood: (1) Wood produced by broad-leaved trees, such as maple, oak, ash, elm; the xylem of woody dicotyledons; characterized by the presence of vessels. (2) Any tree having hardwood as defined under (1).

Haustorium (hôs-tō′rĭ-ŭm): (1) In parasitic vascular plants, a specialized outgrowth from the stem or root that penetrates the living host tissue and absorbs foods or other materials. (2) A specialized fungus hypha that invades a host cell.

Head: inflorescence of sessile or nearly sessile flowers on a very short or flattened floral branch.

Heartwood: nonliving and commonly darker-colored wood surrounded by sapwood.

Hemicellulose (hĕm′ĭ-sĕl′ŭ-lōs): polysaccharides resembling cellulose but more soluble and less complex. Found particularly in cell walls.

Hemiparasite: a parasitic plant containing chlorophyll and therefore partly self-sustaining, as in mistletoe.

Herb (ûrb; hûrb): a nonwoody plant—annual, biennial, or perennial—whose aerial portion is relatively short lived (in the temperate zone, only a single growing season).

Herbaceous: a term referring to any nonwoody plant.

Herbals: early botanical works of the sixteenth and seventeenth centuries. They described many of the plants known at the time, with especial attention to their medicinal and other uses. Some contained attempts at classification.

Herbivore (hûr′bĭ-vŏr): a plant eater.

Herbivorous (hûr-bĭv′ŏ-rŭs): feeding upon plants; opposed to carnivorous.

Heteroecious (hĕt′ĕr-ē′shŭs): a term referring chiefly to rust fungi that have different stages of the life cycle on two unlike hosts.

Heterosporous (hĕt′ĕr-ŏs′pŏ-rŭs): having two kinds of spores, microspores and megaspores. These may differ in size, but more importantly differ in the kind of prothallus (male or female) they produce.

Heterostyly (hĕt′ĕr-ŏ-stī′lĭ): the existence of long and short styles in different flowers of the same species.

Heterothallism (hĕt′ĕr-ŏ-thăl′iz′m): the condition in which the two kinds of gametes that fuse, forming a zygote, are derived from different plants of the same species. Self-fertilization is not possible. (*See* **Homothallism.**)

Heterotrophic (hĕt′ēr-ō-trŏf′ĭk): organisms that must obtain some or all of their food from external sources. In general, applied to animals and non-photosynthetic plants.

Heterozygous (hĕt′ēr-ō-zī′gŭs): the condition that exists when the genes for a given character on the homologous chromosomes are unlike. (*See* **Homozygous.**)

Hexose (hĕk′sōs): a 6-carbon sugar, such as glucose or fructose.

Hilum (hī′lŭm): (1) the central part in a starch grain, surrounded by layers of starch; (2) the scar on the seed left by the stalk that attached the seed to the placenta.

Holdfast: (1) flattened disklike tip of a tendril, used in attachment; (2) basal part of an algal thallus that attaches it to a solid object; may be unicellular or composed of a mass of tissue.

Homologous chromosomes: chromosomes that associate in pairs in the first stage of meiosis; each member of the pair is derived from a different parent.

Homology (hō-mŏl′ō-jī): similarity due to common origin.

Homosporous (hō-mŏs′pōr-rŭs): having but one kind of spore.

Homothallism (hō′mō-thăl′iz′m): the condition in which both kinds of gametes are borne on the same thallus and self-fertilization is possible. Single individuals can form fertile zygotes (*See* **Heterothallism.**)

Homozygous (hō′mō-zī′gŭs): the condition that exists when the genes for a given character on homologous chromosomes are alike. An organism may be homozygous for one or several genes or, rarely, for all genes. (*See* **Heterozygous.**)

Hormone (hôr′mōn): a specific organic substance produced in one part of an organism and moving to and affecting reactions in other parts; in plants, usually a growth hormone.

Humidity: dampness. Relative humidity is the water vapor content of air as a percentage of the saturation content at the same temperature.

Humus (hū′mŭs): a complex mixture of incompletely decomposed organic materials in the soil.

Hybrid (hī′brĭd): the progeny of a cross between two individuals differing in one or more genes.

Hybrid vigor: the increased vigor that frequently is demonstrated by the progeny of a cross between inbred lines; or between unrelated forms, varieties, or species.

Hybridization (hī′brĭd-ĭ-zā′shŭn): the process of crossing individuals of unlike genetic constitution.

Hydrocarbon (hī′drō-kär′bŏn): an organic compound composed of hydrogen and carbon.

Hydrogen bond: a weak bond between a hydrogen atom attached to one oxygen or nitrogen atom and another oxygen or nitrogen atom. Important in water, proteins, and chromosomes.

Hydrolysis (hī-drŏl′ĭ-sĭs): a process in which a complex compound is digested into one or more simpler compounds through a reaction with water.

Hydrophyte (hī′drō-fīt): a plant that grows wholly or partly submerged in water.

Hydroponics (hī′drō-pŏn′ĭks): growth of plants in solutions containing essential elements.

Hyphae (hī′fē): threadlike structures that compose the plant body of a fungus.

Hypocotyl (hī′pō-kŏt′ĭl): the portion of the axis of an embryo or seedling situated between the cotyledons and the radicle.

Imbibition (ĭm′bĭ-bĭsh′ŭn): absorption of water and swelling of colloidal materials because of the adsorption of water molecules onto the internal surfaces of the materials.

Immunity: freedom from infection because of resistance.

Imperfect stage: the stage in the life cycle of a fungus in which only asexual reproduction occurs.

Inbreeding: the breeding of closely related plants or animals. In plants, it is usually brought about by self-pollination.

Incomplete dominance: the condition that results when two different alleles together produce an effect intermediate between the effects of these same genes in the homozygous condition.

Incomplete flower: a flower lacking one or more of the four kinds of floral organs: sepals, petals, stamens, or pistil.

Indehiscent (ĭn′dē′hĭs′ĕnt): remaining closed at maturity, as many fruits.

Independent assortment: the segregation of two or more pairs of genes and their distribution into the gametes independently of one another—that is, under the influence of chance factors only.

Indole-acetic acid (IAA): a widely distributed plant growth hormone of the auxin type. (*See* **Auxin** and **Hormone.**)

Inducer: a substance that brings about derepression of an operator gene by inactivating the repressor produced by a regulator gene.

Indusium (ĭn-dū′zĭ-ŭm): an epidermal outgrowth covering the sorus in ferns.

Inferior ovary: *See* **Floral tube.**

Inflorescence (ĭn′flō-rĕs′ĕns): a flower cluster; the arrangement of flowers on the floral axis.

Integument (ĭn-tĕg′ū-mĕnt): one, or sometimes two,

outer layers of the ovule, which develop into the seed coat.

Intercalary (ĭn-tûr′kȧ-lĕr′ĭ) **meristem:** a meristematic region between two partly differentiated tissue regions.

Intercellular layer: the wall layer, composed chiefly of pectic compounds, that is common to two adjoining cells and lies between the primary walls; also termed the middle lamella.

Interfascicular cambium (ĭn′tēr-fȧ-sĭk′ū-lēr kăm′bĭ-ŭm): a cambium developing between vascular bundles.

Internode: the region of the stem between any two nodes.

Interphase: the stage between two mitoses.

Invertase (ĭn-vûr′tās): *See* **Sucrase.**

Ions: atoms or groups of atoms carrying negative or positive charges. Ions are formed by the ionization (dissociation) of molecules, forming two or more ions.

Irregular: term applied to a flower in which the members of some or all of the floral whorls are unequal. Most irregular flowers can be divided longitudinally into two equal halves in only one plane, a condition termed zygomorphic.

Irritability: the ability of living protoplasm to respond to external stimuli.

Isogamy (ī-sŏg′ȧ-mĭ): a type of sexual reproduction in algae and fungi in which the gametes (or gametangia) are alike in size.

Isotope (ī′sȯ-tōp): a form of an element having the same chemical properties but a different atomic weight from other forms of the element.

Kilocalorie (kcal) (kĭl′ō-kăl′ȯ-rĭ): a unit of heat measure equivalent to 1000 gram calories.

Kinetic (kĭ-nĕt′ĭk) **energy:** energy of motion.

Knot: a portion of the base of a branch embedded in the wood.

Krebs cycle: the long, complicated series of reactions that results in the oxidation of pyruvic acid to hydrogen and carbon dioxide. The hydrogen, held on hydrogen carrier molecules, then goes through the oxidative phosphorylation and terminal oxidation processes.

Lamellae (lȧ-mĕl′ē): layers of protoplasmic membranes observed in the protoplast, particularly in and between the grana of the chloroplasts.

Late wood: the later-formed part of a growth ring; the summer wood.

Lateral: on one side of an organ, as opposed to terminal.

Lateral bud: a bud in the axil of a leaf.

Lateral meristems (mĕr′ĭ-stĕm′z): meristems that give rise to secondary tissue; the cambium and cork cambium.

Latex (lā′tĕks): a milky fluid found in certain plants, such as opium poppy, dandelion, milkweed, and Brazilian rubber tree.

Layering: a horticultural method of propagating plants from stems that form roots while still attached to the parent plant.

Leaf: an organ of limited growth, arising laterally and from superficial tissues of a shoot apex. Usually dorsiventral in structure; may be simple or compound. Commonly consists of blade and petiole, but petiole may be absent (leaf sessile) or blade absent and petiole bladelike (phyllode).

Leaf bud: a bud that produces only leaves.

Leaf gap: an interruption in the continuity of the primary vascular cylinder above the departure of the leaf trace or traces. This break is filled with parenchyma tissue.

Leaf mosaic (mȯ-zā′ĭk): the pattern of leaf arrangement formed by leaves presenting a maximum exposure of surface to the sun.

Leaf scar: a scar left on the twig when the leaf falls.

Leaf trace: a vascular bundle that connects the vascular tissues of the stem with those of the leaf.

Leaflet: one of the parts of a compound leaf.

Legume (lĕ-gūm′): (1) a member of the Leguminosae, the pea or bean family; (2) a type of dry fruit developed from a simple pistil and opening along two sides.

Lemma (lĕm′ȧ): lowermost of two bracts enclosing the flower of grasses.

Lenticels (lĕn′tĭ-sĕl′z): small, corky areas on the surfaces of stems, roots, and other plant parts, which allow interchange of gases between internal tissues and the atmosphere.

Liana (lĭ-ăn′ȧ): a large, woody, climbing plant.

Lichen (lī′kĕn): a mutualistic, composite organization of an alga and a fungus.

Lignin (lĭg′nĭn): the principal noncellulosic constituent of wood. It tends to harden and preserve the cellulose.

Linkage: the tendency of two or more genes to be inherited together because they are borne on the same chromosome.

Lipase (lī′pās): an enzyme that converts fats to fatty acids and glycerine.

Lipid (lĭp′ĭd): a general term for the fatty substances, fats and oils. Lipids have a greatly reduced oxygen content as compared to carbohydrates.

Locus (lō′kŭs): the position on a chromosome occupied by a particular gene.

Long-day plant: a plant that flowers when days are longer than its critical light period.

Lysis (lī'sĭs): disintegration, as of a compound by an enzyme, or of a bacterial cell by phage.

Lysogeny (lī-sŏj'eny): state of a bacterial cell in which the genetic material of a phage is carried from one cell generation to another in a noninfectious form.

Male sterility: condition in flowering plants in which pollen is absent or nonfunctional.

Malt: a cereal, usually barley, that has been allowed to sprout and then has been dried and ground. The product contains amylase and other enzymes.

Maltase (môl'tās): an enzyme that hydrolyzes maltose (malt sugar) to glucose.

Mating type: a strain of organisms incapable of sexual reproduction with one another but capable of such reproduction with members of other strains of the same organism.

Megasporangium (mĕg'a-spō-ran'jĭ-ŭm): the structure containing megaspores.

Megaspore (mĕg'a-spōr'): a spore that germinates and forms a female gametophyte.

Megasporophyll (mĕg'a-spō'rŏ-fĭl): a leaf, usually modified, that bears one or more megasporangia.

Meiosis (mī-ō'sĭs): the two divisions during which the chromosomes are reduced from the diploid to the haploid number.

Meiospore (mī'ō-spōr): haploid spore formed in meiosis; considered a part of the sexual process.

Meristem (mĕr'ĭ-stĕm): a region in which cell division continues, commonly for the life of the plant. Region of protoplasmic synthesis and tissue initiation.

Mesocarp (mĕs'ō-kärp): the middle layer (usually fleshy) of the fruit wall of a drupe or other fruit.

Messenger-RNA: RNA produced in the nucleus and moving to the ribosomes where it determines the order of the amino acids in a polypeptide.

Mesophyll (mĕs'ō-fĭl): the thin-walled parenchyma tissue, containing chloroplasts, that composes the bulk of the leaf between the upper and lower epidermis.

Mesophytes (mĕs'ō-fīt'z): plants that grow in environments that are neither very wet nor very dry; growing under conditions of a medium moisture supply.

Metabolism (mĕ-tăb'ō-liz'm): the sum total of the chemical processes that go in in the plant body.

Metaphase (mĕt'a-fāz): a stage in mitosis in which the chromosomes are arranged on a spindle, midway between two poles.

Micron (mī'krŏn): a thousandth part of a millimeter or one twenty-five thousandth of an inch. Symbol μ.

Micropyle (mī'krŏ-pīl): the opening in the integuments of the ovule.

Microsporangium (mī'krō-spō-ran'jĭ-ŭm): a structure containing microspores.

Microspore: a spore that germinates and forms a male gametophyte.

Microsporophyll (mī'krŏ-spō'fĭl): a leaf, usually modified, that bears one or more microsporangia.

Microtubules (mī-krō-tū-būles): very slender, protein tubes in the cytoplasm.

Middle lamella: *See* **Intercellular layer.**

Millimicron: a thousandth part of a micron or a millionth of a millimeter; one twenty-five millionth of an inch. Symbol mμ.

Mitochondria (mī'tŏ-kŏn'drĭ-a): cytoplasmic organelles that contain the enzymes of the Krebs cycle and of oxidative phosphorylation.

Mitosis (mī-tō'sĭs): a process during which the chromosomes become doubled longitudinally, the daughter chromosomes then separating and forming two genetically identical daughter nuclei. Mitosis is usually accompanied by cell division.

Mixed buds: buds that produce both leaves and flowers.

Mole: *See* **Gram molecular weight.**

Monocotyledon (mŏn'ō-kŏt'ĭ-le'dŭn): also, monocot; a flowering plant with one seed leaf or cotyledon.

Monoecious (mŏ-nē'shŭs): a condition in which both staminate and pistillate flowers (or pollen and seed cones of conifers) are borne on the same plant.

Monohybrid (mŏn'ō-hī'brĭd): a cross between two parents differing in only a single gene.

Morphogenesis (môr'fŏ-jĕn'ĕ-sĭs): the morphological and physiological events involved in the development of an organism.

Morphology (môr-fŏl'ō-jī): the study and scientific explanation of the form, structure, and development of plants.

Mosaic (mŏ-zā'ĭk): (1) term designating the mottled or variegated appearance of a leaf resulting from localized failure of chlorophyll formation; usually caused by virus infection; (2) *See* **Leaf mosaic.**

Motile (mō'tĭl): capable of spontaneous movement.

Multiple fruit: fruit composed of a number of closely associated fruits derived from different flowers, these fruits forming one body at maturity, as a pineapple.

Multiple genes: two or more independent pairs of genes working together to produce a cumulative effect upon the same characteristic of the phenotype.

Mutualism: the living together of two or more organisms in an association that is mutually advantageous. (*See also* **Symbiosis.**)

Mycelium (mī-sē'lĭ-ŭm): collective term applied to a mass of hyphae.

Mycology (mī-kŏl'ō-jī): the study of fungi.

Mycorrhiza (mī'kŏ-rī'zà): pl, **Mycorrhizae** (-zē): the combination of the hyphae of certain fungi with the root of a seed plant.

NAD: *See* **Nicotinamide adenine dinucleotide.**

NADP: *See* **Nicotinamide adenine dinucleotide phosphate.**

Naked bud: a bud without bud scales.

Nastic (năs'tĭk) **movements:** a response to external stimuli in which the nature of the reaction is independent of the direction from which the stimulus comes.

Natural selection: the mechanism of evolution by which environmental factors favor the survival and reproduction of those variants of a population that are best adapted. Contrasted with artificial selection practiced by man in the improvement of domestic plants and animals.

Nectary (něk'tà-rĭ): a gland, usually in a flower, that secretes a sugary nectar attractive to pollinators.

Net (apparent) photosynthesis: net absorption of CO_2 in photosynthesis, without correction for CO_2 obtained from respiration of the tissue.

Net venation: the pattern of venation in which the branching veins in the leaf or leaflet form a network.

Neutron: a nuclear body of an atom, without an electric charge. Neutrons change the molecular weight of a chemical but have very little effect on its chemistry.

Nicotinamide adenine dinucleotide (nĭk'ō-těn-am'ĭd ăd'ĕ-nēn dī-nū'klĕ-ŏ-tūd), **(NAD):** hydrogen acceptor molecule in respiration.

Nicotinamide adenine dinucleotide phosphate (NADP): a hydrogen acceptor molecule in photosynthesis.

Nitrification (nī'trĭ-fĭ-kā'shŭn): the conversion of ammonia nitrogen to nitrate nitrogen by two specific groups of bacteria, the nitrifying bacteria.

Nitrifying bacteria: the bacteria that carry on nitrification.

Nitrogen bases: nitrogen-containing compounds (adenine, guanine, thymine, cytosine, and uracil) found in nucleotides.

Nitrogen-fixing bacteria: bacteria living in the soil, or in the roots of legumes, that convert atmospheric nitrogen into nitrogen compounds in their own bodies.

Node: the region of the stem where one or more leaves are attached. Buds are commonly borne at the node, in the axils of the leaves (*See* **Internode.**)

Nodules (nŏd'ūlz): enlargements, or swellings, on the roots of legumes inhabited by nitrogen-fixing bacteria.

Nucellus (nŭ-sĕl'ŭs): tissue in the young ovule in which the embryo sac develops. Commonly considered to be the wall of the megasporangium.

Nuclear (nū'klĕ-ēr) **membrane:** the outer, bounding membrane of the nucleus.

Nucleic acids: a class of large acidic compounds composed of a series of nucleotides. They are of two kinds, DNA and RNA. Concerned with the storage and replication of hereditary information and the synthesis of proteins.

Nucleolus (nŭ-klē'ŏ-lŭs): one or more spherical dark-staining bodies found in the nucleus; rich in RNA and protein.

Nucleoproteins (nū'klĕ-ŏ-prō'tĕ-ĭnz): combinations of nucleotides and proteins.

Nucleotides (nū'klĕ-ŏ-tīd'z): the structural units of DNA and RNA; compounds composed of an organic phosphate, a pentose sugar, and a nitrogen base. The different nucleotides vary in the nature of their nitrogen bases.

Nucleus (nū'klĕ-ŭs): (1) a specialized body within the cell that contains the chromosomes; (2) the central part of an atom, consisting of neutrons and positively charged protons, except for hydrogen, in which the nucleus consists only of a proton.

Nut: a one-seeded fruit in which the fruit wall is hard, stony, or woody at maturity.

Nutation (nŭ-tā'shŭn): the movement of growing parts, such as stem tips, leaves, or flowers, in which an irregularly circular path is traced in space as growth proceeds.

Omnivore (ŏm'nĭ-vōr): an organism that consumes both plant and animal food.

Omnivorous (ŏm-nĭv'ŏ-rŭs): eating both plant and animal food.

Ontogeny (ŏn-tŏj'ĕ-nĭ): the entire development of an individual, from the fertilized egg to the adult stage; the life history of an individual organism.

Oogamy (ŏ-ŏg'à-mĭ): kind of sexual reproduction in which one of the gametes (the egg) is large and non-motile while the other gamete (the sperm) is smaller and motile.

Oogonium (ō'ŏ-gō'nĭ-ŭm): in certain thallophytes, a unicellular female sex organ that contains one or several eggs.

Oospore (ō'ŏ-spōr): in algae and fungi, the spore resulting from the fertilization of an egg cell by a sperm—opposed to zygospore.

Open pollination (pŏl'ĭ-nā'shŭn): natural pollination by selfing or crossing between two more or less related strains.

Operator gene: *See* **Controlling gene.**

Operculum (ŏ-pûr'kŭ-lŭm): in mosses, the lid or cover of the capsule.

Opposite: term applied to leaves or buds occurring in pairs at a node.

Order: a category of classification above the family and below the class; composed of one or more families. The suffix to the ordinal name is usually *-ales.*

Organ: one of the larger, distinct, and visibly differentiated parts of a plant, such as root, stem, leaf, floral organs. An organ is composed of tissues.

Organelle (or'găn-ĕl): discrete bodies in the cell, each kind with specialized structure and function.

Organic compound: a compound containing carbon and usually hydrogen, with or without other elements.

Osmosis (ŏs-mō'sĭs): the diffusion of water through a differentially permeable membrane from a region of higher diffusion pressure of water to one of lower diffusion pressure of water.

Ovary (ō'vȧ-rĭ): the swollen basal portion of a pistil; the part containing the ovules or seeds.

Ovule (ō'vūl): at maturity, a structure composed of embryo sac, nucellus, integuments, and stalk. Following fertilization the ovule develops into the seed.

Oxidation: the loss of electrons from an atom or a molecule; in biology, an energy-releasing process. Typically (1) the removal of hydrogen atoms, together with their electrons, or (2) the addition of oxygen to a compound.

Oxidative phosphorylation (fŏs'fŏ-rĭl-ā'shŭn): the production of ATP from ADP and P during the oxidation of $NADPH_2$ in respiration.

Palea (pā'lē-ȧ): uppermost of two bracts enclosing the flower of grasses.

Paleobotany (pā'lē-ŏ-): the study of the plant life of the geologic past.

Palisade parenchyma (păl'ĭ-sād pȧ-rĕng'kĭ-mȧ): a leaf tissue composed of columnar chloroplast-bearing cells with their long axes at right angles to the leaf surface.

Palmately compound: a compound leaf with the leaflets attached at the tip of the petiole.

Palmately veined: a term that refers to veins arising from the base of the leaf and radiating outward.

Panicle (păn'ĭ-k'l): an open, branching inflorescence, as in oats and other grasses.

Papain (pȧ-pā'ĭn): a protease found in the latex of the fruit and leaves of the papaya plant.

Papilla (pȧ-pĭl'ȧ): a budlike structure associated with reproduction in certain algae.

Parallel venation: the pattern of venation in which the principal veins of the leaf are parallel, or nearly so.

Parasite: an organism that obtains its food from the living tissues of another organism.

Parenchyma (pȧ-rĕng'kĭ-ma): an unspecialized, simple cell or tissue. The cells are isodiametric or sometimes elongated, usually thin-walled, living at maturity, and retaining a capacity for renewal of cell division.

Parthenocarpy (pär'thĕ-nŏ-kär'pĭ): the production of fruits in the absence of fertilization. Parthenocarpic fruits are usually seedless.

Partial veil: in gill fungi, a layer of tissue joining the margin of the cap to the stalk. The remains on the stalk constitute the annulus. Also called inner veil.

Passive water absorption: the absorption of water through the roots as a result of forces originating in the leaves.

Pasteurization (păs'tēr-ĭ-zā'shŭn): treatment of materials at temperatures high enough to kill nonspore-forming bacteria.

Pathogen (păth'ŏ-jĕn): any organism capable of causing disease.

Pectin (pĕk'tĭn): a complex organic compound present in the intercellular layer and primary wall of plant cells. The basis of fruit jellies.

Pectinase: an enzyme that acts upon pectin; also called pectase.

Pedicel (pĕd'ĭ-sĕl): the stem of an individual flower.

Peduncle (pē-dung'k'l): the stem of an inflorescence.

Pellicle: the outer layer of euglenoid cells, composed of plasma membrane and flat, usually elastic, interlocking strips.

Peptide (pĕp'tĭd, also tīd): two or more amino acids linked by peptide bonds.

Peptide bond: a covalent bond between two amino acids.

Perennial: a woody or herbaceous plant living from year to year, not dying after once flowering.

Perfect stage: phase of the life history of a fungus that includes sexual fusion and the spores associated with such fusions.

Perianth (pĕr'ĭ-ănth): collective term applied to the calyx and corolla.

Pericarp (pĕr'ĭ-kärp): the mature, ripened ovary wall.

Pericycle (pĕr'ĭ-sī'k'l): a layer of cells (parenchyma) just outside the primary phloem and inside the endodermis. If the endodermis is lacking, as in many stems, the pericycle cannot be distinguished from the cortex.

Perithecium (pĕr'ĭ-thē'shĭ-ŭm): a spherical or flask-shaped structure in the ascomycetes containing asci.

Permanent wilting percentage: the lower limit of the available capillary water.

Petal: one of the units of the corolla of the flower.

Petiole (pĕt'ĭ-ōl): the stalk of the leaf.

Petrifaction (pĕt'rĭ-făk'shŭn): a type of plant fossil in which the original cellular tissues are retained and are impregnated with mineral compounds.

pH: a symbol denoting the negative logarithm of the

concentration of the hydrogen ion in grams per liter. A measure of acidity or alkalinity of a solution.

Phage (fāj): a virus that attacks and destroys bacteria.

Phagotrophy (făg'ŏ-trŏ'fĭ): ingestion of food in particulate form.

Phenotype (fē'nŏ-tīp): the external, manifest, or visible characters of an organism, as contrasted with its genetic constitution (the genotype).

Phloem (flō'ĕm): one of the two component tissues of vascular tissues; food conducting.

Phosphate (fŏs'fāt): a compound of phosphorus; in general, phosphoric acid.

Phosphoglyceric acid: monophosphates of glyceric acid formed as intermediates in photosynthesis and other metabolic processes.

Phosphorylation (fŏs-fŏr'ĭ-lā'shŭn): the process by which high-energy, organic phosphate compounds are produced, using respiratory or sunlight energy.

Photoautotrophic (fō'tŏ-ô'tŏ-trŏf'ĭk): organisms, such as green plants, able to manufacture their own basic foods with the energy of light. (*See* **Autotrophic.**)

Photoperiod: length of the light period in a 24-hour day.

Photoperiodism (fō'tŏ-pēr'ĭ-ŭd-ĭz'm): the response of plants to the relative lengths of day and night.

Photophosphorylation (fō'tŏ-fŏs'fŏ-rĭl-ā'shŭn): a phosphorylation reaction dependent upon the energy of light absorbed by the chlorophylls.

Photosynthesis (fō'tŏ-sĭn'thĕ-sĭs): the production of sugar from carbon dioxide and water in the presence of chlorophyll, using light energy and releasing oxygen.

Photosynthetic unit: a sheet of 200–300 chlorophyll molecules together with reactive sites at which the light reactions of photosynthesis are completed.

Phototropism (fō-tŏt'rŏ-pĭz'm): a growth movement in response to one-sided illumination.

Phragmoplast (phrăg'mō-plăst): spindle produced in the region of the equatorial cell plate, preliminary to wall formation.

Phycobilins (fīkō'bĭlĭns): a class of bluish green or red pigments that occur in algae.

Phyllode (fĭl'ōd): a flat, expanded petiole replacing the blade of a leaf.

Phylogeny (fĭ-lŏj'ĕ-nĭ): the evolutionary history of a species, genus, or larger group.

Phylum (fī'lŭm): *See* **Division.**

Physiologic races: subdivisions of a variety, alike in structure but unlike in certain physiological, biochemical, or pathological characters.

Physiology: the study of the activities and processes of living organisms.

Phytochrome (fī'tŏ-krōm): a plant pigment associated with responses of plants to light.

Phytoplankton (fītŏ-plăngk'tŏn): plant plankton.

Pinna (pĭn'ȧ): a primary division (leaflet) of a pinnately compound leaf.

Pinnately compound: a compound leaf with the leaflets arranged along the sides of a common axis.

Pinnately veined: the condition of veins arising from each side of a common axis, like the barbs of a feather.

Pistil (pĭs'tĭl): the central organ of the flower, composed of one or more carpels and enclosing the ovules.

Pistillate (pĭs'tĭ-lȧt) **flower:** a flower with one or more pistils but no functional stamens.

Pit: a cavity where the secondary wall fails to form. In the bordered pit the secondary wall arches over the pit cavity.

Pith: the tissue occupying the center of the stem within the vascular cylinder. It usually consists of parenchyma, but other types of cells may also occur.

Placenta (plȧ-sĕn'tȧ): the region or area of the ovary to which one or more ovules (or seeds) are attached.

Placentation (plăs'ĕn-ta'shŭn): arrangement of the placentas and ovules within the ovary.

Plain-sawed wood: wood sawed in such a way that the tangential surface is exposed on the surface of the board.

Plankton (plăngk'tŏn): microscopic or near microscopic floating or weakly swimming organisms of a body of water.

Plant hormone: *See* **Hormone.**

Plant pathology (pȧ-thŏl'ŏ-jĭ): the study of plant diseases.

Plaque (plăk): round, clear area in a bacterial plate culture, the result of bacterial lysis by phage.

Plasma (plăz'mȧ) **membrane:** the outermost layer of the cytoplasm of a cell.

Plasmodesma (plăz'mŏ-dĕz-mȧ; pl., **Plasmodesmata**): minute cytoplasmic threads that extend through openings in the cell walls, connecting the protoplasts of adjacent living cells.

Plasmodium (plăz-mō'dĭ-ŭm): a multinucleate mass of naked protoplasm (without cell walls); the vegetative phase in the life cycle of a slime mold.

Plasmolysis (plăz-mŏl'ĭ-sĭs): shrinkage of the protoplasm from the cell wall as a result of outward diffusion of water into a solution with a low diffusion pressure of water.

Plastids (plăs'tĭd'z): specialized bodies in the cytoplasm that are the sites of such activities as food manufacture and storage.

Plumule (ploo'mūl): the bud of the embryo. It may consist of a shoot apex alone, or of the apex and one or more embryonic leaves. Also applied to the apical bud of a seedling.

Polar nuclei (nū'klĕ-ī): two nuclei, one derived from each end (poles) of the embryo sac, which become centrally located. They fuse at about the time the pollen tube enters the embryo sac.

Pollen grain: a germinated microspore; a partly developed male gametophyte.

Pollen sac: a cavity in the anther that contains the pollen grains.

Pollen tube: a tube formed following germination of the pollen grain.

Pollination (pŏl'ĭ-nā'shŭn): the transfer of pollen from the anther to the stigma of the same or another flower.

Polypeptide: a long chain of amino acids. (*See* **Peptide.**)

Polyploid (pŏl'ĭ-ploid): more than two chromosome sets (diploid) in one cell.

Polysaccharides (pŏl'ĭ-săk'*a*-rīdz): carbohydrates formed of many simple sugars linked together, as starch or cellulose.

Polysome: aggregation or cluster of ribosomes.

Pome: an accessory fruit with a leathery endocarp, as an apple.

Population: a group of sexually reproducing and interbreeding individuals that share the same gene pool.

Prickle: a spinelike superficial outgrowth of the stem.

Primary endosperm nucleus (ĕn'dŏ-spûrm nū'klĕ-ŭs): the result of the fusion of a male gamete and the two polar nuclei; also called triple-fusion nucleus.

Primary growth: growth resulting from the production of cells by the apical meristems and the growth and differentiation of derivatives of these meristems.

Primary tissues: cells derived from the apical and subapical meristems of root and shoot; opposed to secondary tissues derived from a cambium or cork cambium.

Primary wall: (1) the wall layer on either side of the intercellular layer; (2) a collective term applied to two primary walls and the intercellular layer.

Primordium (prī-môr'dī-ŭm): the early or rudimentary stage of an organ.

Procambium: meristematic tissue that produces primary xylem and phloem. It may also give rise to a cambium.

Procaryotic (prŏ-kăr'ĭ-ŏt'ĭk): organisms (blue-greens and bacteria) in which mitosis is absent and the cell organelles are not limited by membranes.

Progametangium (prŏ-găm'ĕ-tăn'gĭ-ŭm): a cell or organ that becomes the gametangium (*which see*).

Prop roots: roots that arise from the stem above soil level.

Prophage: form in which phage is carried in lysogenized bacterial cells.

Prophase: an early stage in mitosis in which the chromosomes are dispersed throughout the nucleus, and are longitudinally doubled.

Proplastids: embryonic plastids that divide and grow.

Protease (prō'tĕ-ās) an enzyme that digests protein.

Protein (prō'tĕ-ĭn): complex organic compounds constructed from amino acids and composed of carbon, hydrogen, oxygen, and nitrogen. Many proteins also contain sulfur. Proteins are one of the three groups of plant foods, and the chief organic component of protoplasm.

Prothallial (prŏ-thăl'ĭ-ăl) **cells:** vegetative, vestigial cells found in the male gametophyte generation of certain vascular plants, as selaginella and pine.

Prothallus (prŏ-thăl'ŭs): in ferns and other lower vascular plants, the free-living or independent gametophyte generation.

Protoderm: meristematic tissue that gives rise to the epidermis.

Proton: a body in the atomic nucleus. It carries a positive electric charge which is balanced by the negative charge of an electron.

Protonema (prŏ-tŏ-nē'm*a*): in mosses, the threadlike or platelike growth from which the conspicuous plant is developed.

Protoplasm (prō'tŏ-plăz'm): the living material of the cell.

Protoplast: the entire contents, both protoplasmic and nonprotoplasmic, of a cell.

Ptomaines (tō'mān'z): decomposition products produced by the action of bacteria upon proteins; popularly but incorrectly believed to cause food poisoning.

Pure culture: an aggregation of microorganisms all of one kind, so segregated that mixture with other species is impossible.

Pyrenoid (pī'rē-noid): a proteinaceous body in the plastids of many algae and of *Anthoceros;* in the green algae, a center of starch accumulation.

Quarter-sawed wood: wood sawed along a radial plane, exposing the radial surface of the wood.

Raceme (r*a*-sēm'): an inflorescence in which the individual flowers are borne on stems (pedicels) along a central axis.

Radial plane: a view of the wood exposed by a longitudinal cut along a radius. A section cut in this plane for microscopic study is a radial section.

Radially symmetrical: term applied to (1) a type of flower that may be divided into two equal halves in more than one longitudinal plane, also called regular flower; (2) a uniform development around a central axis, such as leaves around a stem.

Radicle (răd′ĭ-k'l): the basal end of the embryonic axis, which grows into the primary root.

Radioisotopes: radioactive elements.

Ray flowers: *See* **Disk flowers.**

Receptacle (rĕ-sĕp′tà-k'l): (1) that part of the axis of a flower stalk that bears the floral organs; (2) in certain liverworts, the stalked structures that bear the sex organs.

Recessive: *See* **Dominant.**

Recombination: *See* **Gene recombination.**

Reduction: gain of electrons by an atom or molecule as by addition of hydrogen or removal of oxygen. (*See* **Oxidation.**)

Regular: applied to a flower in which the parts of each whorl are similar in shape; the flower can be divided into two equal halves along more than one longitudinal plane. Also termed actinomorphic.

Regulator gene: *See* **Controlling gene.**

Repression: the inhibition of an operator gene by a regulator gene through the production of a protein repressor. The removal of this inhibition is known as derepression.

Repressor: *See* **Repression.**

Reproduction: the formation of a new individual by nonsexual or sexual methods.

Resin canal: a tubelike cavity, lined with living cells that secrete resin into the canal.

Respiration: an intracellular process in which food is oxidized with release of energy. The complete breakdown of sugar or other organic compounds to carbon dioxide and water is termed aerobic respiration, although the earlier steps are anaerobic.

Rhizoids (rī′zoid′z): slender, root-hairlike filaments that serve in attachment and absorption in many kinds of plants.

Rhizome (rī′sōm): an underground, usually horizontal, stem, the primary shoot of the plant. Distinguished from a root by the presence of nodes and internodes and sometimes buds and scalelike leaves at the nodes. Often thickened and containing accumulated food.

Ribonucleic (rī′bō-nū-klē′ĭk) **acid (RNA):** the nucleic acid found in the nucleoli and in the cytoplasm. Believed to carry a copy of the genetic information contained in the genes and to apply it in the synthesis of specific protein molecules.

Ribose (rī′bōs): a 5-carbon sugar.

Ribosome (rī′bŏ-sōm′): submicroscopic granules in the protoplasm. They contain RNA and protein and are associated with protein formation.

Ring-porous: hardwoods in which the vessels of the early (spring) wood are large in comparison to those of the late (summer) wood.

Ringing: *See* **Girdling.**

RNA: *See* **Ribonucleic acid.**

Root: a portion of the plant axis; distinguished from rhizomes and stolons by absence of nodes and internodes.

Root apex: the meristematic tissue in the terminal part of the root, that is, the root apical meristem. Sometimes used loosely to include the root cap also.

Root cap: a thimble-shaped mass of parenchyma cells over the root apex, protecting it from mechanical injury.

Root hairs: tubular outgrowths of epidermal cells of the root in the zone of maturation.

Root pressure: the pressure developed in roots that causes guttation and exudation from cut stumps.

Runner: slender, elongated prostate aerial branch from a basal node of an erect stem. It forms buds, and may form adventitious roots at the nodes or near the tip, sometimes both.

Samara (săm′à-rà; sà-mā-rà): a dry fruit bearing wings, useful in dispersal, as in a maple.

Saprophyte (săp′rŏ-fīt): an organism that obtains its food from nonliving organic matter.

Saprophytic (săp′rŏ-fīt′ĭk): obtaining food from nonliving, plant or animal remains.

Saprovore: animal feeding on nonliving organic matter.

Sapwood: outer part of the wood of stem or trunk, usually distinguished from the heartwood by its lighter color.

Scarification (skăr′ĭ-fĭ-kā′shŭn): the seed treatment employed to break dormancy by use of abrasives and other methods.

Scion (sī′ŭn): a portion of a shoot used for grafting. Also spelled cion.

Sclerenchyma (sklĕ-rĕng′kĭ-mà) **fibers:** elongated cells with tapering ends and thick secondary walls; usually nonliving at maturity; supporting tissue.

Scutellum (skŭ-tĕl′ŭm): the portion of the cotyledon that, in grasses, absorbs food from the endosperm at germination.

Secondary growth: increase in thickness of root or stem resulting from the formation of secondary tissues by a lateral meristem, chiefly the vascular cambium.

Secondary tissues: tissues produced by the cambium and cork cambium.

Secondary wall: innermost wall layer, formed after elongation has ceased.

Seed: a structure formed by seed plants following

fertilization. In conifers it consists of seed coat, embryo, and female gametophyte (n) storage tissue. Some angiosperm seeds are composed only of seed coat and embryo; others also contain endosperm ($3n$) storage tissue.

Seed coat: the outer layer of the seed, developed from the integuments of the ovule.

Segregation: the separation of the chromosomes (and genes) from different parents at meiosis.

Self-compatibility: the ability to produce fruits and normal seeds following self-pollination.

Self-incompatibility: inability to produce seed and fruit without cross-pollination.

Self-pollination: the transfer of pollen from the anther to the stigma of the same flower or of another flower on the same plant or within a clone.

Sepal: one of the units of the calyx (*See* **Calyx**).

Separation layer: a layer of specialized cells at the base of the leaf or other plant part. The structure of this layer is responsible for abscission of the leaf or other organ.

Septum (sĕp'tŭm): a dividing wall or partition.

Sessile (sĕs'ĭl): without a stalk.

Seta (sē'tà): in mosses, the stalk that supports the capsule.

Sex organs: the archegonia and antheridia.

Sexual reproduction: the fusion of gametes followed by meiosis and recombination at some point in the life cycle.

Sheath (shēth): the base of a leaf that wraps around the stem, as in grasses.

Shoot: (1) collective term applied to the stem and leaves; (2) any young growing branch or twig.

Shoot apex: the meristematic tissue in the terminal part of the shoot; sometimes considered to consist only of the tissue above the youngest leaf primordia, but more commonly considered to extend below this. Essentially synonymous with apical meristem of shoot. (*See* **Apical meristem.**)

Short-day plant: a plant that flowers when days are shorter than its critical light period.

Shrub: a perennial woody plant of relatively low stature, typically with several stems arising from or near the ground.

Sieve (sĭv) **tube:** a vertical series of food-conducting cells of the phloem of the flowering plants; characterized by sievelike openings on the end walls and sometimes on the side walls also.

Sieve-tube element: one of the component cells of the sieve tube of the flowering plants.

Simple fruit: fruit derived from a single pistil, simple or compound; ovary superior or inferior.

Simple leaf: an undivided leaf; opposed to compound.

Single cross: in corn, a cross involving only two inbred strains and one season.

Softwood: (1) wood produced by coniferous trees, such as larch, pine, cedar, spruce, and fir; characterized by absence of vessels; (2) a tree having softwood as characterized under (1).

Solute (sŏ'lūt): a dissolved substance.

Solution: a homogeneous mixture of two or more substances.

Solvent (sŏl'vĕnt): a substance, usually liquid, in which other substances are dissolved.

Soredium (sŏ-rē'dĭ-ŭm): a small group of algal cells and fungous hyphae that, when freed from a lichen thallus, may become a new plant.

Sorus (sō'rŭs): in ferns, a cluster of sporangia .

Species (spē'shĭz; spē'shēz): a group of closely related individuals; the unit of classification.

Spermatium (spûr-māshēăm): in the rust fungi, a spore (of either a plus or a minus strain) that fuses with a receptive hypha.

Spermagonium (spûr-mă-gōnēăm): in the rust fungi, a flask-shaped structure that contains spermatia.

Spike: an elongated inflorescence, resembling a raceme except that the flowers are sessile.

Spikelet: the unit of the compound inflorescence of the grasses; composed of a cluster of one or more flowers and their associated bracts.

Spindle (spĭn'd'l) **fibers:** a group of fibers that extends from the centromeres of the chromosomes to the poles of the spindle or from pole to pole in a dividing cell.

Spine: a sharp-pointed woody structure, usually modified from a leaf or part of a leaf.

Spirillum (spĭ-rĭl'ŭm): a spiral or corkscrew-shaped bacterium.

Spongy parenchyma: a leaf tissue composed of loosely arranged, chloroplast-bearing cells; also called spongy tissue.

Sporangiophore (spŏ-răn'jĭ-ŏ-fōr'): a structure that bears one or more sporangia (spore cases).

Sporangium (spŏ-răn'jĭ-ŭm): a spore case in which spores are formed.

Spore: a minute reproductive body produced by plants. usually unicellular, but sometimes multicellular. The spore is not a reproductive body in bacteria.

Spore mother cell: a diploid ($2n$) cell that undergoes meiosis and produces (usually) 4 haploid cells (spores) or 4 haploid nuclei.

Spore print: the design produced by the falling of spores when the cap of a gill fungus is placed on a flat surface.

Sporophyll (spō'rŏ-fĭl): a modified leaf or leaflike organ that bears sporangia; applied to the stamens and carpels of angiosperms, fertile fronds of ferns, etc.

Sporophyte (spōr′rŏ-fīt) **generation:** the diploid ($2n$) generation of the life cycle.

Spring wood: the early wood of an annual ring.

Stalk: in bryophytes, the region between the foot and the capsule.

Stamen (stā′mĕn): the organ of the flower producing the pollen, composed (usually) of anther and filament.

Staminate (stăm′ĭ-nàt) **flower:** a flower containing stamens but no functional pistil.

Starch: a carbohydrate, insoluble in water, converted to soluble forms by enzymes; the most common storage carbohydrate in plants.

Starch phosphorylase: an enzyme that converts starch to glucose phosphate.

Starch synthetase (sĭn′thē-tās): the primary enzyme concerned in starch synthesis.

Sterigma (stĕ-rĭg′mà; pl., **Sterigmata**): a slender stalk or short hypha from the tips of which conidia or basidiospores are produced.

Stigma (stĭg′mà): the summit of the pistil; the part that receives the pollen grains.

Stipe: a supporting stalk, such as the stalk of a gill fungus or the leafstalk of a fern.

Stipule (stĭp′ūl): an appendage on either side of the basal part of a leaf of some species of plants.

Stock: that part of the stem that receives the scion in grafting.

Stolon (stō′lŏn): (1) A slender, elongated, prostrate subterranean stem, arising originally from a basal node of an erect stem. It forms buds, and may form adventitious roots, at the nodes or at the tip, or by swelling of the apical part, may produce a tuber. (2) Horizontal surface hyphae of some fungi, such as the bread mold fungus.

Stoma (stō′mà; pl., **Stomata**): an opening or pore; an intercellular space between two specialized epidermal cells, the guard cells.

Stratification: the exposure of moist dormant seeds to low temperatures to break the dormant period.

Strobilus (strŏb′ĭ-lŭs): an aggregation of sporophylls, usually conelike in form.

Structural gene: a gene that controls or influences a particular characteristic of an organism by specifying the kinds of amino acids in a polypeptide chain.

Style (stīl): that part of the pistil that connects the ovary and stigma.

Substrate (sŭb-strāt′): substance acted upon by an enzyme .

Succession: the occupation of an area by an orderly sequence of plant communities.

Succulent: a plant with fleshy, water-storing stems or leaves.

Sucrase (sū′krās): an enzyme that hydrolyzes sucrose into glucose and fructose; also called invertase.

Summer wood: the late wood of an annual ring.

Suspensor: (1) a hypha supporting a zygospore; (2) a structure in the embryo of seed plants that pushes the terminal part of the embryo into the endosperm and probably also absorbs nutrients.

Symplast (sĭm′plăst): a system of living cells connected by plasmodesmata.

Symbiosis (sĭm′bĭ-ō′sĭs): the living together in close association of two or more dissimilar organisms. Includes parasitism, in which the association is harmful to one of the organisms; and mutualism, in which the association is advantageous to both.

Synapsis (sĭ-năp′sĭs): pairing of homologous maternal and paternal chromosomes early in meiosis.

Synergids (sĭ-nûr′jĭd′z): two ephemeral cells lying close to the egg in the mature embryo sac of the ovule of flowering plants.

Tangential plane: the view of the wood exposed by a longitudinal cut at right angles to the vascular rays. A section cut in this plane for microscopic study is a tangential section.

Taproot: a stout, tapering main root from which arise smaller, lateral branches.

Taxon: term applied to any taxonomic category.

Taxonomy (tăks-ŏn′ŏ-mĭ): the science of classification, dealing with the arrangement of plants and animals into categories according to their natural relationships.

Teleology (tĕl′ē-ŏl′ŏ-jĭ, tē′lē-): the explanation of natural phenomena on the basis of design or purpose.

Teliospore (tē′lĭ-ŏ-spōr′): in the smuts and rusts, a thick-walled resting spore in which nuclear fusion occurs. It produces basidia upon germination. The teliospores of smuts are formed within hyphal cells; those of rusts are produced terminally.

Telophase (tĕl′ŏ-fāz): the last stage in mitosis, during which the chromosomes become reorganized into two new nuclei.

Template (tĕm′plĭt): a pattern or mold.

Tendril (tĕn′drĭl): a slender, coiling organ of climbing plants, usually a modified leaf or part of a leaf.

Tepal (tĕp′ăl): unit structure of a perianth that is not differentiated into sepals and petals.

Terminal oxidation: the transfer of electrons and hydrogen ions to oxygen, forming H_2O, in respiration; the only use of free oxygen in the process.

Test cross: a cross of a dominant with a homozygous

recessive; used to determine whether the dominant is homozygous or heterozygous.

Tetrad (tĕt′răd): a group of four spores formed from a spore mother cell after meiosis.

Tetraploid (tĕt′rȧ-ploid): twice the usual, diploid, number of chromosomes.

Thallophyte (thăl′ȯ-fīt): a thallus plant; an alga, bacterium, or fungus.

Thallus (thăl′ŭs): a type of plant body that is undifferentiated into root, stem, or leaf.

Thigmonasty (thĭg-mȯ-năs′tĭ): a nastic movement in response to touch.

Thigmotropism (thĭg-mŏt′rȯ-pĭz′m): a growth (twining) movement as a result of contact with a solid object.

Thorn: a branch that has become hard, woody, and pointed.

Tiller: a branch from the axil of a lower leaf, as in grasses.

Tissue: a group of cells generally similar in origin and function.

Tissue culture: growth of cells or tissues from plants or animals in sterile, synthetic media; an important research tool.

Toxins: poisonous substances produced by living organisms. As generally employed, refers to substances that stimulate the formation of antibodies.

Totipotency: the capacity of a cell or a group of cells to give rise to an entire organism.

Tracheid (trā′kė̆-ĭd): an elongated thick-walled conducting and supporting cell of the wood. It has tapering ends and pitted walls without true perforations.

Tracheophyte (trā′kė̆-ȯ-fīt′): a plant with vascular (xylem and phloem) tissue, such as a fern or seed plant.

Transduction: genetic transfer in bacteria by phage particles.

Transfer-RNA: smaller RNA molecules which unite with specific amino acids and align them on messenger-RNA in the formation of polypeptides.

Transformation: a process in bacteria in which free DNA, released from a donor cell, penetrates a recipient cell and brings about recombination.

Translocation: the movement of food, water, or mineral solutes from one part of a plant to another.

Transpiration: the loss of water from plant tissues in the form of vapor.

Transpiration ratio: ratio of water transpired to dry matter produced by a plant; typically 300–500 pounds of water per pound of dry matter.

Transverse plane: the view of wood exposed by cutting at right angles to the long axis. A section cut in this plane for microscopic study is a transverse or cross section.

Tree: a perennial woody plant with a single stem (trunk).

Trophic (trŏf′ĭk) **level:** a step in the movement of energy through an ecosystem.

Tropism (trō′pĭz′m): tropic movement; a response to an external stimulus in which the direction of the movement is usually determined by the direction from which the stimulus comes.

Tube cell: (1) one of the two cells (the other is the generative cell) resulting from the germination of the microspore of flowering plants; a vegetative cell, probably vestigial; (2) one of the cells of the mature male gametophyte of pine.

Tuber (tū′bĕr): A much enlarged portion of a subterranean branch (stolon) provided with buds on the sides and tip.

Turgor (tûr′gŏr): the swollen condition of a cell caused by internal water pressure.

Turgor pressure: the pressure developed within a cell that causes turgor.

Twig: the portion of a woody branch produced during the latest growing season.

Umbel (ŭm′bĕl): a kind of flower cluster (inflorescence) in which the flower stalks arise from the same point, like the ribs of an umbrella.

Unicellular: composed of a single cell.

Unisexual: usually applied to a flower lacking either stamens or pistil. A perianth may be present or absent.

Universal veil: in fleshy fungi, a layer of tissue covering the fruiting body early in development. The basal remnants constitute the volva.

Uredospore (ů-rē′dȯ-spōr): a vegetative, summer spore; one of the spores in the life·cycle of a rust fungus.

Vacuole (văk′ů-ōl): a cavity within the cytoplasm containing a solution of sugars, salts, pigments, etc., together with colloidal materials.

Vacuole membrane: the innermost layer of the cytoplasm; the tonoplast.

Variation: anatomical or physiological differences among individuals of the same species. Variations may be environmental or hereditary.

Variety: a subdivision of a species.

Vascular (văs′kū-lēr) **bundle:** a strandlike portion of the vascular tissue of a plant, composed of xylem and phloem.

Vascular cambium: the meristematic zone or cylinder, one cell thick, that produces secondary phloem and xylem.

Vascular cylinder: a term applied to the vascular tissues and associated ground tissues of stem or root.

Vascular rays: ribbonlike sheets of parenchymatous tissue that extend radially through the wood, across the cambium, and into the secondary phloem. They are always produced by the cambium.

Vascular tissue: tissue composed of xylem and phloem; the conducting tissues of the Tracheophytes.

Vegetable: botanically, any edible part of a plant not formed from a mature ovary or from an ovary and associated parts.

Vegetative: growth, tissues, or processes concerned with the maintenance of the plant body, as contrasted with tissues or activities involved in sexual reproduction.

Vegetative reproduction: (1) in seed plants, reproduction by means other than by seeds; (2) in lower forms, reproduction by vegetative spores, fragmentation, or division of the plant body.

Vein: a vascular bundle forming a part of the framework of the conducting and supporting tissue of a leaf or other expanded organ.

Venation (vē-nā′shŭn): the arrangement of the vascular bundles, or veins, of a leaf.

Vernalization: the sensitizing of a plant to photoperiod by exposure to near-freezing temperatures at an earlier stage of development.

Vessel: a tubelike structure of the xylem, composed of cells placed end to end and connected by perforations.

Vessel element: one of the cells composing a vessel. (Also called vessel member).

Virus (vī′rus): a submicroscopic entity whose genetic material is composed either of DNA or RNA, and which is directly dependent upon a living host cell for its multiplication.

Vitamins: complex organic compounds constructed by green and some nongreen plants and necessary in minute amounts for normal growth.

Vivipary (vī-vĭp′a̍-rĭ): in plants, the growth of the embryo directly into the seedling without cessation of growth.

Volva (vŏl′va̍): remnants of the universal veil at the base of the stalk of a gill fungus.

Water-absorbing power of the cell: the ability of the cell to absorb water because the diffusion pressure of water within the cell is lower than the diffusion pressure of pure water.

Whorl: (1) three or more leaves or branches at a node; (2) a circle of floral organs, such as a whorl of sepals, stamens.

Wood: technically, the xylem, the woody portion of the vascular tissue.

Wood fibers: the supporting cells of the wood resembling tracheids but generally longer, thicker-walled, with more tapering ends and with reduced pits.

Woody plants: trees and shrubs in which increase in diameter of stems and roots continues from year to year.

Xanthophylls (zăn′thŏ-fĭl′z): yellow to orange carotenoid pigments associated with carotenes in the plastids of plant cells.

Xeromorphic (zē′rŏ-môr′fĭk): anatomical modifications of xerophytes.

Xerophyte (zē′rŏ-fīt): a plant that is adapted to survival under conditions of a limited supply of water in the habitat.

Xylem (zī′lĕm): one of the two component tissues of vascular tissue. (*See* **Phloem, Wood**.)

Zooplankton (zō′ŏ-plangk′tŏn): animal plankton.

Zoospore (zō′ŏ-spōr): a motile spore, found among algae and fungi.

Zygomorphic (zī′gŏ-môr′fĭk): *See* **Irregular.**

Zygospore (zī′gŏ-spōr): a spore formed following the union of two gametes (or gametangia) of approximately the same size.

Zygote (zī′gōt): a cell arising from the fusion of equal or unequal gametes. The fusion of unequal gametes results in a zygote called the fertilized egg.

Index